Optochemical Nanosensors

Series in Sensors

Series Editors: Barry Jones and Haiying Huang

Optochemical Nanosensors

Edited by
Andrea Cusano
Francisco J. Arregui
Michele Giordano
Antonello Cutolo

CRC Press
Taylor & Francis Group
Boca Raton London New York

CRC Press is an imprint of the
Taylor & Francis Group, an **informa** business

A TAYLOR & FRANCIS BOOK

CRC Press
Taylor & Francis Group
6000 Broken Sound Parkway NW, Suite 300
Boca Raton, FL 33487-2742

First issued in paperback 2019

ISBN-13: 978-1-4398-5489-1 (hbk)
ISBN-13: 978-0-367-38065-6 (pbk)

Library of Congress Cataloging-in-Publication Data

Optochemical nanosensors / editors, Andrea Cusano ... [et al.].
    p. cm. -- (Series in sensors)
  Includes bibliographical references and index.
  ISBN 978-1-4398-5489-1 (hardback)
  1. Chemical detectors. 2. Optical detectors. 3. Nanoelectromechanical systems. 4. Biosensors. I. Cusano, Andrea.

TP159.C46O68 2012
543--dc23

2012031218

**Visit the Taylor & Francis Web site at
http://www.taylorandfrancis.com**

**and the CRC Press Web site at
http://www.crcpress.com**

Contents

Part I   Basics

Part II   Research Advances

Part III   Applications

Foreword

This book covers the rapidly growing field of optical chemical nanosensing, which is a new and exciting area of research and development within the larger field of optical chemical sensing and biosensing. Generally speaking, sensors are controlling our life to an extent that was not envisioned some 20 years ago and even today is not realized by many. An average-sized car typically has 20–50 sensors that control functions and warrant safety. All of them are usually so small or well hidden that they are virtually invisible and, thus, ignored. The temperature of refrigerators is kept fairly constant via sensors that switch the cooling system. Sensors record acceleration, position, and pressure and cause city lights to be turned on and off automatically. Physical sensors are true sensors in that they function in an unattended fashion, fully reversibly, and over long time spans without a need for maintenance. Malfunctions such as the loss of reversibility become obvious, for example, by the distinct smell of a refrigerator that has not been cooled for 10 days.

All physical sensors are small and mostly produced in large quantities. Chemical sensors and biosensors, by comparison, are much less established. Present day chemical sensors and biosensors are mainly of the electrochemical type, but optical sensors are also on the rise. They will cover more specific needs than physical sensors (which are almost ubiquitous), in particular in the medical field, in environmental sciences, marine sciences, and in bioprocess control, to give a few examples. Some chemical sensors (such as the oxygen sensor in a car's lambda probe in the catalytic converter) are expected to have very long lifetimes, while chemical sensors and biosensors for medical purposes often are not operated for more than 10 h in vivo. The glucose biosensor in an artificial pancreas (which, in my opinion, is the Holy Grail in biosensor research) is a notable exception. Moreover, in vivo sensors often are of the single use type for safety reasons.

Unlike physical sensors, the price of chemical sensors and biosensors in medical sciences is not a major issue, but their performance is extremely critical, not the least because existing sensors perform excellently. Typical examples are the pH electrode, the Clark electrode for oxygen, the many kinds of ion-selective electrodes for blood electrolytes, and the electrochemical biosensors for in vitro assays of glucose, lactate, or cholesterol.

Optical nanosensors will further widen the field. However, the design of such sensors requires new materials, new methods for their characterization, new optical sensing schemes in addition to established ones, new methods for creating nanosized structures, and new methods for their interrogation in a complex environment such as blood. New materials include the various kinds of Cd(II)- and Pb(II)-based quantum dots, fluorescent dots made from silica, semiconducting silicon, or carbon, upconverting nanoparticles (capable of converting IR light into visible light) or photonic crystals.

Such materials are not necessarily used in spherical shape, but are also in the form of nanosized fibers, films, elongated particles (rods), nanotubes, even flowers, or combinations thereof. There is also a particular need for improved methods for the characterization of nanomaterials (sensors included) because conventional methods such as IR, NMR, or mass spectrometry are not easily applicable here. TEM, SEM, SERS, and the like have therefore gained significant importance.

Nanosensors may also be considered as molecular machines that can provide valuable optical information, which, however, has to be read out and processed in a proper way so as to finally result in analytically useful information such as the concentration of a (bio)chemical. As a result, new sensing schemes and spectroscopies are needed to account for the fact that such sensors, because of their weak signals, need quite sophisticated methods, in particular if the response of a single nanosensor is to be detected. Fortunately, the use of nanometer-sized materials enables certain spectroscopic methods to be applied that work best on a micro- or nanoscale, examples being FRET, certain nonlinear spectroscopies, plasmonic resonance, effects of microring resonance, SERS, diffraction of light in photonic crystals, and the like.

Nanosized sensors enable the study of chemical and biochemical processes at a level and in dimensions that may not have been envisioned some 20 years ago. It has enabled and will enable a closer look at the cell and its function with unprecedented temporal and spatial resolution. In fact, imaging of cells with a spatial resolution of 2–10 nm has become true with fluorescent methods such as STED or STORM, spectroscopies that go far beyond the classical resolution of optical microscopy as defined by Abbe's law. However, it should also be emphasized here that the use of nanoparticles (that easily penetrate cells or even the blood/brain barrier) may imply certain risks that have not been investigated in adequate detail so far.

This book reflects ongoing research very well. Aside from new materials and spectroscopies, the progress made in micro- and nanofluidic devices (including lab-on-a-fiber technology) and nanofibers is also impressive. It is good to see that respective research not only has resulted in fundamental studies and new and generally applicable sensing platforms, but can already be applied to practical situations such as security, environmental monitoring, or detecting hazardous substances. Therefore, this book is a most valuable resource for anybody interested in this very exciting technology.

Otto S. Wolfbeis
Institute of Analytical Chemistry, Chemo- and Biosensors
University of Regensburg
Regensburg, Germany

Preface

Nanotechnology and nanoscale materials are new and exciting fields of research. Nowadays, only considering journal and conference papers, close to 100,000 scientific manuscripts are published yearly according to data extracted from ISI Web of Knowledge[SM] and Scopus[TM] databases. The inherent small size and the unique optical, magnetic, catalytic, and mechanical properties of nanostructured materials, which are not found in bulk materials, permit the development of novel devices and applications previously unavailable. One of the earliest implementations of nanotechnology has been the development of improved chemical and biological sensors. These optical chemical nanosensors can be generally defined as devices that transduce a chemical or biological event to an optical signal, having dimensions smaller than 1000 nm.

Remarkable progress has been made in the last few years in the design and fabrication of optical chemical nanosensors and their utilization worldwide. The emergence of this new technology is a continuous challenge since the new advances imply new applications and new questions to solve. The impact that will be made in the coming years by implementing novel sensing principles as well as new measurement techniques is unpredictable.

In fact, considering the broad field of sensors, what is exciting in sensor research and development today? This is a tough question because there are many significant innovations and inventions being made daily, and more than 200 new scientific studies are published every day related to sensors. Nevertheless, what seems clear is that there is an increasing interest in the practical utilization of nanotechnology in the sensors field and that the application of nanosensors to different types of molecular measurements is expanding rapidly. Therefore, micro- and nanotechnology, novel materials, and smaller, smarter, and more effective systems will play an important role in the future of sensors. The possibilities provided by nanosensor technology have just begun and will continue to revolutionize different key fields such as cellular biology or medicine, among others. For instance, further development of delivery techniques and new sensing strategies to enable quantification of an increased number of analytes are required to facilitate the access of medicaments to specific sites in organs and to deliver these drugs at a controlled and sustained rate to the site of action.

There are additional challenges in the sensors field, such as the promise of ubiquitous sensor systems, providing situational awareness at low cost. In order to fulfill this, there must be a demonstrated benefit that is only gained through further miniaturization. For example, new nanowire-based materials that have unique sensing properties can provide higher sensitivity, greater selectivity, and possibly enhance stability at a lower cost, and such advances are necessary to the sensor future. Nanosensors can improve the world through diagnostics in medical applications; improve health, safety, and security for people; and improve environmental monitoring. The seed technologies are now being developed for a long-term vision that includes intelligent systems that are self-monitoring, self-correcting and repairing, and self-modifying or morphing not unlike sentient beings.

The applications of these new optochemical nanosensors can also be extended to food safety, environmental monitoring, or homeland security among others. In fact, this book deals with the detection of bioterrorist threats, food security, virology, and explosive detection, as will be seen in the different chapters.

The book has been organized by topics of high interest. In order to offer a fast read of each topic, every chapter is independent and self-contained. On the other hand, since nanotechnology is an interdisciplinary field, some chapters overlap into others and are in some way related.

The book has been divided into three different parts: Basics, Research Advances, and Applications. In Part I, Fundamentals of Photonics (Chapter 1) is followed by a more specific Fundamentals of Optical Chemical Sensors (Chapter 2). The part concludes with Chapter 3, which provides a review on optochemical sensors. Part II opens with Chapter 4, Photoluminescent Nanosensors. The state of the art of cantilever-based sensors is covered in Chapter 5. Chapter 6 deals with nanostructured surface plasmon resonance sensors while Chapter 7 reviews fiber-optic nanosensors. Chapter 8 focuses on lab-on-fiber technology. The topics of micro/nanofibers for biochemical sensing, optofluidic sensor, and lab-on-chip nanostructured sensors are covered in Chapters 9 through 11, respectively. Chapters 12 through 14 describe photonic crystals, nanomaterials, and nanostructures and linear and nonlinear spectroscopy at nanoscale. Chapter 15 deals with plasmonic nanostructures and nano-antennas for sensing. The part concludes with Chapters 16 and 17, Overcoming Mass Transport Limitations with Optofluidic Plasmonic Biosensors and Particle Trapping and Optical Micro-Ring Resonators for Chemical Vapor Sensing. Part III begins with Chapter 18, which focuses on the detection of bioterrorist threats. This is followed by Chapter 19, which deals with food safety and security. Chapter 20 covers multifunctional fiber-optic nanosensors for environmental monitoring, and Chapter 21 reviews nano-optical sensors for virology. The book concludes with Chapter 22, Nano-Optical Sensors for Explosive Detection.

Editors

Andrea Cusano received his Laurea degree cum Laude in electronic engineering in 1998, from the University of Naples Federico II, Italy. In 2003, he defended his PhD thesis carried out in the Department of Electronic and Telecommunication Engineering at the University of Naples under the guidance of his tutor Professor Antonello Cutolo. His PhD thesis was entitled *Optoelectronic Sensors for Smart Materials and Structures.* In February 2002, while pursuing his PhD, he was engaged as a research fellow in the Department of Engineering at the University of Sannio to support Professor Antonello Cutolo in creating the Optoelectronic Division. From September 2002 to present, he has been professor for several courses in electronics in the Engineering Department at the University of Sannio. In December 2002, he started his service as permanent researcher at the University of Sannio. In December 2005, he won a national competition as associate professor (scientific area ING-INF/01) at the University of Sannio. His research interests are focused on optical sensors and nanotechnology.

Dr. Cusano is the cofounder of the following spin-off companies focused on the development of fiber-optic sensing systems for industrial applications: OptoSmart S.r.l. in 2005, MDTech in 2007, and OptoAdvance in 2011. He has published over 100 papers in prestigious international journals and more than 150 communications in well-known international conferences worldwide. He currently has four international patents in use in prestigious industrial companies (Ansaldo STS, Alenia WASS, OptoSmart, and MDTech) and more than 10 national patents.

Since 2011, he has been the editor in chief of the *Journal of Optics and Laser Technology* (Elsevier). He is also the associate editor of the *Journal of Sensors* (Hindawi), *The Open Optics Journal* (Bentham Science Publishers), *The Open Environmental & Biological Monitoring Journal* (Bentham Science Publishers), *Sensors and Transducers Journal* (IFSA), the *International Journal on Smart Sensing and Intelligent Systems*, and *Photonic Sensors* (Springer Verlag).

He is the coauthor of 30 chapters published in international books and invited papers in prestigious scientific international journals as well as international conferences. Cusano coedits two special issues on optical fiber sensors: "Special Issue on Optical Fiber Sensors" (*IEEE Sensors*, 2008) and "Special Issue on Fiber Optic Chemical and Biochemical Sensors: Perspectives and Challenges Approaching the Nano-Era" (*Current Analytical Chemistry*, Bentham, 2008).

He is also a coeditor of international books such as *Fiber Bragg Grating Sensors: Recent Advancements, Industrial Applications and Market Exploitation*, Bentham e-book, 2011; *Selected Topics on Photonic crystals and Metamaterials*, World Scientific (2011); and *Photonic Bandgap Structures: Technological Platforms for Physical Chemical and Biological Sensing*, Bentham Science Publishers (2012).

Cusano is a member of the technical program committee of several international conferences such as IEEE Sensors, International Conference on Sensing Technologies, the European Workshop for SHM, the European Workshop on Optical Fiber Sensors, the Asian-Pacific Optical Sensors, the International Conference on Applications of Optics and Photonics; and Bragg Gratings, Photosensitivity, and Poling (BGPP).

Cusano is a consultant for important companies of the Finmeccanica Group such as Ansaldo STS and Alenia WASS. He is also a consultant for CERN in Geneva, where he works on the development of innovative optical sensors for high-energy physics applications.

He has received many international recognitions and awards for his efforts in the development of innovative optical sensing systems. He was also principal investigator for many national projects and a consultant in many European projects.

Francisco J. Arregui is a professor at the Public University of Navarre, Pamplona, Spain. He was part of the team that fabricated the first optical fiber sensor by means of the layer-by-layer self-assembly method at Virginia Tech, Blacksburg, Virginia, in 1998. He is the author of about 300 scientific journal and conference publications, most of them related to optical fiber sensors based on nanostructured coatings. Professor Arregui has been an associate editor of *IEEE Sensors Journal*, the *International Journal on Smart Sensing and Intelligent Systems*, and the *Journal of Sensors*. He is also the editor of the book entitled *Sensors Based on Nanostructured Materials*. The *Journal of Sensors* (Hindawi) was founded in 2007 by Professor Arregui. He served as editor in chief of the journal between 2007 and 2011.

Michele Giordano received his master's degree in chemical engineering from the University of Naples Federico II in 1992. In the same year, he started a PhD course in materials engineering at the University of Naples Federico II and completed it in 1995. In 1998, he completed the formative track within the Institute for Composite Materials Technology (ITMC) of the National Research Council (CNR). Also, in 1998 he acquired a part-time position as a researcher at ITMC CNR, where he become a permanent researcher in 2001. From 2003 to present, he has been a lecturer at the University of Naples Federico II teaching a course on composite materials technology. Since 2005, he has served as a senior scientist at the Institute of Composite and Biomedical Materials (IMCB) CNR. In 2005, he cofounded the research spin-off company "OptoSmart," focused on the development of fiber-optic sensor systems. Since 2006, he has been responsible for the Composite Technology Unit of IMCB-CNR, coordinating the corporate CNR *commessa* project Polymers, composite and nanostructures technologies. In 2007, he cofounded a new research spin-off company, MDTech, conducting research in the field of optical systems for biomedical applications.

His research activities are within the area of engineering and materials science. In particular, his research focuses on nano- and macrocomposite materials, mainly polymer based, including multiscale design and processing of multifunctional composite materials, structural health management systems, and thin films engineering for sensing and optoelectronic applications. He is the author of more than 90 JRC journal papers, more than 130 international conference communications, and 8 book chapters.

Antonello Cutolo received his Laurea degree cum Laude in electronic engineering from the University of Naples Federico II in 1978 after a six months stint at the Fiat Research Center working on online characterization of mirrors for high-power laser systems.

After serving for one year in the Italian Air Force, he spent a year in the Department of Applied Mathematical Physics at the University of Copenaghen, working on propagation in nonlinear structures, soliton interaction, and Josephson structures. In particular, he found the theoretical limit of the bandwidth of finite size Josephson oscillators.

In 1982, he worked at the National Laboratories of Frascati (Rome) to build a free electron laser on the Adone storage ring, where he was in charge of the design and construction of the optical resonator.

In 1983, he was appointed researcher in the Department of Electronics at the University of Naples. From 1983 to 1986, he worked at the Photon Research Lab and at the Stanford Linear Accelerator Center (SLAC) of Stanford University (California) and in the Department of Physics at Duke University (North Carolina), where he designed and constructed a set of novel devices for increasing the peak power (cavity dumpers and mode lockers) and the tunability range (broadband output couplers and higher harmonic generators) of a free electron laser.

In 1987, he was appointed associate professor of quantum electronics at the University of Naples and in 1998, full professor of electronics and optoelectronics at the University of Sannio. He has founded and directed two laboratories of optoelectronics finalized to contactless characterization of electronics devices and materials, optical fiber sensors, nonlinear optics, and nanophotonics applications. Many of the results have found practical applications in the industry.

Professor Cutolo has been the main advisor of several projects in both basic and industrial research, which has led to many patents with large and small companies. He has tutored several students working on their PhD program. He has also published several books, more than 300 papers in international technical journals and conference proceedings, and has filed more than 20 patents both in Europe and in the United States.

He started a high-tech company (Optosmart, a spin-off from the University of Sannio and the Italian National Council of Research) that focused on the production of optical fiber sensor arrays for environment, structural health monitoring, railway security, harbor surveillance, and food quality control application.

In addition, he created the Optosonar Consortium finalized to the application of optical fibers to underwater security and monitoring. He has been a member of the scientific committee of Corista (Consortium for Research on Advanced Remote Sensing Systems) and is a member of the scientific committee of Confindustria.

Professor Cutolo has been chairman or co-chair of several national and international technical conferences, and he is the referee of several international scientific publications. His research interests include optoelectronic modulators and switching, optical characterization of semiconductor devices and materials, laser beam diagnostic, and nonlinear optical devices.

Contributors

Elena Alieva
Institute of Spectroscopy
Russian Academy of Sciences
Moscow, Russia

Hatice Altug
Department of Electrical and Computer
    Engineering
and
Photonics Center
and
Department of Materials Science and Engineering
Boston University
Boston, Massachusetts

Francisco J. Arregui
Sensor Research Laboratory
Department of Electrical and Electronic
    Engineering
Public University of Navarra
Pamplona, Spain

Alp Artar
Department of Electrical and Computer
    Engineering
Boston University
Boston, Massachusetts

Euiwon Bae
School of Mechanical Engineering
Purdue University
West Lafayette, Indiana

Francesco Baldini
Institute of Applied Physics
National Research Council
Firenze, Italy

Romeo Bernini
Institute for Electromagnetic Sensing of the
    Environment
National Research Council
Naples, Italy

Arun K. Bhunia
Department of Food Science
Purdue University
West Lafayette, Indiana

Anja Boisen
DTU Nanotech
Technical University of Denmark
Kongens Lyngby, Denmark

Arif Engin Cetin
Department of Electrical and Computer
    Engineering
and
Department of Materials Science and Engineering
Boston University
Boston, Massachusetts

Si Chen
Department of Applied Physics
Chalmers University of Technology
Göteborg, Sweden

John H. Connor
Photonics Center
Boston University
and
Department of Microbiology
Boston University Medical School
Boston, Massachusetts

Marco Consales
Department of Engineering
Optoelectronic Division
University of Sannio
Benevento, Italy

Jesus M. Corres
Sensor Research Laboratory
Department of Electrical and Electronic
    Engineering
Public University of Navarra
Pamplona, Spain

Alessio Crescitelli
Department of Engineering
Optoelectronic Division
University of Sannio
Benevento, Italy

Brian Culshaw
University of Strathclyde
Glasgow, United Kingdom

Andrea Cusano
Department of Engineering
Optoelectronic Division
University of Sannio
Benevento, Italy

Antonello Cutolo
Department of Engineering
Optoelectronic Division
University of Sannio
Benevento, Italy

Zachary J. Davis
DTU Nanotech
Technical University of Denmark
Kongens Lyngby, Denmark

Anna Chiara De Luca
Dipartimento di Scienze Fisiche
Università di Napoli Federico II
Naples, Italy

Luca De Stefano
Institute for Microelectronics and Microsystems
National Research Council
Naples, Italy

Edoardo De Tommaso
Institute for Microelectronics and Microsystems
National Research Council
Naples, Italy

Ignacio Del Villar
Sensor Research Laboratory
Department of Electrical and Electronic
    Engineering
Public University of Navarra
Pamplona, Spain

Anuj Dhawan
Department of Electrical Engineering
Indian Institute of Technology Delhi
New Delhi, India

and

Fitzpatrick Institute for Photonics
and
Department of Biomedical Engineering
Duke University
Durham, North Carolina

Emanuela Esposito
CNR-ICIB "E. Caianiello"
Pozzuoli, Italy

Xudong Fan
Department of Biomedical Engineering
University of Michigan
Ann Arbor, Michigan

Ambra Giannetti
Institute of Applied Physics
National Research Council
Firenze, Italy

Michele Giordano
Institute for Composite and Biomedical
    Materials
National Research Council
Naples, Italy

Javier Goicoechea
Sensor Research Laboratory
Department of Electrical and Electronic
    Engineering
Public University of Navarra
Pamplona, Spain

Min Huang
Department of Electrical and Computer
    Engineering
and
Photonics Center
and
Department of Materials Science and Engineering
Boston University
Boston, Massachusetts

Mikael Käll
Department of Applied Physics
Chalmers University of Technology
Göteborg, Sweden

Stephan S. Keller
DTU Nanotech
Technical University of Denmark
Kongens Lyngby, Denmark

Vinod Kumar Khanna
CSIR—Central Electronics Engineering Research
    Institute
Council of Scientific and Industrial Research
Pilani, India

Valery Konopsky
Institute of Spectroscopy
Russian Academy of Sciences
Moscow Region, Russia

Jose Miguel López-Higuera
Photonic Engineering Group
University of Cantabria
Santander, Spain

Ignacio R. Matías
Sensor Research Laboratory
Department of Electrical and Electronic
    Engineering
Public University of Navarra
Pamplona, Spain

Colette McDonagh
School of Physical Sciences
National Centre for Sensor Research
Dublin City University
Dublin, Ireland

Rachel A. McKendry
London Centre for Nanotechnology
University College London
London, United Kingdom

Mike McShane
Department of Biomedical Engineering
Texas A&M University
College Station, Texas

Carlo Morasso
London Centre for Nanotechnology
University College London
London, United Kingdom

Balaji Nandagopal
Sri Narayani Hospital & Research Centre
Vellore, India

Joseph Ndieyira
London Centre for Nanotechnology and
    Department of Medicine
University College London
London, United Kingdom

and

Department of Chemistry
Jomo Kenyatta University of Agriculture and
    Technology
Nairobi, Kenya

Emanuele Orabona
Institute for Microelectronics
    and Microsystems
National Research Council
Naples, Italy

Giuseppe Pesce
Dipartimento di Scienze Fisiche
Università di Napoli Federico II
Naples, Italy

Marco Pisco
Optoelectronic Division
Department of Engineering
University of Sannio
Benevento, Italy

Mageshbabu Ramamurthy
Division of Biomedical Research
Sri Narayani Hospital & Research Centre
Vellore, India

Ilaria Rea
Institute for Microelectronics
    and Microsystems
National Research Council
Naplcs, Italy

Ivo Rendina
Institute for Microelectronics
    and Microsystems
National Research Council
Naples, Italy

Armando Ricciardi
Department of Engineering
Optoelectronic Division
University of Sannio
Benevento, Italy

Paul B. Ruffin
United States Army Research Development,
    and Engineering Command
Huntsville, Alabama

Giulia Rusciano
Dipartimento di Scienze Fisiche
Università di Napoli Federico II
Naples, Italy

Sathish Sankar
Division of Biomedical Research
Sri Narayani Hospital & Research Centre
Vellore, India

Antonio Sasso
Dipartimento di Scienze Fisiche
Università di Napoli Federico II
Naples, Italy

Gopalan Sridharan
Division of Biomedical Research
Sri Narayani Hospital & Research Centre
Vellore, India

Yuze Sun
Department of Biomedical Engineering
University of Michigan
Ann Arbor, Michigan

Mikael Svedendahl
Department of Applied Physics
Chalmers University of Technology
Göteborg, Sweden

Maria Tenje
Department of Measurement
　Technology and Industrial Electrical
　Engineering
Lund University
Lund, Sweden

Cosimo Trono
Institute of Applied Physics
National Research Council
Firenze, Italy

Joel Villatoro
Institut de Ciencies Fotoniques
Barcelona, Spain

Tuan Vo-Dinh
Fitzpatrick Institute for Photonics
and
Department of Biomedical Engineering
and
Department of Chemistry
Duke University
Durham, North Carolina

Manuel Vögtli
London Centre for Nanotechnology
University College London
London, United Kingdom

Dorota Wencel
School of Physical Sciences
National Centre for Sensor Research
Dublin City University
Dublin, Ireland

Ahmet Ali Yanik
Department of Electrical and Computer
　Engineering
and
Photonics Center
and
Department of Materials Science and
　Engineering
Boston University
Boston, Massachusetts

Stuart (Shizhuo) Yin
The Pennsylvania State University
University Park, Pennsylvania

Carlos R. Zamarreño
Sensor Research Laboratory
Department of Electrical and Electronic
　Engineering
Public University of Navarra
Pamplona, Spain

Part I

Basics

Part I

Basic

1

Fundamentals of Photonics

Brian Culshaw and Jose Miguel López-Higuera

CONTENTS

1.1 Introduction

Although the study of light began in ancient Greece in 500 BC, very few new contributions were proposed until Galileo (born in 1564), Willebrord Snell van Royen and Descartes (born in 1621) independently discovered the law of refraction. A remarkable development from this simple law was carried out by Pierre Fermat in 1657, which gave an alternative point of view to the phenomenon of refraction.

In the same year that Galileo died (1642), Newton was born. This started a remarkably creative period in history of science. Newton did not have a particularly clear view about the nature of light. He believed that light was corpuscular in nature and that the particles traveled from the object to the eye as a stream of projectiles. These ideas reduced light propagation to processes of reflection and refraction of light to a collision problem [1]. Christian Huygens (in 1690) considered light to be a form of wave that travels from the source to the observer. His theory explained the reflection and refraction phenomena of light. It was then that the controversy between the wave and particle theories of light started. Newton recognized the difficulties in reconciling experimental data with some kind of corpuscular properties. He and some others anyway continued to believe in the corpuscular nature of light up to his death in 1727. Since then, other researchers such as Euler, Young, Fresnel, and Huygens suggested light as a wave in motion and developed a theory that explained the phenomena of optical interference and diffraction (late eighteenth and early nineteenth centuries).

Fizeau (in 1849) measured, precisely, the speed of light and Foucault (in 1850) demonstrated that the speed of light is less in water than in air by the ratio of their refractive indices.

The observation of the polarization light was crucial in leading to the concept that light consists of transverse fields. But there was the problem of what is the medium through which the waves are traveling. James Clerk Maxwell made central contributions to understanding the nature of light and in the early 1860s, using analogies with fluid flow, he discovered that electromagnetic disturbances propagated at the speed of light through the vacuum [2]. Maxwell did not hesitate to identify light as electromagnetic waves. In the following year, the Theory of Electromagnetism was completely rewritten, and Maxwell's equations of the electromagnetic field were revealed for the first time. This theory was validated in the late 1880s when Heinrich Hertz demonstrated that electromagnetic waves had all the properties of light: linear propagation, reflection, refraction, and polarization. The speed of propagation of waves was measured by the product of the wavelength and their frequency, and this turned out to be exactly the speed of light (1889).

Planck (born in 1858) introduced the concept of quantization in order to account for the spectrum of black-body radiation in 1900. He invoked the idea that light was emitted from a radiation body in discrete packets rather than continuously as a wave. Einstein (1879) published his famous paper on the theory of relativity in 1905 in which light plays a central role in the new understanding of the nature of space and time. This consequently made the question of the nature and existence of the ether completely redundant. The speed of light assumes the role of an absolute limiting speed. In his paper about the photoelectric effect [3], Einstein suggested that for some purposes, it will be more appropriate to consider light consisting of particles rather than waves, and Millikan demonstrated Einstein's prediction in 1916.

In 1927, Davisson and Germer showed that electrons have wave-like properties too. This concept had been proposed by de Broglie as early as in 1924, and it proved to be the inspiration behind Schrödinger's formulation of wave mechanics. The new foundations of physics in wave and quantum mechanics were already established. They showed that the explanation of these phenomena required that light was made of quantized packets of energy called photons, a concept that was proposed for the first time by G. Newton Lewis in 1926.

The introduction of the concept of photons caused a dilemma in the theory of light. Today it is believed that light can be viewed either as wave or as a particle, depending on the kind of interaction. This view of the wave–particle duality of photons is most commonly accepted nowadays: wavelike and particle-like properties seem to be complementary aspects of photons.

Photons and electrons appear to behave either as particles or as waves to us only because of the limitations of our modes of thought. We have been conditioned to think in terms of the behavior of objects such as sticks, stones, and waves on water, the understanding of which has been necessary for us to survive, as species, at our particular level of things [4].

Today, photonics is understood as the set of techniques and scientific knowledge that are applied to the generation, propagation, control, amplification, detection, storage, and processing of signals of the optical spectrum, along with their technologies and derived uses. It can be deduced that the field of photonics can be divided into several areas and that, in many cases, electronics and photonics overlap. There are cases where electrons control photons and there are others in which photons control electrons so a "complementary harmony" can exist between electronics and photonics [5].

Within the field of photonics, there are two areas in which the aforementioned complementary harmony between both fields is very clear and contributes to enhance the behavior that could now be possibly achieved together: optoelectronics and photonics sensing, more popularly known as optical sensors area [6].

Optoelectronics is a diverse and wide-ranging subject covering the many facets of the study of the interaction of light with materials. We have endeavored to encapsulate the principal features of the topic in Figure 1.1. In the context of optically based chemical nanosensor systems, the principal question is how will minute samples of the chemical of interest (i.e., to which the sensor should respond) influence the optoelectronic properties of some form of structure to produce a change therein that can be detected and uniquely related to the triggering chemical species.

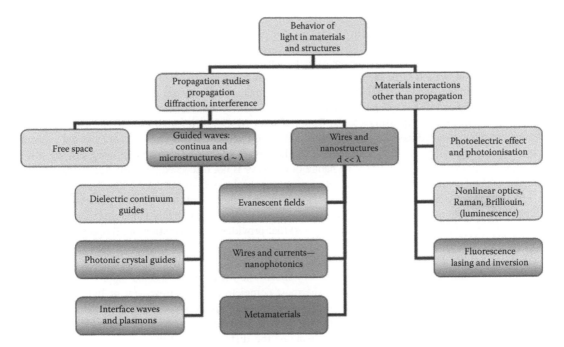

FIGURE 1.1 What is optoelectronics? The dark (metamaterials) areas are "new" to nanophotonics while the dark- and gray-shaded areas represent concepts evolved for guided wave optics that are also applicable to nanophotonic structures, sometimes with appropriate modifications. The blue concepts are well established.

The presence of our sample can influence the optical properties of the detecting structure through three generic mechanisms:

- The material itself (i.e., the sample volume) forms part of the structure and changes in the effective refractive index distribution within the structure with consequent modifications to the optical properties of the structure. These refractive index changes should be viewed in a very general sense to include not only the real and imaginary part of the index itself but also any nonlinear higher order terms.
- Some secondary interface material forms an integral part of the structure, which selectively reacts chemically with the material of interest and consequently changes its refractive index. Again, this should be viewed in the most general possible sense to include linear and higher order terms.
- A secondary interface material can be modified in response to the species of interest through first order phenomena, which may include, for example, changes in temperature, levels of ionization, or changes in dimension, which typically in turn will also involve index changes. In this case, both the influence of the primary interaction and of the consequent index implications must be taken into account in analyzing the optical responses of the host structure.

Figure 1.1 endeavors to encapsulate the conceptual tools required to undertake this analysis assuming, of course, that the relevant modeling processes can be incorporated. The aim of this chapter is to highlight the principles of most of these conceptual tools with additional background provided through the second chapter focusing particularly on florescence and related phenomena. Figure 1.1 also attempts to highlight the principal phenomena of interest (the graded boxes in the diagram) and also to highlight the "new" optoelectronics, which is particularly pertinent for the understanding of these small-scale optical chemical sensing systems.

1.2 Nature of Light

In general, it can be said that light or the optical radiation can be interpreted as an electromagnetic wave phenomenon that follows the same theoretical principle of other electromagnetic radiations. Light propagates in two mutually coupled vector waves as any electromagnetic field does: an electric and a magnetic field wave. This interpretation of light can be called *electromagnetic optics*. However, in some cases it is possible to describe many phenomena using a scalar wave theory or wave function and, then, light treatment can be done using what is called the *wave optics*. When light propagates around or through objects whose dimensions are much greater than the wavelength, it can be modeled as a scalar wave optic but with the wavelength infinitesimally small (whereby the nature of the wave is not discerned), and the light behaviors can be described by mean of rays following a set of geometrical rules or by means of what is called *ray optics*.

Electromagnetic optics provides the most complete treatment of classical optics phenomena, but in some cases the phenomena are quantum in nature (and cannot be classically explained). In such cases, a much more complex treatment (*quantum optics*) that provides an explanation of all optics phenomena is required [7].

In summary, as illustrated in Figure 1.2, ray optics is the simplest one and can be understood as the limit of wave optics when the wavelength is very short. The wave optic is a scalar approximation of the electromagnetic optics; the latter provides the most complete classical treatment of light, and finally the theory of quantum optics provides an explanation of all optical phenomena. The different models are founded in some basic postulates and have certain limits of application. Ray optics explains the cinematic of the propagation. Wave optics explains the diffraction and the interference phenomena. Electromagnetic optics makes it possible to have an exact analysis of the phenomena of classic optics, including the effect of energy at the interfaces. The propagation of light in any medium, including optical waveguides, is governed by Maxwell's equations and can be described by specifying the evolution of the associated electric and magnetic field vectors in space and time. Finally, quantum optics (the most complex one) explains all the optical phenomena that we know, including the interaction of light with matter.

1.2.1 Basic Properties of Light

1.2.1.1 Optical Spectrum

Up to a century ago, light was understood as the visible phenomena for the human eye and only the waves included from 380 to 750 nm band were associated with light. However, after Maxwell's discoveries the electromagnetic spectrum of light goes from the deep ultraviolet (100 nm) to the very far infrared (1 mm) and, hence, the visible part of the light is only a very narrow band included inside it.

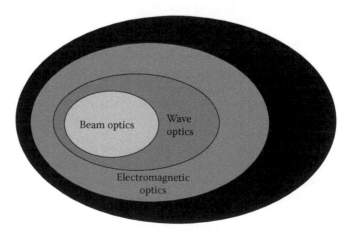

FIGURE 1.2 Illustration of the theories for the treatment of the light ordered (left to right) following their historical development: ray optics, wave optics, electromagnetic optics, and quantum optics, being the later the more complex.

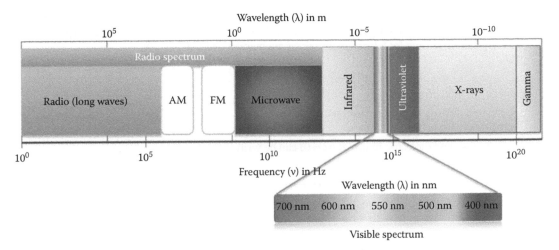

FIGURE 1.3 Electromagnetic spectrum of light in function of the wave wavelength λ, and function of their frequency, ν. It can also be represented in function of the photon energy ($h\nu$) or in function of the wave number λ^{-1}.

The infrared (0.75 μm–1 mm) band is usually subdivided into four regions: near infrared (750 nm–3 μm), middle infrared (3–6 μm), far infrared (6–15 μm), and extreme infrared (15 μm–1 mm). Inside this band, the very interesting band of the terahertz is included. The ultraviolet is known as the region of the electromagnetic spectrum that covers the range between 100 and 380 nm, and it can be subdivided into three bands: near (320–380 nm), middle (280–320 nm), and far ultraviolet (100–280 nm), which overlaps with the x-ray.

By using the facts that speed of light in the vacuum and the photon energy can be written as $c = \lambda f$ and hf, respectively, as can be observed in Figure 1.3, the light spectrum can be expressed in function of the wave wavelength λ, the wave frequency f, and/or in function of the energy of light photons.

1.2.1.2 Propagation of Light

Maxwell demonstrated the symmetrical interdependence of electric and magnetic fields: a changing electric field, even in free space, gives rise to a magnetic field and a changing magnetic field gives rise to an electric field. That is, each of the fields gives rise to the other, which enables the possibility that the two fields mutually sustain each other and then they can propagate as a wave solution (Figure 1.7) of the well-known Maxwell equations given by [8]

$$\nabla \times E + \frac{\partial B}{\partial t} = 0$$

$$\nabla \times H - \frac{\partial D}{\partial t} = J \tag{1.1}$$

$$\nabla \cdot D = \rho$$

$$\nabla \cdot B = 0$$

where E and H are the electric and the magnetic field vectors, respectively. These two field vectors describe an electromagnetic field, D and B include the effects of the field on the matter and are called the electric displacement and the magnetic induction, ρ and J are the electric charge and current densities, and may be considered as the sources of the fields E and H. Nonzero solutions of Maxwell's equations can be obtained even when ρ = 0 and J = 0, given what are known as electromagnetic waves that are able to propagate with a given velocity [7].

Despite the important behaviors of the light that can be deducted from the study of the Maxwell equations, simple sinusoidal solutions of the equations will be considered here to match the purpose of this book. They can be written for a lossy dielectric of attenuation coefficient α [9]:

$$E(x,y,z,t) = E(x,y) \cdot e^{-(\alpha/2)z} e^{j(\omega t - kz)}$$

$$H(x,y,z,t) = H(x,y) \cdot e^{-(\alpha/2)z} e^{j(\omega t - kz)} \tag{1.2}$$

and for a no loss-less dielectric material ($\alpha = 0$) can be written as

$$E_x(z,t) = E_o e^{j(\omega t - kz)}$$

$$H_y(z,t) = H_o e^{j(\omega t - kz)} \tag{1.3}$$

As illustrated in Figure 1.4, these two equations describe a wave propagating in the +z direction with a transversal electric field, E_x, oscillating sinusoidally in the xz plane, and the magnetic field, H_y, oscillating in the yz plane.

It can be observed that when the traversal fields are mutually orthogonal and both are in phase, the two fields mutually sustain each other. The frequency of the wave is given by $f = \omega/2\pi$ and its wavelength by $\lambda = 2\pi/k$, where ω and k are known as the angular frequency wavenumber or propagation constant, respectively. The speed of light can then be obtained as $c = f\lambda = \omega/k$. Also the speed in the free space can be written as $c_0 = (\varepsilon_0 \cdot \mu_0)^{-1/2}$, where ε_0 and μ_0 are the electric permittivity and the magnetic permeability, respectively. It was found that the speed of light in the vacuum is $c_0 = 2.99792458 \times 10^8$ m·s^{-1} exactly [10].

The free-space symmetry of Maxwell's equations is retained for media in which $\rho = 0$ (are electrically neutral) and $J = 0$ (not conduct electric currents), which is a typical case of optical materials of interest for this chapter such as dielectrics and, hence, the speed of light can be written as

$$c = (\varepsilon_0 \cdot \varepsilon_r \cdot \mu_0 \cdot \mu_r)^{-1/2} = c_0 \cdot (\varepsilon_r \cdot \mu_r)^{-1/2} \tag{1.4}$$

where ε_r and μ_r are the relative permittivity and permeability, respectively, which are measures of the enhancement of the electric and magnetic fields, respectively, by the presence of the dielectric media. c_0 and v are the speed of light in the vacuum and the optical media, respectively. The relative permeability and permittivity are defined as: $\mu_r = \mu/\mu_0$ and $\varepsilon_r = \varepsilon/\varepsilon_0$. Note that when $\mu_r = 1$ (diamagnetic material at the frequency of the light: $\mu \approx \mu_0$) and $\varepsilon_r > 1$, which is the case of the most common optical media, the speed of light is decreased in the factor of $\varepsilon_r^{1/2}$, which is known as refraction index of the medium $n = \varepsilon_r^{1/2}$, and hence the velocity of light can be written as

$$c = \frac{c_0}{n} \tag{1.5}$$

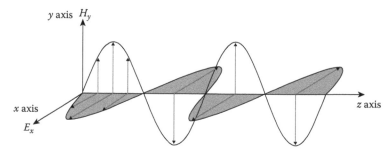

FIGURE 1.4 Sinusoidal electromagnetic wave of light illustration.

These results connect the optical behaviors of the media with its atomic structure. This permits to consider that the medium provides an enhancement of the electric field because that field displaces the atomic electrons from their equilibrium position with respect of the nuclei, which produces an additional field and thus a magnification of the original field. To understand this, it must be noted that to obtain the field vectors solutions it was necessary to introduce the material influence into Maxwell's equations by means of the electric displacement, D, and the magnetic induction, B, vectors and their relationship with E and H fields that depends on the medium nature [6]. This is

$$D = \bar{\varepsilon}E = \varepsilon_0 \cdot E + \Pi; \quad B = \bar{\mu} \cdot H = \mu_0 \cdot H + M \tag{1.6}$$

where
  Π is the induced polarization in the medium
  $\bar{\varepsilon}$ and $\bar{\mu}$ are the dielectric and permeability tensors respectively

In general, both can be complex being included in the imaginary part of ε the gain or loss factors of the optical mediums. With these facts in mind, some additional facts concerning the velocity of light and its interdependence with the material and the frequency of the electromagnetic fields can be useful in understanding several phenomena of interest for the purpose of this book. They are as follows: (i) the value of the refractive index possessed by the material is clearly dependent on the way in which the electromagnetic field of the propagating wave interacts with the atoms and molecules of the optical material; (ii) associated with the binding of electrons in atoms, resonant frequencies can be found, what the index frequency dependency through the dielectric permittivity frequency dependency $\varepsilon(f)$. The latter drives to the conclusion that the speed of light is frequency (wavelength) dependent, and thus it explains the phenomenon of the optical dispersion, which is important when nonmonocromatic light (real light) travels along real optical media.

Moreover, using the plane wave solution of Maxwell's equation, it can be found that E and H are related by

$$\frac{E}{H} = \frac{k}{\left(\varepsilon_0 \cdot \varepsilon_r \cdot \omega\right)} = \frac{1}{\left(\varepsilon_r^{1/2} \cdot \varepsilon_0 \cdot c_0\right)} = Z = \frac{Z_0}{n} \tag{1.7}$$

where
  Z is the impedance of the medium to electromagnetic wave propagation (has the dimensions of ohms)
  $Z_0 = 377 \ \Omega$ is the impedance in the free space
  n is the refraction index of the optical medium [4]

1.2.1.3 Energy and Power of Light

An electromagnetic wave is able to "store" energy in itself. It can be given by

$$U = \frac{1}{2} \cdot \left(\varepsilon E^2 + \mu H^2\right) = \varepsilon E^2 = \mu H^2 \tag{1.8}$$

What suggests the energy "stored" in each of the two fields is the same? As the light electromagnetic wave is propagating at a speed, it is a carrier of energy that fluxes in the direction of the light propagation given by

$$v \cdot U = P = E \times H \tag{1.9}$$

where P is the Poynting vector, which is the vectorial product of the electric by the magnetic fields. This vector is orthogonal to both former vectors being its sense given by the right-handed screw rotation law. That is, in isotropic mediums P lies parallel to k and hence the flux of optical energy through the unit of area "travels" in the same direction as that of light propagation. The time-averaged value of P represents the intensity, I, or irradiance of the light given in power per unit area ($W \cdot m^{-2}$). When both electric and magnetic fields are mutually orthogonal and are in phase, it is proportional to the square of the electric or magnetic fields. This is given by

$$I = \langle P \rangle = \langle E_0 \sin(\omega t) \cdot H_0 \sin(\omega t) \rangle = \frac{1}{2} E_0^2 Z^{-1} \tag{1.10}$$

Light

FIGURE 1.5 Photonic mill image. The four blades inside a high-vacuum glass chamber rotate when light strikes on the vanes. The higher the light intensity, the higher the speed of the rotation blades. (Courtesy of the coauthors.)

This, in real terms, provides a direct measure of E or H amplitudes. However, when H and E are not in phase, the intensity of the light will be different. In the particular case in which their difference of phase is $\pi/2$, the intensity equals zero, which represents that no energy is transported being an example of great interest for the proposed aims of this book: the evanescent waves.

The optical power, $P_0(t)$, flowing into an area, A, normal to the direction of propagation of light is the sum of all intensity contribution from all area, A. Given in unit of Watts, it is

$$P_0(t) = \int_A I(r,t)dA \qquad (1.11)$$

and hence, the optical energy collected in a given time interval is the integral of the optical power over a time interval.

An illustrative evidence that light carries linear energy is the forced rotation of vanes or blades in a vacuum when one side is exposed to a light beam (Figure 1.5). The plane blades of this "photonic mill" are dark on one side, and light that reflects the photons on the other side. When the incident light collides with the blades it provokes their rotation, being the angular velocity proportional to the intensity of the incident light. The higher the intensity, the higher the rotational speed. For more details, see for instance page 6 of Ref. [4].

1.2.1.4 Momentum of Light

As an electromagnetic wave carries linear momentum, p, it provokes a *radiation pressure*, or pressure exerted upon object surfaces from which the photons reflect or scatter. When the light is totally reflected, the radiation pressure (the power flux density divided by the speed of light) is doubled than that if it is absorbed. The linear momentum density, m_d, per unit volume, is given by the vectorial product of the electric displacement and the induction vectors

$$m_d = D \times B \qquad (1.12)$$

which in free space is proportional to the Poynting vector according to the expression [7]

$$m_d = c^{-2} \cdot P \qquad (1.13)$$

FIGURE 1.6 Nanoparticle optical trapping process illustration. The force applied to the nanoparticle by the light provokes its displacement to a place in which the optical field is more intense and traps it.

It must be recalled that a photon is like a typical particle in some aspects but very different in others. In particular, a photon has zero rest mass, which enables it to propagate at the speed of light (only objects with zero mass are able to propagate at the velocity of light) but presents a momentum given by [11]

$$p = h \cdot \frac{f}{c} = \frac{h}{\lambda} \tag{1.14}$$

where h is the Plank constant. According to Newton's second law, the net force acting on an interaction object is equal to the rate change on its momentum. That is,

$$F = \frac{\Delta p}{\Delta t} \tag{1.15}$$

which can also be written as the variation of the impulse, $Ip = \Delta p$, with the time.

Due to the vector nature of p, the total momentum is given by the vector sum of the momentums of all photons included inside the beam. So, the effect of the energetic particles (photons) striking the object in the direction of the traveling wave is that the interaction object is pulled by a force equal to the speed (change with the time) of the total light vector momentum. Based on this principle, it is possible to move and to trap very tiny nanoparticles by the concentration of the light beam (because the force is proportional to the Poynting vector) in a part of the space (Figure 1.6). These are the basics of the optical tweezers, which are used to trap and manipulate for instance biological species.

1.2.1.5 Polarization of Light

Natural light or sunlight is totally depolarized and incoherent. On the other hand, a monochromatic light wave is fully polarized and coherent. Real light is partially polarized, which can be measured by means of a parameter known as the degree of polarization (DoP). When the DoP is equal to one, the field is called fully polarized, and when the DoP is equal to zero, the x and y components of the electromagnetic field are uncorrelated, and then the light is unpolarized. For a better understanding, let us consider that when the x and y electric field components of the propagating electric field (as a plane wave) are in the direction z, $E(z, t)$, can be written as

$$E_x = e_x \cos(\omega t - kz + \varphi_x) \quad \text{and} \quad E_y = e_y \cos(\omega t - kz + \varphi_y) \tag{1.16}$$

As kz can be written as $\omega z/c$, E_x and E_y are periodic functions of $(t - z/c)$ that oscillate at the pulsation ω as illustrated in Figure 1.7a. The E_x and E_y are, in fact, the parametric equations of the ellipse defined as

$$\left[\frac{E_x}{e_x}\right]^2 + \left[\frac{E_y}{e_y}\right]^2 - 2\cos\varphi\left(\frac{E_x \cdot E_y}{e_x \cdot e_y}\right) = \sin^2\varphi \tag{1.17}$$

The angle between the major ellipse axis and the x axis, δ, and the ratio of the minor to the major ellipse axis are defined as

$$\tan a\delta = \left[\frac{2h}{1-h^2}\right]\cdot\cos\varphi \quad \text{and} \quad \sin 2\psi = \left[\frac{2h}{\left(1+h^2\right)}\right]\cdot\cos\varphi \tag{1.18}$$

where $\varphi = \varphi_y - \varphi_x$ and $h = e_y/e_x$ as illustrated in Figure 1.7.

It can be observed that the resulting electric field (fixed a time) rotates at ω angular velocity with the propagating z axis, which is also happening (in a fixed propagation z position) with the time. An observer who follows the field vector approaching him (Figure 1.7) will see the vector rotate and change its amplitude according to an ellipse on the xy plane.

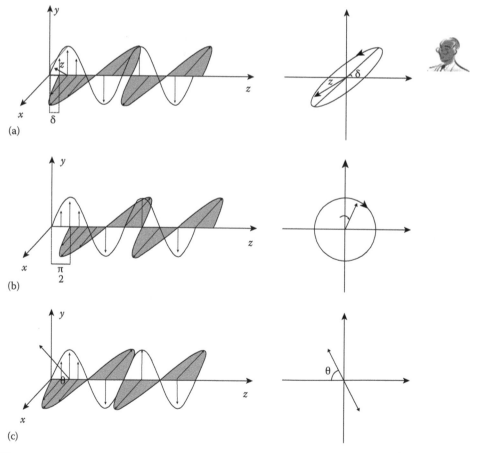

FIGURE 1.7 Light wave polarization illustration, which is a function of the amplitudes and phase differences between the two orthogonal electric field components. Considering that light is approaching the observer, he sees a vector whose (a) amplitude follows the values of an ellipse rotating counterclockwise: left helical polarization; (b) the amplitude is constant but it rotates clockwise: right circular polarization; and (c) the amplitude changes from a positive to a negative value, but it follows a straight line with a constant angle from one of the axes: linear polarization.

If the rotation is clockwise, the light is *right elliptically polarized*; on the contrary, if the electric field rotates counterclockwise, the wave is *left elliptically polarized*. The stay of polarization of the wave is determined by the orientation and shape of the ellipse.

It must also be noted that the polarization ellipse could be degenerated in two particular cases: a circle and a straight line. The light will be circularly polarized when both amplitude fields, e_x and e_y, are equal and to be in phase-quadrature $\varphi = \varphi_y - \varphi_x = (2m + 1)\pi/2$. So *right or left circularly polarized light* can also be observed (Figure 1.7b) when the electric field vector rotates clockwise or counterclockwise, respectively. The light will be *linearly polarized, LP*, when either $\varphi = \varphi_y - \varphi_x = m/\pi$ or $e_x = 0$ or $e_y = 0$, where m is a positive or negative integer (Figure 1.7c).

To have a clear and identifiable polarization state, it is necessary that the two components (E_x and E_y) maintain a constant phase and amplitude relationship. On the contrary, when the phase of the waves changes with the time and this change is chaotic, it is impossible to discern a polarization state. The light is then called *unpolarized*. However, sometimes the light has statistical preferences for a particular plane of polarization because of some anisotropy in the propagating medium. The light is then called *partially polarized*. When the medium where the light propagates presents some directionality, the polarization estate of the light will interact with it, which could be a very useful attribute for a wide set of applications.

Finally, it must be said that similar results can be obtained using the magnetic field instead of the electric field to describe polarization attributes of the light. However, the latter is commonly used because the effect of the electric field on the electrical charges within the atoms tends to be more direct than that of the magnetic field.

1.3 Some Light Phenomena for Optochemical Sensors

1.3.1 Light in Optical Interfaces: Reflection, Refraction, Total Internal Reflection

Here an optical interface is considered as a structure composed of two different dielectric media as shown in Figure 1.8. The study of the reflection and refraction (transmission) and other related phenomena such as total internal reflection, evanescent waves, optical tunneling, and Plasmon resonances are the main aims of this subsection. As a sharp boundary between two media, there are simple relationships that must be obeyed between the fields on the two sides: (i) the components of the fields E and H parallel to the separation plane are equal on the two sides; (ii) the normal components of the D and B must be continuous.

As already illustrated in Figure 1.8, suppose that a plane light wave with a wave vector k_i an electric field amplitude E_o is incident on a plane surface separating of two of different media of refraction indexes n_1 and n_2. The angle of incidence between the incidence vector k_i and the normal to the separation plane is θ_I. After striking the interface surface, a reflected and a transmitted (refracted) wave will be observed. Two cases will be studied: (a) when E is normal to the incident plane, it is denoted by $\perp$(TE waves) and

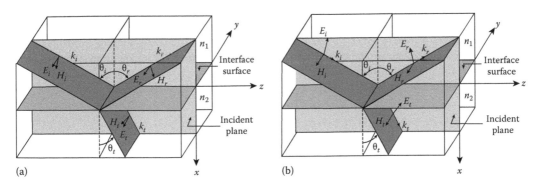

FIGURE 1.8 Light wave approaching the interface between two different optical mediums of refraction index n_1 and n_2 and the effects of the reflection and refraction of light. (a) TE wave (electric field normal to the incidence plane) and (b) TM wave (electric field parallel to the incidence plane).

(b) when electric vector E lies in the incident plane defined by k_i and the normal vector to the separation plane is denoted by $\parallel$(TM waves).

Two solutions that represent plane waves travelling in z direction, in a non-absorbing media, can be obtained from Maxwell's equations [8]:

$$E_x = E_{ox} \exp\left(\omega t - kr + \phi_x\right)$$

$$E_y = E_{oy} \exp\left(\omega t - kr + \phi_y\right) \tag{1.19}$$

Considering the boundary conditions, the reflected, transmitted, and the incident waves at the boundary $(x = 0)$ must present identical components for any time and x, hence the law of reflection and Snell–Descartes' law of refraction can be obtained as

$$n_1 x \sin\theta_i = n_1 x \sin\theta_r = n_2 y \sin\theta_t \tag{1.20}$$

where θ_r and θ_t are the reflected and transmitted angles, respectively (with the normal of the interface surface). According to Fresnel's theory, the relationship between the reflected, E_{or}, and the transmitted, E_{ot}, wave amplitudes and the incident one, E_{oi}, in both cases the TE or perpendicular to the incident plane (sometimes known as S wave) and TM or parallel to the incident plane (sometimes known as P wave) can be obtained as follows.

For TE waves:

$$r_s = \frac{E_{or}}{E_{oi}} = \frac{k_{1x} - k_{2x}}{k_{1x} + k_{2x}}; \quad t_s = \frac{E_{ot}}{E_{oi}} = \frac{2k_{1x}}{k_{1x} + k_{2x}} \tag{1.21}$$

For TM waves:

$$r_p = \frac{E_{or}}{E_{oi}} = \frac{n_1^2 \cdot k_{2x} - n_2^2 \cdot k_{2x}}{n_1^2 \cdot k_{1x} + n_2^2 \cdot k_{2x}}; \quad t_p = \frac{E_{ot}}{E_{oi}} = \frac{2n_1^2 k_{2x}}{n_1^2 k_{2x} + n_2^2 k_{1x}}$$

where k_{1x} and k_{2x} are the normal (x component) of the wave vectors in medium 1 and 2, respectively.

As expected, the Fresnel reflection coefficients r_s and r_p vary as a function of the angle of incidence. The reflectance, defined as reflected or the transmitted and the incident power ratios, respectively, of the TE wave $R_s = |r_s|^2$ is always greater than the reflectance of the TM wave $R_p = |r_p|^2$ except at normal incidence $\theta_1 = 0$ and the grazing incidence ($\theta_1 = 1/2\pi$). Furthermore, the Fresnel reflection coefficient vanishes (for TM polarization) for a particular angle of incidence, which is known as *Brewster's angle* θ_B:

$$\theta_B = \tan^{-1}\left(\frac{n_2}{n_1}\right) \tag{1.22}$$

This suggests that if a plane wave with a mixture of TE and TM waves is incident on the plane interface between two dielectric media at the Brewster angles, the reflected radiation is linearly polarized with the electric field vector perpendicular to the plane of incidence (TE).

For the purposes of this book, it will be interesting to examine the corresponding phenomena provoked by an incident electromagnetic radiation on the surface of an absorbing medium such as a metal. In these cases, the index of refraction of absorbing media 2 is complex, $(n_a - jn_b)$; at normal incidence, the reflection coefficients become

$$r_s = r_p = \frac{1 - \left(n_a - jn_b\right)}{1 + \left(n_a - jn_b\right)} \tag{1.23}$$

Then, the reflectance at normal incidence and the phase shift upon reflection can be described as follows:

$$Rs = Rp = \frac{\left(1 - n_a^2\right) + n_b^2}{\left(1 + n_a^2\right) + n_b^2} \quad \varphi = \tan^{-1}\left(\frac{2n_b}{1 - n_a^2 - n_b^2}\right) \tag{1.24}$$

which for cases (such as metals) in which n_a tends to zero or $(n_a - jn_b)$ tends to infinitum, the reflectance approaches unity and then at the optical frequencies in which this occurs, this interface can be used to fabricate optical mirrors.

When $n_1 > n_2$ and the incident angle θ_{i1} reach a critical value $(\theta_c = \sin^{-1}(n_2/n_1))$, called critical angle, the refracted wave propagates parallel to the interface. Then, for $\theta_{i1} > \theta_c$ the energy of the light must be totally reflected regardless of the polarization state of the electric field vector E, phenomenon that is called *total internal reflection* [10]. The latter is the basis for dielectric wave guiding in layered structures as the integrated optical waveguides or optical fibers.

When total internal reflection occurs, the incident wave will be totally reflected from the surface but the Fresnel transmission coefficients, t_S and t_P, vanish. This drives to the conclusion that the light energy is totally reflected but the electromagnetic fields still penetrate into the second medium without transporting any energy. That is, the normal component of Poynting's vector vanishes, and the power flow is parallel to the boundary surface. In this case, the electric field vector decreases exponentially as the distance (x) from the interface surface increases. Equations show that the transmitted wave actually propagates parallel to the boundary surface. Such a wave has important applications in optical waveguides and optical sensors.

The standing wave that decays exponentially away from the interface into the low refraction index material on the interface is called *evanescent wave*, which is able to propagate parallel to the boundary surface (see Figures 1.13 and 1.14). For the purpose of this book, evanescent fields are very useful to interact with the object species to be detected and for this reason, in addition to its amplitude, a key parameter is the depth of penetration of the exponentially decaying evanescent field. This is defined as the distance from the surface over which the electric field of the standing wave disturbance decays e^{-1} of its initial value at the interface.

If the total reflection occurs $(\theta_{i1} > \theta_c)$ from a plane wave, but a second interface exists within the region of the evanescent wave (Figure 1.9), the total reflection can be frustrated, and appears the phenomenon of *optical tunneling* whereby a wave is partially transmitted through it would be forbidden by geometrical optics [10].

As illustrated in Figure 1.9, when a low-wide layer (of thickness comparable to or less than the optical wavelength) is sandwiched between media of higher index of refraction, because of the finite thickness of the air gap, a small amount of light tunnels through the air gap and transmits into the third medium by means of the resonant tunneling phenomena. This can be useful for sensing purposes.

Confined propagation of electromagnetic radiation can also exist at the interface between two semi-infinite homogeneous media, provided that the dielectric constants of the media are opposite in sign. These modes, whose amplitudes decrease exponentially in the two directions normal to the interface, are commonly called *surface plasmon waves* [6]. This is based on the fact that the electron plasma contributes to the negative dielectric constant of metals when the optical frequency is lower than the plasma frequency.

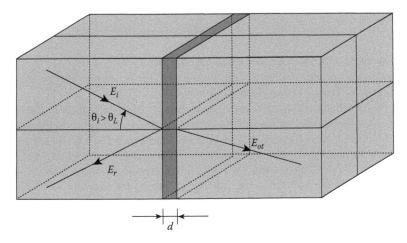

FIGURE 1.9 Optical tunneling of a wave through a very thin layer media, the thickness of which is approximately equal to $(k\beta)^{-1}$. The tunneled wave amplitude, E_{ot}, is observed considerably attenuated.

Surface plasmon waves can also be excited when the interface between two transparent media is coated with a thin layer, typically dozens of nanometers of a metal, enabling the light to penetrate into the metal layer. For $(\theta_{i1} > \theta_c)$, all light is reflected except for a range of angles where the momentum matching conditions of the light are such that surface plasmon waves on the interface between the metal and the lower refractive index is excited.

Since the metal layer is lossy at the optical frequencies, the surface light wave is rapidly damped and, as a consequence, the observed reflectivity of the metal drops to a low level. The effect is maximized when the component of the propagation vector of the incident light along the interface is exactly equal to the propagation vector of the interface plasmon. Its electric field decays exponentially, with a similar penetration depth, into the lower refractive index medium in the same way as an evanescent field does. However, its surface field intensity is typically an order of magnitude or more higher than the pure equivalent evanescent field case. This fact can be successfully exploited for sensing. By placing chemical or biological species on the metal surface layer, a strong interaction between the light and the measurand species can be obtained. That is, by using surface plasmon resonance-based techniques instead of standard evanescent field approaches, optical sensors dotted with higher sensitivities can be obtained.

1.3.2 Interference Light

To point out the main parameters that need to be taken into account to obtain interference patterns and their fundamental behaviors to be used for the purposes of this book, a basic review will be carried out in this section.

First of all, it must be stated that as the interference of light depends on the phase relationship between the superposed waves, it can neither be explained using ray optic approaches nor using the intensity of the light.

Using the linear properties of the wave equation, when two or more monochromatic waves are simultaneously present in the same part of the space and time, the total wave function is the sum of the individual wave functions. So when two monochromatic waves with complex amplitudes E_1 and E_2 are superposed, the result is a monochromatic wave of the same frequency that has a complex amplitude (sum of vectors):

$$E(r) = E_1(r) + E_2(r) \tag{1.25}$$

The intensity of the interference wave can be written as

$$I = |E|^2 = |E_1 + E_2|^2 = |E_1|^2 + |E_2|^2 + E_1^* E_2 + E_1 E_2^* \tag{1.26}$$

That is, the interference wave results:

$$I = I_1 + I_2 + 2(I_1 I_2)^{1/2} \cos\varphi \tag{1.27}$$

where $\varphi = \varphi_1 - \varphi_2$ is the difference of phase of the two waves.

Thus, the interference pattern is the sum of the intensities of the waves plus a factor of interference that depends on the phase difference between the phase waves.

To optimize the interference pattern requires (i) to control the polarization of the two waves (i.e., waves with the same linear polarization), (ii) to control the phase difference independent of the time that drives to use coherent beams, and (iii) equal intensity values ($I_1 = I_2$). With these basic conditions in the space in which the phase difference will reach $\varphi = 0$ or a multiple value of 2π, the intensity will be maximum (four times the individual intensity, $I = 4I_1$). However, in the space areas in which the phase difference takes the value $\varphi = \pi$ or $(2\mu + 1)\pi$, μ being an integer, the intensity will be minimum ($I = 0$ in the ideal case). It must also be remarked that this strong dependence of the intensity on the phase difference can be used as the basic principle to measure phase differences (with very high sensitivity) by detecting the light intensity changes.

For sensing purposes, based on the interference theory, interferometer devices can be constructed. These devices basically split a wave into two waves using an optical splitter, delays them by mean of two optical paths, recombine them in an optical combiner, and detect the intensity of the output light. One of the optical paths is fully isolated (i.e., φ_1 = constant) and the other is rearranged to be phase-dependent with the measurand, M (i.e., $\varphi_2 = g(M)$); then detecting the intensity changes, around a working point, the phase changes can be measured and hence the measurand value, M, can be deducted. Three main types of interferometer devices are commonly used for sensing purposes: the *Mach–Zender*, the *Michelson*, and the *Sagnac interferometers* [6].

Finally, it must be noticed that light interference cannot be observed with unpolarized light since the random fluctuation nature of the wave phases causes random φ values uniformly distributed between 0 and 2π, and then the interference factor goes to zero ($\cos \varphi = 0$).

1.3.3 Light Wave Resonances

Devices able to confine and store light at given frequencies can be called optical resonators. These frequencies are commonly named resonance ones and are determined by their optical materials and architectures. These resonators can be understood as devices with optical feedback in which light is stored by means of a recirculation approach such as: multiple reflections from mirrors; traveling through closed-loop waveguides; re-circulating the light through surface layers or the surface of cylinders, spheres, disks, reflection from periodic structures like short period gratings and by means of proper re-arranges of defects on photonic crystal materials. Using the current technologies, these devices can be micro and/or nano sized which is of interest for the purposes of this book.

Due to their frequency selectivity, resonators can be used as optical filters and as storing of energy in active optical mediums to provoke (for instance) both continuous and pulsed lasing of light. These devices can also be used to provoke places on the space in which the density of light, at the resonance frequency, will be very high and then, very good interaction of light and tiny species can be obtained as a way of their detection and even their identification. It is a very interesting technique for chemical and biological sensing approaches.

Two key parameters characterize an optical resonator: (i) the volume occupied by the confined optical mode/s or modal volume, V, or capacity of spatial storage of energy and (ii) the *quality factor*, Q, which represents a measure of their temporal storage capacity of energy.

Historically the first and useful resonator proposed was the very well-known Fabry–Pérot etalon, composed basically of two parallel mirrors ending in a media of index n and length d. As it is the simplest one and the results can be translated to get general ideas about the topic, we cover this only briefly here.

The light resonant standing waves will be established at the resonant frequencies and associated resonant wavelengths given respectively by (no losses)

$$f_{or} = \frac{m(c_0/n)}{2d}; \quad \lambda_{or} = \frac{2d}{m} \tag{1.28}$$

where m is an integer $m = 1, 2, 3$ and then the adjacent resonance frequencies are separated by a constant frequency difference, Δf_o, also known as the free spectral range, that can be defined as (see Figure 1.9)

$$\Delta f_{or} = \frac{(c_0/n)}{2d} \tag{1.29}$$

In real resonators the optical losses and gains (if the optical medium is active) must be considered and then, very restrictive condition of the optical waves existing inside the resonator is relaxed (net losses) or enhanced (net gain). If r is the reflectivity of the mirrors in a media with no losses, the intensity of the light in the resonator can be written as [7]

$$I = \frac{I_{max}}{\left[1 + (2\mathbf{F}/\pi)2\sin^2\left(\pi f_0/f_{or}\right)\right]} \tag{1.30}$$

where f_0 is the optical frequency and **F** is the finesse of the resonator defined as

$$\mathbf{F} = \left[\frac{\pi |r|^{1/2}}{(1 - |r|)} \right] \tag{1.31}$$

The spectral width δf of each resonator modes can be approximately written as

$$\delta f = \frac{\left[(c_0/n)/2d \right]}{\mathbf{F}} = \frac{\Delta f_{or}}{\mathbf{F}} \tag{1.32}$$

This suggests that the finesse is a measure of the selectivity (in frequency) of the resonant modes (or each band-pass filter). The higher the finesse, the higher the selectivity or the narrower the band-pass filter, δf. However, if losses are considered and α_r is the effective overall distributed-loss coefficient (cm^{-1}) composed of the sum of the distributed medium, α_s, and mirrors α_{m1}, and α_{m2} losses respectively ($\alpha_r = \alpha_s + \alpha_{m1} + \alpha_{m2}$), then the finesse can be approached by $\mathbf{F} = \pi/d\alpha_r$. Then, as occurs with the electric resonant circuits, the optical resonator will be less selective. The higher the optical losses, the lower the frequency resonator selectivity (Figure 1.10b).

In general, the quality factor, Q, is used as a typical factor to characterize resonant circuits [7]. It can be defined as

$$Q = 2\pi \left(\frac{Stored\ Energy}{Energy\ loss\ per\ cycle} \right) \tag{1.33}$$

It can be related to the resonant frequency, f_{or} the spectral width, δf, of the individual resonator modes and finesse factor, F, by the following equation [7]:

$$Q = \left(\frac{f_{or}}{\Delta f_{or}} \right) \cdot \mathbf{F} \tag{1.34}$$

It must be noticed that as the resonator frequencies, f_{or}, are typically much greater that the mode spacing, Δf_{or}, the quality factor of the optical resonator is greater than their finesse factor. Resonant cavities can be used in detection tasks or optical sensors to enhance their sensitivity to the measurand species. For instance, in gas sensors the gases (i.e., methane) can be properly introduced into the resonant cavity of a laser optical resonator.

Resonator architectures in 1D, 2D, and 3D dimensions can be realized to confine and store light. The higher the confinement, the higher the density of modes inside the cavity. So the density of modes per

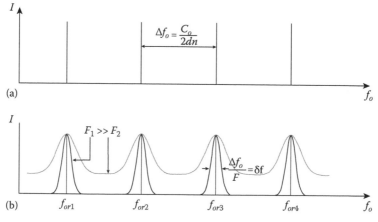

FIGURE 1.10 Illustration of a typical Fabry–Pérot resonator response: (a) the ideal case and (b) the real case in which this multiwave interferometer reflection function is represented versus the finesse.

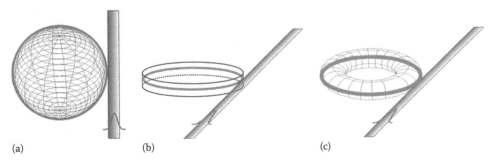

FIGURE 1.11 Three of the nanoresonator structures with higher quality factor Q: (a) nanosphere, (b) nanodisk, and (c) nanotoroid.

unit of area per unit of frequency, M_{2D}, and density of modes per unit of volume per unit of frequency, M_{3D}, in 2D and 3D resonators, respectively, can be approached by

$$M_{2D}(f_0) = \frac{4\pi f_0 n^2}{c_0^2} \quad \text{and} \quad M_{3D}(f_0) = \frac{8\pi f_0^2 n^3}{c_0^3} \tag{1.35}$$

So the density of modes increase lineally in 2D and quadratically with the optical frequency in 3D resonators respectively.

It must be noticed that despite the fact that these mode density approaches are obtained for square and cubic geometries, respectively, they can also be applicable for arbitrary geometries observing the condition that the resonator sizes are large in comparison with the wavelength. It must also be interpreted that these density of modes play here a similar role to the one that the density of quantum states in function of the energy plays in semiconductor theory.

When one or more of spatial dimensions of size of an optical resonator approach the optical wavelength, they are commonly called microresonators. Then these devices present large spacing of the resonator modes and this fact suggests that, in absence of resonance modes, the emissions of light from sources placed within a one word can be inhibited and, on the other hand, the emission of light into particular modes of a high-Q factor of a one word can be enhanced. These effects can be used to enhance the interaction of light with soft materials such as gases or biological species.

Microresonators can be fabricated using dielectric materials to enhance the quality factor, Q, and to reduce the modal volume defined as spatial integral of the optical energy density of the mode, normalized to its maximum value micro resonators can be fabricated using dielectric materials. Several microresonator architectures such as microspheres, microtoroids, microdisks, micropillars and those generated using photonic crystal materials can be used. Two of the more efficient are those based on microspheres and microtoroids structures which are able to offer quality factors as high and 10^{10} and 10^8, respectively.

1.4 Light Guiding

In some applications, light propagates in free space or through optical materials as "free beams." However in a wide set of situations it is necessary to "conduct" or guide the photons to particular places or with special distributions of light for particular applications such as sensing. These kinds of devices, commonly fabricated using dielectric materials, are called optical waveguides or optical guides. The latter, despite other approach and conditions, basically are integrated by a medium of refractive index n_2 embedded into a medium/s of lower index, let's say n_1, and then the light is trapped and confined inside the lower medium by means of the total internal reflection phenomena. In Figure 1.12 several of the used guided architectures both on integrated optics and on fiber-optic technologies are shown. The first one, the planar or slab guide (a), offers confinement in one direction 1D; however the next three, channel (b), strip (c), and fiber waveguides (d), confine the modes in two directions, 2D.

The field components for a planar waveguide involve only elemental functions which are simpler than those modeling propagation in cylindrical waveguides. However, the mathematical procedures are similar, and it is very instructive to analyze both types of waveguides. The use of one or another model depends on the relation between the dimensions of the waveguide layer d and the wavelength λ. If $d \gg \lambda$; it is possible to use geometric optics in order to obtain the propagation constant of the guided modes in multimode waveguides. The limit of application of this theory is the diffraction that appears when $d \approx \lambda$, in this case, in single-mode waveguides, it is necessary to use EM theory that enables the obtention of the constant of propagation of the guided modes and its transverse distribution of energy. In several extreme cases, for the study of light interactions and the propagation in quantum well lasers in which d is of the order of the de Broglie wavelength, it is necessary to use quantum theory [12]. Well-known methods for solving the scalar wave equation include the finite element, staircase approximation, direct integration, and perturbation methods. Various methods of wave-theory analysis such as WKB, Rayleigh Ritz, power-series expansion, finite element, and staircase approximation can be used for any waveguide profiles.

A waveguide can also be understood as a structure that supports electromagnetic waves that propagate through it. With the proper contour conditions, solutions can be obtained from Maxwell's equations (Equation 1.1). Harmonic plane wave represented by Equations 1.2 and 1.3 for lossy and lossless media are two possible solutions. In the mentioned equations, k can be re-emplaced by the propagation constant of the confined eigen mode $\beta(\omega)$, and $E(x, y)$ represents the transversal distributions of the electric field.

Despite the slab guide, it is not very useful in practical terms; however it can be used as a way to obtain a "flavor" of what happens in 1D wave guiding and to obtain preliminary "visions" of what will happen in 2D guides.

As shown in the dielectric slab waveguide illustrated in Figure 1.12a, the guiding region is defined by the layer with the refractive index n_2 and is deposited onto a substrate with refractive index n_1. The refractive index of the medium above the guiding layer is the region of index n_3. In order to achieve true mode guidance, it is necessary that n_2 be larger than n_1 and n_3. The lower order mode of a symmetric slab waveguide ($n_1 = n_3$) does not have a cut-off frequency, which means that, in principle, this mode can propagate at arbitrarily low frequencies. In contrast, all modes of asymmetric slabs ($n_1 \neq n_3$) become cut-off if the frequency of operation is sufficiently low. Optical waveguides also include leaky modes, and in the case of an inverted slab waveguide with $n_2 > n_1$, n_3 only leaky waves are supported.

It must be mentioned that slab waveguides can support two independent TE and TM modes. Depending on ($k_0^2 n_3^2 - \beta^2$) values, no feasible physical solution ($\beta > k_0 n_1$), confined or guided modes ($k_0 n_3 < \beta < k_0 n_1$), substrate radiation modes ($k_0 n_2 < \beta < k_0 n_1$), and the radiation modes of the waveguides ($0 < \beta < k_0 n_3$) can be obtained. An illustration of the qualitative transversal distribution of the mentioned modes can be seen in Figure 1.13. On the left, the optical energy is radiated through the upper layer losing all optical energy coupled inside the central layer. The traversal field distributions for the other cases included in the figure, single mode ($m = 0$), bi-mode ($m = 1$) and four modes ($m = 4$), are also illustrated. The exponential

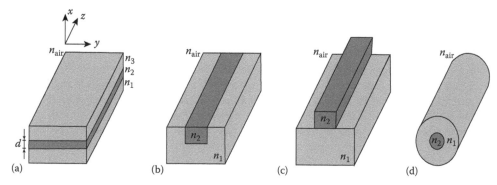

FIGURE 1.12 Four types of optical dielectric waveguides: (a) 1D slab or planar, (b) 2D channel or embedded strip, (c) 2D strip, and (d) fiber-optic waveguide.

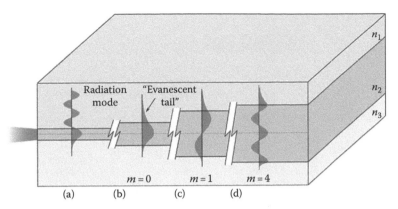

FIGURE 1.13 Illustration of the transversal electric field distribution on a slab waveguide. The launched light (a) radiates through the upper layer, (b) travels in a single mode, (c) travels in two modes, and (d) travels in four transversal modes.

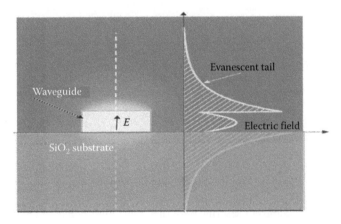

FIGURE 1.14 Illustration of the high evanescent electric field tail supported by a nanoguide.

decrease of the evanescent mode tail in the upper layer must be observed. The extent of the light intensity within the guide is a measure of the confinement factor defined by the ratio of light intensity within the layer to total light intensity. In general terms, the higher the *normalized frequency* value, V, the higher the *mode confinement* and vice versa [13].

As can be observed on the left part of Figure 1.14, in 2D waveguides the light will be confined in two dimensions so, in general, two different field distributions will be supported by the waveguide. On the right part of the figure, the electric field distribution along the vertical axis on the middle of the strip waveguide is represented. It is noticeable that highly enhanced evanescent fields can be created with one word waveguides or nanowires [14]. In Figure 1.14, the nanostrip waveguide was made by silicon on SiO_2 substrates [15]. These kinds of nano-waveguides with very enhanced evanescent mode tail in the upper layer can be used to build approaches to detect chemical and biological species that are the main aim of this book. In Figure 1.15, a schematic illustration of such mentioned devices using optical integrated technologies can be observed. The evanescent field of the mode traveling through the waveguide interacts with the target species provoking behavior modifications or light parameters modulating on the interrogating light. The demodulations of the output light permits to detect and to carry out measurements concerning the mentioned target species.

Coming back to the right part of the Figure 1.12, another way to reach a 2D waveguide is embedding a cylindrical dielectric material with a refractive index n_2 inside other dielectric material of lover refractive index, n_1. If after launching the light core (n_2) the total internal reflection on core-cladding (n_1) interface and the other wave guiding conditions are satisfied, then light can be guided along a large length of this type of waveguide called fiber optic [12]. Single and multimode efficient fiber-optic waveguides have

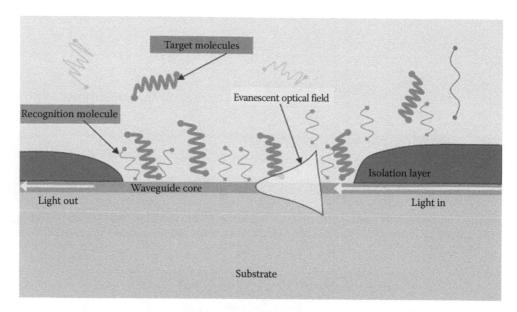

FIGURE 1.15 Illustration of an integrated device for chemical or soft-matter species detection. The upper evanescent tail of the traveling mode interacts with the target species and then the output light includes encoded or modulated information concerning the species. After the light demodulation, the required species information is obtained.

been developed from the famous paper of the Nobel Prize (2009) Charles Kao (co-authored with George Hockham) in 1966 [16]. Thanks to this milestone, today, we enjoy advanced communications by means of the very complex fiber networks around the entire world. However, fiber-optic technologies can also be successfully used for other purposes such as to construct fiber-optic and optical sensors, which is the main aim of this book. Fiber optics are used both as optical channels that communicate the optoelectronic unit with the optical/fiber transducer and also to implement the required transducers.

It must be stated that the complete description of the guided and radiation modes of optical fibers is complicated since there are six-component hybrid fields of great mathematical complexity even if the common simplification that assumes that $n_2 - n_1 \ll 1$ is accepted. Since the refractive index profiles $n(r)$ of most fibers have a cylindrical symmetry, it is convenient to use the cylindrical coordinate system. The components of both electric and magnetic fields are in this case are E_r, E_φ, E_z, H_r, H_φ, and H_z the wave equation involving the transverse components and are very complicated [6] being the solution out-of-the scope of this chapter.

Within the last 20 years new ideas to control and localize light have been reported and widely checked. These approaches are based in a physical effect reported for the first time by Eli Yablonovitch and John, who indecently studied spontaneous emission control and localization of light in novel periodic structures [17,18]. These structures are, today, recognized as photonic crystals or *photonic band gap structures* being widely used both in integrated and in fiber-optic technologies [19]. Photonic crystal waveguides on integrated optics technology have the attractive potential of providing lossless transmission of light around 90° sharp bents and probably is going to be the key to realize integrated devices with low aspect-ration values, which will enable the possibility of obtaining integrated optics devices with a very high scale of integration. Photonic crystal fibers PCF, may, through a lifting of the usual core-cladding index requirements and restrictions, exhibit radically new properties and push the current fiber limits to values that are unknown today.

PCF consists in a collection of optical fibers made out of one single material, which presents a periodic arrangement of air holes across the whole length of the fiber. They can be made by piling up identical hollow tubes of silica in a hexagonal shape, and after the fiber preform is fabricated it is pulled in a conventional fiber drawing tower [20].

PCF, also called microstructured fibers, can operate by means of one of two main working principles: (i) using the index guide principle (total internal reflection as the standard fibers) or (ii) using the photonic band gap effect.

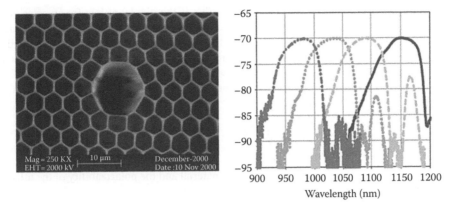

FIGURE 1.16 Illustration of a hollow core photonic crystal fiber. Its pass-band is a function of the structural parameters. (Courtesy of D. Richardson, ORC, U.K.)

The index guide fibers can be classified in several subtypes: The high-numerical aperture HNA, fibers having a central part surrounded by a ring of relatively large holes; large mode area LMA, provided with large dimensions and small effective refractive index contrast to spread only the transverse field; and highly nonlinear HNL, fibers that apply very small core dimensions to provide tight mode confinement [19].

The photonic band gap fibers can be subdivided as low-index LIC, fibers and air-guiding or hollow-core HC, [21] fibers, including the so-called Bragg fibers which present several layers of ring structure constituted by holes/voids around the core as cladding. In these fibers, the optical frequencies that lie within the photonic band gap will not propagate (are prohibited) through the cladding and hence the core–cladding interface acts "like a reflective". The light is then confined and propagated through the hollow core despite the fact that the effective core index is lower than the cladding one. An example of this fiber can be observed in Figure 1.16a. In these kinds of fiber structures, the position of the transmission band-gap is dependent on the holes structure parameters (Figure 1.16b). The possibility to confine important density of photons in the cores enables the consecutions of high light–matter interactions and, hence, offers a structure to be successfully used for sensing of soft matter (gases, liquids) structures, the chemical and biological species by filling properly the hole-core with them [22].

1.5 Light and Matter

The potential interactions between a beam of light and a homogeneous, essentially infinite, volume of material form one of the essential building blocks of our understanding of optoelectronics for chemical sensing (Figure 1.1). These interactions range from the properties of the phase propagation constant as a function of frequency (i.e., dispersion) and the consequent impact thereof on propagation within the material through to nonlinear phenomena involving the incident light being reradiated as a range of new optical frequencies and extend into phenomena including the absorption of light and its re-emergence as heat or ionized atoms. These phenomena are all acutely dependent upon the material through which the light propagates, and consequently we need to establish an underlying framework through which these phenomena may be appreciated.

We start with a classical look at the origins of the dielectric constant embodied within the Clausius–Mossotti equation [23,24]. This equation is derived on a simple basis assuming (correctly) that dielectric properties may be understood through the simple artifact that an electron is connected to an atom through a mass spring damper equivalent mechanism and that the electron moves with respect to the nucleus of the atom in response to the applied input electric field. This in turn induces a time-dependent dipole moment that re-radiates and introduces an effective delay in the ongoing electromagnetic wave. The larger the dipole moment (equivalently the higher the compliance of the effective spring), the greater the "slowing" influence of the re-radiated emission. There are, of course, also resonance effects associated with mass spring damper systems. The displacement (that is the dipole) at low frequencies is in phase with the incident electric field; as the resonance is approached, depending on the quality factor, this displacement increases but changes in phase going through a maximum displacement in quadrature at resonance and then into antiphase (effectively reducing the dielectric constant) above this resonant frequency. The Clausius–Mossotti equation is well known and discussed extensively elsewhere. In its most general form, it can be written as the sum of a number of independent resonators—in effect a quantum mechanical equivalent—and then appears as

$$\varepsilon - 1 = \sum \frac{N\alpha_n}{\left(1 - N\alpha_n/3\right)} \tag{1.36}$$

where N is the volume density of the molecules in the material and the polarizability of the nth electronic oscillator is given by

$$\alpha_n = \frac{\left(q_e^2/m_e\right)}{\left(j\gamma\omega + \left(\omega_0^2 - \omega^2\right)\right)} \tag{1.37}$$

in which q_e is the electronic charge, m_e the electronic mass, γ the damping term in our oscillator, and ω_0 the resonant frequency of a particular oscillator term.

Typically then, the dielectric constant of a material whether solid, liquid, or gaseous follows a relationship of the type shown in Figure 1.17. The index is large at high frequencies, and it decreases after each resonance. Figure 1.17 also shows the imaginary contribution to the index (i.e., the loss terms) that are clearly closely linked to the real terms (formalized in the Kramers–Kronig relationship). These correspond to spectroscopic loss terms in the material and are highly characteristic of specific samples. Indeed the Clausius–Mossotti equation has a remarkably wide range of applicability—exemplified in Refs. [25,26].

In metals, the story is slightly different—or indeed in any conducting material that hosts a free electron density dependent on the material conductivity. The same observation also incidentally applies to plasma in the ionosphere. This lack of restoring force at the molecular level implies that the electrons respond directly to the electric field and in effect makes ω_0 equal to zero. The damping term is then related to electronic collision times in the metal—that is, the classical conductivity. Additionally, this in effect states that the contribution to the index is only from the conduction electrons.

At low frequencies ($\gamma \ll \omega$), this approximates to a dielectric constant that is entirely imaginary. Recall that the refractive index—the parameter of interest in optical systems—is the square root of the dielectric constant, and we immediately see that the index comprises equal real and imaginary components. Since in metals "low frequencies" can extend toward the visible, this observation has, for example, immediate

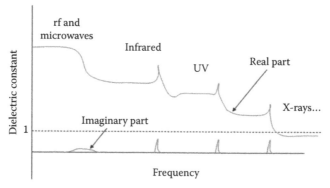

FIGURE 1.17 Very schematic representation of the behavior of dielectric constant vs. frequency with polar, rotational, molecular stretch, electronic, and in the x-ray region, nuclear contributions. The number of terms in the Clausius–Mossotti equation that contribute falls with frequency until eventually we observe dielectric constant <1 at very high frequencies (and sometimes, just close to resonances, as observed in some forms of "slow" light).

implications for the transmission of optical signals in metal films (plasmons), a topic of substantial interest in modern photonics. In addition, the low frequencies in the free electron cloud in the upper atmosphere—the ionosphere are in the tens, maybe 100 MHz range so your satellite TV signal is transmitted without problems while for short wave radio, the MHz signals are reflected from the ionosphere.

At high frequencies (i.e., $\gamma \gg \omega$), the metal refractive polarizability becomes negative—that is, the index is less than unity! Consequently, the phase velocity exceeds that of light in a vacuum (but note not the group velocity [27]). We can also see that here the metal is lossless and consequently totally transparent, so electromagnetic waves pass through it. This critical "high" frequency for most metals is typically in the far ultraviolet to x-ray region, but for the ionosphere microwaves pass through unhindered.

Somewhat inconveniently, the optical region is typically neither "high" nor "low" in terms of its frequency spectrum with respect to the so-called plasma frequency, which defines the critical cut-off point. The properties of an interface comprising a metallic layer between two-phase media have stimulated an entire subdiscipline—the study of surface plasmons sometimes termed "plasmonics," which forms the subject of an entire chapter later in this book. Perhaps though, the most important observation in the context of plasmonics is that the index of the metallic material can still be modeled using the concepts discussed earlier—encapsulated in the appropriate interpretation of Equation 1.37 and that the effective composite wave guide formed by the two dielectric layers and the metal exhibits a relatively slow phase velocity (and so perhaps can contribute toward super-resolution imaging?) and couples into both dielectric media through a short evanescent tail of the order of one wavelength in extent.

The classical view of the dielectric constant can also provide insight into numerous other optical properties of materials. Birefringence may be intuitively understood by observing that in a noncentrosymmetric medium the effective molecular density is different in different (polarization) directions. Consequently, the effective volume of N in the Clausius–Mossotti equation depends on the direction of the applied electric field, and the index is consequently higher in directions corresponding to closer atomic spacing. By the same token, amorphous, noncrystalline materials will not be birefringent with the possible exception of some chiral materials (such as certain sugars), which exhibit what is effectively an omnidirectional, circular birefringence.

The mass spring damper insight also gives some hints on nonlinear phenomena. If we increase the electric field beyond the elastic limit of the "spring," the consequent induced molecular level electric dipole will reradiate at harmonics of the original frequency as well as the fundamental. We can also take this further and speculate that if we make the input electric field high enough, we can, in fact, break the "spring," resulting in photo-ionization [25].

This simplified, quasiclassical approach to the dielectric constant does in fact survive very well under close scrutiny from those preferring the quantum approach. There are, of course, a few things that are different, and we shall explore just some of these.

The energy level model of the atom, or even that of the molecule, are well known and well rehearsed (Figure 1.18). The idea is simple, namely, that an atom or molecule can exist at a number of energy levels at which usually one, the so-called ground state, is stable and the remainder are transient, each with a characteristic lifetime. The basic foundation for this model goes back to the experimental observations of Rydberg on the ratios of the wavelengths for absorption lines in the spectrum from the sun ascribed to hydrogen. The Bohr atom later described this theoretically and while this was a somewhat contrived model with the wisdom of our current hindsight, it still serves as an introduction to the concept of quantized energy levels. At a descriptive level, a single atom has very precisely defined levels associated with it and will absorb energy corresponding to these level differences and thereafter re-emit this energy, almost always as heat, though sometimes—and most important in our context—as light. The distribution of the lifetime of a particular excited state will introduce an equivalent energy distribution in the reradiated emission—be it optical or thermal. When the energy arrives as light—or generally as electromagnetic radiation—the very characteristic absorption spectrum gives us the spectroscopic signature of the atom (Figure 1.19).

If the atom is moving (thermal motion in gases—effectively the normal case for gases and liquids), then we will observe this as a Doppler shift in the absorption spectrum, known as Doppler broadening. Given that the atom is hypothetically "allone", and in fact it will be in a gas and surrounded by distant

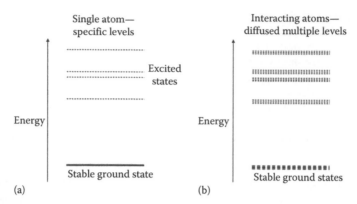

FIGURE 1.18 A simplified energy level model of a single atom—approximating to a gas at low pressure (a) and a group of atoms (b) typified by energy levels in liquids and solids. The energy level diagram's structure is characteristic of a specific element and is unique to that element. Molecules also have similar energy level diagrams, but invariably with significantly more detail therein.

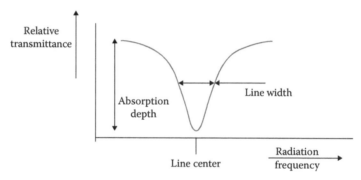

FIGURE 1.19 Absorption line shape characteristic of a gas line. The absorption depth depends on species concentration and the line width depends on the square root of the temperature, which is also a function of the gas pressure.

fellows (so distant that their collision probability is minute) then the velocity distribution will correspond to the thermal motion of the atoms. When the atoms are close to each other, but still not connected as in a liquid or a solid, collisions will become significant and here "close" implies that the collision time is shorter than the reciprocal of the spectral width introduced by the Doppler broadening. When this happens, we can add collision broadening to the effects on the spectral width.

This encapsulates in a very simplistic manner the essential features of gas spectroscopy. The arguments apply to molecules as well as atoms, though at the molecular level there is a much more complex range of resonances available corresponding to rotational and structurally induced changes as well as effects on individual atoms. Measuring the spectral broadening can, though, give a good measurement of temperature changes and at constant pressure or vice versa for pressure changes and a constant temperature. At the nano- and even microscale, however, the concepts of gas spectroscopy become somewhat compromised since there are very few atoms or molecules around and they also interact with each other at much closer proximities than typical in gas spectroscopy.

This brings us to effects in liquids and solids. When the molecules or atoms become close together, the individual energy levels need to adjust their positions into energy bands in order to comply with one of the basic tenets of quantum mechanics, namely, the Pauli exclusion principle. The simplest and most obvious ramification of this is that in contrast with gases the spectral signature of solids comprises very broad lines, which in turn, of course, give us the colors that we see of solid and liquid objects. These very broad lines still have a totally characteristic optical signature reflecting the composition of the material in question, but they are clearly more difficult to distinguish from each other and identify uniquely, particularly in unknown mixtures.

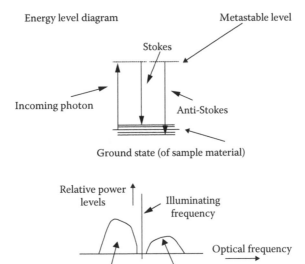

FIGURE 1.20 Raman spectroscopy in solids. An incoming photon is transiently absorbed and re-emitted at a range of energies determined by the optical phonon spectrum of the material through which the light is being transmitted.

These broad lines do though have their applications and a particularly useful one is (Figure 1.20) the Raman spectrum, for transparent media like silica glass fibers, for reflective media, or indeed for small clusters of particles. The essential features are that an incoming photon, in principle of any wavelength, can be absorbed and thereafter reradiated either by dropping to the same level and therefore emerging at the same wavelength, or dropping to one or other range of available levels characteristic of the sample material in question and therefore re-emitting at a different wavelength. The range and the probability distribution of these available wavelengths is very characteristic of a particular medium and is known as the Raman spectrum. The Raman spectrum—that is, the shift in the reemitted radiation is clearly going to be related to the energy level structure of the sample, and in some circumstances, for example, in gases, can directly model the absorption spectrum in a band of lower photon energy than that of the incident photons. There is a probability that the re-emitted photon may emerge at a higher energy (so-called anti-Stokes) though the probability is significantly lower than the probability of emerging at the lower photon energy (Stokes). The ratio though of the probabilities at a given energy from the input photon (i.e., of the Stokes and anti-Stokes radiation) is a unique function of temperature. This characteristic has been used extensively in distributed fiber-optic temperature measurements for over 20 years. The energy difference between the original photon and the re-emitted photon remains within the solid and manifests itself as an optical phonon in the form of heat.

There is another closely related phenomenon, namely, Brillouin scatter, which involves the emission of acoustic rather than optical phonons. Stimulated Brillouin scatter in which an optically induced acoustic phonon manifests itself as an ultrasonic traveling wave along the core of an optical fiber has been utilized extensively as, in effect, a means for measuring acoustic velocity through the Doppler shift induced by the traveling wave reflecting the incident light. The acoustic wavelength is known through the wavelength of the incident light, and consequently the acoustic velocity can be determined from the offset frequency in the reflection. In fibers, this offset frequency is typically 10–20 GHz, and the acoustic velocity depends on the longitudinal strain that the fiber experiences and also the temperature of the environment in which it operates. Provided the temperature effect can be compensated, this also presents an extremely useful tool for distributed measurement of strain fields.

Fluorescence is another widely used phenomenon that depends fundamentally on these very characteristic diffuse energy levels and that has been applied extensively in chemical sensing and especially in microscopy. Indeed, this phenomenon is so important that it is explored in detail in the next chapter.

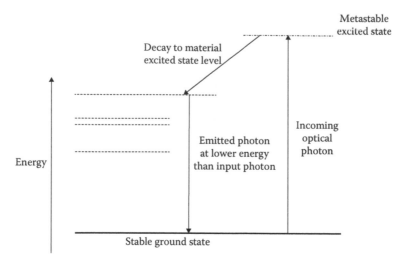

FIGURE 1.21 The basic elements of fluorescence. An incoming photon is absorbed into a metastable state, which relaxes into one of the (photon-emitting) excited levels characteristic of the host material. This state itself then relaxes to emit a photon that is characteristic of the host material spectrum. The remaining energy of the incoming photon is typically dissipated as heat.

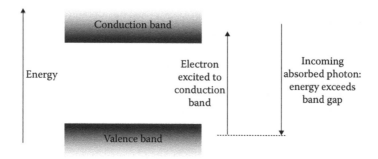

FIGURE 1.22 The photoelectric effect—an absorbed photon that needs to have sufficient energy to overcome the band-gap of the target material creates a conduction electron (hole pair in semiconductors), which in turn produces a detectable increase in the target material conductivity.

The process is slightly different from Raman scatter (Figure 1.21) involving the need for a metastable pump level. Fluorescence is also the basis of laser action, again covered extensively and comprehensively elsewhere, more specifically in this context for optically pumped laser architectures. Here, provided the pumping is sufficiently strong to create a population inversion, the mirrors provide the feedback that selects one or more very specific wavelengths from the fluorescent spectrum and provides amplification. Indeed in the laser we see a very simple example of the interaction between structure and quantum levels, which is at the heart of much of nanophotonics.

The remaining important mechanism involving light—acting as photons—and materials is the photoelectric effect. This was first observed a century or so ago and is essentially (Figure 1.22) a very useful feature that photons above a threshold energy can produce free electrons in the conduction band of an otherwise insulating material or—equivalently—can significantly increase the conductance of an illuminated material.

The effect has been exploited extensively in a huge range of photodetecting materials of which probably the most important are semiconductor devices using silicon and III–V compounds. Silicon as a single element indirect band gap material has by far the most advanced technological base from which devices may be realized but is restricted in the range of absorption properties available. In other words, there is only one band gap in the silicon stable, and the fact that this gap is indirect makes it unsuitable for simple

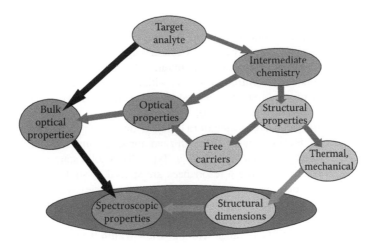

FIGURE 1.23 An overview of the factors influencing the response of nanochemical photonic sensors to a target analyte. The sensor can respond directly via combined "bulk" and structurally dimensionally influenced effects or indirectly through an intermediate chemical process that can change the direct spectral properties of the sensor or the behavior of the structural artifact in which the sensor is configured.

light emission devices (silicon photonics is changing some of that). In contrast, the III–V stable can be tailored through appropriate combinations of indium, gallium, phosphorous, and arsenic to produce a huge range of band gap options, which, among other things, facilitate detection in the all-important near-infrared endemic within fiber communication systems. The III–V stable also exhibits direct band gap behavior so sources including lasers and super-radiant diodes are relatively straightforward to realize. These sources also have the immensely beneficial feature of electrical rather than optical pumping.

In the context of our optochemical nanosensors, these optical properties of materials provide a basis for spectroscopic identification involving linear or nonlinear processes and, by implication, a range of mechanisms through which these optical properties may be modulated through external chemical influences. Furthermore, semiconductor devices are by far the most convenient and most common approaches to sources and detectors to utilize in such sensor systems. The focus in this chapter is, however, predominantly on the nanodomain through which structural properties of the material of which our sensor is fabricated also influence its optical properties [28].

We have glimpsed this by implication in our brief discussion of the mode selectivity in lasers where the spacing of the mirrors determines the actual frequency at which the laser oscillates and selects this from the band possibilities offered through the fluorescence process. There are also other basic techniques and tools available through which the spectroscopic (i.e., dielectric constant) properties of materials may be influenced. Of these, perhaps the most evident are thermal effects, which clearly affect the density of atoms in the structure and will also have an impact on the stiffness of the "spring." Another relevant tool for chemical sensing is the induced free carrier concentration that can occur in certain types of intermediate materials and that will, in turn, change the effective index through adding a plasma into the solid material. We have attempted to encapsulate these phenomena in Figure 1.23, and we will explore some of the implications in more detail in the following section.

1.6 Optics and Structures

We have become very much accustomed to the concept that optical devices are built on scales very much larger than a wavelength, and we have developed an instinctive understanding of most of the consequences thereof. However, the past couple of decades have witnessed technological improvements, enabling component machining at the nano-meter scale with dimensions and tolerances much less than an optical wavelength [29]. Indeed, the nanooptical chemical sensors that are the subject of this book fall predominantly into this category. This then changes the analytical framework from, at one extreme, very much guided-wave devices with dimensions up to tens, hundreds, or thousands of wavelengths to localized devices with dimensions comparable to or much less than a wavelength. In many ways, there are parallels to the evolution of microwaves half a century ago from the copper waveguide, through microstrip into the circuit element regime, which, among other things, facilitates mobile phones and Wi-Fi. Optoelectronics is following a similar evolution, with the flow of electric currents that determine the properties of active microwave systems replaced by the materials interactions that determine the properties of even macroscale optical systems [30].

The interaction between the optical resonator and a lasing medium exemplifies much of what is involved here. As already mentioned, for a resonator of length L comprising a uniform material of index n, the cavity resonances are spaced by a frequency $c/2 \cdot L \cdot n$. Typically, a solid state laser cavity has a length ranging from centimeters upward, so these resonances are spaced by multiples of the order of GHz. Consequently, there are tens or even hundreds of resonant frequencies within the gain curve of the medium, but typically the lasing processes would select but a few around the gain maximum.

If we repeat the exercise for a semiconductor laser with chip dimensions in the region of $100\,\mu m$, then the mode spacings move to the 100 GHz range with but a few within the medium gain curve. Selection processes [31], like distributed feedback, facilitate nominal single frequency operation of semiconductor lasers, so additional control is most certainly possible (Figure 1.24).

Of course, when we get to the very very small, the potential well distribution that determines the quantum mechanical wave function also becomes influenced by the boundary conditions so that in effect the optical properties of extremely small samples of materials become modified. However, since the quantum mechanical wave functions are determined by periodicities linked to the periodicity of the atoms and molecules in the materials themselves, this is significantly below the wavelength of light. So here the quantum dot approach becomes necessary and the structure itself is more akin to a wire than a cavity. It is, however, an unusual "wire" since its properties are dependent on its dimensions; its behavior, therefore, is somewhat different though with the correct approach can be modeled. Indeed, we now have circuit elements rather than waves and cavities and an analysis in terms of inductance resistance and capacitance becomes, with care, achievable [30].

The diffraction grating is another important building block in the micro and nano toolbox and indeed is the essential structure that determines band gap phenomena, whether in the context of the electronic band structure, the transmission characteristics of waveguides made with diffraction gratings as the guiding wall (e.g, photonic crystal fibers), or indeed the properties of a microwave post filter. Figure 1.25

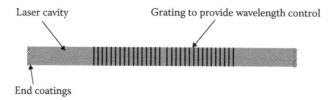

FIGURE 1.24   The principle of distributed feedback wavelength control in semiconductor (and some fiber) lasers. The grating is tightly coupled to the active region of the laser and is phase shifted in the center to turn the grating into a resonator, as indicated schematically in the figure. Typically one or both ends are antireflection coated.

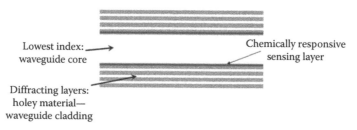

FIGURE 1.25   Bragg diffraction used as the cladding geometry in optical waveguides. The figure also incorporates one possible position (of many) for a chemically active interface that will modify the transmission characteristics of the guide in response to a specific chemical stimulus. With the correct choice of material, the layer need only be nanometers thick to have significant impact.

illustrates the conceptual framework for the so-called Bragg waveguide. In the context of our nano-optical chemical sensors, clearly the properties of this waveguide can—as indicated—be modified by the inclusion of reactive surface layers on the internal surface of the optical core or indeed within the grating elements themselves.

The so-called photonic crystal also offers immense flexibility in waveguide design. For example, the numerical aperture of a conventional dielectric guide can effectively be increased by introducing a carefully structured hole distribution within the cladding. This, in turn enables light to go round much tighter corners than would otherwise be feasible.

The potential variations on this basic theme are immense, enabling control of birefringence and—back to the original context—the inclusion, in principle, of chemically sensitive zones on the interface between the air or, more generally, the surface of the photonic structure involved, and the dielectric whether in core or the cladding. Again though, we are considering a wave-based structure rather than the nanoscale phenomenon associated with the extremely small.

Finally, we have already mentioned the surface plasmon. This has been used in optical sensors for many years, initially in the Kretschmann configuration shown in Figure 1.26. This shows a very characteristic dip in the transmission when the incident angle phase propagation constant matches into the surface plasmon at the dielectric interface, and this dip moves predictably and repeatedly as the index differential across the interface changes. This basic configuration has also been used in waveguide form, again with due attention to phase matching, to realize waveguide polarizers for which the design is optimized to minimize sensitivity to overlay index values or waveguide sensors for which the design is optimized by reverse. Going back to our Clausius–Mossotti equation and the refractive index of a metal, we can easily show that the propagation wavelength (which equals the skin depth) in a metal is extremely small. Consequently, the availability of a nanoscale precision machining does offer the availability to make resonators based on surface plasmons using periodic structures for which—in principle—the sensitivities to changes in loading index can be significantly increased.

There is, however, the significant attenuation associated with any surface plasmon propagation phenomenon (the skin depth phenomenon again) so the useful distances that can be addressed are extremely small, though more than adequate for nanoscale devices. The metal waveguide interface though does have another important manifestation in the realization of the scanning near field optical microscope (Figure 1.27). These enable scanned images with resolutions in the tens of nanometers to be obtained albeit at a speed limited by the power that can be collected and the achievable size of the available illumination spot. The scanning near field optical microscope, SNOM, has however, made substantive and

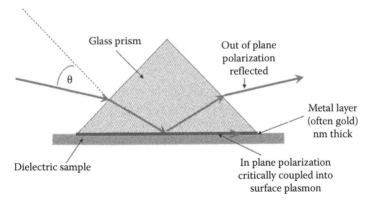

FIGURE 1.26 The essence of surface plasmons, which are often used in nanophotonic devices, is embodied in the so-called Kretschmann configuration sketched here. At a very specific angle q, the reflected light intensity dips corresponding to the p polarization excitation of the surface plasmon mode in the metal dielectric interface. The exact coupling conditions depend on the prism index, the film type (i.e., index) and thickness, and on the sample index. This principle, in various manifestations, has been incorporated into numerous optically based chemical sensor systems and can be extremely sensitive to sample index variations.

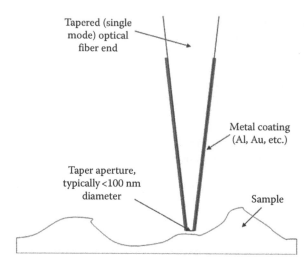

FIGURE 1.27 The principle of the SNOM. The metal coating confines the optical field to within the taper at subwavelength dimensions, somewhat like a coaxial transmission line. The radiation at the end couples to the sample to produce, in reflection, an image with a resolution determined by the diameter of the tip.

useful contributions to imaging science and has also been used in biochemical probes to, for example analyze chemical structure within the dimensions of a biological cell.

This very brief introduction has but scratched the surface of the undeniably fascinating array of effects that can be introduced by combining material properties with structural dimensions, particularly, though not entirely, for optical systems with dimensions measured in nanometers. At these dimensions, our concepts of interacting the quantum mechanical wave function with large numbers of atoms become challenged and our intuitions for the wave treatment of optical systems no longer apply. Indeed we are, on this scale, approaching structural artifacts more akin to optical circuit elements than wave interaction systems. That said, the availability of precise structural definition for dimensions of the order of a small fraction of an optical wavelength also opens up a whole array of synthetic materials with controlled optical properties, of which the so-called photonic crystal is the dominant example. Artifacts fabricated on either the micro- or nanoscale exhibit properties depending upon the optical properties of the material constituents as well as their dimensions. Consequently, changes in these optical properties produce a change in the properties of the structure, which can be observed using a suitable optical interrogation scheme. Such structures are clearly naturals for chemical sensing.

1.7 Some Final Observations

Photonics is the electronics of the twenty-first century.

Is it—well maybe? As a tool there is still much to understand and even more to realize. The technology of producing what we now perceive as "very small" is but in its infancy. But microcircuits were beyond anyone's imagination in 1912. Eli. Yablonovitch's concept of "circuit analysis in metal-optics" may well be laying the foundation for a new philosophy in optics—a new approach to design and realization, not to mention performance ambition somewhat akin to the evolution of microwaves through waveguide to microstrip to microcircuit? There is a potential social need—the challenges of population expansion, health, responsible energy, clean water, reliable food supply will become more and more pressing. And photonics does offer some tantalizing prospects. Virtually all our energy arrives as photons. We have already witnessed a significant communications and data manipulation revolution, which could not have happened without light, especially lasers. And one of the most important and most rapidly evolving roles of light is the physical and chemical characterization of the microworld—and here many of the concepts that form the theme of this book will find their future niche.

So whether photonics actually will be "the electronics of the twenty-first century" remains to be seen, but there is no doubt that as the concepts that we have very briefly outlined here become more understood and more practical, photonics will most certainly have a central role to play in our technological evolution. Chemical sensing will certainly be a very important contributor to this.

Acknowledgment

This work was cosupported by the Spanish Ministry of Education through the project TEC2010-20224-C02-02.

REFERENCES

1. I. Newton, *Opticks*, Dover publications, Inc., New York, 1954.
2. T. Lamb, J. Bourriau, *Colour: Art & Science*, Cambridge University Press, Cambridge, U.K., 1995.
3. A. Einstein, Über einen die Erzeugung und Verwandlung des Lichtes betreffenden heuristischen Gesichtspunkt, *Annalen der Physik*, 17, 132–148, 1905.
4. A. Rogers, Essentials of photonics, 2nd edn., CRC Press, New York, 2009.
5. J.M. López-Higuera, *Optical Sensors*, Dept. of Publications, University of Cantabria, Cantabria, Spain, 1998.
6. J.M. López-Higuera, *Handbook of Optical Fibre Sensing Technology*, John Wiley & Sons, Chichester, U.K., 2002.
7. B.E.A. Saleh, M.C. Teich, *Fundamentals of Photonics*, 2nd edn., Wiley-Interscience, New York, 2006.
8. A. Yariv, *Optical Electronics in Modern Communications*, Oxford University Press, London, U.K., 1997.
9. Y.B. Band, *Light and Matter*, John Wiley & Sons, Chichester, U.K., 2006.
10. S.G. Lipson, H. Lipson, D.S. Tannhauser, *Optical Physics*, 3rd edn., Cambridge University Press, Cambridge, U.K., 1995.
11. J.S. Walker, *Physics*, 2nd edn., Prentice Hall, Pearson Educational International, Upper Saddle River, NJ, 2004.
12. T. Okoshi, *Optical Fibers*, Academic Press, New York, 1982.
13. J.M. Senior, *Optical Fibers Communication*, 3rd edn., Prentice Hall, Pearson Educational International, Edinburgh, U.K., 2009.
14. E. Mazur, *Nonlinear Optics at the Nanoscale*, Invited tutorial at Optoel'11, Santander, Spain, 2011. His talk presentation can be see in the website: http://www.teisa.unican.es/optoel2011/index.php?option=com_content&vie=article&id=56&Itemid=69&lang=es
15. S. Janz, Food and water safety: Silicon photonic wire waveguide sensors detect the presence of pathogens and chemicals, *Photonics Spectra (Features)*, 2008.
16. K.C. Kao, G.A. Hockham, Dielectric-fibre surface waveguides for optical frequencies, *Proceedings of IEE*, 113, 1151–1158, 1966.
17. E. Yablonovich, Inhibited sponatneous emission in solid-state physics and elelctronics, *Physical Review Letters*, 58, 2059–2062, 1987.
18. S. John, Strong localization of photons in certain disordered dielectric superlattices, *Physical Review Letters*, 58, 2486–2489, 1987.
19. A. Bjarklev, A.S. Bjarklev, J. Broeng, *Photonic Crystal Fibers*, Kluwer Academic Publishers, Boston, MA, 2003.
20. J.C. Knight, J. Broeng, T.A. Birk, P.St.J. Russell, Photonic band gap guidance in optical fibers, *Science*, 282, 1476–1478, 1998.
21. T.M. Monro, D.J. Richardson, N.G.R. Broderick, P.J. Bennett, Holey optical fibers: An efficient modal model, *Journal of Lightwave Technology*, 17, 1093–1102, 1999.
22. P. Russel, *Keeping a Tight Focus on Matter*, Invited tutorial at Optoel'11, Santander, Spain, 2011.
23. R.P. Feynman, R.B. Leighton, M. Sands, *The Feynmann Lectures on Physics*, Addison Wesley, Reading, MA, Vol. 2, 1964.
24. C. Kittel, *Introduction to Solid State Physics*, 8th edn., John Wiley & Sons, New York, 2005.
25. W.T. Metaferia, *Nano-Photonics and the Clausius–Mossotti Relation*, DM Verlag Dr. Muller Aktiengesellschaft & Co., Saarbrücken, KG, 2010.

26. P. Mclman, R.W. Davies, Application of the Clausius–Mossotti equation to dispersion calculations in optical fibers, *Journal of Lightwave Technology*, 3, 1123–1124, 1985.
27. J.R. Pierce, *Almost All about Waves*, Original published by MIT, Cambridge, MA, 1976, reissued by Dover Books, 2006.
28. V.A. Markel, T.F. George, *Optics of Nanostructured Materials*, John Wiley & Sons, New York, 2000.
29. E. Yablonobitch, Photonic crystals: Semiconductors of Light, *Scientific American*, 285, 34–41, 2001.
30. M. Staffaroni, J. Conway, S. Vedantam, J. Tang, E. Yablonovitch, *Circuit Analysis in Metal-Optics*, 22pp. Available on line at http://arxiv.org/ftp/arxiv/papers/1006/1006.3126.pdf, January 2012.
31. V.N. Astratov, J.S. Culshaw, R.M. Stevenson, D.M. Whittaker, M.S. Skolnick, T.F. Krauss, R.M. De La Rue, Resonant coupling of near-infrared radiation to photonic band structure waveguides, *Journal of Lightwave Technology*, 17, 2050–2057, 1999.

2

Fundamentals of Optical Chemical Sensors

Francesco Baldini, Ambra Giannetti, and Cosimo Trono

CONTENTS

2.1 Introduction

In optical chemical sensors, the light delivered to the sensing probe is modulated either directly by the parameter being investigated or by a chemical transducer or a biochemical recognition element, the optical properties of which are changed by the interaction with the parameter under study. Absorption and fluorescence are definitely the most exploited optical properties in chemical sensing; however, chemiluminescence, bioluminescence, and Raman scattering are at the basis of many chemical and biochemical sensors. A large class of sensors is represented by sensors based on the change in the refractive index induced by the analyte under investigation on a sensing layer. This change in the refractive index can give rise to a shift in the surface plasmon resonance, may modify the evanescent wave of an optical waveguide, can cause a modulation of the interference spectrum between optical beams, or may alter the optical resonance in optical resonant structures. Generally speaking, the sensors are amplitude-modulation sensors; this means that the intensity of the light is the measured parameter associated with the analyte concentration, but both the wavelength shift and the modulation of the optical signal in the time or frequency domain are exploited as forms of optical transduction.

2.2 Fluorescence

Fluorescence can be used in optical sensors for detecting a chemical substance, thanks to different approaches. Three main cases can be distinguished as follows:

1. The substance being investigated is fluorescent.
2. The substance is not fluorescent, but can be labeled with a fluorophore.
3. The substance interacts with a fluorophore, causing a variation in the emission of fluorescence.

In the first two cases, the equation that links the fluorescence intensity to the concentration is given by

$$I(\lambda_{em}) = kI(\lambda_{exc})\psi(\lambda_{exc}, \lambda_{em})\varepsilon(\lambda_{exc})lc \qquad (2.1)$$

where

$I(\lambda_{exc})$ and $I(\lambda_{em})$ are the intensities of excitation and emission radiation, respectively
ψ and ε are the quantum yield and the absorption coefficient
l is the optical path
c is the concentration
k is a constant that depends on the optical setup and on the configuration of the probe

This equation hypothesizes low absorbance values by the fluorophore. Moreover, if other chromophores are present in the solution under test, a decrease in the fluorescence, caused by the absorption of the excitation light (primary inner filter effect) or of the emission light (secondary inner filter effect) by these chromophores, can be observed. It is apparent that in this case, the previous equation is no longer valid, but that corrective terms are necessary.

In the latter case, of particular interest is the phenomenon known as fluorescence "quenching," in which the fluorescence intensity decreases as a consequence of the interaction with the substance (quencher) under test, which can thus be detected [1]. The interaction with the analyte can either be direct or mediated by a receptor. The main mechanisms of quenching are photoinduced electron transfer (PET), intramolecular change transfer (ICT), intersystem crossing, and energy transfer [2].

In PET-based sensors, the fluorophore is generally bound to a receptor via a spacer, and an electron transfer takes place between the receptor and the excited fluorophore in the absence of the analyte. The transfer is possible in the two directions (from fluorophore to receptor and vice versa), depending on the oxidation and reduction potentials. The result is that, following the electron transfer, the fluorophore returns to the ground state via charge recombination without any emission of photons. In the presence of the analyte bound to the receptor, the oxidation and/or reduction potentials change, the electron transfer becomes energetically not feasible, and, consequently, the fluorescence emission is observed. With the use of this approach, reliable optical sensors for sodium [3] and potassium [4] have been realized by using suitable ionophores as receptors. PET has also been used to study the conformational dynamics of macromolecules [5]. It is important to observe that the PET is an on–off process with the appearance of the fluorescence emission in the presence of the interaction of the analyte with the receptor. This implies that the change in fluorescence is not accompanied by a spectral shift.

In ICT, the fluorophore and the receptor are bound without the presence of any spacer, thus forming a single structure inside which the charge transfer takes place when the interaction with the analyte occurs. The resulting molecule generally has electron donating and withdrawing groups at its opposite ends, and a dipole is created when the charge transfer takes place. This fact explains reasonably well the reason why ICT-based sensors are quite affected by the polarity of the solvent, which can alter the electronic characteristics of the dipole. Contrarily to what occurs in PET, both states of the ICT-based structure are fluorescent, and the interaction with the analyte implies changes of intensities and spectral shifts.

A decrease in the fluorescence intensity may also be due to an energy transfer from the fluorophore in the excited state to another molecule. The most popular energy transfer in chemical sensing is the Förster resonance energy transfer (FRET). In FRET, the interaction is between a fluorophore and the so-called acceptor, the absorption spectrum of which, modulated by the chemical species under investigation, overlaps the emission spectrum of the fluorophore [6,7]. In the presence of a strong closeness between fluorophore and acceptor, the overlapping of the emission and absorption spectra enables the energy absorbed by the fluorophore to be transferred to the acceptor without any emission of photons. The efficiency of this energy transfer η is expressed, according to the Förster theory [8], by

$$\eta = \frac{R_0^6}{R_0^6 + R^6} \qquad (2.2)$$

where

R is the distance between the fluorophore and the acceptor
R_0 is a constant that depends on the fluorophore/acceptor pair and constitutes the distance for which the efficiency is 0.5

Typical values of R are between 10 and 100 Å.

Different schemes can be used in optical sensing with FRET:

1. The acceptor is an absorber, the absorption of which is a function of the investigated analyte. Therefore, the amount of the emitted fluorescence is modulated by the analyte, and the fluorescence intensity in the presence of the acceptor I is given by

$$\frac{I}{I_0} = 1 - \eta \qquad (2.3)$$

where I_0 is the fluorescence intensity in the absence of the acceptor.

2. The acceptor is another fluorophore. In this case, the immediate advantage is an increase in the distance between the wavelength of the excitation beam and the wavelength of the measured emitted fluorescence, which can be a critical parameter in the design of an optical fiber sensor.

The very short distance between the donor and the acceptor makes FRET very useful for studying protein interactions. One interesting application in nanosensing is offered by the molecular beacon, an oligonucleotidic sequence labeled at the two side ends, with the donor on one side and the acceptor, an absorber, on the other side. The sequence is designed in order to form a hairpin structure (stem-loop), which brings the fluorophore and the quencher so close that the energy transfer occurs and no light emission takes place. In the presence of the target, which is the complementary sequence, the hybridization opens the stem-loop structure and the oligonucleotide sequence turns on with the emission of fluorescence, no longer quenched by the acceptor, and quite distant from the fluorescent label (Figure 2.1). Therefore, at a molecular level, the structure acts as a beacon that turns on in the presence of the target. This structure, designed as a molecular beacon, is proposed to monitor the efficiency of drugs at an intracellular level. Another energy transfer mechanism is provided by the Dexter interaction. In the Dexter energy transfer, a bilateral electron exchange takes place between the donor, and the acceptor provided that there is an overlapping between their electronic clouds. This condition implies a very short distance between the donor and the acceptor (less than 10 Å). Unlike the six-power dependence of Förster energy transfer, the reaction rate constant of the Dexter energy transfer exponentially decays as the distance between these two elements increases.

An important general distinction can be made between static quenching, when the quencher interacts with the fluorophore at the ground state forming a nonfluorescent complex (FQ), and dynamic quenching, when the quencher interacts with the fluorophore in the excited state, causing its return to the ground state, without the emission of fluorescence.

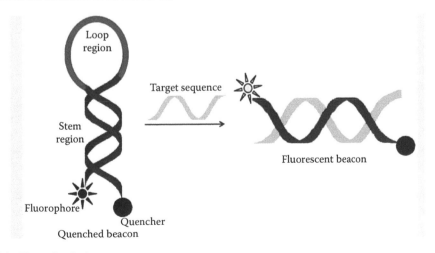

FIGURE 2.1 The molecular beacon.

In both cases, the relationship between fluorescence intensity I and the concentration of quencher [Q] is

$$\frac{I}{I_0} = \frac{1}{1 + K[Q]} \qquad (2.4)$$

where

I$_0$ is the fluorescence in the absence of the quencher

K is a constant equal to the dissociation constant of the chemical reaction that leads to the formation of the nonfluorescent complex in the case of static quenching or is the Stern–Volmer constant in the case of dynamic quenching

The Stern–Volmer constant is equal to the product between the diffusion-controlled rate constant k_q and the fluorescent lifetime τ_0 of the excited state of the fluorophore in the absence of the quencher. The equation is clearly valid in the absence of inner filter effects.

In the case of dynamic quenching, it is more opportune to consider time-dependent decay [9] instead of intensity. In the presence of an interaction with the excited state, the lifetime of the fluorophore is decreased: the higher the concentration of the quencher, the greater the decrease in the lifetime. This is not the case with static quenching, in which the lifetime of the fluorophore is unaffected by a change in the concentration of the quencher. Typical fluorescence decay times fall within the range between 2 and 20 ns, while phosphorescence decay times are in the 1 μs–10 s range.

According to Stern and Volmer, the relationship between the decay time and the concentration is

$$\frac{\tau}{\tau_0} = \frac{1}{1 + K_{sv}[Q]} \qquad (2.5)$$

where

K_{sv} is the Stern–Volmer constant

τ and τ_0 are the decay times of the excited state of the fluorophore in the presence and absence of the quencher, respectively.

Lifetime can be measured either in the time domain or in the frequency domain. In the former case, the fluorophore is excited with a narrow pulse, and the fluorescence decay is monitored. In the latter case, a modulated excitation is used: the fluorescence emission is still modulated at the same frequency, but is decreased in amplitude and is phase-shifted.

The advantage of this approach lies mainly in the fact that there is less dependence on loading or photobleaching the chemical transducer fixed at the end of the optical fibers, which is one of the greatest drawbacks in intensity-modulated chemical sensors. Moreover, fewer problems arise from eventual fluctuations or drifts in the source intensity or photodetector sensitivity, which, on the contrary, heavily affect the intensity of modulated sensors. Furthermore, in the case in which several species interact with the reagent, causing the emission of fluorescences characterized by different decay times, these can be detected simultaneously with the use of time-resolution instrumentation.

Up to a few years ago, the utilization of this technique was thwarted by the need for expensive and cumbersome optoelectronic instrumentation (lasers, rapid detection systems, etc.). At present, the advent of fast and powerful light sources, such as emitting diodes and laser diodes, at wavelengths compatible with the fluorophores, make it possible to realize compact and quite inexpensive optoelectronic units [10–12]. Moreover, the synthesis of new fluorophores characterized by a longer lifetime has made it possible to design lifetime-based sensors working in the microsecond range. All these possibilities make this approach one of the most promising for optical fiber chemical detection. The only drawback, which is intrinsic to the properties of the fibers, is related to the limitation in the length of the optical link due to fiber dispersion: in fact, this approach cannot be proposed in the case of sensor networks.

In lifetime-based sensors, a noteworthy intrinsic correction of the fluctuations coming from the optoelectronic system (e.g., source or detector drift) and from the optical components (filters, fibers, etc.) is performed by the dual lifetime referencing (DLR) [13–15]. In this case, two fluorophores having the same excitation wavelength and very different decay times are combined in the sensing layer. The fluorophore characterized by the shorter lifetime is the indicator, which is sensitive to the investigated parameter, whereas the other one, characterized by the longer lifetime, is independent of the concentration of the parameter. The other condition on making use of DLR is that the emission spectra of the two fluorophores overlap so as to make possible the use of a single photodiode. In this way, it is possible to show that the phase angle of the detected signal, which comes from the superposition of the two fluorophores, depends only on the ratio of the two fluorescence amplitudes, and is, therefore, independent of the optical/optoelectronic components of the sensor and also of the light losses in the optical path.

2.3 Chemiluminescence and Bioluminescence

The production of visible light from a compound excited by a chemical reaction is called chemiluminescence. A chemiluminescent sensor is based on a monitoring of the rate of production of photons, with the light intensity depending on the concentration of a limiting reactant that is involved in the reaction. Bioluminescence is a special case of chemiluminescence, in which the emission of light comes from a living organism and involves a protein (generally an enzyme) in the reaction [16]. Low detection limits can be achieved with this approach, which has the advantage of the absence of any optoelectronic source, since the light is directly generated by chemical/biochemical reactions. In this way, light background effects are drastically reduced, since all the light that reaches the photodetector is modulated by the analyte (e.g., no scattered light directly from the source to the photodetector, as in absorption-based sensors, or interference from the excitation light, as in fluorescence-based sensors). Bioluminescence-based and chemiluminescence-based sensors have been developed with the aim of combining the sensitivity of light-emitting reactions with the convenience of sensors. Several analytical luminescence systems including immobilized reagents with or without the use of optical fibers as optical carriers have been described in the literature. More recently, chemi- and bioluminescence detections have also been proposed for the development of biochips and microarrays [17–19].

The chemiluminescent reactions of greatest interest are those involving hydrogen peroxide (H_2O_2). In this case, the luminescence process is associated with an enzyme-catalyzed reaction that produces H_2O_2. Luminol, lophine, lucigenin, or oxalate derivatives are some of the synthetic compounds that can produce light when oxidized in the presence of H_2O_2 [20–22].

The chemiluminescent reaction of luminol (5-amino-2,3-dihydro-1,4-phthalazinedione) with H_2O_2 in protic solvents (e.g., water) occurs under alkaline conditions in the presence of a catalyst or a cooxidant (Figure 2.2). Transition metal cations (Cr^{3+}, Mn^{4+}, Fe^{2+}, Fe^{3+}, Co^{2+}, Ni^{2+}, Cu^{2+}, Hg^{2+}), either free or complexed to inorganic or organic ligands (e.g., heme-containing proteins, such as horseradish peroxidase), catalyze the luminol chemiluminescence oxidation. The emission spectrum of luminol shows a maximum at 425 nm.

Chemiluminescent reactions are oxidoreduction processes that are generally catalyzed by enzymes. When the oxidoreduction process is generated during an electrochemical reaction, the phenomenon is called electrochemiluminescence or electrogenerated chemiluminescence (ECL) [23]. Among many organic and inorganic ECL systems, ECL based on ruthenium complexes, particularly tris(2,2'-bipyridine)rutheniumII ($Ru(bpy)_3^{2+}$), has received considerable attention [24,25].

Bioluminescence has been found in a wide range of the major groups of organisms, from bacteria and protists to squid and fishes, with numerous phyla in between. Nevertheless, only the firefly and the

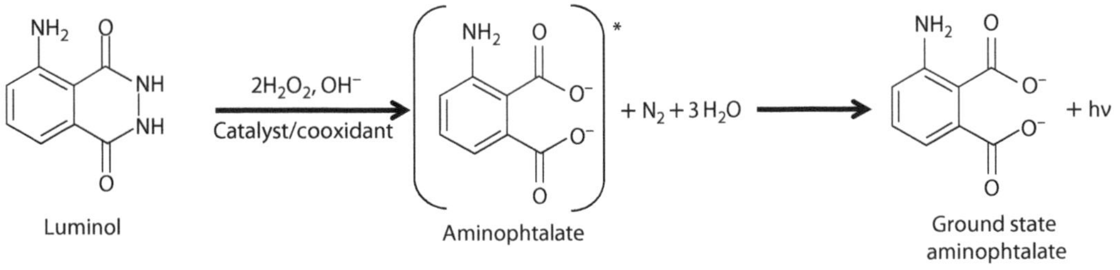

FIGURE 2.2 Chemiluminescent reaction of luminol.

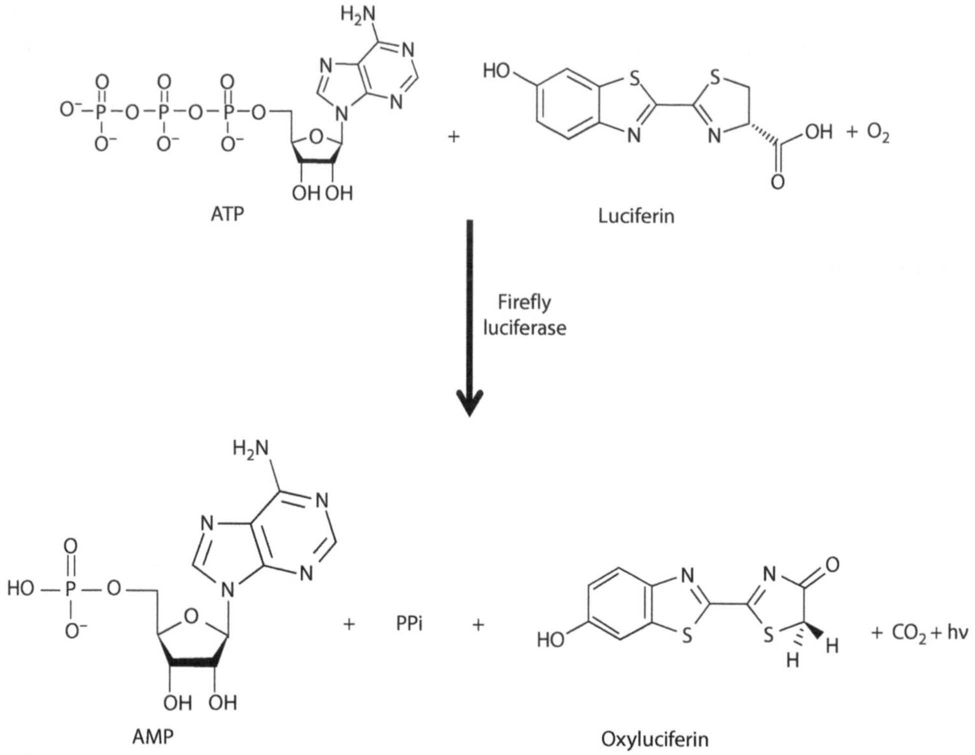

FIGURE 2.3   The firefly luciferase bioluminescence reaction.

marine bacteria bioluminescent systems have actually been evaluated for analytical purposes. Analytical applications of bioluminescence are related mainly to the detection of ATP with the firefly luciferase [26,27] and of NADH with certain marine bacteria systems [28,29]. In some other bioluminescent organisms, luciferase is not involved in the production of light, which is produced directly by a protein-luciferin complex called photoprotein. In this complex, luciferin is tightly or covalently bound to the protein, and the energy is released in the form of light and generally in the presence of H^+ or Ca^{2+} ions.

Firefly luciferase catalyzes the emission of light in the presence of ATP, Mg^{2+}, and firefly luciferin (Figure 2.3). The emission has a maximum peak at 560 nm.

The substrate of the bacterial luciferase reaction includes reduced flavin mononucleotide ($FMNH_2$), O_2, and a long-chain aldehyde (R-CHO) (Figure 2.4). In vitro, decanal is generally used as the aliphatic aldehyde, and the light emission has a maximum peak at 490 nm.

Chemiluminescence and bioluminescence are considered to be very powerful analytical tools, since hydrogen peroxide, ATP, or NAD(P)H can be directly measured. However, theoretically, any compounds or enzyme involved in a reaction that generates or consumes these metabolites can also be measured by one of the appropriate light-emitting reactions.

2.4 Absorption

When a monochromatic beam passes through an absorbing medium, that is homogeneous, a fraction of it is absorbed. Lambert's law (also known as Lambert–Bouguer's law) expresses the dependence of the absorption on the thickness of the medium:

$$I = I_0 e^{-\alpha x} \ln\left(\frac{I}{I_0}\right) = -\alpha x \tag{2.6}$$

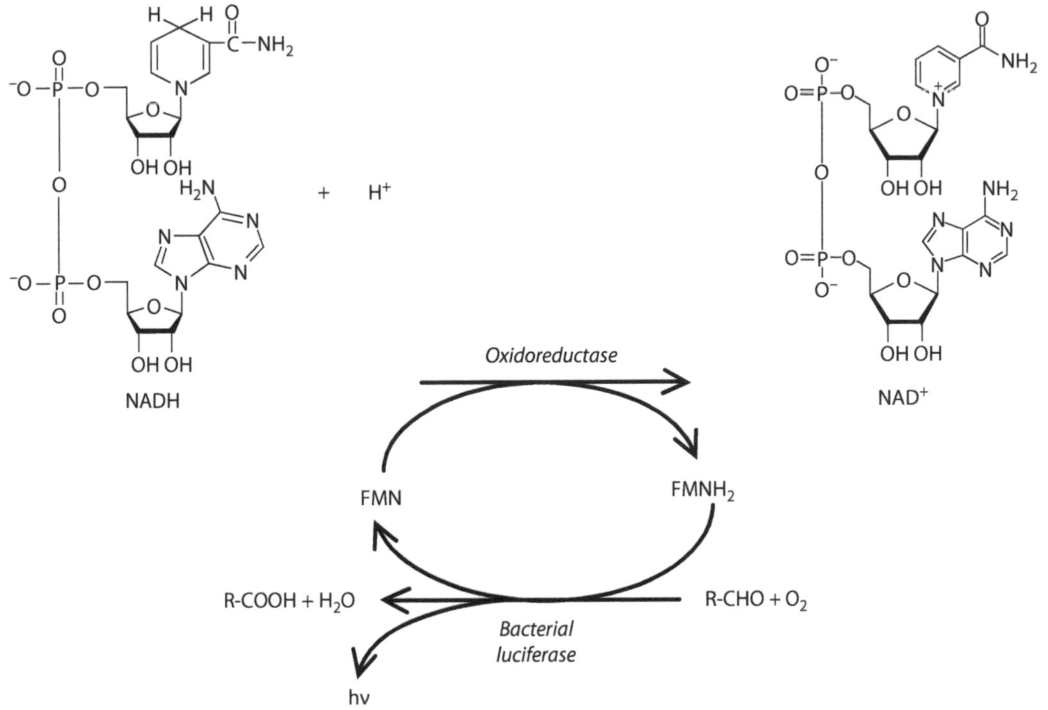

FIGURE 2.4 Bacterial bioluminescent reaction for NADH or NADPH detection.

where
  I_0 is the intensity of the incident radiation (corrected by considering the losses due to the reflection at the medium interface)
  I is the intensity of the radiation, which crosses a thickness x of the absorbing medium
  α is the extinction coefficient

In the case of a liquid solution or a gas, we can assume that every single molecule absorbs the same amount of light, independently of its reciprocal distance. The Lambert–Beer (or simply the Beer) law is based on this hypothesis and relates the intensity of the light to the concentration:

$$I = I_0 e^{-\varepsilon c l} \quad \ln\left(\frac{I}{I_0}\right) = -\varepsilon c l \tag{2.7}$$

where
  I_0 and I are the radiation incident on and emanating from the sample, respectively
  c is the concentration
  l is the path length
  ε is a substance-specific function of the wavelength

The units of measurement of c and ε depend on the way in which the concentration of the absorber is expressed. In the case of solutions, c is usually the molar concentration [mol L^{-1}] and ε is the molar absorption coefficient (or molar absorptivity or molar extinction coefficient) [L mol^{-1} cm^{-1}].

In the case of gas, the concentration is given as number density (numbers of molecules per volume unit) and ε (better indicated with the Greek letter σ) is the absorption (or extinction) cross-section and has units of length squared.

TABLE 2.1

Molar Absorption Coefficient, ε', in Correspondence with the Absorption Peak Wavelength for Different Chemical Compounds

	ε (L mol^{-1} cm^{-1})	λ (nm)	Solvent
Bilirubin	55,000	450.80	Chloroform
Chlorophyll a	111,700	417.80	Methanol
Crystal violet	112,000	509.50	Water
Indocarbocyanine (C3)	133,000	544.25	Methanol
Indodicarbocyanine (C5)	200,000	646.00	Methanol
Indotricarbocyanine (C7)	240,000	742.25	Ethanol
Malachite green	148,900	616.50	Water
Nile blue	76,800	626.75	Methanol

The Lambert–Beer law can be also expressed in terms of the absorbance (or optical density) A:

$$A = \log_{10} \frac{I}{I_0} = \varepsilon'cl \tag{2.8}$$

where

$$\varepsilon' = 2.3 \times \varepsilon \tag{2.9}$$

is the decadic molar absorption coefficient, but is often called simply the molar absorption coefficient. Table 2.1 shows typical values of the decadic molar absorption coefficient.

Clearly, if more than one absorbing species is present, the contribution of each of them must be considered, and Equation 2.3 becomes

$$A = \log_{10} \frac{I_0}{I} = \sum \varepsilon'_i c_i \cdot l \tag{2.10}$$

The relation between transmittance T $(=I/I_0)$ and the absorbance is given by

$$A = \log_{10} \frac{1}{T} \tag{2.11}$$

The simplicity of the Lambert Beer law lies in the linearity between absorbance and concentration. The Lambert–Bouguer law has very few exceptions, which is not the case with the Lambert–Beer law, the application of which requires greater attention. As mentioned earlier, this law is based on the hypothesis that every single molecule absorbs the same amount of light, independently of its reciprocal distance, which means that there is no interaction among the different molecules. This can be considered true only in dilute solution. At high concentrations, dimerization or the formation of more complex aggregates takes place, and, usually, the absorbing properties of these aggregates are different from those of the starting molecules. Another effect that occurs in the presence of highly concentrated solutions is the change in the refractive index, which generally takes places for concentrations of the analyte larger than $\sim 10^{-2}$ M. These changes in the refractive index imply different light reflection losses, which modify the transmitted light.

Deviations from the Lambert–Beer's law occur also in the case of chemical reactions involving the absorbing analyte. Examples are solutions of acid–base indicators or electrolytes.

The Lambert–Beer and Lambert–Bouguer laws are strictly valid only if a monochromatic source is used. If a multiwavelength optical beam is used, as often occurs in optical sensors (e.g., in the case of the use of light-emitting diodes as optical sources), the contribution of all the wavelengths must be considered. For example, Equation 2.8 becomes

$$A' = \log \frac{\int I_0(\lambda)d\lambda}{\int I(\lambda)d\lambda} \tag{2.12}$$

where the integral is evaluated on all the wavelengths emitted by the LEDs. Clearly, A' is no longer related in a linear manner to the concentration of the chemical parameter.

From an instrumental point of view, deviations from linearity occur whenever the spectral bandwidth of the incident light cannot be considered narrow in comparison with the absorption bands in the spectrum.

2.5 Raman Scattering (SERS)

Raman scattering [30] is a particular aspect of matter-radiation interaction. A light beam that interacts with matter is partially absorbed and partially scattered. Most scattered light has the same energy as does incident light, but a small fraction of the absorbed energy is utilized for inducing vibrations in the molecules of the absorbing medium. Therefore, part of the scattered light has a smaller energy and, apparently, a smaller frequency ($E = h\nu$). The differences in frequency between incident and scattered light are equal to the vibration frequencies of the matter.

A Raman spectrum is usually a plotting of the intensity of scattered light as a function of the change in frequency of the scattered beam with respect to the incident one. This spectrum is a characteristic of matter and is practically its "fingerprint" in a similar way to the infrared spectrum. A Raman spectrum can be obtained using any excitation wavelength in the UV, visible, or near infrared regions. This aspect has several advantages:

1. It is possible to perform this investigation on aqueous solution, in the case of which an infrared spectrum would not give any information, due to the high absorption of the water;
2. It is possible to choose the optimum excitation wavelength, at which any interference by fluorescence is completely avoided;
3. The possibility of obtaining Raman spectra using a wavelength in the visible/NIR region as the excitation wavelength makes this technique perfectly compatible with optical fibers, which have the lowest attenuation in this region.

The main disadvantage of Raman spectroscopy is the weakness of the Raman signal. This can be blamed on the fact that Raman intensity is proportional to the intensity of the electromagnetic field that is incident on the analyte, and the source of the electromagnetic radiation is generally distant from the analyte. A great improvement is obtained using surface enhanced Raman spectroscopy (SERS), where the Raman signal is enhanced by adsorbing the analyte on substrates suitably modified with metal surfaces. The first results were achieved by using roughened metal surfaces, and with this approach, Raman signal was noticeably amplified [31–34] with an enhancement of the order of 10^5–10^{10}. The reason for this enhancement is mainly due to an electromagnetic effect. The incident radiation excites the plasmons in the metal surface, and the field generated by the excited plasmons is very effective, due to the proximity of the metal surface and the analyte. The roughness of the metal surface plays a fundamental role in this mechanism, since the intensity of the emitted field greatly depends on the shape of the metal layer. Silver layers are generally used; however, gold, copper, or alkali metals can also be used in SERS, because their excitation wavelength is in the near or in the visible regions.

With the progress in the synthesis of nanomaterials and nanoparticles fabrication, metal nanoparticles in solution and substrate-bound nanostructures have been used as supports for SERS, replacing more and more often the roughened metal surfaces and offering better performances [35,36]. Optical fiber sensors have also been developed that make use of direct Raman spectroscopy or SERS. Different techniques are followed in order to manufacture SERS-active metal layers on the fiber tip that have the proper roughness: for example, the depositing of metal films on alumina micro- or nanoparticles fixed at the end of the fiber using a dip-coating process [37], the evaporation of metal island films on the distal end of a fiber, or the depositing of metal film on fiber tips previously roughened by grinding [38,39]. The miniaturization of the optical fiber tip down to a few tens of nanometers by means of a tapering process recently made possible intracellular bioanalysis in single living human cells [40].

2.6 Chemical Sensing by Means of Refractometry

A large class of sensors is represented by sensors based on the change in the refractive index induced by the investigated analyte on a sensing layer. This approach avoids the labeling of the analyte with a luminescent compound and is employed particularly in biosensing. There are both advantages and disadvantages present in the label-based and label-free methods. The main advantage of the direct detection method is its ability to make analyses without the application of labels [41,42]. In fact, label-based biosensors require either multistep detection protocols or delicately balanced affinities of interacting biomolecules for displacement assays, which can cause sensor cross-sensitivity to nontarget analytes. However, for complex sample monitoring, the label-free methods continue to be susceptible to problems such as low sensitivity and increased background due to nonspecific binding. Therefore, in clinical applications in which the sample matrix can be very complex (e.g., whole blood or serum), the use of a label-based system may be preferable [43]. Moreover, any changes in the sensing surface, for example, due to nonspecific adsorption or fouling, give rise to a perturbation in the refractive index, which is the measured parameter in the label-free system.

2.6.1 Surface Plasmon Resonance

Among all the existing technology platforms that measure the RI of a surrounding medium, those based on surface plasmon resonance (SPR) [44] are the most widely used ones, also due to the presence of SPR-based devices on the market (Biacore system, first of all). Moreover, they are very attractive, because they are characterized by one of the best resolutions ever achieved, of the order of about 10^{-7} refractive index unit (RIU) [45].

In the past decade, SPR sensors have been extensively studied [46]. Various configurations of SPR sensors based on prism [47] and grating couplers [48], optical fiber [49], and integrated optical waveguides [50] have been described.

This physical phenomenon is based on a variation in the light reflected by a fine metallic layer as a result of the SPR. On the basis of the quantum theory, the surface plasmon is the particle that takes into account the quantization of the energy levels of the electromagnetic field that propagates along the boundary between a dielectric and a metal medium. At this interface, the resonance takes place when the momentum of the photons in the plane of the metallic layer matches that of the surface plasmon, k_{sp}. This momentum is a function of the complex dielectric function of the metal, $\varepsilon_m(\omega)$ and of the external layer, ε, according to the relationship

$$k_{sp} = \frac{\omega}{c} \sqrt{\frac{\varepsilon_m(\omega)\varepsilon}{\varepsilon + \varepsilon_m}} \qquad (2.13)$$

If the light is incident to the surface with an angle ϑ, then the wave vector of the light parallel to the surface is

$$k_x = \frac{\omega}{c} \sqrt{\varepsilon_g} \sin \vartheta \qquad (2.14)$$

where ε_g is the dielectric constant of the medium in which the incident light is propagating.

For angles that satisfy the condition of total reflection, an evanescent wave penetrates the metallic layer and, if the layer is sufficiently thin, interacts with the surface plasmon wave. A definite value of ϑ exists for which Equations 2.13 and 2.14, match each other. Experimentally, this resonance can be detected (1) with a monochromatic source, by observing the presence of a minimum in the light reflected changing the angle of incidence on the metal/optical guide interface; (2) with a polychromatic radiation, by observing the presence of a minimum in the reflected light as a function of the wavelength. Clearly, in both cases, this minimum also depends on the refractive index of the external medium ε. Therefore, the presence of a chemical species can be detected by following a variation in the refractive index. This technique is a label-free approach that fulfils the requirements of selectivity by designing a suitable selective layer on the metallic surface on the side of the external medium.

A more frequent use of metal nanoparticles has been witnessed in recent years with the creation of surface plasmons confined to well-defined spaces and no further propagating on metallic layers. The properties of these localized surface plasmon resonances (LSPR) depend on the nanoparticle shape: spherical nanoparticles exhibit a single resonance, whereas two resonances are observable in the case of nanorods, and multiple resonances occur in the case of nanoparticles with a triangular shape [51]. Gold and silver are the materials mainly used, since their resonance bands lie in the visible region. With the advent of nanotechnology and thanks to advanced fabrication techniques, arrays of nanoparticles have been increasingly proposed as optical platforms for chemical and biochemical sensing [52–56]. The interactions between the nanoparticles modify the shape and wavelength peaks of the resonance bands with near-field or far-field dipolar interactions that occur for distances among the nanoparticles smaller than or of the order of the wavelengths, respectively. Nanoholes or nanowells [57,58] created in thin metallic layers, as well as nanoparticles immobilized along optical fibers, are also being proposed for LSPR [59]. The use of nanoparticles and nanostructures in LSPR-based sensors opens up new perspectives, with a potential increase in the sensitivity of SPR-based sensors that can make it possible to achieve single-molecule detection [60,61].

2.6.2 Interference-Based Sensing

Generally speaking, an interferometer is an optical device that splits an optical input beam into two separate beams by means of suitable splitting optical element, redirects them by means of mirrors or waveguides, and recombines them on an optical detector. The intensity measured by an optical detector depends on the interference of two beams, that is, on their phase difference. If one of the interferometer's arms is exposed to some external influence that modifies the phase, this influence modifies the intensity read by the detector. Figure 2.5 provides a schematic drawing of a typical example of a Mach–Zender interferometer (MZI): an integrated optical waveguide is separated into two arms that recombine again into the output waveguide; a thin layer, capable of interacting with the investigated analyte, covers one of the two arms, while the other one is taken as reference for the compensation of other undesired effects (i.e., the influence of temperature). The interaction with the analyte induces a refractive index change on the surface of the sensing arm, and this produces an optical path change in this arm and, consequently, a phase shift and an intensity modulation at the interferometer output. Typically, the sensitivity will increase by increasing the length of the sensing arm; however, since the intensity at the output has a sinusoidal trend, the interferometer may be characterized by low sensitivities if the intensity is at the

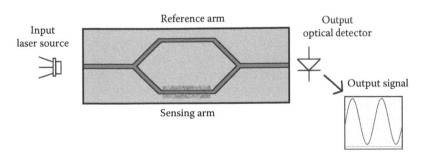

FIGURE 2.5 Mach–Zender integrated optics interferometer.

maximum or minimum value (where the first derivative is zero), and by high sensitivities at half the intensity, in which case the derivative is maximum.

One of the first examples of the application of biosensing consists of an MZI integrated on a silicon substrate with a Si_3N_4 waveguide and etched gratings for input/output coupling [62]. The proposed device can quantitatively and selectively detect a bulk concentration of 5×10^{-11} M of human chorionic gonadotropin (hCG). After this first application, many MZI-based biosensors have been proposed, thanks to improvements in the optical waveguide technologies [63,64] and to the use of new alternative materials [65].

An alternative interferometric technique, called reflectometric interference spectroscopy (RIfS), is based on the interference generated by partial beams reflected by two surfaces of a thin layer (hundreds of nm) that is exposed to the analyte and deposited on a substrate (generally glass) [66]. The interaction between the parameter and the sensing layer causes a change in its optomechanical properties (thickness and/or refractive index). This gives rise to a change in the interference pattern being measured and can be correlated to the concentration of the interacting parameter. In the first prototype, the optical system consisted of a visible light source as the optical source and of a diode array spectrometer to record the whole spectrum. The coupling between the source/detector and the sensing layer was carried out by means of plastic optical fibers; in this way, a sensor array with different sensing layers on the same glass substrate could be implemented. This system was tested for the detection of hydrocarbons, which caused a swelling in the sensing polymeric layer that consisted of polysiloxane films [66], and for the detection of IgG, with the immunoreactions carried out on thin films of synthetic silica [67]. An implemented version of the RIfS makes use of four light-emitting diodes, driven sequentially, which are able to provide sufficient information on the modification of the interference pattern. In this way, a single photodiode was used as the photodetector, instead of the diode array spectrophometer, and this led to a simplification of the whole system. The system was utilized for the determination of volatile organic compounds [68] and was used for a thorough characterization of polymeric sensing layers [69].

2.6.3 Optical Ring Microresonators

Optical ring microresonators are known since more than one decade as optical components for light switching, amplification, and modulation [70]. In a ring resonator, the light is confined thanks to total internal reflection and, due to constructive interference, only well-defined resonant modes can be supported by the resonator. The strong confinement of the light near its surface makes this optical element very promising also for chemical and biochemical sensing [71]. The frequency shift of the resonance modes (the so-called whispering-gallery modes; WGM) caused by the adsorption of the investigated parameters on the surface of the resonators was described for the first time applied to silica microspheres [72]. The excitation of the cavity takes place by means of the evanescent coupling between the microsphere and an optical fiber through which the light emitted by a tunable laser is transmitted. The WGM circulates about the equator of the cavity and the quality factor Q of the resonance is of the order of 10^6–10^8 depending on the quality of the surface of the microspheres. In a silica glass microsphere characterized by a quality factor of 10^8, the presence of a resonance at 1.5 μm corresponds to a lifetime of the resonant wave of 0.1 μs. If a silica glass microsphere with a diameter of 200 μm is considered, the distance that the resonant wave covers is roughly 20 m, with the light within a WGM, which circumnavigates the equator roughly 60,000 times. This gives an idea of the potential sensitivity of the approach, due to the multiple interactions with the external medium along a very long distance. Shift of the order of picometers or less can be easily appreciated with a Q factor of 10^8, which corresponds to a frequency line width of 1 MHz at 1.5 μm, equivalent in terms of wavelength to 1×10^{-14} m. Deposition of layers with atomic thickness on the surface can give a detectable shift of a given resonance frequency. From a theoretical point of view, single-protein detection is also possible [73]. Besides microspheres, other different optical configurations were proposed in the last years. Ring or disk resonators implemented on silicon substrates have the advantage of mass-production thanks to nanofabrication techniques, with the possibility of the implementation of arrays for multianalyte detection [74–77]. In biosensing, where there is generally the necessity of microfluidics for sampling, washing, and mixing, one of the most suitable configurations is constituted by capillary-based resonators [78,79], where the resonance occurs on the transverse section of a glass capillary through which the sample flows.

2.7 Conclusions

In this chapter, the basics of luminescence and absorption chemical sensing and the main refractometric approaches were described. Cantilever-based sensors were not described, since their basic theory is described in detail in Chapter 5.

REFERENCES

1. Yuan, P. and D. R. Walt. 1987. Calculation for fluorescence modulation by absorbing species and its application to measurements using optical fibres. *Anal. Chem.* 59: 2391–2394.
2. Demschenko, A. P. 2009. Mechanism of signal transduction, Chapter 4. In *Introduction to Fluorescence Sensing*, ed. Demschenko, A. P., pp. 249–297. Berlin, Germany: Springer.
3. He, H., Mortellaro, M. A., Leiner, M. J. P., Young, S. T., Fraatz, R. J., and J. K. Tusa. 2003. A fluorescent chemosensor for sodium based on photoinduced electron transfer. *Anal. Chem.* 75: 549–555.
4. He, H., Mortellaro, M. A., Leiner, M. J. P., Fraatz, R. J., and J. K. Tusa. 2003. A fluorescent sensor with high selectivity and sensitivity for potassium in water. *J. Am. Chem. Soc.* 125: 1468–1469.
5. Doose, S., Neuweiler, H., and M. Sauer. 2009. Fluorescence quenching by photoinduced electron transfer: A reporter for conformational dynamics of macromolecules. *Chem. Phys. Chem.* 10: 1389–1398.
6. Jordan, D. M., Walt, D. R., and F. P. Milanovich. 1987. Physiological pH fiber-optic chemical sensor based on energy transfer. *Anal. Chem.* 59: 437–439.
7. Misra, V., Misra, H., Joshi, H. C., and T. C. Pant. 2000. Excitation energy transfer between acriflavine and rhoidamine 6G as a pH sensor. *Sens. Actuat. B* 63: 18–23.
8. Forster, T. 1965. Action of light and organic crystals. In *Modern Quantum Chemistry*, ed. Sinanoglu, O., Part III, pp. 93–137. New York: Academic Press.
9. Szmacinski, H. and J. R. Lakowicz. 1995. Fluorescence lifetime-based sensing and imaging. *Sens. Actuat. B* 29: 16–24.
10. Kieslinger, D., Draxler, S., Trznadel, K., and M. E. Lippitsch. 1997. Lifetime-based capillary waveguide sensor instrumentation. *Sens. Actuat. B* 39: 300–304.
11. Koronczi, I., Reichert, J., Ache, H.-J., Krause, C., Werner, T., and O. S. Wolfbeis. 2001. Sensors for ion detection based on measurement of luminescence decay time. *Sens. Actuat. B* 74: 47–53.
12. Holst, G. and B. Grunwald. 2001. Luminescence lifetime imaging with transparent oxygen optodes. *Sens. Actuat. B* 74: 78–90.
13. Huber, C., Klimant, I., Krause, C., and O. S. Wolfbeis. 2001. Dual lifetime referencing as applied to a chloride optical sensor. *Anal. Chem.* 73: 2097–2103.
14. Von Bueltzingsloewen, C., McEnvoy, A. K., McDonagh, C., MacCraith, B. D., Klimant, I., Krause, C., and O. S. Wolfbeis. 2002. Sol-gel based optical carbon dioxide sensor employing dual luminophore referencing for application in food packaging technology. *Analyst* 127: 1478–1483.
15. Klimant, I., Huber, C., Liebsch, G., Neurauter, G., Stangelmayer, A., and O. S. Wolfbeis. 2001. Dual lifetime referencing (DLR)—A new scheme for converting fluorescence intensity into a frequency-domain or time-domain information. In *New Trends in Fluorescence Spectroscopy*, eds. Valeur, B. and J. C. Brochon, pp. 257–274. Berlin, Germany: Springer Verlag.
16. Roda, A. 2011. A history of bioluminescence and chemiluminescence from ancient times to the present. In *Chemiluminescence and Bioluminescence, Past, Present and Future*, ed. A. Roda, Cambridge, U.K.: Royal Society of Chemistry.
17. Blum, L. J. and S. M. Gautier. 1991. Bioluminescence- and chemiluminescence- based fiberoptic sensors. In *Biosensor Principles and Applications*, eds. Blum, L. J. and P. R. Coulet, pp. 213–247. New York: Marcel Dekker, Inc.
18. Coulet, P. R. and L. J. Blum. 1991. Luminescence in biosensor design. In *Biosensors with Fiberoptics*, eds. Wise, D. L. and L. B. Wingard, Jr., pp. 293–324. Clifton, NJ: The Hummana Press Inc.
19. Blum, L. J. and C. A. Marquette. 2006. Chemiluminescence-based sensors. In *Optical Chemical Sensors*, eds. Baldini, F., Chester, A. N., Homola, J., and S. Martellucci, pp. 157–178. Dordrecht, the Netherlands: Springer.
20. White, E. H. and M. M. Bursey. 1964. Chemiluminescence of luminol and related hydrazides: The light emission step. *J. Am. Chem. Soc.* 86: 941–942.

21. Marquette, C. A. and L. J. Blum. 2006. Applications of the luminol chemiluminescent reaction in analytical chemistry. *Anal. Bioanal. Chem.* 385: 546–554.

22. Aitken, R. J., Buckingham, D. W., and K. M. West. 1992. Reactive oxygen species and human spermatozoa: Analysis of the cellular mechanisms involved in luminol- and lucigenin-dependent chemiluminescence. *J. Cell. Physiol.* 151: 466–477.

23. Huang, H. and J. J. Zhu. 2009. DNA aptamer-based QDs electrochemiluminescence biosensor for the detection of thrombin. *Biosens. Bioelectron.* 25: 927–930.

24. Martin, A. F. and T. A. Nieman. 1997. Chemiluminescence biosensors using tris(2,2′-bipyridyl) ruthenium(II) and dehydrogenases immobilized in cation exchange polymers. *Biosens. Bioelectron.* 12: 479–489.

25. Marquette, C. A. and L. J. Blum. 2008. Electro-chemiluminescent biosensing. *Anal. Bioanal. Chem.* 390: 155–168.

26. Beigi, R., Kobatake, E., Aizawa, M., and G. R. Dubyak. 1999. Detection of local ATP release from activated platelets using cell surface-attached firefly luciferase. *Am. J. Physiol. Cell Physiol.* 276: C267–C278.

27. Cruz-Aguado, J. A., Chen, Y., Zhang, Z., Elowe, N. H., Brook, M. A., and J. D. Brennan. 2004. Ultrasensitive ATP detection using firefly luciferase entrapped in sugar-modified sol-gel-derived silica. *J. Am. Chem. Soc.* 126: 6878–6879.

28. Gautier, S. M., Blum, L. J., and P. R. Coulet. 1990. Multi-function fibre-optic sensor for the bioluminescent flow determination of ATP or NADH. *Anal. Chim. Acta* 235: 243–253.

29. Girotti, S., Ghini, S., Carrea, G., Bovara, R., Roda, A., and R. Budini. 1991. Bioluminescent flow sensor for D-(–)-lactate. *Anal. Chim. Acta* 255: 259–268.

30. Hanlon, E. B., Manoharan, R., Koo, T.-W., Shafer, K. E., Motz, J. T., Fitzmaurice, M., Kramer, J. R., Itzkan, I., Dasari, R. R., and M. S. Feld. 2000. Prospects for in vivo Raman spectroscopy. *Phys. Med. Biol.* 45: R1–R59.

31. Moskovits, M. 1985. Surface-enhanced spectroscopy. *Rev. Mod. Phys.* 57: 783–826.

32. Bello, J. M. and T. Vo-Dinh. 1990. Surface enhanced Raman scattering fiber-optic sensor. *Appl. Spectr.* 44: 63–69.

33. Garrell, R.L. 1989. Surface-enhanced Raman spectroscopy. *Anal. Chem.* 61: 401A–411A.

34. Vo-Dinh, T., Stokes, D. L., Griffin, G.D., Volkan, M., Kim, U.J., and M.I. Simon. 1999. Surface-enhanced Raman Scattering (SERS) method and instrumentation for genomics and biomedical analysis. *J. Raman Spectr.* 30: 785–793.

35. Otto, S. 2006. On the significance of Shalaev's 'hot spots' in ensemble and single-molecule SERS by adsorbates on metallic films at the percolation threshold. *J. Raman Spectr.* 37: 937–947.

36. Surbhi, L., Link, S., and J. H. Naomi. 2007. Nano-optics from sensing to waveguiding. *Nature Phot.* 1: 641–648.

37. Stokes, D. L., Alarie, J. P., and T. Vo-Dinh, Surface-enhanced Raman fiberoptic sensors for remote monitoring. *Proc. SPIE Environ. Monitoring Hazardous Waste Site Remed.* 2504: 552–558.

38. Viets, C. and W. Hill. 1998. Comparison of fibre-optic SERS sensors with differently prepared tips. *Sens. Actuat. B* 51: 92–99.

39. MacDonald, H. L., Jorgenson, R. C., Schoen, C. L., Smith, B. F., Yee, S. S., and K. I. Mullen. 1994. Improving surface-enhanced Raman spectroscopy (SERS) for better sensors. *Proc. SPIE Chem. Biochem. and Environ. Fiber Sens. VI* 2293: 198–208.

40. Scaffidi, J. P., Gregas, M. K., Seewaldt, V., and T. Vo-Dinh. 2009. SERS-based plasmonic nanobiosensing in single living cells. *Anal. Bioanal. Chem.* 393: 1135–1141.

41. Proll, G., Steinle, L., Proll, F., Kumpf, M., Moehrle, B., Mehlmann, M., and G. Gauglitz. 2007. Potential of label-free detection in high-content-screening applications. *J. Chromatogr. A* 1161: 2–8.

42. Fan, X., White, I. M., Shopova, S. I., Zhu, H., Suter, J. D., and Y. Sun. 2008. Sensitive optical biosensors for unlabeled targets: A review. *Anal. Chim. Acta* 620: 8–26.

43. Moreno-Bondi, M. C., Taitt, C. R., Shriver-Lake, L. C., and F. S. Ligler. 2006. Multiplexed measurement of serum antibodies using an array biosensor. *Biosens. Bioelectron.* 21: 1880–1886.

44. Homola, J. 2008. Surface plasmon resonance sensors for detection of chemical and biological species. *Chem. Rev.* 108: 462–493.

45. Piliarik, M. and J. Homola. 2009. Surface plasmon resonance (SPR) sensors: approaching their limits? *Opt. Lett.* 17: 16505–16517.

46. Homola, J. 2003. Present and future of surface plasmon resonance biosensors. *Anal. Bioanal. Chem.* 377: 528–539.
47. Liedberg, B., Lundstrom, I., and E. Stenberg. 1993. Principles of biosensing with an extended coupling matrix and surface plasmon resonance. *Sens. Actuat. B* 11: 63–72.
48. Jory, M. J., Vukusic, P. S., and J. R. Sambles. 1994. Development of a prototype gas sensor using surfaces plasmon resonance on gratings. *Sens. Actuat. B* 17: 1203–1209.
49. Homola, J. 1995. Optical fiber sensor based on surface plasmon excitation. *Sens. Actuat. B* 29: 401–405.
50. Harris, R. D. and J. S. Wilkinson. 1995. Waveguide surface plasmon resonance sensors. *Sens. Actuat. B* 29: 261–267.
51. Stewart, M. E., Anderton, C. R., Thompson, L. B., Maria, J., Gray, S. K., Rogers, J. A., and R. G. Nuzzo. 2008. Nanostructured plasmonic sensors. *Chem. Rev.* 108: 494–521.
52. Nath, N. and A. Chilkoti. 2004. Label-free biosensing by surface plasmon resonance of nanoparticles on glass: Optimization of nanoparticle size. *Anal. Chem.* 76: 5370–5378.
53. Frederix, F., Friedt, J. M., Choi, K. H., Laureyn, W., Campitelli, A., Mondelaers, D., Maes, G., and G. Borghs. 2003. Biosensing based on light absorption of nanoscaled gold and silver particles. *Anal. Chem.* 75: 6894–6900.
54. Fujiwara, K., Watarai, H., Itoh, H., Nakahama, E., and N. Ogawa. 2006. Measurement of antibody binding to protein immobilized on gold nanoparticles by localized surface plasmon spectroscopy. *Anal. Bioanal. Chem.* 386: 639–644.
55. Kim, S. A., Byun, K. M., Kim, K., Jang, S. M., Ma, K., Oh, Y., Kim, D., Kim, S. G., Shuler, M. L., and S. J. Kim. 2010. Surface-enhanced localized surface plasmon resonance biosensing of avian influenza DNA hybridization using subwavelength metallic nanoarrays. *Nanotechnology* 21: 355503.
56. Choi, I. and Y. Choi. 2012. Plasmonic nanosensors: Review and prospect. *IEEE J. Sel. Topics Quantum Electron* 18: 1110–1221.
57. Brolo, A. G., Gordon, R., Leathem, B., and K. L. Kavanagh. 2004. Surface plasmon sensor based on the enhanced light transmission through arrays of nanoholes in gold films. *Langmuir* 20: 4813–4815.
58. Eftekhari, E., Escobedo, C., Ferreira, J., Duan, X., Girotto, E. M., Brolo, A. G., Gordon, R., and D. Sinton. 2009. Nanoholes as nanochannels: flow-through plasmonic sensing. *Anal. Chem.* 81: 4308–4311.
59. Chau, L. K., Lin, Y. F., Cheng, S. F., and J. T. Lin. 2006. Fiber-optic chemical and biochemical probes based on localized surface plasmon resonance. *Sens. Actuat. B* 113: 100–105.
60. Anker, J. N., Hall, W. P., Lyandres, O., Shah, N. C., Zhao, J., and R. P. Van Duyne. 2008. Biosensing with plasmonic nanosensors. *Nat. Mater.* 7: 442–453.
61. Chen, S., Svedendahl, M., Van Duyne, R. P., and M. Käll. 2011. Plasmon-enhanced colorimetric ELISA with single molecule sensitivity. *Nano Lett.* 11: 1826–1830.
62. Heideman, R. G., Kooyman, R. P. H., and J. Greve. 1993. Performance of a highly sensitive optical waveguide Mach-Zehnder interferometer immunosensor. *Sens. Actuat. B* 10: 209–217.
63. Heideman, R. G. and P. V. Lambeck. 1999. Remote opto-chemical sensing with extreme sensitivity: Design, fabrication and performance of a pigtailed integrated optical phase-modulated Mach–Zehnder interferometer system. *Sens. Actuat. B* 61: 100–127.
64. Prieto, F., Lechuga, L. M., Calle, A., Llobera, A., and C. Domínguez. 2001. Optimized silicon antiresonant reflecting optical waveguides for sensing applications. *J. Lightwave Technol.* 19: 75–83.
65. Brucka, R., Melnika, E., Muellnera, P., Hainbergera, R., and M. Lämmerhoferb. 2011. Integrated polymer-based Mach-Zehnder interferometer label-free streptavidin biosensor compatible with injection molding. *Biosens. Bioelectron.* 26: 3832–3837.
66. Gauglitz, G., Brecht, A., Kraus, G., and W. Nahm. 1993. Chemical and biochemical sensors based on interferometry at thin (multi-) layers. *Sens. Actuat. B* 11: 21–27.
67. Brecht, A., Gauglitz, G., and J. Polster. 1993. Interferometric immunoassay in a FIA-system: A sensitive and rapid approach in label-free immunosensing. *Biosens. Bioelectron.* 8: 387–392.
68. Reichl, D., Krage, R., Krummel, C., and G. Gauglitz. 2000. Sensing of volatile organic compounds using a simplified reflectometric interference spectroscopy setup. *Appl. Spectr.* 54(4): 583–586.
69. Dieterle, F., Belge, G., Betsch, C., and G. Gauglitz. 2002. Quantification of the refrigerants R22 and R134a in mixtures by means of different polymers and reflectometric interference spectroscopy. *Anal. Bioanal. Chem.* 374: 858–867.

70. Little, B. E. 2003. Advances in microring resonators. *Proceedings of Integrated Photonics Research Conference (IPR 2003)*, Washington, DC, pp. 165–167.
71. Sun, Y. and X. Fan. 2011. Optical ring resonators for biochemical and chemical sensing. *Anal. Bioanal. Chem.* 399: 205–211.
72. Serpengueze, A., Arnold, S., and G. Griffel. 1994. Excitation of resonances of microspheres on an optical fiber. *Opt. Lett.* 20: 654–656.
73. Vollmer, F. and S. Arnold. 2008. Whispering-gallery-mode biosensing: Label-free detection down to single molecules. *Nat. Methods* 5: 591–596.
74. Chao, C.-Y. and L. J. Guo. 2003. Biochemical sensors based on polymer microrings with sharp, asymmetrical resonance. *Appl. Phys. Lett.* 83: 1527–1529.
75. Yebo, N. A., Lommens, P., Hens, Z., and R. Baets. 2010. An integrated optic ethanol vapor sensor based on a silicon-on-insulator microring resonator coated with a porous ZnO film. *Opt. Expr.* 11: 11859–11866.
76. Armani, A. M. and K. J. Vahala. 2006. Heavy water detection using ultra-high-Q microcavities. *Opt. Lett.* 31: 1896–1898.
77. Gylfason, K. B., Carlborg, C. F., Kazmierczak, A., Dortu, F., Sohlström, H., Vivien, L., Barrios, C. A., van der Wijngaart, W., and G. Stemme. 2010. A packaged optical slot-waveguide ring resonator sensor array for multiplex label-free assays in labs-on-chips. *Lab. Chip* 3: 281–290.
78. White, I. M., Oveys, H., and X. Fan. 2006. Liquid-core optical ring-resonator sensors. *Opt. Lett.* 31: 1319–1321.
79. Zhu, H., White, I. M., Suter, J. D., Zourob, M., and X. Fan. 2007. Integrated refractive index optical ring resonator detector for capillary electrophoresis. *Anal. Chem.* 79: 930–937.

3

Optical Chemical Sensors: A Look Back

Dorota Wencel and Colette McDonagh

CONTENTS

3.1 Introduction and Overview

3.1.1 Introduction and Scope of the Review

Research in optical chemical sensors has been growing steadily over the past 50 years, stretching back to the latter part of the last century. The definition coined by Wolfbeis and coworkers (Cammann et al. 1995, Wolfbeis 2004a) states that *Chemical sensors are miniaturized devices that can deliver real-time and online information on the presence of specific compounds or ions in even complex samples.* Figure 3.1 shows a schematic of the main elements of a chemical sensor, namely, the analyte, the transduction platform, and data processing, leading to the display of analyte concentration.

This review is concerned exclusively with sensors that use optical transduction techniques. In analyzing past literature on this topic, it can be observed that there has been a dramatic increase in output since the late 1990s, which reflects the growing interest in optical biosensors. Furthermore, within this topic, surface plasmon resonance (SPR)-based sensors have played a prominant role. It has been necessary to restrict the scope of this review largely to nonbiological sensors and to limit the optical transduction techniques to that of absorption and luminescence. Hence, sensors using optical properties such as reflectance and refractive index, for example SPR sensors, will be largely excluded. A short section on Raman and surface-enhanced Raman (SERS)-based sensors will be included and will refer to recent key review articles. It is to be noted that Chemiluminescence sensors will be the subject of Chapter 5 while Chapter 8 will provide some background on SPR sensors for biosensor applications while introducing the area of nano-SPR sensors.

In this review, we divide optical chemical sensors into two main categories: direct sensors and indirect or reagent-based sensors. By far, the larger part of the chapter will deal with the latter category and will include both absorption- and luminescence-based sensors for a broad range of analytes such as oxygen (O_2), carbon dioxide (CO_2), pH, cations and anions, volatile organic compounds (VOCs), and gases such as ammonia (NH_3). While it will be assumed that different sensing principles and platforms have been dealt with in detail in Chapter 2, issues in relation to platforms, reagents, encapsulation matrices will be referred to here under specific analyte headings.

Leading on from Chapter 2, the last two parts of this section will provide brief overviews of sensor principles and platforms and sensor reagents and encapsulation matrices. In order to act as a backdrop to the main body of the review, which will deal largely with developments since 2000, we also provide a brief look back to the last century and chart the development of the optical chemical sensors area, dealing mainly with the period 1980–2000. In so doing, we acknowledge the significant contributions

FIGURE 3.1   Principal stages of the operation of a sensor.

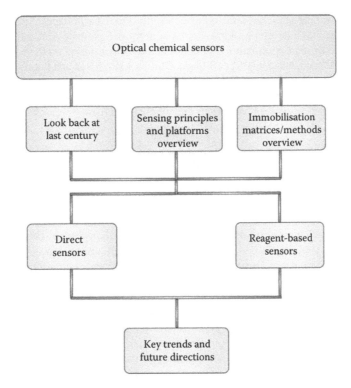

FIGURE 3.2 Structure of review.

of major players in this field who have paved the way for the more recent developments and for the rapid expansion of the field to encompass optical chemical nanosensors, which is the subject of this book. The structure adopted for this review is shown in Figure 3.2.

3.1.2 A Brief Look Back at the Last Century (1980–2000)

Prior to ~1980, the main optical chemical sensors in use were for O_2 and pH and were in the reagent-based sensing category. Indicator strips based on cellulose were in use and pH indicator dyes were immobilized on glass in 1975 (Harper 1975). The birth of optical O_2 sensing can be traced to Kautsky and Hirsch in the 1930s (Kautsky and Hirsch 1931, Kautsky 1939) when they reported O_2-dependent phosphorescence of dyes adsorbed on silica while the first optical O_2 sensor system was described by Bergman (1968). Other early pioneers include Lubbers and Opitz (Opitz and Lubbers 1975) who developed systems for O_2, CO_2, and pH and who coined the word optode, and Hirschfeld (Hirschfeld et al. 1983) who used fiber light guides to measure uranium. Early review articles include those by Kirkbright et al. (1984) and Hirschfeld et al. (1984).

The upsurge of activity in optical chemical sensors from ~1980 onward was partly related to the major advances in optoelectronics, which made available miniaturized light sources such as light emitting diodes (LEDs) and diode lasers and photodiode and CCD detectors. As stated earlier, optical O_2 sensing was one of the main areas where early developments took place. The reader is directed to a comprehensive review of the early developments in optical chemical sensing by Wolfbeis (2004a) in which is summarized the key publications in this area, leading up to the end of the century. From this list, key pioneers in the area can be identified, many of whom are still active. Two very important applications of ion sensing are in the clinical and environmental areas. The main clinical ions are the cations K^+, Na^+, Ca^{2+} while ions such as Hg^{2+}, Pb^{2+}, Cl^-, NO_2, etc. are key environmental analytes. The ion-sensing detection principle of co-extraction goes back to the 1980s when Charlton et al. used a PVC film and valinomycin as an ion carrier to extract K^+ into the film with the same quantity of a red dye, erythrosine whose absorbance was related to the extracted K^+ concentration (Charlton and Fleming 1982). The Seitz group (Saari and Seitz 1983) developed a low-cost system for detection of Al^{3+} and the same group also developed a fluorescence-based ion-pair extraction scheme for Na^+ based on the use of crown ethers as molecular recognition elements (Zhujun et al. 1986). Chromoionophores and fluoroionophores have been and still are used to detect ions such as K^+ (Voegtle 1996). A fluorescence-based scheme for nitrate detection using a polarity-sensitive dye was published in 1995 by the Wolfbeis group (Mohr and Wolfbeis 1995). This scheme can be used for both cations and anions. The main developments in ion sensing during this period are included in the review by deSilva et al. (1997).

3.1.3 Overview of Sensor Principles and Platforms

It is intended that this section will act as a general contextual background for the review of recent sensor developments to be discussed in the remainder of this chapter. The reader is referred to Chapter 2, for a more in-depth treatment of sensor mechanisms and optical platforms.

This review will focus mainly on optical absorption and luminescence transduction methods, the principles of which are well established. In general terms, optical luminescence-based sensing is intrinsically more sensitive than that of absorption. However, not all analytes emit luminescence while many have intrinsic absorption bands which can be detected. Infra-red (IR) spectroscopy has been widely used for the direct detection of gases such as CO_2, NH_3, and methane (CH_4) as the energy-level structure of these gases facilitate transitions in this part of the spectrum (Reyes et al. 2006, de Castro et al. 2007). On the other hand, hydrocarbons and many VOCs have intrinsic absorption in the UV (Velasco et al. 2007). Reagent-based sensing provides more flexibility than these direct-sensing methods in that the absorption or luminescence properties of the reagent can be selected to match available light sources in the visible or IR. Analytes such as O_2 are detected via a collisional quenching mechanism (McDonagh et al. 2001), while ion sensing often involves mechanisms such as co-extraction as discussed earlier (Morf et al. 1989).

As outlined in Section 3.1.1, the sensor transduction platform is an integral part of any optical chemical sensor design and the overall sensor performance is intimately linked to the design and performance of the sensing platform, whether for absorbance- or luminescence-based sensing. The reader is directed to Chapter 2 for a detailed treatment of this topic. An overview is provided here. In the early days of optical chemical sensors research, fiber-optic sensor platforms were widely used and there have been numerous pre-2000 review articles and books published in this area, which are contained in the Wolfbeis (2004a) reference. Both fiber-optic and waveguide-based sensors involve guiding of the light which is confined in a high-refractive-index medium (fiber core or planar waveguide layer). Passive fiber-optic sensors are those in which the sensor response is not linked to a change in the optical properties of the fiber, rather the fiber is used to transport the light to and from the analyte. In the case of active fiber sensing, the fiber is modified in some way, for example, by the encapsulation of an analyte-sensitive dye in the cladding, whereby the fiber participates in the sensor transduction process. Fiber-optic sensors is still an active field and in the last decade, there have been many review articles, particularly in the area of biosensors. These include reviews by Lin (2000) on pH sensors, Marazuela and Moreno-Bondi (2002), Monk and Walt (2004), Bosch et al. (2007), Leung et al. (2007), Orellana and Haigh (2008), and Podoleanu (2010) on biosensors, Vo-Dinh and Kasili (2005) and Del Villar et al. (2008) on nanosensors, Elosua et al. (2006) on VOC sensors, Yang et al. (2002), and Peters (2011) on polymer fiber sensors. In recent years, the range of sensor platforms has broadened out to include planar waveguide platforms, some of which integrate multiple functionalities on a single chip (Lambeck 2006). It could be argued that planar waveguide platforms are more robust and are more compatible with current advanced microfabrication technologies toward integrated microfluidic and lab-on-a-chip systems. However, fiber-optic platforms play a key role in distributed sensing (Culshaw 2004) and also in in vivo biosensing (Utzinger and Richards-Kortum 2003). Many planar waveguide sensors are based on measurement of refractive index (Lambeck 2006). However, there have been some key developments over the last decade or so in absorbance- and luminescence-based systems. Puyol et al. have developed integrated absorbance waveguide systems based on a polyvinyl chloride (PVC) sensor membrane between two curved antiresonant reflecting waveguides (Puyol et al. 2002). Burke et al. in 2004 used an injection-molded waveguide platform incorporating integrated refractive optical elements for NH_3 sensing. Duveneck et al. in 2002 used a multiple-reflection, evanescent excitation system in a fluorescence-based platform, which led to the commercial Zeptosens microarray platform for biomarker discovery. Yimit et al., in 2005, have reviewed thin-film composite optical waveguides for applications such as gas sensing and immunosensing. Over the last decade, the Ligler group (Ligler et al. 2007, Taitt et al. 2008) has developed a suite of fluorescence-based array biosensors which are portable and automated for sensitive detection of multiple targets in applications such as food safety and homeland security. Other relevant reviews of sensor platforms include that of Kuswandi et al. in 2007 on microfluidic platforms, Lambeck (2006) on integrated optical sensor platforms, Law et al. (2004) on refractometric nanowires, and waveguiding platforms by Jiang et al. (2010) and Baldini and Giannetti (2005).

3.1.4 Overview of Sensor Reagents and Immobilization Matrices

When an analyte does not exhibit a convenient spectroscopic absorbance or luminescence, sensing can be achieved indirectly by monitoring the analyte-sensitive response of an intermediate reagent or dye. This is well illustrated by the sensing principle of luminescence-based O_2 sensors where the luminescence intensity and lifetime of O_2-sensitive ruthenium or porphyrin complexes is quenched in the presence of O_2 by a collisional quenching mechanism (McDonagh et al. 2001, O'Mahony et al. 2006). Many optical pH sensors are based on the change in optical absorbance of pH-sensitive indicators such as bromocresol purple (BCP) and bromothymol blue (BTB) (Lobnik 2006), while many fluorescence-based pH sensors use fluorescence of 1-hydroxypyrene- 3,6,8-trisulfonic acid (HPTS) or derivatives thereof, which have absorption and emission bands that are pH-sensitive via protonation and deprotonation processes (Nivens et al. 2002, Kermis et al. 2003). In particular, HPTS can be used in a ratiometric detection mode whereby the ratio of absorption (using dual light sources, e.g., LEDs) or emission bands (using dual photodiodes) can be detected as a self-referenced pH-dependent signal that is immune to interferences such as source or detector drift (Wencel et al. 2009). The principle of fluorescence-based CO_2 sensing is also based on protonation and deprotonation of a pH-sensitive dye using the Severinghaus principle, which involves conversion of CO_2 to carbonic acid (Mills and Eaton 2000, Cajlakovic et al. 2006), and dyes such as fluorescein and HPTS have been used. Sensors for cations often use metal-complexing ligands that are entrapped in molecularly imprinted polymers (MIPS) or in crown ether complexes (Ng and Narayanaswamy 2006). Co-extraction, ion exchange, and chromoionophores and fluoroionophores have also been used for both cations and anions. Looking forward, there is an increasing interest in the use of quantum dots as fluorescent indicators (Costa-Fernandez et al. 2006), particularly for biosensor applications and also in the use of fluorescent nanoparticles (Coto-Garcia et al. 2011).

The reagent is usually immobilized in a solid matrix, for example, a polymer or sol–gel monolith or thin film. The role of the matrix is to entrap the reagent while allowing accessibility of the analyte. The reagent can be physically or covalently entrapped. Covalent immobilization can eliminate reagent leaching and provide more flexibility with regard to matrix structure and morphology. The two main matrix types used in optical chemical sensing are sol–gel and polymer materials. The sol–gel matrix provides a nanoporous, low-density glass-like structure in which reagents are entrapped and into which analyte molecules can diffuse. The sol–gel process is highly tailorable via processing parameters such as sol pH, precursor type and concentration, water content, and curing temperature. The matrix can be rigid and glass-like using precursors such as tetraethoxysilane (TEOS) or it can be more flexible and polymer-like by using the so-called organically modified precursors, for example, methyltriethoxysilane (MTEOS) or propyltriethoxysilane (PTEOS) (MacCraith and McDonagh 2002, Tran-Thi et al. 2011).

Polymers have been widely used as supports for optical chemical sensors since early in the development of this field. They have many desirable properties and compare well with sol–gels as stable, robust supports. There are indications that some reagents are more photostable in sol–gel supports than in polymers (Dunn and Zink 1991). On the other hand, polymers are more stable for applications where high-temperature treatments are required, for example, autoclavation (Voraberger et al. 2001). Some widely used polymer matrices include polystyrene (PS), PVC, poly(methyl methacrylate) (PMMA), and polydimethyl siloxanes (PDMS) (Mohr 2006).

3.2 Direct Spectroscopic Sensors

3.2.1 Introduction

Direct optical sensors measure an intrinsic property of the analyte, for example, some gas sensors measure the intrinsic optical absorbance of the gas while some biological sensors measure the intrinsic fluorescence of specific proteins. While the bulk of this review will be concerned with reagent-mediated optical chemical sensors, a short review of direct sensors is included. Because of the broad definition of this type of sensor and the many transduction techniques that are currently in use, we have selected a limited number of topics under the general heading of spectroscopic sensing. Some recent developments in optical absorption-based sensors will be discussed, which will include non-dispersive infrared (NDIR) sensors and Fourier Transform infrared (FTIR) sensors and IR absorption sensors based on quantum cascade laser (QCL) sources. A short account will be given of progress in direct fluorescence sensors, focusing on a few key applications. A short section on Raman and surface-enhanced Raman spectroscopy (SERS)-related sensing will include some recent developments and reference to review articles.

3.2.2 Absorption-Based Sensors

3.2.2.1 IR Absorption

IR spectroscopy is widely used for detection of gases, for example, CO_2, CO, NO_2, NH_3, and CH_4. The technique in its simplest form consists of illuminating the gas sample, usually contained in a sample cell and measuring the absorption at specific wavelengths which are characteristic of the vibrational modes of the molecule. The fundamental absorption of many molecules occurs in the near and mid-IR from 1 to >10 µm. Gas analysis using direct-sensing methods has applications in key areas such as control of industrial processes such as combustion and in environmental monitoring. NDIR and FTIR sensing are two of the most widely used approaches for IR sensing.

3.2.2.2 NDIR Sensors

The main components of NDIR sensors are a sample chamber, IR light source, often a lamp, a filter which selects the relevant analyte absorption range, and an IR detector. In its simplest form, the absorbed intensity is measured, which is inversely proportional to the analyte concentration according to Beer's Law. These sensors can be used to measure the concentration of a range of organic and inorganic gases but are most commonly used for CO_2 monitoring and many commercial NDIR sensor platforms are on the market for measuring CO_2 for applications such as climate change, automotive emissions, and indoor air quality. NDIR analyzers have been used to monitor CO_2 concentration in the ocean by extracting the gas from a water sample, giving a precision of ~0.05% and an accuracy of ~0.02% (Kaltin et al. 2005). Bandstra et al. (2006) demonstrate high-speed continuous analysis using a gas-permeable hydrophobic membrane contactor, which continuously strips the gas from a flowing stream of seawater with a precision >0.1% and a response time of 6 s. A high-precision spectroscopic system for measuring CO_2 in the harsh environmental conditions of automotive applications was developed, which used a second IR source for internal recalibration of the primary source. The system accommodated the wide temperature and pressure ranges for this application and yielded an accuracy of ~±5% over the complete temperature, pressure, and humidity ranges with a resolution of <15 ppm, a response time of 5 s, and a sensor operation time of >6000 h over 15 years (Frodl and Tille 2006). A study by Pandey (Pandey and Kim 2007) compared different commercial measurement systems for reproducibility and concluded that NDIR sensors are sufficiently reliable to produce highly comparable data, at least in a relative sense. A recent paper reported a CH_4 sensor which used a single light beam (LED) and double wavelengths technology and a pyroelectric detector. The system detected the 5% CH_4 range reliably with sensitivity of 0.5% and was designed for industrial applications (Zhao et al. 2011).

3.2.2.3 FTIR Analysis

FTIR sensing has the advantage of measuring multiple analytes simultaneously, which differentiates it from NDIR analysis. While in the past, sensing systems were complex and bulky, current systems are integrated and compact, offering precise measurement down to single digit ppb levels for many analytes. As with NDIR sensing, the main application is in gas analysis with emphasis on environmental monitoring and also in common with NDIR sensor technology, there are many commercial portable FTIR systems available. The basic FTIR system consists of an interferometer, usually a Michelson, which generates an interferogram from the IR emission of the sample and performs a Fourier Transform to obtain the spectrum. Research over the last decade has concentrated mainly on miniaturization and performance enhancement. Remote-sensing FTIR for air quality monitoring has been reported by Hu et al. (2006). The system displays high resolution and selectivity, requires no sampling and sample preprocessing, and can monitor several analytes simultaneously, in particular VOCs. Advances in chemometrics, computer tomography, and FTIR spectra interpretation are reviewed. Gao et al. (2006) improved the measurement and data analysis technique for monitoring CO and N_2O in the atmosphere, allowing fast and mobile remote detection and identification of trace gas for monitoring of accidental pollution breakout. De Castro et al. (2007) used remote FTIR to characterize low-temperature combustion gases in biomass fuels. A high acquisition speed was necessary in order to evaluate the temporal evolution of gas concentrations and improve the signal-to-noise ratio (SNR). The concentrations of three gases, CO_2, CO, and CH_4, which are the most important carbon-related products of biomass combustion, shrub species in this case, were measured in situ. The results are important in the context of modeling pyrolysis for predicting forest fire behavior. The same authors have also published on using different analysis programs to extract reproducible results from data for ozone and CO monitoring (Briz et al. 2007). FTIR has also been used for analysis of engine exhaust emissions. Reyes et al. (2006) monitored a range of pollutants including CO, NO, SO_2, and NH_3 under different driving conditions using a hybrid car and recently, Xia et al. (2009) used remote FTIR passive sensing to monitor aeroengine exhausts to determine the relative concentrations of some of the aforementioned exhaust gases, which has applications in optimum engine design. A portable FTIR analyzer has been used to measure gas emission from the Yasur volcano in 2005, whereby the relative proportions of SO_2 and HCl measured with time were used to elucidate different explosive events during the eruption (Oppenheimer et al. 2006). In a novel application, field-based FTIR–ATR spectroscopy, in conjunction with chemometric techniques, has been used for rapid qualitative and quantitative analysis of opiates such as morphine and thebaine in commercial poppy cultivation with low limits of detection (Turner et al. 2009). Furthermore, chemometrics was used to identify specific signature peaks which have potential forensic applications in opiate crop analysis. Finally, in a recent paper from Smith et al. (2011), FTIR-derived trace gas measurements of CO_2, CH_4, and CO based on modeling yielded highly accurate results even over large concentration ranges. Open-path FTIR using a forward model coupled to a nonlinear least-squares-fitting procedure was used, where, despite complications arising from absorption line pressure broadening, temperature effects on line width, and convolution of the gas absorption lines and instrument line shape, concentrations for all gases were measured to within 5%. As mentioned at the beginning of this section, much of the recent progress in this area hinges on improved data analysis and modeling.

3.2.2.4 Mid-Infrared Sensing Using Laser Light Sources

We include here a short section on the use of laser light sources, primarily for gas analysis. Laser spectroscopy is the method of choice for many applications such as trace gas analysis because of the higher sensitivity and specificity available arising from the intense and narrow spectral output. Lasers used include tunable diode lasers and quantum cascade lasers (QCLs). Tunable diode laser absorption spectroscopy (TDLAS) systems, using lead salt lasers, have been used to detect ppb of gases such as CH_4 (Werle et al. 2002). However, many of these systems require laser cooling and do not lend themselves to portable systems for use in the field. QCL lasers (Hvozdara et al. 2002, Kosterev and Tittel 2002) operate at near-room temperature and emit in the MIR from about $3\,\mu m$. These wavelengths match well with the fundamental vibrational bands of many gases and other compounds such as VOCs. The first QCL-based system for determination of CO_2 in aqueous phase was reported by Schaden et al. (2004), while a system for measuring trace gases was reported by Kosterev et al. (2000). In recent years, there has been a growing interest in MIR laser-based systems for the analysis of the constituents of exhaled breath. This niche application has huge potential for point-of-care sensing toward diagnosis of a broad range of diseases and medical conditions. As well as the expected gases, N_2, CO_2 and O_2, and H_2O, exhaled breath also contains over 1000 trace organic and inorganic compounds, some of which are indicators for particular diseases or conditions, for example, acetone, which is one of the most studied VOCs in breath and is an indicator for diabetes. Direct analytical tools for breath analysis include gas chromatography, mass spectrometry, and IR absorption. MIR sensing techniques are of interest for this application, particularly since the advent of tunable QCLs that tune very efficiently into signature absorption transitions for many relevant compounds in breath. For information on recent developments in this interesting area, readers are directed to some recent review articles (McCurdy et al. 2007, Davis et al. 2010, Kim et al. 2010, Risby and Tittel 2010, Wen et al. 2010).

3.2.3 Direct Fluorescence-Based Sensors

Direct fluorescence has been used extensively in biomedical applications, in particular using the intrinsic autofluorescence of biological material as a tool for identification of, for example, precancerous development of the epithelium where many human cancers originate. Many applications monitor the fluorescence of nicotinamide adenine dinucleotide (NADH) and flavin adenine dinucleotide (FAD), which are indicators of cell activity. The ratio of NADH and FAD fluorescence is a useful tool for probing cellular metabolism and may be used for early cancer detection. Wang et al. (2009) have reviewed recent literature on using this ratio for the detection of cell death and on the response of NADH lifetime to metabolic perturbation and hypoxic environments. Time-resolved confocal fluorescence spectroscopy was used to study cell metabolism by Wu et al. (Wu and Qu 2006, Wu et al. 2006) who were able to distinguish between cancerous and normal cells by analyzing the NADH lifetime. The intrinsic fluorescence of tryptophan, which is sensitive to the environment, has been used to study protein folding and unfolding, general protein stability, and protein aggregation (Vivian and Callis 2001, Monsellier and Bedouelle 2005, Duy and Fitter 2006).

Remote fluorescence sensing of terrestrial vegetation is used for monitoring plant status and functioning via measurement of photosynthetic efficiency. Such measurements are key to the understanding of dynamic biosphere processes. Measurements can be ground based or mounted in aircraft or satellites. In a recent review of remote sensing of solar-induced chlorophyll fluorescence, Meroni et al. compared different experimental approaches, instrumentation used, methods for estimating the fluorescence, and efforts to correlate the fluorescence to plant physiology (Meroni et al. 2009). Another review article (Malenovsky et al. 2009) reviewed recent satellite spectral remote-sensing techniques for quantitative biophysical and biochemical characteristics of vegetation canopies from measured red and blue-green fluorescence.

Fluorescence has been used to characterize different oils as the fluorescence differs for different compositions. For example, crude petroleum oils have been analyzed using fluorescence lifetime studies (Ryder et al. 2002) and, more recently, petroleum-bearing fluid inclusions have been analyzed in order to discriminate different oil populations in situ (Blamey et al. 2009). Fluorescence has also been used to discriminate between virgin olive oil and sunflower oil (Poulli et al. 2006), and fluorescence spectroscopy in conjunction with artificial neural networks (ANNs) has been used to classify different edible oils (Scott et al. 2003).

3.2.4 Raman and SERS-Based Sensing

Raman spectroscopy, like IR absorption, probes the vibrational energy levels of molecules and can be a useful sensing tool. The technique normally uses laser excitation and good optical filtering to separate the scattered photon from the intense incident beam (Laserna 2001). Although Raman scattering cross sections are many orders of magnitude smaller than fluorescence cross sections, it can still be used for sensing gaseous and liquid analytes. In particular, unlike IR sensing, there is no interference from the vibrational spectrum of water, hence its usefulness in biomedical sensing such as the study of whole cells and tissue, for example, mapping the chemical distribution in cells (Owen et al. 2006), mapping DNA and RNA distribution (Uzunbajakava et al. 2003), and distinguishing between different cancer cells (Puppels 2002, Shafer-Peltier et al. 2002).

For many applications, the use of conventional Raman spectroscopy has been superseded by the SERS technique where the Raman scattering is considerably enhanced, up to a factor of 10^8, from a molecule in close proximity to a nanostructured metal surface. The effect is due to the interaction of the localized surface plasmon resonance at the metal surface with the vibrational levels of the molecule whereby the molecule experiences an electric dipole which is induced by the large electromagnetic field, hence giving rise to a hugely enhanced Raman signal (Stuart et al. 2006a). The main requirements for SERS are the presence of the nanostructured metal surface and the immobilization of the molecule/sample under investigation on or near the surface. Recent advances in nanostructured materials and increased understanding of the SERS mechanism has seen a large growth in the use of the technique in fields such as biomedicine and environmental analysis. The groups of Vo-Dinh and Van Duyne have published a number of review articles on the basics of SERS and highlighting the use of the technique in biomedical applications (Stuart et al. 2006b, Vo-Dinh et al. 2006, Zhang et al. 2006). Nanoparticle-based SERS probes for protein localization ex- and in vivo have been reviewed (Schlucker 2009) and Tripp et al. (2008) have described recent progress in the production of highly reproducible SERS substrates. More recently, two reviews of optical fiber SERS sensors have been published, one describing SERS-active fibers for biomedical applications (Stoddart and White 2009) and one dealing with new high sensitivity fiber platforms (Shi et al. 2009).

3.2.5 Summary

This short section has focused mainly on spectroscopic techniques for optical sensing, both absorption and fluorescence based, with emphasis on the most recent review articles and papers published in the last decade. The largest part of the section was devoted to IR sensing, including NDIR- and FTIR-based sensors, mainly for gas-sensing applications. NDIR-based CO_2 sensing is by far the most widely used method and there are many portable systems on the market. In recent years, portable FTIR sensor systems have also been developed. A short section on direct fluorescence sensing has been included, as also has a section on Raman and SERS sensing.

3.3 Reagent-Based Optical Chemical Sensors

3.3.1 Introduction

As indicated in Section 3.1.4, in the case of reagent-based sensors, sensing can be achieved indirectly by monitoring the analyte-sensitive response of an intermediate reagent or dye which is normally encapsulated in a solid but analyte-permeable matrix. Here we summarize progress over the last decade or so, concentrating on those sensors based on optical absorption and luminescence. Key analytes such as O_2, pH, CO_2, and NH_3 will be dealt with. Recent progress in cation and anion sensing and sensing of VOCs will also be treated. For the most part, our aim is not to give a comprehensive review of all recent publications, as, for analytes such as O_2 and pH, this would be beyond the scope of this chapter due to the large volume of articles published, but to highlight a selection of interesting sensing systems that show enhanced sensor performance for selected analytes. We will focus on the recent advances in sensor materials and sensor performance parameters such as sensitivity, stability, selectivity, and robustness. In many cases, the specific sensor limitations will be also highlighted. There are a number of review articles on optical chemical sensors available for the interested reader (Wolfbeis 2002, 2004b, 2006, 2008, Baldini and Giannetti 2005, McDonagh et al. 2008, Orellana and Haigh 2008).

3.3.2 Optical Oxygen Sensing

3.3.2.1 Introduction

Optical O_2 sensors based on luminescence quenching by O_2 of the emission intensity or excited state lifetime of O_2-sensitive luminophores are the most widely used methods in the development of such optical sensors. In O_2 sensing, luminescence quenching results from collisions between the luminophore in its excited state and the quencher, molecular oxygen, in its ground state. The luminophore returns to the ground state without emission of a photon. If luminescence quenching is purely collisional, the emission intensity and excited-state lifetime are related to the partial pressure of O_2 by the Stern–Volmer equation:

$$\frac{I_0}{I} = \frac{\tau_0}{\tau} = 1 + K_{SV} \cdot pO_2$$

where
  I_0, I, τ_0, and τ are the luminescence intensities and decay times of the luminophore in the absence and
      presence of O_2, respectively
  K_{SV} is the Stern–Volmer constant
  pO_2 is the partial pressure of O_2

The graphical representation of I_0/I or τ_0/τ versus pO_2 is called the Stern–Volmer plot. For an ideal, homogeneous environment, this plot yields a straight line with an intercept at 1 and a slope of K_{SV}, which represents sensor sensitivity. If the luminophore is distributed in a solid matrix such as a polymer or sol–gel, where the entrapped indicator dyes encounter different environmental influences, matrix microheterogeneity will cause different sites to be quenched differently with a resultant downward curve in the Stern–Volmer plot (Bacon and Demas 1987, Carraway et al. 1991, Xu et al. 1994, Eaton et al. 2004). The majority of O_2 sensors exhibit nonlinear Stern–Volmer behavior. Different models were developed and have been used to describe such processes (Lehrer 1971, Kneas et al. 1997, Mills 1999a,b).

For most optical O_2 sensors, a wide variety of hydrophobic O_2-permeable polymeric matrices including sol–gel-derived materials have been used for immobilizing the O_2-sensitive indicator. The hydrophobic nature of such matrices reduces problems associated with dye leaching and quenching by interferents, therefore improving selectivity. They also provide very good mechanical and chemical stability.

The most commonly used O_2-sensitive luminescent indicators belong to the luminescent platinum group of metal complexes with α-diimine ligands, in particular luminescent ruthenium(II) diimine complexes with excited-state lifetimes in the microsecond range and phosphorescent platinum(II) and palladium(II) porphyrins which have longer lifetimes (in the range of 100–1000 μs). Desirable features of all these materials for use in O_2-sensing applications include long excited-state lifetimes, which make lifetime measurements relatively simple and low cost in comparison to short lifetimes (low nanosecond range) of the organic fluorophores, large Stokes shifts, and high quantum efficiency. In addition, they possess intense visible absorptions that simplify sensor design. They exhibit chemical flexibility that allows for the design of systems with desired absorption-emission properties that are compatible with available excitation sources and detectors and for ionic or covalent attachment to the desired solid support. Luminescent indicator dyes should also exhibit very good photostability, so often dye photobleaching has to be minimized in order to achieve good long-term sensor performance. Optical O_2 sensor materials have been reviewed by Amao (2003). Recently, recent progress in luminescence-based O_2 sensors has been described by DeGraff and Demas (2005).

There has been very little activity in the area of development of absorbance-based O_2 sensors that are based on change of color or absorption properties due to the selective absorption of the dye in the presence of O_2. In general, such sensors exhibit poor reversibility or are irreversible and are classified, therefore, as optical probes. They also exhibit limited stability, longer response time, and much lower sensitivity and resolution than luminescence-based O_2 sensors, and therefore these sensors will not be discussed here. However, such irreversible O_2 probes are required in certain applications such as modified atmosphere food packaging. A review has been published by Mills that reports on different colorimetric O_2 indicators suitable for this application (Mills 2005).

3.3.2.2 *Luminescence Intensity-Based O_2 Sensors*

Luminescence intensity-based O_2 sensors have been widely investigated using a range of new solid matrices and O_2-sensitive luminophores. Amao et al. reported on a number of new polymers. O_2 sensors have been fabricated with enhanced sensitivity and photostability using two fluoropolymers as matrices, namely, poly(isobutylmethacrylate-co-trifluoroethylmethacrylate), (poly-IBM-co-TFEM) and poly(styrene-co-2,2,2-trifluoroethyl methacrylate) (poly-styrene-co-TFEM) doped with platinum(II) octaethylporphyrin (PtOEP) and tris(2-phenylpyridine anion) iridium(III) [Ir(ppy)$_3$] complexes (Amao et al. 2000a, 2001). High I_{argon}/I_{oxygen} ratios (which is the measure of sensor sensitivity) of 288 and 296 were observed for PtOEP entrapped in poly-IBM-co-TFEM and in poly-styrene-co-TFEM, respectively, together with good photostability. The intensity decreased by 2% after continuous sensor film irradiation for 24 h. This has been attributed to the presence of the fluoro group in the polymer. Fluorine is the most electronegative of all the chemical elements and it forms a very strong and short bond to carbon, and therefore fluorinated materials exhibit enhanced chemical and thermal stability. [Ir(ppy)$_3$] incorporated in poly-styrene-co-TFEM exhibited lower sensitivity with an I_{argon}/I_{oxygen} ratio of 15.3 and intensity was reduced by 5% after continuous sensor film irradiation for 24 h. Both systems exhibited reasonably linear Stern–Volmer plots. Highly O_2-permeable poly-(1-trimethylsilyl-1-propyne) polymer was also used as a solid support for a porphyrin dye (Amao et al. 2000b). This sensor showed high sensitivity with an I_{argon}/I_{oxygen} ratio of 225, low limit of detection (LOD) (<0.3%), and linear response from 0% to 10% O_2. This system was less photostable as the intensity decreased by 3% when the material was illuminated for 12 h. Different europium(III) complexes immobilized in polystyrene-co-TFEM were also exploited for the fabrication of O_2 sensors. However, they exhibited rather small dynamic range with an I_{argon}/I_{oxygen} of ~2 for the most sensitive system (Amao et al. 2000c). Such systems are suitable for high O_2 concentration detection.

O_2 sensors produced by using poly(ethylene glycol)ethyl ether methacrylate (pPEGMA) polymer with different molecular weights have been demonstrated. PtOEP, palladium(II) (PdOEP), and ruthenium(II) octaethylporphyrin (RuOEP) have been used as O_2-sensitive dyes (DiMarco and Lanza 2000). It was found that by using the polymer with lower viscosity, the O_2 sensitivity can be enhanced. The polymer, when doped with RuOEP dye, exhibited the lowest O_2 sensitivity and chemical stability when compared to PtOEP and PdOEP complexes. PdOEP was the most sensitive system but the least photostable. The sensors also displayed linear Stern–Volmer calibration plots in the range from 0 to 100 Torr O_2 (for PdOEP) and from 0 to 800 Torr O_2 for PtOEP- and RuOEP-based sensor with response time in the range of a few seconds. An increase of up to 600-fold in O_2 sensitivity was found for a series of O_2 sensors when using Pt and Pd octaethylporphyrins immobilized in three different polymers, namely, ethyl cellulose, cellulose acetate butyrate, and PVC (Douglas and Eaton 2002). The difference in sensor performance was correlated with different lifetimes of the porphyrin dyes used together with different O_2 permeability of the polymers. Single exponential decays in the absence of O_2 were observed while in the presence of O_2, the decays became multi-exponential and therefore nonlinear Stern–Volmer plots were observed. This suggests that there was an inhomogeneous distribution of pO_2 in the polymer films. The authors also investigated the influence of humidity on O_2 sensor response (Eaton and Douglas 2002) and found that the O_2 response is significantly decreased when hydrophilic polymers are used as solid matrices as opposed to hydrophobic polymers where the humidity interference was minimized. The cationic platinum metalloporphyrins (Pd(II), Pt(II), and Rh(III)) were also immobilized in cationic-exchange Nafion® membrane (Vasil'ev and Borisov 2002). Phosphorescence decay curves for all the systems were single exponential, which translates to linear Stern–Volmer calibration plots with $r^2 = 0.998$. Values of I_{argon}/I_{oxygen} from 6 to 84 were reported. No significant photobleaching was observed when the sensor membranes were excited at 442 nm for 12 h.

Anti-biofouling coatings have been designed and used in dissolved O_2 (dO_2) sensing (Navarro-Villoslada et al. 2001). Silicone layers loaded with [Ru(II)-tris(4,7-diphenyl-1,10-phenanthroline)] ([Ru(dpp)$_3$]$^{2+}$) and overcoated with a polymer containing phosphorylcholine exhibited reduced adhesion of marine bacteria and thrombocytes. The biofouling-resistant O_2 films exhibited similar performance to uncoated sensors and were validated using the Clark electrode. No statistically significant differences in O_2 readings were observed. Tian and coauthors covalently attached a modified porphyrin complex to the hydrophilic biocompatible polymer poly(2-hydroxyethyl methacrylate)-co-polyacrylamide (PHEMA) and used this material for dO_2 determination (Tian et al. 2010a). The uniform polymer membrane 25 μm in thickness was grafted on quartz surfaces modified with trimethylsilylpropyl acrylate. The sensor solution was deposited on the modified glass and thermal polymerization was performed at 80°C. The sensor exhibited a response time of 50 s, enhanced photostability (less than 1% reduction in intensity over 5 h illumination at 405 nm), and no leaching when compared to the films with physically entrapped dye. However, sensitivity was lower probably due to restricted dye molecule movement in the matrix. The storage stability for the material in ambient in the dark was at least 6 months and in water at least 3 months. The nontoxicity of the new sensor films was also evaluated.

Sol–gel-derived materials have been also used as a solid matrix for the encapsulation of O_2-sensitive indicator dyes. The Bright group reported on high-performance intensity-based sol–gel-derived O_2 sensors (Tang et al. 2003, Tao et al. 2006a). Sol–gel-based O_2 sensor materials were doped with $[Ru(dpp)_3]^{2+}$ and were composed of tetramethoxysilane (TMOS) or TEOS and organically modified silica precursors with varying alkyl chain lengths (C_1–C_{12}). These novel materials outperformed pure TEOS-based O_2 sensors. They exhibited linear Stern–Volmer plots, excellent long-term stability, and much higher O_2 sensitivity; for example, sensor layers composed of TEOS-octyltriethoxysilane (1:1) exhibited four times higher sensitivity than TEOS-based sensors and were stable over an 11-month period (RSD 4%,), whereas the sensitivity of TEOS-based O_2 sensors decreased by 400%. Linear Stern–Volmer plots were also reported for a highly O_2 sensitive system ($I_{N2}/I_{O2} = 35 \pm 4$), which has been demonstrated by using fluorinated ORganically MOdified SILicates (ORMOSIL) as a sensor matrix (Bukowski et al. 2005). The hybrid sol–gel-derived matrix was composed of n-propyltrimethoxysilane and 3,3,3-trifluoropropyl-trimethoxysilane doped with ($[Ru(dpp)_3]^{2+}$ and the higher sensitivity originated from higher O_2 diffusion coefficients within those films. The stability was also very good over a 6-month period with RSD of 2%.

O_2 sensing has also been reported using calcined mesoporous silica spheres in which the tris(bipyridine) ruthenium(II) ($[Ru(bpy)_3]^{2+}$) complex has been incorporated (Zhang et al. 2002). The O_2 sensitivity was dependent on the pore morphology in the sphere, and the response times (2–4 min) were longer than for thin films and most likely related to the relatively large sphere diameter of 0.1–0.2 mm.

Another approach describes an O_2 sensor where a ruthenium (II) complex has been covalently attached to a mesoporous TEOS-based film that was spin coated on glass (Lei et al. 2006). Covalent attachment can increase the sensor stability and eliminate dye leaching. The O_2 sensor exhibited better sensor-to-sensor reproducibility, enhanced sensitivity, and linearity when compared to samples with physically entrapped luminophores.

Other interesting O_2 sensing systems include a report where four different ruthenium(II) and iridium(III) luminescent complexes have been incorporated into transparent metal oxide matrices (Fernandez-Sanchez et al. 2006). The Stern–Volmer constants were ~100 larger than for the same systems based on polystyrene. Furthermore, the response time of <1 s, very good stability over a 9-month period, and possibility of autoclavation and gamma irradiation were reported. The O_2 sensor performance was attributed to the nanoporosity and high total pore volume of these matrices.

3.3.2.3 Luminescence Ratiometric O_2 Sensors

Ratiometric measurement methods have a number of advantages over unreferenced intensity measurements such as immunity to drifts caused by photobleaching effects. However, it can add complexity to the system as two different excitation or emission wavelengths have to be used and not all indicator dyes are compatible with this approach. A fiber-optic ratiometric optical dO_2 sensor was described for use in inter- and intracellular measurements (Park et al. 2005). The phosphorescent O_2-sensitive porphyrin dye was used together with a reference dye, Bodipy 577/618 maleimide, and both were entrapped in a liquid polymer made of PVC, including the plasticizing agent bis(2-ethylhexyl) sebacate. The sensing solution was applied by dip-coating on the submicron-sized fiber tip. The Stern–Volmer plots exhibited reasonable linearity from 0% to 100% O_2, no significant leaching was observed over a 9 h period, and no extensive photobleaching occurred after 1000 pulses of 200 ms with an argon ion laser, 514.5 nm. A low LOD of 14 ppb was reported. Kostov et al. described a ratiometric dual-emission O_2 sensor using a single luminescent dye that possesses both fluorescent and phosphorescent emission (Kostov et al. 2000).

The O_2 detection exploited a modulation of the luminescence at two different frequencies This technique utilizes the large difference between the lifetimes of the two excited states of the luminescent dye [1,2-bis(diphenylphosphino)ethane-Pt{$S_2C_2(CH_2CH_2$-N-2-pyridinium)}](BPh$_4$) where BPh$_4^-$ is a tetraphenylborate anion that was immobilized in cellulose acetate which contained 75% of the plasticizer triethylcitrate. The response time was rather long (3 min), probably due to the film thickness (0.5 mm), and I_{N2}/I_{O2} of ~10 was reported. This autoclavable sensor patch was used for noninvasive monitoring of dO_2 levels in shake flasks during yeast and *Escherichia coli* fermentations (Tolosa et al. 2002). A drift of 3% was reported for tests run for 6 days and sampling every 5 min. Biofouling was not observed. However, an increase of the SNR due to photobleaching was reported after the sensor patch was re-autoclaved and reused. In addition, the sensor exhibited relatively high inaccuracy at higher dO_2 values. The possibility of using quantum dots to provide a reference signal in optical O_2 sensing has also been demonstrated (Jorge et al. 2006, Collier et al. 2011).

3.3.2.4 Phase/Lifetime-Based O_2 Sensors

Trinkel et al. (2000) presented a miniaturized phase fluorometric system that facilitates indirect monitoring of the luminescence lifetime for O_2 trace analysis. The phosphorescent PdOEP was immobilized in the fluoropolymer, Teflon AF. It was the first time that such a highly sensitive O_2 sensor has been reported. The sensor resolution was 0.1 vpm (volume per million) at concentrations up to 20 and <1 vpm over the whole measurement range (0–100 vpm). A linear calibration plot was demonstrated over this concentration range and sensor repeatability was reported to be ±1 vpm over 14 h of measurements. Phase fluorometry was also used in the development of an optical steam sterilizable O_2 sensor for use in bioprocess monitoring applications (Voraberger et al. 2001). The polysulfone and polyetherimide polymers with [Ru(dpp)$_3$]$^{2+}$ as a O_2-sensitive indicator were found to withstand steam sterilization at 135°C. Their physical and chemical properties together with O_2 response did not change after the autoclavation process and therefore recalibration in between sterilization cycles is not required. The O_2 sensor was used in *E. coli* fermentation and its performance matched very well the data obtained using the Clarke electrode. Fiber-optic phase fluorometry detection was also exploited by Papkovsky et al. They used high-impact polystyrene in which platinum(II)-octaethylporphine-ketone (PtOEK) dye has been entrapped (O'Riordan et al. 2000, Papkovsky et al. 2000, 2002). The O_2 sensor performance was characterized over a wide temperature range (−17°C to +30°C) and calibration plots were linear in the range from 0 to 5 kPa. The sensor system has been applied in noninvasive O_2 level monitoring in packaged food samples and in monitoring metabolic activity of living cells.

Sol–gel-derived materials were exploited in the development of a phase fluorometric probe-based dO_2 sensor with a dual LED referencing system for use in waste waters (McDonagh et al. 2001). The O_2-sensitive material composed of [Ru(dpp)$_3$]$^{2+}$ entrapped in an hydrophobic MTEOS-based sol–gel-derived matrix was deposited on disposable PMMA discs. The sensor exhibited excellent SNR and long-term stability. A low LOD of 6 ppb was reported. A wide range of organically modified sol–gel precursors have been used to develop a high-performance O_2-sensing material for breath monitoring applications (Higgins et al. 2008). N-propyltriethoxysilane-based sol–gel film was selected as the optimum material for this application due to sufficient O_2 response in the relevant O_2 concentration range, from 15% to 25%, coupled with negligible humidity interference and short response time of 223 ms for a 1 μm thick film. The sensor film was tested using a commercially available lung simulator and its performance compared well with commercial breath monitoring systems (Burke et al. 2008). It has also been demonstrated that sol–gel materials based on fluorinated precursors in combination with silver nanoparticles minimize drift caused by photobleaching (Estella et al. 2010). A magnetically controlled wireless lifetime-based O_2 sensor for intraocular measurements has been presented (Ergeneman et al. 2008) where a novel iridium complex was entrapped in polystyrene and dip coated on ferromagnetic spheres of 3.25 mm diameter. The size of the spheres needs to be further minimized as spheres below 1 mm in diameter are required for in vivo intraocular measurements. O_2 magnetic sensor macrospheres (MagSeMacs) in the millimeter range spray coated with O_2-sensitive layers have been reported by Mistlberger et al. (2010) and their suitability for use for contactless monitoring of O_2 in a microbioreactor has been presented. Microbeads were also exploited in an O_2 sensor for real-time intracellular O_2 measurements in green plants.

Pt(II)-tetra-pentafluorophenyl-porphyrin was encapsulated in the carboxylated polystyrene microbeads that were injected into the photosynthetically active cells (Schmalzlin et al. 2005). Multifrequency phase modulation was used as a detection technique to differentiate the sensor's O_2-sensitive phosphorescence from the autofluorescence of the plant tissue that arises from chlorophyll. Strong delayed fluorescence of the fullerene C_{70} immobilized in an ORMOSIL matrix and ethyl cellulose has also been exploited in O_2 sensing (Nagl et al. 2007). The sensor was suitable for the detection of O_2 in the ppb range at elevated temperatures. The Stern–Volmer constants were more than one order of magnitude higher than other O_2 sensors reported in the literature. Unfortunately, this sensor exhibited low luminescence brightness at room temperature and high temperature dependence. Recently, a low LOD of 0.015 Pa (25°C) for an O_2 sensor based on palladium(II) complex of 5,10,15,20-*meso*-tetrakis-(2,3,4,5,6-pentafluorophenyl)-porphyrin covalently attached to silica particles with diameter of 5 μm functionalized with amine groups was reported (Borisov et al. 2011). The sensor exhibited linear Stern–Volmer plots in the range from 0.02 to 100 Pa O_2, high photostability, short response time (150–250 ms for a 25 μm thick film), and low temperature dependence.

3.3.2.5 Other Approaches in Optical O_2 Sensing

In 2006, a new visual-based approach to optical O_2 sensing has been reported using two luminescent dyes with different O_2 sensitivities and emission wavelengths. The cyclometalated platinum(II) complex ($\lambda_{em} = 506$ nm, $\tau = 9.2$ μs) and PtOEP ($\lambda_{em} = 646$ nm, $\tau \sim 100$ μs) has been immobilized in ethyl cellulose and coated on glass as two separate layers (Evans et al. 2006). Emission from PtOEP is quenched at low O_2 concentrations due to higher O_2 sensitivity whereas the other complex emits significantly at higher O_2 pressures. This results in a gradual shift in the emission color across the red-yellow-green spectral region with increasing O_2 concentrations allowing for visual color identification. The response profiles of these sensors can be tuned by using multiple luminophore–polymer layers and therefore it is possible to design sensors with different sensitivities and color space response (Evans and Douglas 2006). Sensors exhibiting red-green, red-blue, and red-green-blue color responses have been demonstrated (Evans and Douglas 2009). A similar O_2 sensor based on a combination of a O_2-insensitive cadmium telluride (CdTe) quantum dots ($\lambda_{em} = 550$ nm) and the O_2 indicator [meso-tetrakis (pentafluorophenyl) porphyrinato] platinum(II), ($PtF_{20}TPP$, $\lambda_{em} = 648$ nm) entrapped in the sol–gel matrix have been reported by Wang et al. (2008a). The sensor layers were excited at 395 nm and exhibited a change in color from green to red, which indicated an O_2 pressure reduction. However, the sensor displayed poor stability.

In summary, we present in Table 3.1 a selection of the many papers on optical O_2 sensors which have appeared in the literature since 2000.

3.3.3 Optical Carbon Dioxide Sensing

3.3.3.1 Introduction

The principle of operation of most optical CO_2 sensors is based on the acidic nature of dissolved CO_2 (dCO_2). The sensors utilize pH indicator dyes with pKa between 7.5 and 9.0 that change their absorption or luminescence properties when the pH changes upon exposure to different levels of CO_2. The CO_2 affects the pH of a buffered aqueous medium via carbonic acid formation and its subsequent deprotonation reactions. The number of these probes is limited as only some pH indicators satisfy this pKa requirement and exhibit good photostability.

In the first optical CO_2 sensors, so-called wet sensors, a pH-sensitive dye was dissolved in a sodium bicarbonate solution (Opitz and Lubbers 1975, Zhujun and Seitz 1984, Wolfbeis et al. 1988, Uttamlal and Walt 1995). A gas-permeable, ion-impermeable membrane such as silicone or Teflon was used to prevent leaching of the indicator dye and to avoid proton interference. Single-membrane sensors have also been developed and they rely on the formation of water droplets emulsified in hydrophobic polymers (Heitzmann and Kroneis 1985). The first solid-state CO_2 sensor that does not require the presence of the carbonate buffer was described by Kawabata et al. (1989), and in 1991, Raemer and his coworkers reported on the use of an ion pair and the addition of a lipophilic base, such as tetraoctylammonium

TABLE 3.1

Selection of Papers on Optical O_2 Sensors Published Since 2000

Reference	Detection Method	Comments
Amao et al. (2000a,b)	Luminescence intensity	Europium(III) complex as O_2-sensitive probe
Voraberger et al. (2001)	Phase fluorometry	New autoclavable polymer materials
Kellner et al. (2002)	Luminescence lifetime imaging	Measurement of O_2 distributions in tissues in vitro (planar sensor foils)
Tang et al. (2003)	Luminescence intensity	Stable sol–gel-derived O_2 sensor materials with linear calibration plots
Koo Lee et al. (2010)	2-wavelength (λ) ratiometric	Measurement of intracellular changes of dO_2 in vitro using PEBBLE nanoparticles
Babilas et al. (2005)	Luminescence lifetime imaging	In vivo mapping of O_2 distribution over tumor tissue using transparent planar O_2 sensor
Lei et al. (2006)	Luminescence intensity	Ru(II) complex covalently attached to mesostructured silica
Jorge et al. (2006)	2λ ratiometric	Applications of quantum dots in optical O_2 sensing
Evans et al. (2006)	Visual color identification	Two luminescent dyes with different O_2 sensitivities and emission wavelengths entrapped in one polymer matrix
Nagl et al. (2007)	Luminescence lifetime imaging	Highly sensitive fullerene-based O_2 sensor (ppb range) for sensing at elevated temperatures
Ergeneman et al. (2008)	Frequency-domain lifetime measurement	Magnetically controlled wireless O_2 sensor for intraocular measurements
Achatz et al. (2011)	Luminescence intensity and 2λ ratiometric	NIR excitable O_2 sensor exploiting upconverting nanoparticles in combination with O_2-sensitive iridium(III) complex

hydroxide (TOAOH), both entrapped in a hydrophobic matrix (Raemer et al. 1991), which was later developed by Mills and Chang (1992, 1993, 1994). This approach minimizes the effects of osmotic pressure as there is no aqueous buffer. The hydrophobic support acts as a barrier to ions, so no additional ion-impermeable layer is required. However, a limited shelf life of these sensors has been reported due to the presence of other acidic gases such as SO_2 or NO_2 in the atmosphere that irreversibly protonate the pH indicator. A good review on optical CO_2 sensors is available (Mills 2009).

3.3.3.2 Colorimetric and Unreferenced Luminescence Intensity CO_2 Sensors

As discussed earlier, colorimetric CO_2 sensors utilize pH indicators which have pKa values in the range 7.5–9.0. A colorimetric fiber-optic sensor for in vivo monitoring of gastric CO_2 has been characterized and validated in clinical tests (Baldini et al. 2003). The sensor is composed of cresol red that was incorporated together with TOAOH in ethyl cellulose. The sensor layer was deposited on a transparent Mylar support and overcoated with a gas-permeable silicone coating to prevent pH and ionic strength interference. The absorption band at 590 nm decreased with increasing CO_2 concentrations and a band at 435 nm appeared. The sensor dynamic range extended from 0 to 150 hPa, and resolution of 0.2 hPa at 0 hPa CO_2, accuracy of ±2.5 hPa, and response time <1 min were observed. The sensor performance compared favorably to Tonocap, a commercial instrument currently used for measuring gastric CO_2 (Tonometrics™ Catherers, Datex-Ohmeda, www. gehealthcare.com). Nanostructured metal oxides such as aluminum oxide, silicon oxide, and zirconium oxide embedded in polyvinyl alcohol as CO_2 sensor matrices have also been demonstrated (Fernandez-Sanchez et al. 2007). Colorimetric CO_2-sensitive dyes α-naphtholphthalein (NAF), naphthol blue black (NBB), and calmagite (CMG) were used together with TOAOH. Sensor sensitivity was mainly dependent on the total pore volume and average pore diameter of the metal oxide used and therefore sensor systems exhibited different dynamic ranges and LODs. A dynamic range from 0.6% to 40% (v/v) CO_2 was observed for CMG in aluminum oxide together with a linear response function, LOD of 0.6% (v/v), and response time of <30 s when switching from 0% to 10% CO_2. However, it had to be stored at 4°C. The two other systems were stable even when kept at 25°C and 40°C in ambient air for 2 months. Furthermore, the materials were gamma sterilizable and humidity interference was observed only in the range from 0% to 25% relative humidity for NAF and CMG-based sensor films. A CO_2 sensor based on ion-paired bromothymol blue that has been dissolved in a room temperature ionic liquid (RTIL) matrix has been demonstrated (Oter et al. 2006). The solubility of CO_2 in RTIL such as 1-methyl-3-butylimidazolium tetrafluoroborate is 10–20 times that observed in polymer matrices. However, regeneration of the reagent phase with nitrogen was necessary. An LOD of 1.4% and 10^{-6} M [HCO_3^-] for gas and dissolved phase, respectively, was reported. However, detection of CO_2 in solution suffers from practical limitations. RTILs were also exploited in the development of solid-state CO_2 sensors. Borisov et al. described a range of colorimetric and fluorescent sensors that exploited an emulsion of RTIL in an ion-impermeable polydimethylsiloxane-based silicone matrix (Borisov et al. 2007). Thymol blue (TB) and BTB were used to produce the colorimetric CO_2 sensor. The buffer system comprised the pH indicator, RTIL, 15 vol% water and organic (tetrabutylammonium hydroxide) or inorganic base (sodium hydrocarbonate or sodium phosphate) depending on the system. Such a buffer system was incorporated in a silicone matrix to produce an emulsion-based CO_2 sensor. It was found that the sensor sensitivity is dependent on the nature of the counterion in the RTIL. In addition, certain RTILs such as 1-butyl-3-methylimidazolium tetrafluoroborate suffered from irreversible temperature dependence. The colorimetric sensors exhibited dynamic range from 0% to 100% CO_2 for the BTB-based system (color change from dark blue to yellowish green). The sensor based on TB was much more sensitive with a dynamic range from 0% to 2% CO_2. The optimum fluorescent CO_2 material was based on tosylate RTIL which exhibited linear calibration plots, reproducible response in the temperature range from 10°C to 50°C, response and recovery time of 50 and 320 s, respectively, and moderate photostability. The storage stability was 1 month when the sensor material was stored under 100% relative humidity and 10% CO_2.

A very photostable and highly fluorescent pH indicator, HPTS, has been exploited by Wolfbeis et al. for optical CO_2 sensing. The fiber-optic microsensor with a tip diameter between 20 and 50 μm was produced for high-resolution pCO_2 sensing in marine sediments (Neurauter et al. 2000). HPTS was ion-paired with tetraoctylammonium as the counter ion and was entrapped along with TOAOH in ethyl cellulose. The sensing layer was overcoated with Teflon AF to prevent interference from protons and other ion interferences. The sensor dynamic range extended from 0.05 to 7 hPa pCO_2; response time of <1 min and LOD of 0.04 hPa pCO_2 (60 ppb dCO_2) was reported. The same sensor chemistry was used to fabricate a reservoir-type capillary microsensor for pCO_2 determination (Ertekin et al. 2003). The sensor dynamic range was from 1 to 20 hPa; LOD 1 hPa and response time of 15 s was observed.

Sol–gel-derived materials have also been used in the optical CO_2 development. Dansby-Sparks et al. used an MTEOS-derived sol–gel layer doped with HPTS ion paired with tetraoctylammonium bromide for CO_2 detection in the 0.03%–30% range (Dansby-Sparks et al. 2010). The films were spin coated on glass and thicknesses were in the range from 8 to 30 nm to >1 μm. Films lost sensitivity in <1 week when stored in ambient conditions. However, when stored under vacuum, the storage time extended to at least 1 month. LOD of 0.008% (80 ppm) was reported. In addition, humidity interference was minimized by applying an overcoat layer that was composed of tetraphenylmethyldisilazane.

3.3.3.3 Ratiometric Luminescence-Based CO_2 Sensors

HPTS is very often exploited for ratiometric CO_2 sensing. Depending on the pH, HPTS exhibits two different absorption maxima, one at 404 nm for the acidic form, the other at 455 nm for the basic form, and it exhibits one emission maximum at 515 nm. The presence of dual excitation bands facilitates the use of excitation ratiometric detection.

Ge et al. (2003) reported an autoclavable ratiometric fluorescent dCO_2 sensor. HPTS together with cetytrimethyl ammonium hydroxide (CTMAOH) was incorporated into an ion-impermeable silicone film. An LOD of about 0.03% CO_2, average response and recovery times of 0.7 and 2 min, respectively, were reported. The sensor was insensitive to ion concentrations in the range from 0 to 0.2 M. The sensor was used in continuous measurements of dCO_2 during *E. coli* fermentation. A similar system was reported by Zhu et al. (2006) and was used for CO_2 distribution pattern monitoring in marine sediments. A HPTS ion pair was incorporated in ethyl cellulose along with TOAOH and overcoated with an ion-impermeable silicone layer. The sensor exhibited linear response in the range from 0 to 9 matm and was stable for several weeks when stored in a sealed bag or directly in the marine sediment and for several months when stored at 4°C. A response time of ~2.5 min was observed. A ratiometric dCO_2 sensor based on HPTS ion paired with cetyltrimethylammonium bromide, which has been entrapped in a hybrid sol–gel-based matrix derived from n-propyltriethoxysilane along with the lipophilic organic base, has also been described (Wencel et al. 2010). The sensor spot deposited on a cover slip has been interrogated with a ratiometric optical probe. The sensor was suitable for dCO_2 determination in the range 0%–30% CO_2. The sensor exhibited enhanced temporal stability and sensitivity with LOD of 35 ppb, excellent reversibility, no leaching, and a short response time of 39 s and recovery time of 1.8 min when the sensor was exposed to a 0.2 M (5% of dCO_2) and 1 M (25% of dCO_2) standard solutions of $NaHCO_3$, respectively.

3.3.3.4 Lifetime-Based CO₂ Optical Sensors

A lifetime-based CO_2 sensor with a wide dynamic range from 0% to 100% CO_2 based on resonance energy transfer between a long-lifetime luminescent ruthenium polypyridyl complex whose emission band overlaps the absorption band of a pH-sensitive dye Sudan III has been reported (von Bultzingslowen et al. 2003). The energy transfer caused the modulation of intensity and lifetime of the ruthenium complex and therefore the sensor was interrogated with low-cost phase fluorometric instrumentation. Both dyes together with TOAOH were immobilized in a composite sol–gel/ethyl cellulose hydrophobic matrix. An LOD of 0.06% and a resolution of ±2% were reported. The shelf life of the sensor when kept in darkness was longer than 3 months. The limiting factors were O_2 and humidity cross-sensitivity. A sol–gel-based CO_2 gas sensor employing dual lifetime referencing (DLR) for application in food packaging has been described (von Bultzingslowen et al. 2002). DLR is an internal ratiometric method whereby the analyte-sensitive luminescence intensity signal is converted into the phase domain by co-immobilizing an inert long-lifetime reference luminophore with similar spectral characteristics (Klimant 2003). HPTS was entrapped in a hydrophobic MTEOS-based sol–gel matrix along with CTMAOH and $[Ru(dpp)_3]^{2+}$ as the internal reference. To avoid O_2 cross-sensitivity, a long-lived reference $[Ru(dpp)_3]^{2+}$ was encapsulated in O_2-impermeable polymer nanobeads. The sensor dynamic range extended from 0% to 100% CO_2 with a resolution of ±1% and it exhibited excellent repeatability and stability for at least 50 days. An LOD of 0.08% was measured and cross-sensitivity toward pH and ions was found to be negligible. Furthermore, the sensor output correlated very well with IR spectroscopy that was used as a reference method. Cajlakovic et al. (2006) also exploited the DLR method for CO_2 sensing.

3.3.3.5 Other Approaches in Optical CO₂ Sensing

CO_2 sensors based on the luminescence intensity change of the O_2-insensitive europium(III) complex, tris(thenoyltrifluoroacetonato europium(III) dehydrate ($[Eu(tta)_3]$) (internal reference), in combination with colorimetric CO_2-sensitive indicators TB, phenol red, and cresol red have been reported (Nakamura and Amao 2003). The colorimetric indicators exhibit CO_2-sensitive absorption bands at 600 nm, which decrease with increasing CO_2 concentrations. This in turn causes the increase of the $[Eu(tta)_3]$ intensity at 613 nm when excited at 350 nm. The CO_2 indicators were incorporated in ethyl cellulose together with TOAOH while $[Eu(tta)_3]$ was encapsulated in polystyrene. Both layers were cast on the opposite sides of a glass slide. The TB-based layer was the most sensitive with I_{CO2}/I_{N2} of 15.6. A response time of 4.4 s and recovery time of 36 s were observed when switching between 100% N_2 and 100% CO_2 gases. A similar system based on the colorimetric CO_2 indicator α-naphtholphthalein incorporated in poly(trimethylsilylpropyne) membrane and an internal phosphorescent reference tetraphenylporphyrin has also been demonstrated for use in gas and dissolved phase (Amao and Nakamura 2005, Amao et al. 2005). The storage and operational stability of these sensors has not been reported.

A novel approach to colorimetric CO_2 sensing both in gas and dissolved phase has been reported recently, which exploited the solvatochromic dye Nile red (NR) and ethyl cellulose as a polymer matrix (Ali et al. 2011). The N,N,N′-tributylpentanamidine, TB-PAM was also used which reversibly reacts with

CO_2 and forms a hydrophilic salt in the presence of water. This process is accompanied by a change in the microenvironment polarity and therefore NR was employed to sense these changes. The NR changes both its absorption (color change from brick-red to magenta) and fluorescence properties (from orange to red) when exposed to CO_2. The dynamic range extended from 0% to 100% CO_2 for a gas sensor and from 0 to 1 M solutions of bicarbonate for dCO_2. An LOD of 0.23% and 1.53 hPa and response time of 10 and 25 min was observed for gas and dissolved phase, respectively.

3.3.4 Optical Ammonia Sensing

3.3.4.1 Introduction

NH_3 is a weak base, and one of the main principles of optical measurement of NH_3 relies on the change in optical properties of a pH indicator which is usually entrapped in a gas-permeable hydrophobic matrix. The description of such NH_3 sensors will be discussed in the following section. Due to NH_3 being a weak base, the absorbed/transmitted signal is generally very weak, and hence many of the sensor platforms are based on waveguiding principles. While most of the examples given later rely on changes to optical properties of pH indicators and organic conducting polymers due to protonation/deprotonation, we also include a short section highlighting alternative sensing approaches, for example, direct sensing via reflectance changes due to adsorption of NH_3 on a tin oxide layer.

3.3.4.2 NH₃ Sensors Based on Absorption/Reflectance

A fiber-optic gaseous NH_3 sensor based on the reflectance changes of a Nafion film doped with crystal violet dye was described by Raimundo and Narayanaswamy (2000). The film reacted irreversibly with NH_3. However, a pulse of humid air followed by a pulse of dry air facilitated film recovery greater than 90% at 580 nm. The sensor exhibited a linear range from 0 to 10 ppm and 0 to 50 ppm in dry air, depending on the exposure time. Cross-sensitivity to CO_2 was not observed. However, the film responded to 5 ppm of hydrogen hydrochloride and nitrogen dioxide. The influence of relative humidity on sensor response was also discussed. A NH_3 sensor based on guided wave absorption of BTB was reported (Qi et al. 2001). A composite optical waveguide was fabricated and was composed of a tapered BTB film evaporated onto a potassium ion-exchanged waveguide. The sensor showed a short response time of 1 s and recovery time of 90 s. It was said that a LOD of 1 ppb can be easily realized for the described experimental conditions. The same dye was used in a sensor based on titanium dioxide (TiO_2) which was deposited on an ion-exchange glass waveguide and BTB was spin coated on the TiO_2 film region (Yimit et al. 2003). Such a composite optical waveguide sensor detected NH_3 in the ppt range. Moreno et al. fabricated a fiber-optic evanescent wave absorption-based NH_3 sensor using BCP (Moreno et al. 2005). It was incorporated in a thermoplastic polyurethane resin, Tecoflex® and dip coated on a fiber at 60°C. The response and recovery times when the sensor was exposed to 1% of NH_3 gas in dry environment were <10 min and 5 s, respectively. Humidity and temperature cross-sensitivity had a negligible effect on sensor response. Ten different colorimetric formulations consisting of bromophenol blue (BPB), bromocresol green (BCG), or BCP incorporated in different polymers, namely, poly(vinyl butyral), ethyl cellulose, or PMMA have been evaluated as NH_3 sensors (Courbat et al. 2009). The sensing layers were spin coated on a waveguide and the influence of the polymer matrix, addition of a plasticizer, and nature of the dye on sensor performance was studied. The sensor composed of BPB entrapped in PMMA along with the plasticizer was selected as the optimum composition. It exhibited linear response to NH_3 in the range <2 ppm, very good selectivity, and theoretical LOD of 2 ppb.

A sensing chip has been fabricated by coating a porous Si rugate filter with a composite film composed of chitosan and 2-glycidoxypropyltrimethoxysilane (Shang et al. 2011). The chip was exposed to BTB solution for 1 h to allow for dye molecule adsorption on the film surface. In the presence of NH_3, the BTB was deprotonated and therefore could absorb light reflected from the porous Si chip and a reduction in the reflected light intensity at 610 nm was observed. The sensor performance was characterized by short response time of 15 s (one-layer membrane), very good repeatability (RSD 0.5% for five measurements), and full reversibility. After 1 year of storage at room temperature, the sensor produced 90% of its original response to 160 ppm NH_3.

Silica sol–gel-derived materials have also been exploited in optical NH_3 sensing (Malins et al. 2000a). BCP was entrapped in hybrid sol–gel films composed of TEOS and MTEOS precursors mixed in different ratios and therefore the resulting films exhibited different polarities. The changes in the evanescent absorption at 585 nm were monitored upon sensor exposure to NH_3 gas. It was found that sensor sensitivity decreased with increased sensor film hydrophobicity. The authors also addressed the nonclassical temporal sensor response behavior that was observed in this study. A TEOS-derived solution doped with this dye was also coated on the unclad surface of a fiber (Cao and Duan 2005). The sensor exhibited good reversibility and repeatability. The influence of three different carrier gases, namely, nitrogen, argon, and air on sensor sensitivity and response time was investigated. The shortest response time of 2.7 min and highest NH_3 sensitivity was achieved when air was employed as a carrier gas. The sensor response was also tested at elevated temperatures and a response time of 10 s was observed at 55.5°C. A sol–gel-derived material based on TEOS doped with BCP and coated on a bent optical fiber to produce an absorption-based NH_3 sensor for gas and dissolved phase measurements has been described by Tao et al. (2006b). The NH_3 sensitive sol–gel layer was overcoated with polydimethyl siloxane (PDMS) for aqueous measurements to avoid leaching and cross-sensitivity toward protons and alkali ions. The gas phase sensor displayed a response time of ~1 min and LOD of 13 ppb while the dissolved phase sensor exhibited a 10 min response time and detected NH_3 in water down to 5 ppb. Another NH_3 sensor composed of PMMA doped with BCP exploited the changes of sensing layer refractive index at 1550 nm upon exposure to gaseous NH_3 (Passaro et al. 2007). The sensor exhibited very good linearity, LOD of 4‰, and detected refractive index changes of 8×10^{-5}. The mesostructured material Al-MCM-41 impregnated with BCG has also been evaluated as a NH_3 gas sensor (Chang et al. 2011).

3.3.4.3 NH₃ Sensing Based on Conducting Polymers

Conducting polymers such as polyaniline (PANI) have also been widely used in optical NH_3 sensing as the absorption spectrum is pH dependent. Chemical oxidation was used to produce a PANI film that was deposited on a polyethylene surface (Jin et al. 2001). The PANI film changed its absorption properties and a distinct color change from green to blue was observed upon NH_3 exposure. A response time of 15 s and recovery time of 2 min were reported. The sensor linear response extended from 180 to 18,000 ppm and LOD of 1 ppm (v/v) was observed. The film was fully regenerated by using hydrochloric acid at room temperature. A composite PANI–PMMA-based NH_3 sensor using transmittance measurements was described by Nicho et al. (2001). The changes in the optical properties of the composite film upon NH_3 exposure were studied using a nulling optical-transmittance bridge with a 632 nm laser source. The NH_3 gas could be removed from the film surface by purging with nitrogen and therefore the response was found to be reversible. An LOD from ~10 to 4000 ppm was observed, depending on the conductivity of the film. The PANI/PMMA composite material was also exploited by Airoudj et al. (2008a) to produce an evanescent planar waveguide NH_3 sensor. The sensing layer was deposited on an SU-8 waveguide and changes in transmitted light power upon NH_3 exposure were monitored using a spectrophotometer. The sensing principle was based on the change of the sensing layer refractive index that modified the propagation conditions of the guiding structure thus leading to an increase in the total transmitted light intensity of the SU-8 waveguide. A response time of 6 min was reported. The same authors also presented PANI/epoxy resin composite material that changed its absorbance properties in the presence of NH_3 (Airoudj et al. 2009). The sensor exhibited short response time, was easy to regenerate at room temperature, and possessed good chemical stability. The performance was compared to a pure PANI-based NH_3 sensor. In another paper, they also described PANI deposition on the waveguide using a plasma polymerization technique and evaluated its response toward NH_3 gas sensing (Airoudj et al. 2008b). Other optical NH_3 sensors employing conductive polymers have also been reported (Castrellon-Uribe et al. 2009). PANI was also used to monitor gaseous NH_3 at 1300 nm as the transmission of PANI films at this wavelength increases (Christie et al. 2003). The remote sensing system was based on a standard telecom LED and 100 m of duplex multimode fiber. The sensing film responded to NH_3 at concentrations down to 6 ppm at 50% relative humidity but several hours were required for the sensor output to reach equilibrium. The optical transmittance changes of PANI films in the presence of NH_3 were also reported by Lee et al. (2003).

3.3.4.4 Luminescence-Based NH₃ Sensing

Luminescence-based NH_3 sensing has also been explored. Chen and co-authors immobilized amino-fluorescein (AF) in a range of hybrid organically modified sol–gel-derived films that differed in polarity (Chen et al. 2004). The sensor films were spin-coated on glass slides to produce a dissolved NH_3 sensor. Such sensors exhibited different linear ranges, LOD, response time, and stability, depending on the sol–gel matrix formulation. The fluorescent indicator erythrosin A has been incorporated into a porous plastic optical fiber composed of styrene and diethylene glycol dimethacrylate and used for dissolved NH_3 sensing (Xie et al. 2005). The indicator fluorescence intensity increased in the presence of NH_3 solution. A response time of 22 min, when the NH_3 concentration was varied from 0 to 10^{-3} mol L^{-1}, no leaching, no pH interference, and good reversibility (RSD 2.6%) was observed.

NH_3-sensing sol–gel-derived materials doped with fluorescein have been reported by Persad et al. (2008). Waich et al. presented a fiber-optic microsensor for aqueous NH_3 determination in the range below 100 µg L^{-1} and with an LOD of 0.5 µg L^{-1} (Waich et al. 2007). The sensing layer was composed of a pH-sensitive fluorescent dye, eosin, immobilized in a cellulose ester matrix. The maximum emission wavelength of the dye appeared at 545 nm and shifted to 555 nm in the presence of 1 ppm NH_3 in nitrogen. The indicator was deprotonated by NH_3 and the resulting ammonium ion was stabilized by a cation trap. A protective Teflon layer was deposited on the sensing layer to prevent dye leaching and pH, ionic strength, and alkali ions interferences. The sensor was characterized by very good reversibility and response time of the order of 15 min. The photostability was also investigated and after 105 min of continuous illumination of the sensor exposed to 100 µg L^{-1} NH_3 the intensity was reduced to 95%. The same group presented a DLR trace NH_3 sensor exploiting the resonance energy transfer (RET) process (Waich et al. 2009). Coumarin derivatives were used as donors in RET to pH-sensitive indicators. Two sensing systems were presented. The first employed a coumarin derivative and eosin. The second system was a so-called RET cascade system and employed a coumarin derivative, sulforhodamine and BPB. The dynamic range extended from 1 to 5,000 µg L^{-1} and from 50 to 50,000 µg L^{-1}, depending on the system used. An optical NH_3 sensor employing commercially available cross-linked acrylic ester polymer microparticles in combination with a number of fluorescent dyes such as fluorescein, rhodamine B, eosin, acridine orange, and meso-tetraphenylporphyrine has been evaluated (Takagai et al. 2010). The fluorescein-based sensor was selected as the optimum composition and exhibited a linear response toward NH_3 vapor in the range from 1.0 to 250 ppm and LOD of 0.73 ppm. The sensor sensitivity toward organic amines was also studied.

3.3.4.5 Other Approaches in Optical NH₃ Sensing

NH_3 gas sensing was also realized using two forms of copper(II) complexes that displayed near-infrared (NIR) absorption bands at 735 and 744 nm (Scorsone et al. 2003). Methyl- and ethyl derivatives of a 5-(4%-dialkylaminophenylimino)quinolin-8-one copper(II)perchlorate complex have been immobilized onto the cladding material. An increase in transmitted intensity was observed upon exposure to NH_3. The proposed mechanism of the sensing reaction of these copper(II) complexes with NH_3 was one of ligand exchange. The dissolved NH_3 was monitored using bis(acetylacetoneethylenediamine)tributylphosphin cobalt(III) tetraphenyl complex that was coated on transparent triacetylcellulose membranes (Absalan et al. 2004). The film absorption maximum at 408 nm increased with increasing NH_3 concentration. The sensing mechanism was said to be based on complex formation whereby NH_3 molecules diffuse into the membrane and bind to the cobalt(III) complex. The linear dynamic range extended from 3.3×10^{-4} to 6.9×10^{-3} mol L^{-1}, calculated LOD was 5.0×10^{-5} mol L^{-1}, and response time in the range of 4–6 min was reported. The sensor was fully regenerated in pH 2.0.

Other examples of optical NH_3 sensing described in the literature include mesostructured SBA-3 silica doped with a solvatochromic probe, Reichardt's dye, which was exploited in reversible NH_3 sensor production (Onida et al. 2004). A fiber-optic sensor for dissolved NH_3 determination was presented by Pisco et al. (2006). Tin dioxide (SnO_2) was used as a sensing layer and was deposited on the distal end of the single-mode silica optical fiber. Upon NH_3 exposure, the reflectance increased as a consequence of the interaction between NH_3 molecules and the sensing layer. In another approach, a bent optical fiber has been dip coated with composite sol–gel-derived layer (SiO_2) doped with 25 nm silver nanoparticles (Guo and Tao 2007). Exposure to NH_3 enhanced the attenuation of light power guided through the fiber coated with the composite layer. The feasibility of using a zeolite-coated long-period fiber-grating sensor for gaseous NH_3 detection was also reported by Tang et al. (2011).

3.3.5 Optical pH Sensing

3.3.5.1 Introduction

Most optical pH sensors consist of a pH-sensitive indicator that reversibly interacts with the protons and is immobilized in a proton-permeable solid matrix. Such a matrix dictates the characteristics of the sensor. The pH indicators are typically weak organic acids or bases with distinct optical properties associated with their protonated and deprotonated forms. In most cases, pH sensitivity is based on a color or fluorescence intensity change with a change in the concentration of hydrogen ions (H_3O^+). The pH indicators may be characterized on the basis of their pKa values. For a particular application, the pKa of an indicator should be close to the pH range that the application requires. An approximate definition is that the working range of the pH dye in solution is within ±1.5 pH units of the pKa. It is important to note that the optical technique measures the concentration of the acidic and basic forms of the pH indicator and the pH value is defined in terms of activity (Janata 1987). Therefore, optical pH sensors are cross-sensitive toward ionic strength and this limits their use in practical applications. Ionic strength cross-sensitivity can cause pH errors of up to 1.5 pH units (Edmonds et al. 1988) and depends on the pH indicator charge and its environment or the electrolyte concentration of the sample. The pH indicator can be covalently bound to the polymer or sol–gel matrix (Rottman et al. 1998, Ensafi and Kazemzadeh 1999), which completely eliminates leaching effects. However, often this process involves complex chemistry, and the sensor response of covalently bound indicators is sometimes reduced compared to their performance when physically entrapped. As stated earlier, our focus is on the absorption- and fluorescence-based optical pH sensors described in the literature in the last decade.

As mentioned in Section 3.1.2, the first colorimetric pH strips suitable for continuous pH measurements were developed in the 1970s (Harper 1975). In 1980, Peterson et al. reported on the first fiber-optic absorption-based pH sensor that utilized phenol red (PR) as an indicator dye immobilized in polystyrene microspheres (Peterson 1980) and in 1982, Saari et al. reported on the first fluorescence-based pH sensor using covalently immobilized fluoresceinamine in a cellulose matrix (Saari and Seitz 1982). A review has been published by Lin in 2000, which covers optical pH sensors development in 1990s (Lin 2000).

3.3.5.2 Colorimetric pH Sensors

A colorimetric sol–gel-derived pH sensor was demonstrated, which exhibited a linear response over 3.5 pH units (Lin and Liu 2000). Three pH indicators, namely, BCG, BCP, and PR were co-entrapped in TMOS-based matrix. The conditions that have to be fulfilled in order to design an optical pH sensor that exhibits linear pH response over a broad range were discussed. However, the pH sensor performance was not examined. A wide dynamic range was also achieved by entrapping ethyl violet in a TEOS-based sol–gel matrix. The solution was coated on the unclad core of the fiber (Sharma and Gupta 2004). The sensor exhibited a dynamic range from 2.0 to 12 pH units. Dong et al. (2008) reported on a wide dynamic range pH sensor based on evanescent wave absorption. A sol–gel-based solution was doped with a mixture of three pH dyes, namely, cresol red, BPB, and chlorophenol red, and was deposited on a fiber-optic waveguide. The sensor dynamic range extended from pH 4.5 to 13.0; response time of 5 s and very good reversibility was observed.

Lee et al. (2001) also presented a fiber-optic sol–gel-derived pH sensor based on evanescent wave absorption. The TEOS-based solution doped with BCP or BCG was deposited on an unclad portion of a multimode optical fiber. It was found that by applying multiple sol–gel layers, the pH sensitivity was enhanced. The pKa' values shifted significantly toward higher values upon immobilization (from 6.0 to 8.0 and from 4.4 to 8.0 for BCP and BCG, respectively). A response time of 5 s was reported when the pH was varied from 4.0 to 11.0 and 30 s in the reverse direction. Extensive leaching was observed in alkaline solutions. PR was entrapped in a hybrid sol–gel material composed of TEOS and phenyltriethoxysilane (PhTEOS) (Wang et al. 2003). A small amount of the organically modified silane, PhTEOS, improved phenol red solubility in the material, and therefore leaching was not observed. The sensor exhibited a broad dynamic range from 6.0 to 11.0 pH units, a short response time <20 s, and good stability during continuous usage over 3 months. An optical fiber pH sensor also exploiting PR was used for measuring pH in interstitial fluid (Baldini et al. 2007). The PR was covalently immobilized to the internal wall of a glass capillary that was interrogated with optical fibers. The sensor was pH sensitive in the range from pH 6.0 to 8.0, exhibited a response time ~1 min, and accuracy of 0.07 pH units. The dependence on the ionic strength was also addressed. The sensor was used for in vivo pH measurements on animals. Phenolphthalein and phenol red were covalently immobilized to a diacetylcellulose matrix and poly(vinyl alcohol), respectively (Liu et al. 2004, 2005). A phenolphthalein-based sensor was used for pH determination in the range from 8.0 to 12.0 pH units. Colorimetric pH sensors were also demonstrated using BTB entrapped in mesostructured silica (Miled et al. 2004), neutral red immobilized in layer-by-layer assemblies (Goicoechea et al. 2008), and a mixture of malachite green oxalate and BCG incorporated in triacetylcellulose (Noroozifar et al. 2010).

A colorimetric optical sensor for high and low pH values was also described. Two dyes, victoria blue (VB) and dipicrylamine (DPA) were co-immobilized on triacetylcellulose (Safavi and Bagheri 2003). Two bands were exploited, one at 437 nm and 621 nm related to DPA and VB, respectively. The pKa' values were found to be 1.82 and 11.2 for DPA and VB, respectively. The sensors showed very good reversibility, short response time <1 min, and no leaching was reported. An ANN was used in order to extend the useful sensor working range. ANN was also applied to a sol–gel-derived pH sensor that used BPB as a pH indicator (Suah et al. 2003). As a result, broadening of the limited dynamic range to nearly full range (from pH 2.0 to 12.0) was demonstrated. Capel-Cuevas et al. also exploited ANN and reported on a full-range (pH range from 0 to 14) optical pH sensor array (Capel-Cuevas et al. 2011).

A pH sensor for low pH value determination based on congo red, covalently immobilized on epoxy-activated hydrophilic agarose membrane has also been described (Hashemi and Abolghasemi 2006). The sensor pKa' was determined to be 2.8 and the dynamic range extended from pH 0.5 to 5.0. The same group chemically immobilized a mixture of neutral red and thionin using the same substrate that was suitable for use in a broad pH range from 0.5 to 12.0 (Hashemi and Zarjani 2008). An optical fiber sensor in combination with phenolphthalein was used to monitor the penetration of alkaline solutions into an unsaturated polyester resin that is useful for online health monitoring of polymer composites in a corrosive environment (Gotou et al. 2006).

3.3.5.3 *Luminescence Intensity-Based pH Sensors, Including Ratiometric Methods*

The literature for fluorescence-based pH sensing is dominated by two pH-dependent fluorescent probes, namely, fluorescein and its derivatives and 1-hydroxypyrene-3,6,8-trisulfonic acid (HPTS). These dyes usually absorb in the visible blue region while emission occurs above 500 nm. A composite material doped with fluorescein was fabricated using sol–gel and poly(vinyl alcohol) in order to achieve better material flexibility and stability (Cajlakovic et al. 2002). The intensity-based sensor exhibited a broad calibration plot that was associated with a nonuniform distribution of a pH indicator in the films. Different pKa' values were reported depending on the sensor film formulation and extended from 6.44 to 7.06 at ionic strength of 50 mM. Response time in the range of minutes and no leaching was observed for a 48-h period. The pH error caused by ionic strength interference of up to 0.78 pH units was reported. Fluorescein isothiocyanate was covalently attached to a mesoporous sol–gel film (Wirnsberger et al. 2001).

Weidgans et al. (2004) reported on fluorescent pH sensors with negligible sensitivity to ionic strength. Three different lipophilic fluorescein esters with only one negative charge were synthesized and incorporated in uncharged hydrogel. The hydrogel consists of hydrophobic and hydrophilic domains that facilitate the incorporation of lipophilic fluorescein derivatives with no leaching. The pKa' values were 5.54,

6.97, and 7.28, measured in phosphate buffer with ionic strength adjusted to 100 mM. Sensor membranes exposed to pH 8.0 for 12 h exhibited <4% signal loss. Very good reproducibility (RSD of pKa' values <1.3% for a set of four membranes) was reported. The sensors also exhibited very good photostability. The same hydrogel matrix was used for the entrapment of two other fluorescein derivatives, namely, 2',7'-dihexyl-5(6)-N-octadecyl-carboxamidofluorescein ethyl ester (DHFAE) and 2',7'-dihexyl-5(6)-N-octadecyl-carboxamidofluorescein (DHFA). These dyes exhibited higher pKa' ~ 8.4 and therefore were suitable for pH monitoring in seawater and marine sediments (Schroder et al. 2005). A sensor response time of 90 s was reported for a 6 μm thick membrane when pH was varied from 8.5 to 7.5. Both sensors exhibited good photostability and small temperature dependence. Negligible ionic strength interference was observed for the DHFAE membrane. However, leaching of the indicator was reported. The optical properties of DHFAE facilitated the use of robust dual excitation ratiometric detection. A DLR detection method was used for a DHFA-based sensor and O_2-insensitive long-lived luminescent reference particles were also incorporated in the sensor membrane. In 2007, novel Ph-sensitive coumarin-based fluorescent indicators were described (Vasylevska et al. 2007). These iminocoumarin derivatives are compatible with commercially available LEDs and exhibit moderate to high brightness and excellent photostability. The intensity was reduced by ~10% after 1 h of continuous illumination using a mercury lamp. In order to avoid leaching, the indicators were covalently attached to the surface of amino-modified polymer microbeads. Indicator-loaded microbeads were incorporated into a hydrogel matrix to produce a pH-sensitive material that was virtually independent of the ionic strength due to the uncharged nature of the dyes and the hydrogel used. The sensing materials are suitable for pH determination in different ranges using DLR or dual-emission methods, depending on the dye used. When a mixture of two indicator-loaded microbeads was used, the pH sensor dynamic range extended from pH 1.0 to 11.0. A DLR pH sensor was also reported based on a red excitable fluorescent seminaphthorhodafluor and luminescent inorganic phosphor chromium(III)-doped gadolinium aluminum borate with low temperature sensitivity that acted as a DLR reference (Borisov et al. 2010). Both indicator dye and reference phosphor were excited with an LED at 605 nm. Negligible cross-sensitivity to ionic strength and short response time of 30 s when the pH was varied from 4.5 to 8.5 was reported. Poor storage stability at room temperature and poor photostability of the indicator dye was observed.

Recently, fluorescein was incorporated in a layered double hydroxide (LDH) matrix to produce an optical pH sensor (Shi et al. 2010). The dye, together with a surfactant, was incorporated into the Mg_2Al-LDH matrix using electrophoretic deposition. The highly oriented films were fabricated with thickness in the range from nanometer to micrometer on indium tin oxide substrates. The sensor linear pH range extended from pH 5.0 to 8.5; good reversibility (RSD < 1.5%), storage stability, and short response time of 2 s for film 300 nm thick were reported.

Organometallic ruthenium (II) complexes have been examined for use in luminescence-based pH sensing. However, when using such species for pH sensing, the O_2 cross-sensitivity is often a problem. Malins et al. evaluated a series of protonable ruthenium(II) polypyridyl complexes that were entrapped in sol–gel-based matrices (Malins et al. 2000b). They found that bis(2,2'-bipyridine)[3-(2-phenol)-5-(pyridine-2-yl)-1,2,4-triazole]ruthenium(II) was the most suitable as it exhibited a linear response in the range 3.0–8.0 pH units, resolution of 0.05 pH units, with minimal sensitivity to O_2. The films were dip coated on glass and were stable when kept in air for several months. No leaching was observed. A broad-range pH sensor based on a ruthenium(II) complex consisting of different polypyridyl ligands has been described (Kim et al. 2009). The wide pH range extended from pH 3.0 to 9.0 and was attributed to the multiple steps of protonation and deprotonation. The O_2 cross-sensitivity has not been addressed. Recently, Orellana's group published a paper where they discussed the suitability of Ru(II) polypyridyl complexes as direct pH-sensitive indicators (Tormo et al. 2010). An irreversible excited-state proton transfer process to buffer solution was observed, which makes most of such complexes inappropriate for use as pH indicator dyes.

Lam and others covalently attached a rhenium (I) complex to a sol–gel-matrix derived from trimethoxypropylsilane (Lam et al. 2000). The luminescence intensity decreased with increasing pH values at 484 nm. A lower pKa' was observed in the sol–gel material spin coated on glass (5.80) than in aqueous solution (7.05) and O_2 interference did not seem to affect the response. Lanthanide luminescence-based optical pH sensors have also been reported (Blair et al. 2001, Gunnlaugsson 2001). Recently, pH-sensitive ratiometric europium(III) complexes have been reported that allow for pH measurements in the range from pH 6.0 to 8.0 (Pal and Parker 2008).

A TEOS-derived sol–gel-based pH sensor doped with 10-(4-aminophenyl)-5,15-dimesitylcorrole suitable for pH detection over a wide range was developed by Li et al. (2006). The sensor was excited at 507 nm and the pH-dependent fluorescence was detected at 656 nm. These changes were said to be caused by the change of the dye quantum yield. The sensor linear pH response was in the range from 2.17 to 10.30 pH units. Good photostability was reported as after 3 h of continuous illumination at 507 nm the intensity only reduced to 98%. The sensor also exhibited satisfactory reversibility and reproducibility, and response time of <2 min was reported for the sensor membrane which was 4 μm thick.

Lee et al. immobilized HPTS in an electrostatic layer-by-layer assembly using poly(allylamine hydrochloride) as a polycation and a mixture of HPTS and poly(acrylic acid) as a polyanion (Lee et al. 2000). The multilayers were deposited on glass slides and then were cross-linked at 130°C to induce amide bond formation. The possibility of ratiometric pH sensing using such assemblies was discussed. HPTS was used in the development of robust dual-excitation ratiometric pH sensors for applications in marine sediments (Hulth et al. 2002). The planar pH sensor foils composed of HPTS incorporated in cellulose acetate polymer were used for the detection of two-dimensional (2D) pH distributions in sediments. The sensor exhibited excellent stability, short response time of 5 s, and no leaching. The sensor response was highly dependent on ionic strength and the apparent pKa' was shifted toward lower values (6.4) upon immobilization. Another ratiometric system for use in marine sediments was composed of HPTS covalently attached to poly(vinyl alcohol) membrane (Zhu et al. 2005). The sensor pKa' was 7.06, response time <2 min, and stability of 2 months was reported when the sensor was kept dry at room temperature. The intensity decreased by 5%–9% after 200 pH transitions from pH 6.0 to 8.2. Ionic strength interference was observed. Kermis et al. (2002) immobilized HPTS electrostatically on a basic anion-exchange resin, which was then entrapped in a hydrogel layer and used in bioprocess control monitoring. The apparent pKa' of the immobilized HPTS was found to be 7.7 and, therefore, the linear sensor range extended from pH 6.7 to 8.7. There was no change in the sensor response after 3 weeks of storage in deionized water. The performance also did not change after exposure to 70% ethanol/water that was used as a method of sterilization. A response time of 9 min, leaching, and cross-sensitivity to ionic strength were observed. This ratiometric sensor patch was used in a shake-flask setup to monitor online *E. coli* fermentation processes and the measurements were in agreement with the pH recorded off-line using a pH electrode. In order to improve the performance of the pH sensor, the same authors covalently attached modified HPTS by copolymerization with poly(ethylene glycol) diacrylate (Kermis et al. 2003). Some leaching was still observed but the sensor patch was steam sterilizable and exhibited a short response time of 3 min. The drawback was ionic strength cross-sensitivity. A hybrid sol–gel-derived sensor based on the ratiometric detection of the pH-dependent fluorescence of HPTS, which was ion-paired with hexadecyltrimethylammonium bromide (CTAB) to reduce its solubility in water and hence minimize dye leaching, was also reported (Wencel et al. 2009). The indicator was completely physically encapsulated in a novel hybrid sol–gel matrix composed of 3-glycidoxypropyltrimethoxysilane (GPTMS) and ethyltriethoxysilane (ETEOS), which, due to the degree of organic–inorganic polymerization and cross-linking, resulted in an optimum microstructure that eliminates dye leaching. The sensor displayed a resolution of <0.01 pH units and a short response time of <12 s for a film thickness of 1 μm. The dynamic range of the sensor was from pH 5.0 to 8.0 and temperature and ionic strength interference was also reported.

Hakonen et al. presented a time-dependent nonlinear calibration protocol that compensated for the HPTS-based ratiometric pH sensor drift (Hakonen and Hulth 2008). Sensor accuracy, precision, and lifetime were improved. The sensor repeatability was 0.0044 pH units, which was in close agreement to results obtained with a pH electrode (0.0046).

A different ratiometric scheme using one fluorophore was described by Sanchez-Barragan et al. (2006). It was based on monitoring the fluorescence intensity of the pH-sensitive indicator mercurochrome, another fluorescein-related dye, and the blue excitation light that was reflected by the sensing phase. Both signals were collected by the bifurcated optical fiber and were independent of the excitation light intensity. A H⁺-responsive fluoroionophore based on boron dipyrromethane was covalently immobilized in a mesoporous silica film and exploited in dual-emission ratiometric pH sensing in the range from pH 1.0 to 4.0 (Hiruta et al. 2010).

3.3.5.4 Lifetime-Based pH Sensors

Acridine immobilized in Nafion was evaluated as a lifetime-based pH sensor (Ryder et al. 2003). A dynamic range from 8.0 to 11.0 and cross-sensitivity to ionic strength were reported. Lifetime-based pH sensing based on the luminescent ruthenium(II) complex $[Ru(Ph_2phen)_2DCbpy]^{2+}$ (DCbpy-4,4'-dicarboxy-2,2'-bipyridine) with ~1 μs lifetime incorporated in a copolymer consisting of hydrophobic and hydrophilic domains was reported (Clarke et al. 2000). The sensor film with pKa' ~3.75 was stable in solution for over 2 years with no leaching. The errors associated with O_2 cross-sensitivity were calculated to be below 0.03 pH unit. Similar fiber-optic systems exploiting sol–gel materials as sensor supports were also reported (Goncalves et al. 2008). A system composed of a long-lived europium(III) complex co-entrapped with pH-sensitive BPB in a sol–gel-based layer was also used for lifetime-based pH sensing (Turel et al. 2008). The sensor was suitable for pH detection in the range from 4.0 to 9.5, did not suffer from O_2 cross-sensitivity, but ionic strength interference was reported.

3.3.5.5 Other Approaches in Optical pH Sensing

Organic conductive polymers (OCPs) such as PANI have also been exploited in optical pH sensor development (Lange et al. 2008). The optical changes occur in the NIR red region and have been used to monitor pH and acidic gases. Chemical oxidation was used to produce PANI films that exhibit pH-dependent absorption spectra and, during the protonation–deprotonation reaction, reversibly change color from green to blue (Jin et al. 2000). An apparent pKa of 6.7 was observed with a rapid change in absorbance between pH 5.0 and 8.0. The films could be stored in air for up to 1 month and in water for longer periods. A film thickness of 110 nm was measured and therefore the sensor exhibited a short response time of ~1 s when exposed to 0.1 M HCl and NaOH. Hysteresis was also observed, which was attributed to the inherent nature of the PANI films. A pH sensor based on the PANI–porous Vycor glass nanocomposite was described (Sotomayor et al. 2001). Aniline was polymerized inside the pores of the glass and a glass slide was placed on the distal end of a bifurcated fiber bundle. The polymer changed color from green (acidic form) to blue (basic from) and therefore the reflectance intensity in the range 450–600 nm decreased as pH increased. The sensor exhibited linear response in the range from pH 7.4 to 9.5 and response time of 4.8 and >16 min depending on the slide's thickness. A small influence of temperature and ionic strength on sensor response was observed and there was no leaching. There are a number of other papers available that report on PANI- and other OCPs-based pH sensors (Lindfors et al. 2003, Tsai et al. 2003, Lindfors and Ivaska 2005, Kaempgen and Roth 2006, Ge et al. 2007, Panda and Chattopadhyay 2007). Fluorescent pH-insensitive polystyrene beads coated with PANI were also explored as pH-sensing materials (Pringsheim et al. 2001). The beads maximum fluorescence band overlapped with pH-sensitive PANI absorption spectra and therefore the fluorescence intensity was changed as pH was varied due to an inner filter effect.

Another approach used in the design of optical pH sensors is the photoinduced electron transfer (PET) phenomenon (de Silva et al. 2009). Niu and others explored the use of proton "off–on" behavior of naphthalimide derivative for optical pH sensing (Niu et al. 2004). The N-allyl-4-piperazinyl-1,8-naphthalimide (AMPN) was photo-copolymerized with 2-hydroxyethyl methacrylate and acrylamide on a glass slide that was treated with silanizing agent. The photoinduced energy transfer occurred between the naphthalimide fluorophore and electron-donating NCH_3 group in the piperizine structure. The sensor was excited at 400 nm and exhibited one emission band at 517 nm. The sensor dynamic range extended from pH 4.5 to 9.0 and a response time of 90 s was reported. Cross-sensitivity to ionic strength was not observed but common inorganic ions and organics can interfere. The aforementioned AMPN dye was used in combination with meso-5,10,15,20-tetra-(4-allyloxyphenyl)porphyrin (TAPP) as a ratiometric pH sensor (Niu et al. 2005). The sensor exhibited broad dynamic range from 1.5 to 9.0. The pH PET probes were also described by others (Bojinov and Konstantinova 2007, Bojinov et al. 2008, Georgiev 2011).

Recently, a visual pH sensor was developed based on a dual LED system (Chen et al. 2010). A TEOS-based sol–gel film doped with 5(6)-carboxyfluorescein was deposited on a glass slide. Two LEDs were used in the system—465 nm LED for sensor film illumination and 660 nm LED used as a reference. The fluorescence and reference signal were captured by a CCD camera and appeared as different colors

TABLE 3.2

Examples of Papers on Optical pH Sensors Published Since 2000

Reference	Detection Method	Comments
Malins et al. (2000b)	Luminescence intensity	Ruthenium(III) complex entrapped in sol–gel-based matrix as a pH-sensitive probe
Blair et al. (2001)	Luminescence intensity and ratiometric	Lanthanide complexes immobilized in sol–gel-derived matrix as pH sensors (excitation wavelengths 255–385 nm)
Liebsch et al. (2001)	Time-domain DLR (t-DLR)	New referenced scheme applied to pH sensing—two luminophores with different decay times
Safavi and Bagheri (2003)	Colorimetric	Sensor for high and low pH values exploiting victoria blue (pKa′ = 11.2) and dipicrylamine (pKa′ = 1.82)
Ryder et al. (2003)	Lifetime-based	Acridine immobilized in Nafion® as lifetime-based pH sensor
Weidgans et al. (2004)	Fluorescence intensity	pH sensors with negligible sensitivity to ionic strength (lipophilic fluorescein derivatives embedded in hydrogel)
Niu et al. (2004)	Fluorescence intensity	PET pH sensor based on naphthalimide derivative
Li et al. (2006)	Fluorescence intensity	Amino-functionalized corrole entrapped in sol–gel-based matrix for broad pH range detection
Baldini et al. (2007)	Colorimetric	Phenol red-based fiber-optic pH sensor used for in vivo measurements on animals
Vasylevska et al. (2007)	2λ ratiometric and DLR	Novel pH sensitive coumarin-based fluorescent indicators
Hashemi and Zarjani (2008)	Colorimetric	Mixture of neutral red and thionin, broad pH range from 0.5 to 12.0
Sun et al. (2009)	Upconversion luminescence	First pH sensor exploiting upconverting luminescent lanthanide nanorods in combination with BTB
Borisov et al. (2010)	DLR	Red excitable fluorescent SNARF-DE and luminescent inorganic phosphor as reference. Sensors with intrinsic temperature compensation
Shi et al. (2010)	Fluorescence intensity	Fluorescein incorporated in a layered double hydroxide matrix
Ruedas-Rama et al. (2011)	Lifetime-based	First quantum dot lifetime-based pH nanosensor for intracellular measurements

depending on the sample pH. The sensor exhibited reversible and reproducible response with response time of several seconds. The sensor was validated in rainwater and acid wastewater analysis and compared very well with the results obtained using the standard pH strip and glass electrode.

To summarize this section on pH sensing, Table 3.2 gives a selection of some of the many papers that have been published since 2000.

3.3.6 Cations, Anions, and VOCs

There has been some activity in reagent-based optical sensing for ions. Key developments in the early part of the decade were captured in the review by Wolfbeis (2004a). Much of the recently reported literature on ion sensing relates to non-optical sensor techniques. In recent years, coinciding with the global focus on preserving the environment and on environmental sensing, optical sensing of mercury(II) ion (Hg^{2+}) has probably been more widely reported than other ion species. A sensitive and highly selective colorimetric sensor for Hg^{2+} was reported, which was based on the interaction of Hg^{2+} with 2-mercaptoo-2-thiazoline in plasticized PVC membranes incorporating a proton-selective chromoionophore and lipophilic anion sites. The sensor was used for flow-through determination of Hg^{2+} in river water and had a detection limit of 5×10^{-11} M with a response time of <40 s (Amini et al. 2008). Yari and Abdoli

(2010) also reported a colorimetric sensor for Hg^{2+} which was based on incorporation of the indicator dye 4-phenyl-2,6-bis(2,3,5,6-tetrahydrobenzo[b] [1,4,7] trioxononin-9-yl) pyrylium perchlorate into a sol–gel layer. The detection limit was 1.11×10^{-9} M and was stable, reproducible, and selective. A sensor for Hg^{2+} based on quenching of fluorescence of the porphyrin complex tetra(p-dimethylaminophenyl) porphyrin (TDMAPP) complex in PVC has been reported, which had a LOD of 8.0×10^{-9} mol L^{-1} and was selective, reproducible and reversible (Yang et al. 2009). A fluorescent sensor for Hg^{2+} in water, based on the functionalization of mesoporous silica with the fluorescent chromophore, 1-(4'-hydroxylphenyl)-4-pyrene-2,3-diaza-1,3-butadiene (Py-OH) was reported (Wang et al. 2010), where the Py-OH showed a dramatic increase in fluorescence on exposure to Hg^{2+}. The LOD was 1.7×10^{-7} g mL^{-1}. Other recent reports of optical sensing of Hg^{2+} include Guo et al. (2011), Chen et al. (2011), and Alizadeh et al. (2011).

A colorimetric sensor for uranyl ion was reported, which used Piroxicam as a cation recognition agent and Alizarin as a co-ionophore in PVC membrane (Sadeghi and Doosti 2008). The reported LOD was 6.0×10^{-9} mol L^{-1}. Bulk optode membranes based on thiaglutaric diamide ionophore were developed for silver (I) and copper (II) in aqueous solution where different combinations of ionophore and chromoionophores allowed selective sensing of the two ions (Liu and Qin 2008). Ng et al. (2006) reported a fluorescence sensor for Al^{3+} ions based on MIP, with a LOD of $3.62\,\mu M$. The fluorescence tag was 8-hydroxy-quinoline sulfonic acid. Sensors for Zn^{2+} (Gong et al. 2011) and Fe^{3+} (Su et al. 2010) have also been reported as has a Cl^{-} sensor based on a squaraine rotaxane (Gassensmith et al. 2010).

For optical sensing of VOCs, the reader is referred to the Wolfbeis reviews of 2006 and 2008 as cited earlier. Elosua et al. (2009) published a review of fiber-optic nanosensors for VOCs. The same authors more recently reviewed the area of optical fiber sensors for sick building syndrome. Different optical configurations and sensor types are discussed in the context of detection of relevant analytes such as NH_3, CH_4, and CO_2 (Elosua et al. 2010). Brittle et al. (2011) reported on the use of a range of tetraphenylporphyrin compounds whose absorbance changes in the presence of VOC vapors based on the electron donating/withdrawal strength of specific functional groups. A range of VOC analytes were tested, including acetic acid, butanone, and ethyl acetate. Sensing of toxic metal ions and some VOCs have been discussed in a recent review of optical chemical sensors-based hybrid sol–gel nanoreactors (Tran-Thi et al. 2011).

3.3.7 Multi-Analyte Sensing and Imaging Optical Sensors

3.3.7.1 Introduction

A current trend in the field of optical chemical sensors is the development of multi-analyte sensors. Multiparameter analysis is highly desirable in wide application areas, for example, simultaneous monitoring of O_2, pCO_2, and pH in blood or during the fermentation process. The simultaneous detection of a number of analytes is a challenging task. Cross-sensitivity and spectral overlap can significantly increase the complexity of the sensing system. The availability of luminescent analyte-sensitive probes with optical properties that allow for easy spectral discrimination by using appropriate optical filters is still limited.

Most of the current research has been focused on dual-analyte optical sensors and they will be the subject of the following section. Multi-analyte sensing has been implemented by using a sensor array format or single/two-sensing-layer approach (Nagl and Wolfbeis 2007). This approach allows for the acquisition of several analytical parameters from a single spot. Early work on this approach was demonstrated in 1988 when an O_2 and CO_2 fiber-optic chemical sensor was developed (Wolfbeis et al. 1988). The sensing layers were excited with a single excitation source and appropriate optical filters were employed for wavelength selection of the emitted luminescence. As well as the method of signal separation via spectral distinction using steady-state luminescence intensity, methods based on differences in the excited-state lifetime of luminescent probes can be exploited, such as dual lifetime referencing, rapid lifetime determination, or dual lifetime determination method (Stich et al. 2010).

Imaging using optical sensors is also a very attractive tool in medical and biological application areas and CCD cameras have become very useful in this area. They have been deployed in this field as they facilitate very good temporal and spatial resolution of illuminated areas. An intensity-based, imaging-based system was used for measurements for O_2 distribution in perfused tissue (Rumsey et al. 1988). In 1991, Lakowicz et al. described a lifetime-based fluorescence imaging system (Lakowicz and Berndt 1991, Lakowicz et al. 1992).

In the following sections, we will report on some single-analyte imaging optical sensors and on dual-analyte sensing and imaging luminescent sensors that have been described in the literature in the last decade.

3.3.7.2 Single-Analyte Imaging Optical Sensors

O_2 pH, CO_2, and temperature mapping, using luminescence lifetime imaging, was demonstrated by Liebsch et al. (2000). They used a fast-gateable CCD camera in combination with a pulsed LED array as a light source. The sensor spots were attached at the bottom of the wellplates and the gradients were mapped within ~1 s. The group applied such a system for the first time to measure changes of O_2 distributions in cartilage tissue cross sections where the O_2-sensitive material (PtOEP in prepolymer E4) was spray coated on a polyester support (Kellner et al. 2002). The O_2 gradients were correlated with tissue composition and histological analysis. In 2005, similar system (PtOEP in polystyrene) was exploited using planar sensors for in vitro and in vivo pO_2 surface measurements over normal and tumor tissue (Babilas et al. 2005) and to measure transcutaneous O_2 in humans before, during, and after tourniquet-induced forearm ischemia (Babilas et al. 2008). Holst et al. and Konig et al. also reported on luminescence lifetime imaging with an O_2 sensor composed of a ruthenium complex(II) entrapped in a PVC membrane and ORMOSIL matrices, respectively (Holst and Grunwald 2001, Konig et al. 2005). Polerecky et al. reported on 2D measurements of O_2 distributions in bioirrigated sediments using a luminescence lifetime imaging system. A planar optode has been reported that consisted of an O_2-sensitive platinum(II) meso-tetra (pentafluorophenyl) porphyrin incorporated in polystyrene (Polerecky et al. 2006). Recently, ultra-bright optodes based on a cyclometalated iridium(III) coumarin complex incorporated in polystyrene have been presented (Staal et al. 2011). The thin O_2-sensing films exhibited higher brightness, lower temperature dependence, and improved homogeneity of luminescence lifetime images than sensors based on a ruthenium(II) polypirydyl complex or platinum(II) porphyrin. These sensor films were used in imaging of patterns of O_2 distributions in biofilms. Zhang et al. presented a concept where a dual-emissive, O_2-sensitive boron was synthesized and exploited in the form of nanoparticles (98 nm) as an O_2 sensor for tumor hypoxia imaging (Zhang et al. 2009). This material exhibited green room-temperature phosphorescence and intense blue fluorescence, thus allowing for ratiometric sensing. In this approach, the short-lived fluorescence served as the O_2-insensitive internal reference and oxygenation was monitored by phosphorescence quenching. The linear calibration plot was in the range from 0% to 3% of O_2, which is compatible with hypoxia, and NPs have been employed in ratiometric tumor hypoxia imaging. The oxygen maps showed an excellent contrast between the microvasculature and the tumor tissue.

Referenced time-resolved DLR (t-DLR) pH imaging has been presented where two luminophores with different decay times were used (Liebsch et al. 2001). The two images taken at different time gates were recorded using a CCD camera and the ratio of the images represented a referenced intensity distribution that reflected the pH at each pixel. The t-DLR pH sensor foil was composed of pH-sensitive particles that consisted of a poly(acrylonitrile) core that contained an inert reference [Ru(dpp)$_3$]$^{2+}$ and a hydrogel shell with covalently bound carboxyfluorescein in a polyurethane layer. A t-DLR pH sensor was also fabricated and used for 2D imaging of pH in vivo (Schreml et al. 2011). Fluorescein isothiocyanate was covalently attached to aminocellulose particles while [Ru(dpp)$_3$]$^{2+}$ was entrapped in O_2-insensitive polyacrylonitrile particles and served as an internal reference. Both particles were immobilized in a polyurethane hydrogel matrix. The sensor foils were used for the in vivo luminescent imaging of pH distribution in cutaneous wound healing. The same imaging method was exploited in copper(II) ion determination based on the fluorescence quenching of lucifer yellow in the presence of copper(II) ions (Mayr et al. 2002a).

Imaging pH-sensitive nanotip arrays fabricated on distal faces of coherent fiber-optic bundles was described in 2000 (Liu et al. 2000). A pH-sensitive sensor layer composed of fluorescein isothiocyanate–dextran incorporated in a photopolymer was deposited across the array. Fluorescence images were acquired from such a sensor inserted in a rat liver. A feasibility study was presented on using drop-on-demand microjet printing technology for the design of pH-imaging sensor arrays printed on the surface of an optical fiber image guide (Carter et al. 2006). Another approach was presented by Kreft et al. and described pH-sensitive polyelectrolyte macrocapsules loaded with SNARF-1 dextran which have been used for imaging of intracellular pH values in human breast cancer cells and in fibroplasts (Kreft et al. 2007). A ratiometric imaging ammonium sensor has also been developed using a time-correlated pixel-by-pixel calibration scheme (Stromberg and Hulth 2005). The sensor was used to image the concentration of ammonium ion in soil.

3.3.7.3 Dual-Analyte Optical Sensors

Malins et al. developed a dual-analyte optical sensor platform based on the surface patterning of O_2- and CO_2-sensitive fluorescent dyes incorporated in a sol–gel-derived material (Malins et al. 2000c). Discrete deposition of sensitive zones on a single waveguiding platform was demonstrated. The sensor platform facilitated the efficient excitation of immobilized indicator dyes with a single LED and could be imaged with a CCD camera. Cho et al. presented pin-printed sensor arrays for simultaneous multi-analyte detection (Cho and Bright 2002). O_2- and pH-sensitive sol–gel-based solutions were pin printed in the form of discrete spots on a planar substrate and were based on $[Ru(dpp)_3]^{2+}$ and fluorescein, respectively. The sensor spots were in the order of $100\,\mu m$ in diameter, $1–2\,\mu m$ thick, and were printed at a rate of $\sim$ one sensor element per second. No significant cross talk or interference was observed. Another system involved disposable sensor arrays for fluorescence-based Ca^{2+}, Na^{+}, Mg^{2+}, SO_4^{2-}, Cl^{-}, and Hg^{2+} detection, which were formed on the bottom of a well plate (Mayr et al. 2002b). The fluorescence of the array was imaged using a CCD camera. Walt et al. developed fiber bundle arrays that comprised of a 1000 fused optical fibers (Epstein and Walt 2003, Walt 2010). Such a multi-analyte imaging fiber sensor has been fabricated and was used for simultaneous pH, CO_2, and O_2 monitoring during beer fermentation (Ferguson et al. 1997). Wygladacz et al. exploited imaging fibers that contained microarray wells for fluorescent-based ion sensing (Wygladacz and Bakker 2005). The Na^{+}- and Cl^{-}-sensitive microspheres were deposited on etched fiber bundles and simultaneous Na^{+} and Cl^{-} sensing was demonstrated.

It is well known that optical O_2 sensor response depends on temperature and, therefore, a number of papers have been published that reported on simultaneous sensing of O_2 and temperature using optical sensors. Hradil et al. reported on a technique for temperature correction in pressure-sensitive paint (PSP) applications (Hradil et al. 2002). A lifetime-based PSP system consisted of a sol–gel-based paint doped with an O_2-sensitive and a temperature-sensitive luminophore. The spectral properties of the luminophores were selected to allow LED excitation with a single source and detection by one camera. The temperature phosphor used was manganese-activated magnesium fluorogermanate with lifetime of $\sim$3 ms compared to a lifetime of $\sim$5 μs for the O_2- sensitive $[Ru(dpp)_3]^{2+}$ complex. The large difference in decay times allowed for the O_2- and temperature-dependent lifetime determination without spectral separation. A composite luminescent material has been presented that was composed of O_2-sensitive fluorinated palladium(II) tetraphenylporphyrin and temperature-sensitive dye ruthenium(II) (tris-1,10-phenantroline), both incorporated in polymer microparticles (Borisov et al. 2006). The microparticles were incorporated in a single polymer matrix, allowing for simultaneous O_2 and temperature determination. Both dye luminescences could be separated spectrally or via luminescence decay times. The same temperature probe was combined with an O_2-sensitive fullerene C_{70} to achieve a dual fluorescence sensor for trace O_2 (ppbv range) and temperature (Baleizao et al. 2008). The material consisted of two sensor layers which were combined to suppress any RET from the temperature-sensitive ruthenium(II) complex to C_{70}. The sensor exhibited a temperature operation range between 0°C and 120°C. A combination of luminescent CdSe-ZnS nanocrystals with O_2-sensitive ruthenium(II) complex immobilized in a sol–gel-derived matrix was also explored in simultaneous sensing of O_2 and temperature (Jorge et al. 2008). The quantum dots exhibited a temperature-dependent spectral shift and allowed for self-referenced intensity-based temperature measurements. Another study presented a low-cost plastic optical fiber that was exploited for dual sensing of temperature and O_2 (Chu and Lo 2008). An O_2-sensitive porphyrin dye was immobilized in a sol–gel-based material that was deposited on a fiber end. The temperature-sensing element consisted of epoxy glue that was coated on the side-polished fiber surface. Both indicators were excited with UV LED and the two emission spectra were well resolved.

Luminescent europium(III) complexes have also been exploited as temperature probes (Borisov and Wolfbeis 2006). The complex was incorporated into microbeads and immobilized along with O_2-sensitive beads in a single hydrogel layer that resulted in a dual O_2- and temperature-sensing material. Both indicators were excited with 405 nm LED and their bright luminescence was separated using appropriate interference filters. A europium(III) complex was also exploited as a temperature-sensitive probe in dual sensing of O_2 and temperature by others (Lam et al. 2011). $[Ru(dpp)_3]^{2+}$ was used as an O_2-sensitive probe. The excitation and emission spectra of the two luminophores overlapped significantly and therefore a combination of time-domain and frequency-domain techniques was used to separate the signals.

There are also a number of papers that reported on simultaneous determination of pH and O_2. The pH-sensitive dye carboxyfluorescein was covalently attached to amino-modified poly(hydroxyethyl methacrylate) particles while O_2-sensitive ruthenium(II) complex was incorporated in organically modified sol–gel particles (Vasylevska et al. 2006). Both particles were dispersed in a single-layer hydrogel matrix and placed on the distal end of an optical fiber waveguide. A modified dual luminophore referencing method was used for data acquisition. A polyurethane-based single-layer membrane in which a pH-sensitive lipophilic fluorescein derivative and an O_2-sensitive porphyrin complex were immobilized was described by Schroder et al. (Schroder et al. 2007b). In this work, a novel time-resolved imaging method that combined t-DLR with rapid lifetime determination was exploited and allowed for simultaneous mapping of O_2 and pH in marine sediments. Tian and coauthors used a membrane made from polymerizable pH- and O_2-sensing monomers for simultaneous O_2 and pH sensing (Tian et al. 2010b). Other examples include a dual sensor for time-resolved imaging of CO_2 and O_2 in aquatic systems such as freshwater or seawater sediments (Schroder et al. 2007a), a dual luminescent sensor for imaging of pressure and temperature on surfaces (Stich et al. 2008), and soluble polymeric dual sensor for temperature and pH (Pietsch et al. 2009).

3.3.8 Summary

This section on reagent-mediated optical sensors has provided a comprehensive review of optical absorption and luminescence-based sensors which have been published over the last 10 years. By far, the largest sections covered are those devoted to O_2 and pH sensing, which reflects the large number of papers that are published on these topics. For both these analytes, we have presented a summary in tabular form of key papers published over this period. This section concludes with an account of recent work in the important areas of imaging and multi-analyte sensing.

3.4 Key Trends and Future Perspectives: A View Forward

3.4.1 General Trends

There is a continuing trend toward increased integration and miniaturization in sensor platforms, which builds on recent progress in miniature optoelectronic devices, new developments toward high-density sensor arrays, the availability of fluorescent nanoprobes, and progress in micro- and nano-fluidics. Looking forward to the next chapter and indeed the rest of this book, we give a brief review of recent reports on optical chemical sensors that use nanomaterials, in particular nanoparticles. While the last decade has seen significant growth in the number of papers published in this area, in the next section, we select a number of recent papers, which give a flavor for developments in nano-based optical sensing. Much of the literature on nano-related sensors is in the area of biosensing as nanoparticles have an obvious potential in this area, for example, in the emerging application of intracellular sensing for key analytes such as O_2, pH, and Ca^{2+}. The reader is directed to some useful recent reviews of the range of nanoparticles available and on conjugation strategies for biosensing. A review of multifunctional nanoparticles including magnetic nanoparticles, gold nanoparticles, and quantum dots is given by de Dios and Diaz-Garcia (2010). Nanomaterials, specifically for fluorescence-based biosensing, are reviewed by Zhong et al. (Zhong 2009), bioconjugated silica nanoparticles have been reviewed by the Tan group (Wang et al. 2008b), and the potential of fluorescent core-shell silica nanoparticles is described by the Wiesner group (Burns et al. 2006).

3.4.2 A Selection of Recent Publications in Nano-Based Optical Sensing

A large proportion of recently reported nano-based optical chemical sensors have been for O_2 and pH. In particular, the Wolfbeis group, highlighted in Section 3.1.2 in the context of early developments in optical chemical sensing, has published recently in this area. A nanogel-based ratiometric sensor for intracellular pH has been reported, which is based on FRET inside a polyurethane gel network. Ratiometric detection is facilitated by the use of two fluorophores (Peng et al. 2010). The same group has used upconverting luminescent nanoparticles for O_2 sensing (Achatz et al. 2011), pH sensing (Sun et al. 2009), and NH_3 sensing (Mader and Wolfbeis 2010). For the O_2 sensor, the nanoparticles act as nanolamps that are excited at 980 nm and whose visible luminescence in turn excited an O_2-quenchable iridium complex while the pH and NH_3 sensors are based on an inner filter effect arising from the interaction of the absorption bands of colorimetric probes for pH and NH_3, which interact with the luminescence bands of the nanoparticles. The Klimant group has reported fluorescence-based pH sensors using polymer nanobeads containing fluorescein and HPTS indicators, which have applications in biomedical and marine sensing (Borisov et al. 2009), and also fluorescent pH-sensitive nanospheres, synthesized by emulsion polymerization, for trace NH_3 detection (Waich et al. 2010). Two-photon absorbing semiconducting nanocrystals for ratiometric O_2 detection have been reported, which are based on FRET between the nanocrystals and O_2-sensitive osmium complexes (McLaurin et al. 2009). A CdSe/ZnS quantum dot-based fluorescence lifetime nanosensor for pH has been reported for use in intracellular media (Ruedas-Rama et al. 2011), and a report of the same type of quantum dot used in a ratiometric O_2 sensor has also appeared in the recent literature (Amelia et al. 2011). The Kopelman group, who developed the so-called PEBBLE (photonic explorers for biomedical use with biologically localized embedding) nanoparticles for intracellular sensing (Buck et al. 2004), more recently has reported the development of nontoxic targeted and ratiometric 30 nm diameter nanosensors for O_2, which are synthesized from polyacrylamide hydrogel and contain NIR dyes. The nanoparticles have been used for noninvasive real-time monitoring of O_2 levels in live cancer cells (Koo Lee et al. 2010). A nanosensor for intracellular sensing of Ca^{2+} has been reported recently, which uses the Ca^{2+} indicator Fluo-4 that has been conjugated to dextran and immobilized on the surface of a silica nanoparticle in which is encapsulated a rhodamine reference dye to facilitate ratiometric measurements (Schulz et al. 2011).

3.4.3 Overall Summary of the Chapter

Our primary aim in this chapter was to review progress in optical chemical sensors over the last 10 years or so as a prelude to subsequent chapters, which will focus on nanosensing and the role of nanomaterials in optical chemical sensing. We have concentrated mainly on reagent-based sensing but have included a short section on direct sensors. Throughout the chapter, we have endeavored to direct the reader to all relevant review articles. For reasons stated earlier, we have dealt mainly with optical absorption- and luminescence-based sensors and we have largely excluded optical biosensors from the review. We preface our review by looking back to early developments and we finish with a brief look forward to sensors using nanomaterials and nanosensors, which is the main subject area of this book.

REFERENCES

Absalan, G., M. Soleimani, M. Asadi, and M. B. Ahmadi. 2004. Constructing a new optical sensor for monitoring ammonia in water samples using bis(acetylacetoneethylenediamine)-tributylphosphin cobalt(III) tetraphenylborate complex-coated triacetylcellulose. *Analytical Sciences* 20 (10): 1433–1436.

Achatz, D. E., R. J. Meier, L. H. Fischer, and O. S. Wolfbeis. 2011. Luminescent sensing of oxygen using a quenchable probe and upconverting nanoparticles. *Angewandte Chemie-International Edition* 50 (1): 260–263.

Airoudj, A., D. Debarnot, B. Beche, and F. Poncin-Epaillard. 2008a. A new evanescent wave ammonia sensor based on polyaniline composite. *Talanta* 76 (2): 314–319.

Airoudj, A., D. Debarnot, B. Beche, and F. Poncin-Epaillard. 2008b. New sensitive layer based on pulsed plasma-polymerized aniline for integrated optical ammonia sensor. *Analytica Chimica Acta* 626 (1): 44–52.

Airoudj, A., D. Debarnot, B. Beche, and F. Poncin-Epaillard. 2009. Development of an optical ammonia sensor based on polyaniline/epoxy resin (SU-8) composite. *Talanta* 77 (5): 1590–1596.

Ali, R., T. Lang, S. M. Saleh, R. J. Meier, and O. S. Wolfbeis. 2011. Optical sensing scheme for carbon dioxide using a solvatochromic probe. *Analytical Chemistry* 83 (8): 2846–2851.

Alizadeh, K., R. Parooi, P. Hashemi, B. Rezaei, and M. R. Ganjali. 2011. A new Schiff's base ligand immobilized agarose membrane optical sensor for selective monitoring of mercury ion. *Journal of Hazardous Materials* 186 (2–3): 1794–1800.

Amao, Y. 2003. Probes and polymers for optical sensing of oxygen. *Microchimica Acta* 143 (1): 1–12.

Amao, Y., K. Asai, T. Miyashita, and I. Okura. 2000a. Novel optical oxygen sensing material: Platinum porphyrin-fluoropolymer film. *Polymers for Advanced Technologies* 11 (8–12): 705–709.

Amao, Y., K. Asai, I. Okura, H. Shinohara, and H. Nishide. 2000b. Platinum porphyrin embedded in poly(1-trimethylsilyl-1-propyne) film as an optical sensor for trace analysis of oxygen. *Analyst* 125 (11): 1911–1914.

Amao, Y., Y. Ishikawa, and I. Okura. 2001. Green luminescent iridium(iii) complex immobilized in fluoropolymer film as optical oxygen-sensing material. *Analytica Chimica Acta* 445 (2): 177–182.

Amao, Y., T. Komori, and H. Nishide. 2005. Rapid responsible optical CO_2 sensor of the combination of colorimetric change of alpha-naphtholphthalein in poly(trimethylsilylpropyne) layer and internal reference fluorescent porphyrin in polystyrene layer. *Reactive and Functional Polymers* 63 (1): 35–41.

Amao, Y. and N. Nakamura. 2005. An optical sensor with the combination of colorimetric change of alpha-naphtholphthalein and internal reference luminescent dye CO_2 in water. *Sensors and Actuators B-Chemical* 107 (2): 861–865.

Amao, Y., I. Okura, and T. Miyashita. 2000c. Optical oxygen sensing based on the luminescence quenching of europium(III) complex immobilized in fluoropolymer film. *Bulletin of the Chemical Society of Japan* 73 (12): 2663–2668.

Amelia, M., A. Lavie-Cambot, N. D. McClenaghan, and A. Credi. 2011. A ratiometric luminescent oxygen sensor based on a chemically functionalized quantum dot. *Chemical Communications* 47 (1): 325–327.

Amini, M. K., B. Khezri, and A. R. Firooz. 2008. Development of a highly sensitive and selective optical chemical sensor for batch and flow-through determination of mercury ion. *Sensors and Actuators B-Chemical* 131 (2): 470–478.

Babilas, P., P. Lamby, L. Prantl, S. Schreml, G. Liebsch, O. S. Wolfbeis, M. Landthaler, R. Szeimies, and C. Abels. 2008. Transcutaneous pO_2 imaging during tourniquet-induced forearm ischemia using planar optical oxygen sensors. *Experimental Dermatology* 17 (3): 265–265.

Babilas, P., G. Liebsch, V. Schacht, I. Klimant, O. S. Wolfbeis, R. M. Szeimies, and C. Abels. 2005. In vivo phosphorescence imaging of pO(2) using planar oxygen sensors. *Microcirculation* 12 (6): 477–487.

Bacon, J. R. and J. N. Demas. 1987. Determination of oxygen concentrations by luminescence quenching of a polymer-immobilized transition-metal complex. *Analytical Chemistry* 59 (23): 2780–2785.

Baldini, F., A. Falai, A. R. De Gaudio, D. Landi, A. Lueger, A. Mencaglia, D. Scherr, and W. Trettnak. 2003. Continuous monitoring of gastric carbon dioxide with optical fibres. *Sensors and Actuators B-Chemical* 90 (1–3): 132–138.

Baldini, F. and A. Giannetti. 2005. Optical chemical and biochemical sensors: New trends. *Proceedings of SPIE* 5826: 485–499.

Baldini, F., A. Giannetti, and A. A. Mencaglia. 2007. Optical sensor for interstitial pH measurements. *Journal of Biomedical Optics* 12 (2): 024024.

Baleizao, C., S. Nagl, M. Schaeferling, M. N. Berberan-Santos, and O. S. Wolfbeis. 2008. Dual fluorescence sensor for trace oxygen and temperature with unmatched range and sensitivity. *Analytical Chemistry* 80 (16): 6449–6457.

Bandstra, L., B. Hales, and T. Takahashi. 2006. High-frequency measurements of total CO_2: Method development and first oceanographic observations. *Marine Chemistry* 100 (1–2): 24–38.

Bergman, I. 1968. Rapid-response atmospheric oxygen monitor based on fluorescence quenching. *Nature* 218: 396.

Blair, S., M. P. Lowe, C. E. Mathieu, D. Parker, P. K. Senanayake, and R. Kataky. 2001. Narrow-range optical pH sensors based on luminescent europium and terbium complexes immobilized in a sol gel glass. *Inorganic Chemistry* 40 (23): 5860–5867.

Blamey, N. J. F., J. Conliffe, J. Parnell, A. G. Ryder, and M. Feely. 2009. Application of fluorescence lifetime measurements on single petroleum-bearing fluid inclusions to demonstrate multicharge history in petroleum reservoirs. *Geofluids* 9 (4): 330–337.

Bojinov, V. B. and T. N. Konstantinova. 2007. Fluorescent 4-(2,2,6,6-tetramethylpiperidin-4-ylamino)-1,8-naphthalimide pH chemosensor based on photoinduced electron transfer. *Sensors and Actuators B-Chemical* 123 (2): 869–876.

Bojinov, V. B., D. B. Simeonov, and N. I. Georgiev. 2008. A novel blue fluorescent 4-(1,2,2,6,6-pentamethylpi-peridin-4-yloxy)-1,8-naphthalimide pH chemosensor based on photoinduced electron transfer. *Dyes and Pigments* 76 (1): 41–46.

Borisov, S. M., K. Gatterer, and I. Klimant. 2010. Red light-excitable dual lifetime referenced optical pH sensors with intrinsic temperature compensation. *Analyst* 135 (7): 1711–1717.

Borisov, S. M., D. L. Herrod, and I. Klimant. 2009. Fluorescent poly(styrene-block-vinylpyrrolidone) nanobeads for optical sensing of pH. *Sensors and Actuators B-Chemical* 139 (1): 52–58.

Borisov, S. M., P. Lehner, and I. Klimant. 2011. Novel optical trace oxygen sensors based on platinum(II) and palladium(II) complexes with 5,10,15,20-meso-tetrakis-(2,3,4,5,6-pentafluorphenyl)-porphyrin covalently immobilized on silica-gel particles. *Analytica Chimica Acta* 690 (1): 108–115.

Borisov, S. M., A. S. Vasylevska, C. Krause, and O. S. Wolfbeis. 2006. Composite luminescent material for dual sensing of oxygen and temperature. *Advanced Functional Materials* 16 (12): 1536–1542.

Borisov, S. M., M. Ch Waldhier, I. Klimant, and O. S. Wolfbeis. 2007. Optical carbon dioxide sensors based on silicone-encapsulated room-temperature ionic liquids. *Chemistry of Materials* 19 (25): 6187–6194.

Borisov, S. M. and O. S. Wolfbeis. 2006. Temperature-sensitive europium(III) probes and their use for simultaneous luminescent sensing of temperature and oxygen. *Analytical Chemistry* 78 (14): 5094–5101.

Bosch, E. M., A. J. R. Sanchez, F. S. Rojas, and C. B. Ojeda. 2007. Recent development in optical fiber biosensors. *Sensors* 7 (6): 797–859.

Brittle, S. A., T. H. Richardson, A. D. F. Dunbar, S. M. Turega, and C. A. Hunter. 2011. Tuning free base tetraphenylporphyrins as optical sensing elements for volatile organic analytes. *Journal of Materials Chemistry* 21 (13): 4882–4887.

Briz, S., A. J. de Castro, S. Diez, F. Lopez, and K. Schaefer. 2007. Remote sensing by open-path FTIR spectroscopy. Comparison of different analysis techniques applied to ozone and carbon monoxide detection. *Journal of Quantitative Spectroscopy and Radiative Transfer* 103 (2): 314–330.

Buck, S. M., H. Xu, M. Brasuel, M. A. Philbert, and R. Kopelman. 2004. Nanoscale probes encapsulated by biologically localized embedding (PEBBLEs) for ion sensing and imaging in live cells. *Talanta* 63 (1): 41–59.

Bukowski, R. M., R. Ciriminna, M. Pagliaro, and F. V. Bright. 2005. High-performance quenchometric oxygen sensors based on fluorinated xerogels doped with [Ru(dpp)$_3$]$^{2+}$. *Analytical Chemistry* 77 (8): 2670–2672.

von Bultzingslowen, C., A. K. McEvoy, C. McDonagh, and B. D. MacCraith. 2003. Lifetime-based optical sensor for high-level pCO(2) detection employing fluorescence resonance energy transfer. *Analytica Chimica Acta* 480 (2): 275–283.

von Bultzingslowen, C., A. K. McEvoy, C. McDonagh, B. D. MacCraith, I. Klimant, C. Krause, and O. S. Wolfbeis. 2002. Sol-gel based optical carbon dioxide sensor employing dual luminophore referencing for application in food packaging technology. *Analyst* 127 (11): 1478–1483.

Burke, C. S., J. P. Moore, D. Wencel, A. K. McEvoy, and B. D. MacCraith. 2008. Breath-by-breath measurement of oxygen using a compact optical sensor. *Journal of Biomedical Optics* 13 (1): 014027.

Burke, C. S., L. Polerecky, and B. D. MacCraith. 2004. Design and fabrication of enhanced polymer waveguide platforms for absorption-based optical chemical sensors. *Measurement Science and Technology* 15 (6): 1140–1145.

Burns, A., H. Ow, and U. Wiesner. 2006. Fluorescent core-shell silica nanoparticles: Towards "lab on a particle" architectures for nanobiotechnology. *Chemical Society Reviews* 35 (11): 1028–1042.

Cajlakovic, M., A. Bizzarri, and V. Ribitsch. 2006. Luminescence lifetime-based carbon dioxide optical sensor for clinical applications. *Analytica Chimica Acta* 573: 57–64.

Cajlakovic, M., A. Lobnik, and T. Werner. 2002. Stability of new optical pH sensing material based on cross-linked poly(vinyl alcohol) copolymer. *Analytica Chimica Acta* 455 (2): 207–213.

Cammann, K., E. A. H. Hall, R. Kellner, H. L. Schmidt, and O. S. Wolfbeis. 1995. The Cambridge definition of chemical sensors. In *Proceedings of the Cambridge Workshop on Chemical Sensors and Biosensors*. Cambridge University Press, Cambridge, U.K.

Cao, W. Q. and Y. X. Duan. 2005. Optical fiber-based evanescent ammonia sensor. *Sensors and Actuators B-Chemical* 110 (2): 252–259.

Capel-Cuevas, S., M. P. Cuellar, I. de Orbe-Paya, M. C. Pegalajar, and L. F. Capitan-Vallvey. 2011. Full-range optical pH sensor array based on neural networks. *Microchemical Journal* 97 (2): 225–233.

Carraway, E. R., J. N. Demas, and B. A. DeGraff. 1991. Luminescence quenching mechanism for microheterogeneous systems. *Analytical Chemistry* 63 (4): 332–336.

Carter, J. C., R. M. Alvis, S. B. Brown, K. C. Langry, T. S. Wilson, M. T. McBride, M. L. Myrick, W. R. Cox, M. E. Grove, and B. W. Colston. 2006. Fabricating optical fiber imaging sensors using inkjet printing technology: A pH sensor proof-of-concept. *Biosensors and Bioelectronics* 21 (7): 1359–1364.

Castrellon-Uribe, J., M. E. Nicho, and G. Reyes-Merino. 2009. Remote optical detection of low concentrations of aqueous ammonia employing conductive polymers of polyaniline. *Sensors and Actuators B-Chemical* 141 (1): 40–44.

de Castro, A. J., A. M. Lerma, F. Lopez, M. Guijarro, C. Diez, C. Hernando, and J. Madrigal. 2007. Open-path Fourier transform infrared spectrometry characterization of low temperature combustion gases in biomass fuels. *Infrared Physics and Technology* 51 (1): 21–30.

Chang, Y. C., H. Bai, S. N. Li, and C. N. Kuo. 2011. Bromocresol green/mesoporous silica adsorbent for ammonia gas sensing via an optical sensing instrument. *Sensors* 11 (4): 4060–4072.

Charlton, S. C. and R. L. Fleming. 1982. A colorimetric solid-phase test for potassium. *Clinical Chemistry* 28 (7): 1669–1669.

Chen, X., L. Lin, P. W. Li, Y. J. Dai, and X. R. Wang. 2004. Fluorescent response of sol-gel derived ormosils for optical ammonia sensing film. *Analytica Chimica Acta* 506 (1): 9–15.

Chen, H. X., X. D. Wang, X. H. Song, T. Y. Zhou, Y. Q Jiang, and X. Chen. 2010. Colorimetric optical ph sensor production using a dual-color system. *Sensors and Actuators B-Chemical* 146 (1): 278–282.

Chen, X., Y. Zu, H. Xie, A. M. Kemas, and Z. Gao. 2011. Coordination of mercury(II) to gold nanoparticle associated nitrotriazole towards sensitive colorimetric detection of mercuric ion with a tunable dynamic range. *Analyst* 136 (8): 1690–1696.

Cho, E. J. and F. V. Bright. 2002. Pin-printed chemical sensor arrays for simultaneous multianalyte quantification. *Analytical Chemistry* 74 (6): 1462–1466.

Christie, S., E. Scorsone, K. Persaud, and F. Kvasnik. 2003. Remote detection of gaseous ammonia using the near infrared transmission properties of polyaniline. *Sensors and Actuators B-Chemical* 90 (1–3): 163–169.

Chu, C. S. and Y. L. Lo. 2008. A plastic optical fiber sensor for the dual sensing of temperature and oxygen. *IEEE Photonics Technology Letters* 20 (1–4): 63–65.

Clarke, Y., W. Y. Xu, J. N. Demas, and B. A. DeGraff. 2000. Lifetime-based pH sensor system based on a polymer supported ruthenium(II) complex. *Analytical Chemistry* 72 (15): 3468–3475.

Collier, B. B., S. Singh, and M. McShane. 2011. Microparticle ratiometric oxygen sensors utilizing near-infrared emitting quantum dots. *Analyst* 136 (5): 962–967.

Costa-Fernandez, J. M., R. Pereiro, and A. Sanz-Medel. 2006. The use of luminescent quantum dots for optical sensing. *Trac-Trends in Analytical Chemistry* 25 (3): 207–218.

Coto-Garcia, A. M., E. Sotelo-Gonzalez, M. T. Fernandez-Argueelles, R. Pereiro, J. M. Costa-Fernandez, and A. Sanz-Medel. 2011. Nanoparticles as fluorescent labels for optical imaging and sensing in genomics and proteomics. *Analytical and Bioanalytical Chemistry* 399 (1): 29–42.

Courbat, J., D. Briand, J. Damon-Lacoste, J. Woellenstein, and N. F. de Rooij. 2009. Evaluation of pH indicator-based colorimetric films for ammonia detection using optical waveguides. *Sensors and Actuators B-Chemical* 143 (1): 62–70.

Culshaw, B. 2004. Optical fiber sensor technologies: Opportunities and-perhaps-pitfalls. *Journal of Lightwave Technology* 22 (1): 39–50.

Dansby-Sparks, R. N., J. Jin, S. J. Mechery, U. Sampathkumaran, T. W. Owen, B. D. Yu, K. Goswami, K. Hong, J. Grant, and Z. L. Xue. 2010. Fluorescent-dye-doped sol-gel sensor for highly sensitive carbon dioxide gas detection below atmospheric concentrations. *Analytical Chemistry* 82 (2): 593–600.

Davis, C. E., M. Frank, B. Mizaikoff, and H. Oser. 2010. The future of sensors and instrumentation for human breath analysis. *IEEE Sensors Journal* 10 (1): 3–6.

DeGraff, B. A. and J. N. Demas. 2005. Luminescence-based oxygen sensors. In: *Reviews in Fluorescence 2005*, eds. C. D. Geddes and J. R. Lakowicz, Vol. 125. Springer, New York.

Del Villar, I., I. R. Matias, and F. J. Arregui. 2008. Fiber-optic chemical nanosensors by electrostatic molecular self-assembly. *Current Analytical Chemistry* 4 (4): 341–355.

DiMarco, G. and M. Lanza. 2000. Optical solid-state oxygen sensors using metalloporphyrin complexes immobilized in suitable polymeric matrices. *Sensors and Actuators B-Chemical* 63 (1–2): 42–48.

de Dios A. S. and M. E. Diaz-Garcia. 2010. Multifunctional nanoparticles: Analytical prospects. *Analytica Chimica Acta* 666 (1–2): 1–22.

Dong, S., M. Luo, G. Peng, and W. Cheng. 2008. Broad range pH sensor based on sol-gel entrapped indicators on fibre optic. *Sensors and Actuators B-Chemical* 129 (1): 94–98.

Douglas, P. and K. Eaton. 2002. Response characteristics of thin film oxygen sensors, Pt and Pd octaethylporphyrins in polymer films. *Sensors and Actuators B-Chemical* 82 (2–3): 200–208.

Dunn, B. and J. I. Zink. 1991. Optical-properties of sol-gel glasses doped with organic-molecules. *Journal of Materials Chemistry* 1 (6): 903–913.

Duveneck, G. L., A. P. Abel, M. A. Bopp, G. M. Kresbach, and M. Ehrat. 2002. Planar waveguides for ultra-high sensitivity of the analysis of nucleic acids. *Analytica Chimica Acta* 469 (1): 49–61.

Duy, C. and J. Fitter. 2006. How aggregation and conformational scrambling of unfolded states govern fluorescence emission spectra. *Biophysical Journal* 90 (10): 3704–3711.

Eaton, K. and P. Douglas. 2002. Effect of humidity on the response characteristics of luminescent PtOEP thin film optical oxygen sensors. *Sensors and Actuators B-Chemical* 82 (1): 94–104.

Eaton, K., B. Douglas, and P. Douglas. 2004. Luminescent oxygen sensors: Time-resolved studies and modelling of heterogeneous oxygen quenching of luminescence emission from Pt and Pd octaethylporphyrins in thin polymer films. *Sensors and Actuators B-Chemical* 97 (1): 2–12.

Edmonds, T. E., N. J. Flatters, C. F. Jones, and J. N. Miller. 1988. Determination of pH with acid-base indicators—Implications for optical fiber probes. *Talanta* 35 (2): 103–107.

Elosua, C., C. Bariain, and I. R. Matias. 2010. Optical fiber sensors to detect volatile organic compound in sick building syndrome applications. *The Open Construction and Building Technology Journal* 4: 113–120.

Elosua, C., C. Bariain, I. R. Matias, F. J. Arregui, E. Vergara, and M. Laguna. 2009. Optical fiber sensing devices based on organic vapor indicators towards sensor array implementation. *Sensors and Actuators B-Chemical* 137 (1): 139–146.

Elosua, C., I. R. Matias, C. Bariain, and F. J. Arregui. 2006. Volatile organic compound optical fiber sensors: A review. *Sensors* 6 (11): 1440–1465.

Ensafi, A. A. and A. Kazemzadeh. 1999. Optical pH sensor based on chemical modification of polymer film. *Microchemical Journal* 63 (3): 381–388.

Epstein, J. R. and D. R. Walt. 2003. Fluorescence-based fibre optic arrays: A universal platform for sensing. *Chemical Society Reviews* 32 (4): 203–214.

Ergeneman, O., G. Dogangil, M. P. Kummer, J. J. Abbott, M. K. Nazeeruddin, and B. J. Nelson. 2008. A magnetically controlled wireless optical oxygen sensor for intraocular measurements. *IEEE Sensors Journal* 8 (1–2): 29–37.

Ertekin, K., I. Klimant, G. Neurauter, and O. S. Wolfbeis. 2003. Characterization of a reservoir-type capillary optical microsensor for pCO(2) measurements. *Talanta* 59 (2): 261–267.

Estella, J., D. Wencel, J. P. Moore, M. Sourdaine, and C. McDonagh. 2010. Fabrication and performance evaluation of highly sensitive hybrid sol-gel-derived oxygen sensor films based on a fluorinated precursor. *Analytica Chimica Acta* 666 (1–2): 83–90.

Evans, R. C. and P. Douglas. 2006. Controlling the color space response of colorimetric luminescent oxygen sensors. *Analytical Chemistry* 78 (16): 5645–5652.

Evans, R. C. and P. Douglas. 2009. Design and color response of colorimetric multilumophore oxygen sensors. *ACS Applied Materials & Interfaces* 1 (5): 1023–1030.

Evans, R. C., P. Douglas, J. A. G. Williams, and D. L. Rochester. 2006. A novel luminescence-based colorimetric oxygen sensor with a "traffic light" response. *Journal of Fluorescence* 16 (2): 201–206.

Ferguson, J. A., B. G. Healey, K. S. Bronk, S. M. Barnard, and D. R. Walt. 1997. Simultaneous monitoring of pH, CO_2 and O_2 using an optical imaging fiber. *Analytica Chimica Acta* 340 (1–3): 123–131.

Fernandez-Sanchez, J. F., R. Cannas, S. Spichiger, R. Steiger, and U. E. Spichiger-Keller. 2006. Novel nanostructured materials to develop oxygen-sensitive films for optical sensors. *Analytica Chimica Acta* 566 (2): 271–282.

Fernandez-Sanchez, J. F., R. Cannas, S. Spichiger, R. Steiger, and U. E. Spichiger-Keller. 2007. Optical CO_2-sensing layers for clinical application based on pH-sensitive indicators incorporated into nanoscopic metal-oxide supports. *Sensors and Actuators B-Chemical* 128 (1): 145–153.

Frodl, R. and T. Tille. 2006. A high-precision NDIR CO_2 gas sensor for automotive applications. *IEEE Sensors Journal* 6 (6): 1697–1705.

Gao, M. G., W. Q. Liu, T. S. Zhang, J. G. Liu, Y. H. Lu, Y. P. Wang, X. Liang, Z. Jun, and C. Jun. 2006. Remote sensing of atmospheric trace gas by airborne passive FTIR. *Spectroscopy and Spectral Analysis* 26 (12): 2203–2206.

Gassensmith, J. J., S. Matthys, J. Lee, A. Wojcik, P. V. Kamat, and B. D. Smith. 2010. Squaraine rotaxane as a reversible optical chloride sensor. *Chemistry-A European Journal* 16 (9): 2916–2921.

Ge, C., N. R. Armstrong, and S. S. Saavedra. 2007. pH-sensing properties of poly(aniline) ultrathin films self-assembled on indium-tin oxide. *Analytical Chemistry* 79 (4): 1401–1410.

Ge, X. D., Y. Kostov, and G. Rao. 2003. High-stability non-invasive autoclavable naked optical CO_2 sensor. *Biosensors and Bioelectronics* 18 (7): 857–865.

Georgiev, N. I., V. B. Bojinov, and P. S. Nikolov. 2011. The design, synthesis and photophysical properties of two novel 1,8-naphthalimide fluorescent pH sensors based on PET and ICT. *Dyes and Pigments* 88 (3): 350–357.

Goicoechea, J., C. R. Zamarreno, I. R. Matias, and F. J. Arregui. 2008. Optical fiber pH sensors based on layer-by-layer electrostatic self-assembled neutral red. *Sensors and Actuators B-Chemical* 132 (1): 305–311.

Goncalves, H. M. R., C. D. Maule, P. A. S. Jorge, and J. C. G. Esteves da Silva. 2008. Fiber optic lifetime pH sensing based on ruthenium(II) complexes with dicarboxybipyridine. *Analytica Chimica Acta* 626 (1): 62–70.

Gong, Z. L., B. X. Zhao, W. Y. Liu, and H. S. Lv. 2011. A new highly selective "turn on" fluorescent sensor for zinc ion based on a pyrazoline derivative. *Journal of Photochemistry and Photobiology A-Chemistry* 218 (1): 6–10.

Gotou, T., M. Noda, T. Tomiyama, H. Sembokuya, M. Kubouchi, and K. Tsuda. 2006. In situ health monitoring of corrosion resistant polymers exposed to alkaline solutions using pH indicators. *Sensors and Actuators B-Chemical* 119 (1): 27–32.

Gunnlaugsson, T. 2001. A novel Eu(III)-based luminescent chemosensor: Determining pH in a highly acidic environment. *Tetrahedron Letters* 42 (50): 8901–8905.

Guo, H. and S. Tao. 2007. Silver nanoparticles doped silica nanocomposites coated on an optical fiber for ammonia sensing. *Sensors and Actuators B-Chemical* 123 (1): 578–582.

Guo, L. Q, N. Yin, D. D. Nie, J. R. Gan, M. J. Li, F. F. Fu, and G. N. Chen. 2011. Label-free fluorescent sensor for mercury(II) ion by using carbon nanotubes to reduce background signal. *Analyst* 136 (8): 1632–1636.

Hakonen, A. and S. Hulth. 2008. A high-precision ratiometric fluorosensor for pH: Implementing time-dependent non-linear calibration protocols for drift compensation. *Analytica Chimica Acta* 606 (1): 63–71.

Harper, G. B. 1975. Reusable glass-bound pH indicators. *Analytical Chemistry* 47 (2): 348–351.

Hashemi, P. and M. M. Abolghasemi. 2006. Preparation of a novel optical sensor for low pH values using agarose membranes as support. *Sensors and Actuators B-Chemical* 115 (1): 49–53.

Hashemi, P. and R. A. Zarjani. 2008. A wide range pH optical sensor with mixture of neutral red and thionin immobilized on an agarose film coated glass slide. *Sensors and Actuators B-Chemical* 135 (1): 112–115.

Heitzmann, H. and H. Kroneis. 1985. U.S. Patent 4.557.900.

Higgins, C., D. Wencel, C. S. Burke, B. D. MacCraith, and C. McDonagh. 2008. Novel hybrid optical sensor materials for in-breath O-2 analysis. *Analyst* 133 (2): 241–247.

Hirschfeld, T., J. B. Callis, and B. R. Kowalski. 1984. Chemical sensing in process analysis. *Science* 226 (4672): 312–318.

Hirschfeld, T., T. Deaton, F. Milanovich, and S. Klainer. 1983. Feasibility of using fiber optics for monitoring groundwater contaminants. *Optical Engineering* 22 (5): 527–531.

Hiruta, Y., Y. Ando, D. Citterio, and K. Suzuki. 2010. A fast-response pH optode based on a fluoroionophore immobilized to a mesoporous silica thin film. *Analytical Sciences* 26 (3): 297–301.

Holst, G. and B. Grunwald. 2001. Luminescence lifetime imaging with transparent oxygen optodes. *Sensors and Actuators B-Chemical* 74 (1–3): 78–90.

Hradil, J., C. Davis, K. Mongey, C. McDonagh, and B. D. MacCraith. 2002. Temperature-corrected pressure-sensitive paint measurements using a single camera and a dual-lifetime approach. *Measurement Science and Technology* 13 (10): 1552–1557.

Hu L. P., Y. Li, L. Zhang, L. M. Zhang, and J. D. Wang. 2006. Advanced development of remote sensing FTIR in air environment monitoring. *Spectroscopy and Spectral Analysis* 26 (10): 1863–1867.

Hulth, S., R. C. Aller, P. Engstrom, and E. Selander. 2002. A pH plate fluorosensor (optode) for early diagenetic studies of marine sediments. *Limnology and Oceanography* 47 (1): 212–220.

Hvozdara, L., N. Pennington, A. Kraft, M. Karlowatz, and B. Mizaikoff. 2002. Quantum cascade lasers for mid-infrared spectroscopy. *Vibrational Spectroscopy* 30 (1): 53–58.

Janata, J. 1987. Do optical sensors really measure pH. *Analytical Chemistry* 59 (9): 1351–1356.

Jiang J. F., T. G. Liu, H. W. Li, R. Q. Hui, K. Liu, and Y. M. Zhang. 2010. Review on label-free optical bio-sensing technology based on whisper-gallery-mode. *Spectroscopy and Spectral Analysis* 30 (11): 3076–3080.

Jin, Z., Y. X. Su, and Y. X. Duan. 2000. An improved optical pH sensor based on polyaniline. *Sensors and Actuators B-Chemical* 71 (1–2): 118–122.

Jin, Z., Y. X. Su, and Y. X. Duan. 2001. Development of a polyaniline-based optical ammonia sensor. *Sensors and Actuators B-Chemical* 72 (1): 75–79.

Jorge, P. A. S., C. Maule, A. J. Silva, R. Benrashid, J. L. Santos, and F. Farahi. 2008. Dual sensing of oxygen and temperature using quantum dots and a ruthenium complex. *Analytica Chimica Acta* 606 (2): 223–229.

Jorge, P. A. S., M. Mayeh, R. Benrashid, P. Caldas, J. L. Santos, and F. Farahi. 2006. Applications of quantum dots in optical fiber luminescent oxygen sensors. *Applied Optics* 45 (16): 3760–3767.

Kaempgen, M. and S. Roth. 2006. Transparent and flexible carbon nanotube/polyaniline pH sensors. *Journal of Electroanalytical Chemistry* 586 (1): 72–76.

Kaltin, S., C. Haraldsson, and L. G. Anderson. 2005. A rapid method for determination of total dissolved inorganic carbon in seawater with high accuracy and precision. *Marine Chemistry* 96 (1–2): 53–60.

Kautsky, H. 1939. Quenching of luminescence by oxygen. *Transactions of the Faraday Society* 35: 216–219.

Kautsky, H. and A. Hirsch. 1931. Energy transformations on boundary surfaces. IV. Interaction of excited dyestuff molecules and oxygen. *Chemische Berichte* 64: 2677.

Kawabata, Y., T. Kamichika, T. Imasaka, and N. Ishibashi. 1989. Fiber-optic sensor for carbon-dioxide with a pH-indicator dispersed in a poly(ethylene glycol) membrane. *Analytica Chimica Acta* 219 (2): 223–229.

Kellner, K., G. Liebsch, I. Klimant, O. S. Wolfbeis, T. Blunk, M. B. Schulz, and A. Gopferich. 2002. Determination of oxygen gradients in engineered tissue using a fluorescent sensor. *Biotechnology and Bioengineering* 80 (1): 73–83.

Kermis, H. R., Y. Kostov, P. Harms, and G. Rao. 2002. Dual excitation ratiometric fluorescent ph sensor for noninvasive bioprocess monitoring: Development and application. *Biotechnology Progress* 18 (5): 1047–1053.

Kermis, H. R., Y. Kostov, and G. Rao. 2003. Rapid method for the preparation of a robust optical pH sensor. *Analyst* 128 (9): 1181–1186.

Kim, H. J., Y. C. Jeong, J. Ileo, J. II Rhee, and K. J. Hwang. 2009. A wide-range luminescent pH sensor based on ruthenium(II) complex. *Bulletin of the Korean Chemical Society* 30 (3): 539–540.

Kim, S. S., C. Young, B. Vidakovic, S. G. A. Gabram-Mendola, C. W. Bayer, and B. Mizaikoff. 2010. Potential and challenges for mid-infrared sensors in breath diagnostics. *IEEE Sensors Journal* 10 (1): 145–158.

Kirkbright, G. F., R. Narayanaswamy, and N. A. Welti. 1984. Studies with immobilized chemical reagents using a flow-cell for the development of chemically sensitive fibre-optic devices. *Analyst* 109 (1): 15–17.

Klimant, I. 2003. *Method and Device for Referencing Fluorescence Intensity Signals*, U.S. Patent 6,602,716.

Kneas, K. A., W. Y. Xu, J. N. Demas, and B. A. DeGraff. 1997. Oxygen sensors based on luminescence quenching: Interactions of tris(4,7-diphenyl-1,10-phenanthroline)ruthenium (II) chloride and pyrene with polymer supports. *Applied Spectroscopy* 51 (9): 1346–1351.

Konig, B., O. Kohls, G. Holst, R. N. Glud, and M. Kuhl. 2005. Fabrication and test of sol-gel based planar oxygen optodes for use in aquatic sediments. *Marine Chemistry* 97 (3–4): 262–276.

Koo Lee, Y., E. E. Ulbrich, G. Kim, H. Hah, C. Strollo, W. Z. Fan, R. Gurjar, S. M. Koo, and R. Kopelman. 2010. Near infrared luminescent oxygen nanosensors with nanoparticle matrix tailored sensitivity. *Analytical Chemistry* 82 (20): 8446–8455.

Kosterev, A. A., R. F. Curl, F. K. Tittel, C. Gmachl, F. Capasso, D. L. Sivco, J. N. Baillargeon, A. L. Hutchinson, and A. Y. Cho. 2000. Effective utilization of quantum-cascade distributed-feedback lasers in absorption spectroscopy. *Applied Optics* 39 (24): 4425–4430.

Kosterev, A. A. and F. K. Tittel. 2002. Chemical sensors based on quantum cascade lasers. *IEEE Journal of Quantum Electronics* 38 (6): 582–591.

Kostov, Y., P. Harms, R. S. Pilato, and G. Rao. 2000. Ratiometric oxygen sensing: Detection of dual-emission ratio through a single emission filter. *Analyst* 125 (6): 1175–1178.

Kreft, O., A. M. Javier, G. B. Sukhorukov, and W. J. Parak. 2007. Polymer microcapsules as mobile local pH-sensors. *Journal of Materials Chemistry* 17 (42): 4471–4476.

Kuswandi, B., Nuriman, J. Huskens, and W. Verboom. 2007. Optical sensing systems for microfluidic devices: A review. *Analytica Chimica Acta* 601 (2): 141–155.

Lakowicz, J. R. and K. W. Berndt. 1991. Lifetime-selective fluorescence imaging using an Rf phase-sensitive camera. *Review of Scientific Instruments* 62 (7): 1727–1734.

Lakowicz, J. R., H. Szmacinski, K. Nowaczyk, K. W. Berndt, and M. Johnson. 1992. Fluorescence lifetime imaging. *Analytical Biochemistry* 202 (2): 316–330.

Lam, M. H. W., D. Y. K. Lee, K. W. Man, and C. S. W. Lau. 2000. A luminescent pH sensor based on a sol-gel film functionalized with a luminescent organometallic complex. *Journal of Materials Chemistry* 10 (8): 1825–1828.

Lam, H., G. Rao, J. Loureiro, and L. Tolosa. 2011. Dual optical sensor for oxygen and temperature based on the combination of time domain and frequency domain techniques. *Talanta* 84 (1): 65–70.

Lambeck, P. V. 2006. Integrated optical sensors for the chemical domain. *Measurement Science and Technology* 17 (8): R93-R116.

Lange, U., N. V. Roznyatouskaya, and V. M. Mirsky. 2008. Conducting polymers in chemical sensors and arrays. *Analytica Chimica Acta* 614 (1): 1–26.

Laserna, J. 2001. *An Introduction to Raman Spectroscopy: Introduction and Basic Principles.* Wiley, New York.

Law, M., J. Goldberger, and P. D. Yang. 2004. Semiconductor nanowires and nanotubes. *Annual Review of Materials Research* 34: 83–122.

Lee, S. T., J. Gin, V. P. N. Nampoori, C. P. G. Vallabhan, N. V. Unnikrishnan, and P. Radhakrishnan. 2001. A sensitive fibre optic pH sensor using multiple sol-gel coatings. *Journal of Optics A-Pure and Applied Optics* 3 (5): 355–359.

Lee, Y. S., B. S. Joo, N. J. Choi, J. O. Lim, J. S. Huh, and D. D. Lee. 2003. Visible optical sensing of ammonia based on polyaniline film. *Sensors and Actuators B-Chemical* 93 (1–3): 148–152.

Lee, S. H., J. Kumar, and S. K. Tripathy. 2000. Thin film optical sensors employing polyelectrolyte assembly. *Langmuir* 16 (26): 10482–10489.

Lehrer, S. S. 1971. Solute perturbation of protein fluorescence—Quenching of tryptophyl fluorescence of model compounds and of lysozyme by iodide ion. *Biochemistry* 10 (17): 3254–3263.

Lei, B., B. Li, H. Zhang, S. Lu, Z. Zheng, W. Li, and Y. Wang. 2006. Mesostructured silica chemically doped with Ru-II as a superior optical oxygen sensor. *Advanced Functional Materials* 16 (14): 1883–1891.

Leung, A., P. M. Shankar, and R. Mutharasan. 2007. A review of fiber-optic biosensors. *Sensors and Actuators B-Chemical* 125 (2): 688–703.

Li, C. Y., X. B. Zhang, Z. X. Han, B. Akermark, L. C. Sun, G. L. Shen, and R. Q. Yu. 2006. A wide pH range optical sensing system based on a sol-gel encapsulated amino-functionalised corrole. *Analyst* 131 (3): 388–393.

Liebsch, G., I. Klimant, B. Frank, G. Holst, and O. S. Wolfbeis. 2000. Luminescence lifetime imaging of oxygen, pH, and carbon dioxide distribution using optical sensors. *Applied Spectroscopy* 54 (4): 548–559.

Liebsch, G., I. Klimant, C. Krause, and O. S. Wolfbeis. 2001. Fluorescent imaging of pH with optical sensors using time domain dual lifetime referencing. *Analytical Chemistry* 73 (17): 4354–4363.

Ligler, F. S., K. E. Sapsford, J. P. Golden, L. C. Shriver-Lake, C. R. Taitt, M. A. Dyer, S. Barone, and C. J. Myatt. 2007. The array biosensor: Portable, automated systems. *Analytical Sciences* 23 (1): 5–10.

Lin, J. 2000. Recent development and applications of optical and fiber-optic pH sensors. *Trac-Trends in Analytical Chemistry* 19 (9): 541–552.

Lin, J. and D. Liu. 2000. An optical pH sensor with a linear response over a broad range. *Analytica Chimica Acta* 408 (1–2): 49–55.

Lindfors, T., S. Ervela, and A. Ivaska. 2003. Polyaniline as pH-sensitive component in plasticized PVC membranes. *Journal of Electroanalytical Chemistry* 560: 69–78.

Lindfors, T. and A. Ivaska. 2005. Raman based pH measurements with polyaniline. *Journal of Electroanalytical Chemistry* 580 (2): 320–329.

Liu, Y. H., T. H. Dam, and P. Pantano. 2000. A pH-sensitive nanotip array imaging sensor. *Analytica Chimica Acta* 419 (2): 215–225.

Liu, Z. H., J. F. Liu, and T. L. Chen. 2005. Phenol red immobilized PVA membrane for an optical pH sensor with two determination ranges and long-term stability. *Sensors and Actuators B-Chemical* 107 (1): 311–316.

Liu, Z. H., F. L. Luo, and T. L. Chen. 2004. Phenolphthalein immobilized membrane for an optical pH sensor. *Analytica Chimica Acta* 510 (2): 189–194.

Liu, X. and Y. Qin. 2008. Ion-exchange reaction of silver(I) and copper(II) in optical sensors based on thiaglutaric diamide. *Analytical Sciences* 24 (9): 1151–1156.

Lobnik, A. 2006. Absorption-based sensors. In: *Optical Chemical Sensors*, eds. F. Baldini, A. N. Chester, J. Homola, and S. Martellucci. NATO Science Series Springer, Dordrecht, the Netherlands.

MacCraith, B. D. and C. McDonagh. 2002. Enhanced fluorescence sensing using sol-gel materials. *Journal of Fluorescence* 12 (3–4): 333–342.

Mader, H. S. and O. S. Wolfbeis. 2010. Optical ammonia sensor based on upconverting luminescent nanoparticles. *Analytical Chemistry* 82 (12): 5002–5004.

Malenovsky, Z., K. B. Mishra, F. Zemek, U. Rascher, and L. Nedbal. 2009. Scientific and technical challenges in remote sensing of plant canopy reflectance and fluorescence. *Journal of Experimental Botany* 60 (11): 2987–3004.

Malins, C., T. M. Butler, and B. D. MacCraith. 2000a. Influence of the surface polarity of dye-doped sol-gel glass films on optical ammonia sensor response. *Thin Solid Films* 368 (1): 105–110.

Malins, C., H. G. Glever, T. E. Keyes, J. G. Vos, W. J. Dressick, and B. D. MacCraith. 2000b. Sol-gel immobilised ruthenium(II) polypyridyl complexes as chemical transducers for optical ph sensing. *Sensors and Actuators B-Chemical* 67 (1–2): 89–95.

Malins, C., M. Niggemann, and B. D. MacCraith. 2000c. Multi-analyte optical chemical sensor employing a plastic substrate. *Measurement Science and Technology* 11 (8): 1105–1110.

Marazuela, M. D. and M. C. Moreno-Bondi. 2002. Fiber-optic biosensors—An overview. *Analytical and Bioanalytical Chemistry* 372 (5–6): 664–682.

Mayr, T., I. Klimant, O. S. Wolfbeis, and T. Werner. 2002a. Dual lifetime referenced optical sensor membrane for the determination of copper(II) ions. *Analytica Chimica Acta* 462 (1): 1–10.

Mayr, T., G. Liebsch, I. Klimant, and O. S. Wolfbeis. 2002b. Multi-ion imaging using fluorescent sensors in a microtiterplate array format. *Analyst* 127 (2): 201–203.

McCurdy, M. R., Y. Bakhirkin, G. Wysocki, R. Lewicki, and F. K. Tittel. 2007. Recent advances of laser-spectroscopy-based techniques for applications in breath analysis. *Journal of Breath Research* 1 (1): 014001–014001.

McDonagh, C., C. S. Burke, and B. D. MacCraith. 2008. Optical chemical sensors. *Chemical Reviews* 108 (2): 400–422.

McDonagh, C., C. Kolle, A. K. McEvoy, D. L. Dowling, A. A. Cafolla, S. J. Cullen, and B. D. MacCraith. 2001. Phase fluorometric dissolved oxygen sensor. *Sensors and Actuators B-Chemical* 74 (1–3): 124–130.

McLaurin, E. J., A. B. Greytak, M. G. Bawendi, and D. G. Nocera. 2009. Two-photon absorbing nanocrystal sensors for ratiometric detection of oxygen. *Journal of the American Chemical Society* 131 (36): 12994–13001.

Meroni, M., M. Rossini, L. Guanter, L. Alonso, U. Rascher, R. Colombo, and J. Moreno. 2009. Remote sensing of solar-induced chlorophyll fluorescence: Review of methods and applications. *Remote Sensing of Environment* 113 (10): 2037–2051.

Miled, O. B., D. Grosso, C. Sanchez, and J. Livage. 2004. An optical fibre pH sensor based on dye doped mesostructured silica. *Journal of Physics and Chemistry of Solids* 65 (10): 1751–1755.

Mills, A. 1999a. Response characteristics of optical sensors for oxygen: A model based on a distribution in tau(o) and kappa(q). *Analyst* 124 (9): 1309–1314.

Mills, A. 1999b. Response characteristics of optical sensors for oxygen: Models based on a distribution in tau(o) or kappa(q). *Analyst* 124 (9): 1301–1307.

Mills, A. 2005. Oxygen indicators and intelligent inks for packaging food. *Chemical Society Reviews* 34 (12): 1003–1011.

Mills, A. 2009. Optical sensors for carbon dioxide and their applications. In: *Sensors for Environment, Health and Security*, ed. M.-I Baraton, Vol. 347. Springer, New York.

Mills, A. and Q. Chang. 1992. Modeled diffusion-controlled response and recovery behavior of a naked optical film sensor with a hyperbolic-type response to analyte concentration. *Analyst* 117 (9): 1461–1466.

Mills, A. and Q. Chang. 1993. Fluorescence plastic thin-film sensor for carbon-dioxide. *Analyst* 118 (7): 839–843.

Mills, A. and Q. Chang. 1994. Colorimetric polymer film sensors for dissolved carbon-dioxide. *Sensors and Actuators B-Chemical* 21 (2): 83–89.

Mills, A. and K. Eaton. 2000. Optical sensors for carbon dioxide: An overview of sensing strategies past and present. *Quimica Analitica* 19: 75–86.

Mistlberger, G., K. Koren, S. M. Borisov, and I. Klimant. 2010. Magnetically remote-controlled optical sensor spheres for monitoring oxygen or pH. *Analytical Chemistry* 82 (5): 2124–2128.

Mohr, G. J. 2006. Polymers in optical sensors. In: *Optical Chemical Sensors*, eds. F. Baldini, A. N. Chester, J. Homola, and S. Martellucci. NATO Science Series, Springer, New York.

Mohr, G. J. and O. S. Wolfbeis. 1995. Optical sensing of anions via polarity-sensitive dyes—A bulk sensor membrane for nitrate. *Analytica Chimica Acta* 316 (2): 239–246.

Monk, D. J. and D. R. Walt. 2004. Optical fiber-based biosensors. *Analytical and Bioanalytical Chemistry* 379 (7–8): 931–945.

Monsellier, E. and H. Bedouelle. 2005. Quantitative measurement of protein stability from unfolding equilibria monitored with the fluorescence maximum wavelength. *Protein Engineering Design and Selection* 18 (9): 445–456.

Moreno, J., F. J. Arregui, and I. R. Matias. 2005. Fiber optic ammonia sensing employing novel thermoplastic polyurethane membranes. *Sensors and Actuators B-Chemical* 105 (2): 419–424.

Morf, W. E., K. Seiler, B. Lehmann, C. Behringer, K. Hartman, and W. Simon. 1989. Carriers for chemical sensors—Design-features of optical sensors (optodes) based on selective chromoionophores. *Pure and Applied Chemistry* 61 (9): 1613–1618.

Nagl, S., C. Baleizao, S. M. Borisov, M. Schaeferling, M. N. Berberan-Santos, and O. S. Wolfbeis. 2007. Optical sensing and imaging of trace oxygen with record response. *Angewandte Chemie-International Edition* 46 (13): 2317–2319.

Nagl, S. and O. S. Wolfbeis. 2007. Optical multiple chemical sensing: Status and current challenges. *Analyst* 132 (6): 507–511.

Nakamura, N. and Y. Amao. 2003. Optical sensor for carbon dioxide combining colorimetric change of a pH indicator and a reference luminescent dye. *Analytical and Bioanalytical Chemistry* 376 (5): 642–646.

Navarro-Villoslada, F., G. Orellana, M. C. Moreno-Bondi, T. Vick, M. Driver, G. Hildebrand, and K. Liefeith. 2001. Fiber-optic luminescent sensors with composite oxygen-sensitive layers and anti-biofouling coatings. *Analytical Chemistry* 73 (21): 5150–5156.

Neurauter, G., I. Klimant, and O. S. Wolfbeis. 2000. Fiber-optic microsensor for high resolution pCO(2) sensing in marine environment. *Fresenius Journal of Analytical Chemistry* 366 (5): 481–487.

Ng, S. M. and R. Narayanaswamy. 2006. Fluorescence sensor using a molecularly imprinted polymer as a recognition receptor for the detection of aluminium ions in aqueous media. *Analytical and Bioanalytical Chemistry* 386 (5): 1235–1244.

Nicho, M. E., M. Trejo, A. Garcia-Valenzuela, J. M. Saniger, J. Palacios, and H. Hu. 2001. Polyaniline composite coatings interrogated by a nulling optical-transmittance bridge for sensing low concentrations of ammonia gas. *Sensors and Actuators B-Chemical* 76 (1–3): 18–24.

Niu, C. G., X. Q. Gui, G. M. Zeng, and X. Z. Yuan. 2005. A ratiometric fluorescence sensor with broad dynamic range based on two pH-sensitive fluorophores. *Analyst* 130 (11): 1551–1556.

Niu, C. G., G. M. Zeng, L. X. Chen, G. L. Shen, and R. Q. Yu. 2004. Proton "off-on" behaviour of methylpiperazinyl derivative of naphthalimide: A pH sensor based on fluorescence enhancement. *Analyst* 129 (1): 20–24.

Nivens, D. A., M. V. Schiza, and S. M. Angel. 2002. Multilayer sol-gel membranes for optical sensing applications: Single layer pH and dual layer CO_2 and NH_3 sensors. *Talanta* 58 (3): 543–550.

Noroozifar, M., M. K. Motlagh, S. Bahmanzadeh, and S. Boroon. 2010. A novel optical membrane with extended detection range of pH. *Turkish Journal of Chemistry* 34 (5): 719–730.

O'Mahony, F. C., T. C. O'Riordan, N. Papkovskaia, J. P. Kerry, and D. B. Papkovsky. 2006. Non-destructive assessment of oxygen levels in industrial modified atmosphere packaged cheddar cheese. *Food Control* 17 (4): 286–292.

Onida, B., L. Borello, S. Fiorilli, B. Bonelli, C. O. Arean, and E. Garrone. 2004. Mesostructured SBA-3 silica containing Reichardt's dye as an optical ammonia sensor. *Chemical Communications* (21): 2496–2497.

Opitz, N. and D. W. Lubbers. 1975. New fast-responding optical method to measure pCO_2 in gases and solutions. *Pflugers Archiv-European Journal of Physiology* 355: R120–R120.

Oppenheimer, C., P. Bani, J. A. Calkins, M. R. Burton, and G. M. Sawyer. 2006. Rapid FTIR sensing of volcanic gases released by strombolian explosions at Yasur volcano, Vanuatu. *Applied Physics B-Lasers and Optics* 85 (2–3): 453–460.

Orellana, G. and D. Haigh. 2008. New trends in fiber-optic chemical and biological sensors. *Current Analytical Chemistry* 4 (4): 273–295.

O'Riordan, T. C., D. Buckley, V. Ogurtsov, R. O'Connor, and D. B. Papkovsky. 2000. A cell viability assay based on monitoring respiration by optical oxygen sensing. *Analytical Biochemistry* 278 (2): 221–227.

Oter, O., K. Ertekin, D. Topkaya, and S. Alp. 2006. Room temperature ionic liquids as optical sensor matrix materials for gaseous and dissolved CO_2. *Sensors and Actuators B-Chemical* 117 (1): 295–301.

Owen, C. A., I. Notingher, R. Hill, M. Stevens, and L. L. Hench. 2006. Progress in Raman spectroscopy in the fields of tissue engineering, diagnostics and toxicological testing. *Journal of Materials Science-Materials in Medicine* 17 (11): 1019–1023.

Pal, R. and D. Parker. 2008. A ratiometric optical imaging probe for intracellular pH based on modulation of europium emission. *Organic and Biomolecular Chemistry* 6 (6): 1020–1033.

Panda, B. R. and A. Chattopadhyay. 2007. A water-soluble polythiophene-Au nanoparticle composite for pH sensing. *Journal of Colloid and Interface Science* 316 (2): 962–967.

Pandey, S. K. and K. H. Kim. 2007. The relative performance of NDIR-based sensors in the near real-time analysis of CO_2 in air. *Sensors* 7 (9): 1683–1696.

Papkovsky, D. B., N. Papkovskaia, A. Smyth, J. Kerry, and V. I. Ogurtsov. 2000. Phosphorescent sensor approach for non-destructive measurement of oxygen in packaged foods: Optimisation of disposable oxygen sensors and their characterization over a wide temperature range. *Analytical Letters* 33 (9): 1755–1777.

Papkovsky, D. B., M. A. Smiddy, N. Y. Papkovskaia, and J. P. Kerry. 2002. Nondestructive measurement of oxygen in modified atmosphere packaged hams using a phase-fluorimetric sensor system. *Journal of Food Science* 67 (8): 3164–3169.

Park, E. J., K. R. Reid, W. Tang, R. T. Kennedy, and R. Kopelman. 2005. Ratiometric fiber optic sensors for the detection of inter- and intra-cellular dissolved oxygen. *Journal of Materials Chemistry* 15 (27–28): 2913–2919.

Passaro, V. M. N., F. Dell'Olio, and F. De Leonardis. 2007. Ammonia optical sensing by microring resonators. *Sensors* 7 (11): 2741–2749.

Peng, H. S., J. A. Stolwijk, L. N. Sun, J. Wegener, and O. S. Wolfbeis. 2010. A nanogel for ratiometric fluorescent sensing of intracellular pH values. *Angewandte Chemie-International Edition* 49 (25): 4246–4249.

Persad, A., K. F. Chow, W. Wang, E. Wang, A. Okafor, N. Jespersen, J. Mann, and A. Bocarsly. 2008. Investigation of dye-doped sol-gels for ammonia gas sensing. *Sensors and Actuators B-Chemical* 129 (1): 359–363.

Peters, K. 2011. Polymer optical fiber sensors-a review. *Smart Materials and Structures* 20 (1): 013002–013002.

Peterson, J. I., S. R. Goldstein, R. V. Fitzgerald, and D. K. Buckhold. 1980. Fiber optic pH probe for physiological use. *Analytical Chemistry* 52 (6): 864–869.

Pietsch, C., R. Hoogenboom, and U. S. Schubert. 2009. Soluble polymeric dual sensor for temperature and pH value. *Angewandte Chemie-International Edition* 48 (31): 5653–5656.

Pisco, M., M. Consales, S. Campopiano, R. Viter, V. Smyntyna, M. Giordano, and A. Cusano. 2006. A novel optochemical sensor based on SnO_2 sensitive thin film for Ppm ammonia detection in liquid environment. *Journal of Lightwave Technology* 24 (12): 5000–5007.

Podoleanu, A. G. 2010. Fiber optics, from sensing to non invasive high resolution medical imaging. *Journal of Lightwave Technology* 28 (4): 624–640.

Polerecky, L., N. Volkenborn, and P. Stief. 2006. High temporal resolution oxygen imaging in bioirrigated sediments. *Environmental Science and Technology* 40 (18): 5763–5769.

Poulli, K. I., G. A. Mousdis, and C. A. Georgiou. 2006. Synchronous fluorescence spectroscopy for quantitative determination of virgin olive oil adulteration with sunflower oil. *Analytical and Bioanalytical Chemistry* 386 (5): 1571–1575.

Pringsheim, E., D. Zimin, and O. S. Wolfbeis. 2001. Fluorescent beads coated with polyaniline: A novel nanomaterial for optical sensing of pH. *Advanced Materials* 13 (11): 819–822.

Puppels, G. J. 2002. Preface to the special issue on medical applications of Raman spectroscopy. *Journal of Raman Spectroscopy* 33 (7): 496–497.

Puyol, M., I. Salinas, I. Garces, F. Villuendas, A. Llobera, C. Dominguez, and J. Alonso. 2002. Improved integrated waveguide absorbance optodes for ion-selective sensing. *Analytical Chemistry* 74 (14): 3354–3361.

Qi, Z. M., A. Yimit, K. Itoh, M. Murabayashi, N. Matsuda, A. Takatsu, and K. Kato. 2001. Composite optical waveguide composed of a tapered film of bromothymol blue evaporated onto a potassium ion-exchanged waveguide and its application as a guided wave absorption-based ammonia-gas sensor. *Optics Letters* 26 (9): 629–631.

Raemer, D. B., D. R. Walt, and C. Munkholm. 1991. CO_2 indicator and the use thereof to evaluate placement of tracheal tubes. U.S. Patent 5.005.572.

Raimundo, I. M. and R. Narayanaswamy. 2000. An optical sensor for measurement of gaseous ammonia. *Quimica Analitica* 19: 127–133.

Reyes, F., M. Grutter, A. Jazcilevich, and R. Gonzalez-Oropeza. 2006. Technical note: Analysis of non-regulated vehicular emissions by extractive FTIR spectrometry: Tests on a hybrid car in Mexico city. *Atmospheric Chemistry and Physics* 6: 5339–5346.

Risby, T. H. and F. K. Tittel. 2010. Current status of midinfrared quantum and interband cascade lasers for clinical breath analysis. *Optical Engineering* 49 (11): 111123–111123.

Rottman, C., A. Turniansky, and D. Avnir. 1998. Sol-gel physical and covalent entrapment of three methyl red indicators: A comparative study. *Journal of Sol-Gel Science and Technology* 13 (1–3): 17–25.

Ruedas-Rama, M. J., A. Orte, E. A. H. Hall, J. M. Alvarez-Pez, and E. M. Talavera. 2011. Quantum dot photoluminescence lifetime-based pH nanosensor. *Chemical Communications* 47 (10): 2898–2900.

Rumsey, W. L., J. M. Vanderkooi, and D. F. Wilson. 1988. Imaging of phosphorescence—A novel method for measuring oxygen distribution in perfused tissue. *Science* 241 (4873): 1649–1651.

Ryder, A. G., T. J. Glynn, M. Feely, and A. J. G. Barwise. 2002. Characterization of crude oils using fluorescence lifetime data. *Spectrochimica Acta Part A-Molecular and Biomolecular Spectroscopy* 58 (5): 1025–1037.

Ryder, A. G., S. Power, and T. J. Glynn. 2003. Evaluation of acridine in Nafion as a fluorescence-lifetime-based pH sensor. *Applied Spectroscopy* 57 (1): 73–79.

Saari, L. A. and W. R. Seitz. 1982. pH sensor based on immobilized fluoresceinamine. *Analytical Chemistry* 54 (4): 821–823.

Saari, L. A. and W. R. Seitz. 1983. Immobilized morin as fluorescence sensor for determination of aluminum(III). *Analytical Chemistry* 55 (4): 667–670.

Sadeghi, S. and S. Doosti. 2008. Novel PVC membrane bulk optical sensor for determination of uranyl ion. *Sensors and Actuators B-Chemical* 135 (1): 139–144.

Safavi, A. and M. Bagheri. 2003. Novel optical pH sensor for high and low pH values. *Sensors and Actuators B-Chemical* 90 (1–3): 143–150.

Sanchez-Barragan, I., J. M. Costa-Fernandez, A. Sanz-Medel, M. Valledor, F. J. Ferrero, and J. C. Campo. 2006. A ratiometric approach for pH optosensing with a single fluorophore indicator. *Analytica Chimica Acta* 562 (2): 197–203.

Schaden, S., M. Haberkorn, J. Frank, J. R. Baena, and B. Lendl. 2004. Direct determination of carbon dioxide in aqueous solution using mid-infrared quantum cascade lasers. *Applied Spectroscopy* 58 (6): 667–670.

Schlucker, S. 2009. SERS microscopy: Nanoparticle probes and biomedical applications. *ChemPhysChem* 10 (9–10): 1344–1354.

Schmalzlin, E., J. T. van Dongen, I. Klimant, B. Marmodee, M. Steup, J. Fisahn, P. Geigenberger, and H. G. Lohmannsroben. 2005. An optical multifrequency phase-modulation method using microbeads for measuring intracellular oxygen concentrations in plants. *Biophysical Journal* 89 (2): 1339–1345.

Schreml, S., R. J. Meier, O. S. Wolfbeis, M. Landthaler, R.-M. Szeimies, and P. Babilas. 2011. 2D luminescence imaging of pH in vivo. *Proceedings of the National Academy of Sciences of the United States of America* 108 (6): 2432–2437.

Schroder, C. R., G. Neurauter, and I. Klimant. 2007a. Luminescent dual sensor for time-resolved imaging of pCO_2 and pO_2 in aquatic systems. *Microchimica Acta* 158 (3–4): 205–218.

Schroder, C. R., L. Polerecky, and I. Klimant. 2007b. Time-resolved pH/pO_2 mapping with luminescent hybrid sensors. *Analytical Chemistry* 79 (1): 60–70.

Schroder, C. R., B. M. Weidgans, and I. Klimant. 2005. pH fluorosensors for use in marine systems. *Analyst* 130 (6): 907–916.

Schulz, A., R. Woolley, T. Tabarin, and C. McDonagh. 2011. Dextran-coated silica nanoparticles for calcium-sensing. *Analyst* 136 (8): 1722–1727.

Scorsone, E., S. Christie, K. C. Persaud, P. Simon, and F. Kvasnik. 2003. Fibre-optic evanescent sensing of gaseous ammonia with two forms of a new near-infrared dye in comparison to phenol red. *Sensors and Actuators B-Chemical* 90 (1–3): 37–45.

Scott, S. M., D. James, Z. Ali, W. T. O'Harea, and F. J. Rowell. 2003. Total luminescence spectroscopy with pattern recognition for classification of edible oils. *Analyst* 128 (7): 966–973.

Shafer-Peltier, K. E., A. S. Haka, J. T. Motz, M. Fitzmaurice, R. R. Dasari, and M. S. Feld. 2002. Model-based biological Raman spectral imaging. *Journal of Cellular Biochemistry* 87: 125–137.

Shang, Y., X. Wang, E. Xu, C. Tong, and J. Wu. 2011. Optical ammonia gas sensor based on a porous silicon rugate filter coated with polymer-supported dye. *Analytica Chimica Acta* 685 (1): 58–64.

Sharma, N. K. and B. D. Gupta. 2004. Fabrication and characterization of a fiber-optic pH sensor for the pH range 2 to 13. *Fiber and Integrated Optics* 23 (4): 327–335.

Shi, W., S. He, M. Wei, D. G. Evans, and X. Duan. 2010. Optical pH sensor with rapid response based on a fluorescein-intercalated layered double hydroxide. *Advanced Functional Materials* 20 (22): 3856–3863.

Shi, C., Y. Zhang, C. Gu, B. Chen, L. Seballos, T. Olson, and J. Z. Zhang. 2009. Molecular fiber sensors based on surface enhanced Raman scattering (SERS). *Journal of Nanoscience and Nanotechnology* 9 (4): 2234–2246.

de Silva, A. P., H. Q. N. Gunaratne, T. Gunnlaugsson, A. J. M. Huxley, C. P. McCoy, J. T. Rademacher, and T. E. Rice. 1997. Signaling recognition events with fluorescent sensors and switches. *Chemical Reviews* 97 (5): 1515–1566.

de Silva, A. P., T. S. Moody, and G. D. Wright. 2009. Fluorescent PET (photoinduced electron transfer) sensors as potent analytical tools. *Analyst* 134 (12): 2385–2393.

Smith, T. E. L., M. J. Wooster, M. Tattaris, and D. W. T. Griffith. 2011. Absolute accuracy and sensitivity analysis of OP-FTIR retrievals of CO_2, CH_4 and CO over concentrations representative of "clean air" and "polluted plumes." *Atmospheric Measurement Techniques* 4 (1): 97–116.

Sotomayor, P. T., I. M. Raimundo, A. J. G. Zarbin, J. J. R. Rohwedder, G. O. Neto, and O. L. Alves. 2001. Construction and evaluation of an optical pH sensor based on polyaniline-porous Vycor glass nanocomposite. *Sensors and Actuators B-Chemical* 74 (1–3): 157–162.

Staal, M., S. M. Borisov, L. F. Rickelt, I. Klimant, and M. Kuhl. 2011. Ultrabright planar optodes for luminescence life-time based microscopic imaging of O_2 dynamics in biofilms. *Journal of Microbiological Methods* 85 (1): 67–74.

Stich, M. I. J., L. H. Fischer, and O. S. Wolfbeis. 2010. Multiple fluorescent chemical sensing and imaging. *Chemical Society Reviews* 39 (8): 3102–3114.

Stich, M. I. J., S. Nagl, O. S. Wolfbeis, U. Henne, and M. Schaeferling. 2008. A dual luminescent sensor material for simultaneous imaging of pressure and temperature on surfaces. *Advanced Functional Materials* 18 (9): 1399–1406.

Stoddart, P. R. and D. J. White. 2009. Optical fibre SERS sensors. *Analytical and Bioanalytical Chemistry* 394 (7): 1761–1774.

Stromberg, N. and S. Hulth. 2005. Assessing an imaging ammonium sensor using time correlated pixel-by-pixel calibration. *Analytica Chimica Acta* 550 (1–2): 61–68.

Stuart, D. A., K. B. Biggs, and R. P. Van Duyne. 2006a. Surface-enhanced Raman spectroscopy of half-mustard agent. *Analyst* 131 (4): 568–572.

Stuart, D. A., J. M. Yuen, N. S. O. Lyandres, C. R. Yonzon, M. R. Glucksberg, J. T. Walsh, and R. P. Van Duyne. 2006b. In vivo glucose measurement by surface-enhanced Raman spectroscopy. *Analytical Chemistry* 78 (20): 7211–7215.

Su, B. L., N. Moniotte, N. Nivarlet, G. Tian, and J. Desmet. 2010. Design and synthesis of fluorescence-based siderophore sensor molecules for Fe-III ion determination. *Pure and Applied Chemistry* 82 (11): 2199–2216.

Suah, F. B. M., M. Ahmad, and M. N. Taib. 2003. Applications of artificial neural network on signal processing of optical fibre pH sensor based on bromophenol blue doped with sol-gel film. *Sensors and Actuators B-Chemical* 90 (1–3): 182–188.

Sun, L. N., H. Peng, M. I. J. Stich, D. Achatz, and O. S. Wolfbeis. 2009. pH sensor based on upconverting luminescent lanthanide nanorods. *Chemical Communications* (33): 5000–5002.

Taitt, C. R., L. C. Shriver-Lake, M. M. Ngundi, and F. S. Ligler. 2008. Array biosensor for toxin detection: Continued advances. *Sensors* 8 (12): 8361–8377.

Takagai, Y., Y. Nojiri, T. Takase, W. L. Hinze, M. Butsugan, and S. Igarashi. 2010. "Turn-on" fluorescent polymeric microparticle sensors for the determination of ammonia and amines in the vapor state. *Analyst* 135 (6): 1417–1425.

Tang, X., J. Provenzano, Z. Xu, J. Dong, H. Duan, and H. Xiao. 2011. Acidic ZSM-5 zeolite-coated long period fiber grating for optical sensing of ammonia. *Journal of Materials Chemistry* 21 (1): 181–186.

Tang, Y., E. C. Tehan, Z. Y. Tao, and F. V. Bright. 2003. Sol-gel-derived sensor materials that yield linear calibration plots, high sensitivity, and long-term stability. *Analytical Chemistry* 75 (10): 2407–2413.

Tao, Z. Y., E. C. Tehan, Y. Tang, and F. V. Bright. 2006a. Stable sensors with tunable sensitivities based on class II xerogels. *Analytical Chemistry* 78 (6): 1939–1945.

Tao, S. Q., L. Xu, and J. C. Fanguy. 2006b. Optical fiber ammonia sensing probes using reagent immobilized porous silica coating as transducers. *Sensors and Actuators B-Chemical* 115 (1): 158–163.

Tian, Y., B. R. Shumway, and D. R. Meldrum. 2010a. A new cross-linkable oxygen sensor covalently bonded into poly(2-hydroxyethyl methacrylate)-Co-polyacrylamide thin film for dissolved oxygen sensing. *Chemistry of Materials* 22 (6): 2069–2078.

Tian, Y., B. R. Shumway, A. C. Youngbull, Y. Li, A. K. Y. Jen, R. H. Johnson, and D. R. Meldrum. 2010b. Dually fluorescent sensing of pH and dissolved oxygen using a membrane made from polymerizable sensing monomers. *Sensors and Actuators B-Chemical* 147 (2): 714–722.

Tolosa, L., Y. Kostov, P. Harms, and G. Rao. 2002. Noninvasive measurement of dissolved oxygen in shake flasks. *Biotechnology and Bioengineering* 80 (5): 594–597.

Tormo, L., N. Bustamante, G. Colmenarejo, and G. Orellana. 2010. Can luminescent Ru(II) polypyridyl dyes measure pH directly? *Analytical Chemistry* 82 (12): 5195–5204.

Tran-Thi, T. H., R. Dagnelie, S. Crunairez, and L. Nicole. 2011. Optical chemical sensors based on hybrid organic-inorganic sol-gel nanoreactors. *Chemical Society Reviews* 40 (2): 621–639.

Trinkel, M., W. Trettnak, and C. Kolle. 2000. Oxygen trace analysis utilising a miniaturised luminescence lifetime-based sensor instrumentation. *Quimica Analitica* 19: 112–117.

Tripp, R. A., R. A. Dluhy, and Y. Zhao. 2008. Novel nanostructures for SERS biosensing. *Nano Today* 3 (3–4): 31–37.

Tsai, Y. T., T. C. Wen, and A. Gopalan. 2003. Tuning the optical sensing of pH by poly(diphenylamine). *Sensors and Actuators B-Chemical* 96 (3): 646–657.

Turel, M., M. Cajlakovic, E. Austin, J. P. Dakin, G. Uray, and A. Lobnik. 2008. Direct UV-LED lifetime pH sensor based on a semi-permeable sol-gel membrane immobilized luminescent Eu3+ chelate complex. *Sensors and Actuators B-Chemical* 131 (1): 247–253.

Turner, N. W., M. Cauchi, E. V. Piletska, C. Preston, and S. A. Piletsky. 2009. Rapid qualitative and quantitative analysis of opiates in extract of poppy head via FTIR and chemometrics: Towards in-field sensors. *Biosensors and Bioelectronics* 24 (11): 3322–3328.

Uttamlal, M. and D. R. Walt. 1995. A fiber-optic carbon-dioxide sensor for fermentation monitoring. *Bio-Technology* 13 (6): 597–601.

Utzinger, U. and R. R. Richards-Kortum. 2003. Fiber optic probes for biomedical optical spectroscopy. *Journal of Biomedical Optics* 8 (1): 121–147.

Uzunbajakava, N., A. Lenferink, Y. Kraan, E. Volokhina, G. Vrensen, J. Greve, and C. Otto. 2003. Nonresonant confocal Raman imaging of DNA and protein distribution in apoptotic cells. *Biophysical Journal* 84 (6): 3968–3981.

Vasil'ev, V. V. and S. M. Borisov. 2002. Optical oxygen sensors based on phosphorescent water-soluble platinum metals porphyrins immobilized in perfluorinated ion-exchange membrane. *Sensors and Actuators B-Chemical* 82 (2–3): 272–276.

Vasylevska, G. S., S. M. Borisov, C. Krause, and O. S. Wolfbeis. 2006. Indicator-loaded permeation-selective microbeads for use in fiber optic simultaneous sensing of pH and dissolved oxygen. *Chemistry of Materials* 18 (19): 4609–4616.

Vasylevska, A. S., A. A. Karasyov, S. M. Borisov, and C. Krause. 2007. Novel coumarin-based fluorescent pH indicators, probes and membranes covering a broad pH range. *Analytical and Bioanalytical Chemistry* 387 (6): 2131–2141.

Velasco, E., B. Lamb, H. Westberg, E. Allwine, G. Sosa, J. L. Arriaga-Colina, B. T. Jobson et al. 2007. Distribution, magnitudes, reactivities, ratios and diurnal patterns of volatile organic compounds in the valley of Mexico during the MCMA 2002 & 2003 field campaigns. *Atmospheric Chemistry and Physics* 7: 329–353.

Vivian, J. T. and P. R. Callis. 2001. Mechanisms of tryptophan fluorescence shifts in proteins. *Biophysical Journal* 80 (5): 2093–2109.

Vo-Dinh, T. and P. Kasili. 2005. Fiber-optic nanosensors for single-cell monitoring. *Analytical and Bioanalytical Chemistry* 382 (4): 918–925.

Vo-Dinh, T., F. Yan, and M. B. Wabuyele. 2006. Surface-enhanced Raman scattering for biomedical diagnostics and molecular imaging. *Surface-Enhanced Raman Scattering: Physics and Applications* 103: 409–426.

Voegtle, F., ed. 1996. *Comprehensive Supramolecular Chemistry*, Vol. 2. Pergamon Press, Oxford, U.K.

Voraberger, H. S., H. Kreimaier, K. Biebernik, and W. Kern. 2001. Novel oxygen optrode withstanding autoclavation: Technical solutions and performance. *Sensors and Actuators B-Chemical* 74 (1–3): 179–185.

Waich, K., S. Borisov, T. Mayr, and I. Klimant. 2009. Dual lifetime referenced trace ammonia sensors. *Sensors and Actuators B-Chemical* 139 (1): 132–138.

Waich, K., T. Mayr, and I. Klimant. 2007. Microsensors for detection of ammonia at Ppb-concentration levels. *Measurement Science and Technology* 18 (10): 3195–3201.

Waich, K., M. Sandholzer, T. Mayr, C. Slugovc, and I. Klimant. 2010. A highly flexible polymerization technique to prepare fluorescent nanospheres for trace ammonia detection. *Journal of Nanoparticle Research* 12 (4): 1095–1100.

Walt, D. R. 2010. Fibre optic microarrays. *Chemical Society Reviews* 39 (1): 38–50.

Wang, X. D., X. Chen, Z. X. Xie, and X. R. Wang. 2008a. Reversible optical sensor strip for oxygen. *Angewandte Chemie-International Edition* 47 (39): 7450–7453.

Wang, E. J., K. F. Chow, V. Kwan, T. Chin, C. Wong, and A. Bocarsly. 2003. Fast and long term optical sensors for pH based on sol-gels. *Analytica Chimica Acta* 495 (1–2): 45–50.

Wang, Y., B. Li, L. Zhang, L. Liu, Q. Zuo, and P. Li. 2010. A highly selective regenerable optical sensor for detection of mercury(II) ion in water using organic-inorganic hybrid nanomaterials containing pyrene. *New Journal of Chemistry* 34 (9): 1946–1953.

Wang, H. W., Y. H. Wei, and H. W. Guo. 2009. Reduced nicotinamide adenine dinucleotide (NADH) fluorescence for the detection of cell death. *Anti-Cancer Agents in Medicinal Chemistry* 9 (9): 1012–1017.

Wang, L., W. Zhao, and W. Tan. 2008b. Bioconjugated silica nanoparticles: Development and applications. *Nano Research* 1 (2): 99–115.

Weidgans, B. M., C. Krause, I. Klimant, and O. S. Wolfbeis. 2004. Fluorescent pH sensors with negligible sensitivity to ionic strength. *Analyst* 129 (7): 645–650.

Wen, Z. Y., L. F. Wang, and C. Gang. 2010. Development and application of quantum cascade laser based gas sensing system. *Spectroscopy and Spectral Analysis* 30 (8): 2043–2048.

Wencel, D., B. D. MacCraith, and C. McDonagh. 2009. High performance optical ratiometric sol-gel-based pH sensor. *Sensors and Actuators B-Chemical* 139 (1): 208–213.

Wencel, D., J. P. Moore, N. Stevenson, and C. McDonagh. 2010. Ratiometric fluorescence-based dissolved carbon dioxide sensor for use in environmental monitoring applications. *Analytical and Bioanalytical Chemistry* 398 (5): 1899–1907.

Werle, P., F. Slemr, K. Maurer, R. Kormann, R. Mucke, and B. Janker. 2002. Near- and mid-infrared laser-optical sensors for gas analysis. *Optics and Lasers in Engineering* 37 (2–3): 101–114.

Wirnsberger, G., B. J. Scott, and G. D. Stucky. 2001. pH sensing with mesoporous thin films. *Chemical Communications* (01): 119–120.

Wolfbeis, O. S. 2002. Fiber-optic chemical sensors and biosensors. *Analytical Chemistry* 74 (12): 2663–2677.

Wolfbeis, O. S. 2004a. Optical technology until the year 2000: An historical overview. In: *Optical Sensors: Industrial, Environmental and Diagnostic Applications*, eds. R. Narayanaswamy and O. S. Wolfbeis, Chapter 1. Springer, Berlin, Germany.

Wolfbeis, O. S. 2004b. Fiber-optic chemical sensors and biosensors. *Analytical Chemistry* 76 (12): 3269–3283.

Wolfbeis, O. S. 2006. Fiber-optic chemical sensors and biosensors. *Analytical Chemistry* 78 (12): 3859–3873.

Wolfbeis, O. S. 2008. Fiber-optic chemical sensors and biosensors. *Analytical Chemistry* 80 (12): 4269–4283.

Wolfbeis, O. S., L. J. Weis, M. J. P. Leiner, and W. E. Ziegler. 1988. Fiber-optic fluorosensor for oxygen and carbon-dioxide. *Analytical Chemistry* 60 (19): 2028–2030.

Wu, Y. and J. Y. Qu. 2006. Autofluorescence spectroscopy of epithelial tissues. *Journal of Biomedical Optics* 11 (5): 054023.

Wu, Y., W. Zheng, and J. Y. Qu. 2006. Sensing cell metabolism by time-resolved autofluorescence. *Optics Letters* 31 (21): 3122–3124.

Wygladacz, K. and E. Bakker. 2005. Imaging fiber microarray fluorescent ion sensors based on bulk optode microspheres. *Analytica Chimica Acta* 532 (1): 61–69.

Xia Q., H. F. Zuo, S. C. Li, Z. H. Wen, and Y. H. Li. 2009. Remote passive sensing of aeroengine exhausts using FTIR system. *Spectroscopy and Spectral Analysis* 29 (3): 616–619.

Xie, Z. H., L. Q. Guo, X. H. Zheng, X. C. Lin, and G. N. Chen. 2005. A new porous plastic fiber probe for ammonia monitoring. *Sensors and Actuators B-Chemical* 104 (2): 173–178.

Xu, W. Y., R. C. McDonough, B. Langsdorf, J. N. Demas, and B. A. DeGraff. 1994. Oxygen sensors based on luminescence quenching—Interactions of metal-complexes with the polymer supports. *Analytical Chemistry* 66 (23): 4133–4141.

Yang, Y., J. Jiang, G. Shen, and R. Yu. 2009. An optical sensor for mercury ion based on the fluorescence quenching of tetra(p-dimethylaminophenyl)porphyrin. *Analytica Chimica Acta* 636 (1): 83–88.

Yang, J. S., C. J. Lee, and C. H. Wei. 2002. Fiber-optic chemical sensors: A general review. *Journal of the Chinese Chemical Society* 49 (5): 677–692.

Yari, A. and H. A. Abdoli. 2010. Sol-gel derived highly selective optical sensor for sensitive determination of the mercury(II) ion in solution. *Journal of Hazardous Materials* 178 (1–3): 713–717.

Yimit, A., K. Itoh, and M. Murabayashi. 2003. Detection of ammonia in the Ppt range based on a composite optical waveguide pH sensor. *Sensors and Actuators B-Chemical* 88 (3): 239–245.

Yimit, A., A. G. Rossberg, T. Amemiya, and K. Itoh. 2005. Thin film composite optical waveguides for sensor applications: A review. *Talanta* 65 (5): 1102–1109.

Zhang, P., J. H. Guo, Y. Wang, and W. Q. Pang. 2002. Incorporation of luminescent tris(bipyridine)ruthenium(II) complex in mesoporous silica spheres and their spectroscopic and oxygen-sensing properties. *Materials Letters* 53 (6): 400–405.

Zhang, G., G. M. Palmer, M. Dewhirst, and C. L. Fraser. 2009. A dual-emissive-materials design concept enables tumour hypoxia imaging. *Nature Materials* 8 (9): 747–751.

Zhang, X., N. C. Shah, and R. P. Van Duyne. 2006. Sensitive and selective chem/biosensing based on surface-enhanced Raman spectroscopy (SERS). *Vibrational Spectroscopy* 42 (1): 2–8.

Zhao, Z. J., D. X. Liu, J. L. Zhang, Z. B. Wang, X. Li, and E. M. Tian. 2011. Design of non-dispersed infrared (NDIR) methane gas sensor. *Spectroscopy and Spectral Analysis* 31 (2): 570–573.

Zhong, W. 2009. Nanomaterials in fluorescence-based biosensing. *Analytical and Bioanalytical Chemistry* 394 (1): 47–59.

Zhu, Q. Z., R. C. Aller, and Y. Z. Fan. 2005. High-performance planar pH fluorosensor for two-dimensional pH measurements in marine sediment and water. *Environmental Science and Technology* 39 (22): 8906–8911.

Zhu, Q. Z., R. C. Aller, and Y. Z. Fan. 2006. A new ratiometric, planar fluorosensor for measuring high resolution, two-dimensional pCO_2 distributions in marine sediments. *Marine Chemistry* 101 (1–2): 40–53.

Zhujun, Z., J. L. Mullin, and W. R. Seitz. 1986. Optical sensor for sodium based on ion-pair extraction and fluorescence. *Analytica Chimica Acta* 184: 251–258.

Zhujun, Z. and W. R. Seitz. 1984. A carbon-dioxide sensor based on fluorescence. *Analytica Chimica Acta* 160: 305–309.

Part II

Research Advances

4

Photoluminescent Nanosensors

Mike McShane

CONTENTS

4.1 Introduction

In addition to the advantages of rapid response, small displaced volume, and high spatial resolution that are common to all nanoscale optical devices, luminescence-based nanosensors can also leverage the intrinsic environmental sensitivity of fluorophores/phosphors and the phenomenon of energy transfer. Their compatibility with standard microscopes and other imaging devices makes them even more attractive for biological and biomedical applications. The reduction in size does, however, also involve some loss of performance, particularly in longevity. In this chapter, the fundamental properties and basic scaling laws of nanoscale fluorescence-based sensors will be discussed, followed by a review of examples of nanosensors and an outlook of the future of these promising devices.

The focus of this work will remain on systems that can offer reversible sensing behavior; thus, single-use "detection" approaches, while an important area of development,[1–8] will not be covered herein. We also note that fiber-based devices have been developed using the nanoscale tips of tapered optical fibers.[9–15] While the mode of light delivery is different, the principles of chemo-optical transduction in those devices are the same as what are covered here for particles with sensing chemistry attached to the surface; additional attention will not be given to reviewing the unique aspects of fiber-based systems herein. Instead, emphasis will be placed on nanoscale particles and hollow structures with unique inherent luminescence properties or characteristics well suited to immobilization/encapsulation of sensing materials.

4.2 Fluorescence and Phosphorescence Basics

Photoluminescence is a phenomenon in which an excited molecule absorbs energy in the form of photons (light) and subsequently releases energy in the form of a lower energy photon (longer wavelength light). This behavior is characteristic of certain chemical groups, including benzene rings and double and triple bonds, as well as nanoscale structures such as "quantum dots" (Qdots) wherein quantum effects also give rise to strong luminescence. Collectively, these molecules and materials are referred to as "luminophores."[16]

Photoluminescence has some key features that make it very attractive for sensing purposes. First, the shift in emission wavelength relative to the excitation light results in a very high measurement sensitivity due to the low background contributed by the light source; even *single*-molecule detection is possible.[17] Second, both energetics and kinetics of the emission process depend upon the environment of the luminophores at the instant of excitation and for the time in which they are in the excited state (the "luminescence lifetime"). For example, the solvent properties, temperature, and proximity of other molecules or functional groups may influence luminescence.[16] Thus, the time-, frequency-, and spectral-domain features of luminescence carry information regarding the environment of these groups, making them "optical reporters" of chemical information.[18]

Without delving into details on photoluminescence (a comprehensive treatment may be found in Lakowicz's regularly updated book),[16] here we will focus on some key aspects related to sensing. Specifically, excitation, excited state lifetime and time-dependent emission, quantum yield, brightness, quenching, and energy transfer and the associated basic mathematics will be introduced.

Photons are absorbed by luminophores in proportion to their concentration and relative absorbing power, following the Beer-Lambert mathematical description:

$$I_A(z) = I_0\left(1 - e^{-\varepsilon c z}\right) \approx \varepsilon c z$$

where

$I_A(z)$ is the intensity of light absorbed by the luminophore population up through a sample of depth z
I_0 is the input light intensity
c and ε are the molar concentration and the molar absorption coefficient, respectively

This relation is approximately linear with concentration for dilute solutions.

"Excitation" refers to the promotion of an electron to a higher energy level, which can occur upon absorption of a photon as noted earlier. A population of excited luminophores exhibits an exponential temporal emission profile over which it will emit photons, and the decay time constant τ is a characteristic called the "excited state lifetime." Thus, an excited homogenous luminophore population is typically described as having a time-dependent emission profile following $I_L(t) = I_{L_0} e^{-t/\tau}$ in the time period immediately following the removal of an excitation input. In this equation, I_L is the luminescence intensity and I_{L_0} is the initial (maximum) intensity. Traditionally, "fluorescence" is a shorter-lived process (~1–5 ns) following emission from a singlet state, whereas "phosphorescence" occurs from triplet states accessed by intersystem crossing and decays over much longer times (100 s of nanoseconds to even hours). "Luminescence" covers both terms and also generally includes photo-induced emission from materials exhibiting quantum confinement effects that do not fall into the molecular excitation categories described earlier, such as seen in Qdots.

The quantum yield of a luminophore (φ) is defined simply as the number of photons emitted per photon absorbed. As such, φ is a measure of luminescence efficiency (conversion of excitation photons to emission photons), and values above 0.1 are considered high. When combined together with the molar absorption coefficient (ε), φ characterizes the efficiency of converting input photons to output photons of a different wavelength: the product of φ and ε is called the "brightness." "Quenching" refers to processes that decrease the relative emission from excited luminophores and includes both static and dynamic cases.

Luminescence reporters, aka "indicators," are molecules or supramolecular assemblies that respond to an analyte with a change in their optical properties resulting in a measurable change in luminescence. Typically, there is a decrease or increase in the extinction coefficient, quantum yield, or luminescence lifetime when the indicator binds the target analyte; some indicators exhibit a shift in either excitation or emission spectrum (or both; Figure 4.1).

Dynamic quenching decreases luminescence due to collision between the excited luminophore and some environmental energy acceptor ("quencher"). In collisional quenching, the quencher contacts

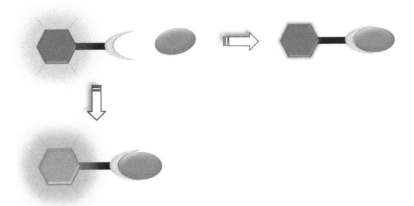

FIGURE 4.1 (Top left) Luminescent indicator, represented as a light absorbing/emitting domain (hexagon) and an analyte-binding region (crescent). Upon binding the target analyte, luminescence intensity may be enhanced or attenuated (top right, shown as attenuation) or spectral properties may be shifted (bottom left, shown as emission color change).

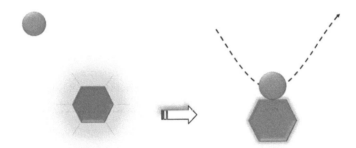

FIGURE 4.2 Luminescent indicator (left) shown to interact with analyte, undergoing dynamic quenching resulting in decreased luminescence intensity and lifetime.

the fluorophores while in the excited state, returning the fluorophore to ground state without photon emission (Figure 4.2).[16] Oxygen is one of the best known collisional quenchers of luminescence; therefore, many luminophores exhibit, to some degree, oxygen sensitivity, and this can be useful in sensing oxygen. The process of collisional quenching is typically characterized by the Stern-Volmer equation $I_L(0)/I_L = \tau_0/\tau = 1 + k_q\tau_0[Q] = 1 + K_D[Q]$. In this equation, $I_L(0)$ and I_L are the luminescence intensities in the absence and presence of the quencher, τ_0 and τ are the lifetimes of the luminophore in the absence and presence of the quencher, k_q is the biomolecular quenching constant, and $[Q]$ is the concentration of the quencher (in this case, molecular oxygen). The Stern-Volmer quenching constant (K_D) is a measure of the quenching efficiency and is calculated as the product of k_q and τ_0.

Resonance energy transfer (RET) refers to the nonradiative energy transfer of excited state energy from an excited state luminophore (the donor, *D*) to another luminophore or absorber (the acceptor, *A*); fluorescence resonance energy transfer (FRET) is the special case when the acceptor is a fluorophore. RET can occur when the following specific conditions hold: (1) the donor and acceptor are in close proximity, typically <10 nm separation; (2) the emission spectrum of the donor overlaps with the absorption spectrum of the acceptor; and (3) the respective emission and absorption dipoles are sufficiently aligned.[16] Thus, RET is a strongly distance-dependent phenomenon that is commonly used to transduce binding between molecules/structures comprising or linked to energy transfer donors and acceptors.

The Förster radius, R_0, is a characteristic value for a D-A pair, defined as the distance at which half the donor molecules are quenched by the acceptor molecules (or, alternatively, the distance at which energy transfer is 50% efficient). The Förster radius is defined as $R_0 = K\sqrt[6]{\kappa^2 n^{-4} \varphi_D J[\lambda]}$, from which the influence of several key properties of the donor and acceptor are evident. *K* is simply a proportionality constant (9.78×10^3), while κ^2 refers to the relative spatial orientation of the dipoles of *D* and *A*, with a range of possible values from 0 (orthogonal dipoles) to 4 (collinear and parallel transitional dipoles). For random orientation (as is usually assumed, especially for solution-phase measurements), κ^2 is 2/3. The remaining terms, n, φ_D, and $J[\lambda]$ correspond to the solvent refractive index, quantum yield of D in the absence of A, and the overlap integral, which measures the degree of overlap between the emission spectrum of D and the absorption spectrum of A, respectively.

In practice, R_0 values vary significantly for different D-A pairs, with values practical for intermolecular energy transfer ranging from 10 to 100 Å.[19] The energy transfer efficiency (E)—the fraction of photons absorbed by D that are transferred to A—is commonly given as the ratio of energy transfer rate $k_T = 1/\tau_D (R_0/r)^6$ to the total decay rate of the donor (τ_D is the natural lifetime of the donor): $E = k_T/\tau_D^{-1} + k_T$. This can be rewritten as $E = R_0^6/R_0^6 + r^6$ to clearly observe the strong distance dependence near R_0. It is important to recognize the impact of energy transfer on observed emission lifetime. The increased rate of de-excitation follows the same strong distance dependence, dropping sharply from the maximum natural lifetime as the sixth order of r.

It is noteworthy that this holds well for dipole-dipole interactions and has been very successful in describing energy transfer between many organic fluorophore pairs.[16,20,21] However, one must appreciate that not all donors and acceptors are "created equal." For example, nanomaterials may be used as donors and/or acceptors. Semiconducting nanocrystals, also called Qdots, exhibit strong photoluminescence ($\varphi = 0.2$–0.7 in buffer, depending upon coatings), which can be tuned in wavelength according to the core crystal size.[22–25] Qdots are very efficient donors, primarily because of the relatively narrow emission profile for low-polydispersity Qdots.[26–30] The large potential spectral separation between excitation and emission is also an advantage. Studies with Qdots have indicated an R_0 that was larger than anticipated from FRET theory. In fact, recent results have suggested that Qdots have a degenerate 2D transition dipole,[31,32] resulting in inaccurate estimates of interaction distances using 1D-1D dipole models.[33] There is also some evidence to suggest that unique surface effects of metal particles used as acceptors may extend the interaction distance to above 200 Å. Changes in dye radiative rate as well as nonradiative Energy Transfer (ET) to the relatively large (few to tens of nanometers diameter) gold nanoparticle acceptors (AuNPs) were observed,[34,35] whereas observations from smaller AuNPs fit the dipole-nanosurface energy transfer (NSET) formalism.[36,37] In this case, the energy transfer efficiency is expressed as

$$E = \frac{N}{N + \left(r/R_0 \right)^4},$$

where

$$R_0 = \left(\frac{0.225 c n^2 \varphi \lambda^2}{(2\pi)^2 \omega_F k_F} \right)^{1/4} \quad \text{and } N \text{ is the number of acceptors proximal to the donor}$$

r is the center-to-center distance between donor and acceptor
κ^2 is the dipole orientation factor
φ is the quantum yield of the donor
n is the refractive index of the surrounding medium
λ is the donor emission wavelength
c is the speed of light in vacuum
ω_F and k_F are the angular frequency and the Fermi wavevector for bulk gold, respectively

The key point is to note a transition from sixth-order dependence on distance to fourth-order, meaning a wider range of distances over which energy transfer will occur (for fixed R_0) as well as decreased sensitivity to distance.

When interactions occur between two nanomaterials, the situation becomes more complex; experimental work on Qdot-AuNP interactions has yielded inconsistent results, ranging from FRET-like behavior to NSET-like behavior.[38–46] Thus, the mathematical descriptions for energy transfer, at this time, are still not settled. Sapsford et al. have recently offered an outstanding review of materials for energy transfer,[19] to which the reader is referred for more details on the basics of energy transfer as well as options for donors and acceptors.

4.3 Calibration Issues

Luminescence is extremely sensitive. However, for it to be used in quantitative measurements (e.g., sensing), appropriate calibration must be performed to relate the measured output property (typically, intensity or lifetime) to the input of interest (i.e., analyte concentration). This raises a critical problem when considering the possibility of distributed sensors that do not maintain a physical connection to the measurement apparatus, as several factors besides analyte concentration may change between measurements: the number of sampled sensors (due to diffusion or leaching); the number of active luminophores (due to bleaching); the efficiency of input light delivery or output light collection (due to motion or other), etc. The key point of note is that a system measuring only intensity will be susceptible to many nonspecific sources of signal change, and therefore, a means of internal calibration or "self-calibration" is essential to achieving a stable quantitative system.

Demchenko has recently described the options for accomplishing this at many levels, from design of reporters to incorporation of additional reference elements.[47] A ratiometric (two-wavelength) excitation or emission response is essential; indeed, this has been used in many nanosensing approaches.[48–57] Ratiometric output may be easily accomplished using combinations of luminophores, including nanomaterials. In comparing the alternatives for intensity-based systems, he concluded that a single reporter exhibiting ratiometric output would yield the most stable self-calibration, primarily because the rate of bleaching or other damage to the indicator would affect all regions of the emission spectrum identically. This could be seen in contrast to a ratiometric system where an organic indicator dye is paired with a Qdot; while the latter would provide a very stable intensity reference, the photobleaching rates for the two luminophores would be different, resulting in a time- and exposure-dependent calibration drift. In cases where such chemistry is not available, luminophores with low bleaching rates should be used and illumination times should be limited to minimize drift. Lifetime analysis should be considered if financially and technically suitable.

4.4 Scaling Laws

Kopelman and Dourado prepared a limited but useful description of the general effects of nanosizing on fluorescent sensors.[58] In their comparison, each "sensor" referred to a particle of radius r, homogeneous and constant concentration of luminophore within the sensor, constant light source power for all sizes, and diffusion-limited behavior. Taking their conclusions, but assuming constant illumination intensity rather than constant power (the latter requires scaling the light source intensity inversely with the particle cross section), the following general set of pros (+) and cons (–) are proposed:

Size Dependence	Property (+)	Property (–)
r^3	Sample volume/invasiveness	
	Absolute detection limit	
	Materials cost	
r^2	Response time	
r	Spatial resolution	Signal-to-noise ratio
$1/r$		Heating
		Leaching
$1/r^2$		Photobleaching
$1/r^3$	Sensitivity	

While these will not hold true for every sensor embodiment, they serve as a basic overview of trends under common conditions. Clearly, nanosizing has some significant advantages if the drawbacks are acceptable or can be overcome by design. Leaching is a time-dependent loss in signal as well as a toxicity risk. Assuming that leaching can be avoided by stable (e.g., covalent) immobilization of sensing chemistry and that local heating under realistic conditions is not deleterious to nanosensor or surrounding environment (i.e., biological cell or tissue), the two other drawbacks of nanosizing remain key issues.

The linearly decreasing SNR and quadratically increasing photobleaching with decreasing particle size constitute a critical "double whammy" to the practical use of nanosensors: to increase signal-to noise ratio, longer exposures or more averages must be taken $\left(SNR \sim \sqrt{nt} \right)$ directly increasing photodestruction in the process. As a result, more sensitive instrumentation and maximally photostable luminophores must be employed to enhance the useful lifetime of nanosensors.

4.5 Nanoscale Sensing Systems

Our review of luminescent nanosensors will begin with "reagentless" systems—those in which a single nano-construct is involved in reversibly responding to a target analyte. A consideration of opportunities to use multicomponent systems will still achieving nanoscale resolution for each sensing element will be provided later.

4.5.1 "Reagentless" Sensors

Reagentless sensors can be categorized several ways. Here, we will primarily differentiate based on the location of the luminescent reporter elements or, stated another way, the position within the construct where the interaction with the analyte occurs to yields a change in luminescence. Therefore, we will focus on (1) surface interactions and (2) embedded systems. The former has the advantage of easy access to the surrounding environment, yielding even faster response times than would be possible for chemistry distributed throughout a nanoscale particle. On the other hand, the smaller volume available at the surface reduces the total amount of indicator that can be packaged in a carrier, limiting signal and resulting in more rapid bleaching or other degradation.

4.5.1.1 Surface-Based Interactions

The simplest platform on which to build a nanoscale luminescent sensor is an inert, nonluminescent nanoparticle: one in which the particle simply serves as a carrier for the sensing chemistry but does not contribute in any substantial way to the response of the sensor to the analyte of interest. The requirements for the particles, in this case, are minimal: they should be stable and amenable to functionalization, allowing immobilization of the sensing chemistry (Figure 4.3).

A step up in complexity is for the nanoparticle to contribute luminescence to the sensor signal in some way, either purely as an internal reference or by acting as an energy transfer donor to other luminophores

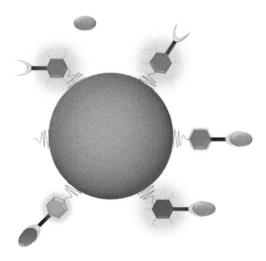

FIGURE 4.3 Sensing chemistry immobilized on an inert nanoparticle carrier. Luminescent indicators are shown as attached by flexible linkers. As described earlier, upon binding the target analyte, luminescence of indicators may be attenuated (right), enhanced, or shifted in spectrum (bottom left).

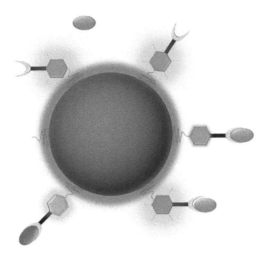

FIGURE 4.4 Sensing chemistry immobilized on a luminescent carrier, which then can contribute reference emission or act as a RET donor for acceptors attached to the surface.

present on the construct (Figure 4.4). In this situation, the particles must have complementary luminescent properties to the sensing reagents: (1) emission that does not substantially overlap with the indicator emission and (2) if serving as a donor, emission should overlap significantly with acceptor absorption.

Examples of luminescent carriers used to provide reference signals include cation and pH sensors. In the first case, Eu-doped polystyrene particles were coated with polyelectrolyte multilayers incorporating cation-sensitive indicators (PBFI, recognizing K[+]).[50] The large Stokes shift and narrow emission of the lanthanide particle provided a well-separated intensity reference signal for normalization of the indictor peak. Similarly, rhodamine-dyed silica beads were functionalized with a reactive silane, then a pH-sensitive luminescent napthalimide was attached.[59] In this particular case, a special spacing layer of silica was deposited to avoid energy transfer between the rhodamine and napthalimide, such that the core was only providing a stable reference signal. It is claimed that a similar approach has yielded ATP and saccharide sensors as well, though the results of these studies have not been made public.[60]

A specific example where Qdots were used as donors or acceptors in a sensor construct involving RET can be found in the work of Nocera.[61] A CdSe/ZnS core/shell dot was chemically conjugated to a squaraine dye, which possesses a pH-dependent absorption spectrum. At lower pH, squaraine shows strong absorption that overlaps with the Qdot emission, allowing efficient RET between the dot and the dye molecule and resulting in a strong red emission from squaraine relative to the Qdot emission. Increasing pH results in decreasing squaraine absorption, disabling RET and yielding a correspondingly higher luminescence from the Qdot (Figure 4.5). The ratio of peaks of the dot and the dye serves

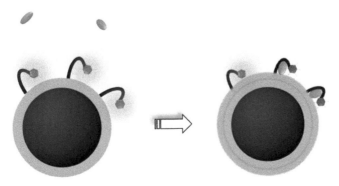

FIGURE 4.5 Energy transfer from core nanoparticle to surface-immobilized sensing chemistry. (Left) Indicator not bound to analyte emits luminescence following energy transfer from core nanoparticle. (Right) Due to analyte binding, indicator changes absorption, resulting in decreased energy transfer and corresponding increase in core luminescence emission.

as a quantitative measurement of pH as opposed to the actual intensity of peaks as is the case with most Qdot (QD) sensors. Thus there is no requirement for calibration or a necessity to excite the sensor at a particular wavelength. However, photobleaching is likely to limit the longevity of the system, as reduced squaraine dye emission with longer exposure time would interfere with pH measurements by shifting the ratiometric calibration curve.

A third category of surface-based nanosensors includes systems in which the nanoparticle behaves as an *active element*. For example, the luminescence of Qdots is a result of the quantum confinement of excitons, leading to a reduction in the band gap energy. Thus, any change in its crystal structure or factors influencing the band gap energy can influence its luminescence (quenching of emission intensity and/or wavelength).[7] There is little selectivity in such responses, which have been observed to result from cation exchange reactions in presence of Ag^{2+} and Pb^{2+} ions. Furthermore, as this involves disruption of the nanocrystal that requires intervention for signal recovery, it is not a practical approach to reversible sensing.

Qdots can, however, be useful as active elements if their surfaces are properly engineered for specific analyte interactions. Because most of the atoms in a nanocrystal are on the surface, chemical modifications (e.g., attachment of ligands) can considerably change luminescence. Functionalization with ligands that are responsive to external factors like temperature, pH, or other chem/bio analytes can endow Qdots with sufficient selectivity to serve as optical probes for the measurement of these interactions (Figure 4.6).

As a practical example, selective sensing toward Zn^{2+} and Cu^{2+} ions has been shown using CdS-capped by cysteine and thioglycerol, respectively.[62,63] Luminescence enhancement was observed in the presence of zinc ions, which was attributed to surface activation. Conversely, quenching resulted from exposure to copper ions, explained by reduction of ions at the surface. Similar strategies of functionalizing Qdots for mercury[64] and anion (cyanide) detection[65,66] have been shown, and this type of Qdot modification remains a new and active area of research. Recent examples of pH,[67] chloride,[68] mercury,[69] and zinc and manganese have been reported.[70] Modified Qdots have also recently been demonstrated as oxygen sensors; oxygen-quenchable pyrenyl ligands added a second oxygen-dependent luminescence signal.[71]

One should be aware that there are several approaches to immobilizing sensor chemistry on such platforms, the most popular being adsorption, direct covalent coupling, physical entrapment, and layer-by-layer (LbL) self-assembly of multilayer films.[72–74] The relative pros and cons of these approaches have been carefully considered in recent reviews.[60,75] However, we note that LbL is of special interest because

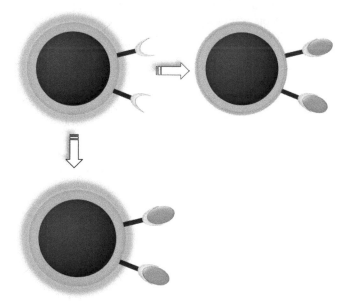

FIGURE 4.6 (Top left) Qdots with surface receptors for target analyte. Binding alters the surface properties, which in turn affect luminescence quantum efficiency (top right, shown as quenching) and/or bandgap energy (shown as shift in emission wavelength).

of the generality and simplicity of the approach to deposit functional nanocomposite films containing encapsulated probes on nanocarriers. No exotic or harsh chemistry is required. Multilayer nanofilms constructed by LbL are attractive for controlling biocatalytic reactions and other interfacial phenomena due to the high surface-to-volume ratio.[76–82] LbL has recently been reviewed extensively,[83,84] and techniques for immobilizing luminescent sensing chemistry within LbL nanofilms have also been described elsewhere.[75] While the reader is referred to these sources for details, the basic categories of useful approaches in this include, briefly, (1) direct electrostatic assembly of anionic and cationic luminescent indicators; (2) covalent attachment of luminescent indicators to charged polymers, followed by incorporation via electrostatic adsorption of the polymers; (3) immobilization of luminescent indicators into preassembled LbL films by electrostatics or precipitation. It is also noteworthy that sensing films can be prototyped by constructing LbL films with sensing materials on macrotemplates and then, once the desired behavior has been achieved, the same process parameters may be used to deposit the same materials onto micro/nanotemplates with confidence that very similar sensing behavior will be observed.[51] This has been demonstrated using glass slides, optical fibers, and nanoparticles. These findings point to the versatility of the approach, as well as its "portability" to the point that functional films can be assembled on nearly any surface using the same process.

4.5.1.2 Embedded Systems

As with the surface-based nanosensors, embedded systems can be categorized by the function of the carrier nanoparticle. Specifically, the nanoparticle may simply serve as an inert carrier with entrapped sensing chemistry, it may contribute luminescence, or the matrix itself may play a key role in the transduction of chemical information. Several of these strategies are highlighted in the subsequent sections, along with specific examples of research-level demonstrations of sensor function in each embodiment. It should be noted that the work of Kopelman on so-called probes encapsulated by biologically localized embedding (PEBBLEs) has largely led the development of nanosensors such as these. PEBBLEs are a class of polymeric or organically modified silica particles (20–500 nm) originally developed as a means of protecting sensing chemistry from the "harsh" internal environment of the intracellular space.[53,85] They have been shown to have advantages when placed in matrices that would otherwise interfere with the selective response to a target analyte (e.g., serum and blood), and techniques for introducing them into living cells have been developed.[53] Various PEBBLE formulations have been demonstrated for response to relevant analytes, and they have different construction depending upon the specific sensing chemistry involved; therefore, they will be discussed categorically by the function of the encapsulating matrix.

As with the surface-based sensing schemes, the simplest platform within which to embed the indicator chemistry is an inert, nonluminescent nanoparticle: the particle simply serves as a carrier for the sensing chemistry (Figure 4.7). The requirements for the particles, in this case, are the same as for surface-immobilized chemistry (stable, functionalizable for immobilization of the sensing chemistry) and have the added requirement of allowing analyte diffusion. In this embodiment, any luminescent

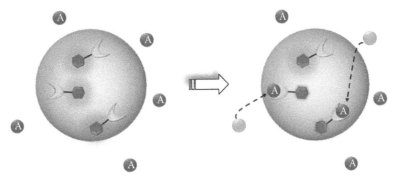

FIGURE 4.7 (Left) Nanosensor with embedded indicators (receptor + luminophore) surrounded by analyte molecules (A). (Right) After equilibration, some analyte molecules bind to the indicators, eliciting a change in luminescence (shown here as quenching).

indicator could be embedded by physical entrapment or covalent attachment to the particle matrix, and analyte molecules diffusing into the matrix would interact with the indicator in a manner similar to what is observed in solution (Figure 4.7). Selective exclusion of potential interferents would be an added benefit. PEBBLEs for pH[86] and calcium,[85] potassium,[87] zinc,[88] chloride,[89] magnesium,[56] iron,[90] copper,[91] sodium,[92] glucose,[93,94] and molecular oxygen have been demonstrated.[55,95-97]

Metal porphyrin compounds are highly sensitive to oxygen; two of the more popular examples of these materials are the Pt(II) and Pd(II) octaethylporphines, both of which possess lifetimes on the order of 100 μs, exhibiting excitation peaks at ca. 375, 500, and 535 nm and emission at 640–660 nm.[98] Pt(II) complexes are highly sensitive to low oxygen levels; however, these complexes are poorly soluble in water, making aqueous application difficult. To overcome this difficulty, surface adsorption to a carrier, usually silicon, organic glassy, cellulose derivatives, or fluoropolymers is common practice; however, the immobilization matrix has been shown to significantly affect O_2 sensitivity,[99,100] and it is understood that the difference in reported sensitivities for various immobilization media are due to the solubility and diffusivity of oxygen in the host matrix.[98] Thus, the encapsulation of the probe is a key factor in affecting sensor performance. Given this observation, the highest O_2 sensitivities with Pt(II) complexes been reported with Pt(II) porphyrin adsorption onto silica-containing matrices.[95,98] It is hypothesized that oxygen molecules adsorb to the surface substrate then rapidly diffuse across the surface; interestingly, silica-containing materials exhibit high oxygen-binding affinities and surface diffusion rates.[98,99] Natively fluorescent polymer nanoparticles have also been demonstrated to be effective as carriers providing reference or donor output.[101]

A key advantage to encapsulating sensor chemistry was shown in the ability to assemble sensors targeting "difficult" analytes by co-localizing multiple chemistries (e.g., probe + ionophore) within the confined volume of the sensor. In such a system, the luminescence readout is indirectly related to the analyte concentration: the indicator responds to a secondary molecule that is itself modulated in proportion to the analyte of interest. For example, a luminescent pH-sensitive probe can be used to indicate the concentration of a metal ion (e.g., K^+) that is bound by a nonluminescent ionophore if the matrix is properly designed (Figure 4.8). PEBBLE sensors based on this concept have been shown to function for univalent (Na^+ and K^+) and divalent (Ca^{2+}, Zn^{2+}, and Mg^{2+}) cations as well as anions (Cl^-).[87,89,97]

Enzymatic sensing, in contrast to the direct or indirect binding effects described earlier, typically relies upon monitoring either a product or co-substrate of a specific reaction between the enzyme and target. A common example is the oxidation of glucose driven by glucose oxidase, described as: glucose + O_2 + H_2O $\xrightarrow{\text{GOx}}$ gluconic acid + H_2O_2.[102] It is sufficient to note that oxygen is a co-substrate with glucose, and both hydrogen peroxide and gluconic acid are produced. Luminescence monitoring of oxygen is typically preferred, as it provides an indirect and fully reversible measure of glucose level. Alternatively, pH or peroxide could similarly be used, though pH shifts are small and peroxide assays are generally slow and nonreversible. Drawbacks specific to using enzymes include (1) changes in activity over time, leading to drifting calibration curves; (2) dependence on local oxygen levels; and (3) consumption of analyte and co-substrates, accompanied with production of by-products. Even if these issues can be overcome, simply creating a system where diffusion and reaction are sufficiently balanced for a sensitive response is a difficult task. Nanoscale enzymatic glucose-sensing particles have also been

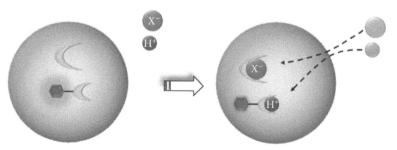

FIGURE 4.8 (Left) Nanosensor with embedded nonluminescent ionophore and luminescent indicators (receptor + luminophore) surrounded by analyte molecules (X^-) and associated protons (H^+). (Right) After equilibration, some analyte molecules bind to the ionophore, and the protons interact with the pH-sensitive indicator (shown here as quenching).

reported in a proof-of-concept study, utilizing a combination of GOx and O_2 indicators.[103] However, performance was inadequate; improvements in stability and control over response are expected with advancements in materials for matrix and diffusion-limiting coatings.

Sensors employing enzymes to drive a reaction that will be monitored with an oxygen indicator, require careful balancing between reaction (consumption) and diffusion (supply) of the co-substrates glucose and oxygen. This balance must be engineered to arrive at a measurable signal change for the expected glucose concentrations. It is well established that the membrane applied to enzyme-based sensors is critical in determining response sensitivity, range, and time.[104–109] When working at the nanoscale, practical realization of transport control requires (1) a deposition method amenable to efficient coating and (2) a highly precise means to control substrate transport, either through precise depositions or by having a wide variety of candidate materials.

Dendrimers—polymer structures that branch isotropically from a single center point, yielding a spherical "particle"[110]—represent a special class of nanostructures that can serve as the matrix for luminescent indicator chemistry.[111–114] Recent work has also seen the demonstration of dendrimers prepared directly from polymerization of indicators such as naphthalene and porphyrin dyes, which have been shown to be useful as pH, ion, and oxygen probes.[115–124]

Another interesting demonstration of dendrimer-based luminescence sensors can be found it the work of Dickson et al., demonstrating metal nanoclusters exhibiting blue photoluminescence; these materials are analogous to Qdots and have been called "gold nanodots" and "silver nanodots."[125,126] Dendrimer-nanodot systems seem to have some potential for sensing applications, as luminescence appears to decrease linearly due to aggregation of the nanoclusters (in an immunoassay)[127] and they would not be expected to exhibit toxicity to biological systems. However, there has yet to be an effective demonstration of reversible sensing employing aggregation of dendrimer-encapsulated nanoparticles.

4.5.2 Sensing with Multimolecular Nanosystems

In contrast to the "reagentless" systems described previously, there are reversible sensing assays involving multiple reagents that can be applied to nanoscale sensing if properly packaged. Some examples of such multimolecular sensing systems will be offered first, followed by a review of relevant encapsulation technology. The classical example of a reversible multiple reagent system that involves luminescence transduction is the competitive binding assay. Epitomized in the work of Schultz toward glucose and galactose analysis,[128–131] this type of system simply uses a RET donor or acceptor-labeled receptor (R) for the target analyte combined with an acceptor- or donor-tagged ligand (L) that will compete with the target analyte for the receptor binding site. Thus, in the absence of analyte, R will only bind L, and energy transfer to the acceptor will be at the maximum value for the system; as analyte is added, some L will be displaced from R, decreasing the energy transfer in proportion to analyte concentration. Traditionally, these assays were based on organic fluorophores or, more recently, fluorescent proteins. However, recent work has elegantly demonstrated the use of nanomaterials in various embodiments of maltose and trinitrotoluene (TNT) sensors.[27,132–134] Figure 4.9 is used to illustrate the general underlying principles of these concepts.

Another interesting approach to RET sensing with Qdots can be seen in multicomponent assays where the distance between the donor and acceptor is modulated in response to the analyte of interest. One representative example of this can be seen in cation recognition via RET between two Qdots tuned to have a spectral overlap.[135] A suspension of crown ether-conjugated Qdots of two different sizes exhibited potassium-dependent energy transfer. This was seen as a result of the aggregation in the presence of K^+ ions due to the guest-host interaction between a potassium ion and two crown ethers. As with the pH sensor described earlier, the ratio of the two peaks can be used to quantitatively calculate the ion concentration; in this case, the photostability is greatly enhanced because the luminescence derives fully from the Qdots. This type of sensor construct has selectivity that depends on the receptor; in this case, selectivity against Cs, Ba, and Mg was high.

Such multimolecular assays rely on the development of nanoscale encapsulating media to enable their use in continuous, reversible sensing applications. Specifically, multimolecular assays, such as the one just described, require that the competing ligands be kept in close proximity to the receptor so they may

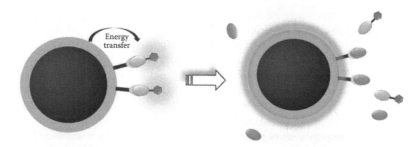

FIGURE 4.9 Cartoon of a generic nanoscale system comprising a luminescent nanoparticle with surface-immobilized receptors paired with a competing ligand with an attached complementary acceptor luminophore (purple oval). (Left) In the absence of analyte (blue oval), the competing ligand binds to the receptor. This brings the acceptor in close proximity to the donor, resulting in efficient energy transfer. (Right) When analyte is present, a fraction of the competing ligands are displaced from receptors. Under these conditions, energy transfer is decreased, yielding a higher emission intensity and longer lifetime from the donor.

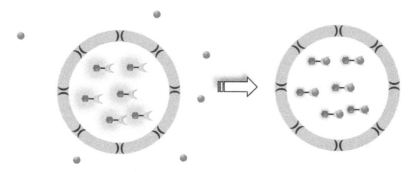

FIGURE 4.10 Hollow carriers with encapsulated sensing chemistry. (Left) Luminescent indicators with high relative emission intensity in the absence of analyte. (Right) Following analyte diffusion, indicators are bound to analyte and decrease emission correspondingly.

continue to compete with the analyte for receptor binding sites by providing a constant total ligand concentration. On the other hand, because of the need for displacement of ligand, these competitive systems do require mobility on the part of either the ligand or the receptor. Freer diffusion will translate into faster equilibration and, hence, shorter response time. Thus, in contrast to the solid/hydrogel matrices described in the previous section, *hollow* structures (Figure 4.10) are more attractive for these systems. A suitable carrier needs to be stable (to avoid leakage), permeable (to analyte, not the sensing reagents), and amenable to fabrication at the nanoscale. Liposomes, polymersomes, and multilayer capsules each have potential to meet these requirements. This subject has been reviewed in other contexts,[75,136–139] so this brief conceptual overview is sufficient before delving into examples of nanoscale multimolecular devices.

4.5.2.1 "X-Somes"

Liposomes are phospholipid-based hollow shells with a hydrophilic interior, and they have been formed from a wide array of natural and synthetic lipids. For sensors, the aqueous interior of the liposome provides a hydrophilic environment that is preferred for many luminescent probes and enzymes, and it also ensures high permeability for hydrophilic species. A drawback of standard liposome technology is the limited stability; to enhance mechanical integrity for long-term sensing applications, liposomes may be crosslinked through use of polymerizable phospholipids, incorporation of hydrophilic polymers, or polymerization of monomer units incorporated into the bilayer. Encapsulation of luminescent probes with liposomes has led to demonstrations of sensors for molecular oxygen,[140] acetylthiocholine chloride,[141] calcium,[142] and pH.[143] *Polymersomes*—synthetic vesicles comprising self-assembling block copolymers—are newer materials that are similar in structure to liposomes and are generally more versatile.[144–147] They have not yet been used effectively for sensing purposes but are useful for encapsulation and, therefore, may impact sensor architectures in the near future.

4.5.2.2 Multilayer Nanocapsules

In the past decade, the multilayer assembly process applied to colloids has been used as the first step in a technique for fabrication of hollow micro/nanocapsules.[75,136,148] Following deposition of multilayer coatings onto the templates, the core material is removed using organic solvents or chemical etching to arrive at hollow capsules.[149–152] The versatility in construction of these tiny capsules and control over their properties makes them attractive for use in sensor applications, especially those in which encapsulation of active molecules and control over transport properties are critical to proper function. A prime example of this can be seen in the recent literature on glucose-sensing "smart tattoo" systems,[153] which can be seen as generic prototypes for similar chemical using optical probes encapsulated in multilayer capsules. Following demonstration of glucose sensitivity in solution-phase, RET-based competitive assay components (apo-glucose oxidase and dextran, labeled with complementary donor-acceptor pair) were encapsulated within a hollow polymeric shell.[154,155] This expected diffusive loading behavior was verified via fluorescence spectroscopy, whereby intensity ratio measurements performed at varying glucose levels revealed a completely reversible response to glucose, matching the observations for the solution-phase assay.[154,155]

4.6 Summary and Conclusions

From this brief introduction to the basics and key examples, it should be evident that the field of luminescent nanosensors is growing rapidly and has a bright future (pun intended) for many applications. With the rapid advancement in nanomaterials, opportunities to incorporate new luminophores are expanding; similarly, new reporter molecules and structures are continuously being developed, as are techniques to integrate sensing elements and carriers. A fairly broad set of examples for ion, pH, oxygen, explosives, saccharides, and other biochemicals has been demonstrated using one or more sensing approaches. What is missing at this point is a penetration of the research and commercial market. Many of the applications for luminescent nanosensors are in the environmental and biomedical areas, which are highly regulated spaces that are very risk-averse. Thus, it is anticipated that adoption by public and private consumers will be slow and limited to niche areas while research and development continues. However, luminescent nanosensors can be very valuable in the hands of experts performing research or specialized testing, and hopefully the breadth of true application of these versatile tools will grow rapidly.

Acknowledgments

Research and writing of this material was funded in part by the National Science Foundation, grant No. 1066928. Any opinions, findings, and conclusions or recommendations expressed in this material are those of the author(s) and do not necessarily reflect the views of the National Science Foundation. Illustrations were prepared with help from Ashvin Nagaraja, whose assistance is gratefully acknowledged.

REFERENCES

1. Mattoussi, H., Palui, G., and Na, H. B. Luminescent quantum dots as platforms for probing in vitro and in vivo biological processes. *Advanced Drug Delivery Reviews* **64**, 138–166, doi:10.1016/j.addr.2011.09.011 (2011).
2. Zhang, C. Y., Yeh, H. C., Kuroki, M. T., and Wang, T. H. Single-quantum-dot-based DNA nanosensor. *Nature Materials* **4**, 826–831 (2005).
3. Huang, C. C. and Chang, H. T. Selective gold-nanoparticle-based "turn-on" fluorescent sensors for detection of mercury(II) in aqueous solution. *Analytical Chemistry* **78**, 8332–8338, doi:10.1021/ac061487i (2006).

4. Lee, S. et al. Polymeric nanoparticle-based activatable near-infrared nanosensor for protease determination in vivo. *Nano Letters* **9**, 4412–4416, doi:10.1021/nl902709m (2009).

5. Welser, K., Grilj, J., Vauthey, E., Aylott, J. W., and Chan, W. C. Protease responsive nanoprobes with tethered fluorogenic peptidyl 3-arylcoumarin substrates. *Chemical Communication (Cambridge)*, 671–673, doi:10.1039/b816637d (2009).

6. Han, C. et al. Urea-type ligand-modified CdSe quantum dots as a fluorescence "turn-on" sensor 99for $CO_3(2-)$ anions. *Photochemical Photobiological Sciences* **9**, 1269–1273, doi:10.1039/c0pp00119h (2010).

7. Kim, J. H. and Chung, B. H. Proteolytic fluorescent signal amplification on gold nanoparticles for a highly sensitive and rapid protease assay. *Small* **6**, 126–131, doi:10.1002/smll.200901635 (2010).

8. Kim, Y. P. et al. Bioluminescent nanosensors for protease detection based upon gold nanoparticle-luciferase conjugates. *Chemical Communication (Cambridge)* **46**, 76–78, doi:10.1039/b915612g (2010).

9. Barker, S. L., Kopelman, R., Meyer, T. E., and Cusanovich, M. A. Fiber-optic nitric oxide-selective biosensors and nanosensors. *Analytical Chemistry* **70**, 971–976 (1998).

10. Barker, S. L., Thorsrud, B. A., and Kopelman, R. Nitrite- and chloride-selective fluorescent nano-optodes and in vitro application to rat conceptuses. *Analytical Chemistry* **70**, 100–104 (1998).

11. Cullum, B. M., Griffin, G. D., and Tuan, V.-D. In *Biomedical Diagnostic, Guidance, and Surgical-Assist Systems III*, San Jose, CA, pp. 35–40, January 21–22, 2001. (SPIE-Int. Soc. Opt. Eng.)

12. Cullum, B. M., Griffin, G. D., and Tuan, V.-D. In *Biomedical Diagnostic, Guidance, and Surgical-Assist Systems IV*, San Jose, CA, pp. 148–154, January 20–21, 2002. (SPIE-Int. Soc. Opt. Eng.)

13. Kasili, P. M., Cullum, B. M., Griffin, G. D., and Vo-Dinh, T. Nanosensor for in vivo measurement of the carcinogen benzo[a]pyrene in a single cell. *Journal of Nanoscience and Nanotechnology* **2**, 653–658 (2002).

14. Dubach, J. M., Harjes, D. I., and Clark, H. A. Fluorescent ion-selective nanosensors for intracellular analysis with improved lifetime and size. *Nano Letters* **7**, 1827–1831, doi:10.1021/nl0707860 (2007).

15. Vo-Dinh, T. and Zhang, Y. Single-cell monitoring using fiberoptic nanosensors. *Wiley Interdisciplinary Reviews: Nanomedicine and Nanobiotechnology* **3**, 79–85 (2011).

16. Lakowicz, J. R. *Principles of Fluorescence Spectroscopy*, 3rd edn. (Springer, New York, 2006).

17. Weiss, S. Fluorescence spectroscopy of single biomolecules. *Science* **283**, 1676–1683, doi:10.1126/science.283.5408.1676 (1999).

18. Slavík, *J. Fluorescent Probes in Cellular and Molecular Biology*. (CRC Press, Boca Raton, FL, 1994).

19. Sapsford, K. E., Berti, L., and Medintz, I. L. Materials for fluorescence resonance energy transfer analysis: Beyond traditional donor-acceptor combinations. *Angewandte Chemie-International Edition* **45**, 4562–4588, doi:10.1002/anie.200503873 (2006).

20. Uzer, T. and Miller, W. H. Theories of intramolecular vibrational-energy transfer. *Physics Reports-Review Section of Physics Letters* **199**, 73–146 (1991).

21. Wu, P. G. and Brand, L. Orientation factor in steady-state and time-resolved resonance energy-transfer measurements. *Biochemistry* **31**, 7939–7947 (1992).

22. Willard, D. M., Carillo, L. L., Jung, J., and Van Orden, A. CdSe-ZnS quantum dots as resonance energy transfer donors in a model protein-protein binding assay. *Nano Letters* **1**, 469–474 (2001).

23. Murphy, C. J. Peer reviewed: Optical sensing with quantum dots. *Analytical Chemistry* **74**, 520 A–526 A, doi:10.1021/ac022124v (2002).

24. Liu, Y. F. et al. Comparison of water-soluble CdTe nanoparticles synthesized in air and in nitrogen. *Journal of Physical Chemistry B* **110**, 16992–17000 (2006).

25. Dabbousi, B. O. et al. (CdSe)ZnS core-shell quantum dots: Synthesis and characterization of a size series of highly luminescent nanocrystallites. *Journal of Physical Chemistry B* **101**, 9463–9475 (1997).

26. Patolsky, F. et al. Lighting-up the dynamics of telomerization and DNA replication by CdSe-ZnS quantum dots. *Journal of the American Chemical Society* **125**, 13918–13919 (2004).

27. Medintz, I. L. et al. A fluorescence resonance energy transfer-derived structure of a quantum dot-protein bioconjugate nanoassembly. *Proceedings of National Academy of Sciences of the United States of America* **101**, 9612–9617 (2004).

28. Ebenstein, Y., Mokari, T., and Banin, U. Quantum-dot-functionalized scanning probes for fluorescence-energy-transfer-based microscopy. *Journal of Physical Chemistry B* **108**, 93–99 (2004).

29. Clapp, A. R. et al. Fluorescence resonance energy transfer between quantum dot donors and dye-labeled protein acceptors. *Journal of the American Chemical Society* **126**, 301–310 (2004).

30. Medintz, I. L., Goldman, E. R., Lassman, M. E., and Mauro, J. M. A fluorescence resonance energy transfer sensor based on maltose binding protein. *Bioconjugate Chemistry* **14**, 909–918 (2003).
31. Brokmann, X., Ehrensperger, M.-V., Hermier, J.-P., Triller, A., and Dahan, M. Orientational imaging and tracking of single CdSe nanocrystals by defocused microscopy. *Chemical Physics Letters* **406**, 210–214 (2005).
32. Chung, I., Shimizu, K. T., and Bawendi, M. G. Room temperature measurements of the orientation of single CdSe quantum dots using polarization microscopy. *PNAS* **100**, 405–408 (2003).
33. van der Meer, B. W., Coker, G., and Chen, S.-Y. S. *Resonance Energy Transfer: Theory and Data.* (VCH Publishers, New York, 1994).
34. Dulkeith, E. et al. Gold nanoparticles quench fluorescence by phased induced radiative rate suppression. *Nano Letters* **5**, 585–589 (2005).
35. Schneider, G. et al. Distance-dependent fluorescence quenching on gold nanoparticles ensheathed with layer-by-layer assembled polyelectrolytes. *Nano Letters* **6**, 530–536 (2006).
36. Yun, C. S. et al. Nanometal surface energy transfer in optical rulers, breaking the FRET barrier. *Journal of American Chemical Society* **127**, 3115–3119 (2005).
37. Jennings, T., Singh, M. P., and Strouse, G. F. Fluorescent lifetime quenching near d = 1.5 nm gold nanoparticles: Probing NSET validity. *Journal of American Chemical Society* **128**, 5462–5467 (2006).
38. Nikoobakht, B., Burda, C., Braun, M., Hun, M., and El-Sayed, M. A. The quenching of CdSe quantum dots photoluminescence by gold nanoparticles in solution. *Photochemistry and Photobiology* **75**, 591–597 (2002).
39. Gueroui, Z. and Libchaber, A. Single molecule measurements of gold-quenched quantum dots. *Physical Review Letters* **96**, 166108-166101–166104 (2004).
40. Wargnier, R. et al. Energy transfer in aqueous solutions of oppositely charged CdSe/ZnS core/shell quantum dots and in quantum dot-nanogold assemblies. *Nano Letters* **4**, 451–457 (2004).
41. Slocik, J. M., Govorov, A. O., and Naik, R. R. Optical characterization of bio-assembled hybrid nanostructures. *Supramolecular Chemistry* **18**, 415–421 (2006).
42. Govorov, A. O. et al. Exciton-plasmon interaction and hybrid excitons in semiconductor-metal nanoparticle assemblies. *Nano Letters* **6**, 984–994 (2006).
43. Jennings, T. L., Schlatterer, J. C., Singh, M. P., Greenbaum, N. L., and Strouse, G. F. NSET molecular beacon analysis of hammerhead RNA substrate binding and catalysis. *Nano Letters* **6**, 13181322 (2006).
44. Carminati, R., Greffet, J.-J., Henkel, C., and Vigoureux, J. M. Radiative and non-radiative decay of a single molecule close to a metallic nanoparticle. *Optics Communications* **261**, 368–375 (2006).
45. Pons, T. et al. On the quenching of semiconductor quantum dot photoluminescence by proximal gold nanoparticles. *Nano Letters* **7**, 3157–3164 (2007).
46. Chang, E. et al. Protease-activated quantum dot probes. *Biochemical Biophysical Research Communications* **334**, 1317–1321 (2005).
47. Demchenko, A. P. The problem of self-calibration of fluorescence signal in microscale sensor systems. *Lab on a Chip* **5**, 1210–1223, doi:10.1039/B507447a (2005).
48. Brown, J. Q., Guice, K. B., Simpson, R. T., and McShane, M. J. Electrostatic self-assembly of nanocomposite hybrid fluorescent sensors. *Proceedings of SPIE—The International Society for Optical Engineering* **5331**, 52–59 (2004).
49. Brown, J. Q. and McShane, M. J. Nanoengineered polyelectrolyte micro- and nano-capsules as fluorescent potassium ion sensors. *IEEE Engineering in Medicine and Biology Magazine* **22**, 118–123 (2003).
50. Brown, J. Q. and McShane, M. J. Core-referenced ratiometric fluorescent potassium ion sensors using self-assembled ultrathin films on europium nanoparticles. *IEEE Sensors Journal* **5**, 1197–1205 (2005).
51. Grant, P. S. and McShane, M. J. Development of multilayer fluorescent thin film chemical sensors using electrostatic self-assembly. *IEEE Sensors Journal* **3**, 139 (2003).
52. Guice, K. B., Caldorera, M. E., and McShane, M. J. Nanoscale internally referenced oxygen sensors produced from self-assembled nanofilms on fluorescent nanoparticles. *Journal of Biomedical Optics* **10**, 064031 (2005).
53. Clark, H. A., Hoyer, M., Philbert, M. A., and Kopelman, R. Optical nanosensors for chemical analysis inside single living cells. 1. Fabrication, characterization, and methods for intracellular delivery of PEBBLE sensors. *Analytical Chemistry* **71**, 4831–4836 (1999).
54. Xu, H., Aylott, J., and Kopelman, R. Sol-gel pebble sensors for biochemical analysis inside living cells. *Abstracts of Papers of the American Chemical Society* **219**, U99–U99 (2000).

55. Xu, H., Aylott, J. W., Kopelman, R., Miller, T. J., and Philbert, M. A. A real-time ratiometric method for the determination of molecular oxygen inside living cells using sol-gel-based spherical optical nanosensors with applications to rat C6 glioma. *Analytical Chemistry* **73**, 4124–4133 (2001).

56. Park, E. J., Brasuel, M., Behrend, C., Philbert, M. A., and Kopelman, R. Ratiometric optical PEBBLE nanosensors for real-time magnesium ion concentrations inside viable cells. *Analytical Chemistry* **75**, 3784–3791 (2003).

57. Lee, Y.-E. K., Kopelman, R., and Smith, R. Nanoparticle PEBBLE sensors in live cells and in vivo. *Annual Review of Analytical Chemistry* **2**, 57 (2009).

58. Kopelman, R. and Dourado, S. Is smaller better? Scaling of characteristics with size of fiber-optic chemical and biochemical sensors. *Chemical, Biochemical, and Environmental Fiber Sensors VIII* **2836**, 2–11 (1996).

59. Doussineau, T., Trupp, S., and Mohr, G. J. Ratiometric pH-nanosensors based on rhodamine-doped silica nanoparticles functionalized with a naphthalimide derivative. *Journal of Colloid and Interface Science* **339**, 266–270 (2009).

60. Doussineau, T. et al. On the design of fluorescent ratiometric nanosensors. *Chemistry—A European Journal* **16**, 10290–10299 (2010).

61. Snee, P. T. et al. A Ratiometric CdSe/ZnS Nanocrystal pH Sensor. *Journal of the American Chemical Society* **128**, 13320–13321, doi:10.1021/ja0618999 (2006).

62. Chen, Y. F. and Rosenzweig, Z. Amino acid and peptide capped CDS quantum dots (QDS) as novel Zn2+ions luminescent sensors. *Abstracts of Papers of the American Chemical Society* **223**, U67–U67 (2002).

63. Moore, D. E. and Patel, K. Q-CdS photoluminescence activation on Zn2+ and Cd2+ salt introduction. *Langmuir* **17**, 2541–2544 (2001).

64. de la Riva, B. S. V., Costa-Fernandez, J. M., Jin, W. J., Pereiro, R., and Sanz-Medel, A. Determination of trace levels of mercury in water samples based on room temperature phosphorescence energy transfer. *Analytica Chimica Acta* **455**, 179–186 (2002).

65. Jin, W. J., Fernandez-Arguelles, M. T., Costa-Fernandez, J. M., Pereiro, R., and Sanz-Medel, A. Photoactivated luminescent CdSe quantum dots as sensitive cyanide probes in aqueous solutions. *Chemical Communications*, 883–885, doi:10.1039/B414858d (2005).

66. Jin, W. J., Costa-Fernandez, J. M., Pereiro, R., and Sanz-Medel, A. Surface-modified CdSe quantum dots as luminescent probes for cyanide determination. *Analytica Chimica Acta* **522**, 1–8, doi:10.1016/j.aca.2004.06.057 (2004).

67. Ruedas-Rama, M. J., Orte, A., Hall, E. A., Alvarez-Pez, J. M., and Talavera, E. M. Quantum dot photoluminescence lifetime-based pH nanosensor. *Chemical Communications (Cambridge)* **47**, 2898–2900, doi:10.1039/c0cc05252c (2011).

68. Wang, Y., Mao, H., and Wong, L. B. Dynamic [Cl-]i measurement with chloride sensing quantum dots nanosensor in epithelial cells. *Nanotechnology* **21**, 55101 (2010).

69. Li, H. B., Zhang, Y., Wang, X. Q., and Gao, Z. N. A luminescent nanosensor for Hg(II) based on functionalized CdSe/ZnS quantum dots. *Microchimica Acta* **160**, 119–123, doi:10.1007/s00604-007-0816-x (2008).

70. Ruedas-Rama, M. J. and Hall, E. A. Multiplexed energy transfer mechanisms in a dual-function quantum dot for zinc and manganese. *Analyst* **134**, 159–169, doi:10.1039/b814879a (2009).

71. Amelia, M., Lavie-Cambot, A., McClenaghan, N. D., and Credi, A. A ratiometric luminescent oxygen sensor based on a chemically functionalized quantum dot. *Chemical Communications (Cambridge)* **47**, 325–327, doi:10.1039/c0cc02163f (2011).

72. Decher, G. Fuzzy nanoassemblies: Toward layered polymeric multicomposites. *Science* **277**, 1232–1237, doi:10.1126/science.277.5330.1232 (1997).

73. Ai, H., Fang, M., Jones, S. A., and Lvov, Y. M. Electrostatic layer-by-layer nanoassembly on biological microtemplates: Platelets. *Biomacromolecules* **3**, 560–564 (2002).

74. Ai, H., Jones, S. A., and Lvov, Y. M. Biomedical applications of electrostatic layer-by-layer nanoassembly of polymers, enzymes, and nanoparticles. *Cell Biochemistry and Biophysics* **39**, 23–43 (2003).

75. McShane, M. J. In *Sensors Based on Nanostructured Materials* (ed. F. J. Arregui), Chapter 8, pp. 253–272, Springer Science + Business Media, New York, NY (2008).

76. Caruso, F., Caruso, R. A., and Mohwald, H. Nanoengineering of inorganic and hybrid hollow spheres by colloidal templating. *Science* **282**, 1111–1114 (1998).

77. Caruso, F., Lichtenfeld, H., Giersig, M., and Mohwald, H. Electrostatic self-assembly of silica nanoparticle—Polyelectrolyte multilayers on polystyrene latex particles. *Journal of the American Chemical Society* **120**, 8523–8524 (1998).

78. Caruso, F. and Mohwald, H. Protein multilayer formation on colloids through a stepwise self-assembly technique. *Journal of the American Chemical Society* **121**, 6039–6046 (1999).

79. Caruso, F. and Schuler, C. Enzyme multilayers on colloid particles: Assembly, stability, and enzymatic activity. *Langmuir* **16**, 9595–9603 (2000).

80. Lvov, Y. and Caruso, F. Biocolloids with ordered urease multilayer shells as enzymatic reactors. *Analytical Chemistry* **73**, 4212–4217 (2001).

81. Fang, M. et al. Magnetic bio/nanoreactor with multilayer shells of glucose oxidase and inorganic nanaparticles. *Langmuir* **18**, 6338–6344 (2002).

82. Stein, E. W. and McShane, M. J. Multilayer lactate oxidase shells on colloidal carriers as engines for nanosensors. *IEEE Transactions Nanobioscience* **2**, 133–137 (2003).

83. McShane, M. J. and Lvov, Y. In *Dekker Encyclopedia of Nanoscience and Nanotechnology* (eds. J. A. Schwarz, C. I. Contescu, and K. Putyera), pp. 1–20. (Marcel Dekker, New York, 2004).

84. Decher, G. and Schlenoff, J. B. *Multilayer Thin Films: Sequential Assembly of Nanocomposite Materials*, p. 543. (Wiley-VCH, Weinheim, Germany, 2006).

85. Clark, H. A., Kopelman, R., Tjalkens, R., and Philbert, M. A. Optical nanosensors for chemical analysis inside single living cells. 2. Sensors for pH and calcium and the intracellular application of PEBBLE sensors. *Analytical Chemistry* **71**, 4837–4843 (1999).

86. Chauhan, V. M., Burnett, G. R., and Aylott, J. W. Dual-fluorophore ratiometric pH nanosensor with tuneable pKa and extended dynamic range. *Analyst* **136**, 1799–1801, doi:10.1039/c1an15042a (2011).

87. Brasuel, M., Kopelman, R., Miller, T. J., Tjalkens, R., and Philbert, M. A. Fluorescent nanosensors for intracellular chemical analysis: Decyl methacrylate liquid polymer matrix and ion-exchange-based potassium PEBBLE sensors with real-time application to viable rat C6 glioma cells. *Analytical Chemistry* **73**, 2221–2228 (2001).

88. Sumner, J. P., Aylott, J. W., Monson, E., and Kopelman, R. A fluorescent PEBBLE nanosensor for intracellular free zinc. *Analyst* **127**, 11–16 (2002).

89. Brasuel, M. G., Miller, T. J., Kopelman, R., and Philbert, M. A. Liquid polymer nano-PEBBLEs for Cl-analysis and biological applications. *Analyst* **128**, 1262–1267, doi:10.1039/b305254k (2003).

90. Sumner, J. P. and Kopelman, R. Alexa Fluor 488 as an iron sensing molecule and its application in PEBBLE nanosensors. *Analyst* **130**, 528–533, doi:10.1039/b414189j (2005).

91. Sumner, J. P., Westerberg, N. M., Stoddard, A. K., Fierke, C. A., and Kopelman, R. Cu+- and Cu2+-sensitive PEBBLE fluorescent nanosensors using DsRed as the recognition element. *Sensors and Actuators B-Chemical* **113**, 760–767, doi:10.1016/j.snb.2005.07.028 (2006).

92. Dubach, J. M., Balaconis, M. K., and Clark, H. A. Fluorescent nanoparticles for the measurement of ion concentration in biological systems. *Journal of Visualized Experiments*, doi:2896 [pii] 10.3791/2896 (2011).

93. Billingsley, K. et al. Fluorescent nano-optodes for glucose detection. *Analytical Chemistry* **82**, 3707–3713, doi:10.1021/ac100042e (2010).

94. Balaconis, M. K., Billingsley, K., Dubach, M. J., Cash, K. J., and Clark, H. A. The design and development of fluorescent nano-optodes for in vivo glucose monitoring. *Journal of Diabetes Science Technology* **5**, 68–75 (2011).

95. Koo, Y.-E. L. et al. Real-time measurements of dissolved oxygen inside live cells by organically modified silicate fluorescent nanosensors. *Analytical Chemistry* **76**, 2498–2505 (2004).

96. Lee, Y. E. et al. Near infrared luminescent oxygen nanosensors with nanoparticle matrix tailored sensitivity. *Analytical Chemistry* **82**, 8446–8455, doi:10.1021/ac1015358 (2010).

97. Brasuel, M. et al. Production, characteristics and applications of fluorescent PEBBLE nanosensors: Potassium, oxygen, calcium and pH imaging inside live cells. *Sensors and Materials* **14**, 309–338 (2002).

98. Han, B.-H., Manners, I., and Winnik, M. A. Oxygen sensors based on mesoporous silica particles on layer-by-layer self-assembled films. *Chemistry of Materials* **17**, 3160–3171 (2005).

99. Lu, X., Manners, I., and Winnik, M. A. Polymer/silica composite films as luminescent oxygen sensors. *Macromolecules* **34**, 1917–1927 (2001).

100. Papkovsky, D. B. and O'Riordan, T. C. Emerging applications of phosphorescent metalloporphyrins. *Journal of Fluorescence* **15**, 569–584 (2005).

101. Wu, C. F., Bull, B., Christensen, K., and McNeill, J. Ratiometric single-nanoparticle oxygen sensors for biological imaging. *Angewandte Chemie International Edition* **48**, 2741–2745, doi:10.1002/anie.200805894 (2009).

102. Wilson, R. and Turner, A. P. F. Glucose oxidase. An ideal enzyme. *Biosensors and Bioelectronics* **7**, 165–185 (1992).

103. Xu, H., Aylott, J. W., and Kopelman, R. Fluorescent nano-PEBBLE sensors designed for intracellular glucose imaging. *The Analyst* **127**, 1471–1477 (2002).

104. Trettnak, W. and Wolfbeis, O. S. Fully reversible fibre-optic glucose biosensor based on the intrinsic fluorescence of glucose oxidase. *Analytica Chimica Acta* **221**, 195–203 (1989).

105. Wolfbeis, O. S., Oehme, I., Papkovskaya, N., and Klimant, I. Sol-gel based glucose biosensors employing optical oxygen transducers, and a method for compensating for variable oxygen background. *Biosensors and Bioelectronics* **15**, 69–76 (2000).

106. Rosenzweig, Z. and Kopelman, R. Analytical properties and sensor size effects of a micrometer-sized optical fiber glucose biosensor. *Analytical Chemistry* **68**, 1408–1413 (1996).

107. Choi, H. N., Kim, M. A., and Lee, W.-Y. Amperometric glucose biosensor based on sol-gel-derived metal oxide/Nafion composite films. *Analytica Chimica Acta* **537**, 179–187 (2005).

108. Wang, B., Li, B., Deng, Q., and Dong, S. Amperometric glucose biosensor based on sol-gel organic-inorganic hybrid material. *Analytical Chemistry* **70**, 3170–3174 (1998).

109. Pandey, P. C., Upadhyay, S., and Pathak, H. C. New glucose sensor based on encapsulated glucose oxidase within organically modified sol-gel glass. *Sensors and Actuators, B: Chemical* **B60**, 83–89 (1999).

110. Imae, T. Structure and functionality of dendrimers. *Kobunshi Ronbunshu* **57**, 810–824 (2000).

111. Crooks, R. M., Lemon, B. I., Sun, L., Yeung, L. K., and Zhao, M. Q. Dendrimer-encapsulated metals and semiconductors: Synthesis, characterization, and applications. *Topics in Current Chemistry* **212**, 81–135 (2001).

112. Goodson, T., Varnavski, O., and Wang, Y. Optical properties and applications of dendrimer-metal nanocomposites. *International Reviews in Physical Chemistry* **23**, 109–150, doi:10.1080/01442350310001628875 (2004).

113. Knoll, W. et al. Nanoscopic building blocks from polymers, metals, and semiconductors for hybrid architectures. *Journal of Nonlinear Optical Physics* **13**, 229–241 (2004).

114. Ibey, B. L. et al. Competitive binding assay for glucose based on glycodendrimer-fluorophore conjugates. *Analytical Chemistry* **77**, 7039–7046, doi:10.1021/Ac0507901 (2005).

115. Grabchev, I., Chovelon, J. M., and Nedelcheva, A. Green fluorescence poly(amidoamine) dendrimer functionalized with 1,8-naphthalimide units as potential sensor for metal cations. *Journal of Photochemistry and Photobiology A* **183**, 9–14, doi:10.1016/j.jphotochem.2006.02.012 (2006).

116. Grabchev, I., Staneva, D., and Betcheva, R. Sensor activity, photodegradation and photostabilisation of a PAMAM dendrimer comprising 1,8-naphthalimide functional groups in its periphery. *Polymer Degradation and Stability* **91**, 2257–2264, doi:10.1016/j.polymdegradstab.2006.04.022 (2006).

117. Ghosh, S. and Banthia, A. K. Towards fluorescence sensing polyamidoamine (PAMAM) dendritic architectures. *Tetrahedron Letters* **43**, 6457–6459 (2002).

118. Finikova, O. et al. Porphyrin and tetrabenzoporphyrin dendrimers: Tunable membrane-impermeable fluorescent pH nanosensors. *Journal of the American Chemical Society* **125**, 4882–4893 (2003).

119. Lebedev, A. Y. et al. Dendritic phosphorescent probes for oxygen imaging in biological systems. *ACS Applied Materials Interfaces* **1**, 1292–1304, doi:10.1021/am9001698 (2009).

120. Sinks, L. E., Roussakis, E., Esipova, T. V., and Vinogradov, S. A. Synthesis and calibration of phosphorescent nanoprobes for oxygen imaging in biological systems. *Journal of Visualized Experiments*, 1731, doi:10.3791/1731 (2010).

121. Sali, S., Grabchev, I., Chovelon, J. M., and Ivanova, G. Selective sensors for-Zn2+ cations based on new green fluorescent poly(amidoamine) dendrimers peripherally modified with 1,8-naphthalimides. *Spectrochimica Acta Part A: Molecular and Biomolecular Spectroscopy* **65**, 591–597, doi:10.1016/j.saa.2005.12.016 (2006).

122. Grabchev, I., Bosch, P., McKenna, M., and Staneva, D. A new colorimetric and fluorimetric sensor for metal cations based on poly(propilene amine) dendrimer modified with 1,8-naphthalimide. *Journal of Photochemistry and Photobiology A* **201**, 75–80, doi:10.1016/j.jphotochem.2008.10.008 (2009).

123. Lei, Y. L., Su, Y. Q., and Huo, J. C. Photophysical property of rhodamine-cored poly(amidoamine) dendrimers: Simultaneous effect of spirolactam ring-opening and PET process on sensing trivalent chromium ion. *Journal of Luminescence* **131**, 2521–2527, doi:10.1016/j.jlumin.2011.06.011 (2011).

124. Brinas, R. P., Troxler, T., Hochstrasser, R. M., and Vinogradov, S. A. Phosphorescent oxygen sensor with dendritic protection and two-photon absorbing antenna. *Journal of American Chemistry Society* **127**, 11851–11862, doi:10.1021/ja052947c (2005).

125. Zheng, J., Nicovich, P. R., and Dickson, R. M. Highly fluorescent noble-metal quantum dots. *Annual Review of Physical Chemistry* **58**, 409–431, doi:10.1146/annurev.physchem.58.032806.104546 (2007).

126. Zheng, J., Petty, J. T., and Dickson, R. M. High quantum yield blue emission from water-soluble Au-8 nanodots. *Journal of the American Chemical Society* **125**, 7780–7781, doi:10.1021/Ja035473v (2003).

127. Triulzi, R. C. et al. Immunoassay based on the antibody-conjugated PAMAM-dendrimer-gold quantum dot complex. *Chemical Communications* **48**, 5068–5070, doi:10.1039/B611278a (2006).

128. Meadows, D. L. and Schultz, J. S. Miniature fiber optic sensor based on fluorescence energy transfer, *Proceedings of SPIE*, **1648**, 202 (1992).

129. Meadows, D. L. and Schultz, J. S. Design, manufacture and characterization of an optical-fiber glucose affinity sensor-based on an homogeneous fluorescence energy-transfer assay system. *Analytica Chimica Acta* **280**, 21–30 (1993).

130. Ballerstadt, R., Dahn, M. S., Schultz, J., and Lange, M. P. Homogeneous affinity fluorescence assay system for galactose monitoring. *Sensors and Actuators, B: Chemical* B**38**, 171 (1997).

131. Ballerstadt, R. and Schultz, J. S. Competitive-binding assay method based on fluorescence quenching of ligands held in close proximity by a multivalent receptor. *Analytica Chimica Acta* **345**, 203 (1997).

132. Medintz, I. L. et al. Self-assembled nanoscale biosensors based on quantum dot FRET donors. *Nature Materials* **2**, 630–638 (2003).

133. Medintz, I. L. et al. Towards the design and implementation of surface tethered quantum dot-based nanosensors. *Material Research Society Symposium Proceedings* **789**, 105–110 (2004).

134. Medintz, I. L. et al. A fluorescence resonance energy transfer quantum dot explosive nanosensor. *Nanobiophotonics and Biomedical Applications II* **5705**, 166–174, doi:10.1117/12.589644 (2005).

135. Chen, C. Y. et al. Potassium ion recognition by 15-crown-5 functionalized CdSe/ZnS quantum dots in H2O. *Chemical Communications*, 263–265, doi:10.1039/B512677k (2006).

136. McShane, M. and Ritter, D. Microcapsules as optical biosensors. *Journal of Materials Chemistry* **20**, 8189 (2010).

137. McShane, M. J. Multilayer nanoengineering techniques for fabrication of opto-chemical probes. In *Proceedings of IEEE Sensors 2003. IEEE International Conference on Sensors*, 2nd edn., Toronto, Ontario, Canada, October 22–24, 2003, Vol. 1, pp. 689–693 (2003).

138. McShane, M. J. Biosensor applications of polyelectrolyte nanofilms and microcapsules. In *Abstracts of Papers, 230th ACS National Meeting*, Washington, DC, August 28–September 1, PMSE-134 (2005).

139. McShane, M. J. In *Topics in Fluorescence Spectroscopy*, Vol. 11 (eds. C. D. Geddes and J. R. Lakowicz). (Springer, New York, 2006).

140. Cheng, Z. L. and Aspinwall, C. A. Nanometre-sized molecular oxygen sensors prepared from polymer stabilized phospholipid vesicles. *Analyst* **131**, 236–243 (2006).

141. Vamvakaki, V., Fournier, D., and Chaniotakis, N. A. Fluorescence detection of enzymatic activity within a liposome based nano-biosensor. *Biosensors and Bioelectronics* **21**, 384–388 (2005).

142. Nguyen, T. and Rosenzweig, Z. Calcium ion fluorescence detection using liposomes containing Alexa-labeled calmodulin. *Analytical and Bioanalytical Chemistry* **374**, 69–74 (2002).

143. Nguyen, T., McNamara, K. P., and Rosenzweig, Z. Optochemical sensing by immobilizing fluorophore-encapsulating liposomes in sol-gel thin films. *Analytica Chimica Acta* **400**, 45–54 (1999).

144. Discher, B. M. et al. Polymersomes: Tough vesicles made from diblock copolymers. *Science* **284**, 1143–1146 (1999).

145. Discher, B. M. et al. Cross-linked polymersome membranes: Vesicles with broadly adjustable properties. *Journal of Physical Chemistry B* **106**, 2848–2854 (2002).

146. Meng, F. H., Engbers, G. H. M., and Feijen, J. Biodegradable polymersomes as a basis for artificial cells: Encapsulation, release and targeting. *Journal of Controlled Release* **101**, 187–198 (2005).

147. Yow, H. N. and Routh, A. F. Formation of liquid core-polymer shell microcapsules. *Soft Matter* **2**, 940–949 (2006).

148. Mohwald, H. From Langmuir monolayers to nanocapsules. *Colloids and Surfaces A: Physicochemical and Engineering Aspects* **171**, 25–31 (2000).

149. Sukhorukov, G., Fery, A., and Mohwald, H. Intelligent micro- and nanocapsules. *Progress in Polymer Science* **30**, 885–897 (2005).
150. Sukhorukov, G. B., Brumen, M., Donath, E., and Mohwald, H. Hollow polyelectrolyte shells: Exclusion of polymers and donnan equilibrium. *Journal of Physical Chemistry B* **103**, 6434–6440 (1999).
151. Sukhorukov, G. B. et al. Stepwise polyelectrolyte assembly on particle surfaces: A novel approach to colloid design. *Polymers for Advanced Technologies* **9**, 759–767 (1998).
152. Sukhorukov, G. B. et al. Multifunctionalized polymer microcapsules: Novel tools for biological and pharmacological applications. *Small* **3**, 944–955 (2007).
153. McShane, M. J. Potential for glucose monitoring with nanoengineered fluorescent biosensors. *Diabetes Technology and Therapeutics* **4**, 533–538 (2002).
154. Chinnayelka, S. and McShane, M. J. Microcapsule biosensors using competitive binding resonance energy transfer assays based on apoenzymes. *Analytical Chemistry* **77**, 5501–5511 (2005).
155. Chinnayelka, S. and McShane, M. J. Glucose sensors based on microcapsules containing an orange/red competitive binding resonance energy transfer assay. *Diabetes Technology and Therapeutics* **8**, 269–278 (2006).

5

Cantilever-Based Sensors

**Maria Tenje, Stephan S. Keller, Zachary J. Davis, Anja Boisen,
Joseph Ndieyira, Manuel Vögtli, Carlo Morasso, and Rachel A. McKendry**

CONTENTS

5.1 Introduction

Cantilever-based sensors present a platform for monitoring chemical reactions and detecting chemical compounds based on nanomechanical effects. Cantilever sensors have become popular because they offer rapid, label-free sensing possibility in a miniaturized platform, readily opening up applications for hand-held and portable devices [1–3]. The sensing ability of the cantilevers is intrinsic since the range of molecules/chemicals that can be detected is not determined by the sensor itself but on the available receptor coatings. The cantilever sensor is simply a transducer of the chemical reactions occurring on its surface.

The so-called *cantilever* is a microstructure that was developed for atomic force microscopy (AFM) in the late 1980s at Stanford University [4,5]. Here, interactions between a sharp tip at the end of the cantilever and the underlying surface are monitored, and information about the surface topography is obtained. This is possible since the interactions result in deflections of the cantilever, which can be monitored very precisely, e.g., by an optical readout system (described in detail in Section 5.3).

Figure 5.1 shows a scanning electron microscope image of a cantilever probe used for AFM, with the sharp tip at the apex. When the cantilever is used as a bio/chemical sensor, there is no need for the sharp tip but the whole lever is the active part of the sensor chip.

By the end of the year 1993, two papers where cantilevers were used as environmental sensors were prepared more or less simultaneously. The first paper to be published describes a measurement of heat evolution from a catalytic reaction $H_2 + O_2 = >H_2O$ by employing a Si cantilever coated with a thin film of the catalyst Pt [6]. The second paper, published only months later, describes the detection of water and mercury vapors, respectively [7] in a closed chamber using Al-coated Si_3N_4 cantilevers. The cantilevers used for measurements in these papers were the probes developed for AFM work. However, the results from this work showed very promising possibilities of applications beyond surface imaging of these microstructures.

It was, however, not until the year 2000 that the research field really started to gain momentum. In that year, Fritz et al. [8] showed that it was possible to detect DNA hybridization on cantilevers, Moulin et al. [9] showed that conformational changes of proteins could be monitored over time, and Raiteri et al. [10] showed that the cantilevers could find applications also within the field of electrochemistry due to their extreme sensitivity to changes at the solid–liquid interface. These publications were the breakthrough of the cantilever-based sensors research field that now involves numerous research groups around the globe resulting in several hundreds of publications every year [11]. The focus of such research is typically centered on developing new areas of applications of the sensors or to further develop the fabrication of the sensors to optimize their performance.

FIGURE 5.1 Scanning electron microscope image of a cantilever used for atomic force microscopy.

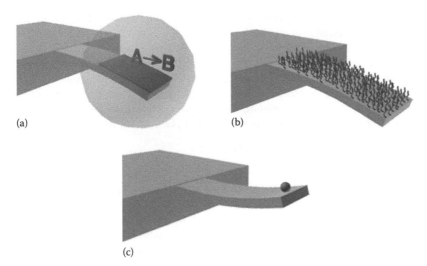

FIGURE 5.2 The three working principles of the cantilever-based sensors: temperature (a), surface stress (b), and mass change (c). They are all described in detail in the following sections.

The working principle of a cantilever sensor relies on one of the three principles:

1. Mechanical expansion or contraction as a result of temperature variations on the cantilever structure or in the close vicinity (Figure 5.2a)
2. Surface stress changes on one of the cantilever surfaces due to surface charge, conformational changes of immobilized molecules, and/or molecular recognition events (Figure 5.2b)
3. Changes in the resonance frequency due to added mass, stiffness of the coating, or changes in the viscosity or density of the surrounding medium (Figure 5.2c)

The three application areas seen in Figure 5.2 can be grouped into the *static* or the *dynamic* mode of operation, where the first relates to observations of cantilever deflections as a result of temperature variations or changes in surface stress and the latter refers to measurements of the cantilever's resonance frequency. It has been realized that the effects are coupled [12,13], but so far no clear methods of how to de-couple them have been presented. We will, therefore, consider the readout methods as three independent applications in this text.

In the following sections, all three operation methods will be discussed in detail with a theoretical approach and discussion on their potentials and limitations. We will also describe the specialized fabrication methods and the tools required to manufacture these small structures (with typical dimensions of only a couple of hundred micrometers), both with focus on classical Si-based sensor chips and the next-generation of polymeric sensor chips. Later in the chapter, the advanced readout methods employed to transduce the mechanical signals will also be described. Sections 5.5 and 5.6 show how best to operate these sensors, and Section 5.7 presents a thorough analysis of results presented in the literature to date.

5.2 Principles of Operation

As described in the introduction, the cantilevers can be applied in two physically different modes of operation, *static mode* (also called bending mode) or *dynamic mode* (also called vibration mode) within three major areas: *temperature sensing*, *surface stress detection*, and *mass measurement*. This section deals with the theory of the three types of applications and discusses how each application can be optimized.

5.2.1 Temperature Sensing

The first application where cantilevers were presented as bio/chemical sensors was actually the detection of heat generated during the catalytic conversion of $H_2 + O_2$ to H_2O over a Pt layer [6]. The work was reported using a 1.5 μm thin and 400 μm long Si cantilever coated with a 400 nm layer of Al. The optical detection system used was capable of resolving cantilever deflections of only 0.1 Å, which meant that this sensor could resolve a temperature variation of only 10^{-5} K when operated at room temperature.

The working principle of using a cantilever as a temperature sensor is simply related to the linear expansion or contraction that all solid materials undergo when exposed to temperature variations [14]. A cantilever that is fabricated in a single material will elongate upon heating.* The increase in length, Δl, is linear with the increase in temperature, ΔT. In Equation 5.1, l_0 describes the initial length of the cantilever and α is the coefficient of linear expansion. Si has a value of $\alpha = 2.6 \cdot 10^{-6}$ K^{-1}.

$$\Delta l = \alpha l_0 \Delta T \tag{5.1}$$

If the cantilever, however, is fabricated in two materials with different *coefficients of linear expansion* (α), then the cantilever will bend. This is because one of the layers expands or contracts more than the other but it is restricted to do so due to the joint structure. The direction and magnitude of deflection is thus determined by the temperature change (ΔT) and the parameters of the cantilever structure, i.e., cantilever length (l) and layer thicknesses (t), as well as the physical properties of the two materials used; Young's modulus (E) and coefficient of linear expansion (α), Equation 5.2.

$$\Delta z = \frac{3l^2\left(\alpha_1 - \alpha_2\right)\Delta T}{t_1 + t_2}\left[\frac{\left(1 + \frac{t_1}{t_2}\right)^2}{3\left(1 + \frac{t_1}{t_2}\right)^2 + \left(1 + \frac{t_1}{t_2}\frac{E_1}{E_2}\right)\left(\frac{t_1^2}{t_2^2} + \frac{t_2}{t_1}\frac{E_2}{E_1}\right)}\right] \tag{5.2}$$

From Equation 5.2, we can calculate the deflection of cantilevers typically used for bio/chemical sensing (1 μm thin and 500 μm long Si cantilevers with a 20 nm Au coating) to be −80 nm/K, which means that these cantilevers can resolve temperature variations of around 6 mK. Not only does it show that cantilevers are highly suitable as miniaturized thermometers but this simple exercise clearly shows the response of the cantilevers to temperature changes of the measurement environment. Small variations in the temperature of the different solutions and buffers used for the bio/chemical assays will result in large artificial signals from the sensors. It is, therefore, critical to always perform *differential measurements* where the deflection of an *active* and a *reference* cantilever is continuously monitored when one wants to analyze the presence of chemicals in solutions. In the final data analysis, the artificial deflections can then be eliminated so that only the specific signal is considered. The concept of reference cantilever is discussed in more detail in Section 5.5.

To take advantage of the high temperature sensitivity of cantilevers, sensors for the field of drug discovery and calorimetry could be developed. Here, temperature changes in the mK (10^{-3} K) range are often measured [15]. Design and fabrication optimization of cantilever sensors is very likely to bring the sensitivity down to an impressive level of μK (10^{-6} K) resolution. In the literature, both cantilevers with more exotic designs such as bi-material cantilevers composed of Ni and Ni-diamond nano-composites [16] as well as more standard designs such as metal coated SiN cantilever [17] have been presented.

5.2.2 Surface Stress Detection

In the second mode of operation, one monitors the *differential surface stress* between the upper and lower sides of the cantilever. This is the most common method of operation of cantilevers for bio/chemical sensing since it is the method where specific chemicals can be identified in a gas or liquid with high specificity and sensitivity. Since the biochemical identification relies on the interaction between a probe molecule and the target, liquid environments are often required so that pH and salt concentrations can be kept constant to avoid the denaturation of the biomolecules. By tailoring the receptive coating of

---

* The thickness and width of the cantilever will naturally increase in proportion to the coefficient of linear expansion and the change in temperature. However, the thickness:width:length ratio of these cantilevers is typically 1:50:500, so it is the elongation that is the dominating effect.

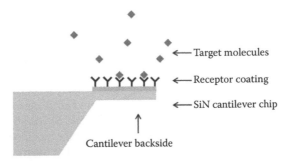

FIGURE 5.3 Schematic drawing of an SiN cantilever chip coated with an Au layer for the immobilization of the receptor coating. The target molecules and the cantilever backside are also marked. The receptors can be referred to as *probes* and the target are sometimes referred to as *the analyte*.

biomolecules applied on one side of the cantilever surface, a large number of analytes can be monitored. Work reported in the literature shows, for example: detection of single base-pair mismatches in DNA hybridization [8,18], RNA [19], detection of pesticides [20], heavy-metal ions [21] in drinking water, detection of drugs in serum [22], and molecular motors [23]. Figure 5.3 shows a schematic drawing of a typical cantilever used in surface stress sensing in order to clarify the nomenclature used in this section. It is the chemical structure of the receptor coating that determines which target molecules the cantilever will react with.

Figure 5.4 shows the raw data of the cantilever deflection from a real measurement. The surface stress signal is generated by the binding of antibiotics in solution (target) to bacterial cell wall peptides (receptor) tethered to the cantilever via Au–thiol linker chemistry. The drug generates a downward bending of the cantilever by −9 nm upon injection of 10 nM vancomycin.

The degree of cantilever deflection can be related to the differential surface stress the cantilever is exposed to via Stoney's equation [24,25]. In Equation 5.3, Δz is the cantilever deflection, $\Delta\sigma$ is the differential surface stress of the cantilever, ν is the Poisson's ratio, E is the Young's modulus of the cantilever material, and l and t are the length and the thickness of the structure, respectively:

$$\Delta z = \frac{3(1-\nu)l^2}{Et^2}\Delta\sigma \tag{5.3}$$

The generated surface stress can in turn be related to the concentration of target molecules present in the solution via the Langmuir adsorption isotherm. However, there is no method available to predict

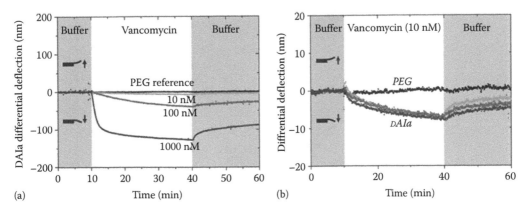

FIGURE 5.4 Raw experimental data to show the nanomechanical bending of cantilevers coated with bacterial cell wall peptide analogues (DAla), upon injection of different concentration of the antibiotic, vancomycin: (a) shows the bending in response to 10, 100, and 1000 nM vancomycin; (b) shows the nanomechanical signals of three independent DAla cantilevers (gray scale), relative to the inert PEG reference coating (black). (Data taken from Ndieyira, J.W. et al., *Nature Nanotechnol.*, 3, 691, 2008.)

the amount of deflection generated by a certain concentration of a specific molecule. This is simply because although several research groups have investigated this mode of cantilever operation over the past 20 years, there still does not exist a complete theoretical model to describe the fundamental origins of the surface stress generated from a molecular recognition event and how it translates into the mechanical deformation of the cantilever, particularly for complex biomolecular interactions.

Nevertheless, recently much progress has been made in addressing this challenging task using model alkane self-assembled monolayer (SAM) coatings with different end groups that respond to changes in pH [26,27]. These SAMs serve as model coatings since their properties can be tuned by simply altering the chain length and end groups. In 2008, Sushko et al. developed the first theoretical model of nanomechanical sensing, combining beam mechanics, mesoscopic soft-matter theory and atomistic modeling to predict the translation of molecular signals in simple SAMs at the nanoscale into the micromechanical response of the entire cantilever device. The model shows that the cantilever response to biochemical changes in its active layer is a complex combination of competing biochemical, elastic, and entropic contributions. Figure 5.5 shows that excellent quantitative agreement was found between experimental measurements (dark gray) and theoretical predictions (light gray). This multiscale model reveals fundamental new physics at the nanoscale and can, in principle, be used to engineer devices with significantly improved biomolecular detection sensitivity *in silico*. Ongoing work is focused on developing the model for more complex biomolecules on cantilevers.

Further progress on understanding the origins of surface stress has been made in work published by Ndieyira et al. on model antibiotic mucopeptides found in drug-resistant hospital "superbugs." It was shown that surface stress depends on the spatial connectivity—percolation—of drug-bound complex [22]. By assuming that the local chemistry and geometric effects responsible for the collective buildup of strain are separable, a general product form for the cantilever response is proposed. The equation is as follows:

$$\Delta\sigma_{eq} = \frac{a \cdot [\text{Van}]}{K_d + [\text{Van}]} \left(\frac{p - p_c}{1 - p_c} \right)^\alpha \quad \text{for } p > p_c \text{ and zero otherwise.} \tag{5.4}$$

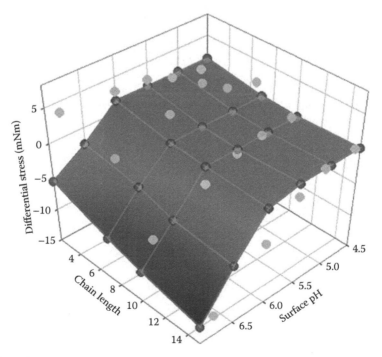

FIGURE 5.5 Three-dimensional plot to directly compare the theoretical (light gray) and experimentally measured (dark gray) surface stress generated by different chain length alkanethiol SAMs in response to different buffer pH environments. (Sushko, M.L. et al., *Adv. Mater.*, 20, 3848, 2008.)

FIGURE 5.6 Schematic illustration of the percolation concept on cantilever arrays where the spatial connectivity of bound drug–target complex on the cantilever is illustrated with a black line, and results in differential downward cantilever bending (compressive surface stress) relative to the neighboring reference cantilever.

The first term is the Langmuir adsorption isotherm, accounting for local molecular binding, and the second term is the power law form describing the large-scale mechanical consequences of stressed network formation. The constant a corresponds to maximum surface stress at full occupancy and K_d is the surface equilibrium dissociation constant on the cantilever. The power law, α, is associated with the elastic interactions between chemically reacted regions on the cantilever.

The buildup of surface stress follows from the connectivity of the chemically transformed network as well as the interactions between nodes of the network. For short-range interactions, there will be a finite percolation threshold, pc, beyond which there will be a connected network, which can produce an apparent bending of the cantilever. On the other hand, for long-range (elastic) interactions, mediated for example, through the silicon substrate, $p_c = 0$.

These findings and underlying concepts have important consequences on the development of improved cantilever devices but may also impact on our understanding of the mechanical mode of action of antibiotics on the outer cell wall surface of real bacteria. Indeed, it is this mechanical rupture event that actually kills bacteria (Figure 5.6).

A further aspect to remember is that the immobilization density of the receptor coating is related to the surface structure of the Au coating [28,29] where the receptor molecules often are immobilized via Au–thiol chemistry. These aspects are further discussed in Sections 5.5 and 5.6.

5.2.3 Mass Detection

Cantilever sensors can also be used to detect and measure changes of their mass. When the cantilever is operated in *dynamic* (mass sensing) mode, molecules are allowed to adsorb on both sides of the structure, which simplifies the application considerably. On the other hand, most resonators can only be operated in air or vacuum due to the large damping effects of any viscous media surrounding the cantilever.

The resonance frequency, f_{res}, of a cantilever is determined by Equation 5.5:

$$f_n = \frac{1}{2\pi} \sqrt{\frac{k_{eff}}{m_{eff}}} \tag{5.5}$$

where
k_{eff} is the effective spring constant
m_{eff} is the effective mass of the cantilever

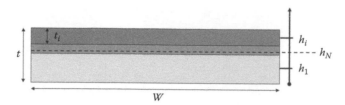

FIGURE 5.7 Cross section of a multilayered cantilever with the position of each layer denoted at h_i and the thickness denoted as t_i. The neutral plane is denoted at h_N.

These two parameters depend on the modal number, the dimensions, and the mechanical properties of the cantilever. The two parameters are also calculated differently depending on whether the masses added onto the structure are allowed to form a homogeneous coating or whether the cantilever is used to detect single particles.

To use the dynamic mode of operation of a cantilever to detect, e.g., explosives in air, the resonance frequency of the cantilever is monitored over time as a homogeneous layer of target molecules are allowed to adsorb onto the structure. In this situation, the resonance frequency can either decrease or increase, depending on whether m_{eff} or k_{eff} changes the most, since the added layer thickness will increase both the effective mass and the effective spring constant at the same time, creating a competitive process.

In order to ascertain how the added layer will affect the mechanical properties of the cantilever, we need to define a neutral plane. The neutral plane is the plane of zero strain within a bent beam with multiple layers, shown in Figure 5.7. In such a joint structure, k_{eff} is found from Equation 5.6 and m_{eff} is found from Equation 5.7:

$$k_{eff} = \frac{C_n^4}{4L^3} \sum_i E_i \left[\frac{wt_i^3}{12} + wt_i(h_i - h_N)^2 \right] \tag{5.6}$$

where
   C_n is the modal number
   L is the length of the beam, and w is the width of the beam
   E is the Young's modulus
   t is the thickness
   h is the position of each respective ith layer

$$m_{eff} = \frac{wL}{4} \sum_i \rho_i t_i \tag{5.7}$$

where ρ is the density of each ith layer.

Tamayo et al. [30] demonstrated theoretically that biological layers absorbing on soft cantilevers, made of the polymer SU-8, will affect the effective spring constant more than the effective mass causing an overall positive frequency shift. Furthermore, compared with standard silicon cantilevers, this positive shift can be much larger than the negative frequency shift predicted by the theory for silicon-based cantilevers.

Another article by Ramos et al. [31] describes the impact of absorbed bacteria cells on a cantilever and demonstrates that bacteria on the free end of the cantilever exhibit decreases in its resonant frequency, whereas bacteria deposited on the base exhibited increases in its resonant frequency. This again demonstrates that the effective spring constant is affected when biological layers are deposited on the base where the strain is concentrated, where the effective mass is affected when biological layers are deposited on the free end, where the displacement is the largest.

If we assume a point mass detection, we can neglect any affects on the cantilever's effective spring constant and thus any shifts in resonant frequency are due to pure mass loading. Due to the extreme sensitivity of the cantilever sensors to added masses and the high resolutions of the measurement equipment used, it has been shown that it is possible to detect mass changes of attograms (10^{-18} g) in air and zeptograms (10^{-21} g) in vacuum with this technique [32,33]. Moreover, by performing measurements at more than one modal number of the cantilever (C_n), it is even possible to detect *where* the single particle is situated on the cantilever [34]. Such accuracy in measurements could present cantilever sensors as more sophisticated mass spectrometers in the future.

When only a point mass is added to the cantilever, the k_{eff} remains unchanged and is again calculated from Equation 5.6. The effective mass is, however, altered and is now found from Equation 5.8:

$$m_{eff+\Delta m} = m_{eff} + \Delta m \left[\frac{z(y_{\Delta m})}{L} \right]^2 \tag{5.8}$$

where

Δm is the mass of the added particle
$z(y_{\Delta m})$ is the position of the mass along the cantilever length
L is the total length of the cantilever

From Equation 5.8, it can be seen that if the particle lands on a nodal point, the effective mass change will be zero and have no effect on changing the f_{res} of the cantilever. Vice versa, the effective mass change will be largest if the particle lands on an antinode. This knowledge can be used to optimize the sensitivity of cantilever sensors operated in dynamic mode by, e.g., only functionalizing some parts of the cantilever to ensure that masses will only bind to the areas of highest sensitivity, to ensure a form of amplification of the mass resolution signal. One could also imagine that different areas of the cantilever were functionalized differently so that multiple target molecules could be detected on a single cantilever.

5.3 Readout Methods

The different modes of operation of the cantilever sensors call for different ways of reading out the generated signals. If resonance frequencies are analyzed, very sensitive electrical equipment is needed, whereas sensitive optical detectors are required when one wants to measure the minimal cantilever deflections in surface stress measurements. The area of application (e.g., medical diagnostics or field-operation) naturally also introduces different requirements for the readout methods used.

5.3.1 External Readout: Optical

The external optical deflection set-up is the most commonly applied method to measure both minimal cantilever deflections generated from surface stress measurements as well as resonance changes of the beams [8,10]. Figure 5.8 shows a schematic drawing of the setup.

In the setup, one uses a laser source and a position-sensitive photodetector (PSD). The laser beam is reflected onto the photodetector via the cantilever beam. Due to geometry, the static cantilever deflection (Δz) can be magnified several orders of magnitudes in this arrangement, simply depending on the angle at which the laser light strikes the cantilever surface. The individual segments on the photodetector ensure that one can specify in which direction the cantilever deflects. The same setup can be used to measure the resonance frequency of the cantilever, typically generated by external actuation [35].

The two limiting factors of this readout method are that one can typically only detect the movement of a single cantilever at a time and that nonspecific signals are easily generated by, e.g., changes in the refractive index of the medium. Moreover, geometric variations among cantilevers within an array and/or misalignment of the laser can result in different signals even for identical assays. These are all aspects one need to take into careful consideration when using external optical readout.

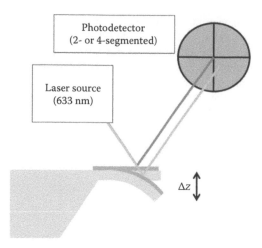

FIGURE 5.8 External optical readout can be used both to detect cantilever deflections as well as cantilever resonance frequencies. In both situations, a laser beam is reflected off the cantilever surface onto a position-sensitive photodetector. As the cantilever bends, the position of the laser beam on the photodetector moves.

There have been a few different approaches to obtaining external optical readout from more than one cantilever, which is critical if one wants to maximize reliability of the measurements by including reference measurements. Researchers at IBM early developed a multiplexing system where an array of eight vertical cavity surface emitting lasers (VCSEL) are aligned to a sensor chip comprising eight cantilevers. Each VCSEL is aligned to an individual cantilever. The VCSELs are switched on one at a time, and their movement is followed on the PSD [36]. Researchers in the United States have developed an optical platform with an array of VCSELs [37]. Here, all VCSELs are switched on at all times, and a charge-coupled device (CCD) camera is used for the optical detection. With this platform, over 100 cantilevers can be monitored simultaneously. Another approach has been to let a single laser source scan across the sensor array [38]. Here, the PSD can also follow the eight cantilever's movements individually. A further advantage is that the scan direction of the laser can easily be adjusted as to scan *along* a cantilever instead of across the whole array. This method has been utilized to obtain the line profile of the cantilever instead of only monitoring the tip deflections [39]. Line profiles can also be obtained by using phase shifting interferometry [40]. When studying the more fundamental aspects of cantilever mechanics, it is important to have an image of the complete cantilever deformation during the assay. This has been made possible by the method of full-field displacement measurements developed in France [41].

To minimize artificial signals, it is critical to use reference cantilevers, which is discussed in Section 5.5. One can also calibrate the system before running an assay. One calibration method is to perform a so-called heat test. This simple test can be used to calibrate the measurement data for mechanical non-uniformity of the sensors as well as slight laser misalignments. The test involves heating the cantilever flow cell by 1°C and monitoring the resulting cantilever deflections. Ideally, all cantilevers in the sensor array should bend equally in response to the temperature change, but typically small variations are seen. A standard deviation for the absolute bending signal of 5% or less is acceptable. Figure 5.9 shows the result for an array of Au-coated Si cantilevers that are subject to a series of heat pulses.

Nonspecific signals will also arise as a result of refractive index change, e.g., from the buffer exchange steps. Before starting the assay, it is, therefore, a good idea to also perform a refractive index test, similar to the heat test described earlier, where the laser is focused on the non-deformable chip body and several different buffers are run through the system. As the buffers are exchanged, the "deflection" signal is monitored. Since the chip body cannot move, it is clear that the signals generated are only artificial and must be subtracted from the raw data at the end of the measurement assay to obtain the true deflection signal. The magnitude and direction of the apparent deflection signals depend on the pH of the exchanged solutions, and typically an increase in the pH of solution results in an apparent downward deflection signal, while a decrease in the solution pH results in an apparent upward deflection signal.

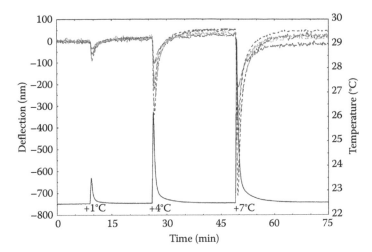

FIGURE 5.9 Monitoring the cantilever deflection with a laser aligned at its free end upon heating to 1°C, 4°C, and 7°C to test the mechanical uniformity and the accuracy of alignment of the cantilevers in the array. Here, Si cantilevers that are 1 μm thin and coated with a 20 nm layer of Au are used.

5.3.2 Integrated Readout: Electrical and Optical

Cantilever-based sensors with integrated readout started to appear in the late 1990s; Integrated readout facilitates realization of compact devices that can operate in even non-transparent environments. The field of self-sensing cantilevers has developed rapidly. Today, several groups around the world are investigating different readout techniques; primarily piezoresistive, piezoelectric, and capacitive readout.

Piezoresistive cantilever sensors have been developed by several groups [42–44] demonstrating the possibility of realizing a dense array of sensors with a compact readout system. A piezoresistive material, like silicon, changes its resistance when strained. This principle works equally well for dynamic and static mode operations. Even highly advanced systems with integrated electronics have been demonstrated [45–47].

Several groups are now implementing piezoelectric readout for cantilever-based sensing. Generally, the cantilevers are coated with a thin layer of piezoelectric material like PZT, which generates an electrical potential when strained. Piezoelectric readout is primarily used in dynamic mode of operation because the potential produced by a static force cannot be maintained by the thin film piezoelectric material.

Capacitive readout is mainly explored for mass detection in nonliquids. Here, typically a cantilever is placed close to a counter electrode and as the cantilever deflects the capacitance of the system changes. The achieved capacitance changes are very small (pF) and capacitive cantilever sensors are, therefore, typically closely integrated with electronics for signal amplification [48–50].

An integrated optical readout has been presented by different groups [51–53]. Just like the electrical methods described earlier, the main advantage is that the complete system can be integrated into a hand-held device. The optical readout methods are most suitable for static operation. Typically, the working principle is to design the cantilever so that it acts both as a mechanical sensor and also as a waveguide. Light can then be passed through the cantilever and collected at the opposite side. The coupling efficiency between the cantilever waveguide and the output waveguide will decrease as the cantilever deflects away from its equilibrium position of perfect alignment. The decrease in optical output intensity can be translated into the readout signal. The working principle is shown schematically in Figure 5.10.

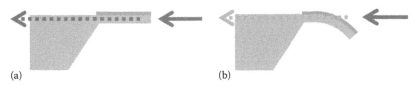

(a)   (b)

FIGURE 5.10 Schematic image of the integrated optical readout principle. The decrease of the intensity of the coupled light is shown as a decrease in the color intensity between image (a) and (b).

5.4 Fabrication

During design and fabrication of a cantilever-based sensor, several requirements have to be considered depending on the sensing mode and the readout method to be used:

1. High cantilever bending for a unit stress applied for surface stress sensing.
2. Low internal damping and a high value of resonance frequency and Q-factor for mass sensing.
3. Control of cantilever geometry with high accuracy since the dimensions have a huge influence on the sensitivity and thus the uniformity of the sensors. For example, precise control of the geometries of reference and measurement cantilevers is crucial to avoid measurement errors.
4. Reflecting cantilever surface with low roughness to avoid light scattering and maximize the intensity of the light reflected during optical readout of the bending.
5. Low initial bending to facilitate alignment of the optical beam and to reduce artificial signals due to changes in temperature, refractive index, etc.

Sensor fabrication requires processing of suspended fragile structures, which can be very challenging. Today, most cantilevers are fabricated in either silicon-based materials or polymers. The processing approaches for the two material categories are typically quite different. In the following, the fabrication of arrays of cantilevers in silicon nitride and the polymer SU-8 are described as these types of devices are the most common ones found in literature. Examples of the two types of cantilevers are shown in Figure 5.11. Furthermore, the discussion is limited to sensors, which are coated on one side with Au to enable external optical readout.

5.4.1 Fabrication of Silicon Nitride Cantilevers

In most cases, silicon-based materials are used for the fabrication of cantilever sensors due to several reasons. The silicon microfabrication is well established and uses technologies initially developed by the integrated circuit (IC) industry in the 1950s. The materials such as silicon (Si), silicon nitride ($Si_{(x-1)}N_x$), and silicon oxide (SiO_2) are also well characterized and stable over time. As a consequence, cantilevers fabricated with theses classical materials can be operated in a large range of temperatures and environmental conditions. Microcantilevers are typically around 1 μm thick and 450–950 μm long. Even extra-sensitive cantilevers with thickness $t = 500$ nm and length $l = 500$ μm are commercially available [54,55], and ultrathin micromechanical devices with $t < 200$ nm have been presented in the literature [56,57].

A typical process for fabrication of silicon nitride cantilevers is illustrated in Figure 5.12 [58].

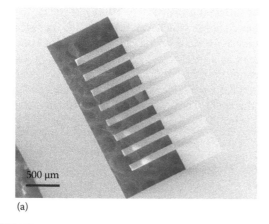
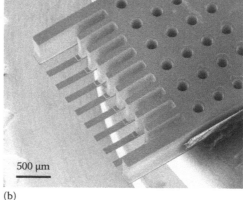

(a)   (b)

FIGURE 5.11   Scanning electron micrographs of silicon (a) and polymer (b) cantilevers. The chips both have an array of eight cantilevers with a length of 500 μm and a width of 100 μm. (Image (a) courtesy Tom Larsen, DTU Nanotech, 2011.)

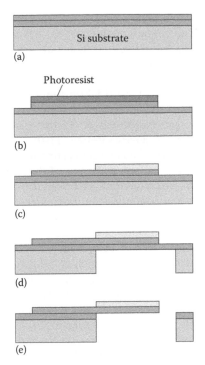

FIGURE 5.12 Fabrication of silicon-based cantilevers by bulk micromachining: (a) substrate preparation by thin film deposition; (b) patterning of cantilevers by photolithography and etching; (c) definition of metal coating; (d) cantilever release by etching through the bulk wafer from the backside; and (e) removal of the etch stop layer.

The approach is called *bulk micromachining* as the suspended structures are defined by three-dimensional etching in and through the silicon substrate. In most cases, a three-layered substrate is prepared (Figure 5.12a). The starting point is a single crystal silicon wafer with a thickness of 350–500 µm. During substrate preparation, a thin film of silicon oxide is grown by low pressure chemical vapor deposition followed by the deposition of a film of silicon nitride. The nitride is the actual device layer, and its thickness defines the final thickness of the devices. The intermediate oxide layer is the so-called *etch stop layer* protecting the device layer during the final release step in order to ensure a well-defined thickness and a highly reflecting surface of the cantilever. The cantilevers are defined by UV photolithography on the front side of the wafer. The photoresist pattern is transferred to the device layer by reactive ion etching (Figure 5.12b). Metal patterns are defined on top of the cantilevers using lift-off (Figure 5.12c). Next, the silicon bulk is etched with potassium hydroxide or by deep-reactive ion etching (Figure 5.12d). Finally, the cantilevers are release by removal of the etch stop layer (Figure 5.12e).

The process results in fully free-standing cantilevers, which are accessible from both sides of the wafer. This is often advantageous. Both sides of the cantilever can easily be inspected, and the cantilever is placed in a liquid or gas flow perpendicular to the cantilever where the etched holes in the wafer serve as microfluidic channels. However, the cantilevers are rather fragile and not well protected. Moreover, etching of the entire bulk of the silicon wafer is time consuming.

5.4.2 Fabrication of Polymer Cantilevers

One of driving factors motivating the introduction of polymers for fabrication of cantilevers is that the Young's modulus typically is two orders of magnitude lower than for traditional Si-based materials. This results in a lower stiffness of the cantilevers and in turn an improvement of the sensitivity in most sensing applications, as discussed in Section 5.2. Another important driving force to investigate fabrication of various polymers is the perspective of reduced material and processing costs. Since the late 1990s, an increasing amount of work on fabrication of polymer-based microcantilevers has been reported. In 1999, Genolet et al. used the negative epoxy photoresist SU-8 to define AFM cantilevers [59]. The Young's modulus of SU-8 is low (around 4 GPa [60]) compared to Si and $Si_{(x-1)}N_x$ (180 and 290 GPa, respectively), which makes it a suitable candidate for micromechanical structures for surface stress measurements [61,62]. In parallel, cantilevers for bio/chemical sensing have been fabricated with other polymers such as polyimide [63,64], parylene [65,66], and TOPAS® [67,68]. Here, we will discuss a typical fabrication process with the material SU-8 illustrated by the process flow in Figure 5.13.

The approach is called *surface micromachining* as the free-standing structures are fabricated by building up layers on the top surface of a carrier substrate. First, a so-called sacrificial layer is applied (Figure 5.13a). A thin film of SU-8 is deposited by spin-coating followed by a baking step to evaporate the solvent. The photosensitivity of SU-8 allows patterning of the cantilever geometries using standard UV-photolithography (Figure 5.13b), which minimizes fabrication time and costs. After definition of the cantilevers in the actual device layer, a 20 nm thin Au film is deposited and patterned using electron-beam evaporation and lift-off (Figure 5.13c). Next, a SU-8 chip body is added on top of the thin polymer film to allow the handling of the cantilevers after device release. For example, subsequent steps of SU-8 photolithography were used for the patterning of a chip body with several hundreds of microns in thickness (Figure 5.13d) [59]. Alternatively, direct integration of SU-8 cantilevers in complete microfluidic systems has been shown [69]. Finally, the cantilevers are released from the front side of the wafer, which is the most critical process step. It means that the sacrificial layer immediately below the cantilever is removed by a selective etch resulting in a cantilever, which is suspended some micrometers above the silicon substrate (Figure 5.13e). Both, wet etching with a liquid [70–73] or dry etching in a plasma [74,75] are possible. Under etch rates are often quite low as the access of the etchant to the sacrificial layer is restricted, resulting in long processing time. Alternatively, antistiction coatings can be used as release layers [76,77].

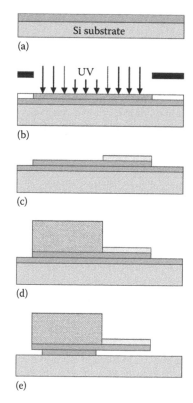

FIGURE 5.13 Fabrication of polymer cantilevers by surface micromachining: (a) deposition of sacrificial layer; (b) patterning of cantilevers by direct UV lithography; (c) definition of metal coating; (d) patterning of chip body; and (e) cantilever release by partial or complete etching of sacrificial layer.

One of the main challenges in using polymer chips for mechanical sensing is the stability of the devices during measurements. Moisture absorption in liquids or degassing in vacuum, results in drift of the output signal [78]. Schmid et al. have shown that a change in air humidity is reflected in a shift of the resonance frequency of SU-8 cantilevers due to water absorption in the material [79]. Other phenomena typically observed for polymers such as creep deformation, ageing, or bleaching can affect the long-term stability of micromechanical devices. To some extent, process optimization can minimize drift and increase time stability of the devices [73].

5.4.3 Comments of Comparison

The state-of-the-art cantilever sensors with length $l = 500\,\mu m$ and width $w = 100\,\mu m$ have a thickness $t = 500\,nm$ for $Si_{(x-1)}N_x$ [58] and $t = 2\,\mu m$ for SU-8 [80]. The performance of these devices should be compared with respect to the requirements stated in the introduction of the section.

For the sensing of surface stress, SU-8 cantilevers are a good alternative to $Si_{(x-1)}N_x$ devices. The theoretical surface stress sensitivity given by Equation 5.3 results in $\Delta z/\Delta\sigma = 36.6\cdot10^{-6}\,m^2/N$ for the polymer cantilevers compared with $\Delta z/\Delta\sigma = 7.6\cdot10^{-6}\,m^2/N$ for the silicon-based cantilevers. Furthermore, a considerable response to charge accumulation upon changes in pH was reported for silicon-based devices [81]. On the other hand, for the measurement of surface stress in liquids, drift due to moisture absorption in the bulk might have an effect on the performance of SU-8 cantilevers. Some authors directly compared polymer and silicon-based cantilevers in surface stress measurements. In general, the performance of polymer devices is comparable to their silicon-based counterparts [82].

For mass sensing, silicon-based cantilevers are preferred. One of the main reasons is the lower internal damping resulting in a higher quality factor during measurements. The mass sensitivity calculated from Equation 5.5 and the dimensions mentioned earlier is $\Delta f_{res}/\Delta m = 21.2\,Hz/ng$ for the $Si_{(x-1)}N_x$ devices compared with $10.0\,Hz/ng$ for the ones fabricated in SU-8.

For sensor operation at elevated temperature (e.g., in calorimetry), silicon-based cantilevers are superior to the ones made of polymer due to the greater temperature stability. Due to the same reason, integration of metal coatings on polymeric structures is more challenging and often results in high initial bending of the cantilevers.

Other important aspects to consider besides the actual sensor performance are the fabrication costs, stability of the chips during handling, and the possibilities to clean the sensors between experiments. In general, fabrication with silicon-based materials is more expensive than with polymers. It is, therefore, not possible to give a simple and direct answer to which type of material is best suited as device material, but it completely depends on the desired application of the sensors.

5.5 Performing Cantilever Assays

The major challenge in obtaining reproducible and specific biological detection when using cantilever sensors lies in ensuring that the properties of the receptor layer is biochemically "active" and uniform for each and every assay. The secret to experimental success lies in a careful pretreatment of the sensor chip with a standardized cleaning and immobilization protocol with linker chemistries to orient the receptor molecules in their active conformation. It is important to appreciate that optimized surface chemistries for cantilever assays, in particular surface stress measurements, can be quite different to those used in conventional sensors, e.g., based on surface plasmon resonance (SPR), since the operating principles are completely different. Here, we present one such successful procedure using Au–thiol chemistry that has been successfully used to detect simple SAMs, DNA, RNA, proteins, and drug–target interactions.

5.5.1 Cantilever Cleaning

First, the cantilevers are cleansed in freshly prepared piranha solution (1:1 ratio of H_2SO_4 and H_2O_2) for 20 min, followed by a thorough rinse in de-ionized water and consecutively pure ethanol. The cantilever chip can then be dried on a hotplate at 70°C for 30 s. This cleaning step is performed to remove any organic contaminants from the chip. Some reports in literature even start with a cleaning step utilizing Aqua Regia to remove the initial Au layer that usually is present on commercial cantilever chips [83]. Once a clean cantilever surface is obtained, a fresh metal layer can be deposited onto the cantilever array. As discussed in Section 5.4, the metal layer typically used comprises a 2 nm adhesion layer (Ti or Cr) followed by a 20 nm film of Au. After metal evaporation, the cantilever array must be sealed under Ar to avoid oxidation of the Au layer. It is advisable to functionalize the array within a few hours after the evaporation.

5.5.2 Cantilever Functionalization

One of the most promising methods to functionalize multiple cantilever arrays with different probe molecules is using microcapillaries [36]. Figure 5.14 shows a schematic representation of such a setup. Each microcapillary array contains assorted probe molecules and due to the confinement of the solutions onto each individual cantilever, cross-contamination can be avoided. This procedure can be performed with sample volumes down to 8 μL, which is an advantage when working with precious and often custom-made probe receptors. The only requirement of this setup is that the microcapillaries are arranged according to the cantilever pitch and size. The cantilevers can be incubated in the microcapillaries filled with alkanethiols for 20 min and washed three times with ethanol and stored in Milli-Q distilled water before use. Alternatively, one can use a complementary microchannel system where a group of cantilevers can be functionalized with a specific probe molecule [84] or microspotting [85], which can not only functionalize much larger arrays but also individually address the upper and lower sides of the cantilever.

5.5.3 Gold Layer Homogeneity

When functionalizing the cantilevers, either by using microcapillaries or microspotting, the binding of the probe molecules onto the cantilever most often relies on the coupling between a thiol moiety of the probe molecule and an Au metal layer. Several groups have studied how the probe molecule immobilization density is related to the deposition method of the metal film [86], the surface roughness [87,88], and the topography of the Au layer [28,29]. A direct relation has not been found, but it is very important to remember the effect of the underlying metal layer since this will greatly affect the final surface stress signal obtained from the sensor. It is also very likely that different probe molecules are affected in different ways, which means that the metal depositions step should be carefully optimized for each individual application.

5.5.4 Reference Cantilevers

Many early experiments reported in the literature used only single cantilevers to detect chemical and biological analytes. However, this is notoriously problematic since biochemical molecules can react nonspecifically with the coating and or the SiO_2 underside of the cantilever. Furthermore, changes in

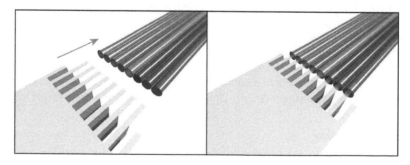

FIGURE 5.14 Schematic representation of glass microcapillaries used for the functionalization of a cantilever array. Each cantilever in the array can be functionalized with different receptors with this configuration.

temperature and refractive index of the buffers used also cause apparent erroneous bending signals that can often mask the biochemically specific signal. To deconvolute the biochemically specific signal, it is essential to use multiple arrays and in situ reference cantilevers coated with an inert layer via differential measurements. When performing a differential measurement, one monitors the deflection/resonance of two identical cantilevers simultaneously during the complete assay. The only difference there should be between the two cantilevers is that one cantilever must react with the target analyte, whereas the other cantilever must be nonresponsive to this molecule. Fritz et al. was the first to develop multiple arrays and differential measurements and was able to detect single nucleotide polymorphisms using two cantilevers coated with different oligonucleotide capture layers [8]. More recently, differential measurements have been used to sense proteins, drugs, and cells with high sensitivity and specificity. However, it is clear that the judicious choice of an appropriate reference coating is of the upper most importance since even small changes in structure can alter the packing density, orientation, and surface energy of the coating. It may not seem so difficult to design a differential measurement but in fact it is not so straightforward, and the most challenging part is to design a perfect irresponsive receptor layer [89].

5.5.5 Liquid Flow, Sample Exchange

Another factor that is important to optimize in the cantilever measurement is the flow rate. Optimized flow rates ensure the efficient exchange of liquids and sufficient steady mass transport of solution materials. The optimum volume of the liquid cell allows fast flow rates to enable perfect liquid exchange to overcome mass transport limitations. The flow rate is especially critical when performing kinetic measurements [90]. Large volume chambers can lead to uncontrollable high flow rates, and they also require large sample volumes, which unnecessarily increases the price of the assay.

5.6 Cantilever Activation

Since the cantilever sensor itself cannot sense any specific analytes, it needs to be activated with a functionalization layer, a so-called *probe layer,* before it becomes useful as a sensing device. The probe layer controls both the specificity (can the cantilever differentiate between species A and species B) as well as the sensitivity (the minimum concentration of species A or species B that can be detected). The functionalization method to employ depends on the mode of operation (static or dynamic) and the environment (air or liquid) that will be used in the assay. It must also be emphasized that the coating layer must be thoroughly characterized, for example, using contact angle, x-ray photoelectron spectroscopy and ellipsometry.

5.6.1 Functionalization Using Gold

Since most commercially available cantilevers are coated with a thin layer of Au on one of its surfaces, it is very convenient to use the Au side for the conjugation of receptors using the well-known Au–thiol chemistry. Care must be taken to first obtain a clean metal surface and to use the cantilevers no more than 24–48 h after preparation, as previously discussed in Section 5.5. The properties of the solvent/buffer will also play an important role, and all layers should be carefully characterized. One must also remember to consider the backside of the cantilever (Si or $Si_{(x-1)}N_x$) and the risk of unspecific binding of the probe molecules and of the analyte on this surface. Here, a few examples from literature are given where specific probe molecules are immobilized onto the cantilever sensor using the Au–thiol chemistry.

In some of the early papers published, DNA oligonucleotides functionalized by a hexyl spacer with a thiol group were conjugated on cantilevers and then used in hybridization studies [8,18]. The conjugation of the nucleic acids was performed using a 10–40 μM solution of the oligonucleotides in triethyl ammonium acetate buffer (50 mM).

Usually, no more than 20 min incubation is necessary and the oligonucleotides are stable for several days at 4°C. This protocol has been exploited even for the conjugation of DNA aptamers [91] and i-motifs [92].

The preparation of cantilevers coated with proteins has been widely studied but is more challenging than DNA due to the delicate tertiary structure of proteins, and considerable care must be addressed to the immobilization protocol in order to maintain biological activity. In some cases, the proteins can be directly adsorbed on the Au using site-specifically engineered thiols or cysteine groups. Backmann et al. prepared cantilevers functionalized with single chain antibody fragments directly linked on Au [93]. These small antibodies are produced artificially and have cysteines that allow the stable binding on the surface.

Protein A, which is produced by *Staphylococcus aureus*, is a protein that is able to bind the Fc region of immunoglobulins and at the same time is spontaneously adsorbed on Au in a stable way. For this reason, protein A is a popular system to create immunosensors with highly oriented antibodies on the surface. Tan et al. [94] used this approach to prepare a cantilever-based immunosensor with the ability to detect small molecules in the part-per-trillion-range. Tan et al. incubated the Au-coated cantilevers in buffer containing 10 µg/mL of protein A for 3 h at 37°C. After a washing step, the immunoglobulin was then attached with an active orientation simply by incubation.

These interesting approaches, however, cannot be employed universally since protein A binds just some classes of immunoglobulin and the standard monoclonal and polyclonal antibodies do not have thiols for the conjugation.

5.6.2 Functionalization Using Self-Assembled Monolayers

One way to "alter" the Au surface into another chemically active surface is to functionalize it with a SAM. The SAM typically consists of short molecules with a thiol moiety at one end and the desired end group unit at the other. Stable SAMs form easily on clean Au surfaces after incubation for about 24 h [95,96]. Since thiols are a very flexible system, many compounds with different chemical groups are commercially available or can be synthesized in a laboratory.

Proteins can be conjugated to SAMs via a number of protocols: for example, an active ester terminating SAM will react with the amino group on a protein though this coupling chemistry will result in randomly oriented proteins since amine groups are typically found all over the protein and so may hinder access to the binding site. The surface active ester can be achieved by immobilizing a carboxylic acid or glutaraldehyde on the cantilever sensor. Wu et al. [97] activated the cantilever surface with 1.5 mM dithiobis—sulfosuccinimidylpropionate in citrate buffer at pH 5 for 5 h at room temperature and then attached an antiprostate-specific antigen antibody dissolved in phosphate buffer. Dutta et al. used a combination of aminoethanethiol and glutaraldehyde to achieve a similar result [98].

It is even possible to reverse the process and activate the proteins with the thiol-containing linker that is then attached on the Au. A SAM presenting an activated carboxylic acid was used for the conjugation of antibodies in a similar way as the previous example [99].

Ndieyira et al. used thiols directly conjugated with tripeptides mimetic of the lipid II from the bacteria cell wall in order to prepare cantilevers for the study of the mechanism of action of antibiotics [22]. Very recently Paoloni et al. published the first paper where click chemistry was employed for the functionalization of cantilevers [100]. In this work, the SAM terminating with an azido group coupled with various alkynes was used for the detection of a nerve-gas from a mixture of solvent vapors.

5.6.3 Functionalization of Silicon

When cantilevers are used as mass sensors, it is not necessary to have an asymmetry between the two sides with respect to functionalization. In this case, the receptors are physisorbed or covalently attached directly on the silicon, typically using SAMs or silanes. Nugaeva et al. adapted a popular protocol used for the conjugation of protein on glass slide using aminopropyltriethoxysilane and glutaraldehyde for the conjugation of protein A followed by antibodies [101].

For a similar purpose, Gupta et al. used the ability of bovine serum albumin (BSA) to be adsorbed on any surface [102]. They adsorbed biotinylated BSA on silicon and then used streptavidin in order to create a versatile platform for the conjugation of antibodies on cantilevers.

5.6.4 Functionalization Using Polymers

Modification of microcantilevers with thin polymer layers is probably the most common approach employed in applications involving chemical vapor sensing. In this case, the major contribution to the response of the cantilevers is due to the swelling of the polymer layer. The adsorption of the analyte molecules into the polymer film results in relatively large differential stresses. Polymers can be conjugated on cantilevers in a number of ways. Battiston et al. [103] used droplets in order to create thick (2–3 μm) layers of polymers on cantilever. These polymers were then used for the detection of different solvents. A similar result can be achieved using an inkjet printer [85]. Many other different approaches of the polymerization of the precursor directly on cantilever [104] or the deposition by spray [105] or spin coating [32] techniques have also been published.

Polymers have even been reported for the conjugation of proteins. Von Muhlen et al. used a carboxy-betaine-derived co-polymer for the conjugation of antibodies on the surface of cantilever using dopamine as the reactive point with silicon [106]. Yan et al. used an Au-coated cantilever functionalization strategy using layer-by-layer nano-assembly of polymers for the conjugation of glucose oxidase [107]. In this work, a negatively charged SAM was adsorbed on Au using thiol chemistry. Subsequently, alternated layers of positively and negatively charged polymers were formed on the surface simply by immersion. On this stable base, the protocol was repeated by substituting the negative polymer with glucose oxidase for a stable conjugation of the enzyme.

Oliviero et al. used a block copolymer for the conjugation of DNA and protein on silicon cantilever and demonstrated that this approach was suitable both for static and dynamic modes [108].

5.6.5 Blocking of Unspecific Bindings

When cantilevers are used for the detection of proteins, it is necessary to consider the fact that the proteins can bind unspecifically both to the lower side (non-Au coated) of the cantilever and to the portions of the upper side (Au coated) that have not been functionalized with the receptor, or nonspecifically to the coating layer. To avoid this problem, many blocking agents such as bovine serum albumin and casein are used [99,109]. Particularly interesting is the PEGylation protocol of the lower side proposed by Backmann et al. [93]. In their work, cantilevers were completely PEGylated using silane chemistry and in a second moment the PEGylated upper side is coated with Au.

As natural oligonucleotides do not have a strong tendency to bind the silicon surface at the concentration typically used in these studies, no blocking agents are necessary.

5.7 Applications and Use of Cantilever Sensors

More than 100 publications, most of them in the past 10 years, have described various applications of cantilever sensors. In this section, some of the most promising applications will be described.

5.7.1 Chemical Sensors

One of the simplest applications of cantilever sensors is the sensing of the pH of a solution. Cantilevers coated with SAMs terminating in an amino- or carboxyl-group can be used as a pH sensor. The functional end groups protonate or deprotonate depending on the pH of the solution and can generate a surface charge that leads the cantilever to bend [110].

Other applications include the detection of ions and chemicals in a liquid solution. For example, the detection of Ca^{2+} ions was reported using cantilever sensors [21]. In another example, cantilevers functionalized with horseradish peroxidase were able to sense hydrogen peroxide [111].

For the detection of various solvent vapors, cantilevers have been coated with different polymer layers [105]. Depending on the analyte molecules, the polymers exhibit characteristic swelling behaviors, and different solvents could, therefore, be discriminated using principal component analysis.

5.7.2 Biomedical Analysis and Drug Discovery

Medical research still relies mostly on assays that require the labeling of targets (e.g. by fluorescence) or multistep preparation procedures (e.g. ELISA). Emerging label-free technologies allow for the investigation of reactions, such as drug-target interactions, in a simple single-step assay.

Different approaches have been reported for the label-free detection of biomolecules on cantilevers. One report demonstrated the specific binding of Taq DNA polymerase to cantilevers coated with DNA aptamers [91]. Another application used single-chain Fv antibody fragments as receptor molecules on cantilevers, which have the specificity to bind different peptides [93]. Cantilevers coated with double-stranded DNA oligonucleotides exhibited the ability to probe the transcription factors SP1 and NF-κB [112]. All these applications demonstrated sensitivities in the nanomolar range or below. The approaches mentioned earlier used SAMs or thiol–Au surface chemistry to attach the receptor molecules onto the cantilever surface. Special thiols that model the bacterial cell wall were used to investigate the binding of the glycopeptide antibiotic vancomycin [22]. In nature, these antibiotics bind to bacterial cell wall precursors, therefore, inhibiting the formation of the bacterial cell wall and lead to lysis of bacteria. Cantilever sensors are a unique tool to study these drug-target interactions as they measure in-plane surface forces due to antibiotic binding, which are thought to be involved in the destabilization of the cell wall of life bacteria.

Some attempts have been made toward the investigation of membrane proteins on cantilever sensors. Membrane proteins are central to many biological processes and are the targets of many drugs. However, measuring interactions with membrane proteins remains difficult. Initial experiments demonstrated the feasibility to form supported bilayers on cantilevers [113]. The subsequent insertion of the pore-forming peptide melittin revealed a change in surface stress. Another research group coated cantilevers with the model transmembrane protein bacteriorhodopsin in liposomes and reported a change in surface stress upon induction of conformational changes in this protein [114]. Subsequent cantilever experiments showed the binding of T5 bacteriophages to liposomes with FhuA receptors from *Escherichia coli* [115].

An innovative cantilever design with an incorporated microfluidic channel allowed for the measurement of the mass and growth rates of single cells [116]. This technology will prove beneficial to the study of cellular responses to different growth factors or drugs.

5.7.3 Medical Diagnostics

Omnipresent concerns about increasing healthcare costs and the urge toward personalized medicine are driving the development of new medical diagnostic devices. Cantilever technology offers a novel tool to detect a multitude of biomarkers in a simple and quick way and positions itself as a promising technology for future medical and point-of-care applications.

The detection of DNA hybridization on a cantilever was amongst the first biological applications of cantilever sensors and are probably the best characterized and understood biological applications on cantilevers [8,18,117–119]. Cantilever surfaces were coated with single-stranded DNA oligomers and upon hybridization of the complementary strand, a surface stress was measured. Thereby, this method is sensitive enough to discriminate between single base mismatches in the DNA strands. This technique was then used to detect mRNA biomarker transcripts in a complex background, which positions the cantilever technology as a promising candidate for a label-free gene expression diagnostics device without the need for target amplification [19].

In the context of applications in gaseous environments, cantilevers coated with different polymers were able to discriminate between breath samples of healthy persons and persons with diabetes or uremia. This opens possibilities for diagnosing some diseases by a noninvasive and simple breath test [120,121]. Furthermore, the specific detection of glucose or ethanol in liquid using cantilever technology could also be used as simple diagnostic or monitoring tests [122,123].

The recent trend toward personalized medicine demands the detection of specific cardiac or cancer biomarkers to make an early diagnosis of the possibility for a heart attack or cancer. The prostate-specific antigen has been detected using cantilever sensors, even in the background of human serum proteins [97]. Additionally, cantilever sensors detected the prostate cancer biomarker, AMACR, directly in patient urine [124]. Other examples include the detection of the cardiac biomarker proteins creatin kinase and myoglobin [99], as well as C-reactive proteins [125].

In a fundamentally different application, cantilevers were coated with a nutritive layer that served as a platform to study the growth of bacteria [126]. As the bacteria divide and proliferate, the resonance frequency of the cantilever decreases. With this method, the growth of *E. coli* could be detected within 1 h, which is a huge improvement compared to current bacterial culture plates. If antibiotics are added to the nutritive layer, this allows quick testing of the bacteria's susceptibility to certain antibiotics and could help to prescribe the correct antibiotic dosages to patients.

5.7.4 Environmental Sensors

Due to their versatility, small size, and robustness, cantilever sensors can be used to monitor several environmental factors. Cantilever-based sensors have been shown to be able to detect toxic and harmful gases that can escape from the laboratory and industrial production units into the environment, such as hydrofluoric acid [127] or hydrogen cyanide [128]. Different cantilever applications recognized small amounts of heavy metals, which can be a major hazard to nature. These include the detection of Pb^{2+} using hydrogel swelling on cantilevers [129] or detection of Cd^{2+} on an antibody-modified cantilever [130] and Hg^{2+} from its affinity to Au [131].

Cantilevers have also been applied to sense explosives like TNT [132] or the nerve agent stimulant dimethyl-methylphosphonate [133]. In the context of biological weapons and pathogenic microorganisms, it has been shown that *Bacillus subtilis* (a stimulant of *Bacillus anthracis*, which causes anthrax) can be captured on cantilevers using selective peptide substrates [134]. Another report demonstrated the feasibility to detect epidemic viruses using cantilevers coated with antiviral antiserum. The technology could measure an exposure to the severe acute respiratory syndrome-associated coronavirus, which caused a serious worldwide epidemic in 2002 [135]. Recently, the detection of *Giardia lamblia* cysts in non-filtered water sources using cantilever sensors was reported [136]. This parasite causes giardiasis, a diarrheal infection with high prevalence in developing countries. This application exemplifies the potential of cantilever sensors to monitor drinking water supplies within a portable device.

5.8 Summary and Conclusions

This chapter has given an introduction to the working principles of these micromechanical devices, both for sensing surface stresses as well as small masses. Theory has been presented on how the structures respond to molecules adsorbing on the surface and interacting with analytes. Two process methods have been discussed, with the focus on how to design and fabricate these miniature structures. It has been shown that it is possible to develop cantilever sensors both using Si-based technologies as well as in polymeric materials. A list of different methods to read out the response signal of the cantilever has been given, with advantages and challenges listed for the different types. We have also presented several tips and tricks that are important to know when using these sensors for analyses. As seen in this chapter, cantilever-based sensors find applications spanning from environmental monitoring out in the field to laboratory-based medical analyses. Today, there are companies retailing complete devices as well as single cantilever chips for analytical assays. We foresee that several more will emerge in the next coming years based on this promising technology.

REFERENCES

1. A. Boisen, S. Dohn, S. S. Keller, S. Schmid, and M. Tenje, Cantilever-like micromechanical sensors, *Reports on Progress in Physics*, **74** (2011) 036101.
2. K. M. Goeders, J. S. Colton, and L. A. Bottomley, Microcantilevers: Sensing chemical interactions via mechanical motion, *Chemical Reviews*, **108** (2008) 522.
3. J. Fritz, Cantilever biosensors, *Analyst*, **133** (2008) 855.
4. G. Binnig, C. F. Quate, and C. Gerber, Atomic Force Microscope, *Physical Review Letters*, **56** (1986) 930.

5. T. R. Albrecht, S. Akamine, T. E. Carver, and C. F. Quate, Microfabrication of cantilever styli for the atomic force microscope, *Journal of Vacuum Science and Technology A—Vacuum Surface Films*, **8** (1990) 3386.

6. J. K. Gimzewski, C. Gerber, E. Meyer, and R. R. Schlittler, Observation of a chemical reaction using a micromechanical sensor, *Chemical Physics Letters*, **217** (1994) 589.

7. T. Thundat, R. J. Warmack, G. Y. Chen, and D. P. Allison, Thermal and ambient-induced deflections of scanning force microscope cantilevers, *Applied Physics Letters*, **64** (1994) 2894.

8. J. Fritz, M. K. Baller, H. P. Lang, H. Rothuizen, P. Vettiger, E. Meyer, H.-J. Güntherodt, C. Gerber, and J. K. Gimzewski, Translating biomolecular recognition into nanomechanics, *Science*, **288** (2000) 316.

9. A. M. Moulin, S. J. O'Shea, and M. E. Welland, Microcantilever-based biosensors, *Ultramicroscopy*, **82** (2000) 23.

10. R. Raiteri, H. J. Butt, and M. Grattarola, Changes in surface stress at the liquid/solid interface measured with a microcantilever, *Electrochimica Acta*, **46** (2000) 157.

11. http://apps.isiknowledge.com, accessed November 22, 2010.

12. A. Subramanian, P. I. Oden, S. J. Kennel, K. B. Jacobson, R. J. Warmack, T. Thundat, and M. J. Doktycz, Glucose biosensing using an enzyme-coated microcantilever, *Applied Physics Letters*, **81** (2002) 385.

13. A. W. McFarland, M. A. Poggi, M. J. Doyle, L. A. Bottomley, and J. S. Colton, Influence of surface stress on the resonance behavior of microcantilevers, *Applied Physics Letters*, **87** (2005) 3.

14. H. D. Young and R. A. Freedman, *University Physics*, Addison-Wesley, New York (1996)

15. J. Comley, Progress in the implementation of label-free detection, *Drug Discovery World*, **Summer** (2008) 77.

16. C. S. Huang, Y. T. Cheng, J. W. Chung, and W. Y. Hsu, Investigation of Ni-based thermal bimaterial structure for sensor and actuator application, *Sensors and Actuators A: Physical*, **149** (2009) 298.

17. N. Lavrik, R. Archibald, D. Grbovic, S. Rajic, and P. Datskos, Uncooled MEMS IR imagers with optical readout and image processing—Art. no. 65421E, in B. F. Andresen, G. F. Fulop, and P. R. Norton, eds., *Infrared Technology and Applications XXXIII*, Orlando, FL, 2007, pp. E5421.

18. R. McKendry, J. Zhang, Y. Arntz, T. Strunz, M. Hegner, H. P. Lang, M. K. Baller et al., Multiple label-free biodetection and quantitative DNA-binding assays on a nanomechanical cantilever array, *Proceedings of the National Academy of Sciences of the United States of America*, **99** (2002) 9783.

19. J. Zhang, H. P. Lang, F. Huber, A. Bietsch, W. Grange, U. Certa, R. McKendry, H. J. Guntgerodt, M. Hegner, and C. Gerber, Rapid and label-free nanomechanical detection of biomarker transcripts in human RNA, *Nature Nanotechnology*, **1** (2006) 214.

20. M. Alvarez, A. Calle, J. Tamayo, L. M. Lechuga, A. Abad, and A. Montoya, Development of nanomechanical biosensors for detection of the pesticide DDT, *Biosensors and Bioelectronics*, **18** (2003) 649.

21. S. Cherian, R. K. Gupta, B. C. Mullin, and T. Thundat, Detection of heavy metal ions using protein-functionalized microcantilever sensors, *Biosensors and Bioelectronics*, **19** (2003) 411.

22. J. W. Ndieyira, M. Watari, A. D. Barrera, D. Zhou, M. Vogtli, M. Batchelor, M. A. Cooper et al., Nanomechanical detection of antibiotic mucopeptide binding in a model for superbug drug resistance, *Nature Nanotechnology*, **3** (2008) 691.

23. W. M. Shu, D. S. Liu, M. Watari, C. K. Riener, T. Strunz, M. E. Welland, S. Balasubramanian, and R. A. McKendry, DNA molecular motor driven micromechanical cantilever arrays, *Journal of the American Chemical Society*, **127** (2005) 17054.

24. G. G. Stoney, The tension of metallic films deposited by electrolysis, *Proceedings of the Royal Society London A—Materials*, **82** (1909) 172.

25. R. J. Jaccodin, and W. A. Schlegel, Measurement of strains at Si-Sio2 interface, *Journal of Applied Physics*, **37** (1966) 2429.

26. M. Watari, J. Galbraith, H. P. Lang, M. Sousa, M. Hegner, C. Gerber, M. A. Horton, and R. A. McKendry, Investigating the molecular mechanisms of in-plane mechanochemistry on cantilever arrays, *Journal of the American Chemical Society*, **129** (2007) 601.

27. M. L. Sushko, J. H. Harding, A. L. Shluger, R. A. McKendry, M. Watari, Physics of nanomechanical biosensing on cantilever arrays, *Advanced Materials*, **20** (2008) 3848.

28. J. Mertens, M. Calleja, D. Ramos, A. Taryn, and J. Tamayo, Role of the gold film nanostructure on the nanomechanical response of microcantilever sensors, *Journal of Applied Physics*, **101** (2007) 034904.

29. M. Godin, P. J. Williams, V. Tabard-Cossa, O. Laroche, L. Y. Beaulieu, R. B. Lennox, and P. Grutter, Surface stress, kinetics, and structure of alkanethiol self-assembled monolayers, *Langmuir*, **20** (2004) 7090.

30. J. Tamayo, D. Ramos, J. Mertens, and M. Calleja, Effect of the adsorbate stiffness on the resonance response of microcantilever sensors, *Applied Physics Letters*, **89** (2006).
31. D. Ramos, J. Tamayo, J. Mertens, M. Calleja, L. G. Villanueva, and A. Zaballos, Detection of bacteria based on the thermomechanical noise of a nanomechanical resonator: Origin of the response and detection limits, *Nanotechnology*, **19** (2008) 035503.
32. M. Li, H. X. Tang, and M. L. Roukes, Ultra-sensitive NEMS-based cantilevers for sensing, scanned probe and very high-frequency applications, *Nature Nanotechnology*, **2** (2007) 114.
33. Y. T. Yang, C. Callegari, X. L. Feng, K. L. Ekinci, and M. L. Roukes, Zeptogram-scale nanomechanical mass sensing, *Nano Letters*, **6** (2006) 583.
34. S. Dohn, S. Schmid, F. Amiot, and A. Boisen, Position and mass determination of multiple particles using cantilever based mass sensors, *Applied Physics Letters*, **97** (2010) 044103.
35. R. Sandberg, A. Boisen, and W. Svendsen, Characterization system for resonant micro- and nanocantilevers, *Review of Scientific Instruments*, **76** (2005) 125101.
36. H. P. Lang, M. Hegner, and C. Gerber, Cantilever array sensors, *Materials Today*, **April** (2005) 30.
37. M. Yue, J. C. Stachowiak, H. Lin, R. Datar, R. Cote, and A. Majumdar, Label-free protein recognition two-dimensional array using nanomechanical sensors, *Nano Letters*, **8** (2008) 520.
38. M. Alvarez and J. Tamayo, Optical sequential readout of microcantilever arrays for biological detection, *Sensors and Actuators B: Chemical*, **106** (2005) 687.
39. J. Mertens, M. Alvarez, and J. Tamayo, Real-time profile of microcantilevers for sensing applications, *Applied Physics Letters*, **87** (2005) 234102.
40. M. Helm, J. J. Servant, F. Saurenbach, and R. Berger, Read-out of micromechanical cantilever sensors by phase shifting interferometry, *Applied Physics Letters*, **87** (2005).
41. F. Amiot, F. Hild, and J. P. Roger, Identification of elastic property and loading fields from full-field displacement measurements, *International Journal of Solids and Structures*, **44** (2007) 2863.
42. A. Boisen, J. Thaysen, H. Jensenius, and O. Hansen, Environmental sensors based on micromachined cantilevers with integrated read-out, *Ultramicroscopy*, **82** (2000) 11.
43. H. T. Yu, X. X. Li, X. H. Gan, Y. J. Liu, X. Liu, P. C. Xu, J. G. Li, and M. Liu, Resonant-cantilever bio/chemical sensors with an integrated heater for both resonance exciting optimization and sensing repeatability enhancement, *Journal of Micromechanics and Microengineering*, **19** (2009) 10.
44. M. Narducci, E. Figueras, M. J. Lopez, I. Gracia, J. Santander, P. Ivanov, L. Fonseca, and C. Cane, Sensitivity improvement of a microcantilever based mass sensor, *Microelectronic Engineering*, **86** (2009) 1187.
45. D. Lange, C. Hagleitner, A. Hierlemann, O. Brand, and H. Baltes, Complementary metal oxide semiconductor cantilever arrays on a single chip: Mass-sensitive detection of volatile organic compounds, *Analytical Chemistry*, **74** (2002) 3084.
46. I. Voiculescu, M. E. Zaghloul, R. A. McGill, E. J. Houser, and G. K. Fedder, Electrostatically actuated resonant microcantilever beam in CMOS technology for the detection of chemical weapons, *IEEE Sensors Journal*, **5** (2005) 641.
47. A. Hierlemann, D. Lange, C. Hagleitner, N. Kerness, A. Koll, O. Brand, and H. Baltes, Application-specific sensor systems based on CMOS chemical microsensors, *Sensors and Actuators B: Chemical*, **70** (2000) 2.
48. S. Ghatnekar-Nilsson, E. Forsen, G. Abadal, J. Verd, F. Campabadal, F. Perez-Murano, J. Esteve, N. Barniol, A. Boisen, and L. Montelius, Resonators with integrated CMOS circuitry for mass sensing applications, fabricated by electron beam lithography, *Nanotechnology*, **16** (2005) 98.
49. J. Arcamone, G. Rius, G. Abadal, J. Teva, N. Barniol, and F. Perez-Murano, Micro/nanomechanical resonators for distributed mass sensing with capacitive detection, *Microelectronic Engineering*, **83** (2006) 1216.
50. E. Forsen, G. Abadal, S. Ghatnekar-Nilsson, J. Teva, J. Verd, R. Sandberg, W. Svendsen et al., Ultrasensitive mass sensor fully integrated with complementary metal-oxide-semiconductor circuitry, *Applied Physics Letters*, **87** (2005) 3.
51. J. W. Noh, R. Anderson, S. Kim, J. Cardenas, and G. P. Nordin, In-plane photonic transduction of silicon-on-insulator microcantilevers, *Optics Express*, **16** (2008) 12114.
52. K. Zinoviev, C. Dominguez, J. A. Plaza, V. J. C. Busto, and L. M. Lechuga, A novel optical waveguide microcantilever sensor for the detection of nanomechanical forces, *Journal of Lightwave Technology*, **24** (2006) 2132.

53. M. Nordström, D. A. Zauner, M. Calleja, J. Hubner, and A. Boisen, Integrated optical readout for miniaturization of cantilever-based sensor system, *Applied Physics Letters*, **91** (2007) 103512.

54. www.concentris.ch, accessed May 06, 2010.

55. www.micromotive.de, accessed May 06, 2010.

56. J. Yang, T. Ono, and M. Esashi, Mechanical behavior of ultrathin microcantilever, *Sensors and Actuators A: Physical*, **82** (2000) 102.

57. D. Ramos, M. Arroyo-Hernandez, E. Gil-Santos, H. D. Tong, C. Van Rijn, M. Calleja, and J. Tamayo, Arrays of dual nanomechanical resonators for selective biological detection, *Analytical Chemistry*, **81** (2009) 2274.

58. L. M. Fischer, C. Pedersen, K. Elkjær, N. N. Noeth, S. Dohn, A. Boisen, and M. Tenje, Development of a microfabricated electrochemical-cantilever hybrid platform, *Sensors and Actuators B: Chemical*, **157** (2011) 321–327.

59. G. Genolet, J. Brugger, M. Despont, U. Drechsler, P. Vettiger, N. F. de Rooij, and D. Anselmetti, Soft, entirely photoplastic probes for scanning force microscopy, *Review of Scientific Instruments*, **70** (1999) 2398.

60. M. Hopcroft, T. Kramer, G. Kim, K. Takashima, Y. Higo, D. Moore, and J. Brugger, Micromechanical testing of SU-8 cantilevers, *Fatigue & Fracture of Engineering Materials and Structures*, **28** (2005) 735.

61. M. Calleja, J. Tamayo, A. Johansson, P. Rasmussen, L. M. Lechuga, and A. Boisen, Polymeric cantilever arrays for biosensing applications, *Sensor Letters*, **1** (2003) 20.

62. J. H. T. Ransley, M. Watari, D. Sukumaran, R. A. McKendry, and A. A. Seshia, SU8 bio-chemical sensor microarrays, *Microelectronic Engineering*, **83** (2006) 1621.

63. X. Wang, K. S. Ryu, D. A. Bullen, J. Zou, H. Zhang, C. A. Mirkin, and C. Liu, Scanning probe contact printing, *Langmuir*, **19** (2003) 8951.

64. A. Gaitas and Y. B. Gianchandani, An experimental study of the contact mode AFM scanning capability of polyimide cantilever probes, *Ultramicroscopy*, **106** (2006) 874.

65. T.-J. Yao, X. Yang, and Y.-C. Tai, BrF3 dry release technology for large freestanding parylene microstructures and electrostatic actuators, *Sensors and Actuators A: Physical*, **97–98** (2002) 771.

66. R. Katragadda, Z. Wang, W. Khalid, Y. Li, and Y. Xu, Parylene cantilevers integrated with polycrystalline silicon piezoresistors for surface stress sensing, *Applied Physics Letters*, **91** (2007) 083505.

67. A. Greve, S. Keller, A. L. Vig, A. Kristensen, D. Larsson, K. Yvind, J. M. Hvam, M. Cerruti, A. Majumdar, and A. Boisen, Thermoplastic microcantilevers fabricated by nanoimprint lithography, *Journal of Micromechanics and Microengineering*, **20** (2010) 015009.

68. www.topas.com, 06.05.2010.

69. A. Johansson, G. Blagoi, and A. Boisen, Polymeric cantilever-based biosensors with integrated readout, *Applied Physics Letters*, **89** (2006) 173505.

70. G. Genolet, New photoplastic fabrication techniques and devices based on high aspect ratio photoresist, PhD thesis, Ecole Polytechnique Fédérale de Lausanne, Lausanne, Switzerland, (2001).

71. L. Dellmann, S. Roth, C. Beuret, G. A. Racine, H. Lorenz, M. Despont, P. Renaud, P. Vettiger, and N. F. de Rooij, Fabrication process of high aspect ratio elastic and SU-8 structures for piezoelectric motor applications, *Sensors and Actuators A: Physical*, **70** (1998) 42.

72. S. Schmid, M. Wendlandt, D. Junker, and C. Hierold, Nonconductive polymer microresonators actuated by the Kelvin polarization force, *Applied Physics Letters*, **89** (2006) 163506.

73. C. Martin, A. Llobera, G. Villanueva, A. Voigt, G. Gruetzner, J. Brugger, and F. Perez-Murano, Stress and aging minimization in photoplastic AFM probes, *Microelectronic Engineering*, **86** (2009) 1226.

74. S. Mouaziz, G. Boero, R. S. Popovic, and J. A. B. J. Brugger, Polymer-based cantilevers with integrated electrodes, *Journal of Microelectromechical System*, **15** (2006) 890.

75. I. G. Foulds, R. W. Johnstone, and M. Parameswaran, Polydimethylglutarimide (PMGI) as a sacrificial material for SU-8 surface-micromachining, *Journal of Micromechanics and Microengineering*, **18** (2008) 075011.

76. M. C. Cheng, A. P. Gadre, J. A. Garra, A. J. Nijdam, C. Luo, T. W. Schneider, R. C. White, J. F. Currie, and M. Paranjape, Dry release of polymer structures with anti-sticking layer, *Journal of Vacuum Science & Technology A: Vacuum, Surfaces, and Films*, **22** (2004) 837.

77. D. Haefliger, M. Nordström, P. A. Rasmussen, and A. Boisen, Dry release of all-polymer structures, *Microelectronic Engineering*, **78–79** (2005) 88.

78. M. Tenje, S. Keller, S. Dohn, Z. J. Davis, and A. Boisen, Drift study of SU8 cantilevers in liquid and gaseous environments, *Ultramicroscopy*, **101** (2010) 596.

79. S. Schmid, S. Kühne, and C. Hierold, Influence of air humidity on polymeric microresonators, *Journal of Micromechanics and Microengineering*, **19** (2009) 065018.

80. S. Keller, D. Haefliger, and A. Boisen, Fabrication of thin SU-8 cantilevers: initial bending, release and time stability, *Journal of Micromechanics and Microengineering*, **20** (2010) 045024.

81. H.-F. Ji, K. M. Hansen, Z. Hu, and T. Thundat, Detection of pH variation using modified microcantilever sensors, *Sensors and Actuators B: Chemical*, **72** (2001) 233.

82. M. Calleja, M. Nordström, M. Alvarez, J. Tamayo, L. M. Lechuga, and A. Boisen, Highly sensitive polymer-based cantilever-sensors for DNA detection, *Ultramicroscopy*, **105** (2005) 215.

83. M. Alvarez, L. G. Carrascosa, M. Moreno, A. Calle, A. Zaballos, L. M. Lechuga, C. Martinez-A, and J. Tamayo, Nanomechanics of the formation of DNA self-assembled monolayers and hybridization on microcantilevers, *Langmuir*, **20** (2004) 9663.

84. M. Nordstrom, M. Calleja, and A. Boisen, Polymeric micro channel system for easy sensitisation of micro-cantilevers, in T. Laurell, J. Nilsson, K. Jensen, D. and J. Harrison, eds., *Micro Total Analysis Systems 2004*, Malmo, Sweden, Vol. 2, 2005, p. 55.

85. A. Bietsch, J. Y. Zhang, M. Hegner, H. P. Lang, and C. Gerber, Rapid functionalization of cantilever array sensors by inkjet printing, *Nanotechnology*, **15** (2004) 873.

86. M. Arroyo-Hernandez, J. Tamayo, and J. L. Costa-Kramer, Stress and DNA assembly differences on cantilevers gold coated by resistive and e-beam evaporation techniques, *Langmuir*, **25** (2009) 10633.

87. N. V. Lavrik, C. A. Tipple, M. J. Sepaniak, and P. G. Datskos, Gold nano-structures for transduction of biomolecular interactions into micrometer scale movements, *Biomedical Microdevices*, **3** (2001) 35.

88. A. G. Hansen, M. W. Mortensen, J. E. T. Andersen, J. Ulstrup, A. Kühle, J. Garnæs, and A. Boisen, Stress formation during self-assembly of alkanethiols on differently pre-treated gold surfaces, *Probe Microscopy*, **2** (2001) 139.

89. M. L. Sushko, Nanomechanics of organic/inorganic interfaces: A theoretical insight, *Faraday Discuss*, **143** (2009) 63.

90. J. Lahiri, L. Isaacs, J. Tien, and G. M. Whitesides, A strategy for the generation of surfaces presenting ligands for studies of binding based on an active ester as a common reactive intermediate: A surface plasmon resonance study, *Analytical Chemistry*, **71** (1999) 777.

91. C. A. Savran, S. M. Knudsen, A. D. Ellington, and S. R. Manalis, Micromechanical detection of proteins using aptamer-based receptor molecules, *Analytical Chemistry*, **76** (2004) 3194.

92. W. M. Shu, E. D. Laue, and A. A. Seshia, Investigation of biotin-streptavidin binding interactions using microcantilever sensors, *Biosensors and Bioelectronics*, **22** (2007) 2003.

93. N. Backmann, C. Zahnd, F. Huber, A. Bietsch, A. Plückthun, H.-P. Lang, H.-J. Güntherodt, M. Hegner, and C. Gerber, A label-free immunosensor array using single-chain antibody fragments, *Proceedings of the National Academy of Sciences*, **102** (2005) 14587.

94. W. Tan, Y. Huang, T. Nan, C. Xue, Z. Li, Q. Zhang, and B. Wang, Development of protein A function-alized microcantilever immunosensors for the analyses of small molecules at parts per trillion levels, *Analytical Chemistry*, **82** (2009) 615.

95. J. Xu and H. L. Li, The chemistry of self-assembled long-chain alkanethiol monolayers on gold, *Journal of Colloid and Interface Science*, **176** (1995) 138.

96. T. Wink, S. J. vanZuilen, A. Bult, and W. P. vanBennekom, Self-assembled monolayers for biosensors, *Analyst*, **122** (1997) R43.

97. G. Wu, R. H. Datar, K. M. Hansen, T. Thundat, R. J. Cote, and A. Majumdar, Bioassay of prostate-specific antigen (PSA) using microcantilevers, *Nature Biotechnology*, **19** (2001) 856.

98. P. Dutta, C. A. Tipple, N. V. Lavrik, P. G. Datskos, H. Hofstetter, O. Hofstetter, and M. J. Sepaniak, Enantioselective sensors based on antibody-mediated nanomechanics, *Analytical Chemistry*, **75** (2003) 2342.

99. Y. Arntz, J. D. Seelig, H. P. Lang, J. Zhang, P. Hunziker, J. P. Ramseyer, E. Meyer, M. Hegner, and C. Gerber, Label-free protein assay based on a nanomechanical cantilever array, *Nanotechnology*, **14** (2003) 86.

100. F. P. V. Paoloni, S. Kelling, J. Huang, and S. R. Elliott, Sensor array composed of "clicked" individual microcantilever chips, *Advanced Functional Materials*, **21** (2011) 372.

101. N. Nugaeva, K. Y. Gfeller, N. Backmann, H. P. Lang, M. Düggelin, and M. Hegner, Micromechanical cantilever array sensors for selective fungal immobilization and fast growth detection, *Biosensors and Bioelectronics*, **21** (2005) 849.

102. A. K. Gupta, P. R. Nair, D. Akin, M. R. Ladisch, S. Broyles, M. A. Alam, and R. Bashir, Anomalous resonance in a nanomechanical biosensor, *Proceedings of the National Academy of Sciences*, **103** (2006) 13362.

103. F. M. Battiston, J. P. Ramseyer, H. P. Lang, M. K. Baller, C. Gerber, J. K. Gimzewski, E. Meyer, and H. J. Guntherodt, A chemical sensor based on a microfabricated cantilever array with simultaneous resonance-frequency and bending readout, *Sensors and Actuators B: Chemical*, **77** (2001) 122.

104. Y. Zhang, H.-F. Ji, G. M. Brown, and T. Thundat, Detection of CrO42-using a hydrogel swelling microcantilever sensor, *Analytical Chemistry*, **75** (2003) 4773.

105. M. K. Baller, H. P. Lang, J. Fritz, C. Gerber, J. K. Gimzewski, U. Drechsler, H. Rothuizen et al., A cantilever array-based artificial nose, *Ultramicroscopy*, **82** (2000) 1.

106. M. G. von Muhlen, N. D. Brault, S. M. Knudsen, S. Jiang, and S. R. Manalis, Label-free biomarker sensing in undiluted serum with suspended microchannel resonators, *Analytical Chemistry*, **82** (2010) 1905.

107. X. Yan, H.-F. Ji, and Y. Lvov, Modification of microcantilevers using layer-by-layer nanoassembly film for glucose measurement, *Chemical Physics Letters*, **396** (2004) 34.

108. G. Oliviero, P. Bergese, G. Canavese, M. Chiari, P. Colombi, M. Cretich, F. Damin et al., A biofunctional polymeric coating for microcantilever molecular recognition, *Analytica Chimica Acta*, **630** (2008) 161.

109. C. Grogan, R. Raiteri, G. M. O'Connor, T. J. Glynn, V. Cunningham, M. Kane, M. Charlton, and D. Leech, Characterisation of an antibody coated microcantilever as a potential immuno-based biosensor, *Biosensors & Bioelectronics*, **17** (2002) 201.

110. M. Watari, J. W. Ndieyira, and R. A. McKendry, Chemically programmed nanomechanical motion of multiple cantilever arrays, *Langmuir*, **26** (2010) 4623.

111. X. D. Yan, X. H. K. Xu, and H. F. Ji, Glucose oxidase multilayer modified microcantilevers for glucose measurement, *Analytical Chemistry*, **77** (2005) 6197.

112. F. Huber, M. Hegner, C. Gerber, H. J. Guntherodt, and H. P. Lang, Label free analysis of transcription factors using microcantilever arrays, *Biosensors Bioelectronics*, **21** (2006) 1599.

113. I. Pera and J. Fritz, Sensing lipid bilayer formation and expansion with a microfabricated cantilever array, *Langmuir*, **23** (2007) 1543.

114. T. Braun, N. Backmann, M. Vogtli, A. Bietsch, A. Engel, H. P. Lang, C. Gerber, and M. Hegner, Conformational change of bacteriorhodopsin quantitatively monitored by microcantilever sensors, *Biophysical Journal*, **90** (2006) 2970.

115. T. Braun, M. K. Ghatkesar, N. Backmann, W. Grange, P. Boulanger, L. Letellier, H. P. Lang, A. Bietsch, C. Gerber, and M. Hegner, Quantitative time-resolved measurement of membrane protein-ligand interactions using microcantilever array sensors, *Nature Nanotechnology*, **4** (2009) 179.

116. M. Godin, F. F. Delgado, S. M. Son, W. H. Grover, A. K. Bryan, A. Tzur, P. Jorgensen et al., Using buoyant mass to measure the growth of single cells, *Nature Methods*, **7** (2010) 387.

117. K. M. Hansen, H. F. Ji, G. H. Wu, R. Datar, R. Cote, A. Majumdar, and T. Thundat, Cantilever-based optical deflection assay for discrimination of DNA single-nucleotide mismatches, *Analytical Chemistry*, **73** (2001) 1567.

118. R. Mukhopadhyay, M. Lorentzen, J. Kjems, and F. Besenbacher, Nanomechanical sensing of DNA sequences using piezoresistive cantilevers, *Langmuir*, **21** (2005) 8400.

119. J. C. Stachowiak, M. Yue, K. Castelino, A. Chakraborty, and A. Majumdar, Chemomechanics of surface stresses induced by DNA hybridization, *Langmuir*, **22** (2006) 263.

120. D. Schmid, H. P. Lang, S. Marsch, C. Gerber, and P. Hunziker, Diagnosing disease by nanomechanical olfactory sensors—System design and clinical validation, *European Journal of Nanomedicine*, **1** (2008) 44.

121. H. P. Lang, A. Filippi, A. Tonin, F. Huber, N. Backmann, J. Zhang, and C. Gerber, Towards a modular, versatile and portable sensor system for measurements in gaseous environments based on microcantilevers, *Proceedings of the Eurosensors XXIII Conference*, **1** (2009) 208.

122. J. H. Pei, F. Tian, and T. Thundat, Glucose biosensor based on the microcantilever, *Analytical Chemistry*, **76** (2004) 292.

123. S. Kim, T. Rahman, L. R. Senesac, B. H. Davison, and T. Thundat, Piezoresistive cantilever array sensor for consolidated bioprocess monitoring, *Scanning*, **31** (2009) 204.

124. D. Maraldo, F. U. Garcia, and R. Mutharasan, Method for quantification of a prostate cancer biomarker in urine without sample preparation, *Analytical Chemistry*, **79** (2007) 7683.

125. K. W. Wee, G. Y. Kang, J. Park, J. Y. Kang, D. S. Yoon, J. H. Park, and T. S. Kim, Novel electrical detection of label-free disease marker proteins using piezoresistive self-sensing micro-cantilevers, *Biosensors Bioelectronics*, **20** (2005) 1932.

126. K. Y. Gfeller, N. Nugaeva, and M. Hegner, Micromechanical oscillators as rapid biosensor for the detection of active growth of *Escherichia coli*, *Biosensors Bioelectronics*, **21** (2005) 528.

127. J. Mertens, E. Finot, M. H. Nadal, V. Eyraud, O. Heintz, and E. Bourillot, Detection of gas trace of hydrofluoric acid using microcantilever, *Sensors and Actuators B: Chemical*, **99** (2004) 58.

128. T. L. Porter, T. L. Vail, M. P. Eastman, R. Stewart, J. Reed, R. Venedam, and W. Delinger, A solid-state sensor platform for the detection of hydrogen cyanide gas, *Sensors and Actuator B: Chemical*, **123** (2007) 313.

129. K. Liu and H. F. Ji, Detection of Pb2+ using a hydrogel swelling microcantilever sensor, *Analytical Sciences*, **20** (2004) 9.

130. S. Velanki, S. Kelly, T. Thundat, D. A. Blake, and H. F. Ji, Detection of Cd(II) using antibody-modified microcantilever sensors, *Ultramicroscopy*, **107** (2007) 1123.

131. X. H. Xu, T. G. Thundat, G. M. Brown, and H. F. Ji, Detection of Hg2+ using microcantilever sensors, *Analytical Chemistry*, **74** (2002) 3611.

132. L. A. Pinnaduwage, A. Wig, D. L. Hedden, A. Gehl, D. Yi, T. Thundat, and R. T. Lareau, Detection of trinitrotoluene via deflagration on a microcantilever, *Journal of Applied Physics*, **95** (2004) 5871.

133. L. A. Pinnaduwage, A. C. Gehl, S. L. Allman, A. Johansson, and A. Boisen, Miniature sensor suitable for electronic nose applications, *Review of Scientific Instruments*, **78** (2007) 055101.

134. B. Dhayal, W. A. Henne, D. D. Doorneweerd, R. G. Reifenberger, and P. S. Low, Detection of *Bacillus subtilis* spores using peptide-functionalized cantilever arrays, *Journal of the American Chemical Society*, **128** (2006) 3716.

135. S. Velanki and H. F. Ji, Detection of feline coronavirus using microcantilever sensors, *Measurement Science and Technology*, **17** (2006) 2964.

136. S. Xu and R. Mutharasan, Rapid and sensitive detection of *Giardia lamblia* using a piezoelectric cantilever biosensor in finished and source waters, *Environmental Science and Technology*, **44** (2010) 1736.

6

Nanostructured Surface Plasmon Resonance Sensors

Mikael Svedendahl, Si Chen, and Mikael Käll

CONTENTS

6.1 Introduction

Early in the 1980s, it was demonstrated that SPRs in thin metal films, so-called propagating plasmons, can be utilized for effective label-free sensing [1,2]. Since then, several companies have commercialized various SPR sensing schemes, and instruments can now be found in many different settings, such as within pharmacology, the food industry, and in medical diagnostics [3–11]. Surface plasmons in metal nanostructures, so-called localized surface plasmon resonances (LSPRs), have been intensively researched over the past decades, and a number of applications have been explored, including enhanced photothermal therapy and solar harvesting, improved imaging techniques, and a number of surface-enhanced molecular spectroscopy methods [12–19]. However, despite the success of the thin film plasmon sensing, the potential label-free LSPR sensing was not investigated until the late 1990s [16,20], although colloidal nanoparticles had been utilized indirectly in various sensing assays based on colorimetric changes induced by aggregation and surface-enhanced Raman scattering (SERS) already since the 1970s [15,21]. During the past decade, the development of refractometric LSPR sensing has focused on nanoparticles immobilized on substrates [22–26]. This trend, which is based on the rapid development of nanofabrication technologies, enables microfluidic solutions to surface functionalization and analyte delivery in analogy with traditional SPR. More important, however, is that the surface-bound nanoparticles can be made in a huge variety of shapes, sizes, compositions, and internal arrangement, which allows for thorough investigation and optimization of both fundamental mechanisms and applications of LSPR sensing.

This chapter is organized as follows: first, the basics of plasmonic refractive index sensing are introduced based on the key properties of plasmons in thin films and nanoparticles. Next, the main biosensing signal transduction parameters are discussed, including how these can be enhanced to achieve single molecule sensitivity. Finally, some current trends within the community are presented as a brief outlook for the coming years of development.

6.2 Plasmonic Sensing: A Brief Background

6.2.1 Propagating Surface Plasmons in Thin Films

One of the main advantages of LSPR sensing compared to traditional SPR thin film sensing is that the former can be performed using a simple UV-VIS spectrophotometer, a piece of equipment available in almost every chemistry laboratory. The reason for the relatively complicated excitation geometries required in SPR sensing is that both energy and momentum conservation has to be fulfilled in the interaction between light and propagating plasmons. Light directly incident from the ambient will not couple energy to the plasmon because the momentum, proportional to the wave number k, of the plasmon is considerably larger than the momentum of light in the ambient dielectric. A nanoparticle, on the other hand, has no translational symmetry, and the LSPR, therefore, has no well-defined k. This means that LSPR excitation, in general, only requires the right photon energy and polarization but no particular magnitude or direction of the incident light momentum.

The relationship between the energy and the momenta of surface plasmons in thin films can be approximated by the dispersion relation for plasmons running along a boundary between two semi-infinite spaces, the metal, and a dielectric ambient. By applying the continuity condition of Maxwell's equations for transverse magnetic waves, one can derive an implicit dispersion relation for the surface plasmon according to the following equation:

$$k(\omega) = \frac{\omega}{c} \sqrt{\frac{\varepsilon_{metal}\varepsilon_d}{\varepsilon_{metal} + \varepsilon_d}} \tag{6.1}$$

Here, ε_{metal} is the frequency dependent and complex dielectric function of the metal, $\varepsilon_d = n_d^2$ is the dielectric constant of the ambient with refractive index n_d, and c is the speed of light in vacuum. By inserting an analytic function or tabulated values for the metal dielectric function, one can calculate the dispersion relation $\omega(k)$. This is done in Figure 6.1 for the case of surface plasmons propagating along a gold–water interface. The figure also shows the dispersion (the light line) of light propagating parallel to the metal interface in water. Since the two curves never cross, it is impossible to convert photons propagating in water to surface plasmons and, at the same time, conserve both energy and momentum. However, if the photons instead propagate in glass, their momenta would increase in accordance with the refractive index and coupling to the plasmon would become possible. This is the idea behind the so-called Kretschmann configuration [27], in which a thin metal film is deposited on a glass prism, and plasmons can be excited at the metal-ambient interface via illumination through the prism.

Figure 6.2A and B illustrates the Kretschmann configuration with the most common excitation and detection geometries of surface plasmons in thin metal films [27]. Incident p-polarized light is coupled via a high refractive index prism to the metal film using an incidence angle above the critical angle. The propagation constant of the light at the prism/thin film interface is thus dependent on the incident angle, which, using Equation 6.1, yields the following condition for surface plasmon polariton (SPP) excitation:

$$k_{inc.} = \frac{\omega}{c} n_{prism} \sin\theta = \frac{\omega}{c} \sqrt{\frac{\varepsilon_{metal}\varepsilon_d}{\varepsilon_{metal} + \varepsilon_d}} = k_{SPP} \tag{6.2}$$

In Figure 6.2B, white light incident via a prism toward the thin metal film is used. The wavelength of the radiation that couples most effectively to the SPP will vary depending on the (fixed) incident angle and

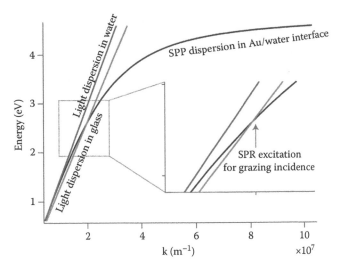

FIGURE 6.1 Dispersion of surface plasmons and light. For any given energy, the surface plasmon at a gold–water interface possess larger momentum than light propagating parallel to the surface in water ($n = 1.33$), but excitation is possible if one utilizes a glass ($n = 1.5$) prism in order to couple light to the plasmon. The figure illustrates this by visualizing that the light-line in the water never crosses the plasmon dispersion curve, while the light line for grazing incidence in glass does.

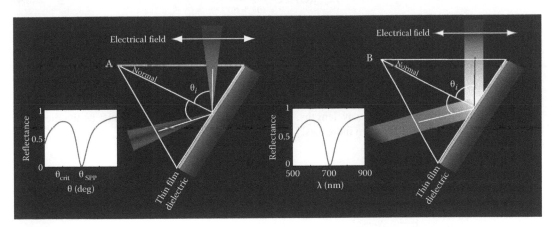

FIGURE 6.2 Detection methods of surface plasmon resonances in thin films. A and B Illustrates angular and wavelength detection using Kretschmann geometry, respectively, and θ_i denotes the light angle of incidence.

the refractive index of the ambient. This is analyzed by studying the decreased reflectivity spectroscopically. The reflectance spectrum shows a dip with a minimum at the SPR-wavelength, which can then be used for tracking the refractive index changes of the ambient.

In Figure 6.2A, the wavelength of the incident light is instead kept constant, while illumination now takes place from multiple incident angles. This yields a range of propagation constants, k, along the metal film. With the light focused on the film, the reflected light is diverging, and an intensity minimum is found at the angle where the energy and momentum coupling conditions are met simultaneously.

An often used alternative to prism coupling in the Kretschmann geometry is to use so-called grating coupling. It is less utilized than the prism coupling methods even though the two provide similar detection limits in the end [28], but it also has some advantages as cheap light sources can be used for excitation [29]. Both spectroscopic and angular detection methods can be used for probing refractive index changes [29–31], just as in the Kretschmann configuration mentioned earlier.

Naturally, monitoring the intensity at a fixed wavelength and incident angle is possible in both grating- and prism-based SPR sensors, although this implementation suffers from a smaller dynamic range as the intensity fluctuations only can be monitored within the width of the resonance.

By exchanging the detector with a CCD camera, this detection method can be used for detection of multiple analytes simultaneously using imaging SPR [32–34]. To overcome the low dynamic range, it is also possible to take images with varying incidence angles or wavelengths, thereby creating a stack of images from which the resonance position can be determined [35,36]. For a much more extensive treatment of thin film SPR sensing, the reader should consult Ref. [37].

6.2.2 Localized Surface Plasmons

The optical response of subwavelength particles is most conveniently described in terms of their dipole polarizability obtained within the quasi-static approximation. Refractometric sensing with spherical noble metal nanoparticles can then be discussed by referring to the Clausius-Mossotti polarizability:

$$\alpha = 4\pi a^3 \frac{\varepsilon_{metal}(\omega) - \varepsilon_d}{\varepsilon_{metal}(\omega) + 2\varepsilon_d} = 4\pi a^3 \left[\frac{\omega_0}{\omega_0^2 - \omega^2 - i\omega\gamma} - \frac{\omega^2 + i\omega\gamma}{\omega_0^2 - \omega^2 - i\omega\gamma} \frac{1 - \varepsilon_d}{1 + 2\varepsilon_d} \right]. \tag{6.3}$$

Here, a is the radius of the sphere, and the dielectric function of the metal has been approximated by the Drude model for free electrons:

$$\varepsilon_{metal}(\omega) = 1 - \frac{\omega_p}{\omega^2 + i\omega\gamma}. \tag{6.4}$$

Here ω_p is the plasma frequency, and γ is a damping parameter. The LSPR frequency is given by $\omega_0 = \omega_p / \sqrt{1 + 2\varepsilon_d}$, which means that the resonance wavelength will red-shift if the refractive index of the surrounding increases. Based on Equation 6.3 and by noting that resonance is achieved when the denominator in the polarizability function is minimized, i.e., when $\text{Re}[\varepsilon_{metal} + 2\varepsilon_d] = 0$, one can also illustrate the refractive index sensitivity as shown in Figure 6.3. Since the dielectric function in the metal becomes more negative for longer wavelengths, the resonance condition will red-shift as the dielectric constant of the surrounding medium increases.

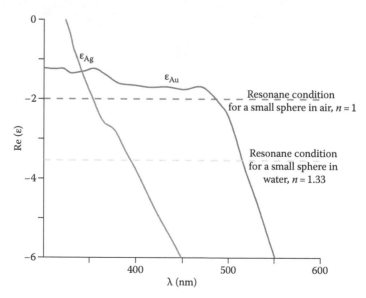

FIGURE 6.3 The real part of the experimentally measured dielectric functions of gold and silver. (From Draine, B.T. and Flatau, P.J., *J. Opt. Soc. Am. A-Opt. Image Sci. Vis.*, 11(4), 1491, 1994; Devoe, H., *J. Chem. Phys.*, 41(2), 393, 1964.) The horizontal lines show the resonance conditions for small spheres in air and water, see Equation 6.3.

The dominant term on the right hand side of Equation 6.3 describes a Lorentzian line shape. This implies that the conduction electrons collectively respond as a damped harmonic oscillator. The resonating surface charges driven by the external field produce an induced field inside and outside of the particle that acts to restore charge neutrality. The restoring force decreases, i.e., the resonance redshifts, if the field lines transverse a medium with a larger dielectric constant. Refractometric plasmonic sensing is thus ultimately a consequence of dielectric screening of surface charges. How sensitive a plasmon is to the environment depends on the ease of electron polarization and the strength of the restoring force.

Similar to small nanoparticles, individual molecules are described by their dipole polarizability. A model that couples the polarizabilities of the nanoparticle and the molecule(s) is, therefore, an appropriate starting point for discussing nanoplasmonic biosensing. In the so-called coupled dipole approximation [38,39], one approximates the nanoparticle and the molecule as point dipoles that are mutually coupled through their induced fields. Solving the coupled dipole equations yields new modified point dipole polarizabilities with modified resonance conditions. In particular, it can be shown that the LSPR frequency of a spherical nanoparticle in the presence of a molecule changes as follows:

$$\tilde{\omega}_0^2 = \omega_0^2 \left(1 - \left(\frac{a_{molecule}}{a_{NP}} \right)^3 \frac{n^2 - n_d^2}{n^2 + 2n_d^2} \left(a_{NP}^3 A \right)^2 \right), \tag{6.5}$$

where

$\alpha_{molecule} = 4\pi a_{molecule}^3 \left(n^2 - n_d^2 \right)/\left(n^2 + 2n_d^2 \right)$ is the polarizability of the molecule

$a_{molecule}$ and a_{NP} are the radii of the molecule and the nanoparticle, respectively

We, here, assume that the molecule is spherical and can be characterized by an effective refractive index, n, and that the whole molecule-particle system is immersed in a medium with refractive index, n_d. The interaction between the dipoles is described by the interaction matrix A. The following three factors control the red shift:

- The volume ratio between the molecule and the particle
- The refractive index contrast between the molecule and the surrounding medium
- The coupling factor—directly related to the local intensity enhancement factor at the location of the molecule

The coupling factor A varies as $1/d^3$, where d is the distance between the two point dipoles, in the simplest approximation. When describing the effect of molecular adsorption on the plasmon resonance, it is thus critical to take the extension of the resonant fields associated with the plasmon into account. In particular, large responses are expected near local "hot spots" with high field enhancement factors, such as at sharp edges or in crevices between particles.

Figure 6.4 illustrates the most common optical setups for LSPR sensing. As mentioned earlier, the excitation conditions for LSPR sensing are much more relaxed than for sensing using propagating plasmons. Dark-field spectroscopy is a popular technique for measuring the response of single nanoparticles [40–43]. As seen in Figure 6.4A, light exciting the nanostructures is incident at high angles, while the scattered light is collected using low numerical aperture optics that is "blind" to the directly transmitted light. Dark-field imaging is also suitable for simultaneous studies of large numbers of isolated particles [41,44], which improves measurement statistics and potentially enable multiplexed measurements of several analytes. The best signal-to-noise ratios in LSPR sensing using high density samples is instead obtained using extinction measurements [45–47]. It can also be advantageous to record specular reflection, since a dense layer of nanostructures behave similar to a homogeneous "metamaterial" surface. In the case of a low interparticle coupling and low reflectance from the substrate supporting the particle layer, the specular reflectance spectrum will show characteristics similar to a single particle scattering spectrum [48].

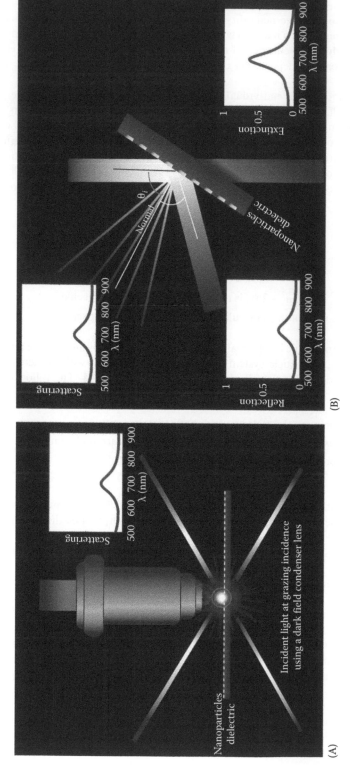

FIGURE 6.4 Illustration of LSPR detection setups. Illustration of LSPR detection setups. (A) Dark-field spectroscopy in an optical microscope. (B) Macroscopic measurements of a layer of supported nanoparticles using extinction, specular reflection, and diffuse scattering.

6.3 An Introduction to Plasmonic Refractive Index Sensing

6.3.1 Bulk Sensing Using Propagating and Localized Plasmons

The simplest form of plasmonic refractometric sensing is to measure the shift of the resonance position as a function of the refractive index of the whole surrounding medium. The measured "bulk" refractive index (RI) sensitivity is also one of the key components that determine the plasmonic signal upon molecular adsorption close to the metal surface.

In order to evaluate the performance of the SPR and LSPR sensing platforms, represented by a 50 nm Au film and a layer of ~100 nm Au nanodisks, respectively, Svedendahl et al. performed a comparative study using essentially identical excitation and detection systems. They also used the same resonance wavelength around 700 nm, in both methodologies, since resonance position is one of the key factors determining the RI sensitivity of plasmonic systems [48]. As expected, the bulk refractive index sensitivity was found to be much higher for the SPR, which exhibited a resonance shift of 3300 nm/RIU, compared with only 180 nm/RIU of the LSPR (Figure 6.5). This difference arises intrinsically from the two plasmonic resonance conditions and can be obtained analyzed analytically by differentiating the resonance conditions with respect to the surrounding refractive index, n [37,49]:

$$\frac{\delta\lambda_{SPR}}{\partial n} = -\frac{2\varepsilon'^2}{n^3\frac{\partial\varepsilon'}{\partial\lambda}}, \quad \frac{\partial\lambda_{LSPR}}{\partial n} = \frac{2\varepsilon'^2}{n\frac{\partial\varepsilon'}{\partial\lambda}} \tag{6.6}$$

where ε' is the real part of the metal dielectric function. However, as stated earlier, the bulk sensitivity is only one parameter determining how useful a plasmonic transducer is. A more appropriate and common measure is the so-called figure of merit defined as $FOM = \frac{\delta\lambda_{SPR}}{\delta n}/FWHM$. Here, $FWHM$ is the full width at half maximum of the resonance [50]. The FOM describes not only the bulk sensitivity but also the *visibility* of a plasmonic resonance, related to the $FWHM$. The latter can be obtained in a similar fashion from the resonance conditions according to [51].

$$FWHM_{SPR} = \frac{4\varepsilon''}{\left|\frac{\delta\varepsilon'}{\partial\lambda}\right|}, \quad FWHM_{LSPR} = \frac{2\varepsilon''}{\left|\frac{\partial\varepsilon'}{\partial\lambda}\right|}, \tag{6.7}$$

for SPR and LSPR, respectively, where ε'' is the imaginary part of the metals dielectric function. The theoretical FOM can then be obtained by combining Equations 6.6 and 6.7, which yields:

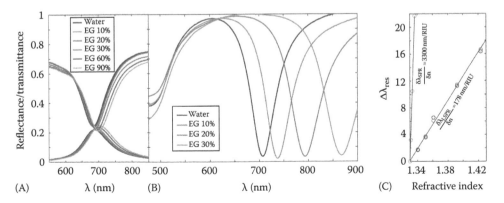

FIGURE 6.5 Bulk refractive index sensitivity measurements with LSPR of Au nanodisk arrays and propagating plasmons in thin Au film. (A) Transmission and reflection spectra for Au nanodisks in contact with water/glycol mixtures growing refractive index, n. (B) Equivalent data for propagating plasmons on a 50 nm thin film. (C) The resonance shifts extracted from (A) and (B). (Reprinted from Svedendahl, M. et al., *Nano Lett.*, 9(12), 4428, 2009. With permission.)

$$FOM_{SPR} = \frac{\varepsilon'}{2n^3\varepsilon''}, \quad FOM_{LSPR} = \frac{|\varepsilon'|}{n\varepsilon''}. \tag{6.8}$$

In the Au film versus Au nanodisk case, the theoretical *FOMs* could be determined to $FOM_{SPR} = 55$ and $FOM_{LSPR} = 12$, respectively, at a resonance position of 700 nm, whereas the experimentally measured values were $FOM_{SPR} = 52$ and $FOM_{LSPR} = 2$. The discrepancy between the theoretical and experimental LSPR *FOMs*, comes from two main assumptions. First, in the theoretical model, one assumes no plasmon radiation damping, i.e., no scattering. This is a poor approximation for the rather large nanodisks investigated here. The second assumption is that the particles are immersed in a homogeneous medium, while they are in reality supported by a high RI substrate. The second implication of this substrate effect will be discussed in the following section.

6.3.2 Substrate Effects

As mentioned earlier, the immobilization of nanoparticles on solid substrates has the immediate drawback that the surface area in contact with the substrate is inaccessible for molecular binding. Furthermore, the substrate also breaks the symmetry, thereby changing the charge and field distribution in and around the particle. This effect generally leads to increased field enhancements close to the high index substrate, thereby decreasing the overlap between the field and the sensing medium. Two of the more common LSPR transducer systems used for biosensing can serve as examples that illustrate these phenomena: nanoholes in thin gold films and gold nanodisks on glass substrates. Brian et al [52] investigated short-range ordered nanohole arrays in thin gold films, a system that support both propagating and localized plasmons [53]. Simulations indicated that the detrimental asymmetry effect mentioned earlier could be decreased substantially by fabricating the nanoarrays on substrates that were index matched to the aqueous sensing medium. Consequently, teflon ($n \approx 1.33$) substrates were used in a comparative study with the glass ($n \approx 1.51$) and TiO₂ ($n \approx 2.4$) substrate. As n_{substr} increased (decreased), the resonance of the nanohole array red (blue) shifted compared with the glass standard. As a consequence of Equation 6.6, one may, therefore, anticipate that the teflon sample would show a lower sensitivity than the TiO₂- and glass-based samples. On the contrary, the opposite trend was found as seen in Figure 6.6. The effective improvement of the sensitivity using the low index substrate was found to be as high as a factor 1.6 and 3, compared with the SiO₂ and TiO₂ samples, respectively. This effect is thus due to the fact that plasmonic field distributions vary strongly with on n_{substr}. As shown in the inset of Figure 6.6A, only 5% of the light intensity overlap with the ambient for a TiO₂ sample compared with 50% for the index matched teflon case.

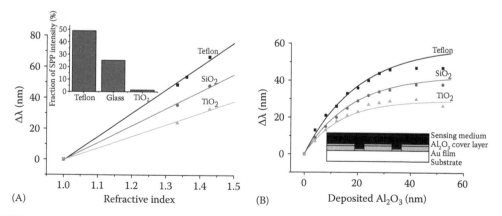

FIGURE 6.6 Effect of the substrate refractive index on nanohole array sensing performance. A symmetric refractive index surrounding the nanohole arrays increases both the bulk (A) and thin film (B) RI sensitivity. Inset (A): fraction of optical intensity at SP resonance that overlaps within the ambient sensing medium. Inset (B): schematic illustration of nanohole array structure. (Reproduced from Brian, B. et al., *Opt. Express*, 17(3), 2015, 2009. With permission.)

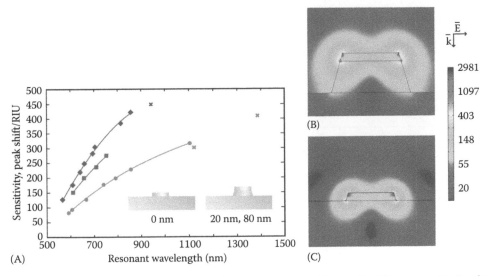

FIGURE 6.7 **(See color insert.)** Reduced substrate effect through nanoparticle elevation. The measured bulk refractive index sensitivities of nanodisks on glass pillars with different height (A) shows the advantage of separating the nanoparticles from the high index substrate. Au nanodisks directly on glass (red) showed smaller sensitivity than nanodisks supported on 20 nm (blue) and 80 nm (green) SiO_2 pillars. The sensitivity enhancement can be understood from the differences in the field distributions between traditional (C) and elevated nanodisks (B). (Reproduced from Dmitriev, A. et al., *Nano Lett.*, 8(11), 3893, 2008. With permission.)

A more direct approach to suppress the influence of the substrate was described by Dmitriev et al., who simply elevated the nanostructures above the substrate by fabricating them on dielectric pillars [54]. The resulting field distributions could be visualized by numerical simulations, as seen in Figure 6.7B and C. With the nanodisk elevated, a larger portion of the plasmonic dipole fields are exposed to the surrounding medium compared with when the disks are attached to the substrate. Bulk refractive index sensitivity measurements for nanodisks on dielectric pillars with different heights revealed that the bulk sensitivity could be doubled with this technique, with the best sensitivity measured with 80 nm tall pillars. Recently, a similar approach based on chemical etching was shown to provide similar advantages also in real biosensing experiments, exemplified by label-free detection of DNA hybridization [55].

6.3.3 Molecular Sensing

As mentioned in the introduction of this section, the bulk sensitivity and the *FOM* are significant parameters for plasmonic sensors. However, as will be evident from this section, it is also crucial to confine this sensitivity to the actual location of the object to be detected. In the case of molecular detection and biosensing, this typically implies that sensitivity should be maximized close to the metal surface, i.e., where molecular attachment takes place.

In a typical sensing experiment, selective adsorption of target molecules to the surface is achieved through functionalization using capture molecules such as antibodies or aptamers.

6.3.3.1 Counting Molecules with Resonance Shifts

The amount of adsorbed biomolecules can be traced by tracking the LSPR wavelength, λ_{res}. The measured resonance shift $\Delta\lambda_{res}$ can be used to estimate the effective refractive index change from the bulk sensitivity through $\Delta\lambda_{res} = \Delta n_{eff}\delta\lambda_{res}/\delta n$. Here, Δn_{eff} is a measure of the effective refractive index change normalized to the approximately exponential decay of the plasmon field according to [24,26]:

$$\Delta n_{eff} = \frac{2}{l_d}\int_0^\infty \left(n(z) - n_{medium}\right)e^{-\frac{2d}{l_d}}\,dz. \tag{6.9}$$

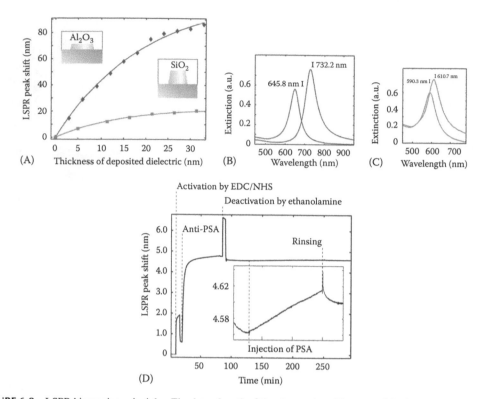

FIGURE 6.8 LSPR biosensing principles. The decay length of the plasmonic gold nanoparticles is probed by deposition of thin dielectric layers. The decay length was found to be $l_d = 38$ nm for the high aspect ratio nanodisks and $l_d = 28$ nm for low aspect ratio disks (A–C). From Equations 6.9 and 6.10, one may estimate the number of prostate-specific antigen protein molecules that correspond to the peak shift in (D). (Reproduced from Chen, S. et al., *Nanotechnology*, 20(43), 434015, 2009. With permission.)

Here, $n(z)$ is the refractive index at distance z from the metal surface; n_{medium} is the ambient refractive index, d is the adsorbed layer thickness, and l_d is the field decay length. The decay length can be estimated experimentally by depositing thin dielectric layers while monitoring the LSPR change. As an example, Figure 6.8A through C shows deposition of alumina and silicon oxide on nanodisks of different diameters and heights.

The calculated refractive index change can be converted to the mass of the protein layer by the following simple formula:

$$\Gamma = \frac{d\Delta n_{layer}}{\delta n / \delta c},$$ (6.10)

where

Γ is the surface mass density

$\Delta n_{layer} = n_{layer} - n_{medium}$

$\delta n / \delta c = 0.182$ cm^3/g is a constant that relates refractive index and mass concentration form most proteins [56]

By utilizing the particles refractive index sensitivity together with the LSPR decay length and the physical dimension of the bio molecules, the amount of proteins adsorbed on the surface of nanoparticles can thus be estimated from Equations 6.9 and 6.10. By using simple algorithms for the peak-shift determination, samples with relatively high density of nanodisks and large sampling area, Chen et al., showed that it is possible to detect average peak-shifts that correspond to less than a single protein molecule per nanoparticle [57].

6.3.3.2 Spatial Sensitivity Confinement

We previously established that the bulk sensitivity is about one order of magnitude larger for thin film plasmons compared with localized plasmons. However, this is also true for l_d, and these two parameters need to be combined in order to effectively compare the two methodologies for mono- and submonolayer biosensing purposes [45,58,59]. Considering the vast differences in *FOM* discussed earlier, one might anticipate that classical SPR is the better alternative for sensing in general. To test whether this is actually the case when it comes to biosensing, we studied biorecognition reactions between streptavidin (SA) and biotin using flat film and nanodisk plasmons [48]. Figure 6.9 shows the corresponding SPR and LSPR shifts versus time during functionalization with biotinylated bovine serum albumin (bBSA), an intermediate step of BSA adsorption that passivates unfunctionalized areas, and the final introduction of SA. Although the LSPR measurements seem to give a smaller overall response, in particular during the functionalization step, the shifts induced by biorecognition, that is, SA binding to bBSA, are surprisingly similar with around 0.59 ± 0.04 nm shifts for both methodologies. These results demonstrate that the *FOM* is not the only factor determining plasmonic sensing capabilities. The decay length is just as important. Thus, in the case of molecular adsorption near the metal surface, the better confinement of the LSPR induced fields leads to a much more efficient use of the bulk refractive index sensitivity. One might also point out that the gold film used in the comparison had a more or less optimum thickness (50 nm), while both the bulk sensitivity, the FWHM, and the field distribution of the LSPR probably could be improved quite substantially by fine-tuning the size, shape, and positioning of the nanoparticles [60–66].

6.3.3.3 Ensemble and Single Particle Assays

By decreasing the number of nanoparticles in a LSPR assay, the sensor can be made extremely small. However, as the number of particles decreases, it becomes necessary to measure in an optical microscope, as illustrated in Figure 6.4A, rather than in a classical macroscopic spectrophotometer. Microscopy measurements can be made in both extinction and dark-field scattering mode, and the relative performance of the two techniques depend on the optical density of the sample and the collection efficiency of the objective [47]. Dahlin et al. recently showed that microextinction measurement from an area of $2 \times 10 \, \mu m^2$ ($\sim$240 nanoparticles) with good SNR [47]. When the sampling area was increased 10 times, the same SNR as in macroextinction experiment was achieved [47]. However, when the particle surface coverage is reduced to the single particle regime, the dark-field scattering methodology is generally favorable.

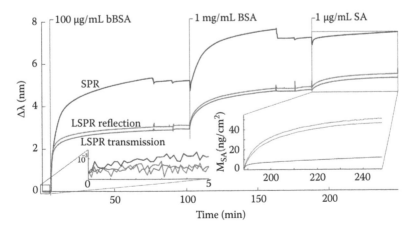

FIGURE 6.9 Biosensing using localized and propagating plasmons. The temporal evolution of the resonance shifts for localized plasmons in Au nanodisks and propagating plasmons on a 50 nm planar gold film. The vertical lines specify the injection of different biomolecules into the sample chamber. The insets show the fluctuations of the resonance wavelength for blank samples (left) and the adsorbed mass after SA injection (right). (Reprinted with permission from Svedendahl, M., Chen, S., Dmitriev, A., and Kället, M., Refractometric sensing using propagating versus localized surface plasmons: A direct comparison, *Nano Letters*, 9(12), 4428–4433. Copyright 2009 American Chemical Society.)

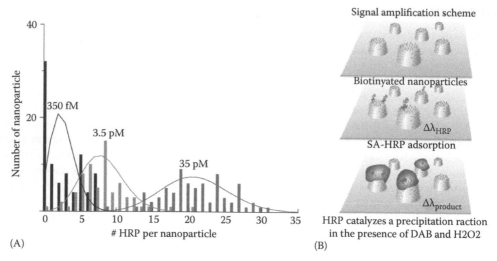

FIGURE 6.10 Illustration of LSPR enhanced ELISA with single molecule sensitivity. Histogram of estimated number of HRP molecules per particle for three concentrations (A) using a precipitation reaction amplifying the refractive index footprint of individual binding events (B). (Reproduced from Chen, S. et al., *Nano Lett.*, 11(4), 1826, 2011. With permission.)

Single or close to single molecule sensitivity is required to detect protein concentrations in the femtomolar range within acceptable reaction times using surface supported nanostructures [41,67]. Plasmonic detection of single biomacromolecules based on structures with ultra-high field-enhancement has been reported using SERS [14] as well as LSPR sensing [43], but the question is whether detection using "hot spots" is robust enough to constitute an analytical alternative to methods based on target amplification, such as polymerase chain reaction, or signal amplification, such as enzyme-liked immunosorbent assay [68]. The basic problem is that the single molecule to be detected must somehow be guided so that it binds at the hotspot only. We recently presented an approach that combines enzymatic amplification with LSPR refractive index sensing using gold nanodisks, a structure that can be expected to exhibit rather uniform enhancement properties [45]. The aim of the study was to develop a simple colorimetric end-point assay with robust single molecule detection capabilities, thus trading binding kinetics for sensitivity. The study was based on dark-field measurements using a variable band-pass liquid crystal tunable filter inserted into the illumination path of the optical microscope [41]. Using stacks of images obtained at varying wavelengths, spectral information from up to a hundred individual nanodisks could be obtained in parallel. As illustrated in Figure 6.10B, Chen et al. studied the interaction between biotin and avidin conjugated to the enzyme horse radish peroxidase, which catalyzes the polymerization of a precipitate in the final step of the assay. These precipitates have a large refractive index, thus enhancing the refractometric footprint of individual binding events by up to two orders-of-magnitude. When the peak-shifts measured for single particles were converted to number of molecules, as in Figure 6.10A, the assay was found to have a sensitivity at or close to the single molecule limit, a conclusion supported by both the statistical signal distribution and computer simulations of the molecular diffusion and binding processes [45]. Although this study is only a model of a real sandwich assay, the method is likely transferrable to more complex sample solutions and may pave the way for highly sensitive diagnostic applications of nanoplasmonic sensing.

6.4 Some Recent Research Developments in Nanoplasmonic Sensing

6.4.1 Improving FOM Using Fano Resonance

The term "Fano resonance" stems from work done by Ugo Fano [69,70] describing the interaction or interference between a discrete resonance and a broad continuum of resonances. The primary effect of this interaction is that the discrete resonance acquires an asymmetric line-shape. Although Fano resonances are abundant in physics, nanoparticle LSP spectra typically consists of symmetric

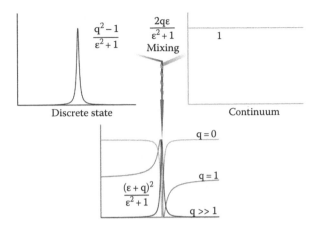

FIGURE 6.11 Illustration of a Fano resonance. Interference (mixing) between a (Lorentzian) discrete state, left, and a continuum, right, may result in various line-shapes depending on the sign and magnitude of the asymmetry parameter q, which is determined by the strength and phase of the interaction.

Lorentzian peaks, although strong plasmon damping due to interband transitions can lead to Fano line-shapes even for simple metal nanostructures in special cases [71]. However, from a nanoplasmonic sensing perspective, the Fano interaction is interesting primarily because it can lead to narrow spectral features with improved *FOM*. The basic idea is to utilize the interaction between a so-called "dark" mode (a discrete dipole-forbidden LSPR) and a "bright" dipolar plasmon. Dark modes are per definition non-radiative, and, therefore, relatively sharp, while "bright" modes suffer from strong radiative damping/broadening and can, therefore, act as the continuum in the Fano interaction. If the two overlap in energy, the Fano interference can lead to a spectral profile with narrow asymmetric peaks or dips with high *FOM*.

Nanoplasmonic Fano interference has been described for a number of different structures, including ordered nanoparticle arrays, asymmetric ring-disk resonators, and nanoparticle multimers. The ring-disk resonator system was investigated by Hao et al. [64,65]. It was shown that the interaction between the elementary ring and disk resonances gives rise to bonding and antibonding modes that are highly sensitive to the physical separation between the resonators and to molecular adsorption in the gaps between them. Hao et al. reported bulk refractive index sensitivities above 1000 nm/RIU and *FOMs* in the range 7–8. Similar sensing characteristics have recently also been reported for certain dark modes in nanocrosses [72]. Fano interferences have also been intensively studied in multimers, such as septa- or heptamers of circular or spherical nanoparticles [73–75]. Such systems have been shown to have high bulk refractive sensitivity and an experimental *FOMs* in the range 5–10 [75,76].

When a narrow dark mode couples to and absorbs light from a broad bright scattering continuum, the resulting scattering, reflectance, or extinction spectrum may exhibit a sharp dip (similar to the $q = 0$ case in Figure 6.11). This phenomenon has been dubbed "plasmon-induced-transparency" in analogy to the so-called electromagnetically induced transparency phenomenon [77,78]. One example is a study by Liu et al. who made structures based on a three-bar system cut out in a planar thin gold film [79]. The "plasmon-induced-transparency" phenomenon was observed in reflectance, with sharp features originating from a narrow quadrupolar dark resonance. These structures showed a refractive index sensitivity of ~600 nm/RIU and a *FOM* around five.

6.4.2 Plasmonic Metamaterials for Sensing

Optical metamaterials are man-made structures designed to exhibit effective optical constants not found in natural materials, the prototypical example being a negative refractive index [80]. In order to exhibit an effective optical constant, the material in principle has to be built up by structures much smaller than the wavelength and the distance between these elementary structures—the "meta atoms"—also has to be much smaller than the design wavelength.

One type of plasmonic metamaterial proposed for sensing applications is the so-called "perfect absorber" structure. A perfect absorber minimizes reflectance and eliminates transmittance through matching of the effective impedance of the metamaterial to the surrounding at a particular design wavelength. "Perfect absorption" might be interesting for sensing in the context of an alternative Figure of Merit, FOM^*, proposed by Becker et al [81]:

$$FOM^* = max \left| \frac{\delta I(\lambda)}{\delta n} \frac{1}{I_0(\lambda)} \right|. \tag{6.11}$$

Here, I_0 is the reference intensity and δI the intensity change upon the refractive index change, δn, of the surroundings. Liu et al. recently reported on a perfect absorber for the infrared consisting of a three layer structure in the form of a gold disk array and a relatively thick gold film separated by a thin MgF_2 spacer [82]. Current distribution plots revealed opposite current directions in the nanodisks and in the metal. This is characteristic for a dark mode, resulting in a large field confinement within the MgF_2 spacer layer. Although the meta-material showed a relatively modest bulk sensitivity of 400 nm/RIU, the potential advantage lies in the possibility to measure the intensity at the nullified reflectance wavelength, which yields a relatively large intensity contrast per RIU. Such a detection scheme could reduce device costs as no expensive spectrometers are used. The authors report a FOM^* of 87, which is about four times larger than for the gold nanorods investigated by Becker et al. [81]

Meta-materials can also be designed to induce so-called super-chiral fields [83], which might be used as ultra-sensitive probes for right- and left-handed versions of molecules, so-called enantiomers. For example, one enantiomer may have useful characteristics in humans, whereas its mirror image perhaps has a detrimental influence. Traditional detection methods for direct measurements of optical activity are quite insensitive, with detection limits in the μg range. However, Hendry et al. recently showed that plasmonic super-chiral fields could dramatically enhance the detection sensitivity of molecular chirality [84]. A planar chiral metamaterial was fabricated by square arrays of left- and right-handed (LH, RH) gammadions, with LSPRs related to strong "left-handed" and "right-handed" chiral fields. These were used to investigate the response from the adsorption of different molecules with various chiral properties and supramolecular structures. Chiral supramolecular structures of the molecules could be measured by following the resonance shifts of "left-handed" and "right-handed" LSPRs and, specifically, by tracking the difference between these shifts. The largest optical activity was found in three β-helical proteins, while α-helical proteins showed almost the same shifts for both gammadions. It was argued that the difference between the LH and RH optical responses was enhanced by six orders of magnitude compared to what could be expected for traditional measurement techniques.

Nanoporous films supporting wave guiding modes serves as the final example of the metamaterial approach to increased plasmonic sensing performances. As the sensing medium thus is incorporated within the metamaterial itself, there might be a better overlap between the sensing volume of the guided mode and the medium to be probed. An interesting illustration of this approach was presented by Kabashin et al. [85], where a metamaterial of closely packed vertically oriented Au nanorods were fabricated and excited through a prism. The authors claim a very high refractive index sensitivity as high and FOM, while the applicability of metamaterials was illustrated through detection of 10 μM biotin (244 Da) [85].

6.4.3 Nanoholes

Nanoholes in thin metal films have received enormous attention since the discovery of the phenomenon extraordinary optical transmission (EOT) by Ebbesen et al. [86]. EOT is primarily due to constructive interference of SPPs propagating between the holes where they can be coupled from/into radiation. EOT, therefore, has a lot in common with grating coupled SPR, although the nanoholes' LSPRs [87,88].

EOT is widely documented for films thicker than ~100 nm. However, it has been argued that the EOT vanishes and the propagating plasmon resonance emerge as a transmission dip (extinction peak) as the film thickness is decreased [53,89]. Spectra from single holes can be interpreted as a superposition of surface plasmon polaritons or as a localized mode derived from the modes of a void [90–93].

Nanohole arrays with only short-range spatial order show sensing characteristics similar to the short-range ordered nanoparticles and can also be used as pores that allow the target molecules to flow through the perforated film surface [94]. This technique provides much faster diffusion kinetics. In a related story, Feuz et al. recently showed that specific binding only to the nanohole walls greatly increased the response to small concentrations of the target analyte in unequilibrated systems [95]. For more details about nanoplasmonic biosensing properties of nanoholes, we refer the reader to recent reviews [96,97].

6.4.4 Plasmonic Sensing for Materials Science

The use of LSPR sensing is of course not restricted to biochemical targets. In a recent series of papers, Langhammer et al., Larsson et al. have investigated the applicability of LSPR sensing for materials science [98–100]. The methodology is based on gold nanodisk arrays with a dielectric spacer layer on top, on which the studied sample material is deposited [98,100]. Changes of the sample material, for example, due to interactions with the surrounding media, can be tracked by the change of the gold nanodisk resonance position and/or width. The spacer layer can be chosen to protect the gold nanodisk transducer, to provide specific surface chemistry or to have an active part in the studied process. Langhammer et al. demonstrated several applications based on this sensing scheme [98], including studies of phase transitions in polystyrene nanospheres and thin polymer films and calorimetric studies of reactions on nanoparticle catalysts and measurements of the kinetics and thermodynamics of hydrogen storage in Pd nanocrystals. The LSPR sensing scheme enables studies of very small nanoparticles that are difficult to probe using standard measurement techniques. In the case of hydrogen storage in Pd nanoparticles, for example, nanoparticles as small as 5 nm in diameter were probed and found to exhibit clear deviations from Pd bulk behavior [98]. This approach has recently been extended to the single particle limit [112,113].

In the case of catalytic reactions, it was demonstrated that oxidation of hydrogen, oxidation of carbon monoxide, and storage and reduction of NO_X on a Pt/BaO nanocatalyst could be followed by LSPR sensing [100].

6.5 Summary and Outlook

This chapter has summarized some of the basic characteristics of SPR sensors. We compared LSPR-based sensors with the more mature SPR technique based on propagating surface plasmons in thin films and argued that although the bulk sensitivity and the figure of merit of the latter is much greater, the nanoparticle biosensing scheme allows a superior sensitivity confinement. The strong field confinement results in a more efficient use of the refractometric sensitivity but also makes the plasmon less sensitive to noise from the surrounding bulk medium, for example, temperature fluctuations and unbound molecules. Another advantage of the LSPRs is the relaxed excitation requirements, compared with the thin film plasmon sensing. This enables a variety of optical setups, from macroscopic reflection and transmission measurements to microextinction and dark-field scattering measurements able to address single nanoparticles. The single particle measurement technique was also shown to enable robust colorimetric single molecule detection when combined with ELISA signal amplification.

Finally, we discussed some of the current trends within the plasmonic sensing community, including Fano resonances, metamaterials, nanohole sensors, and LSPR structures designed for "indirect" sensing of processes interesting for materials science. There are, however, many interesting and important topics in plasmonics that were not addressed. Prominent examples include the interaction of plasmons with other resonances, for example, waveguide, whispering gallery, and Bloch modes [97,101–107], and the possibility to utilize nanostructures (antennas) with directional properties for sensing [108–111]. Indeed, the development of plasmonics is so rapid and diverse that we can expect many significant breakthroughs within the next couple of years!

REFERENCES

1. Liedberg, B., C. Nylander, and I. Lundstrom, Surface-plasmon resonance for gas-detection and biosensing. *Sensors and Actuators*, 1983. **4**(2): 299–304.
2. Nylander, C., B. Liedberg, and T. Lind, Gas detection by means of surface plasmon resonance. *Sensors and Actuators*, 1982. **3**: 79–88.
3. Wassaf, D. et al., High-throughput affinity ranking of antibodies using surface plasmon resonance microarrays. *Analytical Biochemistry*, 2006. **351**(2): 241–253.
4. Navratilova, I. and A.L. Hopkins, Fragment screening by surface plasmon resonance. *ACS Medicinal Chemistry Letters*, 2010. **1**(1): 44–48.
5. Taylor, A.D. et al., Quantitative and simultaneous detection of four foodborne bacterial pathogens with a multi-channel SPR sensor. *Biosensors and Bioelectronics*, 2006. **22**(5): 752–758.
6. Guidi, A. et al., Comparison of a conventional immunoassay (ELISA) with a surface plasmon resonance-based biosensor for IGF-1 detection in cows' milk. *Biosensors and Bioelectronics*, 2001. **16**(9–12): 971–977.
7. Wei, D. et al., Development of a surface plasmon resonance biosensor for the identification of *Campylobacter jejuni*. *Journal of Microbiological Methods*, 2007. **69**(1): 78–85.
8. Ferguson, J.P. et al., Detection of streptomycin and dihydrostreptomycin residues in milk, honey and meat samples using an optical biosensor. *Analyst*, 2002. **127**(7): 951–956.
9. Crooks, S.R.H. et al., Immunobiosensor—An alternative to enzyme immunoassay screening for residues of two sulfonamides in pigs. *Analyst*, 1998. **123**(12): 2755–2757.
10. Nilsson, C.E. et al., A novel assay for influenza virus quantification using surface plasmon resonance. *Vaccine*, 2010. **28**(3): 759–766.
11. Naimushin, A.N. et al., Detection of *Staphylococcus aureus* enterotoxin B at femtomolar levels with a miniature integrated two-channel surface plasmon resonance (SPR) sensor. *Biosensors and Bioelectronics*, 2002. **17**(6–7): 573–584.
12. Kamat, P.V., Photophysical, photochemical and photocatalytic aspects of metal nanoparticles. *Journal of Physical Chemistry B*, 2002. **106**(32): 7729–7744.
13. Kamat, P.V., Meeting the clean energy demand: Nanostructure architectures for solar energy conversion. *Journal of Physical Chemistry C*, 2007. **111**(7): 2834–2860.
14. Xu, H.X. et al., Spectroscopy of single hemoglobin molecules by surface enhanced Raman scattering. *Physical Review Letters*, 1999. **83**(21): 4357–4360.
15. Moskovits, M., Surface-roughness and enhanced intensity of Raman-scattering by molecules adsorbed on metals. *Journal of Chemical Physics*, 1978. **69**(9): 4159–4161.
16. Anker, J. et al., Biosensing with plasmonic nanosensors. *Nature Materials*, 2008. **7**(6): 442–453.
17. Fang, N. et al., Sub-diffraction-limited optical imaging with a silver superlens. *Science*, 2005. **308**(5721): 534–537.
18. Zhang, X. and Z. Liu, Superlenses to overcome the diffraction limit. *Nature Materials*, 2008. **7**(6): 435–441.
19. West, J. and N. Halas, Engineered nanomaterials for biophotonics applications: Improving sensing, imaging, and therapeutics. *Annual Review of Biomedical Engineering*, 2003. **5**: 285–292.
20. Englebienne, P., Use of colloidal gold surface plasmon resonance peak shift to infer affinity constants from the interactions between protein antigens and antibodies specific for single or multiple epitopes. *Analyst*, 1998. **123**(7): 1599–1603.
21. Leuvering, J.H.W. et al., A sol particle agglutination assay for human chorionic gonadotrophin. *Fresenius Zeitschrift Fur Analytische Chemie*, 1980. **301**(2): 132–132.
22. Okamoto, T., I. Yamaguchi, and T. Kobayashi, Local plasmon sensor with gold colloid monolayers deposited upon glass substrates. *Optics Letters*, 2000. **25**(6): 372–374.
23. Nath, N. and A. Chilkoti, A colorimetric gold nanoparticle sensor to interrogate biomolecular interactions in real time on a surface. *Analytical Chemistry*, 2002. **74**(3): 504–509.
24. Malinsky, M.D. et al., Chain length dependence and sensing capabilities of the localized surface plasmon resonance of silver nanoparticles chemically modified with alkanethiol self-assembled monolayers. *Journal of the American Chemical Society*, 2001. **123**(7): 1471–1482.
25. Himmelhaus, M. and H. Takei, Cap-shaped gold nanoparticles for an optical biosensor. *Sensors and Actuators B: Chemical*, 2000. **63**(1–2): 24–30.

26. Haes, A.J. and R.P. Van Duyne, A nanoscale optical biosensor: Sensitivity and selectivity of an approach based on the localized surface plasmon resonance spectroscopy of triangular silver nanoparticles. *Journal of the American Chemical Society*, 2002. **124**(35): 10596–10604.

27. Kretschmann, E. and H. Raether, Radiative decay of non radiative surface plasmons excited by light. *Zeitschrift Fur Naturforschung Part A—Astrophysik Physik Und Physikalische Chemie*, 1968. **A 23**(12): 2135.

28. Piliarik, M. and J. Homola, Surface plasmon resonance (SPR) sensors: Approaching their limits? *Optics Express*, 2009. **17**(19): 16505–16517.

29. Piliarik, M. et al., Compact and low-cost biosensor based on novel approach to spectroscopy of surface plasmons. *Biosensors and Bioelectronics*, 2009. **24**(12): 3430–3435.

30. Vukusic, P.S., G.P. Bryanbrown, and J.R. Sambles, Surface-plasmon resonance on gratings as a novel means for gas sensing. *Sensors and Actuators B: Chemical*, 1992. **8**(2): 155–160.

31. Cullen, D.C., R.G.W. Brown, and C.R. Lowe, Detection of immuno-complex formation via surface-plasmon resonance on gold-coated diffraction gratings. *Biosensors*, 1987. **3**(4): 211–225.

32. Hickel, W., B. Rothenhausler, and W. Knoll, Surface plasmon microscopic characterization of external surfaces. *Journal of Applied Physics*, 1989. **66**(10): 4832–4836.

33. Rothenhausler, B. and W. Knoll, Surface–plasmon microscopy. *Nature*, 1988. **332**(6165): 615–617.

34. Nelson, B.P. et al., Surface plasmon resonance imaging measurements of DNA and RNA hybridization adsorption onto DNA microarrays. *Analytical Chemistry*, 2001. **73**(1): 1–7.

35. Brockman, J.M., B.P. Nelson, and R.M. Corn, Surface plasmon resonance imaging measurements of ultrathin organic films. *Annual Review of Physical Chemistry*, 2000. **51**: 41–63.

36. Andersson, O. et al., Gradient hydrogel matrix for microarray and biosensor applications: An imaging SPR study. *Biomacromolecules*, 2008. **10**(1): 142–148.

37. Homola, J., Surface plasmon resonance based sensors. in: *Springer Series on Chemical Sensors and Biosensors*, ed. O.S. Wolfbeis. 2006, Berlin, Germany: Springer.

38. Draine, B.T. and P.J. Flatau, Discrete-dipole approximation for scattering calculations. *Journal of the Optical Society of America a-Optics Image Science and Vision*, 1994. **11**(4): 1491–1499.

39. Devoe, H., Optical properties of molecular aggregates. I. Classical model of electronic absorption and refraction. *Journal of Chemical Physics*, 1964. **41**(2): 393.

40. Rindzevicius, T. et al., Plasmonic sensing characteristics of single nanometric holes. *Nano Letters*, 2005. **5**(11): 2335–2339.

41. Bingham, J.M. et al., Localized surface plasmon resonance imaging: Simultaneous single nanoparticle spectroscopy and diffusional dynamics. *Journal of Physical Chemistry C*, 2009. **113**(39): 16839–16842.

42. Raschke, G. et al., Biomolecular recognition based on single gold nanoparticle light scattering. *Nano Letters*, 2003. **3**(7): 935–938.

43. Mayer, K.M. et al., A single molecule immunoassay by localized surface plasmon resonance. *Nanotechnology*, 2010. **21**(25): 255503.

44. Nusz, G.J. et al., Dual-order snapshot spectral imaging of plasmonic nanoparticles. *Applied Optics*, 2011. **50**(21): 4198–4206.

45. Chen, S. et al., Plasmon-enhanced colorimetric ELISA with single molecule sensitivity. *Nano Letters*, 2011. **11**(4): 1826–1830.

46. Dahlin, A.B., J.O. Tegenfeldt, and F. Hook, Improving the instrumental resolution of sensors based on localized surface plasmon resonance. *Analytical Chemistry*, 2006. **78**(13): 4416–4423.

47. Dahlin, A.B. ct al., High-resolution microspectroscopy of plasmonic nanostructures for miniaturized biosensing. *Analytical Chemistry*, 2009. **81**(16): 6572–6580.

48. Svedendahl, M. et al., Refractometric sensing using propagating versus localized surface plasmons: A direct comparison. *Nano Letters*, 2009. **9**(12): 4428–4433.

49. Miller, M.M. and A.A. Lazarides, Sensitivity of metal nanoparticle surface plasmon resonance to the dielectric environment. *Journal of Physical Chemistry B*, 2005. **109**(46): 21556–21565.

50. Sherry, L.J. et al., Localized surface plasmon resonance spectroscopy of single silver triangular nanoprisms. *Nano Letters*, 2006. **6**(9): 2060–2065.

51. Kvasnicka, P. and J. Homola, Optical sensors based on spectroscopy of localized surface plasmons on metallic nanoparticles: Sensitivity considerations. *Biointerphases*, 2008. **3**(3): FD4–FD11.

52. Brian, B. et al., Sensitivity enhancement of nanoplasmonic sensors in low refractive index substrates. *Optics Express*, 2009. **17**(3): 2015–2023.
53. Sannomiya, T. et al., Investigation of plasmon resonances in metal films with nanohole arrays for biosensing applications. *Small*, 2011. **7**(12): 1653–1663.
54. Dmitriev, A. et al., Enhanced nanoplasmonic optical sensors with reduced substrate effect. *Nano Letters*, 2008. **8**(11): 3893–3898.
55. Otte, M.A. et al., Improved biosensing capability with novel suspended nanodisks. *Journal of Physical Chemistry C*, 2011. **115**(13): 5344–5351.
56. Defeijter, J.A., J. Benjamins, and F.A. Veer, Ellipsometry as a tool to study adsorption behavior of synthetic and biopolymers at air-water-interface. *Biopolymers*, 1978. **17**(7): 1759–1772.
57. Chen, S. et al., Ultrahigh sensitivity made simple: Nanoplasmonic label-free biosensing with an extremely low limit-of-detection for bacterial and cancer diagnostics. *Nanotechnology*, 2009. **20**(43): 434015.
58. Jain, P.K., W.Y. Huang, and M.A. El-Sayed, On the universal scaling behavior of the distance decay of plasmon coupling in metal nanoparticle pairs: A plasmon ruler equation. *Nano Letters*, 2007. **7**(7): 2080–2088.
59. Murray, W.A., B. Auguie, and W.L. Barnes, Sensitivity of localized surface plasmon resonances to bulk and local changes in the optical environment. *Journal of Physical Chemistry C*, 2009. **113**(13): 5120–5125.
60. Sonnichsen, C. et al., Drastic reduction of plasmon damping in gold nanorods. *Physical Review Letters*, 2002. **88**(7): 077402.
61. Evlyukhin, A.B. et al., Detuned electrical dipoles for plasmonic sensing. *Nano Letters*, 2010. **10**(11): 4571–4577.
62. Henzie, J., M.H. Lee, and T.W. Odom, Multiscale patterning of plasmonic metamaterials. *Nature Nanotechnology*, 2007. **2**(9): 549–554.
63. Sonnefraud, Y. et al., Experimental realization of subradiant, superradiant, and fano resonances in ring/disk plasmonic nanocavities. *ACS Nano*, 2010. **4**(3): 1664–1670.
64. Hao, F. et al., Symmetry breaking in plasmonic nanocavities: Subradiant LSPR sensing and a tunable Fano resonance. *Nano Letters*, 2008. **8**(11): 3983–3988.
65. Hao, F. et al., Tunability of subradiant dipolar and fano-type plasmon resonances in metallic ring/disk cavities: Implications for nanoscale optical sensing. *ACS Nano*, 2009. **3**(3): 643–652.
66. Dondapati, S.K. et al., Label-free biosensing based on single gold nanostars as plasmonic transducers. *ACS Nano*, 2010. **4**(11): 6318–6322.
67. Sheehan, P.E. and L.J. Whitman, Detection limits for nanoscale biosensors. *Nano Letters*, 2005. **5**(4): 803–807.
68. Rissin, D.M. et al., Single-molecule enzyme-linked immunosorbent assay detects serum proteins at sub-femtomolar concentrations. *Nature Biotechnology*, 2010. **28**(6): 595–599.
69. Fano, U., Sullo spettro di assorbimento dei gas nobili presso il limite dello spettro d'arco. *Nuovo Cimento*, 1935. N. S. **12**: 154–161.
70. Fano, U., Effects of configuration interaction on intensities and phase shifts. *Physical Review*, 1961. **1**(6): 1866.
71. Pakizeh, T. et al., Intrinsic Fano interference of localized plasmons in Pd nanoparticles. *Nano Letters*, 2009. **9**(2): 882–886.
72. Verellen, N. et al., Plasmon line shaping using nanocrosses for high sensitivity localized surface plasmon resonance sensing. *Nano Letters*, 2011. **11**(2): 391–397.
73. Hentschel, M. et al., Transition from isolated to collective modes in plasmonic oligomers. *Nano Letters*, 2010. **10**(7): 2721–2726.
74. Fan, J.A. et al., Self-assembled plasmonic nanoparticle clusters. *Science*, 2010. **328**(5982): 1135–1138.
75. Mirin, N.A., K. Bao, and P. Nordlander, Fano resonances in plasmonic nanoparticle aggregates. *Journal of Physical Chemistry A*, 2009. **113**(16): 4028–4034.
76. Lassiter, J.B. et al., Fano resonances in plasmonic nanoclusters: Geometrical and chemical tunability. *Nano Letters*, 2010. **10**(8): 3184–3189.
77. Zhang, S. et al., Plasmon-induced transparency in metamaterials. *Physical Review Letters*, 2008. **101**(4): 047401.
78. Alzar, C.L.G., M.A.G. Martinez, and P. Nussenzveig, Classical analog of electromagnetically induced transparency. *American Journal of Physics*, 2002. **70**(1): 37–41.
79. Liu, N. et al., Planar metamaterial analogue of electromagnetically induced transparency for plasmonic sensing. *Nano Letters*, 2010. **10**(4): 1103–1107.

80. Smith, D.R., J.B. Pendry, and M.C.K. Wiltshire, Metamaterials and negative refractive index. *Science*, 2004. **305**(5685): 788–792.
81. Becker, J. et al., The optimal aspect ratio of gold nanorods for plasmonic bio-sensing. *Plasmonics*, 2010. **5**(2): 161–167.
82. Liu, N. et al., Infrared perfect absorber and its application as plasmonic sensor. *Nano Letters*, 2010. **10**(7): 2342–2348.
83. Tang, Y.Q. and A.E. Cohen, Optical chirality and its interaction with matter. *Physical Review Letters*, 2010. **104**(16): 163901.
84. Hendry, E. et al., Ultrasensitive detection and characterization of biomolecules using superchiral fields. *Nature Nanotechnology*, 2010. **5**(11): 783–787.
85. Kabashin, A.V. et al., Plasmonic nanorod metamaterials for biosensing. *Nature Materials*, 2009. **8**(11): 867–871.
86. Ebbesen, T. et al., Extraordinary optical transmission through sub-wavelength hole arrays. *Nature*, 1998. **391**(6668): 667–669.
87. Degiron, A. et al., Optical transmission properties of a single subwavelength aperture in a real metal. *Optics Communications*, 2004. **239**(1–3): 61–66.
88. Rindzevicius, T. et al., Nanohole plasmons in optically thin gold films. *Journal of Physical Chemistry C*, 2007. **111**(3): 1207–1212.
89. Braun, J. et al., How holes can obscure the view: Suppressed transmission through an ultra thin metal film by a subwavelength hole array. *Physical Review Letters*, 2009. **103**(20): 203901.
90. Park, T.-H. et al., Optical properties of a nanosized hole in a thin metallic film. *ACS Nano*, 2008. **2**(1): 25–32.
91. Alaverdyan, Y. et al., Optical antennas based on coupled nanoholes in thin metal films. *Nature Physics*, 2007. **3**(12): 884–889.
92. Prikulis, J. et al., Optical spectroscopy of nanometric holes in thin gold films. *Nano Letters*, 2004. **4**(6): 1003–1007.
93. Alegret, J., P. Johansson, and M. Käll, Green's tensor calculations of plasmon resonances of single holes and hole pairs in thin gold films. *New Journal of Physics*, 2008. **10**: 105004.
94. Jonsson, M.P. et al., Locally functionalized short-range ordered nanoplasmonic pores for bioanalytical sensing. *Analytical Chemistry*, 2010. **82**(5): 2087–2094.
95. Feuz, L. et al., Improving the limit of detection of nanoscale sensors by directed binding to high-sensitivity areas. *ACS Nano*, 2010. **4**(4): 2167–2177.
96. Jonsson, M.P. et al., Nanoplasmonic biosensing with focus on short-range ordered nanoholes in thin metal films. *Biointerphases*, 2008. **3**(3): FD30–FD40.
97. Stewart, M.E. et al., Nanostructured plasmonic sensors. *Chemical Reviews*, 2008. **108**(2): 494–521.
98. Langhammer, C. et al., Indirect nanoplasmonic sensing: Ultrasensitive experimental platform for nano-materials science and optical nanocalorimetry. *Nano Letters*, 2010. **10**(9): 3529–3538.
99. Langhammer, C., I. Zoric, and B. Kasemo, Hydrogen storage in Pd nanodisks characterized with a novel nanoplasmonic sensing scheme. *Nano Letters*, 2007. **7**(10): 3122–3127.
100. Larsson, E.M. et al., Nanoplasmonic probes of catalytic reactions. *Science*, 2009. **326**(5956): 1091–1094.
101. Shopova, S.I. et al., Plasmonic enhancement of a whispering-gallery-mode biosensor for single nanoparticle detection. *Applied Physics Letters*, 2011. **98**(24): 243104.
102. Vollmer, F. and S. Arnold, Whispering-gallery-mode biosensing: Label-free detection down to single molecules. *Nature Methods*, 2008. **5**(7): 591–596.
103. Min, B.K. et al., High-Q surface-plasmon-polariton whispering-gallery microcavity. *Nature*, 2009. **457**(7228): 455–458.
104. White, I.M., J. Gohring, and X.D. Fan, SERS-based detection in an optofluidic ring resonator platform. *Optics Express*, 2007. **15**(25): 17433–17442.
105. Stewart, M. et al., Quantitative multispectral biosensing and 1D imaging using quasi-3D plasmonic crystals. *Proceedings of the National Academy of Sciences of the United States of America*, 2006. **103**(46): 17143–17148.
106. Dahlin, A.B. et al., Synchronized quartz crystal microbalance and nanoplasmonic sensing of biomolecular recognition reactions. *ACS Nano*, 2008. **2**(10): 2174–2182.

107. Jonsson, M.P., P. Jonsson, and F. Hook, Simultaneous nanoplasmonic and quartz crystal microbalance sensing: Analysis of biomolecular conformational changes and quantification of the bound molecular mass. *Analytical Chemistry*, 2008. **80**(21): 7988–7995.
108. Shegai, T. et al., Angular distribution of surface-enhanced Raman scattering from individual Au nanoparticle aggregates. *ACS Nano*, 2011. **5**(3): 2036–2041.
109. Shegai, T. et al., Unidirectional broadband light emission from supported plasmonic nanowires. *Nano Letters*, 2011. **11**(2): 706–711.
110. Shegai, T. et al., A bimetallic nanoantenna for directional colour routing. *Nature Communications*, 2011. **2**: 481.
111. Shegai, T. et al., Directional Scattering and Hydrogen Sensing by Bimetallic Pd-Au Nanoantennas. *Nano Letters*, 2012. **12**(5): 2464–2469.
112. Shegai, T. and C. Langhammer, Hydride Formation in Single Palladium and Magnesium Nanoparticles Studied By Nanoplasmonic Dark-Field Scattering Spectroscopy. *Advanced Materials*, 2011. **23**(38): 4409–4414.
113. Liu, N., et al., Nanoantenna-enhanced gas sensing in a single tailored nanofocus. *Nature Materials*, 2011. **10**(8): 631–636.

7

Fiber-Optic Nanosensors

Carlos R. Zamarreño, Jesus M. Corres, Javier Goicoechea, Ignacio Del Villar, Ignacio R. Matías, and Francisco J. Arregui

CONTENTS

7.1 Introduction

Nanostructures can enhance, somehow, the behavior of raw materials due to intimate interactions of these structures at a molecular scale (Yadugiri and Malhotra 2010). This novel point of view alters the properties of the materials that overcome the limits of classical physics toward a quantum physics behavior. Some interesting applications include those related to the interactions of electromagnetic radiations, such as light, with structures ordered at the nanoscale level (Huang et al. 1991, Maier et al. 2001). Structures, such as quantum dots (QDs) (Alivisatos 1996), photonic crystals (Joannopoulos et al. 1997), fullerenes (Kroto et al. 1985), and fiber Bragg gratings (FBGs) (Kersey et al. 1997) present fascinating optical phenomena suitable for the development of interesting applications. One particular approach consists of the fabrication of nanostructured coatings onto or within an optical fiber, in such a manner that light traveling through the optical fiber interacts with the fabricated nanostructures (Arregui et al. 2010). The previously described approach has been widely used for the fabrication of optical fiber sensors, which can produce the same or even better response than that of conventional electronic devices (Cusano et al. 2008b). Moreover, optical fiber configuration overcomes traditional electronic devices in several aspects, such as their immunity to electromagnetic fields, the capability to transmit over large distances, the possibility of remote sensing and easy multiplexing, their light weight and small size, or the utilization of light instead of electrical signals in highly flammable or explosive atmospheres (Culshaw 2004, 2005). Furthermore, the ability to produce micro- and nanosensors with optical fibers allows introducing the sensing tip into biological systems in order to obtain precise measurements of some analytes, such as blood compounds, gases, or pH, which requires the utilization of tough and inert films, especially when they are in contact with tissues. These properties can be exploited for the development of refractometers, pH sensors, chemical sensors, biosensors, and other applications. In spite of the mentioned advantages, the growing demand of optical fiber technology is being led by the development of low-cost optical devices such as LEDs, lasers or spectrometers, which has promoted their utilization in many different areas where the price was the most important factor (Culshaw and Kersey 2008).

Therefore, this chapter will be focused on optical fiber sensors based on nanostructured coatings. The fabrication of the nanostructured coatings, which will modify their optical properties as a function of a target parameter, can be also performed by following different fabrication techniques. Moreover, the variations of the coating properties and hence the changes in the reflected or transmitted spectrum can be monitored by using diverse interrogation schemes (Lee et al. 2009). Henceforward, some of the most important fabrication techniques and interrogation schemes (Lee 2003), as well as the underlying theory of some of these phenomena have been revised in order to understand the behavior of the sensors and improve their sensing characteristics.

7.2 Nanofabrication Techniques

The utilization of nanoparticle-based or nanostructured coatings in modern optical fiber sensors can modify or improve their sensitivity to some selected physical or chemical influences. The fabrication of these coatings requires, in many instances, an adequate control and understanding of nanofabrication techniques in order to tailor the characteristics of the structures, which establishes the final properties of the same (Smietana et al. 2010). Nanofabrication techniques involve a deep multidisciplinary study which has to take into account multiple variables aimed to achieve the optimal performances regarding to the sensitivity, response time, working range, hysteresis, or cross-sensitivity of the final sensing device. In this section, we describe the fundamentals and working principles of some of the most popular nanocoating deposition methods in the context of their applicability to optical fiber sensors.

Traditional nanofabrication techniques used in semiconductor or other industries such as spin-coating (Lee et al. 1998), physical vapor deposition (PVD) (Mattox 2000), and electro-spinning (Doshi and Reneker 1995) are intended to be used with flat substrates. This requires, in most cases, an additional effort to produce uniform or quasi-uniform coatings onto the optical fibers due to their distinctive cylindrical or conical geometry. In particular, spin-coating requires an adaptation of the fibers in order to produce flat surfaces, such as D-shaped or side-polished optical fibers embedded into a V-groove substrate (Nagaraju et al. 2008). PVD and its variants (cathodic arc deposition, thermal evaporative deposition, electron beam deposition physical vapor deposition, pulsed laser deposition, or sputter deposition) also need to apply slight modifications to the technique in order to be used on optical fibers. The utilization of polished optical fibers (Bender et al. 1994, Tien et al. 2006, Dikovska et al. 2010, Yang et al. 2010a), multiple step deposition (Díez et al. 1999), rotating (Jorgenson and Yee 1993, Navarrete et al. 2010), or tilting the fibers during the deposition (Fan and Zhao 2005, Konry and Marks 2005, Jayawardhana et al. 2011) have enabled the utilization of PVD for the fabrication of complex nanostructures onto the optical fibers (Allsop et al. 2010b).

Meanwhile, chemical vapor deposition (CVD) can produce single layer, multilayer, composite, nanostructured, and functionally graded coating materials with structural control at atomic or nanometer scale level and unique structure at low processing temperatures. Furthermore, CVD allows the coating of complex shape engineering components and the fabrication of nanodevices. However, CVD usually requires expensive and highly sophisticated equipment or involves the utilization of volatile or toxic chemical species (Choy 2003), which have been partly substituted by the emerging of novel alternative techniques based on CVD, such as plasma enhanced CVD, photo-assisted CVD, atomic layer epitaxy, electrochemical vapor deposition (EVD), metalorganic CVD, flame-assisted CVD, aerosol-assisted CVD, pyrosol or chemical vapor infiltration that can overcome most of the main traditional CVD issues (Smietana et al. 2004).

Sol–gel dip-coating technology is one of the fastest growing fields of contemporary chemistry owing to its simplicity and versatility (Lukowiak and Strek 2009). The final properties of the obtained coatings depend on several factors that can be controlled in order to establish the materials characteristics: refractive index, porosity, surface area, thickness, thermal stability, and mechanical and photochemical properties (Brinker et al. 1991).The easy way of doping is one of the main advantages of the sol–gel-derived materials and this feature is very important for the production of optical and electrochemical sensors. Sol–gel coatings can also be easily combined with optical fibers or planar waveguides, providing intrinsic evanescent wave sensors due to their excellent properties such as transparency, porosity, and high surface areas (Jerónimo et al. 2007, Zamarreño et al. 2010b). In spite of the undeniable advantages of sol–gel technique, it presents a lack of a precise film-thickness control at the nanometer level, which for some applications is considered as critical.

Chemical self-assembly monolayer (CSAM) can also be easily placed onto optical fibers by means of the adequate functionalization of the optical fiber surface. The combination of SAMs and optical fibers enables a broad range of sensing applications, which mainly depend on the functionalized groups. Among them, we can find pH sensors for bioanalysis based on microstructured optical fibers (Schröder et al. 2010), as well as a wide variety of biosensors (Rindorf et al. 2006) for brucella cells (Liebes et al. 2009), dengue virus (Atias et al. 2009), thrombin (Allsop et al. 2010a), and so forth. However, CSAM coatings are usually limited to one specific monolayer and to only certain compounds (Bhat et al. 2002).

The widely used electrodeposition techniques are constrained by the nonconductive nature of the optical fiber. This requires the fabrication of a prior conductive layer onto the fiber by means of some of the previously described methods (Chopra et al. 1983). Moreover, the utilization of transparent conducting oxides (TCOs) enables the utilization of electrodeposition techniques together with absorption- or fluorescence-measuring techniques. In particular, indium tin oxide (ITO) coatings fabricated onto the optical fiber core, as it is shown in the SEM micrograph of Figure 7.1, has enabled the development of many applications, such as chemical sensing, (Kasik et al. 2009b, Zamarreño et al. 2009b) and biological compound determination (Marks et al. 2002, Konry and Marks 2005, Connor and Ferri 2007, Konry et al. 2008).

Photolithography techniques combined with other fabrication procedures, such as spin-coating, PVD, or CVD can easily produce periodic or quasi-periodic structures onto optical fibers, such as LPGs or FBGs (Bennion et al. 1996, Chen et al. 2005), suitable to be used in the fabrication of optical fiber sensors (Allsop et al. 2010a,b). On the other hand, this technique presents several drawbacks, such as the necessity to use a preexisting mask, which has to be positioned in intimate contact with the substrate in order to avoid diffraction, as well as other factors derived from the fabrication process (Smith 2001, Mack 2008).

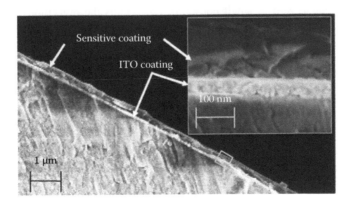

FIGURE 7.1 Cross-section SEM micrograph of an ITO coated optical fiber tip with an electrodeposited sensitive layer onto it.

In contrast, Langmuir-Blodgett (LB) technique enables (James et al. 2005, Arregui et al. 2010), the fabrication of uniform and homogeneous coatings with high-resolution control over the film thickness (1–3 nm per layer) through the proper adjustment of some key parameters, such as concentration, pressure, area, or surface tension. Thus, LB is ideal for waveguide and optical fiber applications (I 1990, Petty 1992). In fact, hundreds of optical fiber sensing applications have been originated within the last decades by the fabrication of nanostructured LB films onto optical fibers (Zhao and Reichert 1991, Ashwell et al. 1999, Ishaq et al. 2005, Veselov et al. 2009). However, this technique is often limited to the utilization of specific chemical species and the fabrication of a few layers due to the formation of defects that are propagated through the structure (James and Tatam 2006). Although less sophisticated than in the case of PVD and CVD, the fabrication of LB films also requires the utilization of some special equipment. An alternative approach is the Layer-by-Layer (LbL) electrostatic self-assembly method (Wang et al. 2005, Decher 1997).

LbL is driven by the electrostatic attraction of oppositely charged materials (Arregui et al. 2002b). This fabrication technique is highly versatile since it allows coating substrates of almost any size or shape, such as an optical fiber, and the utilization of a broad range of materials for the coating formation, including polymers, ceramics, metals, and semiconductors (Arregui et al. 2007). LbL also enables the fabrication of thin–homogeneous coatings with precise thickness by controlling a number of parameters, such as pH, concentration, ionic strength, curing temperature, or ratio between the polyelectrolytes used (Shiratori and Rubner 2000, Ahn et al. 2005, Choi and Rubner 2005, Lee et al. 2007). In contrast to LB, LbL can be performed at room conditions without the necessity of any special apparatus (Arregui et al. 2010). Moreover, LbL permits the molecular-level self-healing of defects that may occur in individual monolayers as additional monolayers are gradually added layer by layer to the substrate during the synthesis process. Furthermore, LbL method is easily scalable, though time-consuming, for multilayer fabrication and industrial applications. Hence, LbL has been extensively exploited in the fabrication of fiber-optic sensors for a large application spectrum as will be detailed in the following sections (Del Villar et al. 2008).

7.3 Optical Fiber Interrogation Schemes

In this section, the basic schemes of fiber-optic nanosensors (FONS) are summarized. In order to increase the sensitivity, various geometries are often used with diverse optical transduction mechanisms such as changes in refractive index, absorption, fluorescence, and surface plasmon resonance (SPR). Optical fibers transmit light on the basis of the principle of total internal reflection (TIR). When this phenomenon occurs, the light is guided through the core of the fiber with slight loss to the surroundings. If the fiber structure is modified (cut, de-cladded, bended, etc.), it is possible to access the traveling light and force it to interact with the external medium. In Figure 7.2, a summary of fiber-optic sensor schematics is shown.

When the incident light is internally reflected at the core/cladding interface of the fiber, its intensity does not abruptly decay to zero. A small portion of light enters the reflecting medium by a fraction of wavelength, far enough for recognition of the different refractive index. This electromagnetic field, called the evanescent wave, has an intensity that decays exponentially with distance, starting at the interface and extending into the medium of lower refractive index. In this way, the evanescent wave interacts with the molecules of the reflecting medium within the penetration depth, producing a net flow of energy across the reflecting surface in the surrounding medium.

In order to interact with the evanescent field, multiple schemes have been proposed as shown in Figure 7.2b–h: tapered optical fiber (TOF) (Corres et al. 2006a), photonic crystal fiber (Rindorf et al. 2006), hollow-core fiber (HCF) (Bravo et al. 2006), long-period fiber gratings (LPGs) (James and Tatam 2003, Chung and Yin 2004). In most of them, the cladding of the optical fiber is modified, using different schemes, along a small section by a sensitive nanomaterial; so any change in the optical or structural characteristic (such a refractive index, thickness, absorption) of the sensitive material causes a change in its transmission properties.

Absorption sensors are governed by the intensity modulation caused by light absorption of the evanescent wave with the result of the attenuation of the guided light in the fiber core (Ramachandran et al. 2002).

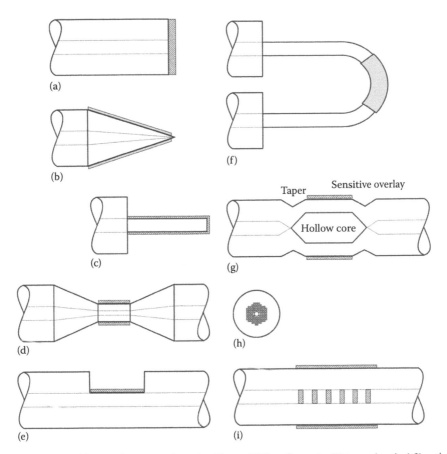

FIGURE 7.2 Summary of fiber-optic sensor schematics (a) nano FP interferometer (b) tapered optical fiber tip, (c) cladding-off fiber, (d) biconical tapered optical fiber, (e) D-shape fiber, (f) U-shaped evanescent wave sensor probe, (g) hollow core fiber, (h) long period grating (LPG), (i) Fiber grating.

The sensitivity depends on the penetration depth into the sensing cladding, which can be calculated by Beer–Lambert law (Allsop et al. 2006, Slavik 2006). For fluorescence sensors, the evanescent light excites a fluorophore in the overlay (usually artificially labeled compounds), and the fluorescence emitted is captured and guided by the fiber. Enzymatic biosensors usually use this type of transduction in which the analyte is biocatalytically converted to a fluorescent product. SPR plays a leading role in real-time affinity biosensing. For SPR sensing, the existence of a dielectric–metal interface is necessary. When light interacts with the interface at a resonance angle, there is a match between the energy of the light photons and the electrons at the metal surface; the transference of energy from light photons to packets of electrons results in the attenuation of the reflected light at the resonance wavelength. The change in the refractive index of the overlay modifies the SPR coupling conditions, and produces a shift in the resonance angle (Marazuela and Moreno-Bondi 2002). Fiber-optic SPR sensors can use different optical structures such as D-shaped, de-cladded, end mirror, or tapered fiber structures. The typically used metals are gold and silver.

In the next paragraphs, some of these interrogation schemes will be described with more detail.

7.3.1 Nanocavities

When the light is guided until an end-cut optical pigtail, the light interacts with the sensing deposition and a nano Fabry–Pérot (FP) cavity is formed between the fiber and the external medium (see Figure 7.2a). The main advantage of this type of sensors is the simple experimental setup as shown in Figure 7.3a. Since the cavity created at the end of fiber is shorter than the coherence length of a LED source, this

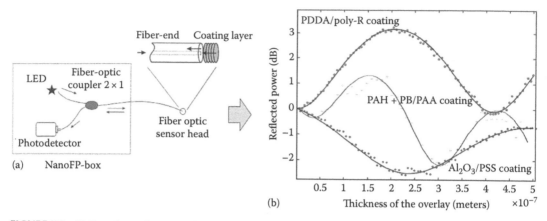

FIGURE 7.3 (a) Experimental setup to detect the reflected optical power by a nano FP cavity. (b) Deposition curves of the control fiber. (Reprinted from Corres, J.M., Arregui, F.J., and Matias, I.R. 2006a, Design of humidity sensors based on tapered optical fibers, *J. Lightwave Technol.*, 24, 11, 4329–4336. © 2006 IEEE.)

permits to avoid the necessity of lasers to monitor the interferometric phenomenon. That is the reason for using this setup in most of the cases (Arregui et al. 1999a,b, 2004, Matias et al. 2003, Del Villar et al. 2005b, 2005d).

The modification of its thickness causes a variation of the reflected optical power as shown in Figure 7.3b, which has been investigated extensively (Arregui et al. 1999a,b, Arregui et al. 2002b, Del Villar et al. 2005b). In addition to this, if an optical property of the sensing dye, such as the refractive index, changes, then a variation in the reflected signal is registered due to the light interference in the FP cavity (Del Villar et al. 2005e). The following expression permits to estimate the thickness of the nanocavity (Arregui et al. 1999b):

$$\varphi = \frac{4n_2 d\pi}{\lambda} \tag{7.1}$$

where
   φ is the phase shift of the optical beam in the cavity formed by the coating with thickness d
   n_2 is the real refractive index of the coating
   λ is the wavelength of the LED

The destructive interference occurs when φ is an odd multiple of π.

Using different coating materials, a grating can be created onto the cut end of the fiber (Arregui et al. 2002). The LbL method can be combined with several sensing materials (James and Tatam 2006, Zamarreño et al. 2007, 2010b, Ma et al. 2010), sensitive to different substances. Using this technique, it is possible to lower the detection limit to nanomolar concentrations. Recently, a very high sensitivity device for biosensing that uses surface-enhanced Raman scattering (SERS) has been developed (Andrade et al. 2010). The optical fiber tip is modified through a "layer-by-layer" procedure to self-assemble silver nanoparticles.

7.3.2 De-Cladded Fibers

Uniformly de-cladded fibers are the predecessors to the development of tapered fiber-optic sensors (Kittidechachan et al. 2008). They are similar to tapered fibers in optical behavior, but the optical power transferred into the evanescent field is usually weak, and hence the penetration depth in the sensing area is low. The reason for the small absorption is that the evanescent absorption coefficient is inversely proportional to the core radius. This translates into a limitation of the sensitivity of the sensor.

To improve the sensitivity, the fiber can be de-cladded and bent. Fiber bending creates higher order modes, which results in coupling more optical power in the evanescent field and having a greater penetration depth. In this case, the increment in the optical power coupled to the evanescent wave depends mainly on the bend radius. There are practical examples involving bending the fiber into a U-Shape probe (Sharma and Gupta 2004). It is also possible to coil a few meters of an optical fiber with modified cladding (Leung et al. 2007). This configuration is useful in applications where distributed sensing is needed.

7.3.3 Tapered Optical Fibers and Side-Polished Fibers

The tapering of the optical fiber allows accessing to the evanescent field in a reflection or transmission scheme. Tapering can be performed by removing the cladding and then tapering the core, or keeping both the core and cladding in place and tapering the entire fiber using heat pulling or chemical etching with hydrofluoric acid (Zamarreño et al. 2007, Kasik et al. 2009, Gonçalves et al. 2010, Martan et al. 2010).

Tapering not only exposes the evanescent field to the surroundings but also increases the evanescent field magnitude and penetration depth (Leung et al. 2007). Tapering enhances the potential of the optical fiber as a sensor: the optical power intensity in the evanescent field of a tapered fiber compared to that of a de-cladded uniform-core fiber in the context of evanescent absorption has been demonstrated to be up to 100 times higher when using adiabatic tapers of $1\mu m$ waist diameter (Díaz-Herrera et al. 2004). Because of that, the vast majority of evanescent fiber-optic biosensors use this optical structure. In addition to this, due to the submicrometer size of optical fiber nanotips, they have much faster response times and are suitable for small analyte volume, for instance for intra-cell monitoring (Vo-Dinh and Kasili 2005).

In a conical tapered tip, the diameter decreases continuously while in a step-etched tapered tip, the diameter abruptly decreases, which results in a high power loss and a low sensitivity (Leung et al. 2007). Conical tips are frequently coated with a metallic nanofilm (thermal vapor deposition of silver, aluminum, or gold), excluding the end part of the fiber, with the purpose of reducing the sensing area and to enable the propagation of the light down the tapered fiber (Uttamchandani and McCulloch 1996).

Conversely, when it is used as a transmission setup, the transmitted optical power through the taper shows an oscillatory behavior with respect to elongation, bend, wavelength, or external refractive index changes (Gonthier et al. 1987). With the focus on sensing, larger amplitude oscillations can be obtained by fabricating tapers where the power is equally divided between two propagating modes. The coupling between the fundamental mode HE11 and the following mode with the same symmetry, HE12, is the main cause of the high oscillations that appear when the refractive index of the external medium is changed (Gonthier et al. 1987). When a thin film is deposited onto a tapered optical fiber, the modes' effective indices change. Those physical parameters, which induce variations in the thickness or refractive index of the film, change the transmitted optical power of the modes. Because of that, those coating thicknesses in the transmission spectrum graph, where the output optical power experiences higher changes, are more sensitive. In Corres et al. (2006), a TOF-based humidity nanosensor has been proposed. In this work, the sensor sensitivity is optimized by adjusting the thickness of the nanofilm with the LbL technique. The coating process is stopped when the slope of the construction curve (Figure 7.4a) is maximal. In Figure 7.4b, the response of the sensor to human breathing is plotted, where the sensitivity enhancement can be seen.

A similar fiber structure consists of polishing the fiber, eliminating the cladding, or at least a part of it (MacDougall et al. 2002, Guo et al. 2004). D-shape fiber or side-polished fibers expose the core that can be coated with a thin film. A metal layer can be used to excite surface plasmon waves. A fiber-optic nanorefractometer fabricated by side-polishing a single-mode fiber has been demonstrated in Slavík and Homola (2001). Sensitivity of 3500 nm/refractive index unit (RIU) has been achieved (Monk and Walt 2004).

Polymer nanowires are a promising alternative for the construction of highly versatile nanosensors. In Gu et al. (2008), nanowires have been used for humidity sensing and NO_2 and NH_3 detection down to subparts-per-million level.

7.3.4 Hollow-Core Fibers and Photonic Crystal Fibers

HCFs consist of air-core fibers where the optical signal is guided mainly by the cladding (Zhu et al. 2004). If a HCF is properly connected between two silica core ones, the optical power can be coupled from the core to the cladding in the first transition, and then, from the cladding to the core in the second transition.

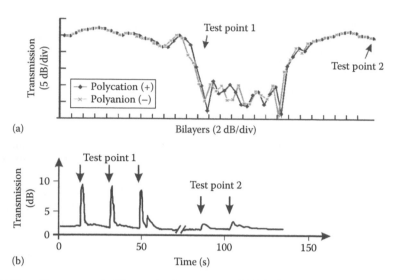

FIGURE 7.4 (a) Deposition curves of a 20 µm waist diameter taper. (b) Experimental response to the human breath, for two coating thicknesses. (Reprinted from Corres, J.M., Bravo, J., Matias, I.R., and Arregui, F.J., Nonadiabatic tapered single-mode fiber coated with humidity sensitive nanofilms, *IEEE Photonics Technol. Lett.*, 18(8), 935–937. © 2006 IEEE.)

An example of this structure is the Multimode-Hollow core-Multimode (MHM) (Bravo et al. 2006) sensor that consists of one short segment of HCF spliced between two standard multimode fibers (MMF). The central part of the structure is coated with the nanofilm. The HCF fiber is composed of a central air hole which acts as the core of the fiber, and a ring of silica around this hole which acts as the cladding. In these devices, the light that is guided in the core of the lead-in MMF can be coupled to the cladding of the HCF due to the tapered region instead of being confined in the air core. The light reaching the HCF section is guided in the core of the lead-in MMF. Then, in the HCF section, the light is guided by the cladding projecting one light ring. Finally, the light is conducted again by the lead-out MMF core. Because the light is guided by the silica cladding in the HCF region, these devices can be used as evanescent-field fiber devices. The main advantage of this structure is that the light is confined in a small area, increasing the effect due to the changes of the optical properties of the media surrounding the segment of HCF fiber.

Microstructured fibers (MFs) (Dong et al. 2005, Liu et al. 2005), holey fibers, or photonic crystal fibers (PCFs), are fibers in which air holes are incorporated within the silica; these cavities increase the interaction with the evanescent field. The size and placement of the air holes provide an extra degree of freedom in controlling light propagation. This kind of microstructured fiber has been employed to detect the presence of toxic gases (Gu 1998, Das and Thyagarajan 2001). There are also hybrid solutions which combine hollow core and holey fibers. In any case, the use of these fibers in future applications is very promising (Lee et al. 2009).

7.3.5 Fiber Bragg Gratings and Long-Period Fiber Gratings

FBGs are optical filters which allow the transmission of some wavelengths and reflect others; this is achieved by introducing a variation in the refractive index of the core of the fiber periodically along a certain length.

The change in refractive index is repeated with a spatial pitch of hundreds of nanometers, and determines the wavelength that will be reflected back. FBG-based sensors are devices normally focused on telecommunications networks and sensors, but measuring the surrounding medium is not possible with standard FBGs. In the work of Gu et al. (2011), a fiber-optic RH sensor based on a thin-core fiber modal interferometer (TCFMI) is proposed. The modal interferometer is used to measure the humidity-induced refractive index change of the nanocoating, whereas the FBG inscribed in the middle of TCFMI is used for the compensation of temperature sensitivity.

However, there is other type of fiber gratings in which the pitch of the refractive variation is much longer (typically between 100 µm to 1 mm) and can be used for evanescent sensing. These devices are known as LPGs (Falate et al. 2006).

In these devices, there is a coupling between the core and co-propagating cladding modes, unlike in FBGs, where there is a coupling between counter-propagating modes. As a result, dips are created in the transmission spectrum. The depth of the attenuation bands can be approximated with this expression (Dong et al. 2005, Kulishov and Azaña 2005):

$$T_i = \cos^2\left(k_i L\right) \tag{7.2}$$

where
  i is the cladding mode order
  k_i is the coupling coefficient
  L is the length of the grating

Using the modified phase-matching condition (Dong et al. 2006), the central resonance wavelength of the attenuation band is given by

$$\beta_{01}(\lambda) + s_0\zeta_{01,01}(\lambda) - \left(\beta_{0j}(\lambda) + s_0\zeta_{0j,0j}(\lambda)\right) = \frac{2\pi N}{\Lambda} \tag{7.3}$$

where
  β_{01} and β_{0j} are the propagation constants of the core and the j cladding modes, respectively
  $\zeta_{01,01}$ and $\zeta_{0j,0j}$ are the self-coupling coefficients of the core and the j cladding modes
  s_0 is the coefficient of the zero-frequency Fourier component of the grating
  Λ is the period of the grating
  N is the diffraction order

If we assume that the overlay of the LPG has a higher refractive index than the cladding, then, as the thickness increases, cladding modes shift their effective index to higher values (Del Villar et al. 2005a,e). In addition, with the increase of the overlay thickness, there is a fading of the attenuation bands (Rees et al. 2002, Del Villar et al. 2005a). For sensing applications, it is clear that during the transitions, the sensitivity is higher. The immediate consequence is that the modification of the parameters of the overlay to set the working point of the sensor at the center of the vanishing zone induces higher wavelength shifts in the transition region (Del Villar et al. 2005e). Using the adequate overlay, the device can be optimized for a maximum wavelength shift and thus a maximum sensitivity. Due to this improved sensitivity to the surrounding medium, LPGs have extended their applicability to measure humidity, pH, and chemical species (Corres et al. 2007a, Cusano et al. 2009), among others, (Han et al. 1998, Kim et al. 2002, Rees et al. 2002, Zhu et al. 2005, Falate et al. 2006, Abi Kaed Bey et al. 2007, Falate et al. 2007, Sakata and Ito 2007, Choi et al. 2008), and it has permitted to improve the sensitivity of previous sensors (Del Villar et al. 2005e).

7.4 Applications

Among all the potential applications related to optical fiber sensors, the monitoring of chemical or biological parameters has been established as a leading topic to these measuring systems. Particularly, optical fiber sensors are under rapid development in environmental monitoring or clinical analysis, where their small size, the flexibility of their presentation, or the capability to perform in vivo and remote measurements is a key factor (Culshaw 2005). In some applications like chemical analysis, agriculture, ecology, or the control of biological systems, it is of great interest and, in some cases, critical to monitor the refractive index, the pH value of the surrounding media, chemical compounds, or biological species (Wolfbeis 2008).

7.4.1 Optical Fiber Refractometers

A fiber-optic sensing structure based on evanescent field, such as tapered optical fibers or de-cladded fibers, can be used as a refractometer. For example, an LPG by itself can be used as a refractometer (James and Tatam 2003, Chung and Yin 2004, Cusano et al. 2005, 2006). When the refractive index of the external medium is lower than that of the cladding, as the external index approaches that of the cladding, the sensitivity of the resonance wavelength to variations of the surrounding medium refractive index (SMRI) is higher (Rees et al. 2002). However, when the SMRI exceeds that of the cladding, the core mode couples to radiation modes (Koyamada 2002) and the dependence of the resonance wavelength on the SMRI is not so important. Instead, the resonance depth is more dependent on the SMRI for values close to the refractive index of the cladding (Rees et al. 2002). In both cases, the sensitivity is higher when the SMRI is similar to that of the cladding.

Using nanotechnology, it is possible to design a device with optimal sensitivity for specific refractive index ranges of the surrounding medium. Once a refractive index for the overlay is chosen, the overlay thickness determines the range of refractive indices where the sensitivity is the highest. This overcomes the limitation of bare LPG-based refractometers to SMRI values close to the cladding refractive index, and improves the sensitivity of bare LPGs in all cases analyzed. The overall wavelength shift is more important for high-order mode resonances. The phenomenon has been experimentally analyzed (James and Tatam 2003, Chung and Yin 2004), obtaining improvements in the sensitivity by up to 10 times (Cusano et al. 2006, Corres et al. 2007). Recently, in Smietana et al. (2010), plasma-deposited diamond-like carbon (DLC) films have been used for the control of the refractive index range and sensitivity. This method has the advantage of biocompatibility, so it can easily find applications in biomedical devices.

In the same line of the refractometer sensitivity optimization but with the advantage of using only a photodetector instead of the usual optical spectrum analyzer, in Del Villar et al. (2004), a fiber-optic photonic-bandgap-based nanorefractometer is analyzed. The design is based on a one-dimensional photonic-bandgap structure with two defects, which originates two defect states inside the bandgap, corresponding to two localized modes. By selecting the adequate parameters, it is possible to determine the variation range of the transmitted power amplitude peak of the localized mode at a specific frequency, obtaining changes of up to 11.2 dB in the optical output power.

Fiber-optic SPR refractometers are generally formed by using a multimode polymer-clad silica fiber with partly removed cladding and an SPR-active metal layer deposited symmetrically around the exposed section of fiber core (Homola 2003). The operating range of SPR-based sensors may be customized by using a thin overlay (Jiří 1997, Chiu and Shih 2008) in the same manner as LPG refractometers. Multimode fiber-optic SPR-sensing devices may suffer from rather low stability because of intermodal coupling and modal noise. On the other hand, SPR sensors based on single-mode (Slavík et al. 1999, Piliarik et al. 2003) or tapered (Monzón-Hernández and Villatoro 2006) optical fiber have been proposed to reduce these problems (Del Villar et al. 2005, Zamarreño et al. 2007), achieving higher resolutions at the cost of a limited operating range. Recently, a novel fiber-optic localized plasma resonance (FO-LPR) sensor has been proposed (Chen et al. 2010b). The exposed surface of a U-shaped optical fiber was modified with self-assembled gold nanoparticles to produce the FO-LPR sensor. In this configuration, a sensitivity of 1.768×10^{-3} RIU was achieved.

In Allsop et al. (2010c), SPR devices were used to detect the variation in the refractive indices of alkane gases with measured sensitivity to index of 8300 dB RIU-1. A minimum concentration of 2% by volume of butane in ethane was achieved using single-mode SPR refractometers.

Lossy mode resonance (LMR)-based optical fiber refractometers are fabricated by using a LMR supporting layer fabricated onto the optical fiber core. These types of resonances present the advantage of being independent with the polarization of lights in contrast to SPR, which only can be generated by TM-polarized light (Del Villar et al. 2010). In addition, these devices can be fabricated by using a wide range of materials from ITO (Zamarreño et al. 2010c), indium oxide (Zamarreño et al. 2010d), TiO_2 (Hernaez et al. 2010a) or polymeric (Zamarreño et al. 2011) coatings fabricated onto optical fibers. Here, a maximum sensitivity of 2.24e-4 RIU per nanometer was achieved.

7.4.2 Optical Fiber pH Sensors

Although pH sensors based on the hydrogen electrode have coped most of the applications for many years, the development of miniaturized optical fiber pH sensors with the previously mentioned advantages associated to the optical fiber has emerged as a good alternative approach and is still a very active research topic.

Fiber-optic pH sensors can take advantage of different optical fiber configurations, such as U-bent optical fiber, cladding removed multimode fiber (CRMMF), plastic optical fibers (POF), microstructured photonic crystal fiber (MPCF) (Russell 2006), hetero-core optical fibers (HCOF), HCF, side-polished fibers, or TOFs, within different pH ranges and using diverse optical structures, as it is summarized in Table 7.1.

TABLE 7.1

Summary of Optical Fiber pH Sensors Based on Nanocoatings

Indicator	Fabrication Technique	Optical Structure	Sensing Mechanism	pH Range	Characteristics	Reference
Methyl orange	Sol-gel	CRMMF	Absorbance	2–10	Broad range	Rayss and Sudolski (2002a)
Bromophenol blue		CRPOF		5–7	Low cost	Alvarado-Méndez et al. (2005)
Methyl red		CRMMF		4.2–6		Dutta et al. (2010)
Acidochrome dye		Reflection-based		4–11	—	Lehmann et al. (1995)
Fluorescenin acrylamide		Side-polished fiber	Fluorescence	6–8	Low leaching	Wallave et al. (2001)
Thymol blue		CRMMF	Absorbance	8–12		Belhadj Miled et al. (2002)
Eosin		MPOF	Fluorescence	1.5–4.5	—	Yang and Wang (2007)
Ethyl violet		CRMMF	Absorbance	2–13	Broad range	Sharma and Gupta (2004)
Neutral red	LbL	U-bent fiber		4–11	—	Surre et al. (2009)
		CRMMF		3–9	High sensitivity	Goicoechea et al. (2008)
	Electrodeposition			3–7	Selective deposition	Zamarreño et al. (2009c)
Ruthenium (II)	Sol-gel	Tapered ends	Fluorescence	2–8	Fluorescence intensity	Gonçalves et al. (2008)
HPTS and Ruthenium (II)		V-taper		6–9	Small samples	Kasik et al. (2010)
HPTS	LbL	Taper		—	Low photobleaching	Goicoechea et al. (2007)
				3–7	Fast response time	Zamarreño et al. (2007)
Phenol red and cresol red	Sol–gel	HCOF	Absorbance	—	—	Seki et al. (2007)
Bromocresol green and cresol red		CRMMF		2–8; 9–13	Broad range	Wu et al. (2010)
Cresol red, bromophenol blue, and chlorophenol red		CRMMF		3–12	High sensitivity	Gupta and Sharma (1997)
		Side-polished fiber		3–13	Small samples	Gupta and Sharma (1998)
		CRMMF		4.5–13	Broad range	Dong et al. (2008)
Cresol red, bromophenol blue, chlorophenol red, and CNTs		CRPOF		3–9	Sensitivity	Alvarado-Méndez (2010)

(*continued*)

TABLE 7.1 (continued)

Summary of Optical Fiber pH Sensors Based on Nanocoatings

Indicator	Fabrication Technique	Optical Structure	Sensing Mechanism	pH Range	Characteristics	Reference
Quantum dots		Tapered ends	Fluorescence	3.5–8.1	—	Maule et al. (2010)
2′,7′-Bis(2-carbonylethyl)-5(6)-carboxyfluorescein		Tapered fibers		5–7	Small samples	Kasik et al. (2009), Martan et al. (2010)
Polymeric thin-film	LbL	CRMMF	Absorbance	—	Fast response	Gui et al. (2010)
Silica Matrix	Sol-gel			7–10.5	No leaching	Rayss and Sudolski (2002b)

Optical fiber-based pH monitoring techniques usually involve the use of colorimetric or fluorescent indicators. These sensors are based on the measurement of the optical properties of the selected indicator, such as absorbance, fluorescent intensity, or fluorescence lifetime, to obtain the pH of the liquid under test. These indicators are, in most cases, immobilized in a matrix (hydrogel or polymeric) (Jerónimo et al. 2007, Korostynska et al. 2007). Many examples have been proposed and demonstrated in literature owing to their simplicity and cost-effective sensing schemes. These include devices based on a single indicator, such as methyl orange (Rayss and Sudolski 2002a), methyl red (Dutta et al. 2010), acidochrome dye (Lehmann et al. 1995), fluorescein acrylamide (Wallave et al. 2001), thymol blue (TB) (Belhadj Miled et al. 2002), ethyl violet dye (Sharma and Gupta 2004), phenol red (PR) (Rovati et al. 2009), neutral red (Goicoechea et al. 2008, Surre et al. 2009, Zamarreño et al. 2009b,c), eosin (Yang and Wang 2007), HPTS (Goicoechea et al. 2007, Zamarreño et al. 2007) and ruthenium(II) (Gonçalves et al. 2008), a mixture of them in order to broaden the measurement range, such as ruthenium(II) and HPTS (Kasik et al. 2010), dipicrylamine and victoria blue (Safavi and Bagheri 2003), PR and cresol red (Seki et al. 2007), bromocresol green and cresol red (CR) (Wu et al. 2010), CR, bromophenol blue and chlorophenol red (Gupta and Sharma 1998, Dong et al. 2008), or adding carbon nanotubes (CNTs) to these indicators (Alvarado-Méndez 2010). An alternative approach can also include different pH-sensitive materials such as QDs (Maule et al. 2010).

Nevertheless, absorption- or fluorescence-based pH sensors have some inherent drawbacks since they are influenced by light intensity fluctuations, temperature, and concentration of the indicator. These drawbacks can be overcome by the use of wavelength-based detection techniques in order to obtain a more accurate and reliable measurement of the pH. Other concerns, such as bleaching or leaching, could be easily avoided by the elimination of the indicator producing a long-term fiber-optic sensor for online pH monitoring applications.

The utilization of films that change the thickness with the pH of the medium (swelling/deswelling), e.g., hydrogels or polyelectrolytes (Hiller and Rubner 2003), can be exploited in the design of robust optical fiber pH sensors based on wavelength monitoring as a function of the coating or external refractive index changes by using some of the interrogation schemes described in Section 7.3. LB and LbL deposition techniques can be used for the thin-film preparation because of the high controllability and repeatability on organic or inorganic nanocoating preparation (Flannery et al. 1999, Del Villar et al. 2010). Moreover, since most of the pH indicators work in the UV-visible range, a fiber-optic pH sensor based on a swellable film that works in the IR range can add a plus to these devices due to their compatibility with fiber-optic sensor networks. Attending to this, PAH/PAA swellable nanostructured thin films have been exploited in the fabrication of optical fiber pH sensors. In Zamarreño et al. (2009a), the PAH/PAA structure was deposited on the end surface of an optical fiber in order to form a FP interferometric nanocavity (see Section 7.3.1). This nanocavity is suitable to be excited using a halogen white light source for the measurement of pH changes, obtaining a wavelength shift of 48 nm between pH 4 and 5, as well as good repeatability and fast response times (180 and 15 s for rise and fall response time, respectively),

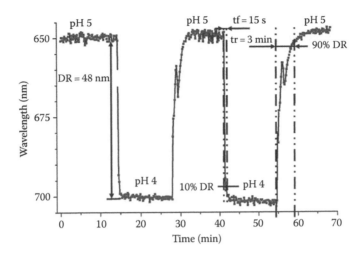

FIGURE 7.5 Dynamic response in wavelength of one of the interferometric maxima detected when the sensor is exposed to pH 5 and 4 steps alternately. (Reprinted from *Sens. Actuators B: Chem.*, 138, Goicoechea, J., Zamarreño, C.R., Matias, I.R., and Arregui, F.J., Utilization of white light interferometry in pH sensing applications by mean of the fabrication of nanostructured cavities, 613–618, Copyright 2009.)

as shown in Figure 7.5 (Goicoechea et al. 2009). In Corres et al. (2007) an LPG-based pH sensor is presented, in which the PAH/PAA nanocoating was deposited on the side-surface of the LPG. Here the variation of the film with the pH produces a shift of the resonant wavelength of the LPG (85 nm), induced by the RI changes, which can be used for the determination of pH in the range 4–7.

In Gu et al. (2009), the fabrication of the PAH/PAA nanocoating onto an HCOF structure is studied, obtaining sensitivities of 0.32 and 0.45 nm/pH unit in the pH ranges 2–7 and 10–7, respectively. An alternative approach is presented in Gui et al. (2010), by using several pairs of porous polymers deposited onto a thin-core optical fiber structure in order to obtain faster response times (20 and 15 s for rise and fall response time, respectively, within the pH range 4.5–5.5).

The PAH/PAA swelling behavior was also exploited in Zamarreño et al., by means of the fabrication of an optical fiber pH sensor based on LMR, achieving a wavelength displacement of 100 nm between pH 3 and pH 6. (Saikia et al. 2009) studied the utilization of silver nanoparticles immersed into a polymeric matrix for the fabrication of a SPR-based pH sensor. Here, the sensitivity of different optical fiber schemes, such as U-bent and U-bent tapered plastic optical fiber, between pH 3.6 and 9.3 was investigated.

7.4.3 Optical Fiber Chemical Sensors

Chemical compounds modify and interfere with natural cycles and can produce a release to the aquatic and terrestrial systems (Oehme and Wolfbeis 1997). Chemical species are essential to many organisms in small doses, whereas high doses or even small doses, in the case of toxic metals, can affect the ecosystem and human health. In this context, the utilization of small, cheap, electromagnetic immune-sensing devices, capable of selectively responding to the presence of trace amounts of these species, is very important and interesting in the fields of medicine, clinical, industrial, or environmental analysis (Malcik et al. 1998, Homola 2003, Wolfbeis 2006, McDonagh et al. 2008, Wolfbeis 2008, Allsop et al. 2009).

Optical fiber chemical sensor design simultaneously accounts for the selection of the specific coating materials (Sevilla III and Narayanaswamy 2003), the appropriate fabrication technique, and the adequate fiber-optic interrogation scheme. These devices are mostly related to the environmental monitoring of air and water pollutants, such as ions, gases, or volatile organic compounds (VOCs), as summarized in Table 7.2, though can be used for a wide range of applications (Jerónimo et al. 2007, Cusano et al. 2008a).

As it was described in the previous section, optical fiber sensors are commonly based on immobilized indicators attached to the optical fiber, which change their color, fluorescence, shape, or optical properties (refractive index) in the presence of the selected analyte. Next, some optical fiber chemical sensing applications that appeared within the last few years will be revised.

TABLE 7.2

Summary of Optical Fiber Chemical Nanosensors

Indicator	Measurand	Fabrication Technique	Interrogation Scheme	Sensing Mechanism	Reference
sPS/aPS	Cations (Na$^+$, K$^+$ and Ca^{2+})	LbL	LPGs	Wavelength shift	Manzillo et al. (2010)
Functionalized carbon dots	Hg^{2+}	Sol–gel	Tapered optical fiber	Fluorescence	Gonçalves et al. (2010)
PVC/ bis(2-ethylhexylsebacate)		LbL	U-shaped optical fiber	Absorbance	Kalvoda et al. (2010)
o-phenylenediamine	Cl$^-$	Electrodeposition	ITO-coated CRMMF		Kasik et al. (2009)
Porphyrin/TiO$_2$	HCl	Sol–gel	MPCF		Huyang et al. (2010)
Pd/Au thin films	Hydrogen	Thermal evaporation	Hetero-Core structured SMF		Luna-Moreno et al. (2011) Monzón-Hernández et al. (2009)
Pd/WO$_3$		Sputtering	Side-polished SMF and MMF		Yang et al. (2010a)
Pd/WO$_3$			SMF end tips and etched MMF		Yang et al. (2010b)
Pd/WO$_3$			MMF end tip		Ou et al. (2010)
Pd		—	MPCF		Minkovich et al. (2006)
			LPGs	Wavelength shift	Kim et al. (2009)
		—	FBGs		Chen et al. (2009)
Bromocresol purple	Ammonia	Sol-gel	Bent optical fiber	Absorbance	Tao et al. (2006)
Tetrakis-(4-sulfophenyl) porphine (TSPP)		LbL	CRMMF		Korposh et al. (2009)
			LPG		Korposh et al. (2010)
ZnO films		Sputtering	Side-polished fiber	Wavelength shift	Dikovska et al. (2010)
Zirconia		LbL	MMF	Absorbance	Galbarra et al. (2005)
SiO$_2$/PDDA + TSPP			CRMMF		Kodaira et al. (2008)
Agarose	Humidity	Sol–gel	CRMMF		Arregui et al. (2003)
			CRMMF	Wavelength shift	Hernaez et al. (2010b)
PAH/PAA thin-films		LbL	CRMMF		Zamarreño et al. (2010b)
Polymer/SiO$_2$		Sol–gel	Fiber end tip	Absorbance	Estella et al. (2010)
PDDA/polyR			HCF		Matias et al. (2007)

TABLE 7.2 (continued)

Summary of Optical Fiber Chemical Nanosensors

Indicator	Measurand	Fabrication Technique	Interrogation Scheme	Sensing Mechanism	Reference
Polyglutamic acid/ polylysine			HCMMF		Akita et al. (2010)
Hydrophobic thin-film coating		Sol–gel dip-coating	LPGs	Wavelength shift	Urrutia et al. (2010)
Polymeric coating			FBGs		Venugopalan et al. (2009)
HPTS	CO_2	Sol–gel	Tapers	Fluorescence	Lo and Chu (2008); Chu and Lo (2009)
Ag thin films	H_2S	Sputtering	Cladding-removed plastic fiber	Absorbance	Angelini et al. (2010)
			CRMMF		Neri et al. (2008)
Thymol blue	Aerosol	Sol–gel	CRPOF		Kulkarni et al. (2010)
Ruthenium(II)	Oxygen		MOFs		Matejec et al. (2008)
Pt/polymer thin films		Sol–gel dip-coating	CRMMF	Luminiscence	Chen et al. (2010a)
Pd and Ag thin films	VOCs	Physical vapor deposition	D-type fiber grating	Wavelength shift	Allsop et al. (2010b)
Functionalized SWCNTs		Langmuir blodguett	Interferometer	Absorbance	Consales et al. (2009) Crescitelli et al. (2008) Consales et al. (2008)
PVP (polyvinylpyrrolidone)		Sol–gel	FBG Fabry-Perot nanocavity	Phase shift	Jesus et al. (2009)
Vapochromic compounds		LbL	Interferometer	Absorbance	Elosua et al. (2006a, 2008, 2009)
Calixarene		Langmuir blodguett	LPGs	Wavelength shift	Topliss et al. (2010)
TiO_2/SiO_2 thin films		Sol–gel	CRMMF	Absorbance	Zaharescu et al. (2008)
Zeolite thin film		Crystallization in-situ	LPGs	Wavelength shift	Zhang et al. (2009)
ITO		Sol–gel	CRMMF		Zamarreño et al. (2010a)
Polymer	TNT	LbL	Fiber end tip	Fluorescence	Ma et al. (2010)

7.4.3.1 Environmental Monitoring

Control and improvement of the quality of drinking water have attracted the attention of specialists as well as consumers all around the world for years. The determination of the presence of different ionic species in water, even at low concentrations, is of great interest in order to prevent pollution and human diseases. As an example, in Manzillo et al. (2010), the utilization of an optical fiber sensor for the detection of different cationic species (K^+, Na^+ and Ca^{2+}) dissolved in water by using an LPG coated with a polymeric thin film is described. The detection of Hg^{2+} was also addressed in several papers. In Kalvoda et al. (2010), a U-shaped optical fiber coated with a polymer/dye thin film is employed,

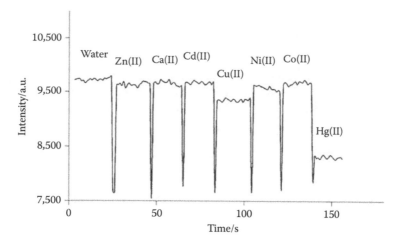

FIGURE 7.6 Steady-state fluorescence quenching (excitation 360 nm and emission 498 nm) of the optic sensing head for different interfering ions at 2.69 M. (Reprinted from *Biosens. Bioelectron.*, 26, Gonçalves, H.M.R., Duarte, A.J., and Esteves da Silva, J.C.G., Optical fiber sensor for Hg(II) based on carbon dots, 1302–1306, Copyright 2010.)

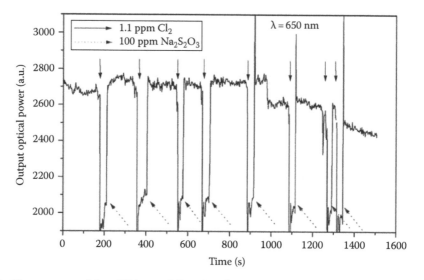

FIGURE 7.7 Time response of the o-PDA-coated fiber detection element to chlorine in water and to the regeneration solution. (Reprinted from *Sens. Actuators, B: Chem.*, 139(1), Kasik, I., Mrazek, J., Podrazky, O., Seidl, M., Aubrecht, J., Tobiska, P., Pospisilova, M., Matejec, V., Kovacs, B., Markovics, A., and Szili, M., Fiber-optic detection of chlorine in water, 139–142, Copyright 2009b.)

whereas in Gonçalves et al. (2010), are employed, TOFs coated with functionalized carbon dots by means of the sol–gel technique. Measurements of the quenching of the fluorescence attributed to Hg^{2+} and cross-sensitivity with other cations are represented in Figure 7.6.

Moreover, detection of chlorine concentration in water is crucial in order to guarantee its quality for bathing and human consumption. A novel approach is presented in Kasik et al. (2009) by means of a commercially available absorption transducer, o-phenylenediamine (o-PDA), electrochemically immobilized onto ITO-coated fiber-optic substrates. Results achieved are shown in Figure 7.7, with a lower detection limit of 0.14 ppm, which makes the detection of chlorine in water within the hygienic limits feasible.

Relative humidity (RH) monitoring also stands up as an important area of investigation and has become an interesting research field in environmental sensing. Optical fiber RH sensors can be easily developed by using a porous gel, such as a silica xerogel deposited onto optical fiber end tips (Estella et al. 2010) or an agarose hydrophilic gel deposited onto CRMMFs (Arregui et al. 2003, Hernaez et al. 2010b). Different approaches are described in Urrutia et al. (2010) and Venugopalan et al. (2009), by means of the deposition of hydrophobic and polymeric coatings onto LPGs and FBG, respectively, using the sol–gel deposition technique. An alternative humidity sensor employs LbL nanostructured PAH/PAA coatings fabricated onto ITO-coated optical fibers, achieving a sensitivity of 5.4 RH% per nm (Zamarreño et al. 2010b). LbL nanostructured films were also fabricated inside HCFs, obtaining fast response times (~300 ms) suitable to be used for breath monitoring (Matias et al. 2007).

Another important concern in environmental monitoring consists of the gas emission control (Dikovska et al. 2007, Yang et al. 2008, Culshaw 2010). Among them, greenhouse effect has to be taken into account. The immobilization of the HPTS fluorescent molecule onto an optical fiber taper via a sol–gel technique enables the determination of CO_2 concentration in atmosphere (Lo and Chu 2008, Chu and Lo 2009), with high sensitivity (ratio of fluorescence intensities ~26) compared with previous devices. Oxygen concentration measurement has been addressed in Matejec et al. (2008), using a ruthenium(II)-complex thin film deposited into a MPCF (see Figure 7.8a), with an excellent response time (see Figure 7.8b). Analogously, in Chen et al. (2010a), the quenching of fluorescence produced by oxygen is monitored. This approach consisted of Pt(II) containing polymer sensing films immobilized onto an optical fiber, showing fast response times (~0.2 s). Preliminary results on optical fiber aerosol-sensing devices have been also accomplished by immobilizing the indicator TB in a sol–gel matrix (Kulkarni et al. 2010).

The development of sensing systems for environmental and human safety monitoring is continuously gaining more attention, thanks to the growing consciousness of the importance of preventing hazardous situations, such as the presence of H_2S at low concentrations. In this context, fiber-optic technologies can be successfully employed for the development of highly sensitive and selective devices to be used in indoor and outdoor permanent monitoring. An example is given in Neri et al. (2008), by means of cladding removed glass fibers with sputtered Ag thin films as transducers. An analogous application can be found for the detection of small quantities and continuous monitoring of H_2S in the field of cultural heritage conservation in museums by using Ag-coated POFs (Angelini et al. 2010). The presence of gaseous hydrogen chloride is addressed in Huyang et al. (2010), by means of the utilization of Porphyrin/TiO_2 thin films deposited in the holes of MPOFs.

Great interest is concentrated on research and development of a reliable hydrogen sensor due to global warming, and the demand of clean and renewable energy, where H_2 has been presented as a clean and possibly inexhaustible energy source, is increasing. Within this context, optical fiber sensors, with Pd-based thin films typically used as transducers, have experienced a rapid development. Latest results are summarized in Table 7.2 by using different interrogation schemes, such as side-polished fibers (Yang et al. 2010a), hetero-core structured SMF with Pd/Au thin films (Monzón-Hernández et al. 2009,

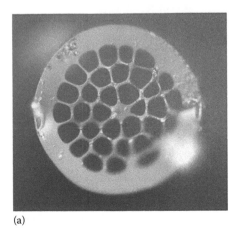

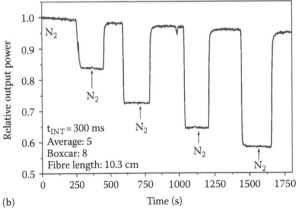

(a)   (b)

FIGURE 7.8 (a) Photo of the cross section of the prepared MSF1 (fiber diameter of 125 μm, hole diameter ~13 μm, and maximum hole pitch ~16 μm) and (b) response time curve of the MOF modified with MTES sensing layers to contact with streams of N2 or O2. (Reprinted from *Mater. Sci. Eng. C*, 28(5–6), Matejec, V., Mrázek, J., Hayer, M., Podrazký, O., Kaňka, J., and Kašík, I., Sensitivity of microstructure fibers to gaseous oxygen, 876–881, Copyright 2008.)

Luna-Moreno et al. 2011) (see Figure 7.9), LPGs (Kim et al. 2009), FBGs (Chen et al. 2009), PCFs (Minkovich et al. 2006) (see Figure 7.10), or Pd/WO3 thin films deposited onto SMF (Yang et al. 2010b) and MMF (Ou et al. 2010) end tips.

Detection and monitoring of ammonia is of enormous importance since it is an air pollutant but also acquires great interest in agriculture and many industrial processes, such as the production of fertilizers,

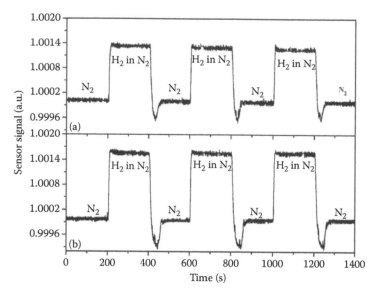

FIGURE 7.9   (a) and (b) Characteristic response of sensors with 8 nm Pd/Au coated optical fibers exposed to three consecutive cycles from a pure nitrogen atmosphere to a mixture 2% and 4% H2 in N2, respectively. (Reprinted from *Sens. Actuators, B: Chem.*, 136(2), Monzón-Hernández, D., Luna-Moreno, D., and Martínez-Escobar, D., Fast response fiber optic hydrogen sensor based on palladium and gold nano-layers, 562–566, Copyright 2009; *J. Fluoresc.*, 21, Aydogdu, S., Ertekin, K., Suslu, A., Ozdemir, M., Celik, E., and Cocen, U., Optical CO$_2$ sensing with ionic liquid doped electrospun nanofibers, 1–7, Copyright 2010.)

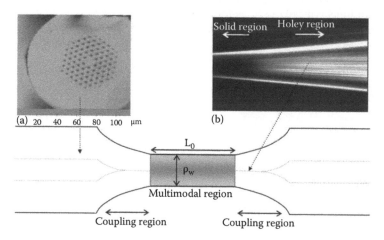

FIGURE 7.10   Images of the cross section of a tapered MOF (a) used to fabricate the tapers and of the expanding zone of the taper (b). The draw illustrates a tapered MOF. The shadowed area represents the gas-permeable thin film. L_0 is the length of the solid multimodal section and ρ_w is the taper waist diameter. (Reprinted from *Opt. Express*, 14(18), Minkovich, V.P., Monzón-Hernández, D., Villatoro, J., and Badenes, G., Microstructured optical fiber coated with thin films for gas and chemical sensing, 8413–8418, Copyright 2006; Zamarreño, C.R., Hernaez, M., Del Villar, I., Matias, I.R., and Arregui, F.J. 2010a, Sensing properties of ITO coated optical fibers to diverse VOCs, *24th Eurosensors Conference*, Linz, Austria, September 5–8, 2010, p. 653; *Appl. Opt.*, 46(13), Dikovska, A.O., Atanasov, P.A., Stoyanchov, T.R., Andreev, A.T., Karakoleva, E.I., and Zafirova, B.S., Pulsed laser deposited ZnO film on side-polished fiber as a gas sensing element, 2481–2485, Copyright 2007, with permission from Elsevier.)

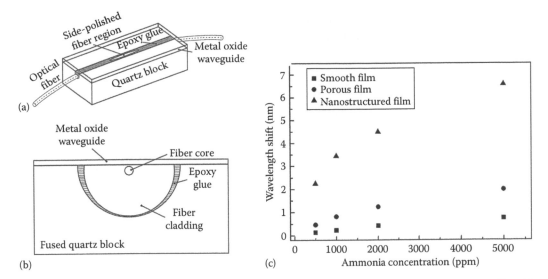

FIGURE 7.11 (a) A schematic, (b) cross section of the side-polished fiber sensor element and (c) wavelength shifts of the smooth, porous, and nanostructured ZnO sensor elements in dependence of the ammonia concentration. (Reprinted from *Sens. Actuators, B: Chem.*, 146(1), Dikovska, A.O., Atanasova, G.B., Nedyalkov, N.N., Stefanov, P.K., Atanasov, P.A., Karakoleva, E.I., and Andreev, A.T., Optical sensing of ammonia using ZnO nanostructure grown on a side-polished optical-fiber, 331–336, Copyright 2010, with permission from Elsevier.)

plastics, explosives, pulp and paper, oil refinery, and power generation. Here, low-cost, simple, highly sensitive, and small-sized nanostructured optical fiber sensors are presented as an opportunity for ammonia detection. In Tao et al. (2006), a sensor capable of detecting ammonia concentrations of 13 and 5 ppb in air and water, respectively, by using an U-shaped optical fiber coated with a bromocresol purple-based nanocoating is developed. The sensitivity of ZnO thin films sputtered onto side-polished optical fibers to gaseous ammonia is represented in Figure 7.11 (Dikovska et al. 2010). LbL technique was used for the fabrication of ammonia-sensitive thin films composed by zirconia (Galbarra et al. 2005) and a porphine-based compound. The porphine-based coating was fabricated onto a CRMMF (Korposh et al. 2009) and LPGs (Korposh et al. 2010), achieving a maximum sensitivity of 1 ppm. An alternative approach was studied in Kodaira et al. (2008), by employing a previous PDDA/SiO$_2$ porous structure in order to improve the reactivity of the device.

7.4.3.2 Volatile Organic Compound Sensing

A growing area of interest is the detection, monitoring, and analysis of VOCs since they can be breathed and in some cases, such as aromatic VOCs, cause adverse health effects. VOCs are commonly used as ingredients in household products or in industrial processes and normally get vaporized at room temperature. Moreover, VOCs are also present in synthetic products as paints, some foods and beverages. In some special workplaces, such as in the chemical industries, is of vital importance to monitor the concentration of these vapors to preserve the health of the workers, and also assure that concentrations do not surpass the safe levels. The emerging of optical fiber sensors offers novel and interesting approaches that can overcome some of the traditional disadvantages of the traditional sensors (Elosua et al. 2006b). In Allsop et al. (2010b), Allsop et al. demonstrated the utilization of a D-shaped optical fiber with a Ag/Pt thin-film grating onto it in order to sense different types of alkanes (methane, ethane, butane, and propane), achieving a minimum concentration detection of 2% of butane in ethane. In the same manner, in Topliss et al. (2010), the detection of several VOCs by the deposition of calixarene-based coatings onto LPGs via the LB technique is studied. The sensing capabilities of these coatings to diverse VOCs show maximum sensitivity for Toluene. In Zaharescu et al. (2008), the detection of toluene by applying different TiO$_2$/SiO$_2$ thin films onto CRMMF, obtaining a detection limit of 0.07 vol.% of toluene in nitrogen is also

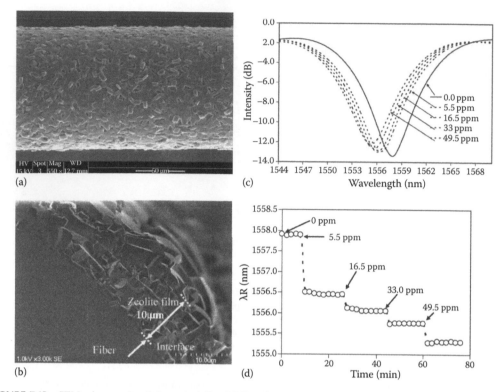

FIGURE 7.12 SEM micrographs of the optical fiber LPG surface (a) and cross section (b) coated with the zeolite film. Displacement of the transmission spectrum produced in response to variations of isopropanol vapor concentration in air (c) and temporal response of the resonance wavelength to concentration variations (d). (Reprinted from *Sens. Actuators, B: Chem.*, 135(2), Zhang, J., Tang, X., Dong, J., Wei, T., and Xiao, H., Zeolite thin film-coated long period fiber grating sensor for measuring trace organic vapors, 420–425, Copyright 2009, with permission from Elsevier.)

studied. LPGs coated with porous zeolite thin films, represented in Figure 7.12a and b, were employed in Zhang et al. (2009), showing high sensitivities in the range of ppm and ppb for isopropanol and toluene, respectively (see Figure 7.12c and d).

Moreover, the sensing properties of dual-resonance, ITO-coated CRMMFs to diverse VOCs has been addressed in Zamarreño et al. (2010a), showing maximum sensitivity to ethanol (26 nm wavelength shift between 0 and 500 ppm.). An optical fiber sensing device for carboxylic acid species in water is also presented in Jesus et al. (2009). This device consisted of PVP (polyvinylpyrrolidone) composite materials fabricated onto SMF end tips combined with a FBG detection system. The results obtained showed a resolution of 0.2% L/L in the 0.6%–3.3% concentration range and fast recovery times of the sensor between 6 and 4 s. Moreover, the configuration used in this work presented potential advantages, such as the fact that it operates in the third optical fiber communication window (1550 nm).

Consales et al. presented the fabrication of low-finesse FP interferometers on the end tips of optical fibers using LB technique, as represented in Figure 7.13a. The utilization of these coatings, based on cadmium arachidate (CdA) and single-walled carbon nanotubes (SWCNTs), enabled the detection of NO_2 as well as different VOCs, as represented in Figure 7.13b (Consales et al. 2008, 2009, Crescitelli et al. 2008). Ethanol, as well as other VOCs have been monitored by means of the immobilization of vapochromic compounds onto the optical fiber end tips in a FP-like optical fiber sensing scheme (Elosua et al. 2006a, 2008).

7.4.4 Optical Fiber Biological Sensors

Nanobiosensors are sensing devices that have biologically active structures attached to the transducer for the detection of the target analyte. The active components can be bacterial cells, enzymes, antibodies, and other bioreceptor proteins. In the last few years, advances have been made in lowering

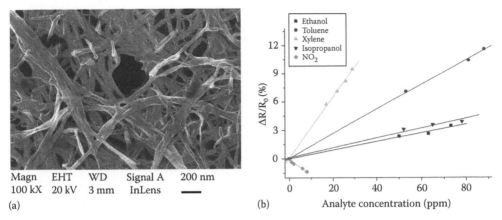

FIGURE 7.13 (a) SEM micrographs of CdA/SWCNT nanocomposite multilayers composed of 10 monolayers, realized by using a SWCNT filler content percentage of 75 wt.% and (b) comparison between the sensor characteristic curves obtained when the multilayer films are subjected to five different chemicals. (Reprinted from *Sens. Actuators, B: Chem.*, 138(1), Consales, M., Crescitelli, A., Penza, M., Aversa, P., Veneri, P.D., Giordano, M., and Cusano, A., SWCNT nano-composite optical sensors for VOC and gas trace detection, 351–361, Copyright 2009; *Sens. Actuators, B: Chem.*, 115(1), Tao, S., Xu, L., and Fanguy, J.C., Optical fiber ammonia sensing probes using reagent immobilized porous silica coating as transducers, 158–163, Copyright 2006, with permission from Elsevier.)

detection limits, increasing sensitivity, selectivity, and in some cases, reversibility as well as the shelf life and long-term stability (Monk and Walt 2004). Now, it is possible to detect microbial pathogens and toxins in minutes rather than days. Biosensors have applications for food-borne contaminants, and waterborne contaminants, as well as infectious disease pathogens. Also, the miniaturization of the sensing schemes, mainly using tapered tips, has enabled monitoring of individual biological living cells (Vo-Dinh and Kasili 2005). This has lead to a better understanding of the microdynamics of living systems.

There are two types of biosensors (Marazuela and Moreno-Bondi 2002): The first type is catalytic biosensors. Usually, the biologically active component is one type of enzyme, although a microorganism, a subcellular organelle, or a tissue slice can also be the biocatalyst. In this type of sensors, a steady-state concentration of a transducer-detectable molecule must be achieved. The concentration of the analyte is related to the rate of formation of a detectable product that is monitored. The second type is affinity biosensors. In these sensors, the receptor molecule is commonly an antibody, but also can be a nucleic acid, or a hormone receptor. The binding of the bioreceptor and the target molecule produces change of a physicochemical parameter measured by the transducer. Mimics of proteins and cells on nanofilms are under development for the enhancement of long-term stability of biosensors. Many research groups are investigating techniques for patterning biomolecules at a molecular level (Holliger and Hoogenboom 1995, Jelinek and Kolusheva, Velasco-Garcia 2009).

7.4.4.1 Catalytic Biosensors

Although advances have been done in the development of substitutes of biologically active components, enzymes are still indispensable tools in biosensing. The high sensitivity, the fast response, the large number of selective reactions that are implicated, and the broad range of analytes that can be detected are among the main reasons of the utilization of enzymes.

Glucose sensing has been always a priority and, currently, due to the high incidence of diabetes, the development of a continuous and implantable glucose sensor remains as a fundamental objective. Both glucose oxidase (GOx) and glucose dehydrogenase have been employed in fiber-optic nanosensors. According to the procedure used in Lenarczuk et al. (2001), glucose and oxygen (O_2) in the presence of enzyme GOx are transformed into glucolactone and hydrogen peroxide (H_2O_2). This last chemical species can be detected using a redox reaction with Prussian blue. The LbL structure $PAH + PB^+/GOx^-$ has been successfully applied in Del Villar et al. (2006). In the presence of glucose and O_2, the sensor

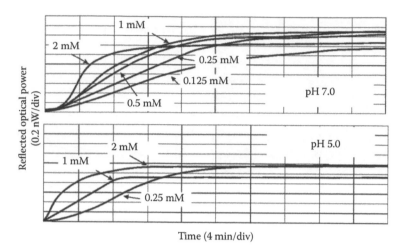

FIGURE 7.14 Reflected optical power for different concentrations of glucose at constant pHs of 5 and 7. (Reprinted from Del Villar, et al., *Opt. Eng.*, 45, 104, 2006. With permission from SPIE.)

generates H_2O_2 as a product of the catalytic reaction. Then, the H_2O_2 oxidizes Prussian white to Prussian blue. A change in the reflected optical power is then produced, as shown in Figure 7.14.

7.4.4.2 Affinity Nanobiosensors

Antibodies have the exceptional capacity to bind specifically the analyte of interest (antigen). When there is an excess of immobilized antibodies with respect to the antigen, the measured signal is directly proportional to the amount of antigen. Other assay formats can be also used in biosensing, like competitive assay, which is based on the competition between a labeled analyte and the analyte in the sample. In this assay, the number of binding sites is limited. Also, in order to decrease the possibility of interference, the sandwich assay can be used. In this assay, there are two types of antibodies which have to bind to the respective sites in the antigen. As in the direct assay, the antigens are incubated with an excess of antibodies. Then, the antigen–antibody composite is incubated with a secondary labeled antibody that has affinity for a second antigenic site. Finally, the transduced signal is proportional to the analyte concentration. Due to the high sensitivity needed, the most commonly used techniques for the evaluation of the antigen–antibody interactions are evanescent wave and SPR. In addition, the higher the affinity of the antibody for the analyte, the better the sensitivity achieved.

Nanodeposition on the cladding of LPGs was achieved for the first time in 2002 (Rees et al. 2002), starting the usage of these devices for biosensing. In Kim et al. (2006), the reflective LPG configuration is selected for the fabrication of an immunosensor and self-assembled polyelectrolyte layers and the immobilization of immunoglobulin G (IgG) is used for sensor fabrication. The sensor's fringe position shift induced by specific antigen–antibody binding has been observed, and nonspecific binding can be minimized with binding block. The binding of antigens to the receptor causes a wavelength shift of 0.1 nm. The temperature cross-sensitivity of the sensor used in this experiment has been measured to be approximately 0.09 nm/°C. The thickness and refractive index of the polymer film can be controlled by varying processing parameters to optimize the refractive index of the composite film, which allows the sensor to operate in the more sensitive range. In Yang et al. (2008), the reflective configuration has been used with the particularity that the region where the phase shift occurs is tapered, resulting in an improved sensitivity.

In Corres et al. (2008), a biconical tapered optical fiber is coated using the LbL electrostatic self-assembly technique with an anti-gliadin antibodies (AGAs)-sensitive nanofilm in order to aid the diagnosis of the celiac disease. Power changes up to 6 dB have been shown with concentrations in the range 1–15 ppm. In Figure 7.15, the dynamic response of the sensor is represented.

In Cullum et al. (2000), a nanobiosensor for the measurement of biomarkers, which are associated with exposure to polycyclic aromatic hydrocarbons is shown.

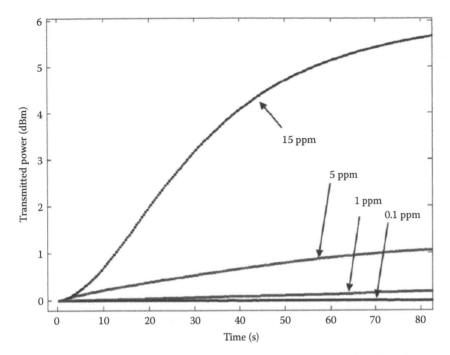

FIGURE 7.15 Dynamic response of sensor T2 after the injection of antibodies. (Reprinted from *Sens. Actuators, B: Chem.*, 135(1), Corres, J.M., Matias, I.R., Bravo, J., and Arregui, F.J., Tapered optical fiber biosensor for the detection of anti-gliadin antibodies, 166–171, Copyright 2008, with permission from Elsevier.)

In Fehr et al. (2003), a nanosensor for monitoring the dynamics of glucose in individual cells is analyzed using fluorescence resonance energy transfer (FRET). The same technique has been applied to maltose detection using QDs as highly fluorescent donors (Medintz et al. 2003).

Conjugates of CdSe/ZnS core/shell QDs-human IgG (QD-IgG) has been prepared in Zhang et al. (2010). The fiber-optic nanosensor demonstrated an important reduction of the detection limit (5 ng/mL with respect to the 10 ng/mL of conventional fluorescein isothiocyanate) and the possibility of multi-analyte assay capability.

In Petrosova et al. (2007), an Ebola virus immunosensor has been studied. In Figure 7.16, the sensing scheme is shown. For the detection of antibodies directed against antigens of the Ebola virus strains, the virus antigens are immobilized using a photoactivable electrogenerated poly(pyrrole-benzophenone) film deposited upon an indium tin oxide (ITO) modified conductive surface fiber-optic. The obtained results reported an increase of one order of magnitude respect standard ELISA test has been reported.

The Group of (Vo-Dinh 2008) has achieved advances in the detection of toxic compounds throughout a single cell using nanobiosensors (see Figure 7.17). Carcinogen polycyclic aromatic hydrocarbons such as benzo[α]pyrene tetrol and benzo[α]pyrene have been detected using antibodies attached to the tip of a tapered optical fiber.

In Zheng and Li (2010), the same structure has been used for the detection of the general cancer biomarker telomerase. The scheme can be seen in Figure 7.18. The nanotip immobilized with a specific antibody demonstrated the successful detection of the telomerase overexpression in cancer cells, providing a potential method for cancer detection.

Diagnosis of genetic diseases and DNA sequencing has promoted the need of genetic nanosensors. In Long et al. (2011), detection of DNA has been achieved using QDs and total internal reflection fluorescence. Apart from the low detection limit of this sensor (3.2 atto-molar of bound target DNA), the self-assembled alkanethiol monolayer is highly stable; it maintains its performance after more than 30 cycles. In Figure 7.19, the schematic of this biosensor is shown.

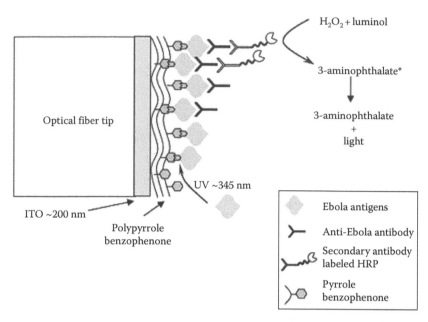

FIGURE 7.16 The biosensor scheme describing the various steps involved in the immunoassay using ITO–poly(pyrrole-benzophenone)-coated optical fibers for the detection of anti-Ebola virus in sera samples. (Reprinted from *Sens. Actuators, B: Chem.*, 122(2), Petrosova, A., Konry, T., Cosnier, S., Trakht, I., Lutwama, J., Rwaguma, E., Chepurnov, A., Mühlberger, E., Lobel, L., and Marks, R.S., Development of a highly sensitive, field operable biosensor for serological studies of Ebola virus in central Africa, 578–586, Copyright 2007, with permission from Elsevier.)

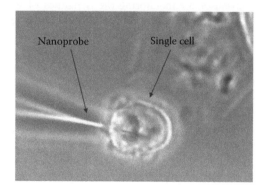

FIGURE 7.17 Photograph of a fiber-optic nanobiosensors inserted into a single cell. (Reprinted from *Spectrochim. Acta Part B: Atomic Spectrosc.*, 63, Vo-Dinh, T., Nanosensing at the single cell level, 95–103, Copyright 2008, with permission from Elsevier.)

7.4.4.3 Other Applications

Fiber-optic sensors are also well suited for the measurement of a wide variety of physical parameters. Among them, mechanical parameters like force, pressure, path length, velocity, acceleration, rotation rate, vibration, flow rate, filling level, bending, and mechanical strain up to breakage, electric and magnetic fields, as well as temperature or sound have been measured using optical fiber sensors (Strobel et al. 2009). However, these applications have been rarely addressed by the utilization of nanostructured coatings (García Moreda et al. 2006).

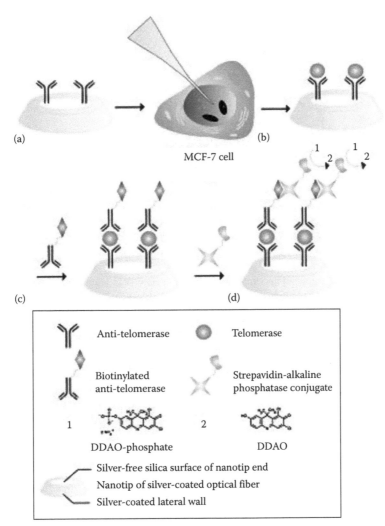

FIGURE 7.18 Single-cell telomerase detection by optical fiber nanobiosensor. (a) Fiber-end after the anti-telomerase adhesion. (b) Fiber-end after the cell-telomerase harvest. (c) Fiber-end after the biotinylated anti-telomerase conjugation. (d) Fiber end after the addition of streptavidin-alkaline phosphatase conjugate fluorescent molecule. (Reprinted from *Biosens. Bioelectron.*, 25(6), Zheng, X.T. and Li, C.M., Single living cell detection of telomerase over-expression for cancer detection by an optical fiber nanobiosensor, 1548–1552, Copyright 2010, with permission from Elsevier.)

7.5 Conclusions

The association of nanocoatings with optical fiber enables advantages of optical fiber (small size, high sensitivity, large bandwidth, immunity to electromagnetic interference, light weight, and ease of implementation of multiplexed and distributed sensors) to be combined with the interesting properties of novel structures ordered at the nanoscale level. In this chapter, we have reviewed some of the most recent applications in optical fiber sensors based on nanostructured coatings covering a wide range of interrogation schemes and fabrication techniques. Thus, in order to achieve an optimal sensor response, a precise selection of the specific coating materials, the appropriate fabrication technique, and the adequate fiber-optic interrogation scheme is required.

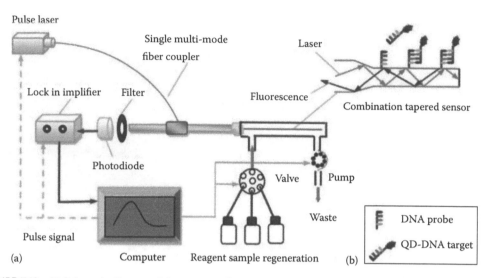

FIGURE 7.19 (a) Schematic diagram of the structure of evanescent wave DNA biosensor and (b) schematic diagram of combination-tapered fiber sensor and mechanical schematic of DNA detection. (Reprinted from *Biosens. Bioelectron.*, 26(5), Long, F., Wu, S., He, M., Tong, T., and Shi, H., Ultrasensitive quantum dots-based DNA detection and hybridization kinetics analysis with evanescent wave biosensing platform, 2390–2395, Copyright 2011, with permission from Elsevier.)

7.6 Future Trends

The results presented in this chapter reveal the potentiality of the integrated optical fiber sensing approach and the perspectives for future applications. Particularly, the integration of novel sensitive nanocoatings combined with advanced sensing configurations and optical fiber structures such as TOFs, HCOF, or MPCFs presents interesting opportunities in the development of novel fiber-optic chemo- and bionanosensors in the next few years.

Acknowledgment

This work was supported by the Spanish Ministry of Education and Science TEC2010–17805 Research Grant.

REFERENCES

Abi Kaed Bey, S.K., Sun, T., and Grattan, K.T.V. 2007, Optimization of a long-period grating-based Mach-Zehnder interferometer for temperature measurement, *Optics Communications*, 272(1), 15–21.

Ahn, J.S., Hammond, P.T., Rubner, M.F., and Lee, I. 2005, Self-assembled particle monolayers on polyelectrolyte multilayers: Particle size effects on formation, structure, and optical properties, *Colloids and Surfaces A: Physicochemical and Engineering Aspects*, 259(1–3), 45–53.

Akita, S., Sasaki, H., Watanabe, K., and Seki, A. 2010, A humidity sensor based on a hetero-core optical fiber, *Sensors and Actuators, B: Chemical*, 147(2), 385–391.

Alivisatos, A.P. 1996, Semiconductor clusters, nanocrystals, and quantum dots, *Science*, 271(5251), 933–937.

Allsop, T., Dubov, M., Martinez, A., Floreani, F., Khrushchev, I., Webb, D.J., and Bennion, I. 2006, Bending characteristics of fiber long-period gratings with cladding index modified by femtosecond laser, *Journal of Lightwave Technology*, 24(8), 3147.

Allsop, T., Nagel, D., Neal, R., Davies, E.M., Mou, C., Bond, P., Rehman, S., Kalli, K., Webb, D.J., Calverhouse, P., Mascini, M., and Bennion, I. 2010a, Aptamer-based surface plasmon fibre sensor for thrombin detection, *Biophotonics: Photonic Solutions for Better Health Care II*, April 12, 2010–April 16, 2010, p. 77151C.

Allsop, T., Neal, R., Davies, E.M., Mou, C., Bond, P., Rehman, S., Kalli, K., Webb, D.J., Calverhouse, P., and Bennion, I. 2010b, Low refractive index gas sensing using a surface plasmon resonance fibre device, *Measurement Science and Technology*, 21(9), 094029 (9pp).

Allsop, T., Neal, R., Mou, C., Brown, P., Saied, S., Rehman, S., Kalli, K., Webb, D.J., Sullivan, J., Mapps, D., and Bennion, I. 2009, Exploitation of multilayer coatings for infrared surface plasmon resonance fiber sensors, *Applied Optics*, 48(2), 276–286.

Allsop, T. et al. 2010c, Low refractive index gas sensing using a surface plasmon resonance fibre device, *Measurement Science and Technology*, 21, 9, 094029. *Doi:10.1088/0957-0233/21/9/094029*.

Alvarado-Méndez, E. 2010, pH biosensor with plastic fiber optic doped with carbone nanotubes used sol-gel technique, Proc. SPIE 7839, 78391C. doi:10.1117/12.867005.

Alvarado-Méndez, E., Rojas-Laguna, R., Andrade-Lucio, J.A., Hernández-Cruz, D., Lessard, R.A., and Aviña-Cervantes, J.G. 2005, Design and characterization of pH sensor based on sol–gel silica layer on plastic optical fiber, *Sensors and Actuators B: Chemical*, 106(2), 518–522.

Andrade, G.F.S., Fan, M., and Brolo, A.G. 2010, Multilayer silver nanoparticles-modified optical fiber tip for high performance SERS remote sensing, *Biosensors and Bioelectronics*, 25(10), 2270–2275.

Angelini, E., Grassini, S., Mombello, D., Neri, A., Parvis, M., and Perrone, G. 2010, Plasma modified POF sensors for in situ environmental monitoring of museum indoor environments, *Applied Physics A: Materials Science and Processing*, 100(3), 975–980.

Arregui, F.J., Ciaurriz, Z., Oneca, M., and Matias, I.R. 2003, An experimental study about hydrogels for the fabrication of optical fiber humidity sensors, *Sensors and Actuators, B: Chemical*, 96(1–2), 165–172.

Arregui, F.J., Dickerson, B., Claus, R.O., Matias, I.R., and Cooper, K.L. 2002a, Polymeric thin films of controlled complex refractive index formed by the electrostatic self-assembled monolayer process, *IEEE Photonics Technology Letters*, 13(12), 1319–1321.

Arregui, F.J., Latasa, I., Matias, I.R., and Claus, R.O. 2004, An optical fiber pH sensor based on the electrostatic self-assembly method, *Proceedings of IEEE Sensors*, Toronto, Canada, 2003, p. 107.

Arregui, F.J., Liu, Y., Matias, I.R., and Claus, R.O. 1999a, Optical fiber humidity sensor using a nano Fabry-Perot cavity formed by the ionic self-assembly method, *Sensors and Actuators B: Chemical*, 59(1), 54–59.

Arregui, F.J., Matias, I.R., and Claus, R.O. 2002b, Optical fiber sensors based on nanoscale self-assembly, in *Transducing Materials and Devices*, ed. Bar-Cohen, Y., October 30, 2002 through November 1, 2002, SPIE, Bellingham, WA, p. 17.

Arregui, F.J., Matías, I.R., and Claus, R.O. 2007, Optical fiber sensors based on nanostructured coatings fabricated by means of the layer-by-layer electrostatic self-assembly method, *EWOFS 2007: Third European Workshop on Optical Fibre Sensors*, Naples, Italy, July 4–6, 2007, 66190F.

Arregui, F.J., Matias, I.R., Corres, J.M., Del Villar, I., Goicoechea, J., Zamarrenoa, C.R., Hernáez, M., and Claus, R.O. 2010, Optical fiber sensors based on layer-by-layer nanostructured films, *24th Eurosensors Conference*, Linz, Austria, September 5–8, 2010, p. 1087.

Arregui, F.J., Matias, I.R., Liu, Y., Lenahan, K.M., and Claus, R.O. 1999b, Optical fiber nanometer-scale Fabry–Perot interferometer formed by the ionic self-assembly monolayer process, *Optics Letters*, 24(9), 596–598.

Ashwell, G.J., Skjonnemand, K., Roberts, M.P.S., Allen, D.W., Li, X., Sworakowski, J., Chyla, A., and Bienkowski, M. 1999, Surface plasmon resonance and nonlinear optical studies of Langmuir-Blodgett films of a betaine dye, *Colloids and Surfaces A: Physicochemical and Engineering Aspects*, 155(1), 43–46.

Atias, D., Liebes, Y., Chalifa-Caspi, V., Bremand, L., Lobel, L., Marks, R.S., and Dussart, P. 2009, Chemiluminescent optical fiber immunosensor for the detection of IgM antibody to dengue virus in humans, *Sensors and Actuators, B: Chemical*, 140(1), 206–215.

Aydogdu, S., Ertekin, K., Suslu, A., Ozdemir, M., Celik, E., and Cocen, U. 2010, Optical CO_2 sensing with ionic liquid doped electrospun nanofibers, *Journal of Fluorescence*, 21, 1–7.

Belhadj Miled, O., Ben Ouada, H., and Livage, J. 2002, pH sensor based on a detection sol-gel layer onto optical fiber, *Materials Science and Engineering C*, 21(1–2), 183–188.

Bender, W.J.H., Dessy, R.E., Miller, M.S., and Claus, R.O. 1994, Feasibility of a chemical microsensor based on surface plasmon resonance on fiber optics modified by multilayer vapor deposition, *Analytical Chemistry*, 66(7), 963–970.

Bennion, I., Williams, J.A.R., Zhang, L., Sugden, K., and Doran, N.J. 1996, UV-written in-fibre Bragg gratings, *Optical and Quantum Electronics*, 28(2), 93–135.

Bhat, R.R., Fischer, D.A., and Genzer, J. 2002, Fabricating planar nanoparticle assemblies with number density gradients, *Langmuir*, 18(15), 5640–5643.

Bravo, J., Matías, I.R., Villar, I.D., Corres, J.M., and Arregui, F.J. 2006, Nanofilms on hollow core fiber-based structures: An optical study, *Journal of Lightwave Technology*, 24(5), 2100.

Brinker, C.J., Frye, G.C., Hurd, A.J., and Ashley, C.S. 1991, Fundamentals of sol-gel dip coating, *Thin Solid Films*, 201(1), 97–108.

Chen, T., Buric, M.P., Xu, D., Chen, K.P., Swinehart, P.R., and Maklad, M. 2009, All-fiber low-temperature hydrogen sensing using a multi-functional light source, *20th International Conference on Optical Fibre Sensors*, Edinburgh, Scotland, October 5–9, 2009, 750317(4pp).

Chen, R., Farmery, A.D., Obeid, A., and Hahn, C.E.W. 2010a, A cylindrical-core fibre optic oxygen sensor based on Pt (II) complexes immobilized in a polymer matrix, *4th European Workshop on Optical Fibre Sensors*, Porto, Portugal, September 8–10, 2010, 76531U.

Chen, C., Tsao, T., Li, W., Shen, W., Cheng, C., Tang, J., Jen, C., Chau, L., and Wu, W. 2010b, Novel U-shape gold nanoparticles-modified optical fiber for localized plasmon resonance chemical sensing, *Microsystem Technologies*, 16(7), 1207–1214.

Chen, X., Zhou, K., Zhang, L., and Bennion, I. 2005, Simultaneous measurement of temperature and external refractive index by use of a hybrid grating in D fiber with enhanced sensitivity by HF etching, *Applied Optics*, 44(2), 178–182.

Chiu, M. and Shih, C. 2008, Searching for optimal sensitivity of single-mode D-type optical fiber sensor in the phase measurement, *Sensors and Actuators B: Chemical*, 131(2), 596–601.

Choi, H.Y., Park, K.S., and Lee, B.H. 2008, Photonic crystal fiber interferometer composed of a long period fiber grating and one point collapsing of air holes, *Optics Letters*, 33(8), 812–814.

Choi, J. and Rubner, M.F. 2005, Influence of the degree of ionization on weak polyelectrolyte multilayer assembly, *Macromolecules*, 38(1), 116–124.

Chopra, K.L., Major, S., and Pandya, D.K. 1983, Transparent conductors-A status review, *Thin Solid Films*, 102(1), 1–46.

Choy, K.L. 2003, Chemical vapour deposition of coatings, *Progress in Materials Science*, 48(2), 57–170.

Chu, C. and Lo, Y. 2009, Highly sensitive and linear optical fiber carbon dioxide sensor based on sol-gel matrix doped with silica particles and HPTS, *Sensors and Actuators, B: Chemical*, 143(1), 205–210.

Chung, K.W. and Yin, S. 2004, Analysis of a widely tunable long-period grating by use of an ultrathin cladding layer and higher-order cladding mode coupling, *Optics Letters*, 29(8), 812–814.

Connor, D.J. and Ferri, B.A. 2007, The conflict within: resistance to inclusion and other paradoxes in special education, *Disability and Society*, 22(1), 63–77.

Consales, M., Crescitelli, A., Penza, M., Aversa, P., Giordano, M., Cutolo, A., and Cusano, A. 2008, SWCNTs-based nanocomposites as sensitive coatings for advanced fiber optic chemical nanosensors, *Optical Sensors 2008*, Strasbourg, France, April 7–10, 2008, p. 70030E.

Consales, M., Crescitelli, A., Penza, M., Aversa, P., Veneri, P.D., Giordano, M., and Cusano, A. 2009, SWCNT nano-composite optical sensors for VOC and gas trace detection, *Sensors and Actuators, B: Chemical*, 138(1), 351–361.

Corres, J.M., Arregui, F.J., and Matias, I.R. 2006a, Design of humidity sensors based on tapered optical fibers, *Journal of Lightwave Technology*, 24(11), 4329–4336.

Corres, J.M., Bravo, J., Matias, I.R., and Arregui, F.J. 2006b, Nonadiabatic tapered single-mode fiber coated with humidity sensitive nanofilms, *IEEE Photonics Technology Letters*, 18(8), 935–937.

Corres, J.M., del Villar, I., Matias, I.R., and Arregui, F.J. 2007a, Enhanced sensitivity in humidity sensors based on long period fiber gratings, *5th IEEE Conference on IEEE Sensors*, Daegu, Korea, October 2006, p. 193.

Corres, J.M., Del Villar, I., Matias, I.R., and Arregui, F.J. 2007b, Fiber-optic pH-sensors in long-period fiber gratings using electrostatic self-assembly, *Optics Letters*, 32(1), 29–31.

Corres, J.M., Matias, I.R., Bravo, J., and Arregui, F.J. 2008, Tapered optical fiber biosensor for the detection of anti-gliadin antibodies, *Sensors and Actuators, B: Chemical*, 135(1), 166–171.

Corres, J.M., Matias, I.R., del Villar, I., and Arregui, F.J. 2007c, Design of pH sensors in long-period fiber gratings using polymeric nanocoatings, *IEEE Sensors Journal*, 7(3), 455–463.

Crescitelli, A., Consales, M., Cutolo, A., Cusano, A., Penza, M., Aversa, P., and Giordano, M. 2008, Novel sensitive nanocoatings based on SWCNT composites for advanced fiber optic chemo-sensors, *2008 IEEE Sensors, SENSORS 2008*, Leece, Italy, October 26–29, 2009, p. 965.

Cullum, B.M., Griffin, G.D., Miller, G.H., and Vo-Dinh, T. 2000, Intracellular measurements in mammary carcinoma cells using fiber-optic nanosensors, *Analytical Biochemistry*, 277(1), 25–32.

Culshaw, B. 2004, Optical fiber sensor technologies: Opportunities and—Perhaps—Pitfalls, *Journal of Lightwave Technology*, 22(1), 39–50.

Culshaw, B. 2005, Research to reality: Bringing fibre optic sensors into applications, *Optical Fibers: Applications*, eds. Jaroszewicz, L.R., Culshaw, B., and Mignani, A.G., August 31–September 2, 2005, p. 1.

Culshaw, B. 2010, Fibre optic systems for gas detection principles, progress and prospects, *Quantum and Nonlinear Optics*, Beijing, China, October 18–19, 2010, 785302(8pp).

Culshaw, B. and Kersey, A. 2008, Fiber-optic sensing: A historical perspective, *Journal of Lightwave Technology*, 26(9), 1064–1078.

Cusano, A., Giordano, M., Cutolo, A., Pisco, M., and Consales, M. 2008a, Integrated development of chemoptical fiber nanosensors, *Current Analytical Chemistry*, 4(4), 296–315.

Cusano, A., Iadicicco, A., Pilla, P., Contessa, L., Campopiano, S., Cutolo, A., and Giordano, M. 2005, Cladding mode reorganization in high-refractive-index-coated long-period gratings: Effects on the refractive-index sensitivity, *Optics Letters*, 30(19), 2536.

Cusano, A., Iadicicco, A., Pilla, P., Contessa, L., Campopiano, S., Cutolo, A., and Giordano, M. 2006, Mode transition in high refractive index coated long period gratings, *Optics Express*, 14(1), 19–34.

Cusano, A., López-Higuera, J.M., Matias, I.R., and Culshaw, B. 2008b, Editorial optical fiber sensor technology and applications, *IEEE Sensors Journal*, 8(7), 1052–1054.

Cusano, A., Pilla, P., Contessa, L., Iadicicco, A., Campopiano, S., Cutolo, A., Giordano, M., and Guerra, G. 2009, High-sensitivity optical chemosensor based on coated long-period gratings for sub-ppm chemical detection in water, *Applied Physics Letters*, 87(23), 234–105.

Das, M. and Thyagarajan, K. 2001, Wavelength-division multiplexing isolation filter using concatenated chirped long period gratings, *Optics Communications*, 197(1–3), 67–71.

Decher, G. 1997, Fuzzy nanoassemblies: Toward layered polymeric multicomposites, *Science*, 277(5330), 1232–1237.

Del Villar, I., Achaerandio, M., Matías, I.R., and Arregui, F.J. 2005a, Deposition of overlays by electrostatic self-assembly in long-period fiber gratings, *Optics Letters*, 30(7), 720–722.

Del Villar, I., Matias, I.R., and Arregui, F.J. 2008, Fiber-optic chemical nanosensors by electrostatic molecular self-assembly, *Current Analytical Chemistry*, 4(4), 341–355.

Del Villar, I., Matias, I.R., Arregui, F.J., and Claus, R.O. 2004, Fiber-optic nanorefractometer based on one-dimensional photonic-bandgap structures with two defects, *IEEE Transactions on Nanotechnology*, 3(2), 293–299.

Del Villar, I., Matias, I.R., Arregui, F.J., and Claus, R.O. 2005b, ESA-based in-fiber nanocavity for hydrogen-peroxide detection, *IEEE Transactions on Nanotechnology*, 4(2), 187–193.

Del Villar, I., Matías, I.R., Arregui, F.J., and Claus, R.O. 2005c, Fiber-optic hydrogen peroxide nanosensor, *IEEE Sensors Journal*, 5(3), 365.

Del Villar, I., Matias, I.R., Arregui, F.J., and Corres, J.M. 2006, Fiber optic glucose biosensor, *Optical Engineering*, 45, 104–401.

Del Villar, I., Matias, I.R., Arregui, F.J., Echeverría, J., and Claus, R.O. 2005d, Strategies for fabrication of hydrogen peroxide sensors based on electrostatic self-assembly (ESA) method, *Sensors and Actuators B: Chemical*, 108(1–2), 751–757.

Del Villar, I., Matías, I., Arregui, F., and Lalanne, P. 2005e, Optimization of sensitivity in long period fiber gratings with overlay deposition, *Optics Express*, 13(1), 56–69.

Del Villar, I., Zamarreño, C.R., Hernaez, M., Arregui, F.J., and Matias, I.R. 2010, Generation of lossy mode resonances with absorbing thin-films, *Journal of Lightwave Technology*, 28(23), 3351–3357.

Díaz-Herrera, N., Navarrete, M.C., Esteban, O., and González-Cano, A. 2004, A fibre-optic temperature sensor based on the deposition of a thermochromic material on an adiabatic taper, *Measurement Science and Technology*, 15(2), 353–358.

Díez, A., Andrés, M.V., and Cruz, J.L. 1999, Hybrid surface plasma modes in circular metal-coated tapered fibers, *Journal of the Optical Society of America A: Optics and Image Science, and Vision*, 16(12), 2978–2982.

Dikovska, A.O., Atanasova, G.B., Nedyalkov, N.N., Stefanov, P.K., Atanasov, P.A., Karakoleva, E.I., and Andreev, A.T. 2010, Optical sensing of ammonia using ZnO nanostructure grown on a side-polished optical-fiber, *Sensors and Actuators, B: Chemical*, 146(1), 331–336.

Dikovska, A.O., Atanasov, P.A., Stoyanchov, T.R., Andreev, A.T., Karakoleva, E.I., and Zafirova, B.S. 2007, Pulsed laser deposited ZnO film on side-polished fiber as a gas sensing element, *Applied Optics*, 46(13), 2481–2485.

Dong, S., Luo, M., Peng, G., and Cheng, W. 2008, Broad range pH sensor based on sol-gel entrapped indicators on fibre optic, *Sensors and Actuators, B: Chemical*, 129(1), 94–98.

Dong, X., Su, L., Shum, P., Chung, Y., and Chan, C. 2006, Wavelength-selective all-fiber filter based on a single long-period fiber grating and a misaligned splicing point, *Optics Communications*, 258(2), 159–163.

Dong, X., Yang, X., Shum, P., and Chan, C.C. 2005, Tunable WDM filter with 0.8-nm channel spacing using a pair of long-period fiber gratings, *IEEE Photonics Technology Letters*, 17(4), 795–797.

Doshi, J. and Reneker, D.H. 1995, Electrospinning process and applications of electrospun fibers, *Journal of Electrostatics*, 35(2–3), 151–160.

Dutta, S., Basak, S., Kumar, R., and Samanta, P.K. 2010, Fabrication of intensity based fiber optic pH Sensor, *2010 3rd International Nanoelectronics Conference, INEC 2010*, Hong Kong, January 3–8, 2010, p. 370.

Elosúa, C., Bariáin, C., Matías, I.R., Arregui, F.J., Luquin, A., Vergara, E., and Laguna, M. 2008, Indicator immobilization on Fabry-Perot nanocavities towards development of fiber optic sensors, *Sensors and Actuators, B: Chemical*, 130(1), 158–163.

Elosua, C., Matias, I.R., Bariain, C., and Arregui, F.J. 2006a, Development of an in-fiber nanocavity towards detection of volatile organic gases, *Sensors*, 6(6), 578–592.

Elosua, C., Matias, I.R., Bariain, C., and Arregui, F.J. 2006b, Volatile organic compound optical fiber sensors: A review, *Sensors*, 6(11), 1440–1465.

Elosua, C., Perez-Herrera, R.A., Lopez-Amo, M., Bariain, C., Luquin, A., and Laguna, M. 2009, Remote sensing network to detect and identify organic vapours, *20th International Conference on Optical Fibre Sensors*, Edinburgh, Scotland, October 5–9, 2009, 750316(4pp).

Estella, J., de Vicente, P., Echeverría, J.C., and Garrido, J.J. 2010, A fibre-optic humidity sensor based on a porous silica xerogel film as the sensing element, *Sensors and Actuators B: Chemical*, 149(1), 122–128.

Falate, R., Frazão, O., Rego, G., Fabris, J.L. and Santos, J.L. 2006, Refractometric sensor based on a phase-shifted long-period fiber grating, *Applied Optics*, 45(21), 5066–5072.

Falate, R., Frazão, O., Rego, G., Ivanov, O., Kalinowski, H., Fabris, J. and Santos, J. 2007, Bending sensitivity dependent on the phase shift imprinted in long-period fibre gratings, *Measurement Science and Technology*, 18, 3123.

Fan, J. and Zhao, Y. 2005, Direct deposition of aligned nanorod array onto cylindrical objects, *Journal of Vacuum Science and Technology B: Microelectronics and Nanometer Structures*, 23(3), 947–953.

Fehr, M., Lalonde, S., Lager, I., Wolff, M.W. and Frommer, W.B. 2003, In vivo imaging of the dynamics of glucose uptake in the cytosol of COS-7 cells by fluorescent nanosensors, *Journal of Biological Chemistry*, 278(21), 19127.

Flannery, D., James, S.W., Tatam, R.P., and Ashwell, G.J. 1999, Fiber-optic chemical sensing with Langmuir-Blodgett overlay waveguides, *Applied Optics*, 38(36), 7370–7374.

Galbarra, D., Arregui, F.J., Matias, I.R., and Claus, R.O. 2005, Ammonia optical fiber sensor based on self-assembled zirconia thin films, *Smart Materials and Structures*, 14(4), 739–744.

García Moreda, F.J., Arregui, F.J., Achaerandio, M., and Matias, I.R. 2006, Study of indicators for the development of fluorescence based optical fiber temperature sensors, *Sensors and Actuators, B: Chemical*, 118(1–2), 425–432.

Goicoechea, J., Zamarreño, C.R., Matias, I.R., and Arregui, F.J. 2007, Minimizing the photobleaching of self-assembled multilayers for sensor applications, *Sensors and Actuators, B: Chemical*, 126(1), 41–47.

Goicoechea, J., Zamarreño, C.R., Matías, I.R., and Arregui, F.J. 2008, Optical fiber pH sensors based on layer-by-layer electrostatic self-assembled Neutral Red, *Sensors and Actuators, B: Chemical*, 132(1), 305–311.

Goicoechea, J., Zamarreño, C.R., Matias, I.R., and Arregui, F.J. 2009, Utilization of white light interferometry in pH sensing applications by mean of the fabrication of nanostructured cavities, *Sensors and Actuators B: Chemical*, 138(2), 613–618.

Gonçalves, H.M.R., Duarte, A.J., and Esteves da Silva, J.C.G. 2010, Optical fiber sensor for Hg(II) based on carbon dots, *Biosensors and Bioelectronics*, 26(4), 1302–1306.

Gonçalves, H.M.R., Maule, C.D., Jorge, P.A.S., and Esteves da Silva, J.C.G. 2008, Fiber optic lifetime pH sensing based on ruthenium(II) complexes with dicarboxybipyridine, *Analytica Chimica Acta*, 626(1), 62–70.

Gonthier, F., Lapierre, J., Veilleux, C., Lacroix, S., and Bures, J. 1987, Investigation of power oscillations along tapered monomode fibers, *Applied Optics*, 26(3), 444–449.

Gu, X. 1998, Wavelength-division multiplexing isolation fiber filter and light source using cascaded long-period fiber gratings, *Optics Letters*, 23(7), 509–510.

Gu, B., Yin, M.-, Zhang, A.P., Qian, J.-, and He, S. 2009, Low-cost high-performance fiber-optic pH sensor based on thin-core fiber modal interferometer, *Optics Express*, 17(25), 22296–22302.

Gu, B., Yin, M., Zhang, A.P., Qian, J., and He, S. 2011, Optical fiber relative humidity sensor based on FBG incorporated thin-core fiber modal interferometer, *Optics Express*, 19(5), 4140–4146.

Gu, F., Zhang, L., Yin, X., and Tong, L. 2008, Polymer single-nanowire optical sensors, *Nano Letters*, 8(9), 2757–2761.

Gui, Z., Qian, J., Yin, M., An, Q., Gu, B., and Zhang, A. 2010, A novel fast response fiber-optic pH sensor based on nanoporous self-assembled multilayer films, *Journal of Materials Chemistry*, 20(36), 7754–7760.

Guo, S., Albin, S., and Rogowski, R. 2004, Comparative analysis of Bragg fibers, *Optics Express*, 12(1), 198–207.

Gupta, B.D. and Sharma, D.K. 1997, Evanescent wave absorption based fiber optic pH sensor prepared by dye doped sol-gel immobilization technique, *Optics Communications*, 140(1–3), 32–35.

Gupta, B.D. and Sharma, S. 1998, A long-range fiber optic pH sensor prepared by dye doped sol-gel immobilization technique, *Optics Communications*, 154(5–6), 282–284.

Han, Y.G., Lee, J.H., and Lee, S.B. 1998, Discrimination of bending and temperature sensitivities with phase-shifted long-period fiber gratings depending on initial coupling strength, *Electronics Letters*, 34, 1773–1775.

Hernaez, M., Zamarreño, C.R., Del Villar, I., Matias, I.R., and Arregui, F.J. 2010a, Lossy mode resonances supported by TiO2-coated optical fibers, *24th Eurosensors Conference*, Linz, Austria, September 5–8, 2010, p. 1099.

Hernaez, M., Zamarreño, C.R., Fernandez-Valdivielso, C., Del Villar, I., Arregui, F.J., and Matias, I.R. 2010b, Agarose optical fibre humidity sensor based on electromagnetic resonance in the infra-red region, *Physica Status Solidi (C) Current Topics in Solid State Physics*, 7(11–12), 2767–2769.

Hiller, J. and Rubner, M.F. 2003, Reversible molecular memory and pH-switchable swelling transitions in polyelectrolyte multilayers, *Macromolecules*, 36, 11, 4078–4083.

Holliger, P. and Hoogenboom, H.R. 1995, Artificial antibodies and enzymes: Mimicking nature and beyond, *Trends in Biotechnology*, 13(1), 7.

Homola, J. 2003, Present and future of surface plasmon resonance biosensors, *Analytical and Bioanalytical Chemistry*, 377(3), 528–539.

Huang, D., Swanson, E.A., Lin, C.P., Schuman, J.S., Stinson, W.G., Chang, W., Hee, M.R., Flotte, T., Gregory, K., Puliafito, C.A., and Fujimoto, J.G. 1991, Optical coherence tomography, *Science*, 254(5035), 1178–1181.

Huyang, G., Canning, J., Åslund, M.L., Naqshbandi, M., Stocks, D., and Crossley, M.J. 2010, Remote gaseous acid sensing within a porphyrin-doped TiO$_2$ sol-gel layer inside a structured optical fibre, *4th European Workshop on Optical Fibre Sensors*, Porto, Portugal, September 8–10, 2010.

I, R.P. 1990, *Langmuir-Blodgett Films*.

Ishaq, I.M., Quintela, A., James, S.W., Ashwell, G.J., Lopez-Higuera, J.M., and Tatam, R.P. 2005, Modification of the refractive index response of long period gratings using thin film overlays, *Sensors and Actuators, B: Chemical*, 107(2), 738–741.

James, S.W., Ishaq, I., Ashwell, G.J., and Tatam, R.P. 2005, Cascaded long-period gratings with nanostructured coatings, *Optics Letters*, 30(17), 2197–2199.

James, S.W. and Tatam, R.P. 2003, Optical fibre long-period grating sensors: Characteristics and application, *Measurement Science and Technology*, 14, R49.

James, S.W. and Tatam, R.P. 2006, Fibre optic sensors with nano-structured coatings, *Journal of Optics A: Pure and Applied Optics*, 8(7).

Jayawardhana, S., Kostovski, G., Mazzolini, A.P., and Stoddart, P.R. 2011, Optical fiber sensor based on oblique angle deposition, *Applied Optics*, 50(2), 155–162.

Jelinek, R. and Kolusheva, S. 2007. Biomimetic Nanosensors in Nanotechnologies for the Life Sciences, ed. Kumar, C.S.S.R. Wiley-VCH, Vol 8.

Jerónimo, P.C.A., Araújo, A.N., and Conceição B.S.M. Montenegro, M. 2007, Optical sensors and biosensors based on sol-gel films, *Talanta*, 72(1), 13–27.

Jesus, C., Silva, S.F.O., Castanheira, M., González Aguilar, G., Frazão, O., Jorge, P.A.S., and Baptista, J.M. 2009, Interferometric fibre-optic sensor for acetic acid measurement, *20th International Conference on Optical Fibre Sensors*, Edinburgh, Scotland, October 5–9, 2009.

Jiří, H. 1997, On the sensitivity of surface plasmon resonance sensors with spectral interrogation, *Sensors and Actuators B: Chemical*, 41(1–3), 207–211.

Joannopoulos, J.D., Villeneuve, P.R., and Fan, S. 1997, Photonic crystals: Putting a new twist on light, *Nature*, 386(6621), 143–149.

Jorgenson, R.C. and Yee, S.S. 1993, A fiber-optic chemical sensor based on surface plasmon resonance, *Sensors and Actuators B: Chemical*, 12(3), 213–220.

Kalvoda, L., Aubrecht, J., Klepáček, R., and Lukášová, P. 2010, Sensing applications of U-optrodes, *4th European Workshop on Optical Fibre Sensors*, Porto, Portugal, September 8–10, 2010, 765329(4pp).

Kasik, I., Martan, T., Podrazky, O., Mrazek, J., Pospisilova, M., and Matejec, V. 2009a, Local real-time detection of pH using fibre tapers, *Optical Sensors 2009*, Prague, Czech Republic, April 20–22, 2009, p. 73561U.

Kasik, I., Mrazek, J., Martan, T., Pospisilova, M., Podrazky, O., Matejec, V., Hoyerova, K., and Kaminek, M. 2010, Fiber-optic pH detection in small volumes of biosamples, *Analytical and Bioanalytical Chemistry*, 398(5), 1883–1889.

Kasik, I., Mrazek, J., Podrazky, O., Seidl, M., Aubrecht, J., Tobiska, P., Pospisilova, M., Matejec, V., Kovacs, B., Markovics, A., and Szili, M. 2009b, Fiber-optic detection of chlorine in water, *Sensors and Actuators, B: Chemical*, 139(1), 139–142.

Kersey, A.D., Davis, M.A., Patrick, H.J., LeBlanc, M., Koo, K.P., Askins, C.G., Putnam, M.A., and Friebele, E.J. 1997, Fiber grating sensors, *Journal of Lightwave Technology*, 15(8), 1442–1462.

Kim, Y.H., Kim, M.J., Rho, B.S., Choi, H.Y., Park, M.-, Jang, J.-, and Lee, B.H. 2009, Ultra-high sensitive sensitive hydrogen sensor using higher order cladding mode coupled by a palladium-coated long-period fiber grating, *20th International Conference on Optical Fibre Sensors*, Edinburgh, Scotland, October 5–9, 2009, 75030Y.

Kim, Y., Paek, U.C., and Han, W.T. 2002, Fiber length dependence of phase change induced by laser-diode pumping in Yb^3-Al^3 co-doped optical fibers, *IEEE Photonics Technology Letters*, 14(12), 1710–1712.

Kim, D., Zhang, Y., Cooper, K., and Wang, A. 2006, Fibre-optic interferometric immuno-sensor using long period grating, *Electronics Letters*, 42(6), 324–325.

Kittidechachan, M., Sripichai, I., Supakum, W., Thuamthai, S., Angkaew, S., and Limsuwan, P. 2008, Construction and evaluation of the fiber-optic sensor system for chemical vapor detection, *Advance Materials Research*, 55–57, 509–512.

Kodaira, S., Korposh, S., Lee, S.-, Batty, W.J., James, S.W., and Tatam, R.P. 2008, Fabrication of highly efficient fibre-optic gas sensors using SiO_2/polymer nanoporous thin films, *3rd International Conference on Sensing Technology, ICST 2008*, Tainan, Taiwan, November 30–December 3, 2008, p. 481.

Konry, T., Heyman, Y., Cosnier, S., Gorgy, K., and Marks, R.S. 2008, Characterization of thin poly(pyrrole-benzophenone) film morphologies electropolymerized on indium tin oxide coated optic fibers for electrochemical and optical biosensing, *Electrochimica Acta*, 53(16), 5128–5135.

Konry, T. and Marks, R.S. 2005, Physico-chemical studies of indium tin oxide-coated fiber optic biosensors, *Thin Solid Films*, 492(1–2), 313–321.

Korostynska, O., Arshak, K., Gill, E., and Arshak, A. 2007, Review on state-of-the-art in polymer based pH sensors, *Sensors*, 7(12), 3027–3042.

Korposh, S., Batty, W., Kodaira, S., Leeb, S.-, James, S.W., Topliss, S.M., and Tatamc, R.P. 2010, Ammonia sensing using a fibre optic long period grating with a porous nanostructured coating formed from silica nanospheres, *4th European Workshop on Optical Fibre Sensors*, Porto, Portugal, September 8–10, 2010.

Korposh, S., Kodaira, S., Batty, W., James, S.W., and Lee, S. 2009, Nanoassembled thin-film gas sensor II. An intrinsic highly sensitive fibre optic sensor for ammonia detection, *Sensors and Materials*, 21(4), 179–189.

Koyamada, Y. 2002, Numerical analysis of core-mode to radiation-mode coupling in long-period fiber gratings, *IEEE Photonics Technology Letters*, 13(4), 308–310.

Kroto, H.W., Heath, J.R., O'Brien, S.C., Curl, R.F., and Smalley, R.E. 1985, C_{60}: Buckminsterfullerene, *Nature*, 318(6042), 162–163.

Kulishov, M. and Azaña, J. 2005, Long-period fiber gratings as ultrafast optical differentiators, *Optics Letters*, 30(20), 2700–2702.

Kulkarni, A., Lee, J., Nam, J., and Kim, T. 2010, Thin film-coated plastic optical fiber probe for aerosol chemical sensing applications, *Sensors and Actuators, B: Chemical*, 150(1), 154–159.

Lee, B. 2003, Review of the present status of optical fiber sensors, *Optical Fiber Technology*, 9(2), 57–79.

Lee, D., Omolade, D., Cohen, R.E., and Rubner, M.F. 2007, pH-Dependent structure and properties of TiO_2/SiO_2 nanoparticle multilayer thin films, *Chemistry of Materials*, 19(6), 1427–1433.

Lee, B., Roh, S., and Park, J. 2009, Current status of micro- and nano-structured optical fiber sensors, *Optical Fiber Technology*, 15(3), 209–221.

Lee, S.G., Sokoloff, J.P., McGinnis, B.P., and Sasabe, H. 1998, Polymer waveguide overlays for side-polished fiber devices, *Applied Optics*, 37(3), 453–462.

Lehmann, H., Schwotzer, G., Czerney, P., and Mohr, G.J. 1995, Fiber-optic pH meter using NIR dye", *Sensors and Actuators: B. Chemical*, 29(1–3), 392–400.

Lenarczuk, T., Wencel, D., Gb, S., and Koncki, R. 2001, Prussian blue-based optical glucose biosensor in flow-injection analysis, *Analytica Chimica Acta*, 447(1–2), 23–32.

Leung, A., Shankar, P.M., and Mutharasan, R. 2007, A review of fiber-optic biosensors, *Sensors and Actuators B: Chemical*, 125(2), 688–703.

Liebes, Y., Marks, R.S., and Banai, M. 2009, Chemiluminescent optical fiber immunosensor detection of Brucella cells presenting smooth-A antigen, *Sensors and Actuators, B: Chemical*, 140(2), 568–576.

Liu, Y., Yu, Z., Yang, H., Zhang, N., Feng, Q., and Zhang, X. 2005, Numerical optimization and simulation to wavelength-division multiplexing isolation filter consisted of two identical long period fiber grating, *Optics Communications*, 246(4–6), 367–372.

Lo, Y.- and Chu, C.- 2008, Fiber-optic carbon dioxide sensor based on fluorinated xerogels doped with HPTS, *19th International Conference on Optical Fibre Sensors*, Perth, Australia, April 15–18, 2008, p. 70040P.

Long, F., Wu, S., He, M., Tong, T., and Shi, H. 2011, Ultrasensitive quantum dots-based DNA detection and hybridization kinetics analysis with evanescent wave biosensing platform, *Biosensors and Bioelectronics*, 26(5), 2390–2395.

Lukowiak, A. and Strek, W. 2009, Sensing abilities of materials prepared by sol-gel technology, *Journal of Sol-Gel Science and Technology*, 50(2), 201–215.

Luna-Moreno, D., Monzon-Hernandez, D., Calixto-Carrera, S., and Espinosa-Luna, R. 2011, Tailored Pd-Au layer produced by conventional evaporation process for hydrogen sensing, *Optics and Lasers in Engineering*, 49, 693–697.

Ma, J., Kos, A., Bock, W.J., Li, X., Nguyen, H., Wang, Z.Y., and Cusano, A. 2010, Lab-on-a-Fiber: Building a fiber-optic sensing platform for low-cost and high-performance trace vapor TNT detection, *4th European Workshop on Optical Fibre Sensors*, Porto, Portugal, September 8–10, 2010, 76531E.

MacDougall, T.W., Pilevar, S., Haggans, C.W., and Jackson, M.A. 2002, Generalized expression for the growth of long period gratings, *IEEE Photonics Technology Letters*, 10(10), 1449–1451.

Mack, C.A. 2008, Proximity distance, *Microlithography World*, 17(4), 10–11.

Maier, S.A., Brongersma, M.L., Kik, P.G., Meltzer, S., Requicha, A.A.G., and Atwater, H.A. 2001, Plasmonics—A route to nanoscale optical devices, *Advanced Materials*, 13(19), 1501–1505.

Malcik, N., Oktar, O., Ozser, M.E., Caglar, P., Bushby, L., Vaughan, A., Kuswandi, B., and Narayanaswamy, R. 1998, Immobilised reagents for optical heavy metal ions sensing, *Sensors and Actuators, B: Chemical*, 53(3), 211–221.

Manzillo, P.F., Pilla, P., Campopiano, S., Borriello, A., Giordano, M., and Cusano, A. 2010, Self assembling and coordination of water nano-layers on polymeric coated long period gratings as promising tool for cation detection, *4th European Workshop on Optical Fibre Sensors*, Porto, Portugal, September 8–10, 2010, 76531Y.

Marazuela, M.D. and Moreno-Bondi, M.C. 2002, Fiber-optic biosensors—An overview, *Analytical and Bioanalytical Chemistry*, 372(5–6), 664–682.

Marks, R.S., Novoa, A., Konry, T., Krais, R., and Cosnier, S. 2002, Indium tin oxide-coated optical fiber tips for affinity electropolymerization, *Materials Science and Engineering C*, 21(1–2), 189–194.

Martan, T., Pospisilova, M., Aubrecht, J., Mrazek, J., Podrazky, O., Kasik, I., and Matejec, V. 2010, Tapered optical fibres for local pH detection, *Journal of Physics: Conference Series*, 206.

Matejec, V., Mrázek, J., Hayer, M., Podrazký, O., Kaňka, J., and Kašík, I. 2008, Sensitivity of microstructure fibers to gaseous oxygen, *Materials Science and Engineering C*, 28(5–6), 876–881.

Matias, I.R., Arregui, F.J., Corres, J.M., and Bravo, J. 2007, Evanescent field fiber-optic sensors for humidity monitoring based on nanocoatings, *IEEE Sensors Journal*, 7(1), 89–95.

Matias, I.R., Del Villar, I., Arregui, F.J., and Claus, R.O. 2003, Comparative study of the modeling of three-dimensional photonic bandgap structures, *JOSA A*, 20(4), 644–654.

Mattox, D.M. 2000, Physical vapor deposition (PVD) processes, *Metal Finishing*, 98(1), 410–423.

Maule, C., Gonçalves, H., Mendonça, C., Sampaio, P., Esteves da Silva, J.C.G., and Jorge, P. 2010, Wavelength encoded analytical imaging and fiber optic sensing with pH sensitive CdTe quantum dots, *Talanta*, 80(5), 1932–1938.

McDonagh, C., Burke, C.S., and MacCraith, B.D. 2008, Optical chemical sensors, *Chemical Reviews*, 108(2), 400–422.

Medintz, I.L., Clapp, A.R., Mattoussi, H., Goldman, E.R., Fisher, B., and Mauro, J.M. 2003, Self-assembled nanoscale biosensors based on quantum dot FRET donors, *Nature Materials*, 2(9), 630–638.

Minkovich, V.P., Monzón-Hernández, D., Villatoro, J., and Badenes, G. 2006, Microstructured optical fiber coated with thin films for gas and chemical sensing, *Optics Express*, 14(18), 8413–8418.D.J. and Walt, D.R. 2004, Optical fiber-based biosensors, *Analytical and Bioanalytical Chemistry*, 379(7–8), 931–945.

Monzón-Hernández, D., Luna-Moreno, D., and Martínez-Escobar, D. 2009, Fast response fiber optic hydrogen sensor based on palladium and gold nano-layers, *Sensors and Actuators, B: Chemical*, 136(2), 562–566.

Monzón-Hernández, D. and Villatoro, J. 2006, High-resolution refractive index sensing by means of a multiple-peak surface plasmon resonance optical fiber sensor, *Sensors and Actuators B: Chemical*, 115(1), 227–231.

Nagaraju, B., Varshney, R.K., Pal, B.P., Singh, A., Monnom, G., and Dussardier, B. 2008, Design and realization of a side-polished single-mode fiber optic highsensitive temperature sensor, *Photonics, Devices, and Systems IV*, August 27–29, 2008, p. 71381H.

Navarrete, M.-, Díaz-Herrera, N., González-Cano, A., and Esteban, O. 2010, A polarization-independent SPR fiber sensor, *Plasmonics*, 5(1), 7–12.

Neri, A., Parvis, M., Perrone, G., Grassini, S., Angelini, E., and Mombello, D. 2008, Low-cost fiber optic H$_2$S gas sensor, *2008 IEEE Sensors, SENSORS 2008*, Leece, Italy, October 26–29, 2008, p. 313.

Oehme, I. and Wolfbeis, O.S. 1997, Optical sensors for determination of heavy metal ions, *Mikrochimica Acta*, 126(3–4), 177–192.

Ou, J., Yaacob, M.H., Campbell, J.L., Kalantar-zadeh, K., and Wlodarski, W. 2010, H$_2$ sensing performance of optical fiber coated with nano-platelet WO$_3$ film, *24th Eurosensors Conference*, Linz, Austria, September 5–8, 2010, p. 1204.

Peterson, I. R. 1990, Langmuir-Blodgett Films, *Journal of Physics D: Applied Physics*, 23(4), 379–395.

Petrosova, A., Konry, T., Cosnier, S., Trakht, I., Lutwama, J., Rwaguma, E., Chepurnov, A., Mühlberger, E., Lobel, L., and Marks, R.S. 2007, Development of a highly sensitive, field operable biosensor for serological studies of Ebola virus in central Africa, *Sensors and Actuators, B: Chemical*, 122(2), 578–586.

Petty, M.C. 1992, Possible applications for Langmuir-Blodgett films, *Thin Solid Films*, 210–211, PART 2, 417–426.

Piliarik, M., Homola, J., Maníková, Z., and Čtyroký, J. 2003, Surface plasmon resonance sensor based on a single-mode polarization-maintaining optical fiber, *Sensors and Actuators B: Chemical*, 90, 1–3, 236–242.

Ramachandran, S., Wang, Z., and Yan, M. 2002, Bandwidth control of long-period grating-based mode converters in few-mode fibers, *Optics Letters*, 27(9), 698–700.

Rayss, J. and Sudolski, G. 2002a, Fibre optic sensor for broad range pH detection, in *Optoelectronic and Electronic Sensors V*, ed. Kalita, W., Rzeszów, Poland, June 5–8, 2002, p. 32.

Rayss, J. and Sudolski, G. 2002b, Ion adsorption in the porous sol–gel silica layer in the fibre optic pH sensor, *Sensors and Actuators B: Chemical*, 87(3), 397–405.

Rees, N.D., James, S.W., Tatam, R.P., and Ashwell, G.J. 2002, Optical fiber long-period gratings with Langmuir—Blodgett thin-film overlays, *Optics Letters*, 27(9), 686–688.

Rindorf, L., Jensen, J.B., Dufva, M., Pedersen, L.H., Høiby, P.E., and Bang, O. 2006, Photonic crystal fiber long-period gratings for biochemical sensing, *Optics Express*, 14, 18, 8224–8231.

Rovati, L., Fabbri, P., and Pilati, F. 2009, Development of a low-cost pH sensor based on plastic optical fibers, *2009 IEEE Instrumentation and Measurement Technology Conference, I2MTC 2009*, Singapore, May 5–7, 2009, p. 1662.

Russell, P.S.J. 2006, Photonic-crystal fibers, *Journal of Lightwave Technology*, 24(12), 4729–4749.

Safavi, A. and Bagheri, M. 2003, Novel optical pH sensor for high and low pH values, *Sensors and Actuators, B: Chemical*, 90(1–3), 143–150.

Saikia, R., Buragohain, M., Datta, P., Nath, P., and Barua, K. 2009, Fiber-optic pH sensor based on SPR of silver nanostructured film, *1st International Conference on Transport And Optical Properties Of Nanomaterials, ICTOPON 2009*, Allahabad, India, January 5–8, 2009, p. 249.

Sakata, H. and Ito, H. 2007, Optical fiber temperature sensor using a pair of nonidentical long-period fiber gratings for intensity-based sensing, *Optics Communications*, 280(1), 87–90.

Schröder, K., Csaki, A., Latka, I., Henkel, T., Malsch, D., Schuster, K., Schneider, T., and Zopf, D. 2010, Microstructured optical fiber with homogeneous monolayer of plasmonic nanoparticles for bioanalysis, *4th European Workshop on Optical Fibre Sensors*, Porto, Portugal, September 8–10, 2010.

Seki, A., Katakura, H., Kai, T., Iga, M., and Watanabe, K. 2007, A hetero-core structured fiber optic pH sensor, *Analytica Chimica Acta*, 582(1), 154–157.

Sevilla III, F. and Narayanaswamy, R. 2003, Optical chemical sensors and biosensors, in *Comprehensive Analytical Chemistry*, ed. Alegret, S., Elsevier, Amsterdam, the Netherlands, Chapter 9, pp. 413–435.

Sharma, N.K. and Gupta, B.D. 2002, Fabrication and characterization of U-shaped fiber-optic pH probes, *Sensors and Actuators B: Chemical*, 82(1–1), 89–93.

Sharma, N.K. and Gupta, B.D. 2004, Fabrication and characterization of a fiber-optic pH sensor for the pH range 2 to 13, *Fiber and Integrated Optics*, 23(4), 327–335.

Shiratori, S.S. and Rubner, M.F. 2000, pH-dependent thickness behavior of sequentially adsorbed layers of weak polyelectrolytes, *Macromolecules*, 33(11), 4213–4219.

Slavik, R. 2006, Extremely deep long-period fiber grating made with CO/sub 2/laser, *IEEE Photonics Technology Letters*, 18(16), 1705–1707.

Slavík, R. and Homola, J. 2001, Novel spectral fiber optic sensor based on surface plasmon resonance, *Sensors and Actuators B: Chemical*, 74(1–3), 106–111.

Slavík, R., Homola, J., and Čtyroký, J. 1999, Single-mode optical fiber surface plasmon resonance sensor, *Sensors and Actuators B: Chemical*, 54(1–2), 74–79.

Smietana, M., Bock, W.J., Szmidt, J., and Pickrell, G.R. 2010, Nanocoating Enhanced Optical Fiber Sensors, *Ceramic Transactions*, 222, 275–286.

Smietana, M., Szmidt, J., Dudek, M., and Niedzielski, P. 2004, Optical properties of diamond-like cladding for optical fibres, *Diamond and Related Materials*, 13(4–8), 954–957.

Smith, H.I. 2001, Low cost nanolithography with nanoaccuracy, *Physica E: Low-Dimensional Systems and Nanostructures*, 11(2–3), 104–109.

Strobel, O., Seibl, D., Lubkoll, J., and Rejeb, R. 2009, Fiber-optic sensors—An overview, *ICTON 2009: 11th International Conference on Transparent Optical Networks*, Island of São Miguel, Azores, Portugal, June 28–July 2, 2009.

Surre, F., Lyons, B., Sun, T., Grattan, V., O'Keeffe, S., Lewis, E., Elosua, C., Hernaez, M., and Barian, C. 2009, U-bend fibre optic pH sensors using layer-by-layer electrostatic self-assembly technique, *Journal of Physics: Conference Series*, 178, 012046. Doi: 1088/1742–6596/178/012046.

Tao, S., Xu, L. and Fanguy, J.C. 2006, Optical fiber ammonia sensing probes using reagent immobilized porous silica coating as transducers, *Sensors and Actuators, B: Chemical*, 115(1), 158–163.

Tien, C., Hung, C., Chen, H., Liu, W., and Lin, S. 2006, Magnetic sensor based on side-polished fiber Bragg grating coated with iron film, *INTERMAG 2006—IEEE International Magnetics Conference*, San Diego, CA, May 8–12, 2006, p. 528.

Topliss, S.M., James, S.W., Davis, F., Higson, S.P.J., and Tatam, R.P. 2010, Optical fibre long period grating based selective vapour sensing of volatile organic compounds, *Sensors and Actuators B: Chemical*, 143(2), 629–634.

Urrutia, A., Rivero, P.J., Goicoechea, J., Arregui, F.J., and Matías, I.R. 2010, Humidity sensor based on a long-period fiber grating coated with a hydrophobic thin film, *4th European Workshop on Optical Fibre Sensors*, Porto, Portugal, September 8–10, 2010, p. 765320.

Uttamchandani, D. and McCulloch, S. 1996, Optical nanosensors—Towards the development of intracellular monitoring, *Advanced Drug Delivery Reviews*, 21(3), 239–247.

Velasco-Garcia, M. 2009, Optical biosensors for probing at the cellular level: A review of recent progress and future prospects, *Seminars in Cell and Developmental Biology*, 20, 27.

Venugopalan, T., Yeo, T.L., Basedau, F., Henke, A.S., Sun, T., Grattan, K.T.V., and Habel, W. 2009, Evaluation and calibration of FBG-based relative humidity sensor designed for structural health monitoring, *20th International Conference on Optical Fibre Sensors*, Edinburgh, Scotland, October 5–9, 2009, p. 750310.

Veselov, A., Thür, C., Chukharev, V., Guina, M., Lemmetyinen, H., and Tkachenko, N.V. 2009, Photochemical proper-
     ties of porphyrin films covering curved surfaces of optical fibers, *Chemical Physics Letters*, 471(4–6), 290–294.

Vo-Dinh, T. 2008, Nanosensing at the single cell level, *Spectrochimica Acta Part B: Atomic Spectroscopy*,
     63(2), 95–103.

Vo-Dinh, T. and Kasili, P. 2005, Fiber-optic nanosensors for single-cell monitoring, *Analytical and Bioanalytical
     Chemistry*, 382(4), 918–925.

Wallave, P.A., Elliott, N., Uttamlal, M., Holmes-Smith, A.S., and Campbell, M. 2001, Development of a
     quasi-distributed optical fibre pH sensor using a covalently bound indicator, *Measurement Science and
     Technology*, 12(7), 882–886.

Wang, Z., Heflin, J.R., Stolen, R.H., and Ramachandran, S. 2005, Analysis of optical response of long period
     fiber gratings to nm-thick thin-film coatings, *Optics Express*, 13(8), 2808–2813.

Wolfbeis, O.S. 2006, Fiber optic chemical sensors and biosensors: A view back, in *Optical Chemical Sensors*,
     Springer, Dordrecht, the Netherlands, 222, 17–44.

Wolfbeis, O.S. 2008, Fiber-optic chemical sensors and biosensors, *Analytical Chemistry*, 80(12), 4269–4283.

Wu, S., Cheng, W., Qiu, Y., Li, Z., Shuang, S., and Dong, C. 2010, Fiber optic pH sensor based on mode-filtered
     light detection", *Sensors and Actuators, B: Chemical*, 144(1), 255–259.

Yadugiri, V.T. and Malhotra, R. 2010, 'Plenty of room'—Fifty years after the Feynman lecture, *Current
     Science*, 99(7), 900–907.

Yang, M., Liu, H., Zhang, D., and Tong, X. 2010a, Hydrogen sensing performance comparison of Pd layer
     and Pd/WO$_3$ composite thin film coated on side-polished single- and multimode fibers, *Sensors and
     Actuators, B: Chemical*, 149(1), 161–164.

Yang, J., Sandhu, P., Liang, W., Xu, C.Q. and Li, Y. 2008, Label-free fiber optic biosensors with enhanced sen-
     sitivity, *IEEE Journal of Selected Topics in Quantum Electronics*, 13(6), 1691–1696.

Yang, M., Sun, Y., Zhang, D., and Jiang, D. 2010b, Using Pd/WO$_3$ composite thin films as sensing materials for
     optical fiber hydrogen sensors, *Sensors and Actuators, B: Chemical*, 143(2), 750–753.

Yang, X.H. and Wang, L.L. 2007, Fluorescence pH probe based on microstructured polymer optical fiber,
     Optics Express, 15(25), 16478–16483.

Zaharescu, M., Barau, A., Predoana, L., Gartner, M., Anastasescu, M., Mrazek, J., Kasik, I., and Matejec, V.
     2008, TiO$_2$-SiO$_2$ sol-gel hybrid films and their sensitivity to gaseous toluene, *Journal of Non-Crystalline
     Solids*, 354, 2–9, 693–699.

Zamarreño, C.R., Bravo, J., Goicoechea, J., Matias, I.R., and Arregui, F.J. 2007, Response time enhancement of
     pH sensing films by means of hydrophilic nanostructured coatings, *Sensors and Actuators, B: Chemical*,
     128(1), 138–144.

Zamarreño, C.R., Goicoechea, J., Matias, I.R., and Arregui, F.J. 2009a, Amplitude interference immune pH
     sensing devices based on white light interferometry, *2008 MRS Fall Meeting: Polymer-Based Smart
     Materials—Processes, Properties and Application*, Boston, MA, December 2–5, 2008, p. 121.

Zamarreño, C.R., Goicoechea, J., Matías, I.R., and Arregui, F.J. 2009b, Laterally selective adsorption of pH
     sensing coatings based on neutral red by means of the electric field directed layer-by-layer self assembly
     method, *Thin Solid Films*, 517(13), 3776–3780.

Zamarreño, C.R., Hernaez, M., Del Villar, I., Matias, I.R., and Arregui, F.J. 2010a, Sensing properties of ITO
     coated optical fibers to diverse VOCs, *24th Eurosensors Conference*, Linz, Austria, September 5–8,
     2010, p. 653.

Zamarreño, C.R., Hernaez, M., Del Villar, I., Matias, I.R., and Arregui, F.J. 2010b, Tunable humidity sensor
     based on ITO-coated optical fiber, *Sensors and Actuators, B: Chemical*, 146(1), 414–417.

Zamarreño, C.R., Hernáez, M., Del Villar, I., Matías, I.R., and Arregui, F.J. 2010c, ITO coated optical fiber
     refractometers based on resonances in the infrared region, *IEEE Sensors Journal*, 10(2), 365–366.

Zamarreño, C.R., Hernáez, M., Del Villar, I., Matías, I.R., and Arregui, F.J. 2011, Optical fiber pH sensor based
     on lossy-mode resonances by means of thin polymeric coatings, *Sensors and Actuators, B: Chemical*,
     155(1), 290–297.

Zamarreño, C.R., Hernáez, M., Matías, I.R., and Arregui, F.J. 2009c, Fiber-optic pH sensors fabrication based
     on selective deposition of neutral red, *IEEE Sensors 2009 Conference—SENSORS 2009*, Christchuch,
     New Zealand, October 25–28, 2009, p. 845.

Zamarreño, C.R., Sanchez, P., Hernaez, M., Villar, I.D., Fernandez-Valdivielso, C., Matias, I.R., and Arregui,
     F.J. 2010d, Dual-peak resonance-based optical fiber refractometers, *IEEE Photonics Technology Letters*,
     22(24), 1778–1780.

Zhang, J., Tang, X., Dong, J., Wei, T., and Xiao, H. 2009, Zeolite thin film-coated long period fiber grating sensor for measuring trace organic vapors, *Sensors and Actuators, B: Chemical*, 135(2), 420–425.

Zhang, Y., Zeng, Q., Sun, Y., Liu, X., Tu, L., Kong, X., Buma, W.J., and Zhang, H. 2010, Multi-targeting single fiber-optic biosensor based on evanescent wave and quantum dots, *Biosensors and Bioelectronics*, 26(1), 149–154.

Zhao, S. and Reichert, W.M. 1991, Modeling of fluorescence emission from cyanine-dye-impregnated Langmuir-Blodgett films deposited on the surface of an optical fiber, *Thin Solid Films*, 200(2), 363–373.

Zheng, X.T. and Li, C.M. 2010, Single living cell detection of telomerase over-expression for cancer detection by an optical fiber nanobiosensor, *Biosensors and Bioelectronics*, 25(6), 1548–1552.

Zhu, Y., Shum, P., Bay, H.W., Chen, X., Tan, C.H., and Lu, C. 2004, Wide-passband, temperature-insensitive, and compact π-phase-shifted long-period gratings in endlessly single-mode photonic crystal fiber, *Optics Letters*, 29(22), 2608–2610.

Zhu, Y., Shum, P., Chen, X., Tan, C.H., and Lu, C. 2005, Resonance-temperature-insensitive phase-shifted long-period fiber gratings induced by surface deformation with anomalous strain characteristics, *Optics Letters*, 30(14), 1788–1790.

8

Lab on Fiber Technology and Related Devices

**Andrea Cusano, Marco Consales, Marco Pisco, Alessio Crescitelli,
Armando Ricciardi, Emanuela Esposito, and Antonello Cutolo**

CONTENTS

8.1 Introduction

In the year 1966, Nobel Prize winner C. K. Kao concluded his study on "Dielectric-Fiber Surface Waveguides for optical frequencies" [1] envisaging that "a fiber of glassy material (…) represents a possible practical optical waveguide with important potential as a new form of communication medium" [1]. He was talking about the low losses optical fibers (or a "dielectric-fiber waveguide (…) with a circular cross-section" [1]) as presently used in optical fiber communication systems.

In the following years, the optical fibers have literally revolutionized the telecommunication industry by providing higher performance, more reliable telecommunication links with ever-decreasing bandwidth cost.

A beneficial side effect of this revolution is the high volume production of optoelectronic and fiber optic components and the diffusion worldwide of a true information superhighway built of glass. In parallel with these developments also fiber-optic sensor technology has been a major user of technology associated with the optoelectronic and fiber-optic communication industry.

It is remarkable that in the last decades during which the optical fiber has driven the communication revolution, fiber optic has been conceived mainly as communication medium. Perhaps for this reason, the fiber optic industry has focused its efforts on a small set of materials and structures able to provide light guidance in the fiber core through total internal reflection in the transparency range of silica glass. Also, the existing optical fiber devices and components (i.e., gratings, polarizing filters, isolators) are basically constituted by silica glass and their functionalities are related to the silica glass properties (and the induced variations).

In the development of photonic systems both for communication and sensing applications, several out-of-fiber optical components, ranging from light sources, modulators, polarizers, up to photodetectors, are currently employed. A significant technological breakthrough would be the integration and development of these components and devices "all in fiber."

Moreover, the integration of advanced functional materials at micro- and nanoscale, exhibiting the more disparate properties, combined with suitable transduction mechanisms is the key for the development of highly integrated and multifunctional technological platforms completely realized in a single optical fiber. This achievement would be the cornerstone of a new photonics technological revolution that would lead to the definition of a novel generation of micro- and nanophotonic devices "all-in-fiber."

The optical fibers are well suited to support this revolution also in virtue of the dynamicity and versatility offered by the optical fiber technology that has shown further developments in the last years in producing specialty fibers. A large set of special optical fibers have been, indeed, proposed, ranging from new classes of optical fibers such as microstructured optical fibers (MOFs) [2,3] and plastic optical fibers (POFs) [4,5], to specific fibers such as double core optical fibers and D shaped fibers and many others constructed for specific applications.

Optical fiber technology thus constitutes a valuable platform, that combined with the new concept of "Lab on Fiber," would enable the implementation of sophisticated autonomous multifunction sensing and actuating systems integrated all in optical fibers with unique advantages in terms of miniaturization, light weight, cost-effectiveness, robustness, and power consumption. Multifunctional labs, able to exchange and manipulate information or to fuse sensorial data, could be realized in a single optical fiber, providing auto diagnostics features as well as new photonic and electro-optic functionalities useful in strategic sectors such as optical processing, environment, life science, safety, and security.

From this perspective, we refer to "Lab on Fiber" technology as the basis for the development of a technological world completely included in a single optical fiber where several structures and materials at nano- and microscale are constructed, embedded, and connected all together to provide the necessary physical connections and light–matter interactions useful to provide a wide range of functionalities and unparalleled performances [6].

In the last decade, many research groups have focused their efforts towards the design and the development of all fiber-integrated devices starting with the assessment of technological steps for their fabrication [7–17]. Two main lines can be traced in the "Lab on fiber" roadmap: the first one relies on the local micro- and nanostructuring of optical fibers in order to increase light–matter interaction providing the basis for functional material integration, while the second was mainly directed to find suitable deposition techniques for the integration of functional materials at micro- and nanoscale in well-defined geometries and shapes.

This chapter reviews the strategies, the main achievements, and related devices in the "Lab on Fiber" roadmap discussing perspectives and challenges that lie ahead.

We first report the research results concerning the development of several technological platforms, implementing the "Lab on Fiber" concept carried out at University of Sannio in our labs as well as in other prestigious research centers. Obviously, many other research groups have contributed to this new technological vision here not reported for brevity.

Finally, we focus the attention on new trends involving novel interesting and high potentiality fabrication strategies ranging from advanced multimaterials stack and drawing technique up to the use of nanotechnologies including standard lithographic tools as well as new nanoimprinting approaches.

8.2 Labs on Fiber

8.2.1 Microstructured Fiber Bragg Gratings

Microstructured fiber Bragg gratings (MSFBGs) can be considered as a simple example of the "Lab on Fiber" concept developed around a well-known optical fiber component that can be considered as a valuable technological platform for the development of highly functionalized devices and components integrated in optical fibers.

MSFBGs constitute an emerging class of fiber Bragg gratings (FBGs), which can be classified in two large categories [8]: one relies on FBGs writing in MOFs, while the second involves standard FBGs where localized structural defects at microscale are properly created within the hosting fiber by post processing techniques [8,18]. Within the first class of MSFBGs, another emerging approach deserves mention and is based on grating formation within optical fiber nanowires and microwires [19], providing a variety of interesting properties, including large evanescent field, flexibility, configurability, high confinement, robustness, and compactness. Nonetheless, here we focus our attention mostly on local MSFBGs, referring to the review paper [8] for the analysis and details on the first category of MSFBGs.

Basically, a locally MSFBG consists in a FBG with single or multiple localized defects breaking the grating periodicity opening up a plenty of possibilities ranging from photonic band-gap engineering up to surrounding refractive index (SRI) sensitization and functional material integration [7,18].

The first demonstration of spectral modification induced by post-processing of UV written FBGs in conventional fibers was proposed by Canning et al. [20]. Some years later, photonic band-gap engineering of fiber gratings was demonstrated by thermal posttreatments using localized heating [21,22].

An alternative approach was proposed in 2005 [23] and later theoretically analyzed [24]. The principle of operation relies on a partial or complete stripping of the cladding layer to introduce a local SRI sensitization of the core mode via evanescent-wave interaction as sketched in Figure 8.1a. This in turn produces a local change in its effective index able to form a distributed phase-shift controlled through the optical properties of the surrounding medium. The main spectral effect is the formation of a band gap inside the stop-band of the grating, similarly to the effect observed in phase-shift FBGs [25,26]. A typical spectral response is plotted in Figure 8.1b and compared with the pristine grating response.

The spectral position of the defect state inside the stop-band is related to the phase delay introduced by the perturbed region, which is affected by the geometric features of the local defect and the SRI [27]. As the SRI changes, a consequent modification of the effective RI and thus of the phase delay occurs, leading to a red shift in the defect state.

Similar behavior occurs when the defect length and the stripping depth are changed. This means that, by a suitable defect design, the spectral tailoring of the final device can be achieved. Moreover, by acting on the SRI, it is possible to actively modulate the defect state enabling all in fiber evanescent-wave sensors and tunable spectral filters.

However, due to the fiber weakening caused by the local cladding stripping, it is obvious that particular attention is required at *the fabrication stage*.

To date, several techniques capable to locally microstructure the optical fibers have been proposed in order to introduce a selective and spatially encoded SRI sensitization of the guided light for the development of MSFBGs. Successively to the first method capable to sensitize to SRI the in-fiber guided light [23], more advanced lithographic and not-lithographic methods were proposed [28–30]. The first method basically included a masking procedure followed by UV laser micromachining at 193 and 248 nm and wet chemical etching [28], while the second relied on the use of the electrical arc discharge (EAD) eventually followed by wet chemical etching [29,30].

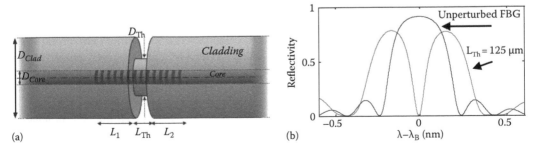

FIGURE 8.1 (a) Schematic of a local MSFBG. (b) Reflected spectrum from a 125 μm MSFBG compared with a standard Bragg spectrum. (Cusano, A. et al., *Proc. SPIE*, 8001, 800122, 2011.)

Following this research line, a different example of locally SRI sensitized fiber involving micro-slot bypass tunnel across optical fiber was successfully provided using tightly focused femtosecond laser inscription and chemical etching [31].

Since the first demonstration of locally cladding stripped MSFBGs, it was clear that the local SRI sensitization and the spectral properties of the final device could be efficiently used for *sensing applications involving refractive index measurements*. The SRI, indeed, affects the phase-shift occurring in the thinned region, leading to a fine tuning of the defect state within the stop-band spectrum [24]. Moreover, the wavelength shift of the defect state due to SRI changes is strongly dependent on the etching length and depth. In particular, as the etching length increases, a larger phase-shift is induced leading to higher SRI sensitivity. On the other side, as the diameter of the thinned region increases, reduced phase-shift is induced in the perturbation region, leading to a lower SRI sensitivity. A similar dependence can be found in the case of strong MSFBGs. However, due to their larger grating bandwidth, strong MSFBGs exhibit a slightly higher sensitivity than the standard grating. Also, theoretical studies reveal that defect state sensitivities can be enhanced through the integration of proper overlays whereas their thickness values and RIs can be properly selected to tune the SRI range with the highest sensitivity [32].

The FBG sensitization to the external refractive index can be exploited for the realization of novel *multiparameter sensing applications* and *wavelength selective filters* as far as multiple defects are accomplished along the grating structure. More defects along the grating permit the formation of more defect states inside the band gap where their spectral distribution depends on a proper combination of the geometrical and physical features of all thinned regions.

In this regard, *chirped MSFBGs* have been proposed as versatile platforms to realize all-fiber single grating WDM tunable filters [33]. Differently from uniform gratings, here multi-defect enables the formation of multi-defect states with exclusive relationship between the defects spectral positions and the corresponding thinned region features and local SRIs. Nevertheless, the main limitation related to the multi-defect configuration relies on the ripple involving the whole chirped MSFBG spectrum since more defects are added.

Finally, strain and thermal characterizations demonstrated that such devices can be used for temperature self-referenced strain measurements. At the same time, suitable combination of strain and temperature actuation enables the possibility to realize tunable narrow band filters for telecommunication applications where the defect state bandwidth can be easily tailored.

Further methods of FBG sensitization to the SRI as well as devices based on MSFBGs engineered for sensing applications can be found in Ref. [34].

8.2.2 Nano-Coated Long-Period Gratings

Also long-period fiber gratings (LPGs) can be considered as a valuable technological platform to be used for the development of highly functionalized Lab on Fiber devices and components. Thanks to the inherent SRI sensitivity of these devices, LPGs only requires proper functional materials integration to open a plenty of new possibilities for both telecommunication and sensing applications without the necessity of local structuring of the host fiber.

Indeed, LPGs are devices consisting of a periodic perturbation of a single mode fiber core and are basically in-fiber diffraction gratings generating light coupling form the core mode to co-propagating cladding modes that probe the external medium in close proximity of the cladding surface via evanescent wave [35]. As a result of the mode coupling process an LPG transmission spectrum shows several attenuation bands or dips related to the different excited cladding modes. LPGs are inherently sensitive to SRI changes and therefore have become increasingly popular devices for the implementation of chemical sensors and biosensors [35]. However, bare LPGs are scarcely sensitive in aqueous environment and they lack of any chemical selectivity or biological affinity.

The behavior is very different when a thin layer of sub-wavelength thickness (ranging in hundreds of nanometers) and with higher refractive index than the cladding is deposited onto the latter [36–40] (see Figure 8.2a). The HRI overlay draws the optical field towards the external medium extending its evanescent wave. As a result there is an increased sensitivity of the device to the SRI changes. Moreover the HRI overlay is a waveguide itself and allows mode propagation depending on its thickness, refractive

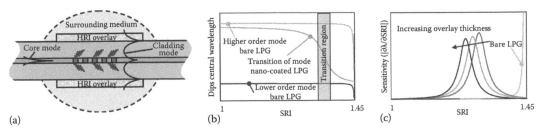

FIGURE 8.2 (a) Schematic of an LPG coated by an HRI overlay. (b) Mode transition in coated LPG. (c) Sensitivity characteristic of the coated LPG.

index (RI) and SRI. For a given material (fixed RI) and overlay thickness, when the SRI is varied in a certain range the lowest order cladding mode is gradually and completely sucked into the overlay. Its effective index becomes close to the overlay RI leaving a vacancy in the effective index distribution of the cladding modes (see Figure 8.2b). At the same time all the higher order cladding modes effective indices shift to recover the previous effective indices distribution. This is reflected through the phase matching condition in the shift of each attenuation band toward the next lower one. In the middle of this modal transition the attenuation bands can exhibit a sensitivity of thousands of nanometers per refractive index unit (RIU) [41,42]. The sensitivity characteristics of the coated LPG are drastically modified compared to the bare device (see Figure 8.2c). In fact, HRI overlays at nanoscale exhibit a resonant-like SRI sensitivity tailored around the desired SRI by changing the overlay thickness. The SRI sensitivity of LPGs can be optimized at the desired working point (ambient index) through the deposition of a high refractive index (HRI) layer by acting on overlays properties [42,43].

Several approaches for depositing coatings of sub-μm thickness onto the surface of LPGs have been examined, such as Langmuir-Blodgett (LB) [44], electrostatic self-assembly (ESA) [37], and dip-coating [45,46] techniques. Recently, in order to speed up the deposition of functional nanocoatings on the surface of LPGs, it has been proposed a novel approach for the deposition, based on the alternate deposition of silica nanoparticles (with a diameter in the range of 40–50 nm), using a Layer by Layer method [47].

Spectral features shifts as high as thousands of nanometers for a unitary change of SRI can be easily obtained and therefore LPGs coated by HRI functional layers have been successfully exploited for pH, humidity, and chemical and biological sensing [48–52].

In particular, the coated device can be used as biochemical sensor if the overlay surface is properly functionalized in the way to specifically concentrate target molecules producing localized refractive index changes.

We proposed two approaches for the surface functionalization of HRI overlays. In the first one, schematized in Figure 8.3a, the hydrophobicity of the HRI overlay (polystyrene) is used to adsorb a protein monolayer of bovine serum albumin (BSA) subsequently modified to covalently link an antibody [53]. In the second one, schematized in Figure 8.3b, a secondary ultrathin layer of poly(methyl methacrylate-co-methacrylic acid) (PMMA-co-MA) is deposited on a primary PS layer to provide a caroboxyl-containing surface minimizing at the same time its impact on the optical design of the device [54]. For this second approach we developed an original solvent/non-solvent strategy for the correct formation of the polymer multilayer structure.

Polymeric coated LPGs working in transition mode have been also used as highly sensitive devices for the monitoring of nanoscale phenomena occurring at the polymer/water interface [55]. In particular, nanoscale layers (about 320 nm) of semicrystalline syndiotactic and amorphous atactic polystyrene (sPS and aPS), have been deposited by dip coating onto LPGs to tune the devices at the transition point. Experimental results demonstrate the polymers capability to orient water molecules in proximity of their surfaces. The sPS and aPS interactions with water have been continuously monitored and then compared demonstrating the higher capability of the crystalline phase of sPS to orient water nanolayers. Moreover, the high sensitivity of the coated LPGs was used to monitor the effect of disorder induced on the interfacial water molecular arrangement by cations (sodium, Na^+, potassium, K^+, and calcium, Ca^{2+}, ions) with different sizes and electrical charge [55].

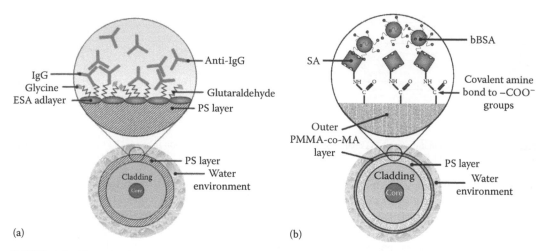

FIGURE 8.3 (See color insert.) (a) Sketch of coated LPG surface with the different biological entities used in the experiment. (b) Sketch of coated LPG surface with the different biological entities used in the experiment.

The experimental results show that, thanks to the water–polymer interaction, sPS-coated LPGs could be successfully employed as high-sensitivity cation sensors. In fact, the monitoring of the disorder induced by cations on the coordinated water layer leads to high sensitivities, in terms of detected RI change for unitary variation of concentration ($\sim 7.80 \cdot 10^{-4}$ RIU·mM^{-1} for Na$^+$ ions, $\sim 9.00 \cdot 10^{-4}$ RIU·mM^{-1} for K$^+$ ions, and $\sim 1.07 \cdot 10^{-3}$ RIU·mM^{-1} for Ca^{2+} ions) [55].

8.2.3 Evanescent-Wave Long-Period Fiber Grating within D-Shaped Optical Fibers

D-shaped optical fibers, due to the planar side, are particularly suited to optical fiber structuring processes (for the reduced aspect ratio) as well as to material integration. They constitute a very interesting platform because they allow the access to light in the core by removal of a thin cladding layer. D-fiber etching was adopted for SRI sensitive gratings [56]. In addition, Brigham Young University researchers demonstrated how to etch an arbitrary length of D-fiber to remove the core of the fiber and then replace the core section with another optical material [57]. The same group proposed, for the first time, a surface relief fiber Bragg grating realized on the flat surface of a D-shaped optical fiber and exploited such a device for the realization of various sensor typologies such as high temperature sensors [58], strain sensors [59], and chemical sensors [60].

An evanescent-wave LPFG (EWLPFG) was already recently proposed for biosensing applications [61] by using a side-polished fiber in combination with periodically patterned photoresist using photolithography. A uniform resist layer was deposited on the side-polished fiber by spin coating and, successively, patterned using a photolithographic process. To investigate the biosensing performance of such a device, further functionalization of the side-polished glass fiber substrate was carried out. The basic idea is very interesting, but the proposed fabrication technique shows certain limitations mainly due to the restricted overlay types to be adopted (only photoresist). Furthermore, the final device has to be properly functionalized for the specific application requiring thus a post-processing stage. Finally, photoresist cannot be easily functionalized for biosensing, severely restricting the versatility of the proposed approach.

To overcame this limitation, we recently proposed a different fabrication approach for the realization of a EWLPFG by periodically patterning a polymeric overlay (polystyrene) deposited onto a "sensitized" D-fiber via UV laser microstructuration [62]. The use of the D-shaped optical fiber as fiber substrate simplifies the realization procedure because it enables to avoid the need to block the fiber in a bulk material for deep side-polishing processes.

As schematically reproduced in Figure 8.4, the D-fiber was made SRI sensitive by means of wet chemical etching based on hydrofluoric acid and, after reaching the desired SRI sensitization of the fiber platform, to induce a periodic modulation of the guided light along the fiber, a uniform and HRI coating was

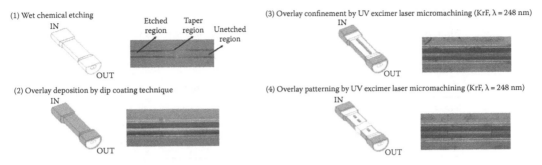

FIGURE 8.4 Technological steps for the fabrication of the EWLPFG within a D-shaped optical fiber.

deposited all around the etched fiber. The HRI overlay was successively confined only in correspondence of the core layer on the flat surface of the structure and periodically patterned by means of laser micromachining technique to induce the LPFG-like modulation of the core refractive index by evanescent wave. Finally, because the main peculiarity of the LPFGs is their spectral SRI sensitivity, a spectral characterization of the realized device versus SRI changes was carried out. SRI sensitivity around the water refractive index of about 700 and 625 nm/RIU for the first and the second attenuation band, respectively, was pointed out. The configuration here presented resulted much more sensitive—more than 1 order of magnitude- of standard LPFG characterized approximately by the same period (SRI sensitivity in water ~8 to 19 nm/RIU) [63] and a bit smaller than those of the best two LPFG configurations from this point of view: the HRI-coated LPFG (>1000 nm/RIU) [53] and the LPFG written in a microstructured optical fiber with optimized geometry (~1500 nm/RIU) [64]. By considering that no optimization of the device parameters was carried out at this stage for the proposed EWLPFG, we believe in the possibility to further increase the achievable SRI sensitivity in water, reducing the gap with the other configurations. Finally, it is important to stress about the main advantage of the reported fabrication technique that relies on its flexibility: materials of different natures could be selected and opportunely deposited onto the flat surface of the D-fiber in order to realize self-functionalized evanescent-wave LPFGs suitable for specific applications, while the sensitivity can be tuned by acting directly on the D-fiber during the chemical etching stage.

8.2.4 Nanostructured Functional Materials for "Lab on Fiber" Components

As shown in previous subsections, the definition of a technological environment for the realization of functional "Lab on Fiber" requires also the capability to integrate, pattern, and functionalize advanced materials with the specific properties at micro- and nanoscale onto and within optical fiber platforms.

Different optical fiber devices have been proposed to be suitably integrated with several functionalized overlays. In all investigated cases, the basic principle is to take advantage from the changes in the optical properties of the sensitive overlays induced by chemical interaction with target analytes to produce modulated light signals. Also, a common feature adopted in all proposed schemes relies on the use of sensitive overlays at nano- or sub-wavelength scale.

8.2.4.1 Carbon Nanotubes Integration

Since their discovery [65], carbon nanotube (CNTs) have been extensively studied as nanostructured materials for many nanoscience applications due to their unique, outstanding, and useful physical characteristics. In particular, the unique morphology of CNTs (their peculiar hollow structure, nanosized morphology, and high surface area) confers them the amazing capability to reversibly adsorb molecules of numerous environmental pollutants undergoing a modulation of their electrical, geometrical, and optical properties, such as resistivity, dielectric constant, thickness, so that CNTs-based chemical sensors offer the possibility of excellent sensitivity, low operating temperature, rapid response time, and sensitivities to various kinds of chemicals [66].

The peculiarity of SWCNTs to change their optical properties due to the adsorption of environmental pollutants was for the first time demonstrated in 2004 [67], when nanoscale CNTs-based overlays

TABLE 8.1

Main Results So Far Achieved with Opto-Chemical Nanosensors Obtained
by the Integration of CNTs with Optical Fibers

Chemical Detection	Room Temperature			Cryogenic Temperatures
	In Air		In Water	
Target	Hydrocarbons, NO_2	Alcohols	Toluene	Hydrogen
Detection limit	$\sim$300 ppb	$\sim$1 ppm	<1 ppm	<1%
Response times	$\sim$10 min	<1 min	$\sim$15 min	$\sim$4 min

were integrated with standard optical fiber for the realization of a fiber-optic sensor for chemical sensing purposes. In that case, very thin films of single-walled (SWCNT) bundles were transferred on fiber facet by means of the molecular engineered Langmuir-Blodgett (LB) technique and used as sensitive coatings for the development of volatile organic compounds (VOCs) optical chemo-sensors.

Since then, different configurations involving buffered and not buffered [68] SWCNTs overlays, as well as CNTs-based nanocomposites layer [69], were successfully integrated on the tips of standard fibers by the LB methods.

The realized SWCNTs-based chemical nanosensors demonstrated excellent sensing capabilities against volatile organic compounds and other pollutants in different environments (air and water) and operating conditions (room temperature and cryogenic temperatures). The main achieved results are summarized in Table 8.1 [66–70].

The integration of CNTs with optical fiber technology aimed at the development of all-in-fiber devices has been the subject of intense research activities. In particular, CNTs received a great deal of interest in the field of photonics since 2004, when, for the first time, they were proposed and exploited for the realization of saturable absorber (SA) [71,72]. CNTs-based SA demonstrated numerous key advantages compared to other SA such as small size, ultra-fast recovery time, polarization insensitivity, high optical damage threshold, mechanical and environmental robustness, chemical stability, tunability to operate at wide range of wavelength bands, and compatibility to optical fibers. Using CNT-based SA, the group of Prof. Yamashita, at the University of Tokyo, successfully realized femtosecond fiber pulsed lasers at various wavelength as well as very short cavity fiber laser having high repetition rates. To this aim, different techniques were exploited for CNTs integration (in the form of a few micrometer-thick layer) either on optical fiber ends or on the flat surface of D-shaped optical fibers, such as spray method [71,72], chemical vapor deposition method [73] or novel optical methods based on the use of light [74]. Besides the saturable absorption, several kinds of CNTs-based nonlinear optical devices were also proposed, exploiting their high third-order nonlinearity. Just to name a few, all-optical wavelength converter based on the nonlinear polarization rotation as well as four-wave mixing-based wavelength converter were realized by using CNTs-coated D-shaped optical fiber [75].

8.2.4.2 Near-Field Opto-Chemical Sensors Based on SnO₂ Particle Layers

Metal oxides are widely used as sensitive materials for electrical gas sensors in environmental, security, and industrial applications. Particularly, tin dioxide was one of the first considered and is still the most frequently used material for these applications, thanks to its excellent sensing performance in presence of small amounts of a wide range of gases [76]. The principle of operation of such gas sensors relies upon a change of electrical conductivity of the sensing material as a consequence of the adsorption of the analyte molecules onto the sensor surface [77]. The main disadvantage of these conductometric sensors is the high working temperature, necessary to increase their sensitivity, which in turn leads to power wastage.

In order to exploit the sensing properties of tin dioxide, we proposed, a few years ago, the integration of tin dioxide particle layers onto the end facet of a single mode optical fiber, as schematically represented in Figure 8.5a, in order to measure the optical response of tin dioxide-based materials to environmental changes, instead of the electrical one [78,79]. For the deposition of tin dioxide films upon

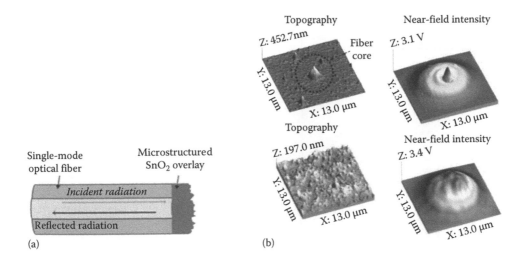

FIGURE 8.5 (a) Schematic view of the near-field opto-chemical sensors based on particle layers sensing configuration. (b) Atomic Force Microscopy topographic images of a flat and microstructured sensitive layers deposited upon the fiber tip in correspondence of the fiber core and corresponding optical near field simultaneously collected by the near field scanning microscope probe in the same region.

the distal end of the optical fiber, we employed the electrostatic spray pyrolysis (ESP) technique, adapted to work onto optical fiber substrates, followed by an annealing procedure. The integration of particle layers onto the optical fiber distal end by ESP revealed itself very simple and inexpensive. Additionally, it allowed us to tailor the structural properties of the film such as crystalline size, thickness, and porosity by changing the deposition process parameters, like the metal chloride concentration, the solution volume, and the substrate temperature [80].

Very interesting results have also been obtained by using SnO_2 particles layers as sensing elements onto the optical fiber end. In fact, this novel sensing scheme turned out to be able to efficiently detect chemicals traces at sub-ppm concentrations either in vapor phase or for water monitoring applications [81].

The layer morphology demonstrated to be a key aspect for optical sensing because it is able to significantly modify the optical near-field profile emerging from the film surface. In fact, local enhancements of the evanescent wave contribute occurs leading to a strong sensitivity to surface effects induced as a consequence of analyte molecule interactions [82].

Since the overlay topography determines the near-field properties, the effect of such process variable and the influence of post-processing thermal treatment on the overlay morphology were investigated by the authors; the obtained results were collected in some recent reports [83]. In particular, AFM-SNOM characterization of the fabricated sensing samples revealed that layers with surface features of the order of the radiation wavelength demonstrated a strong capability to generate and to modify the optical near-field emerging from the overlay (see Figure 8.5b).

The tin dioxide-based near-field sensors exhibited surprising capabilities of detecting the tested aromatic hydrocarbons in air at room temperature, with sensors resolutions of the order of few tens of ppb. The obtained sensitivities turned out to be an order of magnitude higher than those obtained with the sensors in the standard configuration. This demonstrates that the excellent results achieved are not only due to the high surface area of the sensing overlays but could be explained by the fact that the interaction between analytes molecules and sensitive material occurs mainly on the SnO_2 surface by means of the optical near field, with a significant enhancement of the evanescent part of the field.

This is also confirmed by the good results obtained with the proposed near-field fiber-optic sensors in case of ammonia detection in water, at room temperature, where limits of detection below 0.1 ppm were obtained, combined with good reproducibility attitudes and fast responses [81].

Recently, we have been engaged to investigate the possibility to exploit fiber-optic sensing technology as novel fiber-optic humidity sensors to be applied in high-energy physics (HEP) applications and in particular in experiments actually running at the European Organization for Nuclear Research (CERN).

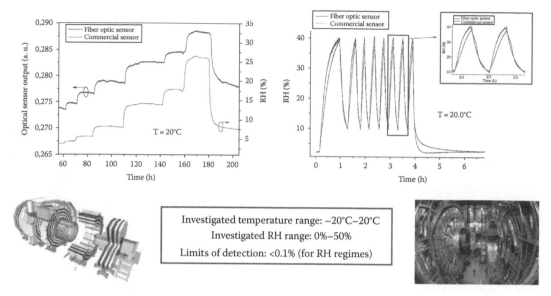

FIGURE 8.6 Fiber-optic humidity sensors at the CERN.

The technologies currently in use at CERN for RH monitoring are mainly based on polymer capacitive sensing elements with on-chip integrated signal conditioning that needs at least three wires for each sensing point [84]. This means that complex cablings are required to control the whole volume inside the tracker with enough RH sensors. In addition, these devices are critically damaged when subject to high radiations doses, not being designed with radiation hardness characteristics. Optical fiber sensors provide many attractive characteristics among which reduced size and weight, immunity to electromagnetic interference, passive operation (intrinsically safe), water, and corrosion resistance.

In addition, modern fiber fabrication technologies allow obtaining fibers with low to moderate levels of radiation induced attenuation (RIA). In this regard, it is worth noting that SMF-28 optical fibers have been selected in the Compact Muon Solenoid (CMS)—one of the four experiments installed in the LHC accelerator—for data transfer and thus have addressed the issues related to their radiation hardness.

In this scenario, near-field fiber-optic sensors based on particle layers of tin dioxide have been employed to perform the monitoring of low values of relative humidity RH at low temperatures. The RH sensing performance of fabricated probes was analyzed during a deep experimental campaign carried out in the laboratories of CERN, in Geneva.

The fiber-optic sensor performance has been investigated at 20°C and 0°C by step changing the RH and waiting until the equilibrium condition inside the climatic chamber was reached [85].

Good agreement was observed between humidity measurements provided by the optical fiber sensors and commercial polymer-based hygrometers at 20°C and 0°C, with limits of detection for low RH regimes below 0.1%.

The dynamic features of the SnO_2-based sensor have also been investigated at 20°C by rapidly and successively changing the RH humidity inside the climatic chamber from approximately 2% to 40% and from 40% to 2%. The obtained results are reported in Figure 8.6 and have been compared with those provided by the commercial electronic sensors (which are known to exhibit a rather fast response time). A good agreement has been found between the responses of fiber-optic and commercial sensors, although the former exhibits a slightly higher response time as demonstrated by the fact that the maximum values of RH it provides are slightly lower than those provided by the electrical counterpart [85].

8.2.5 Functional Material Integration in Microstructured Optical Fibers

MOFs are highly valuable for the design of advanced all-in-fiber components with specific and often unique characteristics. Due to their peculiar composite structure, the MOFs offer a high degree of freedom in their design, whereas the holey structure potentially allows for the integration with specific materials in order to confer it additional functionalities. Several MOFs have been proposed and a wide class of devices have been designed and fabricated since [2,3,9–11].

Specifically, MOFs constitute a fascinating technological platform for sensing applications. They have the potential to dramatically improve the performance of fiber-optic sensors, since a significant portion of the guided light can be located in holes within the fiber in order to maximize the overlap with the material to be sensed.

For example, a large number of optical fiber refractometers based on FBGs have been proposed. Typically, all-in-fiber refractometers rely on the effective refractive index variations of modes whose evanescent wave content is in direct contact with a surrounding refractive index (SRI) to measure. While uniform FBGs in conventional step-index fiber need appropriate treatments to reach SRI sensitization [34], the employment FBGs inscribed in solid-core MOFs easily allows an intrinsic interaction between the guided optical electromagnetic field and the medium inserted into the holes. The enhancing of this interaction can be achieved by a convenient design of the host MOF.

In the past year, Phan Huy et al. investigated the RI response of FBGs in MOFs with different design and determined the wavelength shifts of the Bragg resonance versus the RI of the liquid inserted into the fiber channels. They first explored the RI sensitivity of FBGs written in 1-ring 6-hole fiber and 2-ring 6/12-hole fiber [86]. Successively, in a more recent paper, the same group presented a uniform FBG photowritten in an MOF with a suspended core fiber, usually called steering-wheel fiber (SWF), for further enhancing the RI sensitivity [87]. Due to the small core area of these suspended core fibers (about $9\,\mu m^2$), the electromagnetic field is less confined within the core of the SWF and it extends farther into the holes; hence, a greater interaction between any liquid filling the holes and the evanescent part of the field is possible. This explains the increased sensitivity and, hence, the better resolution obtained with the SWF at low RI values.

An optical fiber refractometer is a key sensor for chemical and biological sensing. In fact, refractometers find application in chemical and biological sensing where refractive index sensors are able to supply information on chemical species as well as biological targets concentrations. In addition, refractometers often offer transduction principles employed in more complex chemical and biological sensing configurations in which a sensitive material undergoes a refractive index change via target molecules absorption [88].

Experimentally, chemical and biological sensing have been demonstrated using absorption spectroscopy in MOFs [89,90] and captured fluorescence-based sensing in liquid-filled hollow-core MOFs [91,92], side excited MOFs [93,94], and double-clad and multi-core (liquid filled) MOF [95]. Both in-fiber excitation and fluorescence recapturing within a liquid-filled, solid-core MOF have been also demonstrated [96].

For example, Ritari et al. studied gas sensing in air-guiding PBFs by filling them in turn with hydrogen cyanide and acetylene at low pressure and measuring absorption spectra [97]. Smolka et al. studied the sensing performances of the MOFs for absorption and fluorescence measurements by infiltrating a few nanoliter of a dye solution directly in the holey structure [92,94,98].

In a wider perspective, the microstructuration of the cladding as well as the presence of the air core [2,3], allow us to develop new functionalized devices by filling the MOF holes with functional materials. Filling fibers and repeating bulk experiments is attractive because structured optical fibers can be considered cells with micro bulk volumes extended over long lengths in a diffraction-free environment, potentially leading to much stronger and larger interactions. Most of these involve straightforward filling of the holes using a pressure system, whether it is gaseous, liquid, metal, or indeed solid powder; some involve reactions within the channel "cells" [99].

Other interesting work involves inserting liquid crystals within structured photonic crystal and photonic band-gap fibers to make various components [100–102], and metal filling to generate metal-assisted mode propagation [103]. Also by manipulating the properties of the in-fiber integrated materials, the guiding features of the MOF itself can be properly changed in order to develop new tunable photonic devices [91,100,101,104].

The integration between the MOFs and specific sensing materials can enable the development of novel opto-chemical sensors, exploiting the change of the guiding properties of the HOFs due to the presence of the sensitive material within the holey structure. In our previous research activities, we integrated

SWCNTs with hollow-core optical fibers (HOFs) to realize sensing probes basically composed by a piece of HOF, spliced at one end with an SOF and covered and partially filled with SWCNTs at the other termination. The sensing capability of the HOF sensors was first investigated in 2006 by exposure to traces of Tetrahydrofuran [105]. After this feasibility analysis, following studies were aimed at investigating the far-field transmission characteristic of SWCNTs-HOF probes with different deposition conditions in order to study the influence of the deposition process on the guiding properties of the final structure [106]. Successively, we demonstrated the sensing capability of the proposed sensors toward volatile organic compounds (VOCs) [107].

Selective filling of cladding holes or the cores of optical fibers is an important step forward in the realization of composite structures. The selective filling can be achieved, whether by plugging or by sealing the holes [91,108–112]. A simple example based on blocking a larger hole with polymer is the filling of a core of a photonic band gap with one liquid and the cladding with another [113]. Controlled splicing was recently used to report the selective filling of holes with three dyes (selected to luminescence in the blue, green, and red), inserted around the core of a regular structured optical fiber and [114]. The laser dyes after excitation generate white light, propagating in the fiber core as blue, green, and red light and are trapped [99].

The accessibility to the air channels of PCF has opened up endless possibility for functionalization of the channel surfaces. In particular, there has been a growing interest in imparting the functionality of surface-enhanced Raman scattering (SERS) in MOFs for sensing and detection. For a wide survey on sensors based on MOFs for SESR see Refs. [115,116]. Gold and silver nanoparticles have been extensively used for SERS because they provide very large Raman enhancement and can be easily synthesized in solution. For example, a special MOF characterized by four big air holes inserted between the solid silica core and the photonic crystal cladding holes has been employed for SERS detection [117]. The gold nanoparticles, serving as the SERS substrate, have been either coated on the inner surface of the holes or mixed in the analyte solution infiltrating in the holes via capillary effect. The SERS signal generated from the probe lighting was collected into the Raman spectrometer to efficiently detect a testing sample of solution of Rhodamine B (RhB). Similarly, SERS characterization with silver nanoparticles (SNPs) coated on the inner wall of MOFs using high-pressure chemical deposition technique [118] has been also demonstrated. Yan et al. presented a HOF SERS probe with a layer of gold nanoparticles coated on the inner surface of the air holes at one end of the fiber. HOFs allowed the direct interaction of light excitation with the analyte infiltrated in the core of the optical fiber. In addition, Du's group reported a SERS-full-length active solid core MOF platform with immobilized and discrete silver nanoparticles [119–121]. The sensing probe requires a uniform incorporation of SERS active nanostructures inside the hollow core along the entire fiber length with controlled surface coverage density, but in this configuration the entire length of the PCF can be utilized for SERS detection in forward propagating mode by increasing the measured SERS signal due to the increased interaction path [115].

8.2.6 Self-Assembling of Nanostructured Materials onto Optical Fibers

The self-assembly of nanostructured materials holds promise as a low-cost, high-yield technique with a wide range of scientific and technological applications [122].

Among the various nanodeposition techniques, the layer-by-layer (LBL) electrostatic self-assembly has been often employed to integrate nanomaterials with the optical fiber. LBL is able to build up nanometric coatings on a variety of different substrate materials such as ceramics, metals, and polymers of different shapes and forms, including planar substrates, prisms, convex, and concave lenses. The deposition procedure for all cases is based on the construction of molecular multilayers by the electrostatic attraction between oppositely charged poly-electrolytes in each monolayer deposited, and involves several steps. The versatility of LBL method for the synthesis of materials permits the application of this technique to design or fabrication of different structures on the tip or the cladding of the optical fiber.

Optical fiber sensors based on nanoFabry-Perots, microgratings, tapered ends, biconically tapered fibers, long period gratings, or photonic crystal fiber have made possible the monitoring of temperature, humidity, pH, gases, volatile organic compounds, H_2O_2, copper, or glucose [12,52,123]. For example, as shown in Ref. [124], a reflective sensors based on the deposition on the tip of optical fiber as well as a coated LPG can work as refractometers, chemical sensors, and biosensors.

8.2.7 Surface Plasmon Resonance Optical Fiber Sensors

The application of the surface plasmon resonance (SPR) phenomenon to the sensing of chemical and biological quantities has attracted the attention of researchers for the extraordinary sensitivities that can be obtained. Plasmons indeed are exceptionally sensitive reporters for chemical phenomena that influence the refractive index of the local environment of a probe [125–127]. SPR sensors based on optical fibers are in turn particularly attractive for chemical and biological sensing for the intrinsic advantages associated with the use of fiber-optic technology. The challenge in exploiting SPR phenomena for optical fiber sensors relies on the difficulty in the implementation of a coupling method for surface Plasmon excitation in the optical fiber. In spite of this, several SPR fiber sensor configurations have been proposed in the last decades. Two very recent reviews collect the relevant results obtained in micro- and nanostructured optical fiber sensors and wide space is reserved to SPR fiber sensors [13,128].

These include symmetrical structures, such as a simple metal-coated optical fiber with and without the cladding layer [129,130], tapered fiber [131,132], structures based on specialty fibers, such as D-shape fiber and one-side metal coated fibers with and without remaining cladding [133,134], and structures with modified fiber tips with flat or angled structures [135–137]. Each of them may contain either overlayer or multilayer structures. The overlayer, which is a layer deposited on the metal layer, can be useful in tuning the measurable range for a fiber-optic SPR sensor [135–139]. Many SPR fiber sensor systems have also been developed with structures that have been further modified, via the use of several types of gratings, hetero core fibers, nano-holes, and so on. More details and also a more complete survey on these configurations can be found in Ref. [13].

Some of the recently developed novel structures include the use of various types of fiber gratings in SPR fiber sensors. Some approaches employ various gratings to couple light from the core mode to various cladding modes and to provide the phase matching needed to excite SPR on the surface of an optical fiber. In one, He et al. suggested an SPR fiber sensor adopting cascaded long period gratings (LPGs) [140]. Tang et al. also employed LPG to SPR fiber sensor with self-assembled gold colloids on the grating portion, which is intended for use in sensing the concentration of a chemical solution and for the label-free detection of bimolecular binding at a nanoparticle surface [141]. An SPR fiber sensor with an FBG for SPP excitation was described by Nemova and Kashyap [142]. An SPR sensor using FBG was also proposed recently, in which HOFs were adopted [143]. Other types of gratings have also been investigated, including tilted fiber gratings [144] and metallic Bragg gratings [145]. Many types of SPR sensors based on PCFs have been proposed [13,126,127,146–149]. For example, in 2008, Hautakorpi et al. proposed and numerically analyzed an SPR sensor based on a three-hole microstructured optical fiber with a gold layer deposited on the holes.

8.3 Toward Advanced Lab on Fiber Configurations

Although the mentioned works, described in the previous sections, led to all in-fiber devices and can be viewed within the same "Lab on Fiber" concept, their main limitations lie in the impossibility of providing fiber micro- and nanostructuring and functional material definition at sub-wavelength scale in a controlled manner. This also prevents the exploitation of many novel and intriguing phenomena that are at the forefront of scientific optical research, involving light manipulation phenomena and excitations of guided resonances in photonic crystals [150,151] and quasi-crystals [152–155] as well as combined plasmonic and photonic effects in hybrid metallo-dielectric structures [156].

This means that more sophisticated fabrication strategies and technologies need to be used to create the necessary technological scenario for advanced Lab on Fiber components and devices.

8.3.1 Multimaterial Fibers: Macro-to-Micro Approach

In the past years, a prestigious group at MIT has been also actively involved in research activities devoted to the integration on optical fibers of conducting, semiconducting, and insulating materials thus giving rise to specific multimaterial fibers each defined by its own unique geometry and composition.

The technological strategy in achieving this goal exploited a preform-based fiber-drawing technique consisting of the thermal scaling of a macroscopic multi-material preform.

Key to this approach is the identification of materials that can be co-drawn and are capable of maintaining the preform geometry in the fiber and the prevention of axial- and cross-sectional capillary breakup. The following general conditions, highlighted by the same authors, are needed in the materials used in this process:

- At least one of the fiber materials needs to support the draw stress and yet continuously and controllably deform (thus it should be amorphous in nature and resist devitrification, allowing for fiber drawing with self-maintaining structural regularity).
- All the materials (vitreous, polymeric or metallic) must flow at a common temperature.
- The materials should exhibit good adhesion/wetting in the viscous and solid states without cracking even when subjected to rapid thermal cooling.

All these requirements lead inevitably to a reduction of available materials to integrate in all-in-fiber devices by the MIT approach. Specifically, the class of materials employed to fabricate multimaterial fibers are chalcogenide glasses and polymeric thermoplastics. In producing optoelectronic fiber devices, metals need to be co-drawn along with the glasses and polymers. In addition, only low-melting-point metals or alloys are suitable for the thermal-drawing process.

Although the approach proposed at MIT suffers some limitations on the class of materials to be used and on the geometries implementable (necessarily longitudinally invariant), it demonstrated great potential.

In fact, several distinct fibers and fiber-based devices were realized by this approach [14,157–165]. Photodetecting fibers, that behave as photodetectors with sensitivity to visible and infrared light at every point along its entire length, were obtained from a macroscopic preform consisting of a cylindrical semiconductor chalcogenide glass core, contacted by metal conduits encapsulated in a protective cladding [157,159,160]. Another example of multimaterial fiber is a hollow-core photonic band-gap transmission line surrounded with a thin temperature-sensitive semiconducting glass layer, which is contacted with electrodes to form independent heat-sensing devices [162,163].

An omnidirectional reflecting multilayer structure, surrounded with metallic electrodes, has been also employed to allow simultaneous transport of electrons and photons along the fiber [162]. A solid nanostructured fiber consisting of a chalcogenide glass core and a two-dimensional photonic crystal cladding has been proposed to generate supercontinuum light at desired wavelengths when the highly nonlinear core is pumped with a laser [163]. Finally, multimaterial piezoelectric fibers have been recently proposed and obtained by thermally drawing a structure made of a ferroelectric polymer layer, spatially confined and electrically contacted by internal viscous electrodes and encapsulated in an insulating polymer cladding. The fibers show a piezoelectric response and acoustic transduction from KHz to MHz frequencies [164,165].

8.3.2 Nano-Transferring Approach

Alternatively to the thermal scaling, a promising nanofabrication method has been proposed by an excellent group at the Harvard University, in Massachusetts [15], enabling nanoscale metallic structures created with electron-beam lithography to be transferred to unconventional substrates (small and/or nonplanar) that are difficult to pattern with standard lithographic techniques. The approach uses a thin sacrificial thiol-ene film that strips and transfers arbitrary nanoscale metallic patterns from one substrate to another, and was successfully exploited to transfer a variety of gold and silver patterns and features to the facet of an optical fiber.

Using this novel transfer technique, a bidirectional fiber based probe was already proposed and demonstrated for in situ surface-enhanced Raman scattering detection [166]. One facet of the probe features an array of gold optical antennas designed to enhance Raman signals, while the other facet of the fiber is used for the input and collection of light. The array of nanoscale optical antennas was first fabricated on a planar silicon wafer by means of electron-beam lithography, and subsequently lifted-off, stripped from the substrate, and transferred to a silica fiber end facet.

Simultaneous detection of benzenethiol and 2-[(E)-2-pyridin-4-ylethenyl]pyridine was demonstrated through a 35 cm long fiber [166].

Although the method is said by the authors to be immediately applicable to other metals, such as platinum and palladium, which bind to thiols strongly and to silicon weakly, some limitations occur for a simple extension to other materials for which a proper selection of substrates and design of interfacial chemistry have to be performed. In particular, eligible materials must (1) be able to be patterned on a flat substrate without forming covalent bonds with the substrate and (2) have a surface functionality that allows the material to be stripped. In addition, the adhesive strength of nanoparticles to the fiber end facet is too low to endure even some kinds of cleaning processes and too many intermediate and high precision fabrication steps need to be carried out, from the nanoscale pattern fabrication performed on planar substrates to its transfer onto the facets of optical fibers.

The same group at Harvard University recently proposed a further integrated approach for fabricating and transferring patterns of metallic nanostructures to the tips of optical fibers [16]. The process combines the technique of nanoskiving (e.g., the thin sectioning of patterned epoxy nanoposts supporting thin metallic films to produce arrays of metallic nanostructures embedded in thin epoxy slabs) with manual transfer of the slabs to the fibers facet.

Specifically, an ultramicrotome equipped with a diamond knife is used to section epoxy nanoscale structures coated with thin metallic films and embedded in a block of epoxy. Sectioning produces thin epoxy slabs, in which the arrays of nanostructures are embedded, that can be physically handled and transferred to the fiber tip.

For the transfer, the floating slab is positioned on the surface of a water bath, and is then captured with the tip of the fiber by holding the fiber with tweezers and pressing down manually on the slabs. As the slab is driven under the surface of the water using the tip, it wraps itself around the tip. Withdrawing the tip from the water bath, the slab results irreversibly attached to the tip. Finally, air plasma is used to etch the epoxy matrices, thus leaving arrays of metallic structures at the nanoscale on the fiber facet.

The proposed technique, so far demonstrated to transfer several nanostructure typologies such as gold crescents, rings, high-aspect-ratio concentric cylinders, and gratings of parallel nanowires, is said by the authors to enable the realization of optical components useful for many applications, including sensing based on localized surface plasmon resonances (LSPRs), label-free detection of extremely dilute chemical and biological analytes using surface-enhanced Raman scattering (SERS), optical filters diffraction gratings, and nose-like chemical sensors.

However, it still presents two main limitations: the first is related to the weak adhesion of the structures to the glass surface (that are attached to the fiber through Van der Waals forces); the second is the possible formation of defects that can occur due to both mechanical stresses of sectioning, combined with the intrinsic brittleness of evaporated films and the compressibility of the epoxy matrix, (mainly fracture of individual structures and global compression in the direction of cutting) and to the folding of the slabs during the transfer procedure (formation of crease running across the core of the fiber).

Based on the same transfer technique, in 2009 a highly sensitive refractive index and temperature monolithic silicon photonic crystal (PC) fiber tip sensor was also proposed [167,168]. The PC slab was firstly fabricated by reactive ion etching (RIE) on standard wafers in a silicon foundry, and successively it was released, transferred, and bonded to the facet of a single-mode fiber using a micromanipulator and focusing ion beam (FIB) tool. More specifically, the PC is cut by the ion-beam to a size moderately larger ($\sim$30 μm) than the fiber core ($\sim$9 μm), so that the PC completely covers the core and also allows for some misalignment during assembly. Then the micromanipulator needle tip is welded to the PC by Pt deposition using a combination of ion-beam and e-beam to cover the top and sides of the tip. The PC is then completely released from the wafer by either ion-milling of the remaining connecting sections or by pulling the PC apart. During this operation, the holes act as perforations that allow the PC to easily break free. The slab is then transferred to the fiber and aligned to the core of the cleaved fiber facet. The PC is bonded to the fiber tip at several points using Pt deposition, and finally, the micromanipulator needle tip is cut away by ion-milling to completely release the device. The realized PC fiber tip sensors demonstrated very high sensitivity to refractive index and temperature and have the attractive feature of returning a spectrally rich signal with independently shifting resonances that can be used to extract multiple physical properties of the measurand and distinguish between them.

Among the transfer methods, the UV-based nanoimprint and transfer lithography (NITL) technique proposed by Scheerlinck et al. as flexible, low-cost, and versatile approach for defining submicron metal patterns on optical fiber facets [169] also deserves mention. This technique was applied for the fabrication of optical probes for photonic integrated circuits based on a waveguide-to-fiber gold grating coupler [169].

8.3.3 Micro- and Nanotechnologies Directly Operating on Optical Fiber Substrates

An alternative approach to achieve advanced lab on fiber configurations could be the use of micro- and nanotechnologies directly operating on fiber substrates avoiding thus the critical transferring step.

This means that well-assessed technologies for both deposition (spin coating, dip coating, sputtering, evaporation, etc.) and sub-wavelength machining (Focused Ion Beam, Electron Beam Lithography, Reactive Ion Etching, etc.) have to be adapted to operate on optical fibers taking into account the particular geometry of the substrates.

The management, assessment, and conjunction of advanced micro- and nanomachining techniques, enriched by the technological capability to integrate, pattern, and functionalize advances materials at micro- and nanoscale onto and within micro- and nanostructured optical fibers can be envisioned as the key aspect to define a technological environment for Lab on Fiber implementation and advanced functional devices realization.

Following this technological strategy, valuable advanced fiber-optic devices have been realized in the recent years. Fiber-top cantilevers have been realized directly onto the fiber top by means of nanomachining techniques such as focused ion beam [170] and femtosecond laser micromachining [171]. Similarly, ferrule-top cantilevers have been machined by ps-laser ablation on the top of a millimeter-sized ferrule that hosts an optical fiber [172].

Both kinds of miniaturized cantilevers fabricated directly onto the fiber tip are easily exploitable in sensing applications. Light coupled in from the opposite end of the fiber allows us to remotely detect the tiniest movement of the cantilever. Consequently, any event responsible for that movement can be detected. The vertical displacements of the cantilever can be measured by using standard optical fiber interferometry and starting from this configuration, applications for chemical sensing [173], refractive index measurements [174], and atomic force microscopy [175] have been reported since.

Dhawan et al. in 2008, instead, proposed the use of FIB milling for the direct definition of ordered arrays of apertures with sub-wavelength dimensions and submicron periodicity on the tip of gold-coated fibers [176].

The fiber-optic sensors were formed by coating the prepared tips of the optical fibers with an optically thick layer of gold via electron beam evaporation, and then using FIB milling to fabricate the array of sub-wavelength apertures. Interaction of light with sub-wavelength structures such as the array of nanoapertures in an optically thick metallic film leads to the excitation of surface Plasmon waves at the interfaces of the metallic film and the surrounding media, thereby leading to a significant enhancement of light at certain wavelengths. The realized plasmonic device was proposed for chemical and biological detection. The spectral position and magnitude of the peaks in the transmission spectra depend on the refractive index of the media surrounding metallic film containing the nanohole array, enabling the detection of the presence of chemical and biological molecules in the localized vicinity of the gold film.

An e-beam lithography nanofabrication process was instead employed by Lin et al. enabling the direct patterning of periodic gold nanodot arrays on optical fiber tips [177]. EBL lithography was preferred with respect to FIB because FIB milling of gold layers results in unwanted doping of silica and gold with gallium ions [177]. A cleaved fiber is firstly coated by a nanometer thick gold layer by vacuum sputtering methods (a few nm thick Cr layer is also used as an adhesion layer) followed by the deposition of an electron beam resist. To this aim a "dip and vibration" coating technique was exploited enabling to achieve a uniform thickness coating layer (at list in an area large enough to cover the optical fiber mode) on the fiber facet. Then the EBL process was used to create the two dimensional nanodot array pattern in the e-beam resist that was successively transferred to the Au layer by RIE etch with Argon ions. The remaining ZEP was finally striped by dipping the fiber tip in the resist developer. The localized surface plasmon resonance of the E-beam patterned gold nanodot arrays on optical fiber tips was then utilized for biochemical sensing [178].

8.3.3.1 Hybrid Photonic-Plasmonic Crystal on the Fiber Facet

Here we report the design, fabrication, and characterization of a two-dimensional hybrid metallo-dielectric PC nanostructure directly realized on the end face of a standard single mode optical fiber [6].

The device is fabricated by means of conventional nanoscale deposition and fabrication/patterning (EBL) tools typically exploited for planar devices and here properly adapted to enable direct in-fiber operations.

In particular, as schematically reproduced in Figure 8.7a, the fabrication procedure consists of three main technological steps: (1) dielectric overlay deposition with flat surface over the fiber core, (2) dielectric overlay nanopatterning, and (3) metallic superstrate deposition. A positive tone electron beam resist (ZEP 520A, Zeon Chemicals) and gold have been selected as dielectric and metallic materials, respectively.

One of the peculiarities and main innovations of this approach relies on the capability to deposit ZEP layers onto the fiber facet with controlled and uniform thickness over the fiber core by using a modified) spin-coating technique (patent pending). Specifically, this technique enables the direct deposition of ZEP overlays onto the cleaved end of standard optical fibers, with controllable thickness ranging from 100 to 400 nm and flat surface areas nearly 50 μm in diameter large (see Figure 8.7b).

The schematic representation of the metallo-dielectric fiber tip device is shown in Figure 7.8a. It consists of a 200 nm thick ZEP layer (refractive index $n \sim 1.54$) patterned with square lattices of holes and covered with a 40 nm thick gold film deposited on both the ridges and grooves.

If the PC structure is illuminated in out-of-plane configuration as the case of single mode fiber illumination in the paraxial propagation regime, hybrid plasmonic–photonic resonances are expected to be excited due to the phase matching condition between the scattered waves and the modes supported by the hybrid PC slab [179].

A numerical analysis of this structure has been carried out by means of the commercial software COMSOL Multiphysics (RF module) based on the finite element method. Following the same approach of [152], by exploiting the crystal translational and mirror symmetries, the computational domain can be reduced to one quarter of unit cell, terminated with perfectly electric conducting (PEC) and vertical

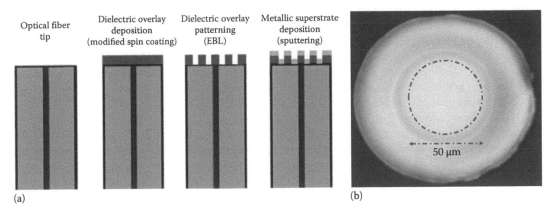

| Optical fiber tip | Dielectric overlay deposition (modified spin coating) | Dielectric overlay patterning (EBL) | Metallic superstrate deposition (sputtering) | |

50 μm

(a) (b)

FIGURE 8.7 (a) Schematics of the main technological steps for the fabrication of the hybrid metallo-dielectric PC based fiber tip device. (b) Microscope image of the ZEP overlay deposited by spin coating on the fiber facet.

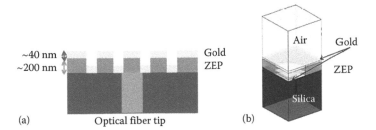

~40 nm   Gold
~200 nm ZEP

Air   Gold

ZEP

Silica

(a) Optical fiber tip   (b)

FIGURE 8.8 (a) Schematic of the hybrid metallo-dielectric PC structure integrated on the optical fiber tip. (b) 3D view of ¼ of unit cell (computational domain).

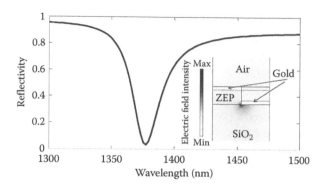

FIGURE 8.9 Simulated reflectivity of the hybrid metallo-dielectric structure: the ZEP and gold layer thicknesses are 200 and 40 nm, respectively. The PC period is a = 900 nm and the holes radius is 450 nm. (Inset) Electric field intensity distribution at the reflectivity minimum.

perfectly magnetic conducting (PMC) boundary conditions, placed two by two in the opposite walls. The resulting structure (shown in Figure 8.8b) supports a transverse-electromagnetic (TEM) wave emulating the normally incident plane wave. The Lorentz–Drude model was used for modeling Au in the IR wavelengths.

Figure 8.9 shows the numerically retrieved reflectance spectrum for a Pc structure exhibiting period a = 900 nm and the holes radius r = 225 nm corresponding to a filling factor (radius to period ratio) r/a = 0.25. The high reflectivity (>85%) base line is interrupted by a resonance dip centered at 1378 nm with a Q-factor as high as 53. In the inset of Figure 8.9 the cross section of the electric field distribution corresponding to the resonant mode evaluated at the resonance wavelength is shown.

According to PC theory, the resonant wavelength can be tailored for the specific application by a proper choice of the device parameters, such as lattice period, filling factor, dielectric, and gold thickness.

The designed 2D PC hybrid metallo-dielectric structure was fabricated on the end face of a standard single-mode fiber with core and cladding diameters of 9 and 125 μm, respectively. With reference to Figure 8.7a, the fiber facet was firstly overcoated with a 200 nm thick layer of ZEP by means of the modified spin-coating technique. The microscope top view of the spinned fiber is shown in Figure 8.7b: the concentric circles of different colors are attributed to the ZEP film thickness variation due to edge bead. However, a uniform thickness area of approximately 50 μm in diameter around the center of the optical fiber has been estimated, which is sufficient to cover the active optical area corresponding to the fiber core. With the purpose of estimating the obtained layer thickness, a hole was drilled onto the coated fiber tip by an excimer laser, operating at a wavelength of 248 nm; then the layer thickness was measured via an optical profilometer (Veeco, Wyko 9100 NT) and was found to be around 200 ± 10 nm.

The square lattice, consisting of a 100 × 100 μm² matrix of circular holes, was written on the covered fiber tip, by using a Raith 150 e-beam lithography system. Finally, a 40 nm gold layer was sputtered on the fiber tip to realize the photonic–plasmonic device (manuscript in preparation). The SEM image (top view) of the fabricated device is shown in the Figure 8.3a together with a zoomed-in image of the holey structure (Figure 8.10a).

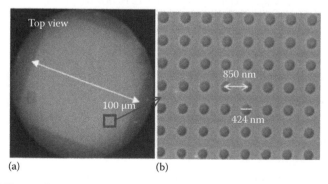

(a)   (b)

FIGURE 8.10 (a) SEM image of the patterned fiber and (b) zoomed detail of (a).

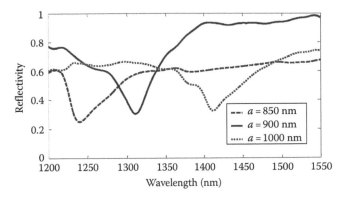

FIGURE 8.11 Measured reflectivity of hybrid metallo-dielectric PC-based structures characterized by a period $a = 850$ (dashed line), $a = 900$ nm (solid line), and $a = 1000$ nm (dotted line).

The fabrication process follows almost ordinary lithographic techniques, allows rapid prototyping with a 90% yield and is able to produce robust and reusable sensors.

Optical reflectance spectra have been measured using as reference mirror a second optical fiber terminated with a 160 nm thick gold film.

The measured reflectance is shown in Figure 8.11 (solid line). A resonance dip centered at 1311 nm was found with a Q-factor of about 23. Comparing the experimental reflectivity with the numerical counterpart it is possible to notice a considerable blue shift (67 nm), together with a reduction of both the visibility and the Q-factor. These discrepancies can be attributed to the fabrication tolerances and mainly to the numerical analysis carried out in the hypothesis of plane wave excitation assuming an infinite structure in the periodicity plane.

In order to demonstrate the tunability of our device via a proper crystal design, we also fabricated other two samples with different periods (850 and 1000 nm) and same filling factor ($r = 425$ nm and $r = 500$ nm). Since the resonant wavelengths are directly related to the lattice period, a red shift is expected in the case of higher lattice period while a blue shift should occur in the opposite case. Indeed, we experimentally observed a red shift of approx. 100 nm for $a = 1000$ nm and a blue shift of approx. 70 nm for $a = 850$ nm (see Figure 8.11).

Preliminary results, here reported, open up very intriguing perspectives for the development of a novel generation of fiber optic micro and nano devices for specific applications.

In order to demonstrate the flexibility of the proposed technological platform, acoustic detection tests have been also carried out.

Indeed, in light of the typical low Young's modulus of polymeric materials (such as the ZEP constituting the dielectric of the aforementioned hybrid crystal), significant variations in the geometrical characteristics of the patterned dielectric slab are expected in response to an applied acoustic pressure wave, hence promoting a consequent shift of the resonant wavelength.

Preliminary experiments have been carried out using the sample characterized by a period $a = 900$ nm (e.g., the one whose reflectance spectrum is shown in Figure 8.11). A schematic representation of the exploited setup is shown in Figure 8.12.

An acoustic tone of frequency f was generated by an audio speaker and then launched onto the fiber facet. The sample was interrogated by means of a tunable laser (Thotlabs—INTUN TL1300-B) locked at the wavelength corresponding to the steepest slope of the resonance right edge. The optical power reflected by the fiber tip is modulated as a consequence of the resonance shift induced by the applied acoustic wave, depending on the effective strain acting on the hybrid structure. The modulated reflected power is delivered through a circulator to a photodetector whose electrical response is proportional to the intensity of the applied wave. Finally the electric signals are amplified and stored via a PC.

Furthermore, in order to gather information about the actual incident acoustic pressure, a reference microphone was placed in close proximity of the fiber sensor.

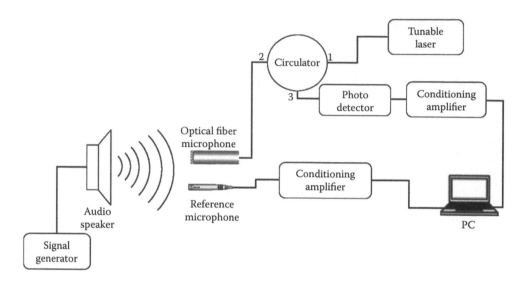

FIGURE 8.12   Schematic representation of the experimental setup exploited for the preliminary acoustic measurements.

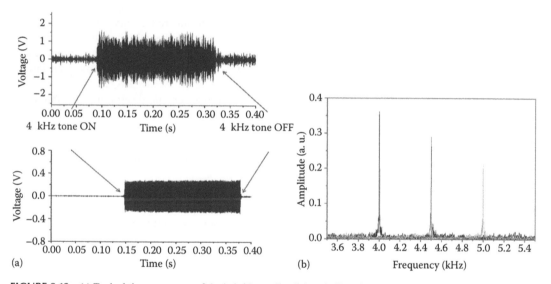

FIGURE 8.13   (a) Typical time responses of the hybrid metallo-dielectric fiber tip device (top) and reference microphone (bottom) to a 4 KHz acoustic pressure pulse with duration 200 ms. (b) FFT spectra obtained by the optical fiber sensor in response to three acoustic tones with increasing frequency (4, 4.5, and 5 kHz).

In Figure 8.13a (upper curve), the typical time response of the hybrid crystal-based device to a 4 kHz acoustic tone with duration of about 230 ms is reported. For comparison, the response of the reference microphone is also reported in Figure 8.13a (lower curve).

Data clearly reveal the capability of the fiber sensor to detect the incident acoustic wave along with the good agreement with the reference device. As evident, the electrical signal is delayed in respect with the optical counterpart, the delay being due to the slightly longer distance at which the reference microphone is located from the acoustic source.

In addition it is worth noting that, although a relatively high noise level is visible in Figure 8.13a (the standard deviation of the sensor signal—σ_{noise}—in absence of acoustic wave is nearly 0.1 V), that can be attributable to the instability of the utilized tunable laser, the response of the metallo-dielectric fiber facet device was found to be more than an order of magnitude higher than the noise level.

In Figure 8.13b the FFT spectra obtained by the same optical fiber microphone in response to three acoustic tone with different frequencies (4, 4.5, and 5 kHz) are also shown.

The ability of the optical device to discriminate among them is clearly evident. The different amplitude of the sample response to the three tones is due to the different amplification the optical signal undergoes during the electronic processing.

Finally, we point out that these results are only preliminary and no efforts were made to optimize the hybrid photonic–plasmonic structure with the aim of maximizing the sensitivity for the specific applications. Hence, further optimization margins exist by properly tailoring the crystal design, in terms of lattice tiling, period, filling fraction, and dielectric and metallic layer thickness.

It is worth noting that interesting perspectives can also be foreseen for affinity-based biological and chemical sensing applications. In these cases, however, as the goal is generally to detect the adsorption of a thin layer of biological material occurring at the device surface (both on the ridges and inside the grooves), an extremely high surface sensitivity is required for the device in order to be able to detect even slight local refractive index changes occurring at the gold surface. To achieve this aim, further configurations are currently under investigation and design, involving two different approaches: a proper tailoring of the geometrical features of the photonic crystal-based device and the use of HRI overlay which represents a useful approach commonly exploited to draws the optical field towards the external medium, hence, improving its interaction with the surroundings [34,40].

8.4 Conclusions

Since the optical fiber has been conceived as a reliable and high performance communication medium, in less than half century, optical fiber technology has driven the communication revolution and nowadays represents one of the key elements of the global communication network.

In the last decade, two significant advancements have characterized optical fiber technology:

- The development of novel optical fiber platforms (MOF, POF, HOF, …)
- The development of multifunctional and multimaterial devices integrated in optical fiber

All these collective efforts can be viewed as concurring to the "Lab on Fiber" realization. The basic idea of the Lab on Fiber concept relies on the development of a new technological world completely included in a single optical fiber where several structures and materials at nano- and microscale are constructed, embedded, and connected all together to provide the necessary physical connections and light-matter interactions useful to provide a wide range of functionality for specific communication and sensing applications.

The research efforts carried out in this direction in the last decade have confirmed the potentiality of the "Lab on Fiber" concept through feasibility analysis devoted to identify and define technological scenarios for the fabrication of highly functionalized all fiber components, devices, and systems for different strategic sectors.

In the future scenario, it can be envisioned that a novel generation of fiber-optic micro- and nanodevices for sensing and communication applications could arise through the concurrent addressing of the issues related to the different aspects of the Lab on Fiber global design which requires highly integrated and multidisciplinary competencies ranging from optoelectronics photonics to physics, material science, biochemistry up to micro- and nanotechnologies. Moreover, a highly integrated approach involving continuous interactions of different backgrounds is required to optimize each single aspect with a continuous feedback that would enable the definition of overall and global design.

In fact, since their first appearing in 1966, optical fibers continue to extraordinarily revolutionize the technological evolution of our planet.

REFERENCES

1. K. C. Kao and G. A. Hockham, Dielectric-fibre surface waveguides for optical frequencies, *Proc. IEEE*, 113, 1151, 1966.
2. P. Russell, Photonic crystal fibers, *Science*, 299, 358–362, January 2003.
3. J. C. Knight, Photonic crystal fibres, *Nature*, 424, 847–851, August 2003.
4. O. Ziemann, J. Krauser, P. E. Zamzow, and W. Daum, *POF Handbook: Optical Short Range Transmission Systems*, Springer-Verlag, Berlin, Germany, 2008.
5. M. A. van Eijkelenborg, M. C. J. Large, A. Argyros, J. Zagari, S. Manos, N. A. Issa, I. Bassett, S. Fleming, R. C. McPhedran, C. M. de Sterke, and N. A. P. Nicorovici, Microstructured polymer optical fibre, *Opt. Exp.*, 9 (7), 319–327, 2001.
6. A. Cusano, M. Consales, M. Pisco, A. Crescitelli, A. Ricciardi, E. Esposito, A. Cutolo, Lab on fiber technology and related devices, part I: A new technological scenario; Lab on fiber technology and related devices, part II: The impact of the nanotechnologies, *Proc. SPIE*, 8001, 800122, 2011.
7. A. Cusano, D. Paladino, and A. Iadicicco, Microstructured fiber Bragg gratings, *J. Lightwave Technol.*, 27, 1663, 2009.
8. A. Cusano, M. Giordano, A. Cutolo, M. Pisco, and M. Consales, Integrated development of chemoptical fiber nanosensors, *Curr. Anal. Chem.*, 4 (4), 296–315, 2008.
9. P. St. J. Russell, Photonic-crystal fibers, *J. Lightwave Technol.*, 24 (12), 4729–4749, 2006.
10. J. Canning, Fibre gratings and devices for sensors and lasers, *Laser Photonics Rev.*, 2, 275, 2008.
11. B. J. Eggleton, C. Kerbage, P. S. Westbrook, R. S. Windeler, and A. Hale, Microstructured optical fiber devices, *Opt. Exp.*, 9 (13), 698–713, 2001.
12. F. J. Arregui, *Sensors Based on Nanostructured Materials*, Springer, New York, 2009.
13. B. Lee et al., Current status of micro- and nano-structured optical fiber sensors, *Opt. Fiber Technol.*, 15, 209–221, 2009.
14. A. F. Abouraddy et al., Towards multimaterial multifunctional fibres that see, hear, sense and communicate, *Nat. Mater.*, 6, 336, 2007.
15. E. J. Smythe, M. D. Dickey, G. M. Whitesides, and F. A. Capasso, A technique to transfer metallic nanoscale patterns to small and nonplanar surfaces, *ACS Nano*, 3, 59–65, 2009.
16. D. J. Lipomi, R. V. Martinez, M. A. Kats, S. H. Kang, P. Kim, J. Aizenberg, F. Capasso, and G. M. Whitesides, Patterning the tips of optical fibers with metallic nanostructures using nanoskiving, *Nano Lett.*, 11 (2), 632–636, 2011.
17. D. Iannuzzi et al., Monolithic fiber-top sensor for critical environments and standard applications, *Appl. Phys. Lett.*, 88, 053501, 2006.
18. A. Cusano and D. Paladino, Developments and applications of microstructured fiber Bragg gratings, in *Optical Fiber Communication Conference*, OSA Technical Digest (CD) (Optical Society of America, 2011), Paper OTuC1, Los Angeles, California, USA.
19. G. Brambilla, Optical fibre nanowires and microwires: A review, *J. Opt.*, 12, Article ID 043001, 2010.
20. J. Canning and M. G. Sceats, π-phase-shifted periodic distributed structures in germanosilicate fibre by UV post-processing, *Electron. Lett.*, 30 (16), 1344–1345, 1994.
21. M. Janos and J. Canning, Permanent and transient resonances thermally induced in optical fibre Bragg gratings, *Electron. Lett.*, 31 (12), 1007–1009, 1995.
22. D. Uttamchandani and A. Othonos, Phase shifted Bragg gratings formed in optical fibres by post-fabrication thermal processing, *Opt. Commun.*, 127 (4–6), 200–204, 1996.
23. A. Iadicicco, A. Cusano, S. Campopiano, A. Cutolo, and M. Giordano, Microstructured fiber Bragg gratings: Analysis and fabrication, *Electron. Lett.*, 41 (8), 466–468, 2005.
24. A. Cusano, A. Iadicicco, D. Paladino, S. Campopiano, A. Cutolo, and M. Giordano, Micro-structured fiber Bragg gratings. Part I: Spectral characteristics, *Opt. Fiber Technol.*, 13 (4), 281–290, 2007.
25. R. Zengerle and O. Leminger, Phase-shifted Bragg-grating filters with improved transmission characteristics, *J. Lightwave Technol.*, 13 (12), 2354–2358, 1995.
26. L. Wei and J. W. Y. Lit, Phase-shifted Bragg grating filters with symmetrical structures, *J. Lightwave Technol.*, 15 (8), 1405–1410, 1997.
27. A. Cusano, A. Iadicicco, S. Campopiano, M. Giordano, and A. Cutolo, Thinned and micro-structured fiber Bragg gratings: Towards new all fiber high sensitivity chemical sensors, *J. Opt. A Pure Appl. Opt.*, 7 (12), 734–741, 2005.

28. A. Iadicicco, S. Campopiano, D. Paladino, A. Cutolo, and A. Cusano, Micro-structured fiber Bragg gratings: Optimization of the fabrication process, *Opt. Exp.*, 15 (23), 15011–15021, 2007.

29. A. Cusano, A. Iadicicco, D. Paladino, S. Campopiano, and A. Cutolo, Photonic band-gap engineering in UV fiber gratings by the arc discharge technique, *Opt. Exp.*, 16 (20), 15332–15342, 2008.

30. D. Paladino, A. Iadicicco, S. Campopiano, and A. Cusano, Not-lithographic fabrication of micro-structured fiber Bragg gratings evanescent wave sensors, *Opt. Exp.*, 17 (2), 1042–1054, 2009.

31. K. Zhou et al., Refractometer based on micro-tunnel in fiber Bragg grating by chemically assisted femtosecond laser processing, *Opt. Exp.*, 15, 15848–15853, 2007.

32. A. Cusano, A. Iadicicco, D. Paladino, S. Campopiano, A. Cutolo, and M. Giordano, Micro-structured fiber Bragg gratings. Part II: Towards advanced photonic devices, *Opt. Fiber Technol.*, 13 (4), 291–301, 2007.

33. M. Pisco, A. Iadicicco, S. Campopiano, A. Cutolo, and A. Cusano, Structured chirped fiber Bragg gratings, *J. Lightwave Technol.*, 26 (12), 1613–1625 (2008).

34. A. Cusano, A. Cutolo, and M. Giordano, Fiber Bragg gratings evanescent wave sensors: A view back and recent advancements, in *Sensors, Book Series: Lecture Notes in Electrical Engineering*, Springer, Berlin, Germany,, Vol. 21, pp. 113–152, 2008.

35. S. W. James and Tatam, R. P., Optical fibre long-period grating sensors: Characteristics and application, *Meas. Sci. Technol.*, 14, R49–R61, 2003.

36. S. W. James, N. D. Rees, G. J. Ashwell, and R. P. Tatam, Optical fibre long period gratings with Langmuir Blodgett thin film overlays, *Opt. Lett.*, 9, 686–688, 2002.

37. I. Del Villar, M. Achaerandio, I. R. Matias, and F. J. Arregui, Deposition of overlays by electrostatic self assembly in long-period fiber gratings, *Opt. Lett.*, 30, 720–722, 2005.

38. I. Del Villar, I. R. Matias, F. J. Arregui, and P. Lalanne, Optimization of sensitivity in long period fiber gratings with overlay deposition, *Opt. Exp.*, 13, 56–69, 2005.

39. P. Pilla, A. Iadicicco, L. Contessa, S. Campopiano, A. Cutolo, M. Giordano, and A. Cusano, Optical Chemo- Sensor based on long period gratings coated with δ form Syndiotactic Polystyrene, *IEEE Photonics Technol. Lett.*, 17, 1713–1715, 2005.

40. A. Cusano, A. Iadicicco, P. Pilla, L. Contessa, S. Campopiano, A. Cutolo, and M. Giordano, Cladding modes re-organization in high refractive index coated long period gratings: Effects on the refractive index sensitivity, *Opt. Lett.*, 30, 2536–2538, 2005.

41. A. Cusano, A. Iadicicco, P. Pilla, L. Contessa, S. Campopiano, A. Cutolo, and M. Giordano, Mode transition in high refractive index coated long period gratings, *Opt. Exp.*, 14 (1), 19–34, 2006.

42. A. Cusano, P. Pilla, M. Giordano, and A. Cutolo, Modal transition in nano-coated long period fiber gratings: Principle and applications to chemical sensing, in *Advanced Photonic Structure for Biological and Chemical Detection*, Ed. Fan, X., Springer, New York, pp. 35–75, 2009.

43. P. Pilla, A. Cusano, A. Cutolo, M. Giordano, G. Mensitieri, P. Rizzo, L. Sanguigno, V. Venditto, and G. Guerra, Molecular sensing by nanoporous crystalline polymers, *Sensors*, 9 (12), 9816–9857, 2009.

44. N. D. Rees, S. W. James, R. P. Tatam, and G. J. Ashwell, Optical fiber long-period gratings with Langmuir-Blodgett thin-film overlays, *Opt. Lett.*, 27 (9), 686–688, 2002.

45. A. Cusano, P. Pilla, L. Contessa, A. Iadicicco, S. Campopiano, A. Cutolo, M. Giordano, and G. Guerra, High sensitivity optical chemosensor based on coated long-period gratings for sub-ppm chemical detection in water, *Appl. Phys. Lett.*, 87 (23), 234105, 2005.

46. Z. Gu and Y. Xu, Design optimization of a long-period fiber grating with sol–gel coating for a gas sensor, *Meas. Sci. Technol.*, 18 (11), 3530–3536, 2007.

47. D. Viegas, J. Goicoechea, J. L. Santos, F. M. Araújo, L. A. Ferreira, F. J. Arregui, and I. R. Matias, Fiber optic long period grating sensors with a nanoassembled mesoporous film of SiO_2 nanoparticles, *Sensors*, 9, 519–527, 2009.

48. J. M. Corres, I. R. Matias, I. Villar, and F. J. Arregui, Design of pH sensors in long-period fiber gratings using polymeric nanocoatings, *IEEE Sens. J.*, 7, 455–463, 2007.

49. M. Konstantaki, S. Pissadakis, S. Pispas, N. Madamopoulos, and N. A. Vainos, Optical fiber long-period grating humidity sensor with poly(ethylene oxide)/cobalt chloride coating, *Appl. Opt.*, 45 (19), 4567–4571, 2006.

50. D. Viegas, J. Goicoechea, J. M. Corres, J. L. Santos, L. A. Ferreira, F. M. Araújo, and I. R. Matías, A fiber optic humidity sensor based on a long-period fiber grating coated with a thin film of SiO_2 nanospheres, *Meas. Sci. Technol.*, 20 (3), 034002, 2009.

51. Cusano, A., Iadicicco, A., Pilla, P., Cutolo, A., Giordano, M., and Campopiano, S., Sensitivity characteristics in nanosized coated long period gratings, *Appl. Phys. Lett.*, 89, 201116/1–201116/3, 2006.

52. S. James and R. Tatam, Fiber optic sensors with nano-structured coatings, *Pure Appl. Opt.*, 8, S430–S444, 2006.

53. P. Pilla, P. Foglia Manzillo, V. Malachovska, A. Buosciolo, S. Campopiano, A. Cutolo, L. Ambrosio, M. Giordano, and A. Cusano, Long period grating working in transition mode as promising technological platform for label-free biosensing, *Opt. Exp.*, 17, 20039–20050, 2009.

54. P. Pilla, V. Malachovska, A. Borriello, A. Buosciolo, M. Giordano, L. Ambrosio, A. Cutolo, and A. Cusano, Transition mode long period grating biosensor with functional multilayer coatings, *Opt. Exp.*, 19 (2), 512–526, 2011.

55. P. F. Manzillo, P. Pilla, A. Buosciolo, S. Campopiano, A. Cutolo, A. Borriello, M. Giordano, and A. Cusano, Self assembling and coordination of water nano-layers on polymer coated long period gratings: Toward new perspectives for cation detection, *Soft Mater.*, 9 (2), 238–263, 2011.

56. G. Meltz, S. J. Hewlett, and J. D. Love, Fiber grating evanescent wave sensors, *Proc. SPIE*, 2836, 342–350, 1996.

57. D. J. Markos, B. L. Ipson, K. H. Smith, S. M. Schultz, and R. H. Selfridge, Controlled core removal from a D-shaped optical fiber, *Appl. Opt.*, 42, 7121–7125, 2003.

58. T. L. Lowder, K. H. Smith, B. L. Ipson, A. R. Hawkins, R. H. Selfridge, and S. M. Schultz, High-temperature sensing using surface relief fiber bragg gratings, *IEEE Photonics Technol. Lett.*, 17, 1926–1928, 2005.

59. R. H. Selfridge, S. M. Schultz, T. L. Lowder, V. P. Wnuk, A. Mendez, S. Ferguson, and T. Graver, Packaging of surface relief fiber bragg gratings for use as strain sensors at high temperature, *Proc. SPIE*, 6167, 616702.1–616702.7, 2006.

60. T. L. Lowder, J. D. Gordon, S. M. Schultz, and R. H. Selfridge, Volatile organic compound sensing using a surface relief d-shaped fiber bragg grating and a polydimethylsiloxane layer, *Opt. Lett.*, 32 (17), 2523–2525, 2007.

61. H. S. Jang, K. N. Park, J. P. Kim, O. J. Kwon, Y.-G. Han, and K. S. Lee, Sensitive DNA biosensor based on a long-period grating formed on the side-polished fiber surface, *Opt. Exp.*, 17, 3855–3860, 2009.

62. G. Quero et al., Evanescent-wave LPFG in D-fiber by periodically patterned overlay, *Proceedings of SPIE*, 7653, Paper 7631G-1, September 8–10, 2010—Best Student Paper Award at EWOFS 2010.

63. X. Shu, L. Zhang, and I. Bennion, Sensitivity characteristics of long-period fiber gratings, *J. Lightwave Technol.*, 20, 255–266, 2002.

64. L. Rindorf, and O. Bang, Highly sensitive refractometer with a photonic-crystal-fiber long-period grating, *Opt. Lett.*, 33, 563–565, 2008.

65. S. Iijima, Helical microtubules of graphitic carbon, *Nature*, 354, 56–58, 1991.

66. M. Penza, G. Cassano, P. Aversa, F. Antolini, A. Cusano, A. Cutolo, M. Giordano, and L. Nicolais, Alcohol detection using carbon nanotubes acoustic and optical sensors, *Appl. Phys. Lett.*, 85, 2378–2381, 2004.

67. M. Consales, S. Campopiano, A. Cutolo, M. Penza, P. Aversa, G. Cassano, M. Giordano, and A. Cusano, Carbon nanotubes thin films fiber optic and acoustic VOCs sensors: Performances analysis, *Sens. Actuators B*, 118, 232, 2006.

68. M. Consales, A. Cutolo, M. Penza, P. Aversa, G. Cassano, M. Giordano, A. Cusano, Carbon nanotubes coated acoustic and optical VOCs sensors: Towards the tailoring of the sensing performances nanotechnology, *IEEE Trans. Nanotechnol.*, 6, 601–612, 2007.

69. M. Consales, A. Crescitelli, M. Penza, P. Aversa, P. Delli Veneri, M. Giordano, and A. Cusano, SWCNT nano-composite optical sensors for VOC and gas trace detection, *Sens. Actuators B*, 138, 351–361, 2009.

70. M. Consales, A. Crescitelli, S. Campopiano, A. Cutolo, M. Penza, P. Aversa, M. Giordano, and A. Cusano, Chemical detection in water by single-walled carbon nanotubes-based optical fiber sensors, *IEEE Sens. J.*, 7 (7), 1004 (2007).

71. S. Y. Set et al., Laser mode locking using a saturable absorber incorporating carbon nanotubes, *J. Lightwave Technol.*, 22 (1), 51–56, January 2004.

72. S. Y. Set et al., Ultrafast fiber pulsed lasers incorporating carbon nanotubes, *J. Sel. Top. Quantum Electron.*, 10 (1), 137–146, January/February 2004.

73. S. Yamashita et al., Saturable absorbers incorporating carbon nanotubes directly synthesized onto substrates and fibers and their application to mode-locked fiber lasers, *Opt. Lett.*, 29 (14), 1581–1583, July 2004.

74. K. Kashiwagi and S. Yamashita, Optically manipulated deposition of carbon nanotubes onto optical fiber end, *Jap. J. Appl. Phys.*, 46 (40), L988–L990, October 2007.

75. Y. W. Song et al., Carbon nanotube mode lockers with enhanced nonlinearity via evanescent field interaction in D-shaped fibers, *Opt. Lett.*, 32 (2), 148–150, January 2007.

76. G. Sberveglieri, Recent developments in semiconducting thin-film gas sensors, *Sens. Actuators B*, 23, 103, 1995.

77. M. Batzill and U. Diebold, The surface and materials science of tin oxide, *Prog. Surf. Sci.*, 79, 47, 2005.

78. M. Pisco, M. Consales, S. Campopiano, A. Cutolo, R. Viter, V. Smyntyna, M. Giordano, and A. Cusano, A novel opto-chemical sensor based on SnO_2 sensitive thin film for ppm ammonia detection in liquid environment, *J. Lightwave Technol.*, 24 (12), 5000–5007, 2006.

79. A. Cusano, M. Consales, M. Pisco, P. Pilla, A. Cutolo, A. Buosciolo, R. Viter, V. Smyntyna, and M. Giordano, Opto-chemical sensor for water monitoring based on SnO_2 particle layer deposited onto optical fibers by the electrospray pyrolysis method, *Appl. Phys. Lett.*, 89, 111103, 2006.

80. E. H. A. Diagne and M. Lumbreras, Elaboration and characterization of tin oxide-lanthanum oxide mixed layersprepared by the electrostatic spray pyrolysis technique, *Sens. Actuators B*, 78 (1–3), 98, 2001.

81. A. Buosciolo, M. Consales, M. Pisco, M. Giordano, and A. Cusano, Near-field opto-chemical sensors, in *Optical Fiber New Developments*, Ed. C. Lethien, INTECH, Vukovar, Croatia. 2009.

82. A. Cusano, P. Pilla, M. Consales, M. Pisco, A. Cutolo, A. Buosciolo, and M. Giordano, Near field behavior of SnO_2 particle-layer deposited on standard optical fiber by electrostatic spray pyrolysis method, *Opt. Exp.*, 15 (8), 5136–5146, 2007.

83. A. Buosciolo, M. Consales, M. Pisco, A. Cusano, and M. Giordano, Fiber optic near-field chemical sensors based on wavelength scale tin dioxide particle layers, *J. Lightwave Technol.*, 26 (20), 3468–3475, 2008.

84. M. Fossa and P. Petagna, Use and calibration of capacitive RH sensors for the hygrometric control of the CMS tracker, CMS NOTE2003/24. Cern, Geneve, Switzerland, September 2003.

85. M. Consales et al., Fiber optic humidity sensors for high-energy physics application at CERN, *Sens. Actuators B*, 159 (1), 28, 66–74, November 2011.

86. M. C. Phan Huy, G. Laffont, Y. Frignac, V. Dewynter-Marty, P. Ferdinand, P. Roy, J.-M. Blondy, D. Pagnoux, W. Blanc, and B. Dussardier, Fibre Bragg grating photowriting in microstructured optical fibres for refractive index measurement, *Meas. Sci. Technol.*, 17 (5), 992–997, 2006.

87. M. C. Phan Huy, G. Laffont, V. Dewynter, P. Ferdinand, P. Roy, J.-L. Auguste, D. Pagnoux, W. Blanc, and B. Dussardier, Three-hole microstructured optical fiber for efficient fiber Bragg grating refractometer, *Opt. Lett.*, 32 (16), 2390–2392, 2007.

88. M. Pisco, A. Ricciardi, S. Campopiano, C. Caucheteur, P. Mégret, and A. Cusano, Time delay measurements as promising technique for tilted fiber Bragg grating sensors interrogation, *IEEE Photonics Technol. Lett.*, 21 (23), 1752–1754, December 1, 2009.

89. L. Rindorf, P. E. Hoiby, J. B. Jensen, L. H. Pedersen, O. Bang, and O. Geschke, Towards biochips using microstructured optical fiber sensors, *Anal. Bioanal. Chem.*, 385, 1370–1375, 2006.

90. C. M. B. Cordeiro, M. A. R. Franco, G. Chesini, E. C. S. Barretto, R. Lwin, C. H. B. Cruz, and M. C. J. Large, Microstructured-core optical fibre for evanescent sensing applications, *Opt. Exp.*, 14, 13056–13066, 2006.

91. Y. Huang, Y. Xu, and A. Yariv, Fabrication of functional microstructured optical fibers through a selective-filling technique, *Appl. Phys. Lett.*, 85, 5182, 2004.

92. S. Smolka, M. Barth, and O. Benson, Highly efficient fluorescence sensing with hollow core photonic crystal fibers, *Opt. Exp.*, 15, 12783, 2007.

93. J. B. Jensen, P. E. Hoiby, G. Emiliyanov, O. Bang, L. H. Pedersen, and A. Bjarklev, Selective detection of antibodies in microstructured polymer optical fibers, *Opt. Exp.*, 13, 5883–5889, 2005.

94. S. Smolka, M. Barth, and O. Benson, Selectively coated photonic crystal fiber for highly sensitive fluorescence detection, *Appl. Phys. Lett.*, 90, 111101, 2007.

95. S. O. Konorov, A. M. Zheltikov, and M. Scalora, Photonic-crystal fiber as a multifunctional optical sensor and sample collector, *Opt. Exp.*, 13, 3454–3459, 2005.

96. S. V. Afshar, S. C. Warren-Smith, and T. M. Monro, Enhancement of fluorescence-based sensing using microstructured optical fibres, *Optics Express*, Vol. 15, Issue 26, pp. 17891–17901, 2007.

97. T. Ritari, J. Tuominen, H. Ludvigsen, J. Petersen, T. Sørensen, T. Hansen, and H. Simonsen, Gas sensing using air-guiding photonic bandgap fibers, *Opt. Exp.*, 12 (17), 4080–4087, 2004.

98. Y. Ruan, T. C. Foo, S. Warren-Smith et al., Antibody immobilization within glass microstructured fibers: A route to sensitive and selective biosensors, *Opt. Exp.*, 16 (22), 18514–18523, 2008.

99. J. Canning, Structured optical fibres and the application of their linear and non-linear properties, in *Selected Topics in Photonic Crystals and Metamaterials*, Eds. A. Andreone, A. Cusano, A. Cutolo, and V. Galdi, World Scientific, Singapore, pp. 389–452, 2011.

100. T. Larson, J. Broeng, D. Hermann, and A. Bjarklev, Thermo-optic switching in liquid crystal infiltrated photonic bandgap fibres, *Electron. Lett.*, 39, 1719–1720, 2003.

101. T. Larson, A. Bjarklev, D. Hermann, and J. Broeng, Optical devices based on liquid crystal photonic bandgap fibres, *Opt. Exp.*, 11, 2589–2596, 2003.

102. M. Haakestad, M. Alkeskjold, M. Nielsen, L. Scolari, J. Riishede, H. Engan, and A. Bjarklev, Electrically tuneable photonic bandgap guidance in a liquid-crystal-filled photonic crystal fibre, *IEEE Photonics Technol. Lett.*, 17, 819–821, 2005.

103. J. Hou, D. Bird, A. George, S. Maier, B. T. Kuhlmey, and J. C. Knight, Metallic mode confinement in microstructured fibres, *Opt. Exp.*, 16, 5983–5990, 2008.

104. G. Christian, P. Domachuk, V. Ta'eed, E. Mägi, J. Bolger, B. Eggleton, L. Rodd, and J. Cooper-White, Compact tunable microfluidic interferometer, *Opt. Exp.*, 12, 5440–5447, 2004.

105. A. Cusano, M. Pisco, M. Consales, A. Cutolo, M. Giordano, M. Penza, P. Aversa, L. Capodieci, and S. Campopiano, Novel opto-chemical sensors based on hollow fibers and single walled carbon nanotubes, *IEEE Photonics Technol. Lett.*, 18 (22), 2431–2433, 2006.

106. M. Pisco, M. Consales, M. Penza, P. Aversa, M. Giordano, A. Cutolo, and A. Cusano, Photonic bandgap modification in hollow optical fibers integrated with single walled carbon nanotubes, *Microwave Opt. Technol. Lett.*, 51 (11), 2729–2732, 2009.

107. M. Pisco, M. Consales, A. Cutolo, M. Penza, P. Aversa, and A. Cusano, Hollow fibers integrated with single walled carbon nanotubes: Bandgap modification and chemical sensing capability, *Sens. Actuators B Chem.*, 129 (1), 163–170, 2008.

108. C. Kerbage, R. S. Windeler, B. J. Eggleton, P. Mach, M. Dolinski, and J. A. Rogers, Tunable devices based on dynamic positioning of micro-fluids in micro-structured optical fiber, *Opt. Commun.*, 204, 179–184, 2002.

109. C. Martelli, P. Olivero, J. Canning, N. Groothoff, B. Gibson, and S. Huntington, Micromachining structured optical fibers using focused ion beam milling, *Opt. Lett.*, 32, 1575–1577, 2007.

110. S. Yiou, P. Delaye, A. Rouvie, J. Chinaud, R. Frey, G. Roosen, P. Viale, S. Février, P. Roy, J. Auguste, and J. Blondy, Stimulated Raman scattering in an ethanol core microstructured optical fiber, *Opt. Exp.*, 13, 4786–4791, 2005.

111. K. Nielsen, D. Noordegraaf, T. Sørensen, A. Bjarklev, and T. P. Hansen, Selective filling of photonic crystal fibres, *J. Opt. A Pure Appl. Opt.*, 7, L13–L20, 2005.

112. L. Xiao, W. Jin, M. S. Demokan, H. L. Ho, Y. L. Hoo, and C. Zhao, Fabrication of selective injection microstructured optical fibers with a conventional fusion splicer, *Opt. Exp.*, 13 (22), 9014–9022, 2005.

113. C. de Matos, C. M. B. Cordeiro, E. M. dos Santos, J. S. Ong, A. Bozolan, and C. H. B. Cruz, Liquid-core, liquid-cladding photonic crystal fibers, *Opt. Exp.*, 15, 11207–11212, 2007.

114. J. Canning, M. Stevenson, T. K. Yip, S. K. Lim, and C. Martelli, White light sources based on multiple precision selective micro-filling of structured optical waveguides, *Opt. Exp.*, 16 (20), 15700–15708, 2008.

115. Y. Han and H. Du, Photonic crystal fiber for chemical sensing using surface-enhanced Raman scattering, in *Photonic Bandgap Structures*, Eds. M. Pisco, A. Cusano, A. and Cutolo, Bentham Science Publisher, Oak Park, IL, eISBN: 978-1-60805-448-0, to be published in 2012.

116. X. Yang, C. Shi, R. Newhouse, J. Z. Zhang, and C. Gu, Hollow-core photonic crystal fibers for surface-enhanced Raman scattering probes, *Int. J. Opt.*, Hindawi publishing corporation, Vol.2011, Article ID 754610, pp. 11, doi:10.1155/2011/754610, 2011.

117. H. Yan, J. Liu, C. Yang, G. Jin, C. Gu, and L. Hou, Novel index-guided photonic crystal fiber surface-enhanced Raman scattering probe, *Opt. Exp.*, 16 (11), 8300–8305, 2008.

118. A. Amezcua-Correa, J. Yang, and C. E. Finlayson et al., Surface enhanced Raman scattering using microstructured optical fiber substrates, *Adv. Funct. Mater.*, 17, 2024–2030, 2007.

119. M. K. Khaing Oo, Y. Han, R. Martini, S. Sukhishvili, and H. Du, Forward-propagating surface-enhanced Raman scattering and intensity distribution in photonic crystal fiber with immobilized Ag nanoparticles, *Opt. Lett.*, 34 (7), 968–970, 2009.

120. M. K. Khaing Oo, Y. Han, J. Kanka, S. Sukhishvili, and H. Du, Structure fits the purpose: Photonic crystal fibers for evanescent-field surface-enhanced Raman spectroscopy, *Opt. Lett.*, 35 (4), 466–468, 2010.

121. Y. Han, S. Tan, M. K. Khaing Oo, D. Pristinski, S. Sukhishvili, and H. Du, Towards full-length accumulative surface-enhanced Raman scattering-active photonic crystal fibers, *Adv. Mater.*, 22 (24), 2647–2651, 2010.

122. G. Whitesides, J. Kriebel, and B. Mayers, Self-assembly and nanostructured materials, *Nanostruct. Mater.*, Springer, New York, pp. 217–239, 2009.

123. F. Arregui et al., Optical fiber sensors based on Layer-by-Layer nanostructured films, *Procedia Eng.*, 5, 1087–1090, 2010.

124. I. D. Villar, I. R. Matias, and F. J. Arregui, Fiber-optic chemical nanosensors by electrostatic molecular self- assembly, *Curr. Anal. Chem.*, 4 (4), 341–355(15), October 2008.

125. J. Homola, S. S. Yeea, and G. Gauglitz, Surface plasmon resonance sensors: Review, *Sens. Actuators B Chem.*, 54 (1–2), 3–15, January 25, 1999.

126. J. Homola, Surface plasmon resonance sensors for detection of chemical and biological species, *Chem. Rev.*, 108 (2), 462–493, 2008.

127. M. E. Stewart, C. R. Anderton, L. B. Thompson, J. Maria, S. K. Gray, J. A. Rogers, and R. G. Nuzzo, Nanostructured plasmonic sensors, *Chem. Rev.*, 108, 494–521, 2008.

128. S. Roh, T. Chung, and B. Lee, Overview of the characteristics of micro- and nano-structured surface plasmon resonance sensors, *Sensors*, 11, 1565–1588, 2011.

129. A. K. Sharma and B. D. Gupta, Fibre-optic sensor based on surface plasmon resonance with Ag–Au alloy nanoparticle films, *Nanotechnology*, 17 (1), 124–131, 2006.

130. M. Kanso, S. Cuenot, and G. Louarn, Sensitivity of optical fiber sensor based on surface plasmon resonance: Modeling and experiments, *Plasmonics*, 3, 49–57, 2008.

131. E. M. Yeatman, Resolution and sensitivity in surface Plasmon microscopy and sensing, *Biosens. Bioelectron.*, 11, 635–649, 1996.

132. J. Homola, I. Koudela, and S. Yee, Surface plasmon resonance sensor based on diffraction gratings and prism couplers: Sensitivity comparison, *Sens. Actuators B Chem.*, 54 (1), 16, 1999.

133. S.-F. Wang, M.-H. Chiu, J.-C. Hsu, R.-S. Chang, and F.-T. Wang, Theorctical analysis and experimental evaluation of D-type optical fiber sensor with a thin gold film, *Opt. Commun.*, 253, 283–289, 2005.

134. M.-H. Chiu and C.-H. Shih, Searching for optimal sensitivity of single-mode D-type optical fiber sensor in the phase measurement, *Sens. Actuators B*, 131, 1120–1124, 2008.

135. R. Slavik, J. Homola, and J. Ctyroky, Miniaturization of fiber optic surface Plasmon resonance sensor, *Sens. Actuators B*, 51, 311–315, 1998.

136. W. J. H. Bender, R. E. Dessy, M. S. Miller, and R. O. Claus, Feasibility of a chemical microsensor based on surface plasmon resonance on fiber optics modified by multilayer vapor deposition, *Anal. Chem.*, 66 (7), 963–970, 1994.

137. M. Piliarik, J. Homola, Z. Manikova, and J. Ctyroky, Surface plasmon resonance sensor based on a single-mode polarization-maintaining optical fiber, *Sens. Actuators B*, 90, 236–242, 2003.

138. M.-H. Chiu, C.-H. Shih, and M.-H. Chi, Optimum sensitivity of single-mode Dtype optical fiber sensor in the intensity measurement, *Sens. Actuators B*, 123, 1120–1124, 2007.

139. R. Slavik, J. Homola, J. Ctyroky, and E. Brynda, Novel spectral fiber optic sensor based on surface plasmon resonance, *Sens. Actuators B*, 74, 106–111, 2001.

140. Y.-J. He, Y.-L. Lo, and J.-F. Huang, Optical-fiber surface-plasmon-resonance sensor employing long-period fiber gratings in multiplexing, *J. Opt. Soc. Am. B*, 23 (5), 801–811, 2006.

141. J.-L. Tang, S.-F. Cheng, W.-T. Hsu, T.-Y. Chiang, and L.-K. Chau, Fiber-optic biochemical sensing with a colloidal gold-modified long period fiber grating, *Sens. Actuators B*, 119, 105–109, 2006.

142. G. Nemova and R. Kashyap, Fiber-Bragg-grating-assisted surface plasmonpolariton sensor, *Opt. Lett.*, 31 (14), 2118–2120, 2006.

143. G. Nemova and R. Kashyap, Modeling of plasmon-polariton refractive-index hollow core fiber sensors assisted by a fiber Bragg grating, *J. Lightwave Technol.*, 24 (10), 3789–3796, 2006.

144. T. Allsop, R. Neal, S. Rehman, D. J. Webb, D. Mapps, and I. Bennion, Characterization of infrared surface plasmon resonances generated from a fiber-optical sensor utilizing tilted Bragg gratings, *J. Opt. Soc. Am. B*, 25 (4), 481–490, 2008.

145. W. Ding, S. R. Andrews, T. A. Birks, and S. A. Maier, Modal coupling in fiber tapers decorated with metallic surface gratings, *Opt. Lett.*, 31 (17), 2556–2558, 2006.

146. B. Gauvreau, A. Hassani, M. F. Fehri, A. Kabashin, and M. Skorobogatiy, Photonic bandgap fiber-based surface plasmon resonance sensors, *Opt. Exp.*, 15 (18), 11413–11426, 2007.

147. M. Hautakorpi, M. Mattinen, and H. Ludvigsen, Surface-plasmon-resonance sensor based on three-hole microstructured optical fiber, *Opt. Exp.*, 16 (12), 8427–8432, 2008.

148. A. Hassani and M. Skorobogatiy, Design of the microstructured optical fiber-based surface plasmon resonance sensors with enhanced microfluidics, *Opt. Exp.*, 14 (24), 11616–11621, 2006.

149. A. Hassani, B. Gauvreau, M. F. Fehri, A. Kabashin, and M. Skorobogatiy, Photonic crystal fiber and waveguide-based surface plasmon resonance sensors for application in the visible and near-IR, *Electromagnetics*, 28, 198–213, 2008.

150. S. G. Johnson, S. H. Fan, P. R. Villeneuve, J. D. Joannopoulos, and L. A. Kolodziejski, Guided modes in photonic crystal slabs, *Phys. Rev. B*, 60, 5751–5758, 1999.

151. S. Fan and J. D. Joannopoulos, Analysis of guided resonances in photonic crystal slabs, *Phys. Rev. B*, 65, 235112, 2002.

152. A. Ricciardi, I. Gallina, S. Campopiano, G. Castaldi, M. Pisco, V. Galdi, and A. Cusano, Guided resonances in photonic quasicrystals, *Opt. Exp.*, 17 (8), 6335–6346, 2009.

153. M. Pisco, A. Ricciardi, I. Gallina, G. Castaldi, S. Campopiano, A. Cutolo, A. Cusano, and V. Galdi, Tuning efficiency and sensitivity of guided resonances in photonic crystals and quasi-crystals: A comparative study, *Opt. Exp.*, 18 (16), 17280–17293, 2010.

154. A. Ricciardi, M. Pisco, I. Gallina, S. Campopiano, V. Galdi, L. O' Faolain, T. F. Krauss, and A. Cusano, Experimental evidence of guided-resonances in photonic crystals with aperiodically ordered supercells, *Opt. Lett.*, 35 (23), 3946–3948, 2010.

155. A. Ricciardi, M. Pisco, A. Cutolo, A. Cusano, L. O'Faolain, T. F. Krauss, G. Castaldi, and V. Galdi, Evidence of guided resonances in photonic quasicrystal slabs, *Phys. Rev. B*, 84, 085135, 2011.

156. X. Yu, L. Shi, D. Han, J. Zi, and P. V. Braun, High quality factor metallodielectric hybrid plasmonic–photonic crystals, *Adv. Funct. Mater.*, 20 (12), 1910–1916, 2010.

157. S. D. Hart, G. R. Maskaly, B. Temelkuran, P. Prideaux, J. D. Joannopoulos, and Y. Fink, External reflection from omnidirectional dielectric mirror fibers, *Science*, 296, 510–513, April 2002.

158. M. Bayindir, F. Sorin, S. Hart, O. Shapira, J. D. Joannopoulos, and Y. Fink, Metal-insulator-semiconductor optoelectronic fibres, *Nature*, 431, 826–829, October 2004.

159. M. Bayindir, A. F. Abouraddy, F. Sorin, J. D. Joannopoulos, and Y. Fink, Fiber photodetectors codrawn from conducting, semiconducting and insulating materials, *Opt. Photonics News*, 15, 24, 2004.

160. K. Kuriki, O. Shapira, S. D. Hart, G. Benoit, Y. Kuriki, J. Viens, M. Bayindir, J. D. Joannopoulos, and Y. Fink, Hollow multilayer photonic bandgap fibers for NIR applications, *Opt. Exp.*, 12, 1510–1517, 2004.

161. M. Bayindir, O. Shapira, D. Saygin-Hinczewski, J. Viens, A. F. Abouraddy, J. D. Joannopoulos, and Y. Fink, Integrated fibers for self monitored optical transport, *Nat. Mater.*, 4, 820–824, 2005.

162. M. Bayindir, A. F. Abouraddy, J. Arnold, J. D. Joannopoulos, and Y. Fink, Thermal-sensing fiber devices by multimaterial codrawing, *Adv. Mater.*, 18, 845–849, 2006.

163. M. Bayindir, A. F. Abouraddy, O. Shapira, J. Viens, D. Saygin-Hinczewski, F. Sorin, J. Arnold, J. D. Joannopoulos, and Y. Fink, Kilometer-long ordered nanophotonic devices by preform-to-fiber fabrication, *IEEE J. Sel. Top. Quantum Electron.*, 12, 1202, 2006.

164. S. Egusa, Z. Wang, N. Chocat, Z. M. Ruff, A. M. Stolyarov, D. Shemuly, F. Sorin, P. T. Rakich, J. D. Joannopoulos, and Y. Fink, Multimaterial piezoelectric fibres, *Nat. Mater.*, 9, 643, 2010.

165. F. Sorin and Y. Fink, Multimaterial fiber sensors, *Proc. SPIE*, 7653, 765305, 2010.

166. E. J. Smythe, M. D. Dickey, J. Bao, G. M. Whitesides, and F. Capasso, Optical antenna arrays on a fiber facet for in situ surface-enhanced Raman scattering detection, *Nano Lett.*, 9 (3), 1132–1138, 2009.

167. I. W. Jung, B. Park, J. Provine, R. T. Howe, and O. Solgaard, Monolithic Si photonic crystal slab fiber tip sensor, in *Proceedings of the IEEE Photonics Society, International Conference on Optical MEMS and Nanophoton*, Hilton Clearwater, Florida, USA, pp. 77–78, 2009.

168. I. W. Jung, B. Park, J. Provine, R. T. Howe, and O. Solgaard, Photonic crystal fiber tip sensor for precision temperature sensing, in *Proc. IEEE Lasers and Electro-Optics Society (LEOS) Annual Meeting Conference*, Antalya, Turkey, pp. 761–762, 2009.

169. S. Scheerlinck et al., Metal grating patterning on fiber facets by UV-based nano imprint and transfer lithography using optical alignment, *J. Lightwave Technol.*, 27 (10), 1415–1420, May 15, 2009.

170. D. Iannuzzi et al., Fibre-top cantilevers: Design, fabrication and applications, *Meas. Sci. Technol.*, 18, 3247, 2007.

171. A. A. Said et al., Carving fiber-top cantilevers with femtosecond laser micromachining, *J. Micromech. Microeng.*, 18, 035005, 2008.

172. G. Gruca, S. de Man, M. Slaman, J. H. Rector, and D. Iannuzzi, Ferrule-top micromachined devices: Design, fabrication, performance, *Meas. Sci. Technol.*, 21, 094033, 2010.

173. D. Iannuzzi, S. Deladi, M. Slaman, J. H. Rector, H. Schreuders, and M. C. Elwenspoek, A fiber-top cantilever for hydrogen detection, *Sens. Actuators B*, 121, 706–708, 2006.

174. C. J. Alberts, S. de Man, J. W. Berenschot, V. J. Gadgil, M. C. Elwenspoek, and D. Iannuzzi, Fiber-top refractometer, *Meas. Sci. Technol.*, 20, 034005, 2009.

175. D. Iannuzzi, S. Deladi, J. W. Berenschot, S. de Man, K. Heeck, and M. C. Elwenspoek, Fiber-top atomic force microscope, *Rev. Sci. Instrum.*, 77, 106105, 2006.

176. A. Dhawan, M. D. Gerhold, and J. F. Muth, Plasmonic structures based on subwavelength apertures for chemical and biological sensing applications, *Sens. J. IEEE*, 8, 942–950, 2008.

177. Y. Lin, Y. Zou, Y. Mo, J. Guo, and R. G. Lindquist, E-beam patterned gold nanodot arrays on optical fiber tips for localized surface plasmon resonance biochemical sensing, *Sensors*, 10 (10), 9397–9406, 2010.

178. Y. Lin, Y. Zou, and R. G. Lindquist, A reflection-based localized surface plasmon resonance fiber-optic probe for biochemical sensing, *Biomed. Opt. Exp.*, 2, 478–484, 2011.

179. D. Rosenblatt, A. Sharon, and A. A. Friesem, Resonant grating waveguide structures, *IEEE J. Quantum Electron.*, 33 (11), 2038–2059, 1997.

9

Micro/Nanofibers for Biochemical Sensing

Joel Villatoro

CONTENTS

Subwavelength-diameter optical fibers, also called optical micro/nanofibers (MNFs), represent a relatively new way of guiding light and are promising for the development of novel sensing platforms. An MNF has diameters ranging from a few hundred nanometers to a few microns and can be fabricated, e.g., by heating and pulling bulk glass or conventional optical fibers. The dimensions of a micro/nanofiber are smaller than or comparable to the wavelength of the light they typically guide. Owing to this, MNFs exhibit propagation properties different from those of conventional optical fibers, e.g., the power of the guided mode propagating along the MNF has a considerable part in the evanescent wave which may reach 100% in some cases. This property is crucial for optical sensing since it makes possible a stronger light–analyte interaction, which in turn leads to MNF-based sensors with higher sensitivity and better detection limits. MNFs are flexible and have an extraordinary mechanical strength for which they can be bent down to micrometric radii. This makes possible the development of ultracompact resonators and other sensors in which the amount of sample necessary to carry out the sensing task can be on the order of micro- or nanoliters. MNFs can be doped or coated with functional materials or films to increase the sensitivity and to provide them the specificity and selectivity to a target gas, chemical or biological parameter. In addition, MNFs are compatible with the modern nanotechnology trends, which may lead to the development of novel ultrahigh-sensitive nanosensor modalities for detection or/and monitoring various chemical or biological quantities. This chapter is devoted to these promising applications.

9.1 Introduction

Since some decades ago it was known that communications with higher bandwidths could be achieved at optical frequencies but the right transmission medium did not exist. Thus, in the 1960s, inspired by submicron-diameter fibers for transmission of images researches became interested in the use of thin, unclad optical fibers as a transmission medium [1]. Such fibers were considered an alternative to free space since the later represented many drawbacks for optical communications. For single-mode transmission, a requisite for long-haul communications, an unclad optical fiber must be smaller than the wavelength it guided, thus posing serious technical challenges. For example, it was extremely difficult to fabricate, and protect properly, kilometers of subwavelength-diameter fiber (SDF). Efficient injection of light into a tiny fiber also posed serious problems. Another problem was surface contamination since anything that touched the fiber would scatter the guided light giving rise to considerable transmission losses which were undesirable for long-haul transmission. Thus, the idea of using SDFs was considered impractical and was soon abandoned. The research was then focused on single-mode, clad optical fibers, which solved the problems of size and surface contamination.

Clad fibers were gradually deployed all over the world and became crucial in local area networks and communications systems. With the growing use of optical fibers in these applications the demand of reliable all-fiber devices such as directional couplers also grew. In a coupler, fiber tapers (fibers with a millimeter-length section in which their diameter decreases down to a certain value and then it increases to the initial fiber diameter) are essential. Due to the relevance of fiber tapers for the development of passive devices, in the 1980s and the 1990s, some research groups analyzed their optical properties in detail [2,3]. The first evidence of fiber tapers with submicrometer diameters was reported by Bilodeau et al. [4]. However, it was Bures and Ghosh who first demonstrated theoretically that the power density of the evanescent wave in a tapered fiber could be significantly enhanced if the optical fiber was tapered down to subwavelength dimensions [5]. At the beginning of the 2000s some groups demonstrated experimentally the potential of micrometer diameter tapered fibers for nonlinear optics, optical sensing, quantum optics, and for the development of miniature devices [6–8].

In 2003, interest in SDFs (also called optical, photonic, or optical fiber nanowires) was rekindled by Liming Tong et al. in an inspiring paper published in *Nature* [9]. Fibers with diameters ranging from ten to several hundred nanometers with high size uniformity with lengths of several millimeters were fabricated with a two-step drawing process. It was demonstrated that glass SDFs represented a simple but effective medium for guiding and controlling photons at the nanoscale and were an alternative to other nanometer diameter waveguides (e.g., plasmonic- or photonic-crystal-based waveguides) which were more difficult to fabricate. The inconvenient extremely thin fibers for the aforementioned communications were proposed as advantages for the development of novel optical sensors and other microphotonic devices with subwavelength-width structures. Some interesting features of subwavelength-diameter fibers were also revealed and their potential applications in both fundamental and applied research sounded promising. The work by Tong et al. motivated the research on the fabrication and applications of other nanometer-sized waveguides made of other materials such as oxides or polymers [10–15].

To launch light into short sections of free-standing SDFs, evanescent wave coupling via van der Waals attraction was proposed [9]. One of the associated problems of this approach was the need of precise positioners and visualization instruments, as well as high insertion losses and long-term stability issues due to mechanical coupling. To overcome these issues optical fiber nanotapers (also called fiber nanotapers or simply nanotapers) were introduced by Brambilla and collaborators [16]. A fiber nanotaper consists of a standard optical fiber with a section thinned down to nanometric dimensions (the taper waist). In a nanotaper, the fundamental mode of the untapered section of the fiber is adiabatically transformed into a guided mode of the thinnest part, i.e., the nanofiber, and back, resulting in a highly efficient coupling of light into and out of the nanofiber. In addition to their enhanced evanescent waves fiber nanotapers exhibit low loss and configurability. This makes it possible to devise ultracompact interferometers and resonators with different geometries (even in three dimensions), among other functional devices, see for example [17,18].

Although the evanescent waves of any waveguide can be accessed by removing the cladding, partially or totally, in optical micro/nanofibers (this is the term most research groups preferred since it embraces SDFs, nanotapers and fiber with diameters larger than the wavelength) the evanescent waves are much stronger and more open to the environment. If the micro/nanofiber (MNF) diameter is much smaller than the wavelength it guides, the percentage of power in the evanescent wave approaches 100%. Other advantages of micro/nanofibers are their low transmission loss, in many cases it is negligible, and their configurability. For these reasons MNFs are considered ideal candidates for the development of highly sensitive chemical and biological sensors [17–20]. Strong evanescent waves make possible a stronger light–analyte interaction, which in turn leads to sensors with higher sensitivity, faster response times, and better detection limits even when the interaction length is short, on the order of a few hundred micrometers. MNFs are compatible with the modern nanotechnology trends, which may lead to the development of innovative sensor modalities. In this chapter we discuss these promising applications as well as the challenges that have to be faced before MNF-sensors become a real alternative to other nanoscale sensors, or a commercial reality.

9.2 Fabrication and Properties of Micro/Nanofibers

There exist different approaches to fabricate nanometer diameter waveguide; the fabrication method basically depends on the material the waveguide is made of and the diameter of interest. The waveguide diameter in turn depends on the target application. Single crystalline oxide nanofibers (nanowires), for example, are typically synthesized or grown chemically [10–14]. This technique allows the fabrication of rectangular or semicylindrical nanofibers with different indices of refraction (typically high). The diameter of such nanowires is typically between a few tens to a few hundred nanometers but their lengths are only a few hundred microns [10–14]. The manipulation and coupling of light in and out of such ultrasmall waveguides is difficult which imposes the use of sophisticated imaging optics [10–12]. Chemically synthesized nanowires guide light with high attenuation, although they can be useful for some sensing applications as demonstrated in [10–14].

9.2.1 Techniques for Fabricating Glass Nanofibers

Free-standing silica or polymer micro/nanofibers can be fabricated by drawing conventional telecommunications optical fiber, bulk glass, or solvated polymer [9,15,21]. Fiber tapers, on the other hand, are typically fabricated by heating and pulling or by etching optical fibers [16,22,23]. The advantage of fiber tapers over free-sanding SDFs is that their fabrication is more controlled since it can be monitored in real time. The handling and manipulation of nanotapers is simple, in fact, they can be carried out with conventional macroscopic tools and equipment. In addition, fiber nanotapers have natural connection to fiber pigtails with standard dimensions for which they can be easily connected to light sources and spectrometers or power meters which simplifies substantially any experiment. The schematic representation of a fiber taper and micrographs of free-standing subwavelength- or micron-diameter fibers are shown in Figure 9.1. As it can be seen the fiber diameter is uniform and the surface smooth.

To fabricate MNFs or fiber nanotapers a flame torch produced by a controlled mixture of butane (or hydrogen) and oxygen, a miniature ceramic microheater, or a CO_2 laser can be used as heating sources [16,17,22]. All these heating sources reach a temperature high enough to soft the glass. The election of the heating source basically depends on the nanofiber diameter or on the application of interest. Flames, for example, are prone to turbulence and can fluctuate because of slight bursts in the gas flow. To fabricate nanotapers with a CO_2 laser imposes the used of focusing optics and a precise alignment system. On the other hand, ceramic microheaters are more stable which leads to better control over taper losses and shape and ensures high reproducibility [22].

9.2.2 Guiding Properties and Evanescent Wave

Micro/nanofibers, independently of the material they are made of, guide light by the total internal reflection effect. In general, there is a high index contrast between the material surrounding the waveguide (air, aqueous environment or low-index polymer) and that it is made of (glass, silicon, polymer, or oxide). Thus, to analyze the guiding and modal properties of MNFs one needs the exact solutions of Maxwell's equations. Rigorous analysis has been carried out by several authors; see for example [24]. For many applications, optical sensing in particular, the single-mode operation regime is of interest as well as the fraction or percentage of power propagating in the evanescent wave (η_{EF}). To do so, it is necessary to calculate the spot size of the light propagating in an MNF, which is strongly dependent on the fiber diameter through the V factor or normalized frequency [17,24]:

$$V = \frac{2\pi}{\lambda} rNA \qquad (9.1)$$

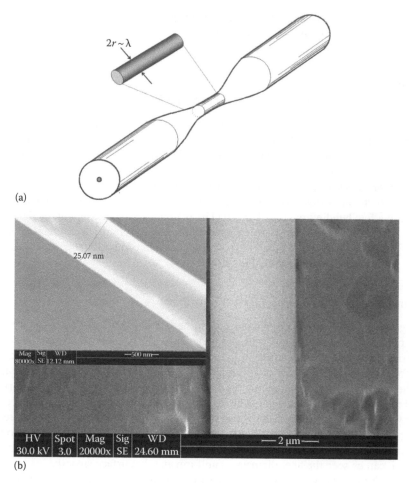

FIGURE 9.1   (a) Illustration of a fiber taper, highlighting the thinnest part (micro/nanofiber). $2r$ refers to the taper waist or MNF diameter and λ to the wavelength of the guided light. (b) SEM images of a 750 nm and a 3 μm diameter fiber.

In Equation 9.1, λ is the wavelength of the guided light, r the MNF radius, and NA is the numerical aperture. The latter is a well-known definition in fiber optics, i.e., $NA = \sqrt{n_f^2 - n_e^2}$, n_f and n_e, respectively, being the refractive indices of the MNF and the external medium. In analogy with optical fibers, for $V < 2.405$ an MNF experiences single-mode operation. As the MNF diameter decreases (thus V) the fundamental mode becomes less and less bound to the MNF and the beam spot size expands until it becomes much larger than r [17,24]. Ultimately, the extension of the evanescent field and η_{EF} depend on the ratio λ/r. Figure 9.2 shows η_{EF} as a function of the normalized wavelength λ/r for two different environments that surround a glass nanofiber [19]. It can be seen that η_{EF} increases monotonically for increasing λ/r and reaches 100% for $\lambda/r \sim 5$ when the external medium is Teflon (refractive index of 1.31). However, when the MNF is in air, η_{EF} reaches 100% for $\lambda/r \sim 10$. The results shown in Figure 9.2 suggest the strong influence of the external environment on η_{EF}.

   The sensitivity of a sensor based on evanescent wave interactions is enhanced by increasing the interaction length or the power of the evanescent field. Due to the growing demand of miniaturization it is always desirable to have compact sensors without losing sensitivity. In conventional waveguides it is not always possible to have full control of the power in the evanescent waves, thus sensors with long interaction lengths are devised. With optical MNFs the situation is different since one can control their diameter, and therefore, the power in their evanescent waves. Note from Figure 9.2 that strong light–analyte interactions are possible even if the MNF is protected.

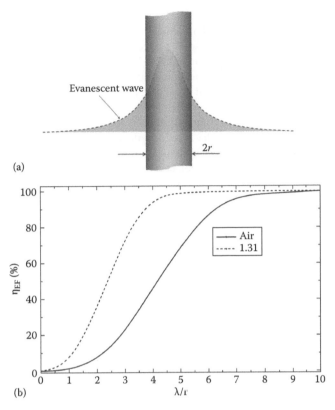

FIGURE 9.2 (a) Illustration of the evanescent wave of the fundamental mode in an MNF. (b) Percentage of power propagating in the evanescent wave as a function of the normalized wavelength (λ/r). For the calculations the nanofiber is assumed to be made of glass and the surrounding medium is assumed to be air (solid line) or Teflon (dotted line).

9.2.3 Guiding Limitations and Degradation

Although it is apparent that the sensitivity of MNF-based sensors can be tailored or enhanced by just making the fiber thinner and thinner, there are two important factors that have to be taken into account. Theoretically there is no cutoff of the fundamental mode for a cylindrical waveguide; however, a nanofiber has a threshold diameter for effective light guidance [25,26]. This threshold is caused by minute diameter fluctuations or nonuniformities in the waveguide that strongly affect the guided mode. Theoretical calculations presented in [25] showed that the transmission loss, T, in a tapered fiber with diameter $2r$ ($<\lambda$) depends on the characteristic length of the fiber diameter variation, L, by the expression [25]

$$T \cong \exp\left[-\frac{0.687L}{\sqrt{(2r(2r_0 - 2r))}} \exp\left(\frac{0.27\lambda^2}{(2r)^2} \right) \right] \tag{9.2}$$

In Equation 9.2, $2r_0 - 2r$ refers to variations in the fiber diameter. L can be taken as the length of the MNF or thinnest part (waist) of the taper. For diameters comparable to or much larger than the wavelength of the guided light the term $2r_0 - 2r$ will be close to zero and the transmission loss is negligible. However, when $2r/\lambda$ is much less than 1, an insignificant small nonuniformity of the nanofiber can completely kill the propagating mode [25]. Thus, the transmission losses of extremely thin fibers can be too high making them useless for many applications and even the guidance in nanofibers with certain diameters can be impossible. Hartung and collaborators found out that for efficient transmission, the minimum nanofiber diameter ($2r$) is related to the wavelength of the light it guides by the simple expression [26]

$$2r \cong 0.24\lambda \tag{9.3}$$

Thus, if the external medium is air, an MNF must have a diameter $2r \sim 370\,\text{nm}$ for $\lambda = 1550\,\text{nm}$ or $2r \sim 96\,\text{nm}$ for $\lambda = 400\,\text{nm}$ to guide light with minimal losses. These threshold values are consistent with a mode distribution where η_{EF} is about 97% ($\lambda/r \sim 8$), see Figure 9.2.

Another issue of nanotapers, and probably nanofibers in general, is the fact that they degrade quickly when they are exposed to air, thus making necessary an adequate protection. During the fabrication of glass nanofibers water vapor is originated since glass is heated to a high temperature (1000°C, or more). This results in nanocracks on the nanofiber surface that may grow with time. Cracks and dust particles cause scattering of the guided light resulting in substantial and irrecoverable transmission losses [27–30]. In addition, cracks make the MNF fragile and brittle, thus complicating its handling and manipulation. In Figure 9.3 the transmission loss as a function of time, measured at $\lambda = 1550\,\text{nm}$, of a $1\,\mu\text{m}$-diameter fiber is shown. It can be seen that the losses increase exponentially with time. The results shown in the figure suggests that subwavelength diameter fibers have a lifetime of hours. To avoid these issues an adequate protection of the nanofiber is therefore essential. This is well known by manufacturers of telecom optical fibers who protect the fiber immediately after drawing it. In the case of MNFs the protection must warrant good mechanical strength and must introduce minimal insertion losses without sacrificing the strong evanescent waves (their main advantage for sensing). To this end, MNFs are partially or totally embedded in low-index polymers such as Teflon (refractive index of 1.31), hydrophobic aerogel (refractive index of 1.05) or PDMS (refractive index of 1.402). The embedding process is carried out by depositing the protecting polymer over a substrate (wafer) or by covering completely the MNF with such polymers [27–30]. The protection is done immediately after fabricating the MNF.

Figure 9.3 shows that by partially and totally embedding MNFs in Teflon the degradation issues are substantially or totally eliminated. The figure also shows some photographs of a MNF protected with Teflon layers [28]. In Figure 9.3b, for example, the whole wafer is shown while supercontinuum light was

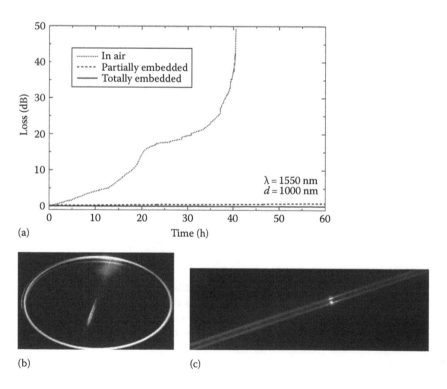

(a)

(b)   (c)

FIGURE 9.3 (a) Transmission loss as a function of time in a $1\,\mu\text{m}$ diameter tapered fiber when it is in air, partially or totally embedded in Teflon. The wavelength of the optical source was $1.55\,\mu\text{m}$. (b) and (c) are photographs of some MNFs protected with Teflon (refractive index of 1.31).

injected to the MNF. Figure 9.3c shows the scattering induced by a dust particle in an embedded MNF, situation that must be avoided. It is important to point out that embedded MNFs have the advantage that semi-planar structures can be obtained, thus making the MNF-based sensors more versatile and easier to handle. In addition, microfluidics channels can be fabricated or incorporated on the protected fibers. All these advantages combined with the long-term stability make protected MNFs really promising for sensing applications.

9.3 Biochemical Sensors Based on Micro/Nanofibers

As discussed in the previous section the fundamental mode in a MNF has a considerable amount of power in the evanescent wave even if it is embedded in a protecting material. This is advantageous for sensing since the sensitivity of protected MNFs is not sacrificed. Free-standing MNFs and nanotapers offer several possibilities for optical sensing. For example, by measuring the spectral properties of the transmitted light the concentration or presence of an analyte (sample) on the MNF surface can be obtained. This technique is known as evanescent wave spectroscopy, a simple but powerful technique. Typically, a wavelength or a range of wavelengths is or are absorbed by the analyte. Evanescent wave spectroscopy can be carried out directly without the need of intermediate coatings. Another possibility of sensing with free-standing MNFs is via evanescent wave attenuation or absorption. In this case the MNF needs to be coated with an intermediate layer or material whose optical properties change when it is exposed to a certain gas or chemical parameter. The changes experienced by the intermediate material make the intensity of the propagating beam to change. The measurements are typically carried out at a fixed wavelength and the transmission changes are easily monitored with a photodetector.

9.3.1 Sensing with Straight or Freestanding Micro/Nanofibers

One of the first demonstrations of MNF for gas sensing was reported by the present author and a collaborator [31]. A single-mode optical fiber was tapered down to a diameter of 1 μm, and then it was immediately coated with a 4 nm-thick palladium film over 2 mm. A palladium film was selected as an intermediate layer since it has the ability to selectively absorb hydrogen. When the film is exposed to hydrogen it is converted to a palladium hydride film whose optical properties are different from those of a hydrogen-free Pd film. The hydration of palladium causes that both the real and imaginary parts of the palladium refractive index to change [32,33]. This in turn causes attenuation changes of the evanescent waves of the propagating mode. The sensor response of the described device to consecutive exposures to hydrogen as well as the observed transmission as a function of hydrogen concentration are shown in Figure 9.4 along with an illustration of the device. The wavelength of the light source was 1550 nm. The response time of the sensor was found to be ~10 s which was between 3 and 5 times faster than that of other optical hydrogen sensors and about 15 times faster than that of some electrical nano hydrogen sensors reported until those days. The rapid response of the device was attributed to the ultra low volume of palladium that was rapidly filled with hydrogen.

Based on theoretical calculations showing that a single spherical nanoparticle in the vicinity of a MNF can induce strong scattering [34], a novel approach to sense hydrogen with MNF coated with nanoparticles was demonstrated by Monzón-Hernández et al. [35]. To achieve hydrogen-sensitive nanoparticles, MNFs were first coated over a few millimeters with a PdAu layer in island form. The islands were achieved by depositing an ultrathin layer (thickness in the 2–4 nm range) at low deposition rate (0.1 Å/s). Then the islands were grown with a thermal annealing process consists of exposing the samples in an air atmosphere at 200°C during 1 h [35]. The addition of gold to palladium enhances the absorption of hydrogen and the annealing process segregates and reshapes the islands allowing the formation of individual islands (in fact nanoparticles) that are well separated from the rest [35]. Figure 9.5 shows an illustration and microscope images of a MNF decorated with PdAu nanoparticles as well as the observed transmission changes in a 10-μm-diameter and a 5-μm-diameter MNF when they are exposed

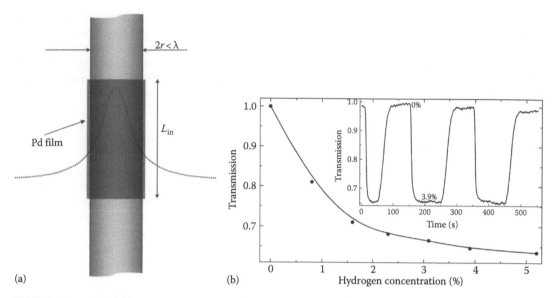

FIGURE 9.4 (a) Illustration of an MNF coated with a Pd layer over a length L_{in}. The dotted line represents the evanescent wave. (b). Transmission as a function of hydrogen concentration observed in a 1 μm-diameter fiber coated over 2 mm with a 4 nm-thick Pd layer. The inset shows the transmission as a function of hydrogen concentration of a sensor after repeated exposures to 3.9% of hydrogen in nitrogen. The measurements were carried out at 1550 nm.

successively to 2% or 4% of hydrogen in nitrogen (used as a carrier gas). The transmission changes were more prominent in the thinner fiber due to its stronger evanescent wave as discussed here. A pulsed-like behavior in the transmission can also be noted. This behavior was attributed to physical changes of the PdAu nanoparticles when they are exposed to hydrogen [35]. The response of MNFs decorated with nanoparticles was found to be around 5 s, considerable faster than many other hydrogen sensors that exploit phenomena at the nanoscale.

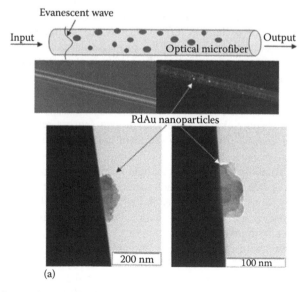

FIGURE 9.5 (a) Illustration and images of an MNF decorated with nanoparticles. The latter become visible when supercontinuum light is launched to the MNF. The closed-up images of the nanoparticles were taken with a transmission electron microscope.

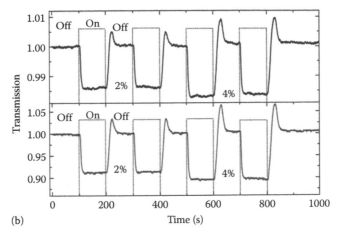

(b)

FIGURE 9.5 (continued) (b) Transmission as a function of time observed in a 10 μm- (top plot) and 5 μm-diameter fiber (bottom plot) when they were exposed to 2% and 4% hydrogen concentration. In both cases the length of the decorated zone was 2 mm and λ = 1550 nm. The on/off operation of the hydrogen valve is shown with dotted lines. (Reprinted from *Sens. Actuators B*, 151, Monzón-Hernández, D. et al., Optical microfibers decorated with PdAu nanoparticles for fast hydrogen sensing, 219, Copyright 2010, with permission from Elsevier.)

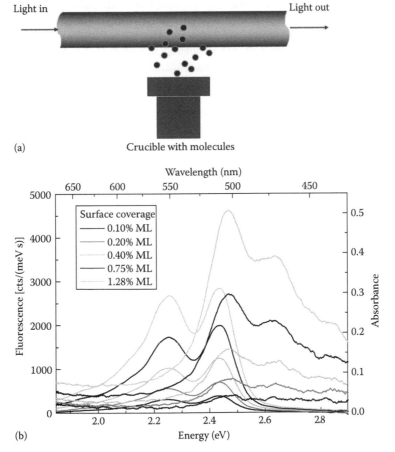

FIGURE 9.6 (a) Schematic representation of a nanofiber exposed to molecules. (b) Absorption and fluorescence spectra of surface-adsorbed PTCDA molecules during deposition. The surface coverages are indicated. (Plot reprinted from Stiebeiner, A. et al., Ultra-sensitive fluorescence spectroscopy of isolated surface-adsorbed molecules using an optical nanofiber, *Opt. Express*, 17, 21704, 2009. With permission of Optical Society of America.)

The results reported in [14,34,35] suggest that optical sensors with small footprints and high sensitivity can be developed by simply decorating MNF with suitable nanoparticles. In theory, a single nanoparticle on a nanofiber can induce detectable scattering losses for which sensors with sensitivity down to single particle (or molecule) seems feasible [34].

Chemical sensing by means of evanescent wave spectroscopy in a nanofiber was reported in [36,37]. The schematic diagram of the experiment is shown in Figure 9.6. A optical fiber, single mode for the 450–600 nm range, was tapered down to 320 nm and exposed to 3,4,9,10-perylenetetrac arboxylic dianhydride molecules (PTCDA) at ambient conditions. The molecules were deposited on the taper waist by heating to 325°C a crucible with PTCDA crystals placed beneath the thinnest section of the fiber. The sublimated molecules were adsorbed on the nanofiber and the corresponding absorption was measured with a conventional spectrometer. The length of the nanofiber that was covered with the PTCDA molecules was only 1 mm. With the same set up the authors measured the fluorescence spectra which were obtained by excitation of the molecules with a diode laser emitting at 406 nm. It is important to point out that evanescent wave spectroscopy by means of a nanofiber is several orders of magnitude more sensitive than that of conventional methods based on free beam absorption.

Polymer nanowires doped with certain materials are also good candidates for chemical and humidity sensing [38–40]. The materials to fabricate such nanowires include poly-(methyl methacrylate) (PMMA), polystyrene (PS), polyacrylamide (PAM), and polyaniline/polystyrene (PANI/PS). To make these nanowires sensitive to a certain gas or a chemical parameter, functional materials are added to the solvated polymers before drawing the nanowires. The advantage in this case is that sensors with detection limits (this refers to the minimum quantity change in the parameter being sensed that can be detected by the sensor) are as low as sub-parts-per-million level can be developed [38–40]. Moreover, the response time (this is defined as the time for a sensor to reach 90% of the signal change) of polymer nanowires sensors is between 1 and 2 orders of magnitude faster than those of conventional sensors. In Figure 9.7 the schematic diagram of a 350 nm-diameter PANI/PS

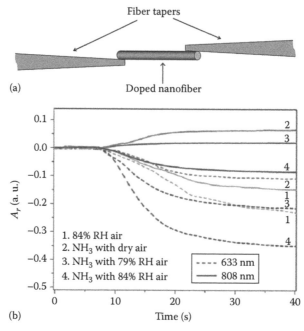

FIGURE 9.7 (a) Schematic diagram of a doped nanowire coupled to tapered fibers in both extremes. (b) Time-dependent response of a 370 nm diameter nanowire exposed to (1) 84% RH air, (2) 5 ppm NH3, (3) 5 ppm NH3 with 79% RH air, and (4) 5 ppm NH3 with 84% RH air respectively, which were simultaneously monitored using two lasers with wavelengths of 633 and 808 nm. (Plot reprinted from Gu, F. X. et al., Polyaniline/polystyrene single-nanowire devices for highly selective optical detection of gas mixtures, *Opt. Express*, 17, 11230, 2009. With permission of Optical Society of America.)

fiber connected to tapered fibers is shown. Upon being exposed to gas mixtures of NH_3 (ammonia) or humidity, the coupling efficiency between the PANI/PS nanowire and the fiber taper changes. This results in transmission changes that are easily measured with a simple transmission measurement setup consisting, e. g., of a laser and a photodetector. The plot in Figure 9.7 shows the time response of the referred nanowire to different NH_3 gas concentration or relative humidity (RH) measured at two different wavelengths.

From the aforementioned results it is clear that free-standing MNFs and nanotapers with or without intermediated layers can be used for a myriad of sensing applications. They can also be decorated with different nanoparticles or nanolayers sensitive to a specific gas or chemical parameter. The main advantages of MNFs for gas or chemical sensing are the fast response times and the low detection limits (subparts per billion) that can be achieved due to the rapid diffusion of the parameter being sensed (gases or chemical parameters) within the nanometer material.

9.3.2 Sensing with Microfiber Resonators

MNFs offer many more possibilities for optical sensing due to their configurability and large evanescent waves. The mode propagating in two adjacent MNFs can overlap and couple [41–43]. Surface attraction forces (Van der Waals and electrostatic) keeps the fibers together even if they do not have identical dimensions or shape or if they are made of different materials [41–43]. Due to this property all-MNF interferometers [44–49] can be devised as well as resonators in different configurations [50–68]. Figure 9.8 illustrates the coupling between two identical MNFs as well as a resonator built by simply coiling a microfiber onto itself [44,45]. The loop resonator shown in Figure 9.8 was created by manipulating the fiber with translation and rotation stages under an optical microscope [45]. The shape and diameter of all-MNF resonators can be controlled while they are created. Thus all-MNF resonators can be an alternative to other resonators (e.g., based on microspheres or capillaries) which are more complex [69,70].

In a resonator, a light wave circumnavigates a ring of radius R and returns in phase. The resonant length, λ_r, is given by [70]

$$\lambda_r = \frac{2\pi R n_e}{m} \quad (9.4)$$

In Equation 9.4, n_e is the effective refractive index (RI) experienced by the optical resonant mode and m is the integer number of orbital wavelengths λ_r/m. Due to the circulating nature of the resonant mode in a resonator, light interacts repeatedly (up to millions of time) with an analyte deposited on its surface via

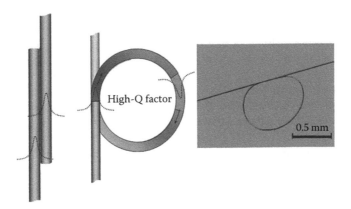

FIGURE 9.8 Schematic representation of the coupling between two adjacent and an all-MNF resonator. The arrows indicate the circulation of the light and the dotted lines the guided mode. The photograph shows a resonator built with 0.9 μm-diameter fiber. (Photograph reproduced from Sumetsky, M. et al., The microfiber loop resonator: Theory, experiment, and application, *J. Lightw. Technol.*, 24, 242, 2006. With permission. Copyright 2006 IEEE.)

evanescent wave effects. This is equivalent to having a sensor with an extremely long effective interaction length, which is determined by

$$L_e = \frac{Q\lambda_r}{2\pi n_e} \tag{9.5}$$

Q is called the resonator quality factor (Q-factor), which represents the number of round trips that the light wave can make along the resonator. The Q-factor ranges from 10^4 to 10^8, depending on the resonator configuration [70]. Therefore, despite a small physical size, in a resonator the effective interaction length is on the order of a few tens of centimeters.

The transmission spectrum of a resonator exhibits narrow dips whose line width ($\delta\lambda_r$) or FWHM, depends on the Q-factor as $\delta\lambda_r = \lambda_r/Q$. Due to the high Q-factors, in most resonators $\delta\lambda_r$ is very narrow, on the order of picometers. When a resonator is used for sensing applications the maximum Q-factor (sensitivity) is searched. The sensitivity of the resonance wavelength to changes of radius (ΔR) or changes of effective refractive index (Δn_e) is [71]

$$\frac{\Delta\lambda_r}{\lambda_r} = \frac{\Delta R}{R} + \frac{\Delta n_e}{n_e} \tag{9.6}$$

Let us analyze the potential of a resonator for optical sensing. Let us assume that the resonator radius does not change, that the Q-factor is $\sim 10^6$, $\lambda_r \sim 1500\,\text{nm}$, and $\Delta n_e \sim 10^{-6}$ (this change can be caused, e.g., by the presence of a 1 nm protein monolayer on the resonator surface). Under these conditions the shift of the resonance dips will be $\Delta\lambda_r \sim 1.5\,\text{pm}$, which is comparable with $\delta\lambda_r$, and therefore, easily detectable. Due to these remarkable properties resonator are appealing for optical sensing. In fact, resonators are capable of detecting single molecule (provided they are large) and tens of nanometer-sized single particle or virus [71–74]. By periodically pattern a molecular surface monolayer on the resonator surface any kind of label-free, large or small molecule is, in principle, detectable, as demonstrated by Dominguez-Juarez et al. [75].

Coupling the light in and out of certain resonators typically imposes alignment constraints that may limit the implementation of functional sensors [69–75]. In all-MNF resonators, the alignment is not an issue since light is guided by the fiber itself. Moreover, MNF-based resonators can be supported or embedded in low-index polymers to make them mechanically and optically stable [62–68]. Monolithic MNF resonators can also be implemented since microfibers can be spliced or glued with spheres [58–61]. Even if they are packaged or protected all-MNF resonators reach considerable high quality or Q-factors. Therefore, all-MNF resonators can be good candidates for biosensing applications due to their flexibility in their design.

An example of an alignment-free resonator built with polystyrene microspheres attached to a MNF was reported by Gregor et al. [60]. Microspheres made from chemically grown polystyrene with diameters of 31 and 50 μm were attached on a 1.2 μm-diameter fiber [60]. Then a tunable diode laser emitting near 780 nm was used to excite the resonances of the polystyrene spheres. This simple system enabled determination of gas concentration. Polystyrene spheres experience changes in their index of refraction and diameter when they are exposed to helium due to changes of temperature suffered by the material that constitutes the sphere. Figure 9.9 shows the relative shift in the resonances as well as the temperature of the spheres for different helium/argon mixtures (argon is used as a carrier gas). These experiments suggest that microspheres made of other gas-sensitive materials attached to MNF can be used for the detection of specific gasses and eventually of other chemical parameters provided that they change the resonator's diameter or refractive index.

Another example of a refractometric sensor based on MNF resonator was reported in [57]. A 480 μm diameter resonator built with 2.3 μm diameter fiber was implemented. The resonator was immersed in water and then different amounts of ethanol were added to change the refractive index in a controlled manner. The resonator transmission spectra were measured for each mixture or refractive index. Figure 9.10 shows the observed dips and their position for different quantities of ethanol in water. The sensitivity of this simple resonator was found to be around 1.8×10^{-5} which is really high for an aqueous environment.

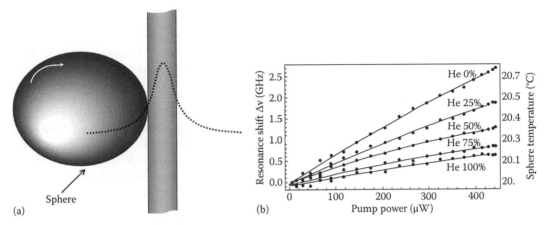

FIGURE 9.9 (a) Schematic representation of a gas-sensitive sphere attached to a MNF. (b) Resonance shifts induced by laser heating for volume mixtures of helium and argon. The indicated pump power is the power in the taper taking coupling and transmission losses into account. (Reprinted with permission from Gregor, M. et al., An alignment-free fiber-coupled microsphere resonator for gas sensing applications, *Appl. Phys. Lett.*, 96, 231102, 2010. Copyright 2010, American Institute of Physics.)

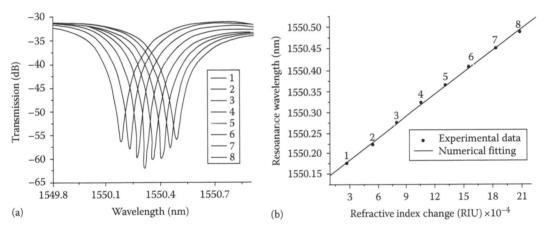

FIGURE 9.10 (a) Transmission spectra over 1 nm of a 480 μm diameter resonator for different refractive indices. (b) Position of the dip as a function of the refractive index change. The eight peaks were obtained by adding 5 μL of ethanol to 500 μL of water. (Reprinted from Guo, X. and Tong, L. M., Supported microfiber loops for optical sensing, *Opt. Express*, 16, 14429, 2008. With permission of Optical Society of America.)

All-MNF resonators are capable of resolving RI changes on the order of 10^{-5}, and even higher. Thus, they are promising for detecting or analyzing biomolecular interactions or for other biosensing applications. The referred resolution is sufficient for monitoring molecular bindings, chemical or biochemical reactions since these can be detected as minute refractive index changes [76–78]. For biosensing applications, MNF resonators must be coated with biolayers, i.e., thin layers that that contain probe molecules which selectively identify or interact with a target molecule [79]. This functionalization of the surface will provide a MNF resonator the required specificity, an important feature of any sensing platform. However, the biocoatings or biolayers must not affect the resonator's sensing capabilities or optical qualities. Another important issue is the integration of MNF resonators with microfluidic systems to create fully integrated biosensing chips. The fuctionalization of MNF resonators as well as their integration with microfluidics systems present challenges of substantial importance for the sensor community.

9.4 Conclusions and Outlook

Subwavelength-diameter optical fibers (or micro/nanofibers) manufactured from optical fibers, bulk glass or polymers exhibit outstanding optical and mechanical properties. In glass nanofibers, for example, the propagating mode has a considerable amount of its energy in the evanescent wave (approaching 100% in some cases). MNFs exhibit low transmission losses and can be configured in different geometries due to their robustness and flexibility. This allows the development of ultracompact resonators and interferometers. That is why MNFs are considered ideal candidates for sensing applications. In general the interaction length required to carry out a sensing task with MNFs is on the order of a few hundreds micrometers. Thus, the amount of sample required for the sensing task can be on the order of micro or nanoliters. Due to these properties, the detection limit of a nanofiber-based sensor can be as low as parts per billion. MNF-based sensors are promising for single molecule, single virus or single particle detection.

A free-standing MNF and nanotaper represent the simplest configuration for optical sensing. The strong evanescent waves in such MNFs can interact directly with monolayers or molecules present on the fiber surface. Such interactions result in detectable changes in the MNF transmission. The absorption of monolayers on a MNF surface, for example, can be detected by measuring the transmission spectra. This type of experiments has demonstrated the power and versatility of MNFs for evanescent wave spectroscopy in which the achievable sensitivity is orders of magnitude higher than that of conventional surface spectroscopy. An MNF can also be doped with functional materials that make the coupling between MNFs or their transmission spectra to change when the nanofiber is exposed to a certain gas or chemical parameter. In this case, the detection limits and time response are, respectively, in the order of parts per billion and milliseconds due to the nanometric dimensions of the waveguide.

Free-standing MNFs and nanotapers can also be coated with polymeric or metallic nanoparticles or nanolayers permeable to specific gases or chemical parameters. The optical properties of such nanoparticles or nanolayers change when they are exposed to the target chemical parameter. Such changes perturb the guided mode via evanescent-wave interactions giving rise to detectable transmission changes in the MNF or nanotaper. One of the main advantages in this case is the fast response time of the sensors due to the short interaction length and the nanometric dimensions of the intermediate particles or layers. It has been demonstrated that some gas sensors based on MNFs exhibit much faster response time that others based on waveguides coated with similar gas-permeable layers.

All-MNF resonators exhibit high Q-factors, thus they can be a real alternative to more complex resonators that are fabricated by lithographic methods or that require a critical alignment. MNFs can be spliced or glued with microspheres; in addition, all-MNF resonators can be embedded without sacrificing their performance. Therefore, the implementation of monolithic or packaged all-MNF resonators, something essential for real-world applications, is promising. Many research groups are working in this direction; thus it would not be a surprise to see MNF resonators in the market in the near future.

For biosensing applications MNF resonators must be coated with biolayers, i.e., thin layers that are sensitive to biological targets such as virus, bacteria, or antigens. Such biolayers will provide a MNF resonator the required specificity, an important feature of any sensing platform intended for applications in complex environments such as blood, saliva, or urine samples. The functionalization process is a necessary step for the development of label-free biosensors based on MNF resonators. The coating of the resonator surface must not affect the device's sensing capabilities or optical properties. Another important issue is the integration of MNF resonators with microfluidic sample processing systems (Lab-on-a-Chip systems) to create fully integrated biosensing chips. Such chips must contain several resonators each of them funtionalized to detect a specific biological target. Presently, multiplexed or multi-analyte detection in a single chip is a must. When these steps are carried out, MNF-based sensors will have a significant impact in crucial areas such as medical diagnostics, food analysis, and environmental monitoring, among others. The functionalization of all-MNF resonators, the multiplexed detection, and their integration with microfluidics represent outstanding challenges for the sensor community. Several groups are working with vigor on this direction and thus, all the aforementioned issues will soon be overcome.

REFERENCES

1. Kapany, N. S., High-resolution fibre optics using sub-micron multiple fibres, *Nature,* 184, 881, 1959.
2. Lacroix, S. et al., Tapered monomode optical fibers: Understanding large power transfer, *Appl. Opt.*, 25, 4421, 1986.
3. Love, J. D. et al., Tapered single-mode fibres and devices Part 1: Adiabaticity criteria, *IEE Proc.*, 138, 343, 1991.
4. Bilodeau, F. et al., Compact, low-loss, fused biconical taper couplers: Overcoupled operation and anti-symmetric supermode cutoff, *Opt. Lett.*, 12, 634–636, 1987.
5. Bures, J. and Ghosh, R., Power density of the evanescent field in the vicinity of a tapered fiber, *J. Opt. Soc. Am. A*, 16, 1992, 1999.
6. Birks, T. A., Wadsworth, W. J., and Russell, P. St. J., Supercontinuum generation in tapered fibers, *Opt. Lett.*, 25, 1415, 2000.
7. Cai, M., Painter, O., and Vahala, K. J., Observation of critical coupling in a fiber taper to a silica-microsphere whispering-gallery mode system, *Phys. Rev. Lett.*, 85, 74, 2000.
8. Kakarantzas, G. et al., Miniature all-fiber devices based on CO_2 laser microstructuring of tapered fibers, *Opt. Lett.*, 26, 1137, 2001.
9. Tong, L.M. et al., Subwavelength-diameter silica wires for low-loss optical wave guiding, *Nature,* 426, 816, 2003.
10. Law, M. et al., Nanoribbon waveguides for subwavelength photonics integration, *Science,* 305, 1269, 2004.
11. Sirbuly, D. J. et al., Optical routing and sensing with nanowire assemblies, *Proc. Natl. Acad. Sci. USA*, 102, 7800, 2005.
12. Sirbuly, D. J. et al., Multifunctional nanowire evanescent wave optical sensors, *Adv. Mater.*, 19, 61, 2007.
13. Sirbuly, D. J. et al., Biofunctional subwavelength optical waveguides for biodetection, *ACS Nano*, 2, 255, 2008.
14. Sirbuly, D. J., Létant, S. E., and Ratto, T. V., Hydrogen sensing with subwavelength optical waveguides via porous silsesquioxane-palladium nanocomposites, *Adv. Mater.*, 20, 4724, 2008.
15. Zhu, H., Wang, Y., and Li, B., Tunable refractive index sensor with ultracompact structure twisted by poly(trimethylene terephthalate) nanowires, *ACS Nano*, 3, 3110, 2009.
16. Brambilla, G., Finazzi, V., and Richardson, D., Ultra low loss optical fiber nanotapers, *Opt. Express*, 12, 2258, 2004.
17. Brambilla, G. et al., Optical fiber nanowires and microwires: Fabrication and applications, *Adv. Opt. Photon.*, 1, 107, 2009.
18. Tong, L. and Sumetsky, M., *Subwavelength and Nanometer Diameter Optical Fibers*, Springer, Berlin, Germany, 2010.
19. Brambilla, G., Optical fibre nanotaper sensors, *Opt. Fiber Technol.*, 16, 331, 2010.
20. Zhang, L., Lou, J., and Tong, L., Micro/nanofiber optical sensors, *Photon. Sens.*, 1, 31, 2010.
21. Tong, L. et al., Photonic nanowires directly drawn from bulk glasses, *Opt. Express*, 14, 82, 2006.
22. Ding, L., Ultralow loss single-mode silica tapers manufactured by a microheater, *App. Opt.* 49, 2441, 2010.
23. Zhang, E. J., Sacher, W. D., and Poon, J. K. S. Hydrofluoric acid flow etching of low-loss subwavelength-diameter biconical fiber tapers, *Opt. Express*, 18, 22593, 2010.
24. Tong, L., Lou, J., and Mazur, E. Single-mode guiding properties of subwavelength-diameter silica and silicon wire waveguides, *Opt. Express*, 12, 1025, 2004.
25. Sumetsky, M. How thin can a microfiber be and still guide light? *Opt. Lett.*, 31, 870, 2006.
26. Hartung, A., Brueckner, S., and Bartelt, H. Limits of light guidance in optical nanofibers, *Opt. Express*, 18, 3754, 2010.
27. Xu, F. and Brambilla, G., Preservation of micro-optical fibers by embedding, *Jpn. J. Appl. Phys.*, 6675, 2006.
28. Lou, N. et al., Embedded optical micro/nano-fibers for stable devices, *Opt. Lett.* 35, 571, 2010.
29. Xiao, L. et al., Stable low-loss optical nanofibres embedded in hydrophobic aerogel, *Opt. Express*, 19, 764, 2011.
30. Chuo, S. M. and Wang, L. A., Propagation loss, degradation and protective coating of long drawn microfibers, *Opt. Commun.*, 12, 2825, 2011.

31. Villatoro, J. and Monzón-Hernández, D., Fast detection of hydrogen with nano fiber tapers coated with ultra thin palladium layers, *Opt. Express*, 13, 5087, 2005.

32. Lewis, F. A., *The Palladium Hydrogen System*, Academic Press, London, U.K., 1967.

33. Flanagan, T. B. and Oates, W. A., The palladium–hydrogen system, *Annu. Rev. Mater. Sci.*, 21, 269, 1991.

34. Wang, S., Pan, X., and Tong, L., Modeling of nanoparticle-induced Rayleigh–Gans scattering for nano-fiber optical sensing, *Opt. Commun.* 276, 293, 2007.

35. Monzón-Hernández, D. et al., Optical microfibers decorated with PdAu nanoparticles for fast hydrogen sensing, *Sens. Actuators B*, 151, 219, 2010.

36. Warken, F. et al., Ultra-sensitive surface absorption spectroscopy using sub-wavelength diameter optical fibers, *Opt. Express*, 15, 11952, 2007.

37. Stiebeiner, A. et al., Ultra-sensitive fluorescence spectroscopy of isolated surface-adsorbed molecules using an optical nanofiber, *Opt. Express*, 17, 21704, 2009.

38. Gu, F. et al., Polymer single-nanowire optical sensors, *Nano Lett.*, 8, 2757, 2008.

39. Gu, F. X. et al., Polyaniline/polystyrene single-nanowire devices for highly selective optical detection of gas mixtures, *Opt. Express*, 17, 11230, 2009.

40. Zhang, L. et al., Fast detection of humidity with a subwavelength-diameter fiber taper coated with gelatin film, *Opt. Express*, 17, 13349, 2008.

41. Huang, K., Yang, S., and Tong, L., Modeling of evanescent coupling between two parallel optical nanowires, *Appl. Opt.* 46, 1429, 2007.

42. Wang, G. H. et al., Theoretical investigation of nanowaveguide-based optical coupler using mode expansion transfer matrix, *Microw. Opt. Technol. Lett.* 52, 1123, 2010.

43. Hong, Z. et al., Coupling characteristics between two conical micro/nano fibers: Simulation and experiment, *Opt. Express*, 19, 3854, 2011.

44. Sumetsky, M., Dulashko, Y., and Hale, A., Fabrication and study of bent and coiled free silica nanowires: Self-coupling microloop optical interferometer, *Opt. Express,* 12, 3521, 2004.

45. Sumetsky, M. et al., The microfiber loop resonator: Theory, experiment, and application, *J. Lightw. Technol.,* 24, 242, 2006.

46. Li, Y. and Tong, L., Mach–Zehnder interferometers assembled with optical microfibers or nanofibers, *Opt. Lett.*, 33, 303, 2008.

47. Themistos, C. et al., Characterization of silica nanowires for optical sensing, *J. Lightw. Technol.,* 27, 5537, 2009.

48. Wu, P. H., Sui, C. H., and Ye, B. Q., Modelling nanofiber Mach-Zehnder interferometers for refractive index sensors, *J. Modern Opt.,* 56, 2335, 2009.

49. Kou, J. et al., Microfiber-probe-based ultrasmall interferometric sensor, *Opt. Lett.*, 35, 2308, 2010.

50. Sumetsky, M. et al., Optical microfiber loop resonator, *Appl. Phys. Lett.*, 86, 161108, 2005.

51. Jiang, X. S. et al., Demonstration of optical microfiber knot resonators, *Appl. Phys. Lett.*, 88, 223501, 2006.

52. Xu, F., Horak, P., and Brambilla, G., Optimized design of microcoil resonators, *J. Lightw. Technol.,* 25, 1561, 2007.

53. Xu, F., Horak, P., and Brambilla, G., Conical and biconical ultra-high-Q optical-fiber nanowire microcoil resonator, *Appl. Opt.*, 46, 570, 2007.

54. Xu, F. and Brambilla, G., Manufacture of 3-D microfiber coil resonators, *Photon. Technol. Lett.*, 19, 1481, 2007.

55. Sumetsky, M., Optimization of optical ring resonator devices for sensing applications, *Opt. Lett.* 32, 2577, 2007.

56. Shi, L. et al., Simulation of optical microfiber loop resonators for ambient refractive index sensing, *Sensors*, 7, 689, 2007.

57. Guo, X. and Tong, L. M., Supported microfiber loops for optical sensing, *Opt. Express*, 16, 14429, 2008.

58. Pal, P. and Knox, W. H., Fabrication and characterization of fused microfiber resonators, *Photon. Technol. Lett.*, 21, 766, 2009.

59. Wang, P. et al., Fusion spliced microfiber closed-loop resonators, *Photon. Technol. Lett.*, 22, 1075, 2010.

60. Gregor, M. et al., An alignment-free fiber-coupled microsphere resonator for gas sensing applications, *Appl. Phys. Lett.*, 96, 231102, 2010.

61. Yan, Y. Z. et al., Packaged silica microsphere-taper coupling system for robust thermal sensing application, *Opt. Express*, 19, 5753, 2011.

62. Vienne, G., Li, Y.H., and Tong, L. M., Effect of host polymer on microfiber resonator, *IEEE Photon. Technol. Lett.,* 19, 1386, 2007.
63. Xu, F. and Brambilla, G., Embedding optical microfiber coil resonators in Teflon, *Opt. Lett.,* 32, 2164, 2007.
64. Xu, F., Horak, P., and Brambilla, G., Optical microfiber coil resonator refractometric sensor, *Opt. Express,* 15, 7888, 2007.
65. Xu, F. et al., An embedded optical nanowire loop resonator refractometric sensor, *Opt. Express,* 16, 1062, 2008.
66. Xu, F. and Brambilla, G., Demonstration of a refractometric sensor based on optical microfiber coil resonator, *Appl. Phys. Lett.,* 92, 101126, 2008.
67. Jung, Y. et al., Embedded optical microfiber coil resonator with enhanced high-Q, *IEEE Photon. Technol. Lett.* 22, 1638, 2010.
68. Birks, T. and Xia, L., High finesse microfibre knot resonators made from double-ended tapered fibres, *Opt. Lett.,* 36, 1098, 2011.
69. Jokerst, N. et al., Chip scale integrated microresonator sensing systems, *J. Biophoton,* 2, 212, 2009.
70. Sun, Y. and Fan X., Optical ring resonators for biochemical and chemical sensing, *Anal. Bioanal. Chem.,* 399, 205, 2011.
71. Vollmer, F. and Arnold, S., Whispering-gallery-mode biosensing: Label-free detection with sensitivity down to single molecules, *Nat Meth.,* 5, 591, 2008.
72. Armani, A. M. et al., Label-free, single-molecule detection with optical microcavities, *Science,* 317, 783, 2007.
73. Vollmer, F., Arnold, S., and Keng, D., Single virus detection from the reactive shift of a whispering-gallery mode, *Proc. Natl. Acad. Sci. USA,* 105, 20701, 2008.
74. Zhu, J. et al., On-chip single nanoparticle detection and sizing by mode splitting in an ultrahigh-Q microresonator, *Nat. Photon.,* 4, 46, 2010.
75. Dominguez-Juarez, J. L., Kozyreff, G., and Martorell, J., Whispering gallery microresonators for second harmonic light generation from a low number of small molecules, *Nat. Commun.,* 2, 254, 2011.
76. Leung, A., Shankar, M. P., and Mutharasan, R., A review of fiber-optic biosensors, *Sens. Actuators B,* 125, 688, 2007.
77. Bosch, M. E. et al., Recent development in optical fiber biosensors, *Sensors,* 7, 797, 2007.
78. Fan, X. et al., Sensitive optical biosensors for unlabeled targets: A review, *Anal. Chim. Acta,* 6, 8, 2008.
79. Hunt, H. K., Soteropulos, C., and Armani, A. M., Bioconjugation strategies for microtoroidal optical resonators, *Sensors,* 10, 9317, 2010.

10

Optofluidic Sensors

Romeo Bernini

CONTENTS

10.1 Introduction

Optofluidics is a very promising field that has seen significant improvement over the last few years. Optofluidics is essentially the marriage of photonics and microfluidics [1–4]. Such integration, at micro- and nanoscale, of photonic devices with microfluidic components has the potential to provide compact, effective sensors for lab-on-a-chip tools that can be used to detect, manipulate, and sort cells, virus, and biomolecules in fluidics.

Currently, a typical optical sensor, for chemical or biological applications, can be divided in two main parts, the physical optical sensor and the liquid under analysis. This structure requires a complex and strong integration on the same chip, of microfluidic devices and optical devices. A different approach to this task is to use optofluidic to design innovative sensors. This allows us to realize a full integration between the microfluidic part and the optical part, resulting in a reduction of the device dimensions, and a high interaction efficiency between the light and the substance, which can be very useful in sensing applications.

Several optofluidic devices for sensing and analysis have been demonstrated based on different transduction methods including absorbance, interferometric, and scattering. This chapter reviews the state of the optofluidics sensing devices and techniques for chemical and biochemical analysis.

10.2 Optofluidic Waveguides

The possibility of guiding light through a fluid offers very interesting applications especially in sensing field. The development of optical lab-on-chip requires the integration on the same chip of microfluidic and optical devices waveguides in order to continue to reduce the device size and cost, and improve the sensitivity. In this field optofluidic waveguides can represent an important building block for the realization of more complex devices and systems. In fact, they represent the maximum integration and functionality between waveguides, the key element of the optical structure, and microchannels, the key element of microfluidics.

The first demonstration of optofluidic waveguides was given by Daniel Colladon [5] and Jacques Babinet [6] in Paris in the early 1840s and few years later confirmed by John Tyndall [7]. They show the possibility to light guiding in a jet of water falling from a tank. This was also the first example of guiding of light by refraction, the principle that makes conventional waveguides possible. Several years later, optical waveguides with a liquid core were explored for enabling light deflectors [8] or long-haul optical communications [3]. However, only in the last years a great effort has been devoted to the development novel and high efficient integrated optofluidic waveguides.

Conventional solid core waveguides are based on total internal reflection confinement mechanism, which requires a high refractive index core surrounded by low refractive indexes claddings. However, this approach cannot be easily replicated in optofluidic sensing applications. In fact, the fulfillment of the total internal reflection (TIR) condition, when low refractive indices of aqueous solutions (1.33) are used as core materials, is a very hard task while considering that common solid materials employed for microelectronic and microfluidic fabrication exhibit higher refractive indexes (1.4–3.5). In order to overcome this problem, many techniques have been proposed in order to realize optofluidic waveguides.

The most simple way to realize a hollow waveguide is to use thin film metal cladding in order to confine the light into the core [9–11]. Metal-coated waveguides have the advantages of guiding light through any core liquid regardless of refractive index and, in principle, collecting light in a very large solid angle. However, typically, these waveguides are characterized by high propagation losses at optical frequencies that make them ill-suited for sensing applications.

Taking into account this problem, in the past years a great effort has been devoted to develop optofluidic waveguides based on TIR. In fact, one of the main advantages of TIR waveguides is that, neglecting the absorption and imperfection effects, they are lossless. In order to achieve the TIR condition when the core material is a liquid, it is necessary to find or develop cladding materials with very low refractive indices. A possible approach is to use a class of fluorinated polymers, such as Teflon AF, that are optically transparent and exhibit refractive index of 1.29–1.31, slightly lower than that of water (1.33). Liquid core fibers consisting of either Teflon AF tubes or glass capillaries coated internally with Teflon AF have been fabricated and used in sensing applications since 1990s [12,13]. More recently, the fabrication, at the chip-scale, of liquid core waveguides by spin-coating a Teflon AF solution onto silicon, glass, or PDMS substrates has been demonstrated [14,15]. A different fabrication method of Teflon AF-coated optofluidic waveguides by selective coating of PDMS channels has been proposed [16]. However, the use of Teflon AF as cladding poses strong challenges because fluorinated polymers are nonpolar, contain no reactive chemical functionality, thus the adhesion to substrate depends primarily on physical rather than chemical interactions. In particular, Teflon AF has poor adhesion to commonly used substrates, requiring additional adhesion-promoting fabrication steps. Furthermore, for the same reason, the surface of Teflon AF offers little possibility for chemical functionalization, as might be required for many biosensor applications.

Nanoporous dielectrics represent an alternative to fluorinated polymers to form low refractive index cladding. Nanoporous materials are formed by introducing nanosized air pores into a solid matrix material. The refractive index of the composite material is a weighted average of the refractive index of air and that of the host material. This peculiar structure offers a great flexibility in terms of obtainable refractive indexes for the cladding that can be easily tuned by varying the degree of porosity. In addition, they exhibit excellent adhesion to substrates and the high surface area typically associated with these materials could be utilized for increasing the binding density of sensor probes with a sensitivity enhancement. The first attempt to use nanoporous dielectrics for the fabrication of liquid core waveguides was performed using the sacrificial porogen/template method [17]. In this approach the nanoporosity is introduced by forming a nanocomposite film comprising a matrix material and a thermally labile species (porogen) followed by a high-temperature heating step. However, these films, with their rigid structures, regularly suffer from large residual thermal stresses which may initiate cracking and buckling, especially when thick films, as for liquid core waveguides, are desired. In order to address this issue, an alternative method of preparation of nanoporous dielectrics based on the deposition of dielectric nanoparticles wherein the porosity of the film arises from the voids formed between individual nanoparticles has been proposed [18]. This configuration minimizes the thermal stresses during the curing/calcination temperature cycle enabling the formation of thick (as thick as 3.6 µm), crack-free films with refractive indexes as low as 1.047. Using this approach two dimensional liquid core waveguides have been successfully realized by spin-coating with low refractive index layer silicon microchannels [18].

Recently, the use of water-ice as potential relatively low refractive index (1.309) cladding has been proposed [19]. However, ice requires that appropriately low temperature is maintained. Typically, liquid core waveguide based on TIR have been fabricated with large core diameters of 200–500 μm. Single-mode waveguides are difficult to achieve due to the very small required core thicknesses (1–2 μm). In order to overcome some limitations of liquid-core/solid-cladding waveguides, liquid-core/liquid-cladding waveguides have been proposed and fabricated using both glass capillary technology [20], and planar geometry by replica molding polymeric techniques [21–23]. These waveguides are microfluidic devices in which the light is confined inside a high refractive index liquid (the core) by a low refractive index liquid (the cladding), both liquids flowing laminarly. Because L2 waveguides are dynamic, many optical properties (refractive index contrast, geometry, etc.) can be easily changed. Furthermore they have optically smooth side walls independently of the channel roughness. The liquid core/liquid cladding guiding concept has been implemented both in one and in two-dimensional geometries. Typically, aqueous solution of CaCl with a high refractive index (1.44–148) was used as the core of these waveguides, while deionized water formed its cladding. However, the choice of liquids poses strong limitations due to the requirement that the refractive index of the liquid "core" exceed that of the liquid "cladding." In particular, light could not be guided in water (n = 1.33) or in low refractive index aqueous solutions without the addition of dissolved species to increase the refractive index of aqueous core. These limitations may preclude its application for biosensing where water is the natural medium for biological processes, as the dissolved species may undesirably interfere with the sensor probes and analytes. In addition, these waveguides are expected to be susceptible to vibrations and the diffusion of analyte molecules of interest into the cladding regions [3]. This issue can be circumvented by using hybrid approaches [24] or by substituting the liquid cladding with a gas (air) [25]. Liquid core/air claddings eliminate both the refractive index problem and diffusion phenomena. A plasmonic liquid–liquid (PLL) waveguide using Au nanoparticles (NPs) colloidal solution has been also proposed [26]. In this waveguide a colloidal solution with suspended Au NPs becomes an effective metallic cladding layer and water is the core layer.

Another important approach of the waveguide structure able to guide the light in a liquid channel is given by the slot waveguide [27–29]. A slot waveguide is composed by two high refractive index dielectric nanowires separated by a low refractive index gap region of tens of nanometers. In these devices, even if the light is guided by the TIR mechanism, there is a strong enhancement and confinement of the optical field in a small gap filled with low-index material. Usually slot waveguides are based on Silicon-on-Insulator (SOI) technology and realized by defining two silicon wires at a distance of few hundred of nanometer on a silicon dioxide substrate. More recently, slot waveguide based on silicon nitride on silica was also demonstrated [30,31]. Due to the very tight confinement and interaction area, slot waveguides represent a promising approach for waveguide sensing into nanofluidic structure allowing extremely low analyte concentrations. Furthermore, because slot waveguides are single mode they can be used in order to realize more complex interferometric or resonant devices as illustrated in the next sections.

The problem to confine the light in low refractive index materials can be solved using confinement mechanisms different from the TIR one. Recently, dielectric cladding hollow optical waveguide based on bragg reflectors [32] or photonic band gap [33,34] has been proposed and fabricated. The light-guiding mechanism in this new class of devices is fundamentally different from that of traditional waveguides, which rely on total internal reflection and thus demand that the index of refraction of the core material be higher than that of the surrounding material. In these waveguides each layer boundary causes a partial reflection of an optical wave. For a range of wavelength, called the photonic stopband, the multiple reflections combine with constructive interference, and the layers act as a high-quality reflector. This different mechanism affords great potential in terms of design freedom and novel applications. However, its fabrication as integrated device is not easy and requires high periodicity and high index contrast between the cladding layers, which typically necessitates the use of high-index materials like silicon that is absorbing in the visible range, so limiting the range of potential applications. A different way to realize hollow optical waveguide is the use of leaky waveguide such as Anti Resonant Reflecting Optical Waveguide (ARROW) [35–40]. Differently from conventional waveguides, in ARROW waveguides the field is not confined in the core region by total internal reflection but by dielectric cladding layers designed to form high reflectivity Fabry-Perot mirrors. This peculiar structure offers some unique properties that make it very attractive for photonics integrated circuits with applications ranging from telecommunications to sensors. Even though the ARROW modes are leaky, low-loss propagation over large distances can be achieved. Furthermore, singlemode waveguides can be realized also with large core dimensions. These waveguides can be easily designed using very simply optical equations that take into account the interference phenomena in the cladding layers.

10.3 Inteferometric Sensors

Integrated interferometric devices are the most used and sensitive devices for sensing applications [41]. Interferometry is used to detect the phase difference between two or more collimated light beams of a coherent light source. Many optical configurations have been proposed in literature like the conventional interferometer (Mach–Zehnder, Young). Typically, these devices are based on conventional solid core waveguides in which the analyte interacts with the evanescent part of the waveguide mode. In the last years, a great effort has been devoted to find new integration strategies based on optofluidic devices.

An optofluidic Mach–Zehnder interferometer (MZI) has recently been realized using liquid-core ARROW waveguides [42]. The device is based on small core 5 × 10 µm waveguides resulting in quasi-single mode behavior that is an essential feature for an interferometric device. The design consists of an input optical waveguide that is split into two separate interferometer arms by a y-junction before recombining to form the output optical waveguide. Differently from conventional MZI, because the whole device is filled with the liquid under analysis, the two arms must have different lengths resulting in an asymmetric device in order to induce a relative phase shift at the output. However, this asymmetric geometry could induces a strong intensity unbalance of the beam in each arm due to the different waveguide bend losses with a strong degradation of the device's performance due to the low visibility of the interferometer. In order to overcome this problem, optimized geometry in order to minimize the intensity unbalance between the two arms also for high asymmetric Mach–Zehnder configurations has been realized [43]. The device is based on T-branches and 90°-bent waveguides (Figure 10.1). This configuration permits us to obtain very high visibility values with a strong improvement on the device sensitivity.

Optofluidic MZIs have been also developed integrating solid-core waveguides with liquid-core waveguides or microchannels [44,45]. An interesting fabrication technique allowing the seamless integration of solid-core and liquid-core waveguides has been demonstrated and applied to MZIs, for integrated refractometry [44]. The technique allows single-mode liquid-core waveguides to be realized. However, since the structures are index guided, currently only liquids with a refractive index larger than 1.45 can serve as a core medium. An optofluidic Mach–Zehnder interferometer has been used for measuring the

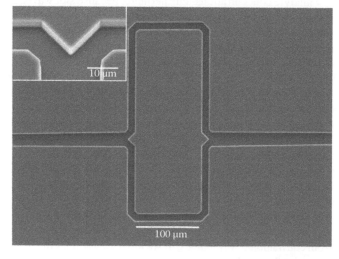

FIGURE 10.1   SEM picture of the optofluidic MZI based on ARROW waveguide. (Reprinted from Testa, G., Huang, Y., Sarro, P.M., Zeni, L., and Bernini, R., High-visibility optofluidic Mach–Zehnder interferometer, *Opt. Lett.*, 35(10), 1584–1586, Copyright 2010. With permission of Optical Society of America.)

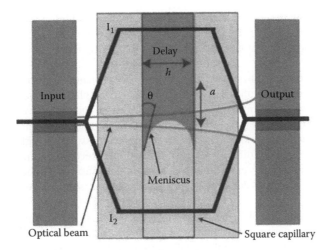

FIGURE 10.2 Schematic representation of the single beam fluid Mach–Zender interferometer. (Reprinted with permission from Domachuk, P., Grillet, C., Taeed, V., Mägi, E., Bolger, J., and Eggleton, B.J., Microfluidic interferometer, *Appl. Phys. Lett.* 86, 024103. Copyright 2005, American Institute of Physics.)

dry/wet mass of a single living cell using. The device consists of the fiber Mach–Zehnder interferometer and the fiber-optical trap, both of which are integrated onto a single chip. In experiment, a single living cell is captured by the fiber-optical trap, and then the cell's refractive index and diameter are simultaneously determined by the spectral shift in response to the buffer modulation [46].

Recently a new class of highly compact tunable interferometric fluid-based devices that achieve phase differences by propagating a beam across a fluid/air discontinuity has been introduced (Figure 10.2) [47]. Using this principle, a single-beam compact microfluidic Mach–Zehnder interferometer has been demonstrated. It relies on the high contrast in refractive index across a fluid meniscus to achieve compactness. Improved design of this interferometer have been proposed to avoids the problems related to non flat surface (meniscus) due to surface tension between liquid and air [48,49]. As reported here, typically, integrated interferometers are based on Mach–Zehnder configuration because it avoids the need for optical mirrors. However, using the flexibility given by optofluidics a Michelson interferometer has been demonstrated. The device is based on droplet microfluidics used to create a droplet grating. The droplet grating is formed by a stream of plugs, of two immiscibile liquids, in the microchannel with constant refractive index variation. It has a real-time tunability in the grating period through varying the flow rates of the liquids and index variation via different combinations of liquids. This optofluidic Michelson interferometer is highly sensitive and is suitable for the measurement of biomedical and biochemical buffer solutions [50].

Optofluidic interferometers have been also developed for real-time and label-free detection and recognition of single viruses and larger proteins [51]. The proposed method makes use of nanofluidic channels in combination with optical interferometry. Elastically scattered light from single viruses traversing a stationary laser focus is detected with a differential heterodyne interferometer and the resulting signal allows single viruses to be characterized individually. Heterodyne detection eliminates phase variations due to different particle trajectories, thus improving the recognition accuracy as compared to standard optical interferometry. This approach is able to resolve nanoparticles of various sizes, and detecting and recognizing different species of human viruses from a mixture. The detection system can be readily integrated into larger nanofluidic architectures for practical applications.

10.4 Resonant Sensors

Optical resonators are an emerging technology for high sensitivity miniaturized biological or biochemical sensing [52]. The high sensitivity of these devices relies on the interaction length between the light and sample that is not limited by the physical length but is related to the number of revolutions of light in the resonator, characterized by the quality factor Q. In order to achieve high performance these structures require an optimized integration of the optical and microfluidic parts. In many cases the design of optical part has been privileged respect to microfluidic ones sacrificing the feasibility and the multiplexing capability of the device.

Microdroplet resonators are the first example of optofluidic resonators and have been extensively studied because they represent the optofluidic versions of solid microsphere resonators [53–55]. They offer a unique integration between the microfluidics and the resonator and exhibit very high Q-factor due to the best surface quality that ensure very low surface scattering losses [56]. Unfortunately, droplets also suffer from several practical limitations, related to the complexity associated with manipulating and controlling droplet structures and the difficult to coupling the light. Recently, some configurations to solve this problem have been proposed by embedding the microdroplets in a liquid medium or in a microfluidic channel and the optical coupling is achieved using a standard fiber-taper coupler [57].

A recent advancement in the architecture of ring resonator sensors is represented by liquid core ring resonator (LCORR) [58]. In LCORR the wall of a fused silica capillary forms a ring resonator. The evanescent field of the whispering gallery mode (WGM) in the core is used to detect the refractive index change near the interior surface. The LCORR configuration achieves dual use of the capillary as a fluidic channel to effectively deliver the sample to the sensor head without sacrificing performance or increasing complexity. This device has been successfully applied in chemical and biochemical detection like cancer biomarkers [59,60].

Using a similar approach integrated silicon optofluidic ring resonator based on ARROW waveguide has been demonstrated [61]. In this device four 90°-bent ARROW waveguides are used to form a rectangular ring resonator and a multimode interference (MMI) liquid core ARROW coupler with a 50:50 splitting ratio acts as coupling element between the ring and the bus waveguide (Figure 10.3).

Optofluidic resonators have been also fabricated using photonic crystals. The optical field inside the holes of photonic crystal is significantly stronger than the evanescent fields of WGM sensors, thereby strengthening the light–matter interactions and enabling a limit of detection on the order of 63 ag total bound mass. Furthermore these devices can be easily multiplexed for simultaneous detection of multiple analytes (Figure 10.4) [62,63].

Recently, optofluidic Fabry-Perot (FP) resonators have been developed using gold coated fiber-based FP [64], fiber–Bragg-grating [65,66] or integrated vertically etched silicon Bragg reflectors [67]. All these devices use the change in optical path length in the cavity when light propagates through different samples to determine the refractive index of the sample. Based on the measured refractive index value, researchers can determine the concentration of solute in a solution or study the relation of living cell's RI with the cell permeability, cell viability, the effective indexes, and sizes of cancerous cells [68]. In order to increase the sensitivity optofluidic Fabry–Perot cavity sensor with integrated flow-through micro-/nanofluidic channels has been proposed [69]. The sensor employs a microsized capillary with

FIGURE 10.3 SEM picture of the optofluidic ring resonator based on ARROW waveguide. (Reprinted with permission from Testa, G., Huang, Y., Sarro, P.M., Zeni, L., and Bernini, R., Integrated silicon optofluidic ring resonator, *Appl. Phys. Lett.* 97, 131110. Copyright 2010, American Institute of Physics.)

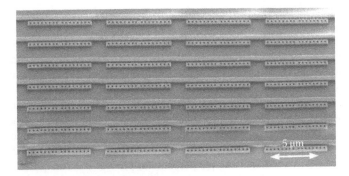

FIGURE 10.4 SEM picture of the photonic crystals nanoscale optofluidic sensor arrays. (Reprinted from Mandal, S. and Erickson, D., Nanoscale optofluidic sensor arrays, *Opt. Express*, 16, 1623–1631,Copyright 2008. With permission of Optical Society of America.)

many built-in micro-/nanosized placed inside a between a FP cavity. This micro-/nanohole arrays enable three-dimensional surface detection which greatly enhances the sensitivity.

10.5 Absorbance Sensors

Absorbance-based techniques are very interesting since they offer the potential to permit a label-free spectral detection and identification of an analyte in a very easy way. However, in this case, the devices miniaturization reduces the interaction length of light with a fluid and thus limits the sensitivity of a device. In order to overcome this problem, optofluidic waveguides has been applied in order to increase the interaction efficiency between the light and fluid. The use of optofluidic waveguides permits a direct optical probing of the liquid sample with a strong advantage in the integration and with increased interaction efficiency. In such waveguides the mode field and the liquid sample can share the same path, in this way leading to a stronger interaction as compared to the solid core waveguides, in which only the evanescent tail of the mode field is sensitive to the surrounding medium.

An example of this approach is the first monolithically integrated rubidium vapor cell using hollow-core ARROW waveguides on a silicon chip. The cells have a volume more than 7 orders of magnitude less than conventional bulk cells. The micrometer-sized mode areas enable high beam intensities over near centimeter lengths, in particular an optical density in excess of 2, and saturation absorption spectroscopy on a chip have been obtained [70].

Analogously, ARROW liquid core waveguides have been used as long path absorbance cell for colorimetric protein assay. In particular, a protein assay based on the method of Bradford, have been used as a simple and accurate procedure for determining concentration of solubilized protein. In this dye-binding assay a differential color change of a dye occurs in response to various concentrations of protein. So measuring the absorbance at a fixed wavelength by comparison to a standard curve provides a relative measurement of protein concentration [71]. However, this approach requires that reagents are premixed using, for example, an additional integrated microfluidic mixer. This problem can be overcome with an appropriate biological surface modification of the inner walls of liquid waveguide.

Using horseradish peroxidase (HRP) attached to the inner walls of a liquid core silicon leaky waveguides an enzymatic sensor for the measurement of hydrogen peroxide has been fabricated and characterized [72]. The core filled with a transparent 0.1 M acetate buffer solution pH 5 containing 50 μM 2,2′-azino-bis(3-ethylbenzthiazoline-6-sulfonic acid) (ABTS) as an enzyme co-substrate. In presence of H_2O_2, a catalytic reaction is carried out with the concomitant oxidation of the ABTS, resulting in the generation of the ABTS+ green colored product. Hence, the absorbance spectrum of the whole hollow core should change with the concentration of H_2O_2 and the reaction could be easily detected. Such a system merges into one single microstructure several functionalities. In fact the core of a hollow waveguide is used, at the same time, to confine the light, as a microfluidic channel and, with the appropriate biological surface modification, as a bioreactor.

Another way to integrate the bioreactor functionality in an optofluidic evanescent wave sensor has been demonstrated using liquid core–liquid cladding waveguides by adopting the differential optical absorption method [73]. In particular, two external light sources with distinctly different wavelengths were coupled to the L2 waveguide. The exponentially decaying evanescent wave interacts with analytes dissolved in the cladding fluids or products formed by in situ chemical reactions at the core–cladding interface. The analyte molecules exhibit distinctly different light absorbance at the two wavelengths during the light–analyte interaction. Therefore, by using the normalized absorbance calculated from the intensity ratio of the two wavelengths instead of the absolute magnitude of either signal, unwanted effects from omnipresent external noise sources can be reduced. Compared to traditional design in which reagents are premixed, continuous operation of the sensor can be achieved in an L2 waveguide evanescent wave sensor because an additional sample preparation process is not required. Also, the proposed approach will reduce errors caused by handling and environmental sources during the sample preparation process.

To further increases the device sensitivity more complex device like optical ring resonators have been used. A 100 μm radius high Q microring resonator integrated with PDMS microfluidic channels was used to enhance the interaction length between evanescent light and a cladding liquid. The absorption spectra of less than 2 nL of N-methylaniline from 1460 to 1610 nm with 1 nm resolution and effective free space path lengths up to 5 mm was successfully measured [74].

REFERENCES

1. Psaltis D., Quake S. R., and Yang C. 2007. Developing optofluidic technology through the fusion of microfluidics and optics. *Nature* 442:381–386.
2. Monat C., Domachuk P., and Eggleton B. J. 2007. Integrated optofluidics: A new river of light. *Nat. Photonics* 1:106—114.
3. Schmidt H. and Hawkins A. R., 2010. *Handbook of Optofluidics*, CRC Press, Boca Raton, FL.
4. Fainman Y., Lee L., Psaltis D., and Yang C. 2009. *Optofluidics—Fundamentals, Devices, and Applications*, McGraw-Hill, New York.
5. Colladon D. 1842. On the reflections of a ray of light inside a parabolic liquid stream. *Compt. Rend.* 15:800–802.
6. Babinet J. 1842. Note on the transmission of light by sinuous canals. *Compt. Rend.* 15:802.
7. Tyndall J. 1854. On some phenomena connected with the motion of liquids. *Proc. R. Inst. Great Britain* 1:446.
8. Taylor G. W. 1972. Liquid optical fibers. *Appl. Optics* 11:786–790.
9. McMullin J. N., Narendra R., and James C. R., 1993. Hollow metallic waveguides in Silicon V-grooves. *IEEE Photon. Technol. Lett.* 5:1080–1082.
10. Miyagi M., Hongo A., and Kawakami S., 1983. Transmission characteristics of dielectric-coated metallic waveguide for infrared transmission: Slab waveguide model. *IEEE J Quantum Electron.* 19:136–145.
11. Miura T., Koyama F., Aoki Y., Matsutani A., and Iga K. 2001. Hollow optical waveguide for temperature-insensitive photonic integrated circuits. *Jpn. J. Appl. Phys.* 40:688–690.
12. Dress P. and Franke H. 1996. A cylindrical liquid-core waveguide. *Appl. Phys. B.* 69:12–19.
13. Altkorn R., Koev I., Van Duyne R. P., and Litorja M., 1997. Low-loss liquid-core optical fiber for low-refractive-index liquids: Fabrication, characterization, and application in Raman spectroscopy. *Appl. Opt.* 36:8992–8998.
14. Datta A., Eom I., Dhar A., Kuban P., Manor R., Ahmad I., Gangopadhyay S., Dallas T., Holtz M., Temkin H., Dasgupta P. 2003. Microfabrication and characterization of teflon AF-coated liquid core waveguide channels in silicon. *IEEE Sens. J.* 3:788–795.
15. Wu C. W. and Gong G. C. 2008. Fabrication of PDMS-based nitrite sensors using Teflon AF coating microchannels. *IEEE Sens. J.* 8:465–469.
16. Cho S. H., Godin J., and Lo Y.-H. 2009. Optofluidic waveguides in Teflon AF-coated PDMS microfluidic channels. *IEEE Photon. Technol. Lett.* 21:1057–1059.
17. Risk W.P., Kim H.C., Miller R.D., Temkin H., and Gangopadhyay S. 2004. Optical waveguides with an aqueous core and a low-index, nanoporous cladding. *Opt Exp.* 12:6446–6455.

18. Korampally V., Mukherjee S., Hossain M., Manor R., Yun M., Gangopadhyay K., Polo-Parada L., and Gangopadhyay S. 2009 Development of a miniaturized liquid core waveguide system with nanoporous dielectric cladding: A potential biosensing platform. *IEEE Sens. J.* 9:1711–1718.

19. Sugiya K., Harada M., and Okada T. 2009. Water-ice chip with liquid-core waveguide functionality. Toward lab on ice. *Lab Chip* 9:1037–1039.

20. Takiguchi H., Odake T., Ozaki M., Umemura T., and Tsunoda K. I. 2003. Liquid/liquid optical waveguides using sheath flow as a new tool for liquid/liquid interfacial measurements. *Appl. Spectrosc.* 57:1039–1041.

21. Dirac H. and Gravesen P. 2001. Realisation and characterisation of all liquid optical waveguides, *Proceedings of the 14th IEEE International Conference on Micro Electro Mechanical Systems*, New York, pp. 459–462.

22. Wolfe D. B., Conroy R. S., Garstecki P., Mayers B. T., Fischbach M. A., Paul K. E., Prentiss M., and Whitesides G. M. 2004. Dynamic control of liquid-core/liquid-cladding optical waveguides. *Proc. Natl Acad. Sci. U.S.A.* 24:12434–12438.

23. Lee K. S., Kim S. B., Lee K. H., Sung H. J., and Kimb S. S. 2010. Three-dimensional microfluidic liquid-core/liquid-cladding waveguide. *Appl. Phys. Lett.* 97:021109.

24. Bernini, R., De Nuccio E., Minardo A., Zeni L., and Sarro P.M. 2008. Liquid-core/liquid-cladding integrated silicon ARROW waveguides. *Opt. Commun.* 281:2062–2066.

25. Lim J. M., Kim S. H., Choi J. H., and Yang S.M. 2008. Fluorescent liquid-core/air-cladding waveguides towards integrated optofluidic light sources. *Lab Chip* 8:1580–1585.

26. Huang H. J., Tsai D. P., and Liu A. Q. 2008. A plasmonic liquid waveguide sensor using nanoparticles for label-free measurement applications. *Twelfth International Conference on Miniaturized Systems for Chemistry and Life Science*, San Diego, CA, pp. 1952–1954.

27. Almeida V. R., Xu Q., Barrios C. A., and Lipson M. 2004. Guiding and confining light in void nanostructure. *Opt. Lett.* 29:1209–1211.

28. Xu Q., Almeida V.R., Panepucci R.R., and Lipson M. 2004. Experimental demonstration of guiding and confining light in nanometer-size low-refractive-index material. *Opt Lett.* 29:1626–1628.

29. Feng N.N., Michel J., and Kimerling L.C. 2006. Optical field concentration in low-index waveguides. *IEEE J. Quantum Electron.* 42:885–890.

30. Barrios C. A., Gylfason K. B., Sánchez B., Griol A., Sohlström H., and Holgado M. 2007. Slot-waveguide biochemical sensor. *Opt. Lett.* 32:3080–3082.

31. Barrios C. A., Sanchez, B., Gylfason K. B., Griol, A., Sohlström H., Holgado M., and Casquel R. 2007. Demonstration of slot-waveguides structures on silicon nitride/silicon oxide platform. *Opt. Express* 15:6846–6856.

32. Lo S. S., Wang M. S., and Chen C. C. 2004. Semiconductor hollow optical waveguides formed by omni-directional reflectors. *Opt. Express* 12:6589–6593.

33. DeCorby R. G., Ponnampalam N., Nguyen H. T., Pai M. M., and Clement T. J. 2007. Guided self-assembly of integrated hollow Bragg waveguides. *Opt. Express* 15:3902–3915.

34. Epp E., Ponnampalam N., Newman W., Drobot B., McMullin J. N., Meldrum A. F., and DeCorby R. G. 2010. Hollow Bragg waveguides fabricated by controlled buckling of Si/SiO2 multilayers. *Opt. Express* 18:24917–24925.

35. Grosse A., Grewe M., and Fouckhardt H. 2001. Deep wet etching of fused silica glass for hollow capillary optical leaky waveguides in microfluidic devices. *J. Micromech. Microeng.* 11:257–262.

36. Goddard N. J., Singh K., Bounaria F., Holmes R. J., Baldock S. J., Pickering L. W., Fielden P. R., and Snook R. D. 1998. Anti-resonant reflecting optical waveguides (ARROWS) as optimal optical detectors for micro TAS Applications. *Proceedings of the μTAS'98 Workshop*, Banff, Canada, Vol. 1, pp. 97–100.

37. Delonges T. and Fouckhardt H. 1995. Integrated optical detection cell based on Bragg reflecting waveguides. *J. Chromatogr. A* 716:135–139.

38. Bernini R., Campopiano S., and Zeni L. 2002. Silicon micromachined hollow optical waveguides for sensing applications. *IEEE J. Sel. Top. Quantum Electron.* 8:106–110.

39. Campopiano S., Bernini R., Zeni L., and Sarro P. M. 2004. Microfluidic sensor based on integrated optical hollow waveguide. *Opt. Lett.* 29:1894–1896.

40. Yin D., Deamer D. W., Schmidt H., Barber J. P., and Hawkins A. R. 2004. Integrated optical waveguides with liquid cores. *Appl. Phys. Lett.* 85:3477–3479.

41. Fan X., White I. M., Shopova S. I., Zhu H., Suter J. D., and Sun Y. 2008. Sensitive optical biosensors for unlabeled targets: A review. *Anal. Chim. Acta* 620:8–26.
42. Bernini, R., Testa G., Zeni L., and Sarro P. M. 2008. Integrated optofluidic Mach–Zehnder interferometer based on liquid core waveguides. *Appl. Phys. Lett.* 93:011106–011109.
43. Testa G., Huang Y., Sarro P.M., Zeni L., and Bernini R. 2010. High-visibility optofluidic Mach-Zehnder interferometer. *Opt. Lett.* 15:1584–1586.
44. Dumais P., Callender C. L., Noad J. P., and Ledderhof C. J. 2008. Integrated optical sensor using a liquid-core waveguide in a Mach-Zehnder interferometer. *Opt. Express* 16:18164–18172.
45. Crespi A., Gu Y., Ngamsom, Hoekstra H. J. W. M., Dongre C., Pollnau M., Ramponi R., van den Vlekkert H. H., Watts P., Cerullo G., and Osellame R. 2010. Three-dimensional Mach-Zehnder interferometer in a microfluidic chip for spatially-resolved label-free detection. *Lab Chip* 10:1167–1173.
46. Song W. Z., Liu A. Q., Swaminathan S., Lim C. S., Yap P. H., and Ayi T. C. 2007. Determination of single living cell's dry/water mass using optofluidic chip. *Appl. Phys. Lett.* 91:223902.
47. Monat C., Domachuk P., Grillet C., Collins M., Eggleton B. J., Cronin-Golomb M., Mutzenich S., Mahmud T., Rosengarten G., and Mitchell A. 2008. Optofluidics: A novel generation of reconfigurable and adaptive compact architectures. *Microfluid. Nanofluid.* 4:81–95.
48. Bedoya A. C., Monat C., Domachuk P., Grillet C., and Eggleton B. J. 2011. Measuring the dispersive properties of liquids using a microinterferometer, *Appl. Opt.*, 50(16):2408–2412.
49. Lapsley M. I., Chiang I. K., Zheng, Y. B. Ding X., Mao X., and Huang T. J. 2011. A single-layer, planar, optofluidic Mach–Zehnder interferometer for label free detection. *Lab Chip* 11:1795–1800.
50. Chin L. K., Liu A. Q., Soh Y. C., Limb C. S., and Lin C. L. 2010. A reconfigurable optofluidic Michelson interferometer using tunable droplet grating. *Lab Chip* 10:1072–1078.
51. Mitra A., Deutsch B., Ignatovich F., Dykes C., and Novotny L. 2010. Nano-optofluidic detection of single viruses and nanoparticles. *ACS Nano* 4:1305–1312.
52. Sun Y. and Fan X. 2011. Optical ring resonators for biochemical and chemical sensing. *Anal Bioanal. Chem.* 399:205–211.
53. Barnes M. D., Ng K. C., Whitten W. B., and Ramsey M. 1993. Detection of single Rhodamine 6G molecules in levitated microdroplets. *Anal. Chem.* 65:2360–2365.
54. Tanyeri M., Nichkova M., Hammock B. D., and Kennedy I. M. 2005. Chemical and biological sensing through optical resonances in microcavities. *Proc. SPIE* 5699:227–236.
55. Tanyeri M. and Kennedy I.M. 2008. Detecting single bacterial cells through optical resonances in microdroplets. *Sens. Lett.* 6:326–329.
56. Zadeh M. H. and Vahala K. J. 2006. Fiber-taper coupling to Whispering-Gallery modes of fluidic resonators embedded in a liquid medium. *Opt. Express* 14:10800–10810.
57. Yu Y. F., Bourouina T., Ng S. H., Yap P. H., and Liu A. Q. 2009. A single droplet carrier with optical detector. *Thirteenth International Conference on Miniaturized Systems for Chemistry and Life Sciences*, Korea, Vol. 1, pp. 979–981.
58. White I. M., Oveys H., and Fan X. 2006. Liquid-core optical ring-resonator sensors. *Opt. Lett.* 31:1319–1321.
59. Sun Y., Liu J., Howard, D. J., Fan X., Frye-Mason G., Ja S.I., and Thompson A. K. 2010. Rapid tandem-column micro-gas chromatography based on optofluidic ring resonators with multi-point on-column detection. *Analyst* 135:165–171.
60. Zhu H., Dale P. S., Caldwell C. W., and Fan X. 2009. Rapid, label-free detection of breast cancer biomarker CA15–3 in clinical human serum samples with opto-fluidic ring resonator sensors. *Anal. Chem.* 81:9858–9865.
61. Testa G., Huang Y., Sarro P. M., Zeni L., and Bernini R. 2010. Integrated silicon optofluidic ring resonator. *Appl. Phys. Lett.* 97:131110.
62. Mandal S. and Erickson D. 2008. Nanoscale optofluidic sensor arrays. *Opt. Express* 16:1623–1631.
63. Mandal S., Goddard J. M., and Erickson D. 2009. A multiplexed optofluidic biomolecular sensor for low mass detection. *Lab Chip* 9:2924–2932.
64. Song W. Z., Zhang X. M., Liu A. Q., Lim C. S., Yap P. H., and Hosseini H. M. 2006. Refractive index measurement of single living cells using on-chip Fabry–Pérot cavity. *Appl. Phys. Lett.* 89:203901.
65. Domachuk, P. Littler I., Cronin-Golomb M., and Eggleton B. 2006. Compact resonant integrated microfluidic refractometer. *Appl. Phys. Lett.* 88:093513.

66. Chin L. K., Liu, A. Q., Lim C. S., Zhang X. M., Ng J. H., Hao J. Z., and Takahashi S. 2077. Differential single living cell refractometry using grating resonant cavity with optical trap. *Appl. Phys. Lett.* 91:243901.

67. St-Gelais R., Masson J., and Peter Y. A. 2009. All-silicon integrated Fabry–Pérot cavity for volume refractive index measurement in microfluidic systems. *Appl. Phys. Lett.* 94:243905.

68. Shao H., Kumar D., and Lear K. L. 2006. Single-Cell detection using optofluidic intracavity spectroscopy. *IEEE Sensors J.* 6(6):1543–1550, December 2006.

69. Guo Y., Li H., Reddy K., Shelar H. S., Nittoor V. R., and Fan X. 2011. Optofluidic Fabry–Pérot cavity biosensor with integrated flow-through micro-/nanochannels. *Appl. Phys. Lett.* 98:041104.

70. Yang W. G., Conkey D. B., Wu B., Yin D. L., Hawkins A. R., and Schmidt H. 2007. Atomic spectroscopy on a chip. *Nat. Photonics* 1:331–335.

71. Bernini R., De Nuccio E., Minardo A., Zeni L., and Sarro P. M. 2007. Integrated silicon optical sensors based on hollow core waveguide. *Proc. SPIE* 6477:647714.

72. Cadarso V. J., Fernandez-Sanchez C., Llobera A., Darder M., and Domınguez C. 2008. Optical biosensor based on hollow integrated waveguides. *Anal. Chem.* 80:3498–3501.

73. Lim J. M., Urbanski J. P., Choi J. H., Thorsen T., and Yang S. M. 2011. Liquid waveguide-based evanescent wave sensor that uses two light sources with different wavelengths. *Anal. Chem.* 83:585–590.

74. Nitkowski A., Chen L., and Lipson M. 2008. Cavity-enhanced on-chip absorption spectroscopy using microring resonators. *Opt. Express* 16:11930–11936.

11

Lab-on-Chip Nanostructured Sensors for Chemical and Biological Applications

Luca De Stefano, Edoardo De Tommaso, Emanuele Orabona, Ilaria Rea, and Ivo Rendina

CONTENTS

11.1 Introduction

Over the last few years, a large demand for very sensitive, highly specific, cost-effective, and rapid methods to detect the concentration of chemical or biological species has been identified in several fields such as clinical diagnostics, drug development, environmental monitoring, food quality control, security, and counterterrorism. In particular, in so-called optical biochips, the miniaturization, parallelization, and integration of optical sensing schemes and measurements of biochemical targets can be achieved [1]. In a lab-on-chip (LOC), all the microfluidics and the optical transducing elements are placed on the same substrate (see Figure 11.1 for a typical LOC architecture). LOC systems are characterized by small dimensions (which imply small sample volumes, in the picoliter–femtoliter range), low cost, short analysis time and, furthermore, they can successfully perform multiplexed analyses within point-of-care (POC) testing schemes.

In this introductory paragraph, we will focus on the sensing element of a LOC; in particular, we will report a brief review on some of the most recently conceived nanostructured transducing elements and architectures that have been implemented in LOC systems.

Multiplexed quantification of three well-characterized cancer biomarkers (the antigens CEA, CA125, and Her-2/*Neu* (C-erbB-2)) from saliva and serum has been performed, exploiting the fluorescence emitted by functionalized nanoparticle quantum dots (QDs) in a microporous agarose bead array [2]. The scheme of detection is shown in Figure 11.2. The agarose beads, loaded into an anisotropically etched silicon chip provided with the proper microfluidics (see Figure 11.3), constitute

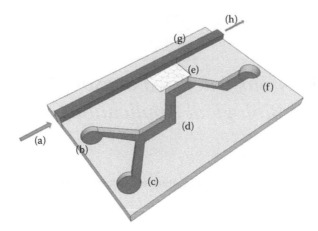

FIGURE 11.1 Typical scheme of a nanostructured, optical lab-on-chip: (a) incoming light; (b) inlet 1; (c) inlet 2; (d) microchannel; (e) nanostructured sensing area; (f) outlet; (g) waveguide; (h) light out.

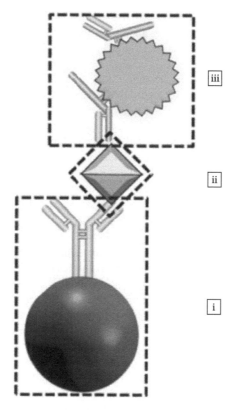

FIGURE 11.2 Scheme of detection for the single element of the QD-based nano-biochip described in text: agarose bead with immobilized antibody (i); target antigen (ii); detecting antibody/QD conjugate (iii). (Adapted from Jesse, V.J. et al., *Biosens. Bioelectron.*, 24, 3622, 2009. With permission.)

the underlying support structure for capturing antibody immobilization. The semiconductor, fluorescent QDs are composed of a CdSe core and a ZnS coat, followed by a polymer passivation layer, resulting in particles ranging from 5 to 20 nm. When the antigen (i.e., the cancer biomarker) links to the antibody bound to the bead, it is in turn recognized by the antibody/QD conjugate. The resulting nano-biochip is characterized by a 30 times signal amplification relative to that of biochips based

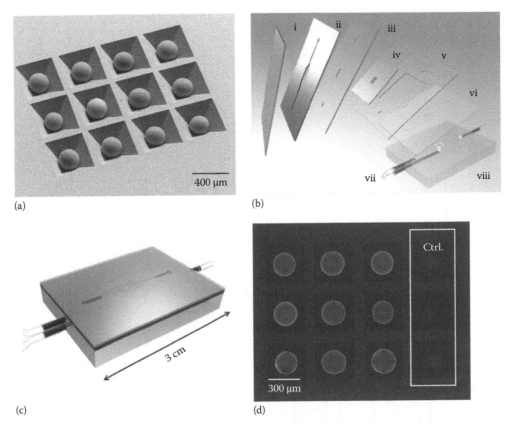

FIGURE 11.3 **(See color insert.)** Details of the QD-based nano-biochip array described in the text. SEM image of beads loaded in anisotropically etched silicon chip (a); sandwich scheme of the chip (iv) fitted between adhesive layer (ii), cover slip (i), laminate layers (iii, v, vi), PMMA base (vii), and inlet and outlet ports (vii) (b); sealed LOC assembly (c); fluorescent image of beads after immunoassay, including negative controls as imaged with CCD camera (d). (Reprinted from Jesse, V.J. et al., *Biosens. Bioelectron.*, 24, 3622, 2009. With permission.)

on standard molecular fluorophores and by a reduction in limit of detection of nearly two orders of magnitude (e.g., 0.02 ng/mL for CEA) relative to the more traditional enzyme-linked immunosorbent assay (ELISA).

Arrays of micro- and nanoscaled surface structures (micro/nanowells, nanotips, nanopyramids etc.) can be fabricated, over relatively large areas and with high reproducibility, by means of selective etching within fused optical fiber bundles [3,4] (see also Figure 11.4). By using the large difference in refractive index between core and cladding of the single fiber of the bundle (which corresponds to a single "pixel" in the array), i.e., making use of large numerical apertures, the pixel diameter can be reduced to submicron sizes. A certain degree of disorder may be introduced in the alignment of the single fibers of the bundle in order to minimize cross talk. Different physical effects can be exploited within this scheme. Raman signal can be strongly enhanced by tuning electromagnetic and chemical interactions between the probed compound and the textured surface of the fiber array (surface-enhanced Raman scattering, SERS) [5] (see also Figure 11.5). The optical fibers of the bundle can also be coated with a metal that serves as an electrode surface; an electrochemical reaction can then be used to immobilize a biorecognition element on the fiber surface or to modulate a luminescence signal that is monitored through the optical fiber itself [6,7]. Nanostructured optical fiber arrays (also in the form of randomly ordered array sensors) can be successfully used as biosensor matrices for genomics, proteomics, and metabolomics; in particular, bead-based DNA sensors have been recently extended to nanoscale fiber arrays [8] (see Figure 11.6) characterized by an ultra-high packing density (up to 4.5×10^6 array elements mm^{-2}). As far as single-molecule detection is concerned, kinetic information from large populations

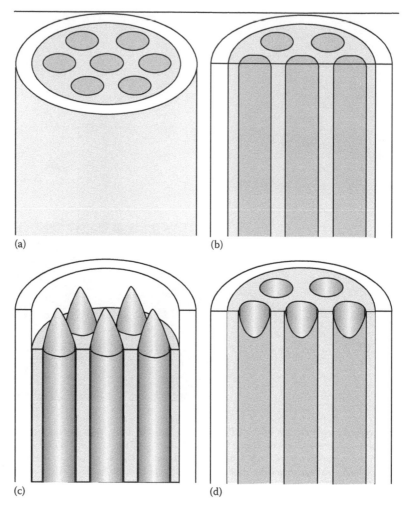

(a) (b) (c) (d)

FIGURE 11.4 Schematic representation of a fiber bundle (a); fiber bundle cross section (b); array of conical tips formed by removal of the cladding (c); array of wells formed by etching of the cores (d). (Reprinted from Deiss, F. et al., *Anal. Bioanal. Chem.*, 396, 53, 2010.)

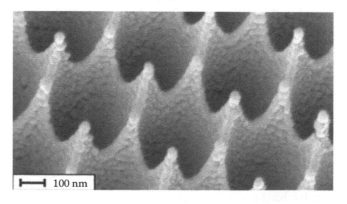

FIGURE 11.5 High-resolution SEM image of the metal coating over an etched fiber bundle used as SERS substrate. (Modified from Deiss, F. et al., *Anal. Bioanal. Chem.*, 396, 53, 2010. With permission.)

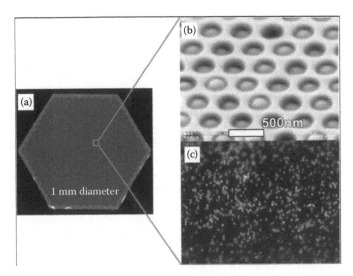

FIGURE 11.6 (a) Optical microscope image of a fiber-based nanoarray comprising 300nm diameter fibers; (b) magnified SEM image of a portion of the array showing nanobeads trapped in etched wells; (c) microscope image of the fluorescent nanobeads in the nanoarray. (Reprinted from Tam, J.M. et al., *Biosens. Bioelectron.*, 24, 2488, 2009. With permission.)

of single-enzyme molecules has been collected simultaneously, making use of the array format of the bundle [9]; the same platform has been used to detect single-molecule DNA hybridization and a detection limit of 1 fmol L^{-1} has been reported [10].

Silicon micro- and nanostructured sensing devices for chemical and biological monitoring present several advantages, such as CMOS compatibility, potential for mass fabrication, and large light–matter interaction strengths [11]. Among the possible technological solutions based on silicon photonics, porous silicon (PSi) sensors [12,13], obtained from electrochemical etching of crystalline silicon in a water–fluorhydric acid–ethanol solution, are characterized by a large surface area (several hundreds of m^2 over cm^3, see Figure 11.7), which can be used as host for a considerable greater number of probe molecules compared to planar, solid surfaces. With a proper modulation of current density and etching time during the electrochemical etching, several optical structures can be made, such as resonant microcavities, Bragg reflectors, Thue-Morse filters, and waveguides [14] whose optical characteristics change when probe–target interaction takes place. Minimum detectable refractive index changes are of the order of 1×10^{-4} [15] and detection limits in the μM–nM range for the concentration of target molecules have been reported in literature [16].

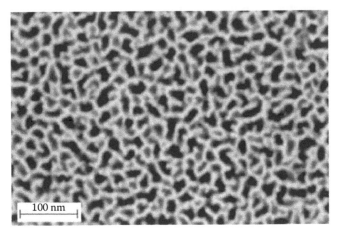

FIGURE 11.7 SEM top view image of a chemical oxidized PSi device, characterized by an average pores diameter of about 20nm. (Adapted from Rea, I. et al., *J. Phys. Chem. C*, 114, 2617, 2010. With permission.)

DNA-probe molecule coverage within the nanostructured internal surface of PSi can be maximized by means of direct synthesis of DNA oligonucleotides inside the porous matrix [17]. More details on PSi fabrication and characterization will be given in succeeding paragraphs.

In silicon waveguide sensors, target molecules can be detected either by attaching fluorophores to the molecules and measuring the fluorescence induced by the evanescent field, or by measuring the phase change determined by exposure to the target molecules within an interferometric scheme. By reducing the waveguide core thickness, the waveguide mode becomes delocalized and field–matter interaction is strongly increased: this is what is obtained in the so-called silicon wire waveguide [18]. In a slot waveguide [19], light is confined in a nanometer-scaled slot of low-refractive-index material sandwiched between two strips of high index material, achieving improved sensitivity if compared to SOI wire waveguides.

Other silicon-based micro- and nanostructured sensors include micro-ring resonators, silicon wave-guide sensors [20]. The operation of silicon micro-ring resonators is based on the optical coupling between a ring-shaped waveguide and one or more linear waveguides patterned on a planar surface, typically an input and an output waveguide. When incoming light has a wavelength that satisfies the resonance conditions, it couples into the micro-ring and continuously recirculates within it. A fraction of this resonant light escapes the micro-ring structure and couples into the output waveguide. The presence of a target analyte over the top surface of the micro-ring (i.e., within the evanescent field) changes the effective refractive index of the mode propagating into the structure, thus causing a shift in resonance wavelength, which can be determined by monitoring the spectrum at the output port. Proper function-alization of the micro-ring surface allows to add selectivity to the sensing system and to detect specific interaction between a bioprobe and its proper target (e.g., protein–ligand, DNA–cDNA interactions). In Figure 11.8, an example of micro-ring resonator chip can be viewed. A drawback of this kind of devices is that, being characterized by very high values of Q factor, even small temperature drifts could cause measurable resonance shifts.

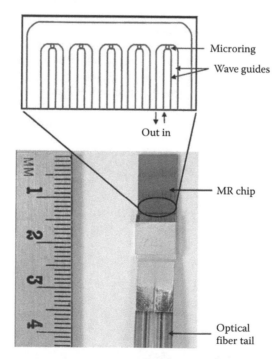

FIGURE 11.8 Five-ringed micro-ring resonator chip. A magnified view of the five rings and the individual input and out-put ports are depicted in the drawing. The input and output waveguides are "U-shaped." A single micro-ring is centered at the base of the "U," vertically coupled to input and output waveguides. (Reprinted from Ramachandran, A. et al., *Biosens. Bioelectron.*, 23, 939. With permission.)

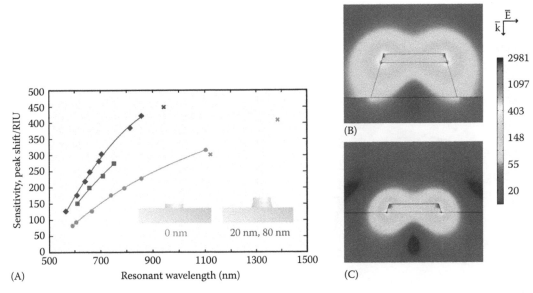

FIGURE 6.7 Reduced substrate effect through nanoparticle elevation. The measured bulk refractive index sensitivities of nanodisks on glass pillars with different height (A) shows the advantage of separating the nanoparticles from the high index substrate. Au nanodisks directly on glass (red) showed smaller sensitivity than nanodisks supported on 20 nm (blue) and 80 nm (green) SiO₂ pillars. The sensitivity enhancement can be understood from the differences in the field distributions between traditional (C) and elevated nanodisks (B). (Reproduced from Dmitriev, A. et al., *Nano Lett.*, 8(11), 3893, 2008. With permission.)

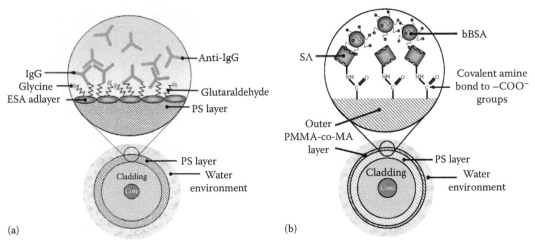

FIGURE 8.3 (a) Sketch of coated LPG surface with the different biological entities used in the experiment. (b) Sketch of coated LPG surface with the different biological entities used in the experiment.

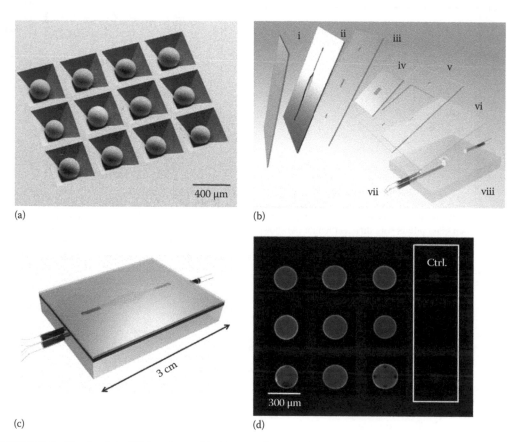

(a)

(b)

(c)

(d)

FIGURE 11.3 Details of the QD-based nano-biochip array described in the text. SEM image of beads loaded in aniso-tropically etched silicon chip (a); sandwich scheme of the chip (iv) fitted between adhesive layer (ii), cover slip (i), laminate layers (iii, v, vi), PMMA base (vii), and inlet and outlet ports (vii) (b); sealed LOC assembly (c); fluorescent image of beads after immunoassay, including negative controls as imaged with CCD camera (d). (Reprinted from Jesse, V.J. et al., *Biosens. Bioelectron.*, 24, 3622, 2009. With permission.)

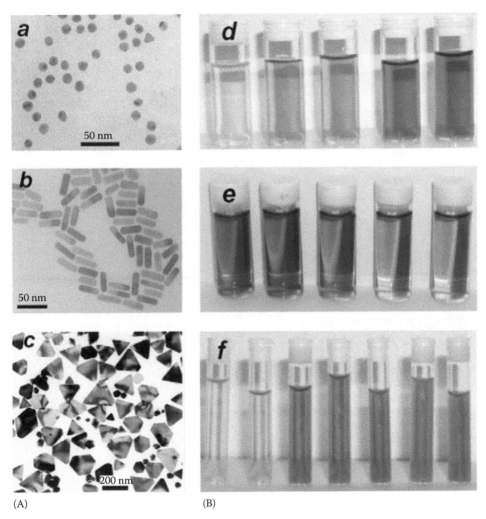

FIGURE 13.3 (A): Transmission electron micrographs of Au nanospheres and nanorods (a, b) and Ag nanoprisms (c, mostly truncated triangles) formed using citrate reduction, seeded growth, and DMF reduction, respectively. (B): Photographs of colloidal dispersions of AuAg alloy nanoparticles with increasing Au concentration (d), Au nanorods of increasing aspect ratio (e), and Ag nanoprisms with increasing lateral size (f). (Reprinted from *Mater. Today*, 7, Liz-Marzán, L.M., Nanometals: Formation and color, 26–31, Copyright 2004, with permission from Elsevier.)

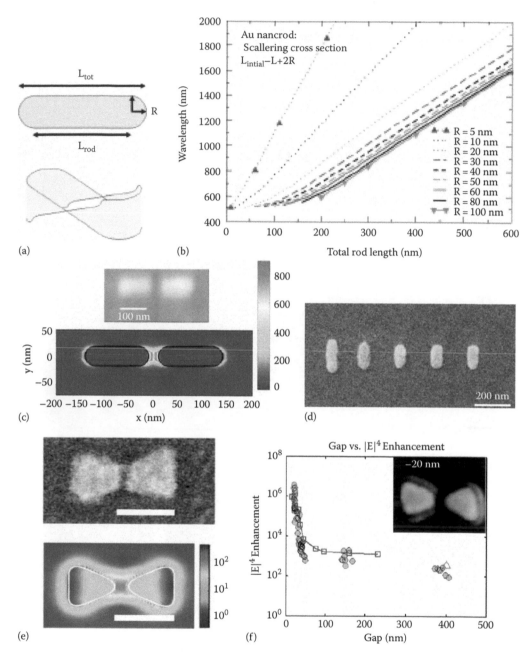

FIGURE 15.2 (a) Schematic of a cylindrical nanorod nano-antenna with hemispherical caps and the surface charge density typical for a dipolar resonance. (b) The dependence of the dipole resonance wavelength on L tot for different R. The resonance wavelength extracted from the far-field scattering is shown. (Reprinted from Bryant, G.W. et al., *Nano Lett.*, 8(2), 631, 2008. With permission.) (c) Dipole nano-antenna, consisting of two rod-shaped nanopillars separated by a small gap: (above) SEM micrograph of a resonant optical dipole antenna; (below) numerical simulation of the total electric field intensity enhancement with respect to the incident intensity. (Reprinted from Cubukcu, E. et al., *Appl. Phys. Lett.*, 89, 093120-1, 2006. With permission.) (d) Yagi-Uda nano-antenna, fabricated by employing electron beam lithography and lift-off. (Reprinted from Novotny, L. and Van Hulst N., *Nat. Photon.*, 5, 83, 2011. With permission.) (e) Bow-tie nano-antenna, consisting of two triangular-shaped nanopillars separated by a small gap between them: (above) SEM micrograph of a gold bow-tie nano-antenna. Scale bar is 100 nm, (below) Finite-difference time-domain calculation of local intensity enhancement. (Reprinted from Kinkhabwala, A. et al., *Nat. Photon.*, 3, 654, 2009. With permission.) (f) Figure showing the effect of spacing between the triangles forming the bow-tie nano-antenna on the electric field enhancement in the region between the nano-triangles. (Reprinted from Schuck, P.J. et al., *Phys. Rev. Lett.*, 94, 017402-1, 2005. With permission.)

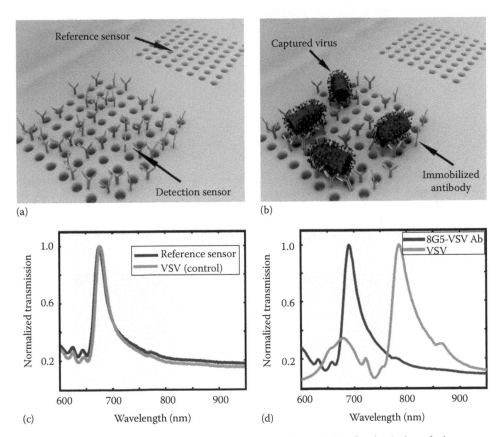

(a) (b)

(c) (d)

FIGURE 16.2 Renderings (not drawn to scale) illustrate (a) before and (b) after incubation of viruses on a reference and a detection sensor with antibodies functionalized on it. (c) Minimal red shifting in resonance frequencies is observed for reference sensors after the VSV incubation, and washing. (d) As a result of accumulation of the VSV, increasing effective refractive index results in a strong red-shifting of the plasmonic resonances (∼100 nm). (From Yanik, A.A. et al., *Nano Lett.*, 10(12), pp. 4962–4969, 2010.).

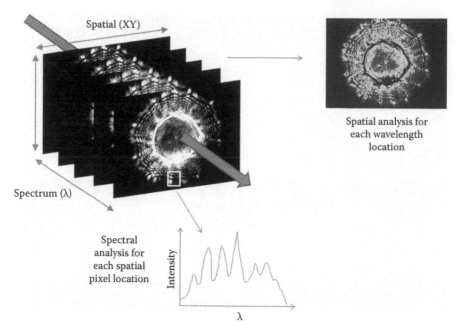

FIGURE 19.5 Schematic diagram showing principles of hyperspectral or multispectral imaging setup. Spatial information is typically recorded as intensity images while the spectral information consists of reflection intensity and generates three-dimensional data set called hyperspectral data cube.

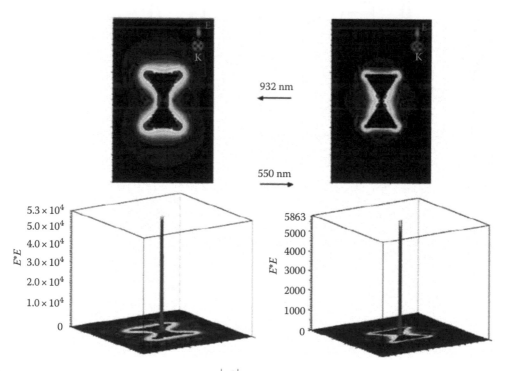

FIGURE 22.10 The calculated enhancing factor of $\left|E_T^2\right|$ or a "tip-to-tip" prism dimer shape nanostructured silver metal surface at the resonant wavelengths of 550 and 932 nm, respectively. The prism has a 60 nm edge dimension, a 2 nm snip, and a thickness of 12 nm. (From Hao, E. and Schatz, G., *J. Chem. Phys.*, 120, 357, 2004.)

11.2 Micro- and Nanofabrication Techniques for Lab-on-Chip

11.2.1 Top-Down Approach in Miniaturizing a Laboratory

As far as micro- and nanofabrication techniques of LOC systems and/or microelectromechanical systems (MEMS), micro-opto-electro-mechanical systems (MOEMS), and nano-electromechanical systems (NEMS) are concerned, we can distinguish between *top-down* and *bottom-up* methods. *Top-down* approach implies the reduction of patterns previously designed on a large scale into micro- and/or nanoscale, while in *bottom-up* techniques, the self-assembling and self-organizing ability of atoms and molecules themselves is exploited in order to obtain micro- and/or nanostructures. When at least one dimension of the patterned channels falls in the nanometric range, we can talk about *nanochannels*. Since such dimensions are comparable with size of molecules dispersed in the handled fluid, new kinds of phenomena have to be taken into account [21].

Among the most used *top-down* methods, *bulk/film machining* is based on more or less standard photolithographic techniques followed by wet or dry etching (for bulk substrate) or by chemical etching (for film etching). The obtained structure can be covered by bonding another wafer on the patterned substrate or film, leading to the formation of micro- or nanochannels. In standard photo-lithographic techniques, usually UV light is used, in combination with a proper mask, to transfer an image of the desired pattern on a photosensitive coating of organic polymer (the *photoresist*) placed over a silicon wafer or a film (see Figure 11.9). The parts of the photoresist exposed to UV can be successively removed, leaving the desired pattern on the substrate (or on the film). The resolution of the technique obviously depends on the wavelength of the radiation used in the exposition of the photo-resist and can hardly be lower than the half of the wavelength itself: e.g., standard UV light (usually around 250 nm in wavelength) guarantees, in improved experimental setups, resolutions of the order of 70–100 nm [22], while the use of radiation of smaller wavelength (extreme UV light and x-rays) is limited by the requirement of special materials in the fabrication of lenses and masks. Resolutions of the order of 10 nm can be achieved by means of electron beam lithography (EBL) [23] and focused ion beam (FIB) lithography [24]. In the first case, a beam of electrons is used to write the pattern on the photoresist, while in the second case, the pattern is written directly on the substrate (or on the film). Interference of two or more coherent beams can be used as well in patterning of nanoscopic features in so-called interferometric lithography (IL) [25], which is especially suitable when periodic and quasi-periodic patterns have to cover relatively large areas with high spatial coherence. Among other maskless approaches in micro- and nanopatterning of substrates and/or films, we recall the direct laser writing (DLW) technique [26–28] (see also Figure 11.10), which allows the reproducible fabrication of micrometric and sub-micrometric features (feature size <40 nm and line widths <200 nm have been reached), both in ceramics and in semiconductors.

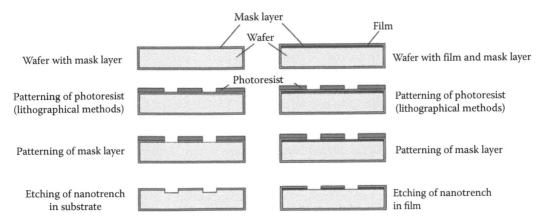

FIGURE 11.9 Bulk/film-machining fabrication process flowchart. (Modified from Mijatovic, D. et al., *Lab. Chip,* 5, 492, 2005. With permission.)

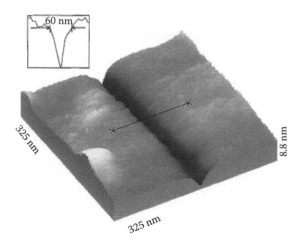

FIGURE 11.10   AFM image of a laser-etched trench in bulk silicon. (Reproduced from Müllenborn, M. et al., *Appl. Phys. Lett.*, 66, 3001, 1995. With permission.)

As far as silicon wet etching is concerned, it is usually performed by means of alkaline solutions and it can lead to both isotropic and anisotropic structures. On the other hand, dry etching can be accomplished not only by means of EBL and FIB techniques, but also by means of reactive ion etching (RIE) and deep reactive ion etching (DRIE) [29].

11.2.2 Porous Silicon: Fabrication and Characteristics

As mentioned in Introduction, PSi is a nanostructured material, which can be successfully used as a sensing element within a LOC system, due to its high specific surface area (up to $600 \, m^2/cm^3$), the possibility to chemically modify the material surface in order to link bioprobes, and its ability to change its electrical and optical properties when the porous matrix interacts with an analyte. Other advantages comprise the low cost of the fabrication technique and the compatibility with micro- and nanoelectronic technologies (thus implying the possibility of integration of PSi devices in hybrid systems).

In Figure 11.11, a typical setup for PSi fabrication is represented and described in detail. The electrochemical etching of crystalline silicon in a HF–ethanol–water solution leads to the following total reaction:

$$Si + 2H^+ + 6F^- + 2h^+ \rightarrow SiF_6^{2-} + H_2$$

Porosity is defined as the ratio of the volume of void over the overall material volume as follows: $\left(P = V_{void}/V_{total} \right)$. The morphology and dimensions of the pores depend on the doping of the starting material: n^+ type silicon leads to the so-called *macroporous* silicon (characterized by a mean pore diameter d greater than $50 \, nm$); p^+ silicon leads to *mesoporous* silicon ($2 \, nm < d < 50 \, nm$), very often used in biological applications; finally, p type silicon leads to the formation of *microporous* silicon ($d < 2 \, nm$) [30]. Once the composition of the etching solution, the doping of the material and, consequently, the mean morphology of the pores are defined, the porosity of a single layer depends directly on the current density of the anodization process, while the thickness of the layer depends on the duration of the etching itself. Since, in virtue of Bruggeman effective medium approximation [31], different porosities correspond to different refractive indexes, with a proper modulation of current density and etching time, a huge variety of multilayered optical structures can be obtained (Bragg and Thue-Morse filters, optical microcavities, optical waveguides etc.) [32] (see also Figure 11.12).

In order to understand the sensing principle of a PSi optical transducer, let us consider, as an example, a PSi Bragg mirror (BM) with a sufficiently narrow stop band. When the porous matrix is exposed to an analyte as a vapor, a gas, or a liquid, it penetrates into the pores, increasing the average refractive index of the overall structure. This causes a red shift in the reflectivity spectrum of the BM

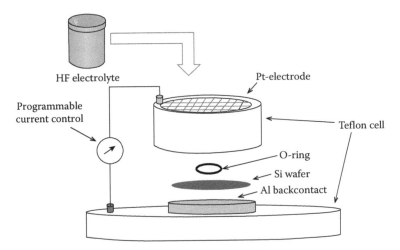

FIGURE 11.11 Typical electrochemical etching setup for PSi fabrication. The silicon wafer itself constitutes the anode: its backside is in contact with an aluminum plate while the front side is sealed with an O-ring and exposed to the anodizing electrolyte. The cathode is made of platinum or any other HF-resistant conductor. The electrolization cell is made of an acid-resistant polymer such as polyvinylidene fluoride (PVDF) or polytetrafluoroethylene (PTFE). (From Rea, I. et al., Porous silicon integrated photonic devices for biochemical optical sensing, in *Crystalline Silicon*, Intech Open Access Publisher, Chapter 12, ISBN: 978-953-307-587-7.)

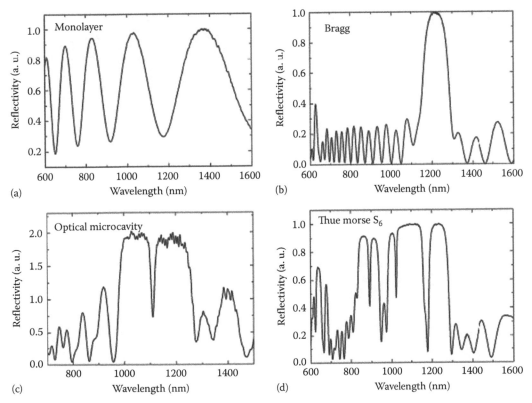

FIGURE 11.12 (a–d) Reflectivity spectra of some of the possible mono- and multilayered optical structures based on PSi. (From Rendina, I. et al., *SPIE Newsroom*, DOI: 10.1117/2.1200801.0982, 2008.)

(see Figure 11.13), which can be related to the concentration of the analyte. In order to give selectivity to this sensing scheme, the surface of PSi has to be chemically modified (*functionalization* process) in order to link biological macromolecules able to detect specific target (*bioprobes*), such as DNA single strand or a protein (antibody or enzyme). Among the various possible functionalization techniques for silicon substrates, particularly efficient is the one based on a first thermal oxidation and on subsequent treatment with aminopropyltriethoxysilane (APTES) and glutaraldehyde (GA) as linkers (see Ref. [33] for more details on the chemical functionalization steps).

11.2.3 Silicon–Glass Anodic Bonding and Other Sealing Methods

Anodic bonding (AB) is widely used when a glass wafer has to be bonded over a silicon substrate (leading, if the substrate has been properly and previously patterned by means of any of the techniques described in precedent paragraph, to the formation of micro- or nanochannels) [34]. AB is based on the use of large electric fields (with voltages ranging from 0.5 up to 2.5 kV) in conjunction with high temperatures (ranging from 200°C to 600°C), according to the scheme described in Figure 11.14: the drift of Na+ ions present in the glass toward the cathode leaves the O^{2-} ions free to react with silicon, leading to the formation of Si–O–Si bonds. Other sealing methods include *polymer bonding*, which is employed for plastic substrates and in which polydimethylsiloxane (PDMS) is very often used; *eutectic bonding* is applied mainly for metallic alloys, e.g., for Au and Si by heating them in a joint such that they diffuse together to form an alloy composition; *direct bonding* exploits the ability of very high polished substrates (such as silicon) to adhere together [21].

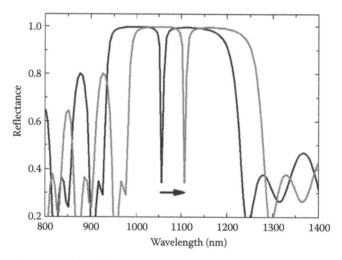

FIGURE 11.13   Schematic representation of the red shift undergone by a PSi BM when exposed to an external analyte.

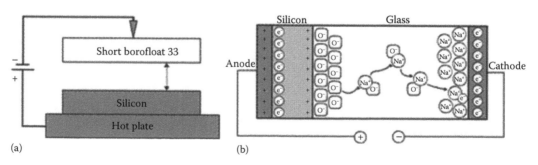

FIGURE 11.14   (a) Schematic of a typical setup used for silicon–glass anodic bonding. (b) Reaction scheme between silicon and glass.

11.3 Why Microfluidics?

"Microfluidics is the science and technology of systems that process or manipulate small (10^{-9}–10^{-18} L) amounts of fluids, using channels with dimensions of tens to hundreds of micrometres" [35]. Why microfluidics is so important in chemical and biological sensing? Biological and chemical targets are almost always transported by a fluid, as well "in vitro" and "in vivo." In "in vitro" microsystems, this happens because the target molecules are most of the time extracted from a liquid (e.g., DNA and cells) and the biochemical reactions on these targets are performed in an aqueous environment. Furthermore, microfluidic devices offer the ability to work with smaller reagent volumes, shorter reaction times, and the possibility of parallel operation. It represents a fundamental tool to integrate entire laboratory onto a single chip, i.e., a LOC. Microfluidic devices exploit their most obvious characteristic, small size, and less obvious characteristics of fluids in microchannels, such as laminar flow. They offer fundamentally new capabilities in the control of concentrations of molecules and cells in space and time. Microfluidic devices are employed in chemistry, biology, genomics, proteomics, pharmaceuticals, biodefense, and other areas where its advantages exceed standard methodologies. Microfluidic field was born exploiting silicon technologies as photolithography successfully applied in microelectronics and MEMS. In fact, first devices were fabricated in silicon and glass, largely used for microprocessors and MEMS devices, but these materials have been largely displaced by plastics [36]. For analyses of biological samples in water, devices fabricated in glass and silicon are usually unnecessary or inappropriate. Silicon, in particular, is expensive, and opaque to visible and ultraviolet light, so cannot be used with conventional optical methods of detection.

11.3.1 Microfluidic Basics

In this paragraph, microfluidic fundamentals as the continuum hypothesis, the laminar flow, and diffusion mechanism are presented.

Although fluids appear continuous in most common applications, they are quantized on the length scale of intermolecular distances. We usually deal with fluids as continuous and use the average fluid properties rather than considering the dynamics of single molecules. This approximation is called continuum hypothesis and states that the macroscopic properties of a fluid is the same if the fluid were perfectly continuous instead of, as in reality, consisting of molecules. In microfluidic volumes, the continuum approximation remains valid. In fact, a picoliter volume of water contains 3×10^{13} molecules, a number large enough to make this hypothesis valid. In nanofluidics, where smaller length scales and less fluid molecules are present, the continuum hypothesis starts to break down. As shown in Figure 11.15, measured quantity on a fluid volume encounters large fluctuations due to the molecular structure of the fluid as its volume decreases. In microfluidic range, steady and reproducible measurements are possible.

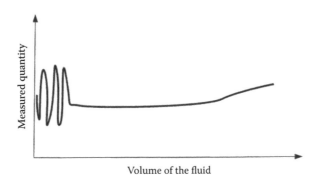

Volume of the fluid

FIGURE 11.15 A sketch of a measured physical quantity of a liquid as a function of the volume probed. It encounters large fluctuations due to the molecular structure of the fluid as its volume decreases. In microfluidic range, steady and reproducible measurements are possible.

The behavior of fluids at the microscale can differ from "macrofluidic" behavior in that factors such as surface tension, energy dissipation, and fluidic resistance start to dominate the system. Reasoning on scaling laws can be useful to anticipate the changes one may expect from miniaturizing a given system. When going from the macroscale to the microscale, an increase of the surface area to volume (SAV) ratio is inevitable. This implies that the surface-related force became dominant with respect to volume-related force. For example, in microfluidics, the capillary force was influenced by the systems more than gravity.

In fluid dynamics, the flow regime is described by the Reynolds Number, a dimensionless parameter given by

$$\mathrm{Re} = \frac{\rho u_0 L_0}{\eta} \tag{11.1}$$

where

ρ and η are the fluid density viscosity
u_0 is the characteristic velocity of the fluid
L_0 is the *hydraulic* diameter

In cylindrical channel, the flow transition value takes place at $\mathrm{Re} = 2300$: $\mathrm{Re} < 2300$ indicates a laminar flow; as Re approaches to 2300, the flow starts to show some turbulence and for $\mathrm{Re} > 2300$, the flow is considered turbulent. In microfluidics, the flow in microchannels is almost always laminar, which means that the velocity of a particle in a fluid stream is not a random function of time and is possible to predict its position in the fluid as function of time. In microfluidics, the Reynolds number is generally <0.1 because of the small size of microchannels. For example, a microchannel with circular cross section with a radius of $25\,\mu m$ filled with water that flows at $100\,\mu m/s$ has a Reynolds number of 5×10^{-6}. As shown in Figure 11.16, one consequence of laminar flow is that two or more streams flowing in contact with each other will not mix except by diffusion [37]. Diffusion is the motion of particles or molecules from regions of high concentration toward regions of low concentration. Diffusive transport is driven by random, thermally Brownian motion of particles such that, given enough time and the absence of external influences, a species will be homogeneously distributed throughout a finite volume. Diffusion can be modeled in one dimension by the equation $d^2 = 2Dt$, where d is the distance a particle moves in a time t, and D is the diffusion coefficient of the particle. For example, a biomolecule with $D = 1 \times 10^{-10}\,m^2/s$ in water goes through few microns in 1 s, but it employs 5×10^6 s for 1 cm. This is not an advantage when mixing is desirable; so to enhance mixing mechanism in microfluidic devices, many schemes have been proposed to maximize the interfaces between solutions to allow diffusion to act quickly [38].

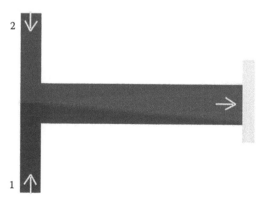

FIGURE 11.16 Two streams flowing in contact will not mix except by diffusion. As the time of contact between two streams increases, the amount of diffusion between the two streams increases.

11.3.2 Modeling and Numerical Calculations

The fluid dynamics obey the Navier–Stokes equations that essentially represent the continuum version of the Newton's law $F = ma$:

$$\frac{\partial \vec{u}}{\partial t} + \vec{u} \cdot \nabla \vec{u} = -\frac{\nabla p}{\rho} + \frac{\mu}{\rho} \nabla^2 \vec{u} \tag{11.2}$$

where u, p, ρ, and μ are the velocity field, the pressure, the density, and dynamic viscosity of the fluid, respectively [39]. To the Navier–Stokes equation, the mass conservation equation must be added:

$$\frac{\partial \rho}{\partial t} + \nabla \cdot (\rho \vec{u}) = 0 \tag{11.3}$$

which for a liquid becomes the incompressibility condition:

$$\nabla \cdot \vec{u} = 0 \tag{11.4}$$

The Navier–Stokes equation is notoriously difficult to solve analytically because it is a nonlinear differential equation. Analytical solutions can however be found in a few, but very important cases. In most cases, the flow can be considered laminar with a parabolic profile since the flow in the microchannel is in the low Reynolds number condition. Usually, boundary conditions for the equations are "no-slip" ($u = 0$) at microchannel walls.

An important class of analytical solutions to the Navier–Stokes equation is the pressure-driven, steady-state flows in channels, also known as Poiseuille Flows. In this flow, the fluid is driven through a long, straight, and rigid channel by imposing a pressure difference between the two ends of the channel. It can be described by the following law:

$$\Delta p = R_{hyd} Q \tag{11.5}$$

where
Δp is the *pressure* difference along the channel
R_{hyd} is the *hydraulic* resistance
Q is the flow *rate*

This law is completely analogous to the Ohm's law $\Delta V = RI$, relating the electrical current I through a wire with the electrical resistance R and an electrical potential difference ΔV along the wire. For example, the hydraulic resistance of microchannel with circular cross section can be calculated by the following equation:

$$R_{hyd} = \frac{8\mu L}{\pi} \cdot \frac{1}{a^4} \tag{11.6}$$

where a and L are the radius and length of the microchannel, respectively. Instead, the hydraulic resistance of microchannel with rectangular cross section can be calculated by the following equation:

$$R_{hyd} = \frac{12\mu L}{1 - 0.63(h/w)} \cdot \frac{1}{h^3 w} \tag{11.7}$$

where L, h, and w are the length, height, and width of the microchannel, respectively [40]. As for electric resistors, it is easy to show, for low Reynolds Number, the law of additivity of hydraulic resistors in a series coupling

$$R = \sum_i R_i \tag{11.8}$$

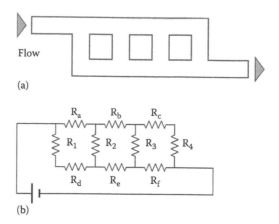

(a)

(b)

FIGURE 11.17   For a given microfluidic network, (a) it is always possible to draw the equivalent electric network (b). Channels with hydraulic resistances R_{hyd} become resistors, flow rates Q become currents, and pumps delivering pressure differences Δp become batteries.

and the law of additivity of inverse hydraulic resistances in a parallel coupling

$$\frac{1}{R} = \sum_i \frac{1}{R_i} \tag{11.9}$$

For a given microfluidic network, it is possible to draw the equivalent electric network. Channels with hydraulic resistances R_{hyd} become resistors, flow rates Q become currents, and pumps delivering pressure differences Δp become batteries (see Figure 11.17). For any given Fluidic network or circuit, one can then apply Kirchhoff's laws:

a. At any node in a microfluidic circuit, the sum of flow rates is zero.

b. The sum of the pressure differences around any closed circuit is zero.

In chemical and biological sensing, the study of the transport of chemical species in bulk liquid phase is fundamental. This is described by the following convection and diffusion equation:

$$\frac{\partial B}{\partial t} + \vec{u} \cdot \nabla B = D\nabla^2 B \tag{11.10}$$

where D (m²/s) is the diffusion coefficient of the chemical species B in bulk phase.

   It also possible to describe the interaction between two chemical species by a first-order Langmuir equation. Hu et al. simulated in their work (Hu et al. [41]), the binding between a chemical species A (mol/m²) bound to a microchannel wall and a second chemical species B (mol/m³) present in a buffer solution as follows:

$$A + B \rightleftarrows C \tag{11.11}$$

$$\frac{\partial C}{\partial t} = k_a A \cdot (B - C) - k_d C \tag{11.12}$$

where
   C (mol/m²) is the complex created by the two molecular species
   k_a is the association rate constant (M⁻¹ s⁻¹)
   k_d is the dissociation rate constant (s⁻¹)

This equation can be used for antigen–antibody [41,42] or protein–ligand reaction [43], or other biochemical interactions [44]. The equilibrium complex concentration C_{eq} can be expressed as

$$C_{eq} = \frac{AB}{A + k_a/k_d} \tag{11.13}$$

where the ratio k_a/k_d represents the affinity constant.

It is crucial to understand the behavior of the complex formation rate $C(t)$ and equilibrium concentration C_{eq} in the microfluidic configuration in order to maximize the sensor response as function of some model parameters.

11.4 Optical Chemical Sensing in Nanostructured Material-Based Lab-on-Chip

The ability to control the porosity and the morphology of the PSi layers, through the value of the current density and the etching time, allows the fabrication of optical transducers, which can be used for chemical and biochemical sensing. Lot of experimental work, exploiting the worth noting properties of PSi in chemical and biological sensing, has been recently reported in literature [45,46]. PSi is an almost ideal material as transducer due to its porous structure, like a natural sponge, having a specific surface of the order of 200–500 m² cm⁻³ [47], so that a very effective interaction with several adsorbates is assured. Moreover, PSi is an available and low-cost material, completely compatible with standard integrated circuit processes, so that it could usefully be employed in the so-called smart sensors [48]. Several different transducer schemes have been proposed, based on changes in capacitance, resistance, reflectivity, and photoluminescence properties of the material when biochemical molecules adsorb to its surface. The electric measurements are relatively straightforward and the control electronics for the device can be easily integrated on a silicon chip. On the other hand, optical measurements can be less straightforward than electric ones but safer in case of flammable vapors.

The sensing mechanism in monitoring of chemical substances is based on the refractive index changes of PSi due to the partial substitution of the air in the pores by the chemicals to be detected: a liquid infiltrates, by capillarity, the pores matrix until the hydrostatic pressure stops the pore filling. In a similar way, gaseous or vapor substances can condensate in the pore due to capillary condensation. The liquid phase has always a refractive index greater than air, so that the average refractive index of the infiltrated PSi increases and this change can be detected in several ways.

11.4.1 Capillary Condensation

The condensation conditions depend not only on the average pore size, distribution, and shape but also on the strength of the interaction between the fluid and the pore walls [49]. Once the pore's shape and the surface chemistry are fixed, a one-to-one correspondence exists between the condensation conditions and the pore diameters, given by the following Kelvin equation:

$$k_B T \rho_l \ln\left(\frac{p_{sat}}{p}\right) = 2\gamma_{lg} \cos\frac{\theta}{R} \tag{11.14}$$

where
ρ_l is the density of the liquid phase
γ_{lg} is the liquid–gas surface tension at temperature T
R is the radius of the pores
p/p_{sat} is the relative vapor pressure into the pore
θ is the contact angle

From the Equation 11.14, it is easy to see that the relative pressure increases with the average radius of pores R.

The sensitivity is a key issue of a sensor, so that several experimental works investigating the sensitivity of the different PSi structures have been reported in literature [50,51]. Ouyang et al. focused their research on the sensitivity of PSi microcavities as function of the material properties, such as pore size, porosity, and number of layers. However, a general scheme to determine the performances of PSi optical sensors has not been proposed yet. In this paragraph, the sensitivities to the pore refractive index changes of two different PSi photonic structures are compared. In particular, a simple model to study the behavior of PSi multilayered structures on exposure to different compounds and to determine their response curve is proposed. The two multilayered structures analyzed are a one-dimensional periodic multilayer, the BM, and an aperiodic multilayer, the Thue-Morse Sequence (TMS). Both the PSi structures are composed of 64 layers, 32 with high (H) refractive index (low-porosity), and 32 with low (L) refractive index (high-porosity). The layers thicknesses are $d_H = \lambda_0/4n_H$, and $d_L = \lambda_0/4n_L$, respectively. The different spatial order of the layers is the only difference between the two structures.

The interaction of PSi sensors with the chemical species induces a variation of effective refractive index of PSi layers, thus a shift of the multilayer reflectivity spectrum. The average refractive index of the PSi layer, n_p, can be determined in the near-infrared range by using the Bruggemann effective medium approximation for a heterogeneous mixture of components (nanocrystalline silicon and pore contents):

$$(1-p)\frac{n_{Si}^2 - n_p^2}{n_{Si}^2 + 2n_p^2} + (1-p-\Lambda)\frac{n_{air}^2 - n_p^2}{n_{air}^2 + 2n_p^2} + \Lambda\frac{n_{ch}^2 - n_p^2}{n_{ch}^2 + 2n_p^2} = 0 \tag{11.15}$$

where

Λ is the layer liquid fraction (LLF), i.e., the volume filled by the chemical species with refractive index n_{ch}

p is the porosity of layer

n_{Si} and n_{air} are the refractive indices of silicon and air, respectively

From Equation 11.15, the relative variation of the refractive index, $\Delta n_p/n_p$ as function of p, Λ, and n_{ch} can be numerically determined. In Figure 11.18a and b, the behavior of $\Delta n_p/n_p$ as function of Λ and n_{ch} for

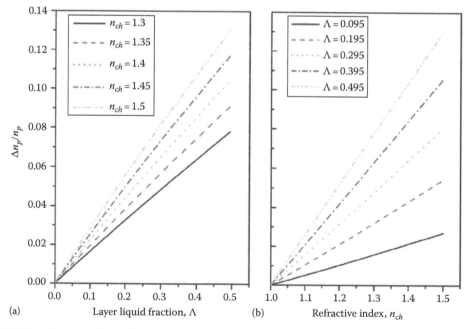

FIGURE 11.18 Relative variation of PSi refractive index layer with $p = 0.5$ as a function of layer liquid fraction for different refractive indices n_{ch} (a); as a function of n_{ch} for different LLFs (b). (From Moretti, L. et al., Periodic versus aperiodic: Enhancing the sensitivity of porous silicon based optical sensors, *Appl. Phys. Lett.*, 90, 191112, 2007. Copyright 2007, American Institute of Physics.)

a PSi layer with $p = 0.5$ is reported: the relative change of the average refractive index of the layer has a linear dependence on the filling factor and the refractive index of the chemical species:

$$\frac{\Delta n_p}{n_p} = c_p (n_{ch} - 1)\Lambda \tag{11.16}$$

The constant c_p depends on the layer porosity. It is well known that the refractive index change, due to the interaction of the PSi multilayers with external agents, preserves the shape of the reflectivity spectrum, so that it is still possible to individuate the resonant characteristics, i.e., the transmittance peaks of the TMS or the high reflectivity stop band of the BM.

The shape of reflectivity spectrum depends on the phase modulation of each layer $\varphi_i = 2\pi n_i d_i/\lambda$; for a couple of layers, the phase modulation is $\varphi = \varphi_L + \varphi_H$. The reflectivity can be factorized as a product of two contributions:

$$R = A\left(\frac{n_L}{n_H}\right) \cdot \Re(\varphi) \tag{11.17}$$

where the function A takes into account the value of the reflectivity due to the refractive index contrast, and $\Re$ is a shape factor due to the different optical paths of the light into the layers. If the reflectivity is simply shifted on a wavelength range without changing the shape during the measurements process, it is possible to write that

$$\Re\left[\varphi(\lambda_r, n_i, d_i)\right] \cong \Re\left[\varphi(\lambda_r + \Delta\lambda_r n_i + \Delta n_i d_i)\right] \tag{11.18}$$

where

  λ_r is the characteristic wavelength used as a reference to measure the spectral shift
  Δn_i is the variation of the layer refractive index due to the interaction of the devices with the chemical species

The equality $\varphi(\lambda_r, n_i, d_i) = \varphi(\lambda_r + \Delta\lambda_r n_i + \Delta n_i d_i)$ can be deduced by the Equation 11.18. By evaluating the variation of φ, an expression for $\Delta\lambda_r$ as function of layer refractive index variations, Δn_L, and Δn_H, can be deduced by the following equation:

$$\Delta\lambda_r = 2\frac{\lambda_r}{\lambda_0}\left(d_L \Delta n_L + d_H \Delta n_H\right) = \frac{\lambda_r}{2}\left(\frac{\Delta n_L}{n_L} + \frac{\Delta n_H}{n_H}\right) \tag{11.19}$$

This formula is a powerful tool in the design of all resonant optical sensors, based on the average refractive index change. Combining the Equation 11.19 with Equation 11.16, it is possible to completely characterize the optical response of whatever PSi multilayer:

$$\frac{\Delta\lambda_r}{\lambda_r} = \frac{(n_{ch} - 1)}{2}\left(c_L \Lambda_L + c_H \Lambda_H\right) \tag{11.20}$$

It is clear that the sensitivity of PSi multilayer depends strictly on the filling capability of the layers. The BM and the TMS have the same response as a function of the LLF. In Figure 11.19 are shown the reflectivity spectra of BM (a) and TMS (b) when unperturbed, and on exposure to a methanol ($n_{ch} = 1.328$) saturated atmosphere.

The BM reflectivity shows a classic photonic band gap centered at the Bragg wavelength 2 $(n_L d_L + n_H d_H) = 712$ nm. This wavelength is a natural candidate as monitor wavelength $\lambda^{\text{B}}\text{M}_r$, since it is simply recognizable after the interaction process. On the other hand, the TMS spectrum shows a more complex photonic band gap structure due to the aperiodic sequence of the layers: three photonic band gaps can be observed in wavelength intervals centered at 640, 890, and 1120 nm, and three resonant transmittance peaks at 894, 1030, and 1184 nm. In this case, it is possible to choose λ_r among the one of the resonant transmittance peaks. In particular, the spectral shift of the resonance transmittance

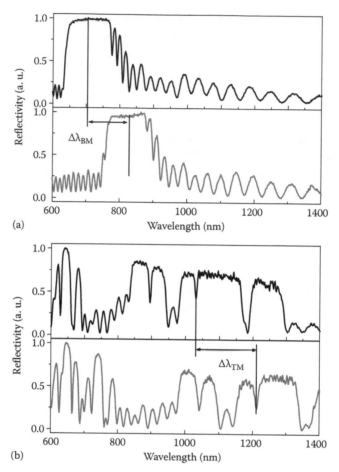

(a)

(b)

FIGURE 11.19 Experimental reflectivity spectra of Bragg multilayer (a, BM indexed) and Thue-Morse multilayer (b, TMS indexed) composed by 64 layers. Bottom plots show the reflectivity spectra after exposure to methanol. λBM and λTMS are the monitor wavelength for Bragg and Thue-Morse multilayer, respectively. (From Moretti, L. et al., Periodic versus aperiodic: Enhancing the sensitivity of porous silicon based optical sensors, *Appl. Phys. Lett.*, 90, 191112, 2007. Copyright 2007, American Institute of Physics.)

peak at 1030 nm (λ_r^{TMS} = 1030 nm) is monitored. In Figure 11.20, using the Equation 11.20, the relative wavelength shift, $\Delta\lambda_r/\lambda_r$ as a function of Λ on exposure to methanol obtained in the case of TMS and BM is reported. The filling can be assumed to proceed uniformly into the entire multilayer stack until the low-porosity layers are completely filled ($\Lambda_L = \Lambda_H = \Lambda$ for $\Lambda < p_L$), then the filling process proceeds only in high-porosity layers [52]: the filling curves in Figure 11.20 show a slope change when $\Lambda = p_L$; the values of c_L and c_H are 0.927 and 0.677, respectively. In Table 11.1, are shown the experimental

TABLE 11.1

Chemical Organic Substances and Its Refractive Index Used in Sensing Experiment

Solvent	n_{ch}	$\Delta\lambda_r^{TMS}$(nm)	Λ^{TMS}	$\Delta\lambda_r^{BM}$(nm)	Λ^{BM}
Methanol	1.328	180	0.745	108	0.599
Pentane	1.358	199	0.761	118	0.599
Isopropanol	1.377	209	0.761	125	0.599
Isobutanol	1.396	215	0.745	127	0.467

Note: $\Delta\lambda_r^{TMS}$ ($\Delta\lambda_r^{BM}$), and Λ^{TMS} (Λ^{BM}) is the wavelength shift and the layer liquid fraction for Thue-Morse (Bragg) structures. The reference wavelength λ_r is 1030 nm for TMS and 712 nm for BM.

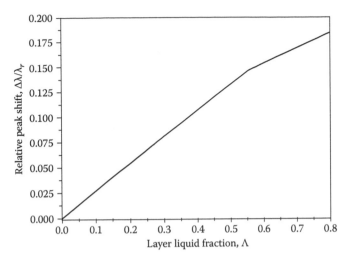

FIGURE 11.20 Calculated relative wavelength shift $\Delta\lambda r/\lambda r$ for BM and TMS as a function of layer liquid fraction Λ after exposure of methanol ($n_{ch} = 1.328$). The two curves coincide. (From Moretti, L. et al., Periodic versus aperiodic: Enhancing the sensitivity of porous silicon based optical sensors, *Appl. Phys. Lett.*, 90, 191112, 2007. Copyright 2007, American Institute of Physics.)

wavelength shifts and the LLFs extrapolated from the curve of Figure 11.20 for several compounds. First of all, it is worth noting that the pores filling is a characteristic parameter of the multilayer sequence, and is almost invariant respect to compound investigated, in agreement with Gurvitsch's rule that states that the volume of liquid adsorbed should be the same for all adsorptives on a given porous solid.

As shown in Figure 11.21 the response curves of both structures, the sensitivities, normalized to the reference wavelengths, are respectively, $S^{TMS} = 0.51(0.05)$ RIU^{-1} (Refractive Index Units) and $S^{BM} = 0.41(0.05)$ RIU^{-1}; the TMS higher sensitivity can be ascribed to higher filling capability. This effect can

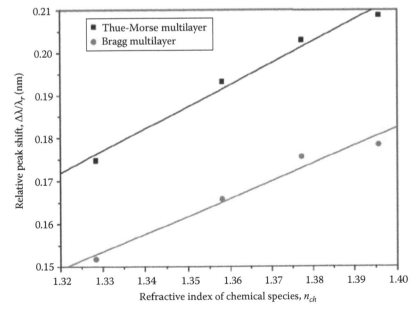

FIGURE 11.21 Experimental response curves of TMS and BM. (From Moretti, L. et al., Periodic versus aperiodic: Enhancing the sensitivity of porous silicon based optical sensors, *Appl. Phys. Lett.*, 90, 191112, 2007. Copyright 2007, American Institute of Physics.)

be explained by considering the number of *L–H* interfaces in the different multilayers. The different spatial order of layers between BM and TMS can be observed:

$$TMS: LHHLHLLHHLLHLHHL\dots$$

$$BM: LHLHLHLHLHLHLHLH\dots$$

It is possible to conclude that the periodic arrangement of the BM induces a greater number of porosity gradients, due to the presence of a greater number of *L–H* interfaces (63), than in TMS (42). These interfaces produce an inhomogeneity in the pore network, which obstacles the propagation of the liquid phase and thus reduces the filling capability of structures. Therefore, after the condensation process or the capillary penetration, the liquid phase finds a more homogeneous pore network in TMS with respect to BM.

11.4.2 Time-Resolved Measurements

Time response is a key issue in most of the chemical sensing applications: devices for homeland security or medical POC monitoring are required faster and faster in order to allow effective countermeasures. PSi optical transducers have response time that depends not only on the physical phenomena involved (i.e., equilibrium between adsorption and desorption in the PSi layer) but also on the geometry of the test chamber and on the measurement procedure, i.e., static or continuous flow mode. In static condition, the response time is mainly determined by the diffusion of the gas into the chamber volume: in fact, when vapor is in contact with the PSi surface, the capillary condensation takes place instantaneously. One way to speed up the optical detection is using some technological solution like the flow injection (FIA). The FIA technique is commonly exploited to perform different chemical and biological diagnosis in a wide range of sensing applications and is based on injecting by a valve of small and well-defined volume of sample into a continuously flowing carrier stream to which appropriate auxiliary reagent streams can be added. The sample disperses and reacts with the components of the carrier in a reactor, forming a species that is sensed by a detector and recorded. A schematic of the flow injection analysis is shown in Figure 11.22. Thus, in contrast to conventional continuous flow procedures, FIA does not rely on complete mixing of sample and reagent. Combined with the inherent exact timing of all events, it is not necessary to wait until all chemical reactions are in equilibrium. These feats, which allow transient signals to be used as the readout, do not only permit the procedures to be accomplished in a very short time, but have opened new ways to perform an array of chemical analytical assays, which are very difficult and in many cases directly impossible to implement in a traditional way. Thus, in FIA it is possible to base the assay on the measurement of metastable compounds that exhibit particularly interesting analytical characteristics. The concept of FIA depends on a combination of three factors: reproducible sample injection volumes, controllable sample dispersion, and reproducible timing of the injected sample through the flow system. The system is ready for instant operation as soon as the sample is introduced. FIA offers several advantages in terms of considerable decrease in sample (normally using 10–50 μL) and reagent consumption, high sample throughput (50–300 samples per hour), reduced residence times (reading time is about 1–40 s), shorter flow injection analysis reaction times (1–60 s),

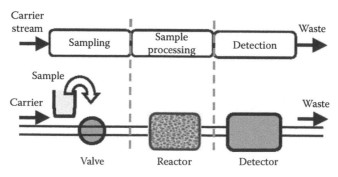

FIGURE 11.22   Scheme of the flow injection analysis principle.

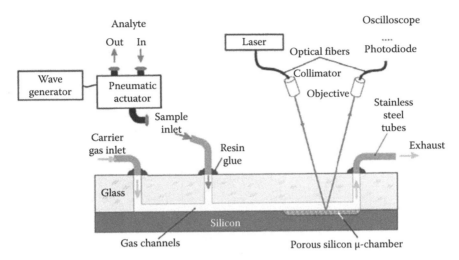

FIGURE 11.23 Scheme of the complete optical microsystem and of the read-out experimental setup. (From De Stefano, L. et al.: An integrated pressure-driven microsystem based on porous silicon for optical monitoring of gaseous and liquid substances. *Physica Status Solidi A Appl. Mater. Sci.* 2007. 204. 1459–1463. Copyright Wiley-VCH Verlag GmbH & Co. KGaA. With permission.)

easy switching from one analysis to another (manifolds are easily assembled and/or exchanged), reproducibility, reliability, low carry over, high degree of flexibility, and ease of automation. Perhaps, the most compelling advantage of the FIA technique is the great reproducibility in the results obtained by this technique, which can be set up without excessive difficulties and at very low cost of investment and maintenance. These advantages have led to an extraordinary development of FIA, unprecedented in comparison to any other technique [53].

Fast optical chemical sensing can thus be obtained by the Psi-based microsystem, whose scheme and read-out experimental setup is shown in Figure 11.23. In this system, the analyte is sent into the reaction chamber by a carrier stream after the valve opening. The interaction between the analyte and the PSi transducer is monitored by time-resolved optical measurements: a laser beam with 633 nm wavelength was used at nearly normal incidence; the signal of a receiving photodetector before, during, and after the exposure to gaseous substances (isopropanol, ethanol, and methanol) has been measured as a function of the time.

The reaction micro-chamber with the PSi on the bottom was produced by a two-step electrochemical etch of a <100> silicon wafer, p⁺ type, using an HF/EtOH (30:70) solution. The first step creates a thick sacrificial layer of high-porosity PSi that is completely removed by rinsing the samples in a NaOH solution. The result is an empty chamber into the silicon substrate with depth of 150 μm. Since the diameter of the PSi area is always about 1 cm, the total volume of the analysis chamber is 12 μL. By using a fresh etching HF/EtOH (30:70) solution, without moving the chip from the electrochemical cell, a second etch step (150 mA/cm² for 45 s) produces PSi monolayer on the bottom of approximately 6 μm thickness and a porosity of 70%. A Borofloat 33 type glass has been chosen as a chip cover. In order to feed gas or liquid substances, inlet and outlet channels have been mechanically drilled in the top glass wafer, on the opposite sides of the porous region. A pneumatic actuator has been used for the injection valve control. We have optimized the frequency of the valve opening (100 mHz) and the pressure of the carrier searching for the fastest time response of the sensor device.

On exposure to the vapors, due to the phenomenon of capillary condensation, the average refractive index of the layer increases, and, as a consequence, the optical thickness of the PSi layer also increases. The result of a time-resolved measurement is shown in Figure 11.24: in the case of Isopropanol, we have obtained a response time to the solvent presence, which is 156 ms, while the signal returns to its original value in shorter time but always in the order of milliseconds ($\tau_{rec} = 24$ ms).

The response time depends not only on the physical phenomena involved (i.e., equilibrium between adsorption and desorption of each substances in the PSi layer) but also on the geometry of the test

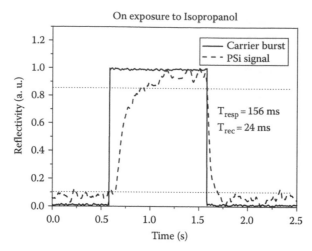

FIGURE 11.24 Time-resolved measurement in case of Isopropanol. The optical response and the time reference of the carrier have been shifted for the sake of clarity. (From De Stefano, L. et al.: An integrated pressure-driven microsystem based on porous silicon for optical monitoring of gaseous and liquid substances. *Physica Status Solidi A Appl. Mater. Sci.* 2007. 204. 1459–1463. Copyright Wiley-VCH Verlag GmbH & Co. KGaA. With permission.)

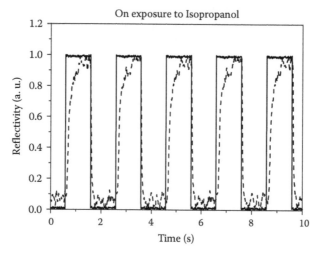

FIGURE 11.25 Repeatability and reversibility of the gas-sensing measurements. (From De Stefano, L. et al.: An integrated pressure-driven microsystem based on porous silicon for optical monitoring of gaseous and liquid substances. *Physica Status Solidi A Appl. Mater. Sci.*. 2007. 204. 1459–1463. Copyright Wiley-VCH Verlag GmbH & Co. KGaA. With permission.)

chamber and on the measurement procedure. In the same experimental way, we have verified the stability and repeatability of the sensor on several pulses of the gas analyte (Figure 11.25).

11.5 Optical Sensing of Chemical and Biological Molecules Based on Porous Silicon

Unfortunately, a PSi sensor cannot discriminate the components of a complex mixture because the sensing mechanism is not selective. Some researchers have chemically or physically modified the Si–H surface sites in order to enhance the sensor selectivity through specific interactions. The common approach is to create a covalent bond between the PSi surface and the biomolecules that specifically recognize the target analytes [54,55]. The reliability of a biosensor strongly depends on the functionalization process: how fast, simple, homogenous, and repeatable it is. This step is also very important for the stability of the sensor: it is well known that "as-etched" PSi has a Si–H terminated surface due to the Si dissolution process, which is very reactive. The substitution of the Si–H bonds with Si–C ones guarantees a much more stable surface from the thermodynamic point of view. Three different PSi surface modification strategies in order to realize an optical biosensor are mentioned here: the target is the fabrication of sensitive label-free biosensors that are highly requested for applications in high-throughput drug monitoring and disease diagnostics; unlabelled analytes require, in fact, easier and faster analytical procedures.

FT-IR spectroscopy (Thermo—Nicholet NEXUS) has been used to compare the different passivation procedures: a pure chemical process based on Grignard Reactives; a photoinduced chemical modification based on the undecenoic organic acid and a passivation method simultaneous to the etching process. In each case, the carboxyl-terminated monolayer covering the PSi surface acts as a substrate for the chemistry of the subsequent attachment of the DNA sequences.

Before the functionalization process, the PSi substrate is immersed in an aqueous ethanol solution, containing millimolar concentration of KOH, for 15 min. This alkaline treatment produces an increase in the porosity of about 15%–20% [56], so improving the infiltration of the biomolecular probes into the pores. The process also removes most of the Si–H bonds from the PSi surface that can be restored by rinsing the PSi device in a low concentration HF-based solution (5 mM) for 30 s.

11.5.1 PSi Functionalization Strategies

The chemical functionalization is based on the ethyl magnesium bromide (CH_3CH_2MgBr) as a nucleophilic agent, which substitutes the Si–H bonds with the Si–C.

The reaction was made at 85°C in an inert atmosphere (Argon) to avoid the deactivation of the reactive, for 8 h. The chip was thus washed with a 1% solution of CF_3COOH in diethyl ether, and then with deionized water and pure diethyl ether. The modified surface chip was characterized by infrared spectroscopy to verify the efficiency of the method. The FT-IR spectrum is shown in Figure 11.26. The characteristic absorption peaks, which identify the presence of organic species, are well evident after the functionalization.

The photoactivated chemical modification of PSi surface was based on the UV exposure of a solution of alkenes that bring some carboxylic acid groups. The PSi chip is pre-cleaned in an ultrasonic acetone bath for 10 min, then washed in deionized water. After dried in N_2 stream, it is immediately covered with 10% N-hydroxysuccinimide ester (UANHS) solution in CH_2Cl_2. The UANHS was synthesized in-house as described in Ref. [54]. This treatment results in covalent attachment of UANHS to the PSi

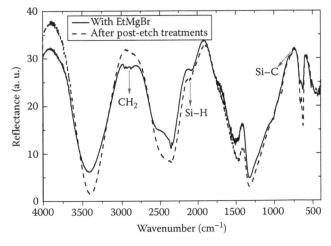

FIGURE 11.26 FT-IR spectra of the PSi monolayer before and after the pure chemical functionalization process based on EtMgBr. (From De Stefano, L. et al., Porous silicon based optical biochips, *J. Opt. A Pure Appl. Opt.*, 8, S540–S544, 2006. With permission IOP.)

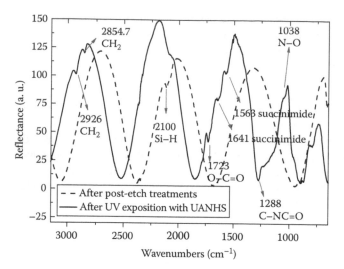

FIGURE 11.27 FT-IR spectra of the PSi monolayer before and after the photoinduced functionalization process based on UV exposure. (From De Stefano, L. et al., Porous silicon based optical biochips, *J. Opt. A Pure Appl. Opt.*, 8, S540–S544, 2006. With permission IOP.)

surface clearly shown in the FT-IR spectrum, shown in Figure 11.27. The chip was then washed in dichloromethane in an ultrasonic bath for 10 min and rinsed in acetone to remove any adsorbed alkene from the surface.

The chemical modification of the PSi surface by directly using a functionalizing agent during the etching process has been studied. Some organic acids (eptinoic and pentenoic acid with concentrations ranging from 0.4 up to 3 M) have been introduced in the electrochemical cell in the presence of a diluted HF solution (HF:EtOH = 1:2). In this case, a current density of 60 mA/cm² was applied to etch an area of 0.07 cm². The FT-IR spectrum is shown in Figure 11.28.

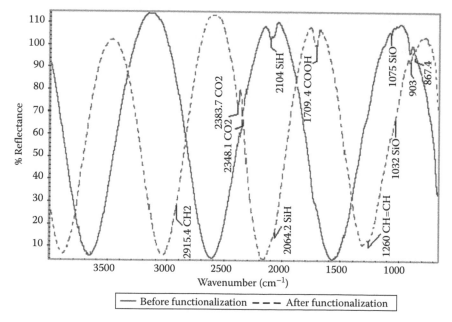

FIGURE 11.28 FT-IR spectra of the PSi monolayer before and after the functionalization during the etching process. (From De Stefano, L. et al., Porous silicon based optical biochips, *J. Opt. A Pure Appl. Opt.*, 8, S540–S544, 2006. With permission IOP.)

11.5.2 Optical Detection of DNA–DNA Hybridization

Among the three procedures experimented, the photoinduced method is the best one due to several reasons: the relaxed reaction conditions (atmospheric pressure and room temperature); the shorter reaction time; and the best reaction yield (largest peaks recorded in FT-IR measurements). This last result is somewhat expected because the reactive considered has a so-called "outgoing group," the succinimide, which promotes its substitution with the ammine group of the DNA probe.

In view of these considerations, each chip with a photochemical-modified surface was incubated, over night, with fluorescent DNA single strand. After the chemical bonding of the labeled *ss*DNA, the chip was observed by the fluorescence macroscopy system. Under the light of the 100 W high-pressure mercury lamp, a high and homogeneous fluorescence on the whole chip surface was found; the fluorescence still remains bright even after two overnight dialysis washings in a HEPES solution and in deionized water, as it can be seen in Figure 11.29a through c. The yield of the chemical functionalization has been studied by spotting different concentrations of the fluorescent *ss*DNA and measuring the fluorescence intensities of the images before and after the washings.

The results shown in Figure 11.30 confirm the qualitative findings of Figure 11.29: the fluorescent intensities decrease but remain of the same order of magnitude. From this graph, it is also possible to estimate the concentration of the DNA probe, which saturates the binding sites available ($\approx 300 \mu M$).

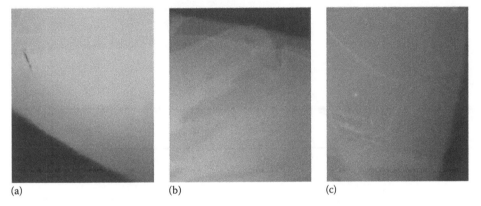

(a)   (b)   (c)

FIGURE 11.29 (a) Fluorescence of the chip surface after the binding of the labeled *ss*DNA; (b) after the overnight dialysis in HEPES solution; (c) after the overnight dialysis in deionized water. (From De Stefano, L. et al., DNA optical detection based on porous silicon technology: from biosensors to biochips, *Sensors*, 7, 214–221, 2007. With permission of MDPI.)

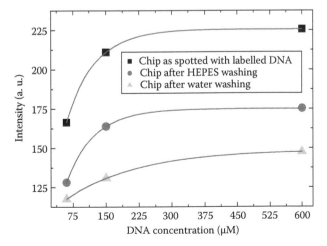

FIGURE 11.30 Fluorescence intensities of the chip surface after the binding of the labeled *ss*DNA and the two overnight dialysis as a function of the cDNA concentration. (From De Stefano, L. et al., DNA optical detection based on porous silicon technology: From biosensors to biochips, *Sensors*, 7, 214–221, 2007. With permission of MDPI.)

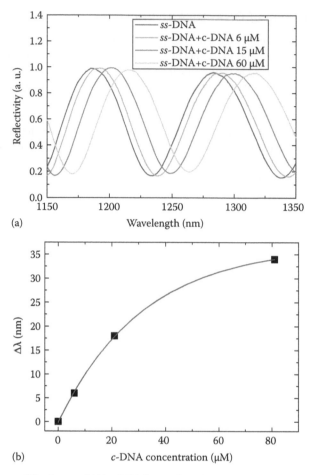

(a)

(b)

FIGURE 11.31 (a) Fringes shifts due to *ss*DNA–*c*DNA interaction. (b) Dose-response curve as a function of the *c*DNA concentration. (From De Stefano, L. et al., DNA optical detection based on porous silicon technology: From biosensors to biochips, *Sensors*, 7, 214–221, 2007. With permission of MDPI.)

The PSi optical biosensors measure the change in the average refractive index of the device: when a bio-recognition event takes place, the refractive index of the molecular complex changes and the interference pattern on output is thus modified.

The label-free optical monitoring of the *ss*DNA–*c*DNA hybridization is simply the comparison between the optical spectra of the PSi layer after the UANHS and probe immobilization on the chip surface and after its hybridization with the *c*DNA. Each step of the chip preparation increases the optical path in the reflectivity spectrum recorded, due to the substitution of the air into the pores by the organic and biological compounds. The interaction of the *ss*DNA with its complementary sequence has been detected as a fringes shift in the wavelengths, which corresponds to a change in the optical path. Since the thickness d is fixed by the physical dimension of the PSi matrix, the variation is clearly due to changes in the average refractive index. In Figure 11.31a, the reflectivity spectra of the PSi layer for different *c*DNA concentration are shown, while in Figure 11.31b, a dose-response curve is shown. A control measurement has been made using a *nc*DNA sequence: a very small shift (less than 2 nm) has been recorded in the reflectivity spectrum with respect to the one obtained after the probe linking. The sensor response has been fitted by a monoexponential growth model, $y = A(1 - e^{-Bx})$, where A is the amplitude and B is the rate constant. The limiting sensitivity, i.e., the sensitivity in the limit of zero ligand concentration, $S = AB$, has been calculated obtaining the value 1.16 (0.04) nm/μM, which corresponds to a limit of detection (LOD) of 0.26 μM for a system able to detect a wavelength shift of 0.1 nm. The LOD is defined as three times the ratio between the standard deviation and the sensitivity according to IUPAC definition [57].

REFERENCES

1. Seitz, P. 2008. Optical biochips. In *Biophotonics*, eds. L. Pavesi, and P.M. Fauchet, pp. 217–237. Berlin, Germany: Springer-Verlag.
2. Jokerst, J. V., Raamanathan, A., Christodoulides, N., Floriano, P. N., Pollard, A. A., Simmons, W. G., Wong, J. et al., 2009. Nano-bio-chips for high performance multiplexed protein detection: Determinations of cancer biomarkers in serum and saliva using quantum dot bioconjugate labels. *Biosens Bioelectron* 24:3622–3629.
3. Deiss, F., Sojic, N., White, D. J., and Stoddart, P. R. 2010. Nanostructured optical fibre arrays for high-density biochemical sensing and remote imaging. *Anal Bioanal Chem* 396:53–71.
4. Ma, Z. Y., Ma, L. Y., and Su, M. 2008. Engineering three-dimensional micromirror arrays by fiber-drawing nanomanufacturing for solar energy conversion. *Adv Mater* 20:3734–3738.
5. White, D. J., Mazzolini, A. P., and Stoddart, P. R. 2008. Nanostructured optical fiber with surface-enhanced Raman scattering functionality. *Optics Lett* 30:598–600.
6. Jin, E. S., Norris, B. J., and Pantano, P. 2001. An electrogenerated chemiluminescence imaging fiber electrode chemical sensor for NADH. *Electroanalysis* 13:1287–1290.
7. Konry, T., Novoa, A., Cosnier, S., and Marks, R. S. 2003. Development of an electroptode immunosensor: Indium tin oxide-coated optical fiber tips conjugated with an electopolymerized thin film with conjugated cholera Toxin B subunit. *Anal Chem* 75:2633–2639.
8. Tam, J. M., Song, L., and Walt, D. R. 2009. DNA detection on ultrahigh-density optical fiber-based nano-arrays. *Biosens Bioelectron* 24:2488–2493.
9. Rissin, D. M., Gorris, H. H., and Walt, D. R. 2008. Distinct and long-lived activity states of single enzyme molecules. *J Am Chem Soc* 130:5349–5353.
10. Li, Z., Hayman, R. B., and Walt, D. R. 2008. Detection of single-molecule DNA hybridization using enzymatic amplification in an array of femtoliter-sized reaction vessels. *J Am Chem Soc* 130:12622–12623.
11. Weiss, S. M., Rong, G., and Lawrie, J. L. 2009. Current status and outlook for silicon-based optical biosensors. *Physica E* 41:1071–1075.
12. Ouyang, H., Christophersen, M., Viard, R., Miller, B. L., and Fauchet, P.M. 2005. Macroporous silicon microcavities for macromolecule detection. *Adv Mater* 15:1851–1859.
13. De Stefano, L., Rendina, I., Rossi, A. M., Rossi, M., Rotiroti, L., and D'Auria, S. 2007. Biochips at work: Porous silicon microbiosensor for proteomic diagnostic. *J Phys Condens Matter* 19:395007–395011.
14. Rendina, I., De Tommasi, E., Rea, I., Rotiroti, L., and De Stefano, L. 2008. Integrated optical biosensors and biochips based on porous silicon technology. *Proc. SPIE* 6898:D8981.
15. DeLouise, L. A., Kou, P. M., and Miller, B. L. 2005. Cross-correlation of optical microcavity biosensor response with immobilized enzyme activity. Insights into biosensor sensitivity. *Anal Chem* 77:3222–3230.
16. De Tommasi, E., De Stefano, L., Rea, I., Di Sarno, V., Rotiroti, L., Arcari, P., Lamberti, A., Sanges, C., and Rendina, I. 2008. Porous silicon based resonant mirrors for biochemical sensing. *Sensors* 8:6549–6556.
17. Rea, I., Oliviero, G., Amato, J., Borbone, N., Piccialli, G., Rendina, I., and De Stefano, L. 2010. Direct synthesis of oligonucleotides on nanostructured silica multilayers. *J Phys Chem C* 114:2617–2621.
18. Densmore, A., Xu, D. X., Waldron, P., Janz, S., Cheben, P., Lapointe, J., Delage, A., Lamontagne, B., Schmid, J.H., and Post, E. 2006. A silicon-on-insulator photonic wire based evanescent field sensor. *IEEE Photon Technol Lett* 18:2520–2522.
19. Almeida, V. R., Xu, Q., Barrios, C. A., and Lipson, M. 2004. Guiding and confining light in void nanostructure. *Optics Lett* 29:1209–1211.
20. Ramachandran, A., Wang, S., Clarke, J., Ja, S. J., Goad, D., Wald, L., Flood, E. M. et al., 2008. A universal biosensing platform based on optical micro-ring resonators. *Biosens Bioelectron* 23:939–944.
21. Mijatovic, D., Eijkel, J. C. T., and van den Berg, A. 2005. Technologies for nanofluidic systems: Top down vs. bottom-up—A review. *Lab Chip* 5:492–500.
22. Alexe, M., Harnagea, C., and Hesse, D. 2004. Non-conventional micro- and nanopatterning techniques for electroceramics. *J Electroceram* 12:69–88.
23. Tseng, A. A., Chen, K., Chen, C. D., and Ma, K. J. 2003. Electron beam lithography in nanoscale fabrication: Recent development. *IEEE T Electron Pa M* 26:141–149.
24. Giannuzzi, L. A. and F. A. Stevie. 2005. *Introduction to Focused Ion Beams*. New York: Springer.
25. Maldovan, M. and E. L. Thomas. 2008. *Periodic Materials and Interference Lithography for Photonics, Phononics and Mechanics*. Weinheim, Germany: Wiley-VCH.

26. Müllenborn, M., Dirac, H., and Petersen, J. W. 1995. Silicon nanostructures produced by laser direct etching. *Appl Phys Lett* 66:3001–3003.
27. Rea, I., Marino, A., Iodice, M., Coppola, G., Rendina, I., and De Stefano, L. 2008. A porous silicon Bragg grating waveguide by direct laser writing. *J Phys Condens Matter* 20:365203–365207.
28. Rea, I., Iodice, M., Coppola, G., Rendina, I., Marino, A., and De Stefano, L. 2009. A porous silicon-based Bragg grating waveguide sensor for chemical monitoring. *Sensor Actuat B Chem* 139:39–43.
29. Madou, M. J. 1997. *Fundamentals of Microfabrication*. Boca Raton, FL: CRC Press.
30. Canham, L. 1997. *Properties of Porous Silicon*. London, U.K.: Inspec.
31. Bruggeman, D. A. G. 1935. Berechnung verschiedener physikalischer Konstanten von heterogenen Substanzen. *Ann Phys* 24:636–679.
32. Rendina, I., De Tommasi, E., Rea, I., Rotiroti, L., and De Stefano, L. 2008. Porous silicon optical transducers offer versatile platforms for biosensors. *SPIE Newsroom*, DOI: 10.1117/2.1200801.0982.
33. Rea, I., Lamberti, A., Rendina, I., Coppola, G., Gioffrè, M., Iodice, M., Casalino, M., De Tommasi, E., and De Stefano, L. 2010. Fabrication and characterization of a porous silicon based microarray for label-free optical monitoring of biomolecular interactions. *J Appl Phys* 107:014513–014516.
34. De Stefano, L., Malecki, K., Della Corte, F. G., Moretti, L., Rea, I., Rotiroti, L., and Rendina, I. 2006. A microsystem based on porous silicon-glass anodic bonding for gas and liquid optical sensing. *Sensors* 6:680–687.
35. Whitesides, G. M. 2006. The origins and the future of microfluidics. *Nature* 442: 368–373.
36. Sia, S. K. and Whitesides, G. M. 2003. Microfluidic devices fabricated in poly(dimethylsiloxane) for biological studies. *Electrophoresis* 24:3563–3576.
37. Beebe, D. J., Mensing, G. A., and Walker, G. M. 2002. Physics and applications of microfluidics in biology. *Annu Rev Biomed Eng* 4:261–286.
38. Stroock, A. D., Dertinger, S. K. W., Ajdari, A., Mezic, I., Stone, H. A., and Whitesides, G. M. 2002. Chaotic mixer for microchannels. *Science* 295:647–651.
39. Squires, T. M. and Quake, S. R. 2005. Microfluidics: Fluid physics at the nanoliter scale. *Rev Mod Phys* 77:977–1026.
40. Bruus, H. 2008. *Theoretical Microfluidics*. New York: Oxford University Press.
41. Hu, G., Gao, Y., and Li, D. 2007. Modeling micropatterned antigen-antibody binding kinetics in a microfluidic chip. *Biosens Bioelectron* 22:1403–1409.
42. Kim, D.R. and Zheng, X. 2008. Numerical characterization and optimization of the microfluidics for nanowire biosensors. *Nano Lett* 8:3233–3237.
43. Yang, C. K., Chang, J. S., Chao, S. D., and Wu, K. C. 2007. Two dimensional simulation on immunoassay for a biosensor with applying electrothermal effect, *Appl Phys Lett* 91:113904.
44. Orabona, E., Rea, I., Rendina, I., and De Stefano, L. 2011. A porous silicon based microfluidic array for the optical monitoring of biomolecular interactions. *SPIE Proc* 8073:807314–807316.
45. Dancil, K. P. S., Greiner, D. P., and Sailor, M. J. 1999. A porous silicon optical biosensor: Detection of reversible binding of IgG to a protein A-modified surface. *J Am Chem Soc* 121, 7925–7930.
46. De Stefano, L., Moretti, L., Rossi, A. M., Rocchia, M., Lamberti, A., Longo, O., Arcari, P., and Rendina, I. 2004. Optical sensors for vapors, liquids, and biological molecules based on porous silicon technology, *IEEE Trans Nanotech* 3:49.
47. Herino, R., Bomchil, G., Barla, K., Bertrand, C., Ginoux, J. L. 1987. Porosity and pore-size distributions of porous silicon layers. *J Electrochem Soc* 134:1994–2000.
48. De Stefano, L., Moretti, L., Rendina, I., Rossi, A. M., and Tundo, S. 2004. Smart optical sensors for chemical substances based on porous silicon technology. *Appl Opt* 43:167.
49. Neimark, V. and Ravikovitch, P. I. 2001. Capillary condensation in MMS and pore structure characterization, *Micropor Mesopor Mater* 44–45:697–707.
50. Ouyang, H. and Fauchet, P. 2005. Biosensing using porous silicon photonic bandgap structures. *Proc SPIE* 6005:600508–600511.
51. Ouyang, H., Striemer, C., and Fauchet, P. M. 2006. Quantitative analysis of the sensitivity of porous silicon optical biosensors. *Appl Phys Lett* 88:163108; Volk, J., Le Grand, T., Barsony, I., Gombkoto, J., and Ramsden, J. J. 2005. Porous silicon multilayer stack for sensitive refractive index determination of pure solvents. *J Phys D* 38:1313–1317.
52. Snow, P. A., Squire, E. K., Russell, P. S. J., and Canham, L. T. 1999. Vapor sensing using the optical properties of porous silicon Bragg mirrors. *J Appl Phys* 86:1781.

53. Ljibisa, R. 1994. *Sensors Technology and Devices.* Boston, MA: Artech House, p. 207.
54. Yin, H. B., Brown, T., Gref, R., Wilkinson, J. S., and Melvin, T. 2004. Chemical modification and micropatterning of Si(100) with oligonucleotides. *Microelectron Eng* 73/74:830–836.
55. Hart, B. R., Letant, S. E., Kane, S. R., Hadi, M. Z., Shields, S. J., and Reynolds, J. G. 2003. New method for attachment of biomolecules to porous silicon. *Chem Commun* 3:322–323.
56. DeLouise, L. A. and Miller, B. L. 2004. Optimization of mesoporous silicon microcavities for proteomic sensing. *Mat Res Soc Symp Proc* 782, A5.3.1–A5.3.7r.
57. IUPAC Compendium of Analytical Nomenclature. 1997. *Definitive Rules*, 3rd edn., Section 10.3.3.3.14: Limit of Detection.
58. Moretti, L. et al. 2007. Periodic versus aperiodic: Enhancing the sensitivity of porous silicon based optical sensors. *Appl Phys Lett* 90:191112.
59. De Stefano, L. et al. 2007. An integrated pressure-driven microsystem based on porous silicon for optical monitoring of gaseous and liquid substances. *Phys Status Solidi A Appl Mater Sci* 204, 1459–1463.
60. De Stefano, L. et al. 2006. Porous silicon based optical biochips. *J Opt A Pure Appl* 8, S540–S544.

12

Photonic Crystals as Valuable Technological Platform for Chemical and Biological Nanosensors

Valery Konopsky and Elena Alieva

CONTENTS

12.1 Introduction

Registration of a bounded optical wave propagating along a surface under investigation is a widely used method among state-of-the-art optical chemical and biochemical sensor techniques [1,2]. Utilizing photonic crystals (PCs) permits to confine and concentrate an optical field in the desirable interaction area between light and (bio)sensing material and, therefore, to enhance the performance of the optochemical sensors. PCs are materials that possess a periodic modulation of their refraction indices (RIs) on the scale of the wavelength of light [3]. The multiple reflection from the periodic RI boundaries in such materials can lead to the destructive interference of the optical waves and to the formation of the bands where light propagation is forbidden.

There are two main PC designs that may be utilized in optochemical sensors (see schematic illustration for 1D examples in Figure 12.1). In the first case, a high-index layer that is in contact with the external medium under study is nanostructured, and the periodic modulation of the RI takes places for the optical waves propagating along the layer. In the second case, the 1D PC structure is a simple periodic multilayer stack, and optical surface waves (SWs) are excited at the external flat layer, which may be a low-index layer (SiO_2 layer, for example). The second case will be described in more detail in the present chapter, while the first case will be outlined in a basic way, for completeness and comparison.

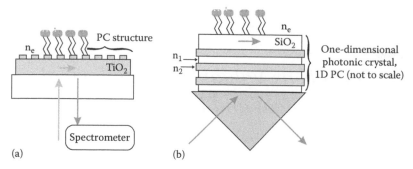

FIGURE 12.1   Outline of two main PC designs. A photonic crystal slab sensor (a) and a sensor based on PC surface waves (b).

12.2 Photonic Crystal Slab Sensors

One possible employment of photonic crystals (PCs) in (bio)sensing may be realized in the form of photonic crystal slabs (PCSs) [4]. The PCS is a high-index guiding layer with a periodically nanostructured surface (Figure 12.1a). The high-index layer supports in-plane waveguide modes, where in-plane propagation is strongly modified by the PC structure. Simple wavelength interrogation of such structures provides useful information about the slab environment since spectral resonances in PCS may be very narrow and sensitive to RI of external media and to adlayer deposition on the slab. Multichannel registration (image-based detection) may be performed in PCS by sequential mechanical scanning of the slab [5,6]. Such PCS sensors also may be ultracompact with a large dynamic range for environment RI monitoring (but with moderate RI detection limit $\Delta n \sim 2 \times 10^{-3} \dots 2 \times 10^{-4}$) [7,8]. A possible limitation for large-scale implementation of PCS sensors may be the cost and difficulties of cleaning a PCS for repeated use due to its highly developed surface. To solve these issues, a cost-effective production method such as microreplication of PC structures on plastic substrates has been developed to generate disposable PCS sensors [9]. Label-free biosensor devices based on a PCS platform are now commercially available [10]. PCS structure with incorporated labels (chromogenic probes) also may be employed for optical sensing of chemical analytes [11].

12.3 Sensors Based on PC Surface Waves

In the next embodiment of PC in optochemical sensors (Figure 12.1b), the wavelength or angle interrogation of optical surface waves, which can exist on the external surface of the multilayer stack, is used. The existence of the optical SWs in PC crystals may be deduced through an analogy between electron waves traveling in the periodic potential of the ordinary crystals and optical waves in PC crystals [12]. In both cases, frequency intervals exist in which wave propagation is forbidden. This analogy may be extended to include surface levels that can exist in band gaps of electronic crystals (Tamm states). In PCs, they correspond to optical SWs with dispersion curves located inside the photonic band gap. Sometimes, these PC SWs are also called Bloch surface waves [13] or optical Tamm states [14].

Optical surface modes in one-dimensional photonic crystals (1D PCs) were studied in the 1970s, both theoretically [15,16] and experimentally [17]. Twenty years later, the excitation of optical SWs in a Kretschmann-like configuration was demonstrated [18,19]. In the recent years, the PC SWs have been used in ever-widening applications in the field of optical sensors [20–24]. In contrast to surface plasmon-polaritons (SPPs), both p-polarized and s-polarized optical surface waves can be used in PC SW sensor applications [25].

The main difference between PC SWs and other surface waves, employed in different sensing techniques, is their unique tunable properties, which permits the design an appropriate PC structure for any desirable wavelength range and excitation angle. Such flexibility is hardly possible for surface waves in other standard techniques. For example, in the surface plasmon resonance (SPR) technique [26], the surface-bound wave is a SPP propagating along a gold or silver surface. In optical waveguide techniques, the bounded wave (waveguide mode) is excited in a high-refractive index dielectric layer either via the frustrated total internal reflection (TIR) from a low-refractive index spacer [27] or via a grating coupler, as that occurring in the integrated-optical waveguide technique [28,29]. In both cases, an evanescent field produced by the optical wave has a penetration depth in water of ~100 nm. Therefore, all the above-mentioned techniques are sensitive not only to surface-bound biomolecular interactions but also to changes in the volume RI caused by variations in liquid temperature and composition. However, to distinguish these two different contributions to detect signal, one needs to record at least two optical waves with different characteristics (e.g., with different penetration depths) simultaneously.

It is a PC SW technique that is able to reliably segregate the volume and the surface contributions from analytes in the detected signals. The reason is the possibility to excite a PC SW with a very large penetration depth in liquid and use this wave as a reference of liquid RI. While in standard waveguides (where an RI of the medium under the waveguide film is larger than the RI of the liquid above the film), it is impossible to obtain a large penetration mode depth and obtain the large penetration depth differences for different modes. Therefore, standard waveguides are limited in their ability to reliably segregate volume and surface contributions from an analyte, because both modes have a similar (small) penetration depth in a liquid (≤100 nm).

A large penetration depth in a liquid (>1 μm) can only be obtained using so-called "reverse" waveguides, where RI of the medium under the waveguide film is smaller than the RI of the liquid above the film [30,31]. 1D PC structures are particularly advantageous as they possess the similar feature (in their band gap regions) as substrates in the "reverse" waveguide, despite the fact that a PC structure consists of media with RIs larger than the RI of the liquid (e.g., seven alternative layers of SiO_2 and Ta_2O_5). The large difference among penetration depths of PC SWs in liquids is the main advantage of biosensors based on PC SWs [22]. This attribute allows for (reliable) segregation of the volume and the surface contributions from an analyte and increases sensitivity of molecule detection.

The next important advantage of PC SWs is the absence of light damping in a metal and, therefore, their large propagation length, which may be higher than the SPP propagation length by more than one order of magnitude. Moreover, such a PC-based biosensor may be also used as a very sensitive Abbe refractometer, because a direct measurement of the critical angle immediately gives the RI n_e of the liquid under investigation [32]. This approach allows one unknown target value, n_e, to become model-independent, ultimately increasing the reliability and integrity of results.

12.4 Experimental Materials and Methods

12.4.1 Photonic Crystal Structure

The desirable 1D PC structure may be theoretically deduced, for example, using a previously described impedance approach [33]. In Section 12.5.4, the dispersion relation for SWs on the external interface of the 1D PC is given in impedance terms. The following 1D PC structure was derived from this dispersion relation and was used in experiments as follows: *substrate*/$(LH)^3L'$/*water*, where L is a SiO_2 layer with thickness $d_1 = 183.2$ nm, H is a Ta_2O_5 layer with $d_2 = 111.2$ nm, and L' is a SiO_2 layer with $d_3 = 341.6$ nm. The SiO_2/Ta_2O_5 seven-layer structure (started and finished by SiO_2 layers) is deposited by magnetron sputtering. The prism and the glass plate substrate are made from BK-7 glass. The RIs of the substrate, SiO_2, Ta_2O_5, and water at λ = 658 nm, are $n_0 = 1.514$, $n_1 = n_3 = 1.47$, $n_2 = 2.1$, and $n_e = 1.331$, respectively.

12.4.2 PC SW Biosensor

All measurements presented here were done on the first commercially available PC SW-based biosensor EVA [34]. This PC SW biosensor with an independent registration of the critical angle of TIR from the liquid is outlined in Figure 12.2. In this figure, a sketch of the biosensor and typical signals from the photodiode array are shown. A laser beam from fiber-coupled diode laser (λ = 658 nm) is sent to the sensor surface through a polarization-maintaining fiber cable (to improve the quality of a beam profile).

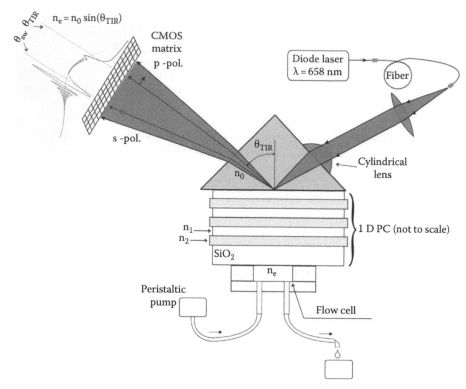

FIGURE 12.2   A sketch of the biosensor based on angle interrogation of a PC SW. The typical reflection profile is shown near the CMOS matrix. Note that this setup also may be used as a critical-angle refractometer, if only θ_{TIR} angle is measured.

The beam is focused by a cylindrical lens so that the excitation angle of one s-polarized PC SW (existing in this 1D PC) structure and TIR angle (in p-polarization) are contained in the convergence angle of the beam.

After reflection from the sensor surface, the reflection profile contains information about the TIR angle (transferred by the p-polarized part of the beam) and about the angle of the PC SW excitation (transferred by the s-polarized part of the beam). Moreover, the sharpness of the reflection near the critical angle and the measurement precision of the liquid RI herein are much higher than those used in standard critical-angle refractometers on uncoated prisms that enhance the RI sensitivity of the biosensor [25,32].

Such types of PC SW sensors also possess 1D spatial selectivity in a direction perpendicular to the plane of the Figure 12.2 (i.e., along the focus line of the cylindrical lens). This fact permits recording of several reactions with an analyte simultaneously if different ligands are deposited on the PC in several linear target bands. In this way, several tests can be performed at once that increases the throughput of the sensor.

12.4.3 Reagents

All biochemicals were purchased from Sigma-Aldrich (Germany) and were used immediately after preparation. 3-Triethoxysilylpropylamine, APTES [molecular weight = 221.37] was used to convert the OH-terminated SiO_2 surface to an NH_2-terminated [35]. Biotinamidohexanoyl-6-aminohexanoic acid N-hydroxysuccinimide ester [biotin-X-X-NHS; molecular weight = 567.7] dissolved in N,N-dimethylformamide [DMF] was used to biotinylate the amino-terminated surface. The streptavidin [molecular weight $M_{str} \sim 60,000$] was deposited on the biotinylated surface. The free biotin [vitamin H; molecular weight $M_b = 244.31$] was used as a test to detect small molecule binding with a streptavidin monolayer. All experiments were carried out in phosphate-buffered saline [PBS; pH = 7.2].

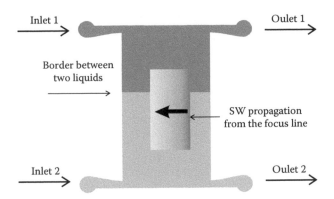

FIGURE 12.3 A flow cell of the biosensor with two inlet tubes and two outlet tubes. Such flow cell design permits movement of the border between two liquids by simply changing the pressure in the inlets.

12.4.4 Sample Preparation

Samples were prepared as follows: first, glass slides were processed by ionized air in a plasma cleaner for 1 min. Next, these precleaned slides (with expected OH bonds on the ultrahydrophilic SiO_2 surface) were immersed in 1% aminosilane solution in 95% acetone/water for 5 min. The glass slides, now with expected NH_2 bonds on the ultrahydrophobic surface, were then dried by argon and desiccated in vacuum for 30 min. To biotinylate the NH_2-terminated surface, the slides were left overnight in solution, where biotin-X-X-NHS was dissolved by DMF to a final concentration of 500 μg/mL. Afterward, the slides with a biotinylated surface were sequentially sonicated and thoroughly rinsed with DMF and PBS to remove any excess of biotin-X-X-NHS.

12.4.5 Flow Cell

The flow cells were made from a glass slide with two or four holes through which glass tubes were fitted to serve as inlet and outlet, respectively. Microstructure of the flow cell was drawn in the glass by glass ablation with a CO_2 laser. The inlet tube was connected to a small tank filled with the solution under investigation. Flow velocity was controlled by a peristaltic pump "Ismatec Reglo Digital." The depth of the flow cell in the glass was determined by the power of the CO_2 laser, by frequency of the laser pulses, and by velocity of the laser beam during the ablation process. A typical depth was from 35 to 50 μm and corresponding flow cell volume was from 3.5 to 5 μL. The dead volume of the flow cell system was ~25 μL.

The construction of the cell with two inlets and two outlets presented in Figure 12.3 permits movement of the border between two liquids by changing the pressure in the inlets. It may be useful if the surface of the PC chip is covered by one target ligand, and several analytes should be tested with this ligand. If the inert reference liquid (e.g., PBS) is sent in one inlet and several analytes under study are sequentially sent to another inlet with sequentially increasing pressure, then all needed information about analytes-ligand interaction may be obtained without chip changing. This is one more way to increase the sensitivity and throughput of the sensor.

12.5 Results

To verify the sensitivity of the biosensor and to compare it with existing label-free methods, we present the unsmoothed experimental data of free biotin binding on the streptavidin monolayer. Initially (see Figure 12.4), we present the buildup of the streptavidin monolayer on the biotinylated surface. Then, the flow cell was rinsed by PBS and biotin was injected into PBS running through the flow cell (see Figure 12.5).

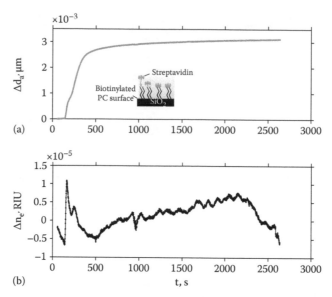

FIGURE 12.4   Immobilization of streptavidin on a biotinylated surface (a) and corresponding changes of RI of the buffer during this injection (b). The measurement time was 1 s per point (no posterior data averaging and smoothing). In the inset, a corresponding process is illustrated.

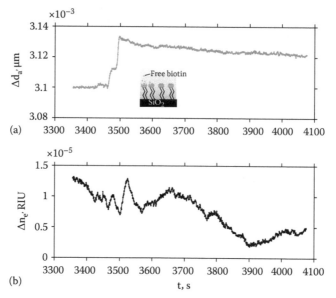

FIGURE 12.5   Adlayer thickness changes during free biotin binding to the streptavidin monolayer (a) and corresponding changes of RI of the buffer during this injection (b). The measurement time was 1 s per point (no posterior data averaging and smoothing). In the color, a corresponding process is illustrated.

12.5.1 Streptavidin Monolayer Deposition

Streptavidin (diluted in PBS to a concentration of $c_{str} = 12\,\mu g/mL$) was run through the flow cell with a volumetric flow rate of $v_{str} = 0.3\,mL/min$. Figure 12.4 illustrates that the increase of the adlayer thickness due to immobilization of streptavidin on a biotinylated surface (a) occurs with kinetics different from those of the RI change of buffer during injection (b). This fact indicates that the volume and surface contributions from an analyte are indeed separated into different registration channels.

12.5.2 Biotin Binding to the Streptavidin Monolayer

The top side of Figure 12.5 presents the adlayer thickness changes observed during free biotin binding to the streptavidin monolayer and the RI changes of the analyte (bottom) during this biotin solution injection. Biotin [in a concentration of $c_{b(low)} = 0.9\,\mu g/mL$] was injected into PBS running through the flow cell with volumetric flow rate of $v_b = 0.4\,mL/min$. It is clear that the biosensor reliably detects the increase in streptavidin monolayer thickness upon free biotin binding.

12.5.3 Data Handling

Data from the CMOS matrix were acquired, processed, and presented using software supplied with the device. Changes of the critical angle position ΔP_0 and the resonance peak positions ΔP_1 on the CMOS matrix were converted to changes of the angle parameters $\Delta\theta_{TIR}$ and $\Delta\theta_{SW}$, respectively.

The RI of the liquid was derived as in that for classical critical-angle Abbe refractometers through the angle of total internal reflection θ_{TIR}. The liquid RI is then given by:

$$n_e = n_0 \sin(\theta_{TIR}), \tag{12.1}$$

where n_0 is the RI of the prism in which the critical angle θ_{TIR} is measured. To derive the changes of the adlayer thickness from the changes of the resonance angle $\Delta\theta_{SW}$ and Δn_e (known from (12.1)), we used the dispersion relation (12.2) derived in [33] and given in the next section.

12.5.4 Dispersion Relation for Surface Waves

The dispersion relation for optical surface waves in 1D PC structure may be obtained through the impedance approach [33]:

$$k_{z(3)}d_3 = \pi M + \arctan\left(\frac{-i\left(Z_{(PC)}^{into} + Z_{(e)}\right)Z_{(3)}}{Z_{(3)}^2 + Z_{(PC)}^{into}Z_{(e)}}\right), \tag{12.2}$$

where
$k_{z(3)} = (2\pi/\lambda)n_3\cos(\theta_3)$
M is a whole number.

This is a general dispersion relation that is valid for both polarizations. If the goal is to obtain the dispersion of the s-polarized optical surface wave, one should use the Z_s impedances in (12.2) and (12.6). Accordingly, to obtain the dispersion of the p-polarized optical surface wave, one must use Z_p impedances in (12.2) and (12.6).

The normal impedance Z, appearing in these equations, is a very useful value if one considers an interaction of light with a plane interface or with a set of plane interfaces (i.e., with a multilayer structure) [36,37]. It is equal to the ratio of the tangential components of the electric field to the magnetic field:

$$Z = \frac{E_{tan}}{H_{tan}}. \tag{12.3}$$

Impedances for the s-polarized wave (in which the electric field vector is orthogonal to the incident plane—TE wave) and for the p-polarized wave (in which the electric field vector is parallel to the incident plane—TM wave) are correspondingly shown:

$$Z_s = \frac{1}{n\cos(\theta)} \quad (\text{for TE wave}) \tag{12.4}$$

$$Z_p = \frac{\cos(\theta)}{n} \quad (\text{for TM wave}). \tag{12.5}$$

The input impedance of the semiinfinite multilayer (i.e., 1D PC) has the next form:

$$Z_{(PC)}^{into} = -\frac{i}{2} \frac{\left(\left(Z_{(2)}^2 - Z_{(1)}^2 \right) \tan(\alpha_2) \tan(\alpha_1) \pm \sqrt{s} \right)}{Z_{(2)} \tan(\alpha_1) + Z_{(1)} \tan(\alpha_2)} \tag{12.6}$$

where $s = -4 Z_{(1)} Z_{(2)} (Z_{(2)} \tan(\alpha_1) + Z_{(1)} \tan(\alpha_2))(Z_{(1)} \tan(\alpha_1) + Z_{(2)} \tan(\alpha_2)) + [(Z_{(2)}^2 - Z_{(1)}^2) \tan(\alpha_1) \tan(\alpha_2)]^2$.

The advantages of the presented dispersion relation (12.2) for PC SWs are its compact structure and its unified form for both polarizations. In addition, a visible physical interpretation of the impedance terms in the current dispersion relation allows it to be easily extended for use with more complicated structures. For example, the addition of an adsorption layer between the layer $_{(3)}$ and the external medium $_{(e)}$ may be taken into account simply by changing the impedance $Z_{(e)}$ by the impedance $Z_{(e)}^{into}$ calculated via the recursion relation (see [33] for details), where impedance of the external medium $Z_{(e)}$ is convoluted with an impedance of the adsorption layer $Z_{(a)}$.

12.6 Discussion

The presented PC SW sensor measures two parameters: the RI of the liquid n_e and the thickness of adsorption layer d_a on the external PC surface.

The sensitivity of a sensor to the liquid RI can be expressed as a product of two terms [26,38]:

$$S_{RI} = \frac{\partial I}{\partial \theta_0} \frac{\partial \theta_0}{\partial n_e} = \frac{\partial R_P}{\partial \theta_0} \frac{\partial \theta_0}{\partial n_e}, \tag{12.7}$$

where the derivative of the detected laser light intensity I as a function of angle θ_0 is taken as the sharpness of the reflection coefficient R_p near the critical angle θ_{TIR} in our case (and for Abbe refractometers).

A high sensitivity of the RI detection in the presented method arises from the fact that the sharpness of the reflection near the critical angle is much higher in our system than both in SPR-based systems and in standard critical-angle refractometers on uncoated prisms. In the standard critical-angle Abbe refractometer $\max(\partial R_s/\partial \theta_0) \cong 130$ [1/rad] for s-polarization and $\max(\partial R_p/\partial \theta_0) \cong 170$ [1/rad] for p-polarization (in the best case scenario). In SPR-based sensors, the maximum sharpness of an SPR curve (near its half-width points) is $\max(\partial R_p/\partial \theta_{SPR}) \sim 15–25$ [1/rad] only (for the gold film in the water at the same wavelength—see, e.g., [39]), while in the presented system $\max(\partial R_p/\partial \theta_0)$ is about 600 [1/rad], that is, 3–5 times greater than in Abbe refractometers without multilayer coating, and more than 20 times greater than in SPR-based sensors.

Such a high value of S_{RI} leads to the decrease of the baseline noise arising from the laser light intensity noise δI because $\delta n_e \sim \delta I/S_{RI}$—see for example [26]. But, unfortunately, other types of δn_e noises, such as a thermodynamic noise of the liquid RI arising from temperature fluctuations, do not depend on the S_{RI} value.

The RI resolution (determined by a RI baseline noise) of the presented PC SW sensor is $\delta n_e \cong 10^{-7}$ RIU/Hz$^{1/2}$. The ultimate RI resolution of the modern SPR sensors is about 2×10^{-7} RIU for the sensors operating in the wavelength range of 750–800 nm (see, e.g., Figure 12.4 from [39]), while the ultimate resolution of the sensors operating near 658 nm (like the presented one) is about 10^{-6} RIU. Therefore, the presented refractometer provides about one order of magnitude improvement in the RI resolution in comparison with SPR sensors.

A baseline noise of adsorption layer thickness measurements in the presented PC SW sensor is $\delta d_a \cong 3 \times 10^{-13}$ m/Hz$^{1/2}$ = 0.3 pm/Hz$^{1/2}$ (corresponding to mass sensitivity $\delta m_a \cong 0.3$ pg/mm^2). The limit of detection is usually defined as three standard deviations from the baseline noise [40].

Such thickness resolution is reached despite the fact that s-polarized PC SW in the presented setup is less localized near the surface compared to SPP waves. The physical reason that compensates the weak localization of PC SWs near the surface and increases sensitivity is the long-range propagation length of PC SWs. The distinguishing feature of this long-range propagation is the fringe pattern near the resonance angle θ_{TIR}. This fringe pattern cannot be theoretically predicted from Fresnel-based calculations

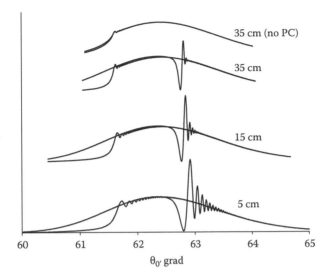

FIGURE 12.6 Angular resonance curves (light gray for s-polarization and dark gray for p-polarization) at different distances from the 1D PC. The fringe pattern near the SW resonance (62.63 grad) is a distinguishing feature of long-range SW propagation. The sharp edge of the curve near the TIR angle (61.54 grad) was specially designed for this 1D PC structure.

of reflection curves since the incident field is approximated as an infinite flat front in these calculations, and the limited propagation of PC SWs is compared to infinity. However, describing the reflection of a focused restricted Gaussian laser beam, the fringe pattern appears when PC SW propagation length becomes much more than a waist of the incident Gaussian beam. One can obtain a qualitative picture of this phenomenon considering destructive interference of two collinear Gaussian beams when one of them (the reflected beam in our case) has much less waist than another (the beam from PC SWs reradiated back to the prism). The waist difference leads to the difference in Rayleigh ranges and to the difference in wavefront curvatures of these beams at some distances. Intersections of constant phase fronts of these beams produce the interference.

The fringe pattern near the resonance dip is illustrated in Figure 12.6 for different distances from the prism with PC. This fringe pattern, observed on the larger-angle side of the resonance dip, is the distinguishing feature of all long-range surface waves, and a similar fringe pattern was observed not only with PC SWs [22] but also with ultra long-range SPPs [21,23]. It may be easily inferred that for very large distances, where wavefront curvatures of both waves (reflected and reradiated from the surface) equalize, the fringe pattern will disappear, and only the resonance dip will be preserved [21]. Indeed, in Figure 12.6, one can see that this fringe pattern is more pronounced on the distance of 5 cm, less pronounced on the distance of 15 cm, and nearly disappears on the distance of 35 cm. This phenomenon is very useful for high precision measurements, because lots of these very sharp fringes are distributed on many CMOS matrix pixels and subpixel registration of a fringe pattern shift becomes possible.

The highest curve in Figure 12.6 is the reflection curve from the prism without multilayer coating (no PC). It is provided to illustrate that the sharpness of the curve near the TIR angle is much higher for our PC structure than the one on the bare surface.

12.7 Conclusions

The development of PC-based biochemical sensors is a fast growing area in recent years. The design flexibility of PC structures permits researchers to devise appropriate sensors for applications in different domains of study. PC sensors can be designed for any optical wavelength and may be used in label-free or label-based detection. The optical surface waves excited on the 1D PC interface are an effective means to guide and concentrate optical waves in the field of interaction between light and sensing material at the external side of the 1D PC.

The exploitation of the 1D PCs as substrates supporting the long-range surface wave propagation permits researchers to

1. Increase the sensitivity of PC SW biosensors to the level $\delta d_a \cong 3 \times 10^{-13}$ m/Hz$^{1/2}$ (that corresponds to mass sensitivity $\delta m_a \cong 0.3$ pg/mm^2),

2. Segregate surface and volume events in biosensing (that may be an important advantage in applications where temperature and composition of the liquid under study vary over a wide range),

3. Enhance the detection of RI variation in the Abbe-like refractometer to the level $n_e \cong 10^{-7}$ RIU/Hz$^{1/2}$,

4. Work with thick target ligands, such as living cells, with thickness up to 1 μm,

5. Obtain 1D spatial selectivity that makes multichannel registration possible and increases throughput of the sensor,

6. Use the same PC chip many times, since a thick final SiO$_2$ layer may be effectively cleaned by some active treatment (e.g., in a plasma cleaner).

REFERENCES

1. Lambeck, P. V. Integrated optical sensors for the chemical domain. *Meas. Sci. Technol.*, 17, R93, 2006.
2. Fan, X. et al. Sensitive optical biosensors for unlabeled targets: A review. *Anal. Chim. Acta*, 620, 8, 2008.
3. Yablonovitch, E. Photonic band-gap structures. *J. Opt. Soc. Am. B*, 10, 283, 1993.
4. Fan, S. and Joannopoulos, J. D. Analysis of guided resonances in photonic crystal slabs. *Phys. Rev. B*, 65, 235112, 2002.
5. Chan, L. L. et al. A label-free photonic crystal biosensor imaging method for detection of cancer cell cytotoxicity and proliferation. *Apoptosis*, 12, 1061, 2007.
6. Chan, L. L. et al. Label-free imaging of cancer cells using photonic crystal biosensors and application to cytotoxicity screening of a natural compound library. *Sens. Actuator B-Chem.*, 132, 418, 2008.
7. Chow, E. et al. Ultracompact biochemical sensor built with two-dimensional photonic crystal microcavity. *Opt. Lett.*, 29, 1093, 2004.
8. Nazirizadeh, Y. et al. Low-cost label-free biosensors using photonic crystals embedded between crossed polarizers. *Opt. Express*, 18, 19120, 2010.
9. Cunningham, B. et al. A plastic colorimetric resonant optical biosensor for multiparallel detection of label-free biochemical interactions. *Sens. Actuator B-Chem.*, 85, 219, 2002.
10. www.srubiosystems.com
11. King, B. H. et al. Internally referenced ammonia sensor based on an electrochemically prepared porous SiO$_2$ photonic crystal. *Adv. Mater.*, 19, 4044, 2007.
12. Kossel, D. Analogies between thin-film optics and electron band theory of solids. *J. Opt. Soc. Am.*, 56, 1434, 1966.
13. Descrovi, E. et al. Guided Bloch surface waves on ultrathin polymeric ridges. *Nano Lett.*, 10, 2087, 2010.
14. Goto, T. et al. Tailoring surfaces of one-dimensional magnetophotonic crystals: Optical Tamm state and Faraday rotation. *Phys. Rev. B*, 79, 125103, 2009.
15. Yeh, P., Yariv, A., and Hong, C.-S. Electromagnetic propagation in periodic stratified media. I. General theory. *J. Opt. Soc. Am.*, 67, 423, 1977.
16. Arnaud, J. A. and Saleh, A. A. M. Guidance of surface waves by multilayer coatings. *Appl. Opt.*, 13, 2343, 1974.
17. Yeh, P., Yariv, A., and Cho, A. Y. Optical surface waves in periodic layered media. *Appl. Phys. Lett.*, 32, 104, 1978.
18. Robertson, W. M. and May, M. S. Surface electromagnetic waves on one-dimensional photonic band gap arrays. *Appl. Phys. Lett.*, 74, 1800, 1999.
19. Robertson, W. M. Experimental measurement of the effect of termination on surface electromagnetic waves in one-dimensional photonic band gap arrays. *J. Lightwave Tech.*, 17, 2013, 1999.

20. Shinn, A. and Robertson, W. Surface plasmon-like sensor based on surface electromagnetic waves in a photonic band-gap material. *Sens. Actuator B-Chem.*, 105, 360, 2005.
21. Konopsky, V. N. and Alieva, E. V. Long-range propagation of plasmon polaritons in a thin metal film on a one-dimensional photonic crystal surface. *Phys. Rev. Lett.*, 97, 253904, 2006.
22. Konopsky, V. N. and Alieva, E. V. Photonic crystal surface waves for optical biosensors. *Anal. Chem.*, 79, 4729, 2007.
23. Konopsky, V. N. and Alieva, E. V. Long-range plasmons in lossy metal films on photonic crystal surfaces. *Opt. Lett.*, 34, 479, 2009.
24. Guo, Y. et al. Real-time biomolecular binding detection using a sensitive photonic crystal biosensor. *Anal. Chem.*, 82, 5211, 2010.
25. Konopsky, V. N. and Alieva, E. V. A biosensor based on photonic crystal surface waves with an independent registration of the liquid refractive index. *Biosens. Bioelectron.*, 25, 1212, 2010.
26. Homola, J. Surface plasmon resonance sensors for detection of chemical and biological species. *Chem. Rev.*, 108, 462, 2008.
27. Cush, R. et al. The resonant mirror—A novel optical biosensor for direct sensing of biomolecular interactions. I. Principle of operation and associated instrumentation. *Biosens. Bioelectron.*, 8, 347, 1993.
28. Tiefenthaler, K. and Lukosz, W. Sensitivity of grating couplers as integrated-optical chemical sensors. *J. Opt. Soc. Am. B-Opt. Phys.*, 6, 209, 1989.
29. Lukosz, W. Integrated optical chemical and direct biochemical sensors. *Sens. Actuator B-Chem.*, 29, 37, 1995.
30. Horvath, R., Lindvold, L., and Larsen, N. Reverse-symmetry waveguides: theory and fabrication. *Appl. Phys. B-Lasers Opt.*, 74, 383, 2002.
31. Horvath, R., Pedersen, H., and Larsen, N. Demonstration of reverse symmetry waveguide sensing in aqueous solutions. *Appl. Phys. Lett.*, 81, 2166, 2002.
32. Konopsky, V. N. and Alieva, E. V. Critical-angle refractometer enhanced by periodic multilayer coating. *Sens. Actuator B-Chem.*, 150, 794, 2010.
33. Konopsky, V. N. Plasmon-polariton waves in nanofilms on one-dimensional photonic crystal surfaces. *New J. Phys.*, 12, 093006, 2010.
34. www.pcbiosensors.com
35. Li, J. et al. Assembly method fabricating linkers for covalently bonding DNA on glass surface. *Sensors*, 1, 53, 2001.
36. Brekhovskikh, L. *Waves in Layered Media*. New York: Academic, 1980.
37. Delano, E. and Pegis, R. Methods of synthesis for dielectric multilayer filters. In E. Wolf (Ed.), *Progress in Optics*, vol. VII, Chapter 2, pp. 67–137 (see pp. 77, 130). Amsterdam, the Netherlands: North-Holland, 1969.
38. Berini, P. Bulk and surface sensitivities of surface plasmon waveguides. *New J. Phys.*, 10, 105010, 2008.
39. Piliarik, M. and Homola, J. Surface plasmon resonance (SPR) sensors: Approaching their limits? *Opt. Express*, 17, 16505, 2009.
40. Analytical Methods Committee. Analytical methods committee. *Analyst*, 112, 199, 1987.

13

Nano-Materials and Nano-Structures for Chemical and Biological Optical Sensors

Ignacio Del Villar, Javier Goicoechea, Carlos R. Zamarreño, and Jesus M. Corres

CONTENTS

13.1 Introduction

Nanotechnology has had an enormous impact in lots of scientific fields and applications, but most of the time we do not have a precise idea about what it is exactly. Sometimes we only imagine one of the simplest approaches: bare nanoparticles. Nevertheless, scientists have worked very hard for many years and the initial idea of simple nanoparticles became obsolete very quickly.

Simple nanoparticles have been improved and engineered and have now become complex and sophisticated pieces of matter, often called nanostructures, allowing the researchers to reach new applications, or improve significantly the existing ones.

The field of nanotechnology applied to chemical and biological opticals is very extensive and there are a lot of scientific disciplines working together. This chapter will demonstrate the main scientific contributions reported in the last years and give an overview of this topic.

Depending on the material, or combination of materials, of each type of nanoparticle or nanostructure, different physical phenomena will be observed, and they will be suitable for specific sensing applications. Consequently, this chapter will present the optical sensing applications of such nanoparticles and nanostructures, grouping them into material families: metallic nanoparticles, semiconductor quantum dots, magnetic nanoparticles, lanthanides, nanotubes, and nanocoatings.

13.2 Nanoparticles

Research developed on the field of nanoparticles has demonstrated how the tiniest pieces of matter have great importance in sensing applications, thanks to numerous phenomena such as localized surface plasmon resonance (LSPR), surface enhanced Raman spectroscopy (SERS) or quantum confinement. Structurally they differ from bulk materials only in their small size, but nanoparticles show dramatically different macroscopic properties due to different phenomena related with their size and shape.

This specific kind of nanomaterials is one of the simplest approaches to nano-structured materials. Nevertheless, this makes it very attractive from the application point of view, since in most of the cases it is a cost-effective solution to many applications where more sophisticated structures are not applicable.

Nanoparticle is a very general term, and involves small particles of almost any material and with different structures, from the simplest simple nanocrystal to complex core-multishell architectures, or chemically surface-modified structures. For the sake of clarity, in the following paragraphs the field of nanoparticles has been structured into different groups in order to sum up the most relevant scientific contributions regarding the underlying phenomena that make each nanoparticle type useful for specific biosensing applications.

13.2.1 Metallic Nanoparticles

Metals have been widely used and studied by the scientific community since ancient times. Nowadays one cannot think of almost any engineering or scientific application without metals. Traditionally they have only been considered from the macroscopic point of view and they have proved to be exceptional materials thanks to their mechanical properties (with application in building, engineering, transport, automotive, aeronautics, etc.) and their electrical and thermal conductivities. Particularly, the electrical properties of some metals have been studied, not only from the electronic-conduction point of view, but also from their interaction with electromagnetic waves. In this last aspect significant progress has been made during the twentieth century, as the undulatory physics was understood and developed.

Moreover, as the fabrication techniques allowed the fabrication of metallic thin-films, new optical phenomena were observed. This is the case of *surface plasmon resonance* (SPR), where a very specific interaction between a metallic thin-film with electromagnetic radiation occurs. This phenomenon consists in a resonant coupling between an incident electromagnetic wave and the surface of a metallic-thin film, where some of the energy of the light is transferred to the surface free electrons of the metal, causing a collective oscillation, or electrical charge wave. Only the resonant modes of the incident light are altered, and resonant condition is very sensitive to changes that involve the optical properties of the metallic thin-film and its dielectric environment. This phenomenon is typically reproduced using the following optical setup, known as Kretschmann configuration, as shown in Figure 13.1.

The surface plasmon wave (SPW) is a TM-polarized charge-density wave; this means that the magnetic vector of the incident electromagnetic wave is perpendicular to the propagation direction of the SPW and parallel to the plane of the metal–dielectric interface. The propagation constant of the SPW

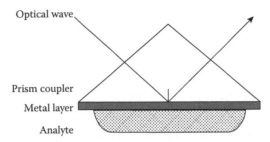

FIGURE 13.1 Typical Kretschmann configuration for the observation of the Surface Plasmon Resonance phenomenon. (Reprinted from *Sens. Actuat. B Chem.*, 54, Homola, J., Sinclair, S.Y., and Gauglitz, G., Surface plasmon resonance sensors: Review, 3–15, Copyright 1999, with permission from Elsevier.)

propagating at the interface between a semi-infinite dielectric and a thin-film metal is given by the following expression (Homola et al. 1999, 3–15):

$$\beta = k\sqrt{\frac{\varepsilon_m n_s^2}{\varepsilon_m + n_s^2}} \tag{13.1}$$

where
  k denotes the free space wave number
  ε_m the dielectric constant of the metal ($\varepsilon_m = \varepsilon_{mr} + i\varepsilon_{mi}$)
  n_s the refractive index of the dielectric (Homola et al. 1999, 3–15)

From Equation 13.1, the SPW can be supported by the structure, provided that $\varepsilon_{mr} < -n_s^2$. There are several metals that fulfill this condition (Ordal et al. 1983, 1099–1119), where gold and silver are the most used materials because their SPR coupling band is located in the visible region of the spectrum. This optical phenomenon have found many applications in the sensor field since the SPR coupling condition is extremely sensitive to variations in the properties of the surrounding medium and the metal layer.

The physical phenomenon changes slightly when the metallic material is distributed into nanoparticles rather than in a continuous thin-film. In this case, the charge-density oscillation does not affect only the nanoparticle surface, but the whole nanoparticle electronic distribution is altered (see Figure 13.2). This particularization of the SPR phenomenon for the case of the nanoparticles is called localized surface plasmon resonance (LSPR).

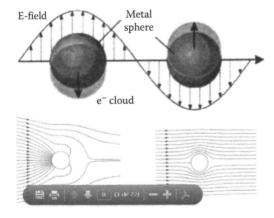

FIGURE 13.2 Schematic view of the LPSR phenomenon. (From Hutter, E. and Fendler, J.H.: Exploitation of localized surface plasmon resonance. *Adv. Mater.* 2004, 16, 1685–16706. Copyright Wiley-VCH Verlag GmbH & Co. KGaA. Reprinted with permission.)

There are many sensing applications for the thin-film SPR setups (Mitchell 2010, 7323–7346; Ho et al. 2012); it is even possible to find commercially available SPR based sensor equipment (ProteOn™ series from Bio-Rad Inc.; Cole Parmer Inc.). Nevertheless, this work is devoted to nanoparticles. Consequently, the focus is on the LSPR-based sensors rather than SPR devices.

13.2.1.1 Synthesis of Metallic Nanoparticles

The synthesis of metallic nanoparticles capable of hosting LSPR is relatively easy and cost effective. Although there are several synthesis techniques available, such as vacuum deposition using templates (Yonzon et al. 2005, 438–448), electron-beam lithography (Marqués-Hueso et al. 2010, 2825–2830), laser ablation (Mafuné et al. 2001, 5114–5120), or electrodeposition (El-Deab et al. 2006, 1792–1798), most metallic nanoparticles are synthesized mostly from metallic-salt solutions with further reduction and stabilization. Since gold and silver are the most used materials for LPSR optical sensing, their synthesis routes have been thoroughly studied and they are well known by the scientific community. Depending on some parameters such as the concentration of the chemicals involved, the presence of surfactants, and the kinetics of the reduction reaction, the characteristics of the final nanoparticles change dramatically. The exact conditions of the reducing reaction induce dramatic changes in the crystallization of the metallic nanoparticles.

In the very first stages of the synthesis reaction, some nucleation seeds are created. Every seed must be protected by a surfactant from aggregation with other surrounding incipient nanoparticles, otherwise uncontrolled agglomeration of metallic particles will lead to massive precipitation. The following stage is the growing of the nanocrystals, where metal ions are reduced directly onto the surface of the seeds. As previously commented, the very precise kinetics of the nucleation and growth steps, and the nature and amount of capping agent, some crystallographic growth directions may be inhibited and other may be improved, yielding an anisotropic growth of the nanocrystal.

The main differences among the synthesis routes are the exact conditions of the seeding and growth stages controlled by the surfactant and concentration of the reactives (Liz-Marzán 2004, 26–31). Figure 13.3 shows different TEM images of gold spherical nanoparticles, nanorods, and silver truncated triangles. As previously introduced, these nanoparticles show a special interaction with the incident light thanks to the LSPR phenomenon, which leads to a very intense coloring of the colloidal dispersions being observed. This optical appearance depends dramatically on the size, shape, and aggregation state of the suspended metallic nanoparticles.

Gold nanoparticle synthesis routes are generally based on an aqueous precursor solution of $HAuCl_4$, and one of the most used surfactants for the synthesis of gold nanoparticles and nanorods is CTAB (hexadecyl cetyltrimethylammonium bromide). Some researchers have demonstrated how a very dilute impurity in the CTAB surfactant can greatly affect nanorod formation. Any imperfection modifies the growth of the rest of the particle, altering its final shape (Smith and Korgel 2008, 644–649). For the case of silver nanoparticles the most used precursor salt is silver nitrate, and most of the authors use poly(vinyl pyrrolidone) (PVP) as capping agent. Some wet-synthesis routes include other metallic ions during the synthesis process resulting in different shape formation, or even bi-layer structures such as the Ag-capped gold nanorods reported in Pérez-Juste et al. (2005, 1870–1901). Figure 13.4 shows the visual aspect and the UV-VIS spectroscopic characterization of several Au nanorod dispersions with different amount of silver present during the synthesis of the nanocrystals. The Ag/Au molar ratio affects dramatically to the size and shape of the final nanoparticles, and this has a direct consequence on the LSPR absorption bands, and also in the visual aspect (color) of the dispersions.

Other synthesis mechanisms have been reported rather than the direct crystallization in the presence of a capping agent. Several authors have proposed an in situ reduction of the metallic ions trapped inside templates or restricted into certain volumes of nanoreactors, such as the pores of a sol-gel (Rivero et al. 2011, 1–7). In such cases the nanoparticles are directly synthesized within the structure of a supporting medium that prevents the aggregation of particles. Other approaches use the charged groups of polyelectrolytes in Layer-by Layer films as nanoreactors for in situ synthesis of silver nanoparticles. Rubner and

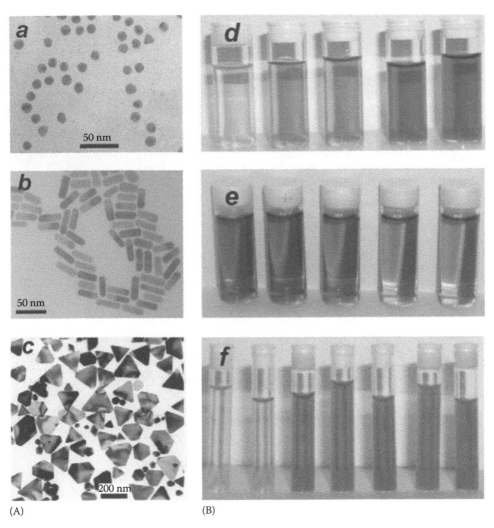

FIGURE 13.3 **(See color insert.)** (A): Transmission electron micrographs of Au nanospheres and nanorods (a, b) and Ag nanoprisms (c, mostly truncated triangles) formed using citrate reduction, seeded growth, and DMF reduction, respectively. (B): Photographs of colloidal dispersions of AuAg alloy nanoparticles with increasing Au concentration (d), Au nanorods of increasing aspect ratio (e), and Ag nanoprisms with increasing lateral size (f). (Reprinted from *Mater. Today*, 7, Liz-Marzán, L.M., Nanometals: Formation and color, 26–31, Copyright 2004, with permission from Elsevier.)

co-workers (Wang et al. 2002b, 3370–3375) demonstrated how the availability of charged functional groups within the LbL film determined strongly the size, shape, and agglomeration of the synthesized Ag NPs (see Figure 13.5).

Apart from the above mentioned colloidal synthesis routes there are other lithographic techniques that allow the creation of metallic nanoparticles with a very controlled spatial distribution, yielding in very repetitive patterns of metallic nanoparticles. One example is the nanosphere lithographic (NSL) technique (Yonzon et al. 2005, 438–448). The NSL process starts by building a self-assembled monolayer of microspheres with a high package ratio. After that, a gold thin-film is evaporated over the system, and the microspheres act as a mask. This intermediate step is called metal film over nanospheres (MFON), and it is very useful in other applications as will be discussed later. If the nanosphere mask is removed, the metallic film remains onto the substrate only in the interstitial regions of the sphere-mask. The result is a well ordered array of gold triangular nanoparticles, as shown in Figure 13.6.

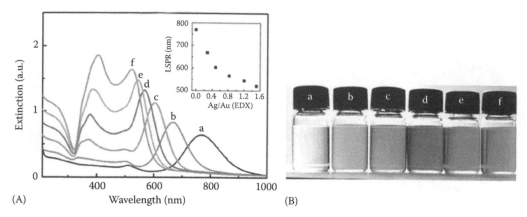

FIGURE 13.4 (A) UV-vis extinction spectra of (a) Au NRs and (b–f) the Au@Ag core/shell nanocrystals with Ag/Au molar ratios of 0.28, 0.49, 0.83, 1.20, and 1.51. The inset in (A) shows the relationship between the longitudinal SPR position and the Ag/Au molar ratio. (B) Photographs of nanocrystal dispersions corresponding to the curves in (A). (Reprinted with permission from Xiang, Y., Wu, X., Liu, D., Li, Z., Chu, W., Feng, L., Zhang, K., Zhou, W., and Xie, S., Gold nanorod-seeded growth of silver nanostructures: From homogeneous coating to anisotropic coating, *Langmuir*, 24(7), 3465–3470. Copyright 2008 American Chemical Society.)

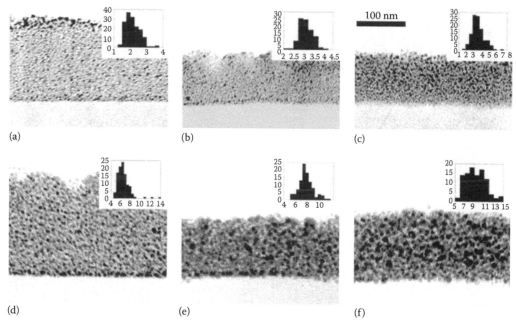

FIGURE 13.5 Cross-sectional TEM images of $(PAHx/PAAx)_{20.5}$ + Ag with multilayers assembled at polyelectrolyte solution pHs of x) (a) 4.5, (b) 3.5, and (c) 2.5. In (a–c) the LbL coatings were dipped once in a silver nitrate solution and further reduced. (d–f) Show the same pH sequence (d) 4.5, (e) 3.5, and (f) 2.5 but this time the LbL films had 5 load/reduction cycles. Nanoparticle diameter histograms (in nanometers) are shown in insets. (Reprinted with permission from Wang, T.C., Rubner, M.F., and Cohen, R.E., Polyelectrolyte multilayer nanoreactors for preparing silver nanoparticle composites: Controlling metal concentration and nanoparticle size, *Langmuir*, 18(8), 3370–3375. Copyright 2002 American Chemical Society.)

13.2.1.2 LSPR-Based Sensors

There are many applications for LSPR. This is possible because the resonant coupling of the light to the collective electronic oscillation strongly depends on the size, shape, aggregation state of the nanoparticles, or variation of their external refractive index. It is important to note that the LSPR

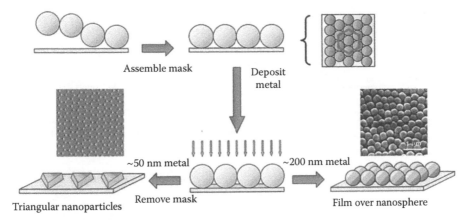

FIGURE 13.6 Scheme of the NSL and MFON nanoparticle fabrication techniques. Firstly a nanosphere mask layer is assembled onto a substrate. After that a metal (typically gold or silver) is evaporated over that mask. If a thin layer of metal is evaporated (bottom left) the mask can be removed leaving a regular triangle shaped nanoparticles pattern over the substrate (NSL). If a thicker layer of metal is evaporated (bottom right) a MFON structure is created. This MFON is very useful in SERS applications as discussed in Section 13.2.1.3. (Reprinted from *Talanta*, 67, Yonzon, C.R., Stuart, D.A., Zhang, X., McFarland, A.D., Haynes, C.L., and Van Duyne, R.P., Towards advanced chemical and biological nanosensors—An overview, 438–448, Copyright 2005, with permission from Elsevier.)

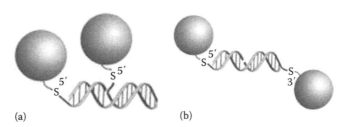

FIGURE 13.7 (a) Head to head alignment of gold nanoparticle probes. (b) Head to tail alignment of the gold nanoparticle probes. The aggregation state of the overall nanoparticle system depends dramatically on the DNA hybridization sequence. (Reprinted with permission from Storhoff, J.J., Elghanian, R., Mucic, R.C., Mirkin, C.A., and Letsinger, R.L., One-pot colorimetric differentiation of polynucleotides with single base imperfections using gold nanoparticle probes, *J. Am. Chem. Soc.*, 120(9), 1959–1964. Copyright 1998 American Chemical Society.)

resonances are extremely sensitive to small changes in the environment, and can produce a measurable signal with interaction of only a few target molecules, if the systems are properly designed and fabricated.

One of the most direct methods to detect the binding of a couple of desired biomolecules is taking advantage of the dramatic change on LSPR resonances in a metallic nanoparticle dispersion when the state of aggregation of the NPs is modified. For example it has been reported to be a simple and effective detector of DNA strand interactions (Alivisatos et al. 1996, 609–611; Mirkin et al. 1996, 607–609; Elghanian et al. 1997, 1078–1081; Storhoff et al. 1998, 1959–1964; 2000, 4640–4650). In these works two gold nanoparticle-colloids are bound to two different non-complementary single strand-DNA chains. When a third oligonucleotide is added to the mixture, the gold colloids are immobilized one close to the other during the hybridization process, forming an aggregate and altering the plasmonic resonant condition (see Figure 13.7). Such change in the aggregation causes a dramatic change of color of the system, perfectly visible to the naked eye, as shown in Figure 13.8. Moreover, the color returns to the initial condition when dehybridization of the DNA oligonucleotides is carried out. Since a small change in the aggregation pattern of the metallic nanoparticles causes a very strong change of the optical appearance of the system, this technique allows differentiation between similar DNA strands with only a difference of one or two nucleotides.

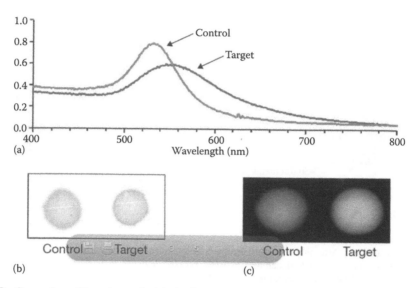

(a)

(b)　　　　　　　　　　　　　　　　　　(c)

FIGURE 13.8 Comparison of detection methodologies for monitoring DNA-AuNP binding to target DNA. (a) UV-visible spectrum of target and control samples. (b) visual aspect of 1-μL spot of each sample. (c) Two spots deposited over a glass optical waveguide. The observed light corresponds to the evanescent light–induced scatter from each sample (1 μL) spotted onto a side-illuminated glass slide. (Reprinted with permission from Macmillan Publishers Ltd., *Nat. Biotechnol.*, Storhoff, J.J., Lucas, A.D., Garimella, V., Bao, Y.P., and Müller, U.R., Homogeneous detection of unamplified genomic DNA sequences based on colorimetric scatter of gold nanoparticle probes, 22(7), 883–887, Copyright 2004.)

Using this approach it is possible to measure the spectral distribution of the scattered light from the nanoparticle aggregates when they are back-illuminated using white light. As this scattered light changes dramatically depending on the aggregation of the nanoparticles, it is possible lower the detection limit of this setup to the zeptomolar range (Storhoff et al. 2004, 883–887).

Using this aggregation concept, multiple biosensors have been reported. For example, metallic nanoparticles can be functionalized with catalytically active DNA molecules, designed to show a high specificity toward Pb^{2+}, Cu^{2+}, Zn^{2+}, or Hg^{2+} (Zhong et al. 2009, 5022–5027). The same concept has been carried out with antibody-antigen specific interactions (Liao and Hafner 2005, 4636–4641) instead of DNA hybridizations to develop aggregation based carbohydrate sensors (Hone et al. 2003, 7141–7144), cysteine sensors, and glutathionic acid sensors (Sudeep et al. 2005, 6516–6517).

Apart from the aggregation based LSPR biosensors, there is another LSPR sensor family based in a different working principle. In this case, the variation of the plasmonic resonance condition is caused by changes on the refractive index of the medium surrounding the metallic nanoparticles. The principle of operation of these sensors is shown schematically in Figure 13.9.

This configuration provides lower sensitivity than the aggregation approaches, but it offers several advantages. This approach is usually open to miniaturation and multiplexing, and some laboratories have combined miniaturized chips with microfluidics to obtain lab-on-a-chip LSPR devices as it is shown in

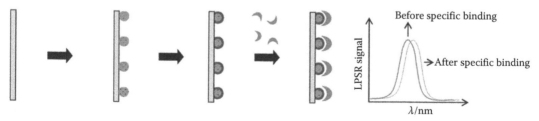

FIGURE 13.9 Schematic representation of the preparation and response of LSPR biosensors based on refractive index changes. (Reprinted from *Nano Today*, 4, Sepúlveda, B., Angelomé, P.C., Lechuga, L.M., and Liz-Marzán, L.M., LSPR-based nanobiosensors, 244–251, Copyright 2009, with permission from Elsevier.)

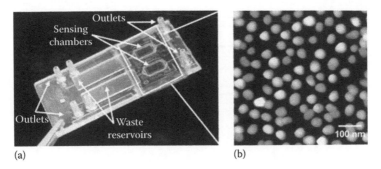

FIGURE 13.10 (a) Photograph of a microfluidic integrated LSPR biosensor. (b) AFM image of the gold nanoparticle film on a quartz substrate. (Reprinted from *Microelectron. Eng.*, 86, Huang, C., Bonroy, K., Reekman, G., Verstreken, K., Lagae, L., and Borghs, G., An on-chip localized surface plasmon resonance-based biosensor for label-free monitoring of antigen-antibody reaction, 2437–2441, Copyright 2009, with permission from Elsevier.)

Figure 13.10 (Huang et al. 2009, 2437–2441). In addition to this, the configuration is very similar to thin-film SPR sensors, which permits most of the applications (biochemical functionalizations, etc.) used in the thin-film sensors to be directly applicable to the LSPR setup. For example, biotinylated gold nanoparticles were used to sense streptavidin, and a detection limit of 0.83 nM was achieved (Nath and Chilkoti 2002, 504–509). Other examples like the detection of Human Serum Albumin using its specific antibody to complete the selective binding reaction gave detection limits up to 10 nM (Fujiwara et al. 2006, 639–644).

The sensitivity of the LSPR to variations in the external refractive index also depends very strongly on the size and shape of the metallic nanoparticles. In Yonzon et al. (2005, 438–448), it is reported that the thickness of metallic nanoislands fabricated by NSL affects dramatically to the sensitivity of the device. NSL nanoparticles of a thickness of 15 nm have three times more dynamic range than 50 nm thickness nanoparticles, as it is shown in Figure 13.11.

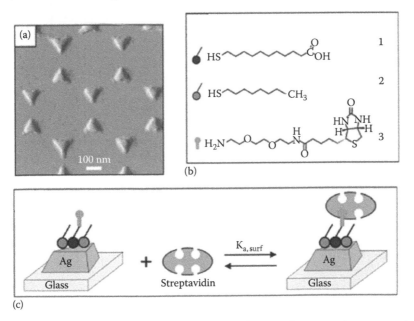

FIGURE 13.11 NSL silver nanoparticles templated using 200 nm diameter polystyrene spheres as template. (a) Shows an SEM image of the Ag NP structure. (b) Shows several thiolated bifunctional molecules used to functionalize the NPs. (c) Scheme of the selective adsorption of streptavidin onto the biotynilated NPs. (Reprinted with permission from Haes, A.J. and Van Duyne, R.P., A nanoscale optical biosensor: Sensitivity and selectivity of an approach based on the localized surface plasmon resonance spectroscopy of triangular silver nanoparticles, *J. Am. Chem. Soc.*, 124(35), 10596–10604. Copyright 2002 American Chemical Society.)

(continued)

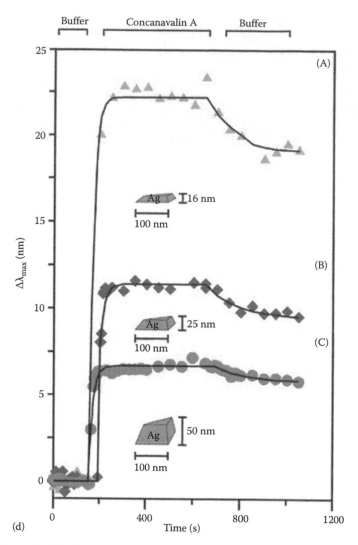

(d)

FIGURE 13.11 (continued) NSL silver nanoparticles templated using 200 nm diameter polystyrene spheres as template. (d) Difference of the response of different NSL structures. The thickness of the Ag nanoislands has a dramatic impact on the sensitivity of the system. (Reprinted from *Talanta*, 67, Yonzon, C.R., Stuart, D.A., Zhang, X., McFarland, A.D., Haynes, C.L., and Van Duyne, R.P., Towards advanced chemical and biological nanosensors—An overview, 438–448, Copyright 2005, with permission from Elsevier.)

The geometry of nanoparticles synthesized using other techniques also affects the LSPR sensitivity of the system to external index variations, as it is shown in Figure 13.12, an effect that is similar to that observed when varying the thickness of metallic NSL nanoparticles.

13.2.1.3 SERS Spectroscopy

One of the most valuable characterization and analysis techniques in chemistry and biology applications is vibrational spectroscopy, in all their forms. This technique provides information about the structure, morphology, and functional groups of the analyzed molecules, since it is based on the excitation of the natural vibrational frequencies of the molecules. Therefore, it is a very useful tool for material identification.

One of the most interesting vibrational techniques is Raman spectroscopy, because of its inherent ability to distinguish between molecules with great structural similarity, such as the isomers glucose and fructose (Yonzon et al. 2004, 78–85). Nevertheless, it has also some drawbacks, such as the high

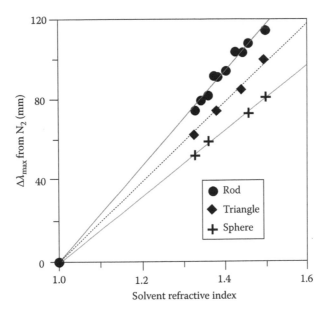

FIGURE 13.12 Shape effect with increasing solvent refractive index in Ag nanoparticles. (Reprinted from *Talanta*, 67, Yonzon, C.R., Stuart, D.A., Zhang, X., McFarland, A.D., Haynes, C.L., and Van Duyne, R.P., Towards advanced chemical and biological nanosensors—An overview, 438–448, Copyright 2005, with permission from Elsevier.)

excitation laser power required, and long acquisition times. This is due to the small normal Raman scattering cross-section of most of the molecules. Nonetheless, there are some special physical conditions in the surroundings of metallic nanoparticles when the LSPR coupling occurs. The resonant oscillation of the electrons of the nanoparticles induce a strong electric field near the surface of the nanoparticle. Consequently, the molecules nearby the surface experiment an enhancement in their Raman scattering cross-section, making more efficient their excitation. Enhancements up to eight orders of magnitude in the Raman scattering emission are typically observed from the molecules surrounding the metallic nanoparticles (Haynes and Van Duyne 2003, 7426–7433), reaching values of 14–15 orders of magnitude of enhancement in some special cases (Kneipp et al. 1997, 1667–1670; Nie and Emory 1997, 1102–1106). Another advantage of SERS is that measurements can take place in aqueous solution, its sensitivity reaches to trace level detection limits, and simultaneous multi-component detection can be achieved. That is why SERS is specially promising in fields such as the pharmaceutical industry, biological applications such as bacteria detection, and the identification of traces and other molecular species.

The very first approaches used highly rough metallic substrates obtained by several oxidation-reduction cycles of the surface of the metal. The nanoscale roughness of the surface of the metallic layer was enough to show a significant enhancement of the Raman scattering spectra. However, such metallic substrates were not optimized at all because the distribution of the electric field enhancement was completely randomly distributed throughout the surface. This made almost impossible to obtain the same SERS spectrum intensity from the same sample, just by using two different metallic substrates. This issue seriously difficult the utilization of SERS for quantitative determination of the chemical species.

In the last years a lot of effort has been devoted to study and optimize the size, shape and spatial distribution of the metallic nanoparticles over the substrate. As an example, Figure 13.13 shows the effect of the electric field enhancement in the gaps between metallic nanoparticles as the distance between them is varied, consequently the Raman scattering is significantly enhanced in these regions when the right configuration is used.

One of the most robust and used substrates for SERS is the so called metal film over nanosphere (MFON). This approach is one intermediate step of the already discussed Nanosphere lithography (NSL) technique for creating LSPR sensor arrays, as it was previously introduced (see Figure 13.6). The nanospheres are placed over the substrate and afterwards coated with a metallic layer by evaporation as it can be seen in Figure 13.14. This yields an artificial and regular nanoparticle distribution ideal for SERS applications.

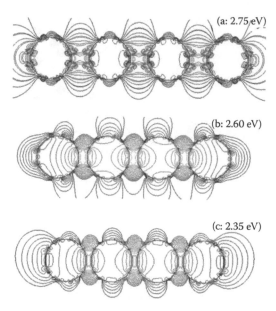

FIGURE 13.13 Electric field lines in a 2D cross-section of a group of four spherical metallic nanoparticles. As the distance between NPs is varied the electrical field enhancement grows in the gaps: (a) 2.75 eV; (b) 2.60 eV; (c) 2.35 eV. (Reprinted with permission from Sweatlock, L.A., Maier, S.A., Atwater, H.A., Penninkhof, J.J., and Polman, A., Highly confined electromagnetic fields in arrays of strongly coupled ag nanoparticles, *Phys. Rev. B Condens. Matter Mater. Phys.*, 71(23), 1–7. Copyright 2005 American Physical Society.)

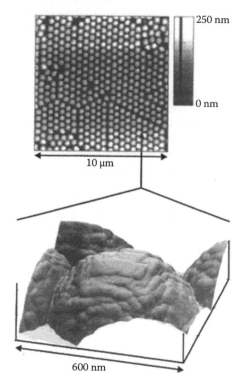

FIGURE 13.14 MFON structure used as substrate for SERS applications. (Reprinted with permission from Dick, L.A., McFarland, A.D., Haynes, C.L., and Van Duyne, R.P., Metal film over nanosphere (MFON) electrodes for surface-enhanced raman spectroscopy (SERS): Improvements in surface nanostructure stability and suppression of irreversible loss, *J. Phys. Chem. B*, 106(4), 853–860. Copyright 2002 American Chemical Society.)

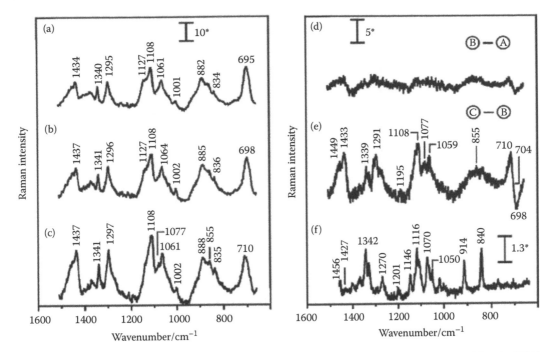

FIGURE 13.15 Three SERS spectra showing detection of glucose in the presence of Bovine Serum Albumin (BSA). (a) Shows the SERS spectrum of the EG3 functinoalization SAM. (b) Shows the spectrum after 1mg/mL of BSA has been injected into de flow cell. (c) Shows the SERS spectrum when a 100mM solution of glucose is injected in the measuring cell. (d) Shows the difference in the SERS spectra from (a) and from (b), where no signal was detected. (e) Represents the difference between (c and b), and shows the SERS spectrum from the glucose molecules adsorbed onto de AgFON substrate. (f) Is the normal Raman spectrum from crystalline glucose, for comparison. (From Haynes, C.L., Yonzon, C.R., Zhang, X., and Van Duyne, R.P.: Surface-enhanced raman sensors: Early history and the development of sensors for quantitative biowarfare agent and glucose detection. *J. Raman Spectrosc.* 2005. 36. 471–484. Copyright Wiley-VCH Verlag GmbH & Co. KGaA. Reprinted with permission.)

SERS has been successfully used for measurement and identification of a biomolecules, toxines, or living organisms.

In the case of the glucose SERS sensors it is important to know that glucose molecules do not have any particular affinity for the silver or gold surfaces. However, in Haynes et al. (2005, 471–484) Haynes and coworkers demonstrate how glucose molecules can be accumulated very close to Ag-based MFON surface using an adequate functionalization. In this case the authors chose a similar Self-Assembled Monolayer (SAM) in a similar way to that used in high-performance liquid chromatography (HPLC) (Blanco Gomis et al. 2001, 173–180). The functionalized AgFON substrate showed an increased chemical stability and furthermore allowed the glucose adsorption near the silver surface, enabling the SERS phenomenon. Figure 13.15 shows the SERS spectra for tri(ethylene glycol) (EG3-OME)-functionalized AgFON substrate.

Other authors report indirect detection of anthrax bacillus using SERS. Another AgFON substrate was used to detect dipicolinate acid (DPA) a chemical agent present in the protective shell of the *bacillus subtilis* spores. The *bacillus subtilis* is a harmless simulant for the *bacillus anthracis*, the pathogen of the Anthrax acute disease. Figure 13.16 shows the characteristic SERS spectrum of a 2.1 × 10⁻¹⁴ M *bacillus subtilis* spore suspension. In this case, the DPA biomarker shows an inherent affinity for the silver surface of the AgFON substrate and therefore the surface functionalization was not necessary. The specificity of the SERS spectra allows to differentiate this specific bacillus respect to other biomolecules, and this is specially important in applications such as biological warfare sensors.

Even more complex organisms can be identified by SERS. Different types of bacteria show different SERS spectra. This occurs because their cell membrane chemistry changes from one case to other. In Jarvis and Goodacre (2003, 40–47) a very simple approach for bacteria identification in water

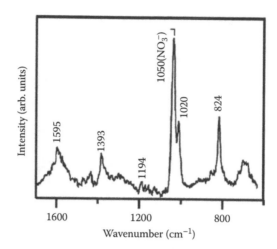

FIGURE 13.16 SERS spectrum of $2.1 \times 10{-14}$ M spore suspension (2.6×10^3 spores in $0.2\,\mu L$, $0.02\,M$ HNO_3) on AgFON; $\lambda_{ex} = 750\,nm$, $P_{ex} = 50\,mW$, acquisition time $= 1\,min$. (Reprinted with permission from Zhang, X., Young, M.A., Lyandres, O., and Van Duyne, R.P., Rapid detection of an anthrax biomarker by surface-enhanced raman spectroscopy, *J. Am. Chem. Soc.*, 127(12), 4484–4489. Copyright 2005 American Chemical Society.)

solution using SERS is reported. The metallic nanoparticles are citrate-capped thermally reduced Ag nanoparticle colloidal suspensions. The colloidal nano-Ag suspensions were mixed with the bacteria samples and, by evaporation of the solvent, the metallic nanoparticles trend to aggregate and dispersant between the bacteria. Figure 13.17.b shows a SEM micrograph, where nanoparticle aggregates are visible over the *Escherichia coli* bacteria. It is in the surrounding of these Ag-NP aggregates where the Raman cross-section enhancement of the molecules takes place, so the shifts observed in the SERS

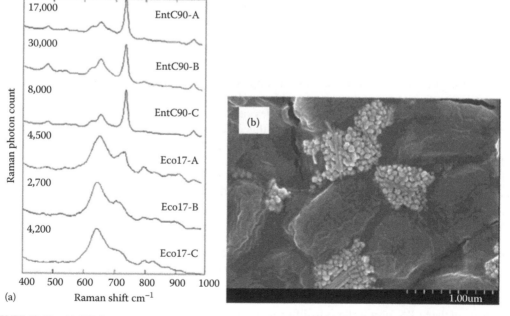

FIGURE 13.17 (a) SERS spectra representing an *Enterococcus sp.* (EntC90) and *E. coli* (Eco17) taken using aggregated colloidal silver solutions from three separate batch preparations (A–C). (b) A SEM image of *E. coli* and colloidal Ag-NP aggregates. The heterogeneity in size and shape of the colloidal silver particles can clearly be seen. (Reprinted with permission from Jarvis, R.M. and Goodacre, R., Discrimination of bacteria using surface-enhanced Raman spectroscopy, *Anal. Chem.*, 76(1), 40–47. Copyright 2003 American Chemical Society.)

spectra are a product of cell wall biochemistry or other chemical components external to the cell. In Figure 13.17a some spectra for *Enterococcus sp.* and *Escherichia coli* are presented, where it is easy to observe the Raman scattering spectral differences, making possible the bacteria identification in a simple and fast manner.

13.2.2 Quantum Dots

13.2.2.1 Quantum Dot Nanoparticles

The term Quantum Dot (QD) does not describe a particular chemical compound or material, but it refers to a very specific physical phenomenon that happens in a particular spatial region of the matter. This phenomenon is the quantum confinement, which consists on a spatial region of semiconducting material surrounded by infinite potential barriers. When an electron-hole pair recombines within a semiconductor material, a defined spatial region is altered. This affected volume is often treated as a particle (quasi-particle) called *exciton*, which has a particular size (delimited by its Bohr radius) and depends strongly on the semiconductor properties and some physical conditions such as the temperature of the system. The quantum confinement happens when any potential barrier limits the exciton into a smaller region than its Bohr radius in any direction of the space. When the confinement affects only one dimension it is possible to speak about quantum wells, while if the confinement occurs in two dimensions it is known as quantum wires. Finally, quantum dots are those regions of one material where the excited states are confined in all spatial directions. The confinement is an interesting phenomenon with a dramatic influence on the distribution of the density of States (DOS) in the confined regions, giving very interesting properties (see Figure 13.18).

There are multiple ways to achieve quantum confinement in any spatial direction of a particular region of a semiconductor. Nevertheless, the most straightforward way to get the three dimensional quantum confinement is through nanoparticles. When a semiconducting nanocrystal is considered, the size of the nanoparticle itself determines the confinement condition. This is possible because the external medium (solvent) acts as infinite potential barrier. Therefore, when the size of the nanocrystal is smaller than the Bohr radius, the electron-hole pair is confined and its DOS is limited to discrete energy states. Consequently some authors refer these QDs as *artificial atoms*. This quantum-distribution of the energy states gives some special properties ideal for optoelectronic devices, such as the following:

- Narrow and intense emission peaks
- Displacement of the allowed energy states due to the confinement condition. Controlling the degree of confinement it is possible to tune the bandgap properties of the QDs
- DOS pile up near the band edge. Consequently, more transitions can contribute to the optical response at the same energy, yielding higher fluorescence efficiencies

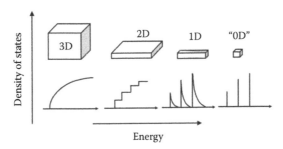

FIGURE 13.18 Scheme that shows the quantization of the Distributions of the Density of States (DOS) in a semiconductor material, when the electron-hole pairs are not confined (3D, or bulk) or confined in one spatial direction (2D, quantum well), in two directions (1D, quantum wire), or in the three directions (0D, or Quantum Dot). (Reproduced with permission from Springer Science+Business Media: Quantum dots for sensing, 2008, Goicoechea, J., Arregui, F.J., and Matias, I.R., ed. Arregui, F.J., Copyright 2008.)

These characteristics confer the semiconductor nanocrystals a highly fluorescent performance, which makes them one of the most serious alternatives to organic fluorophores in applications such as diagnosis imaging. Their use in biological applications has grown dramatically recently, as the QDs have been more easily synthetized, with higher qualities and their surface chemistry has been more understood and controlled.

In the last years the synthesis routes have experienced a significant advance. The most popular synthesis routes involved Cd-chalcogenide semiconductors, due to the availability of chemical precursors and high-quality crystallization routes. The very first works in the 1990s used dry approaches to form the nanocrystals such as Nd:YAG pulsed laser ablation (Koyama et al. 1992, 156–160), or electrospray of the precursors (Danek et al. 1994, 2795–2797). However, those techniques were substituted by wet approaches, which made possible good quality monodisperse nanocrystals using very inexpensive equipment. The chemical precursors were dispersed into a solvent with a surfactant. When the reaction started, the crystals grew from seed points creating sphere-like particles. The degree of quantum confinement could be easily controlled by stopping the crystallization reaction at the desired crystal size and, consequently, allowing the adjustment of the degree of compression of the electron-hole pair within the QD.

The most studied synthesis route uses organic-coordinating solvents such as trioctylphosphine oxide (TOPO), trioctylphosphine (TOP) or hexadeylamine at high temperatures (more than 300°C). These solvents act as the support medium for the reaction, and at the same time stabilize the surface of the nanoparticles, preventing aggregation of the QDs. Murray et al. (1993, 8706–8715) reported one of the first works using this route using $Cd(CH_3)$ as precursor, and achieved a high quantum efficiency and monodispersion of the QDs. Nevertheless, the main drawback of these first synthesis routes was the big hazard of the production process, since precursors were unstable, toxic, and even explosive. In addition to this, they were also expensive. Peng and coworkers (Peng and Peng 2001, 183–184) proposed a new approach using CdO instead $Cd(CH_3)$ as precursor, which lead to new synthesis routes which were more stable with excellent results (efficient and monodisperse nanoparticles), using simple and mild experimental conditions, opening the door for large-scale production of Cd-chalcogenide QDs. There has been a lot of contributions since then, reporting more sophisticated synthesis routes (Ludolph et al. 1998, 1849–1850; Crouch et al. 2003, 1454–1455; Kim et al. 2004a, 93–97; Medintz et al. 2005, 435–446; Costa-Fernández et al. 2006, 207–218) using other semiconductors (Landin et al. 1998, 262–264; Kim et al. 2005, 10526–10532), complex core-shell structures (Bruchez Jr. et al. 1998, 2013–2016; Chan and Nie 1998, 2016–2018), and so forth. Even in the last years some groups have reported water based synthesis routes, in which the organic-coordinating system is replaced by water, at lower temperatures (Ren et al. 2008, 17242–17243; Deng et al. 2009, 434–442; Huang et al. 2011, 348–351; Khatei and Koteswara Rao 2011, 159–164; Zou et al. 2011).

One of the most attractive aspects of the QD nanocrystals is their versatility from the functionalization point of view. Many different approaches have been reported for modifying the surface properties of the QDs, achieving, for example, to change the solubility of the nanoparticles. But the possibilities go much further, permitting us to attach to the surface of the QDs, dyes, functional groups, radio-markers, and the like. This possibility combined with their excellent behavior as high performance fluorophores have boosted their applications especially in bio-sciences.

13.2.2.2 *In Vitro Biological Applications*

Measuring certain target parameters in a biological system is one of the most challenging works that a scientist can think about. Not only from the point of view of the exigent detection limits desired in most of the applications, but because the whole system is an extremely complex environment which can introduce significant errors in the measurement. That is why sometimes it is necessary to carry out the process in a very controlled and reproducible environment outside the living organism. In the in-vitro assays there is the possibility of controlling in a very precise way the measurement conditions and at the same time it is possible to use some characterization techniques impossible to carry out inside a living organism without damaging it.

As it has been previously introduced, one of the most extended applications of the QDs is their use as high performance fluorophores. Although it is possible to see in the bibliography more sophisticated sensing mechanisms, most of the applications are based on highly fluorescent nanoparticles adequately functionalized replacing the organic fluorophores traditionally used in in-vitro bioassays.

This is the case of some immunoassays and DNA-detection applications. The excellent fluorescent efficiency of the QD nanocrystals can lower the detection limit of the well known ELISA (Enzime-Linked Immunosorbent Assay) or WB (Western-Blot) assays. This is possible thanks to a very well developed surface chemistry that allows to functionalize the semiconductor nanocrystal with some specific molecules, such as antibodies. The antibody-antigen interaction is very specific and it allows the detection of small concentrations of antigens or other target biomolecules in a sample. As a result, some immunoassays like ELISA are used currently as diagnosis tool and food quality control due to their good reliability and high throughput.

One of the most used funtionalization for QDs in immunoassay applications is the streptavidin, because of its extraordinary affinity to Biotin, making this a highly selective and strong interaction. Therefore, avidin-conjugated QDs can replace traditional florophores from ELISA assays and lower their detection limit (Goldman et al. 2002a, 407–414; 2002b, 6378–6382).

There are also other advantages of the use of semiconductor QDs. As the wavelength of the emission peak is determined by the size of the nanocrystals, it is possible to use adequately functionalized QDs with different size (and consequently with different colors). This opens the door to optical multiplexing of multiple target molecules in one pot assays (Han et al. 2001, 631–635). In that approach, different color QDs are embedded into polymeric beads and they are differently functionalized, making possible to detect different targets through optical color-multiplexing with an excellent sensitivity and photostability. Moreover, it is possible to excite the different QDs using the same light source due to the wide absorption band of the semiconductor nanocrystals. Such applications were impossible with the conventional organic fluorophores.

A similar approach was reported by Goldman et al. (2004, 684–688). In this work a multiplexed fluoroimmunoassay was carried out for the detection of four different toxins (cholera toxin, ricin, shiga-like toxin one, and staphylococcal enterotoxin B). The whole immunoassay was performed simultaneously in a single well of a microtiter plate using amine-reactive NHS biotin conjugated QDs.

There are other works that use more complicated assays, combining the QDs with other nanoparticles. In Yu et al. (2011, 7804–7811) the authors propose an optical immunoassay for detecting the cyanotoxin microcystin-LR (MCYST-LR) where QDs were functionalized with monoclonal antibodies (mAbs) against MCYST-LR. Also, MCYST-LR functionalized magnetic beads were used as competitor agent. As it is shown in Figure 13.19, the competitor agent is firstly mixed into the sample, and the anti-MCYST-LR functionalized QDs were added afterwards. The QDs interacted with the MCYST-LR free in the sample, and also with the MCYST-LR present in the surface of the magnetic beads. After a magnetic separation, the fluorescence of the QD adsorbed onto the magnetic beads is measured. As in all the competitive assays, the final optical signal is inverse to the MCYST-LR amount in the original sample.

In other works, an alternative sensing mechanism is proposed based on the resonant coupling of the excitation of the QD to a neighbor acceptor. This energy coupling, known as Förster resonance energy transfer (FRET) (Förster 1948, 55), consists of the transference of the excitation energy from a donor fluorophore, in this case the QD, to an acceptor agent placed in the surrounding of the QD. This transference mechanism is only efficient at very short distances (only a few nanometers). Figure 13.20 shows an example of FRET resonance where the emission spectrum of QDs emitting at 655 nm is transferred to an organic fluorophore (Alexa Fluor). When QDs have been conjugated with the Alexa Fluor, part of the excited states is transferred by FRET to this organic dye. In the conjugated form it is possible to observe the two emission peaks, one in 655 nm from the QDs and other around 700 nm from the Alexa Fluor.

One of the most straightforward approaches to a FRET-based sensor is to functionalize the QD emitter with a selective binder for the target analyte, and to perform a competitive assay with an acceptor binding agent (Russ Algar and Krull 2011, 148–154). When the acceptor adsorption is significant, it results

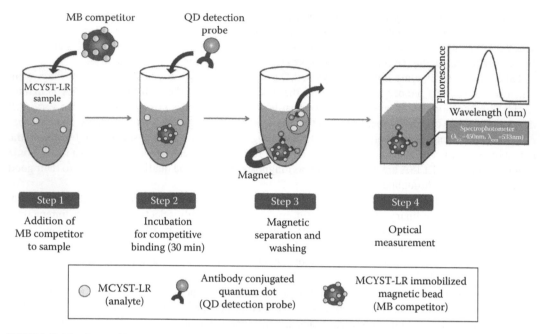

FIGURE 13.19 Competitive fluorescence immunoassay using magnetic nanoparticles as competitive agent. As the analyte concentration rises in the sample, the adsorbed QDs onto the magnetic particles decreases, and so it does the fluorescent signal observed in the magnetically separated sample. (Reprinted with permission from Yu, H.-W., Jang, A., Kim, L.H., Kim, S.-J., and Kim, I.S., Bead-based competitive fluorescence immunoassay for sensitive and rapid diagnosis of cyanotoxin risk in drinking water, *Environ. Sci. Technol.*, 45(18), 7804–7811. Copyright 2011 American Chemical Society.)

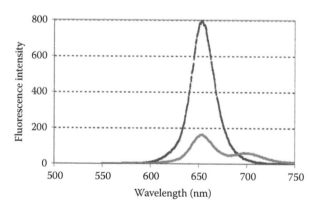

FIGURE 13.20 Emission spectrum of ITK655 QDs (Invitrogen) and the same QDs conjugated with Alexa fluor. It is clearly visible how the QDs emission peak is reduced, transferring part of the energy to the Alexa Fluor, which shows its emission peak near 700 nm. (Reprinted from *Anal. Biochem.*, 357, Nikiforov, T.T. and Beechem, J.M., Development of homogeneous binding assays based on fluorescence resonance energy transfer between quantum dots and Alexa fluor fluorophores, 68–76, Copyright 2006, with permission from Elsevier.)

in a decrease in the fluorescence emission of the QDs, because some of the energy is transferred to the acceptor and further re-emitted or simply dissipated. Furthermore, simply by using the desired surface chemistry for the QDs it is possible to vary the target agent that will cause the conjugation of the acceptor, making possible the FRET phenomenon. Therefore it is possible to see different optical biosensors using this approach, for example, in Figure 13.21 it is shown the sensing mechanism proposed for QD-based detection of histone-modifying enzymes (Ghadiali et al. 2011, 3417–3420).

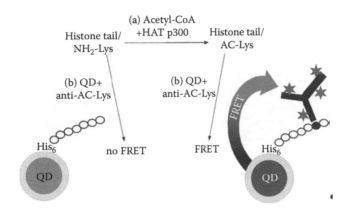

FIGURE 13.21 Scheme of the working principle of a FRET optical sensor of histone-modifying enzymes. (From Ghadiali, J.E., Lowe, S.B., and Stevens, M.M.: Quantum-dot-based FRET detection of histone acetyltransferase activity. *Ann. Chem. Int. Ed.* 2011. 50. 3417–3420. Copyright Wiley-VCH Verlag GmbH & Co. KGaA. Reprinted with permission.)

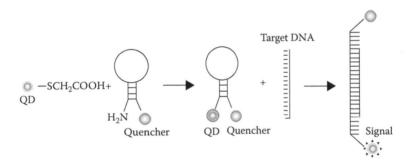

FIGURE 13.22 Schematic view of the working principle of Molecular Beacons (MBs) used to identify target DNA strands. (Reprinted from *Sens. Actuat. B Chem.*, 102, Kim, J.H., Morikis, D., and Ozkan, M., Adaptation of inorganic quantum dots for stable molecular beacons, 315–319, Copyright 2004, with permission from Elsevier.)

A more sophisticated FRET based sensor is used for DNA detection. A selective polynucleotide is sandwiched between two sections of complementary oligonucleotides. This structure is known as Molecular beacon (MB). One oligonucleotide has an emitter (QD) at one extreme, and the other one has a quencher (Kim et al. 2004b, 315–319). As shown in Figure 13.22, the initial state of the MB both the QD and the acceptor are very close, so the FRET mechanism is efficient and the emission from the QD is not observed. But when the sample contains the specific sequence of nucleotides complementary to those in the central part of the MB, a hybridization reaction occurs, and the MB unfolds. This conformational change in the MB destroys the FRET coupling between the QD and the quencher, and consequently its emission becomes visible. (Kim et al. 2004b, 315–319; Sato et al. 2011, 11650–11656; Yeh et al. 2011, 267–270).

Quantum Dots have also caused an enormous impact in the cellular research community. One of the most important advances is the observation of cellular processes almost impossible to see with conventional fluorophores. In this case one of the most interesting properties of the QDs is their outstanding photostability, combined with their high fluorescent efficiency. Therefore QDs have contributed to a significant improvement in 3D cell reconstructed imaging using confocal fluorescence microscopy (Lacoste et al. 2000, 9461–9466; Tokumasu and Dvorak 2003, 256–261). Using different diameter QDs, it is also possible to use different probes inside the same cell, and with the appropriate functionalization it is possible to selectively monitor different parts of a living cell in an in-vitro assay.

Other researchers have used the QDs for cell tracking applications (Derfus et al. 2004, 961–966; Courty et al. 2006, 1491–1495; Ruan et al. 2007, 14759–14766). In such applications the outstanding fluorescence fluorescence efficiency and photostability of QDs make possible even single molecule tracking.

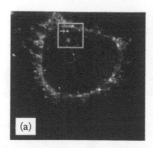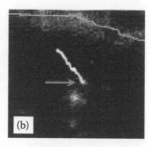

FIGURE 13.23 Direct observation of active transport of endocytosed Tat-QDs inside living cells. (a, b) Directed motion from the cell periphery to an intracellular region adjacent to the cell nucleus. The red box area of image (a) is magnified in image (b). The white line is the trajectory of one Tat-QDs vesicle pointed by the red arrow. The green line shows the plasma membrane boundary of the cell. (Reprinted with permission from Ruan, G., Agrawal, A., Marcus, A.I., and Nie, S., Imaging and tracking of tat peptide-conjugated quantum dots in living cells: New insights into nanoparticle uptake, intracellular transport, and vesicle shedding, *J. Am. Chem. Soc.*, 129(47), 14759–14766. Copyright 2007 American Chemical Society.)

The fluorescence from individual QDs can be observed using fluorescence confocal microscopy and if QDs probes are used with the proper biofunctionalization it is possible to study the internal processes of cells. Figure 13.23 shows the tracking of a single QD inside a cell, which can give very valuable information about the uptake mechanisms, and different transfer operations in different parts of the cell.

13.2.2.3 In Vivo Applications

When QDs are introduced into a living organism it is possible to distinguish two types of applications: non-targeted and targeted imaging. In the non-targeted imaging the QDs are not biofunctionalized. They are simply introduced into a living organism and it is possible to observe their natural accumulation and elimination processes. On the other hand, if the QDs have been surface-modified with a desired biomolecule that selectively attaches to a target region, we speak about targeted imaging.

In vivo optical imaging has been traditionally very limited because of the strong absorption of almost any living organism in the visible region of the spectrum. Nevertheless, there are a couple of windows in the near infrared where living bodies where the absorbance is significantly lower. These windows are in the 700–900 nm region and in the 1200–1600 nm. It is a very difficult task to find organic fluorophores with their emission in such wavelengths, but here is where QDs overcome this limitation.

For example, in Kim et al. (2004a, 93–97) it is reported how NIR CdTe(CdSe) core(shell) type II QDs were synthesized and functionalized with an oligomeric phosphine coating to render them soluble, disperse, and stable in serum. The injection of 400 pmol of near-infrared quantum dots in the lymphatic system of a living animal allowed the sentinel lymph node mapping up to 1 cm deep under the skin in real time using excitation fluency rates of only 5 mW/cm². As it is shown in Figure 13.24, the fluorescence from the QDs accumulated in the sentinel lymph node helped the surgeon to identify the target tissues under complete image guidance.

As previously mentioned, one of the most interesting properties of the QDs is the possibility of color multiplexing of multiple targets. One of the most straight-forward ways to achieve multitarget imaging using QDs is to prepare different diameter QDs with different surface functionalizations. When the QD probes are injected into a living organism, the selective fluorophores accumulate in the targeted areas, making possible the imaging assisted localization of the tissues of interest, such as different carcinogen cells (Jiang et al. 2008, 6–12). QDs can also be encapsulated into polymeric beads to achieve multicolor labels. Using this approach it is possible to use fluorescence-assisted imaging for diagnosis. One example of this is shown in Gao et al. (2004, 969–976), where a living mouse which has been inoculated with three differently colored microbead solutions at three different locations and the injection zones were clearly visible and identified under UV illumination.

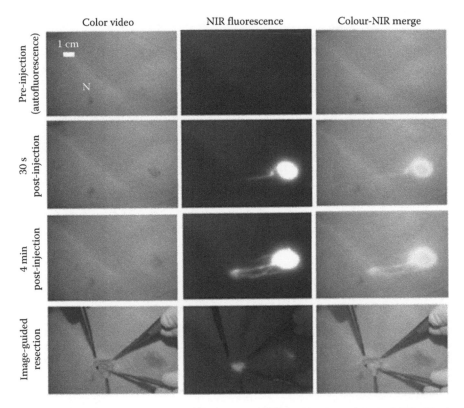

FIGURE 13.24 Images of the surgical field in a pig injected intradermally with 400 pmol of NIR QDs in the right groin. Four time points are shown from top to bottom: before injection (autofluorescence), 30 s after injection, 4 min after injection, and during image-guided resection. (Reprinted with permission from Macmillan Publishers Ltd., *Nat. Biotechnol.*, Kim, S., Lim, Y.T., Soltesz, E.G., De Grand, A.M., Lee, J., Nakayama, A., Parker, J.A. et al., Near-infrared fluorescent type II quantum dots for sentinel lymph node mapping, 22(1), 93–97, Copyright 2004.)

13.2.2.4 Other Sensing Applications

QDs have also been used as optical sensors for other parameters such as ion concentration, temperature, chemical sensor, etc. Sometimes the sensing mechanisms are the same as the ones used in bioassays (previously discussed), with the only difference that the surface chemistry of the QDs is modified properly to be selective to other target. For example there FRET based QD sensing mechanism has been successfully used for applications such as TNT trace detection. As it is shown in Figure 13.25, QDs were functionalized with an anti-TNT single-chain Fv protein (TNB245). Such QDs are firstly conjugated with a TNT simile (trinitrobenzene) labeled with a FRET quencher and therefore the fluorescence of the QDs is not visible. When the complexated QDs are exposed to a TNT containing sample, the TNT molecules binds to the anti-TNT functional coating, displacing the TNB-quencher and causing an increase of the fluorescent signal.

Other sensors have been proposed based on the QD's sensitivity to certain ions or chemical species. For example, the presence of H_2O_2 can alter the mercaptopropionic acid capping layer of the QDs, and induce non-radiant preferent recombination traps in the surface of the nanocrystals, quenching significantly their fluorescent emission. In Hu et al. (2010, 997–1002) the authors show an indirect glucose sensor, using the enzyme glucose-oxidase which catalyzes the oxidation of the glucose and generates hydrogen peroxide as subproduct. As seen in Figure 13.26, the H_2O_2 generated by the enzymatic metabolization of the glucose decreases the emission from QDs.

A similar fluorescence quenching mechanism is purposed by other groups for ion detection. In Ali et al. (2007, 9452–9458) QDs are capped with glutathione (GSH) shells. GSH and its polymeric form, phytochelatin, is a very specialized natural heavy metal ion collector. As a result of specific

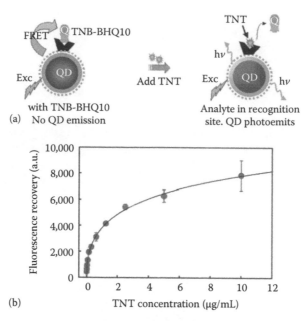

(a)

(b)

FIGURE 13.25 (a, b) FRET based QD TNT sensor. (Reprinted with permission from Goldman, E.R., Medintz, I.L., Whitley, J.L., Hayhurst, A., Clapp, A.R., Uyeda, H.T., Deschamps, J.R., Lassman, M.E., and Mattoussi, H., A hybrid quantum dot—Antibody fragment fluorescence resonance energy transfer-based TNT sensor, *J. Am. Chem. Soc.*, 127(18), 6744–6751. Copyright 2005 American Chemical Society.)

interaction, the fluorescence intensity of GSH-capped QDs is selectively reduced in the presence of heavy metal ions such as Pb^{2+}. The detection limit of Pb^{2+} is found to be 20 nM. The underlying quenching mechanism is, once more, the creation of non-radiative recombination sites at the surface of the QD. There are other ion sensors based on the quenching of the fluorescence of the QDs, such as Ag^+ (Chen and Zhu 2005, 147–153), Cu^+ (Fernández-Argüelles et al. 2005, 20–25) and CN^- (Sarkar et al. 2002, 5045–5047).

Finally other sensors take advantage of intrinsic properties of the QDs for the detection of other parameters such as temperature. Figure 13.27 shows the response of a sensitive coating loaded with QDs to temperature changes (Larrió.n et al. 2009). In this case the sensing principle is based on the alteration of the confinement condition of the nanocrystals in view of the temperature dependency of the electron-hole pair natural size. Consequently, the confinement degree is altered if the size of the electron-hole pair is increased and the size of the nanocrystal is kept constant. This leads to a blue-shift of the fluorescence emission peak of the QDs. Another advantage of the QDs is their extremely small size which makes possible to build the sensitive coatings over the inner holes of a photonic crystal fiber, as it is reported in Larrión et al. (2009).

13.2.3 Magnetic Nanoparticles

13.2.3.1 Introduction

In our present society it is not possible to think about an automobile, a computer, or a factory without the presence of magnetic sensors (Lenz and Edelstein 2006, 631–649). This is explained by the enormous variety of technologies that have been developed for many decades: search-coil sensors, fluxgate sensors, SQUID sensors, magnetoresistive sensors (including anisotropic magnetoresistance (AMR), Giant magnetoresistance (GMR), magnetic tunnel junction sensors (MJT), extraordinary magnetoresistance and ballistic magnetoresistance), giant magnetoimpedance (GMI) sensors, magnetodiodes, magnetotransistors, magnetostrictive sensors, magnetooptic sensors, MEMS based sensors, optical pump, nuclear precession, and overhauser sensors.

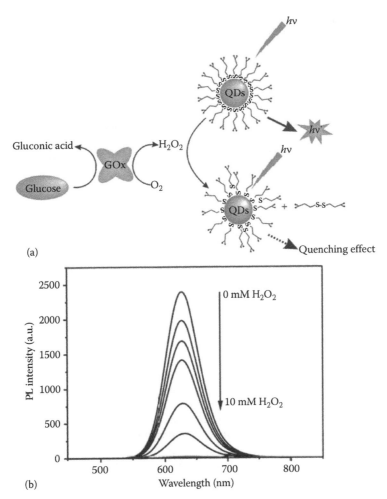

FIGURE 13.26   (a, b) H_2O_2 based QD sensor for the detection of glucose. (Reprinted from *Talanta*, 82, Hu, M., Tian, J., Lu, H.-T., Weng, L.-X., and Wang, L.-H., H2O2-sensitive quantum dots for the label-free detection of glucose, 997–1002, Copyright 2010, with permission from Elsevier.)

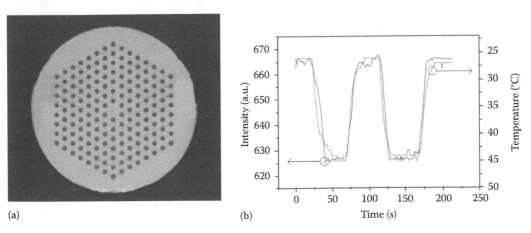

FIGURE 13.27   (a) PCF LMA-20 cross section. (b) Dynamic behavior of an electronic and the PCF QD-based (optical) sensors. The intensity measurement of y-axis is the average of the maximum intensity values between 600 and 640 nm. (Reprinted from Larrión, B. et al., *J. Sensors*, 932471, 2009, doi:10.1155/2009/932471. With permission.)

All the previous technologies can be used in different groups of applications (Lenz and Edelstein 2006, 631–649). In view of the title of this chapter, the focus will be centered on chemical and biological applications, where there is an interaction between magnetic nanoparticles and optics (Pankhurst et al. 2003). The interaction between these two technologies can be understood if we first review the history of physiological imaging. Computed tomography and magnetic resonance imaging (MRI) (Haacke et al. 1999) have been used since many years for visualizing the physiological function of tissue and organs (Mulder et al. 2007, 307–324). However, with the progressive expansion of nanotechnology it has been possible to move towards visualization of cellular or even molecular level (Mulder et al. 2006a, 1–6; Quarta et al. 2007, 298–308). Nowadays the three main techniques used for this last purpose are MRI, Positron emission gtomography (PET) (Phelps 2000, 661–681) and single photon emission computed tomography SPECT (Germano et al. 1995, 2138–2147). Among these techniques, MRI is the most promising one because it has characteristics such as versatility (its performance can be improved by combination with optical detection methods such as fluorescence, luminescence, or even with PET or SPECT), capability to generate 3D images of opaque and soft tissues, high spatial resolution, capability to obtain simultaneously physiological and anatomical information, and because there is no delivery or radiation burden (Mulder et al. 2006a, 1–6; Kim et al. 2009, 372–390). However, MRI has a low sensitivity compared to PET and SPECT. This problem can be solved by adequate selection of MRI contrast agents. In order to obtain a better sensitivity, the contrast agents must catalytically shorten the relaxation times (T1 and T2) of bulk water protons (Berry and Curtis 2003, R198–R206; Mulder et al. 2007, 307–324). T1 is the spin-lattice or longitudinal relaxation time, whereas T2 is the spin-spin or transverse relaxation time. So there are two categories. The first one is referred to as T1 agents because, on a percentage basis, they alter 1/T1 of tissue more than 1/T2 owing to the fast endogenous transverse relaxation in tissue (Caravan 2006, 512–523). Paramagnetic complexes (Gd^{3+}) or manganese (Mn^{2+}) chelates (Lin and Koretsky 1997, 378–388; Caravan 2006, 512–523), are typical T1 contrast agents. Among them, gadolinium diethylenetriaminepentaacetate (Gd-DTPA) is the most widely used T1 contrast agent. The second category is referred to as T2 agents or superparamagnetic iron oxide (SPIO) nanoparticles (Bulte and Kraitchman 2004, 484–499). They accelerate the transverse (T2) relaxation of water protons and exhibit dark contrast. In fact, this group is included in the group of IO nanoparticles (Louie 2010, 3146–3195), that will be explained next. The term SPIO is used to refer to superparamagnetic particles in general, but when their size is very small they are considered as ultrasmall superparamagnetic iron oxide (USPIO). The utilization of either SPIO nanoparticles or paramagnetic nanoparticles is combined in many cases with other materials such as dye doped silica materials, quantum dots, lanthanide compounds, and near infrared (NIR) emitting nanostructures. Most of them allow optical imaging. This permits us to combine the advantages of both methods (optical imaging allows for rapid screening [Kim et al. 2009, 372–390]) and to overcome the limitations of the simple application of MRI. Moreover, it is also possible to use trimodal imaging by combination with PET isotopes such as ^{64}Cu, which adds a third probe (Nahrendorf et al. 2008, 379–387). The combination of different imaging techniques is called multimodal imaging (Mulder et al. 2007, 307–324; Kim et al. 2009, 372–390; Louie 2010, 3146–3195).

However, the aim of the scientists is not only the monitorization of in-vitro or in-vivo assays, but also the control and manipulation of physiological events. Once this objective has been fulfilled, the active treatment of many diseases such as cancer (Corot et al. 2004, 619–625), or neurological ones (Ferrari 2005, 161–171) could become real. Before showing the applications, the different types of sensing elements, their synthesis, and characterization will be explained in the next section.

13.2.3.2 Iron Oxide (IO) and Other Iron Derived Nanoparticles

13.2.3.2.1 Synthesis

The low toxicity (it has been experimentally checked that, upon metabolism, iron ions are added to the body's iron stores and eventually incorporated by erythrocytes as hemoglobin allowing their safe use [Weissleder et al. 1989b, 167–173]), the biodegradability (Sun et al. 2008, 1252–1265), and the fairly good sensitivity of iron when used as contrast agent for MRI, make iron oxide IO nanoparticles an ideal platform for many in-vitro and in-vivo applications. There are also other combinations of iron oxide such as ferrite (Fe_3O_4), (Ma et al. 2009, 1368–1371), FePt (Gao et al. 2007, 11928–11935) or FeCo (Yoon et al. 2005, 1068–1071). FeCo owns a degree of cytotoxicity that prevents its usage without an adequate coating (Sun et al. 2008, 1252–1265). The same occurs for FePt (Vazquez et al. 2008), where further studies regarding its toxicity are required. The iron oxide option has been favored recently due to the controversy generated around the toxicity observed in some patients with pre-existing kidney disease when applying gadolinium based contrast agents (the most typical agents used in the subgroup of paramagnetic nanoparticles) (Idée et al. 2008, 77–88; Shellock and Spinazzi 2008, 1129–1139). Certain conditions of administration of this product have led to nephrogenic system fibrosis (NSF) (Louie 2010, 3146–3195).

Several techniques are used for the chemically synthesization of iron oxide nanoparticles: co-precipitation, thermal decomposition, microemulsion, and hydrothermal synthesis (Lu et al. 2007a, 149–154), and more sophisticated ones such as laser pyrolysis (Tartaj et al. 2002, 4556–4558) or chemical vapor deposition (Willard et al. 2004, 125–170).

Among them, co-precipitation method is the most extended one and it has been commercialized (Ferridex and Combidex). This method permits the simultaneous precipitation of Fe^{2+} and Fe^{3+} ions in aqueous media, which is environmentally harmless (Kim et al. 2009, 372–390). For the sake of its applicability in biological applications the nanoparticles are coated with hydrophilic polymers. Critical parameters that determine the properties of the particle formation are the solution pH value and the coating material, which serves as a surfactant (Sun et al. 2008, 1252–1265). However, Ferridex, Combidex, and other commercialized components such as Resovist, and AMI-288/gerumoxytrol presented a low specificity (Park et al. 2009, 1553–1566). This poor specificity comes from polydisperse and poor crystalline nature of these products. This limits its applicability to domains such as cancer tumor detection. To solve this question, much effort has been devoted towards the development of targeted IO nanoparticle based probes for tumor diagnostics or other applications (Neuberger et al. 2005, 483–496; Torchilin 2006, 1532–1555). In addition to this, thermal decomposition method has permitted researchers to synthetically control the important features of these probes, such as size, magnetic dopants, magneto-crystalline phases, and surface states. These innovative nanoparticle probes exhibit superior magnetism and MR contrast effects which have been shown to be better than that of conventional magnetic nanoparticle MR contrast agents (Jun et al. 2008, 5122–5135).

13.2.3.2.2 Surface Coating

In most of the cases the nanoparticles are synthesized in nonpolar inorganic solvents. If this technology is to be applied for in-vitro or in-vivo applications, there are several issues to consider. First, the probes are required to be soluble in aqueous buffers (Mulder et al. 2007, 307–324). Second, even though the nanoparticles present magnetic properties, they have a tendency to agglomerate as a result of their high surface energy. Finally it is critical to avoid its uptake by the Reticuloendothelial system (RES). A common failure in targeted systems is caused by the opsonization of the particles on entry into the bloodstream, rendering the particles recognizable by the body's major defense system (Berry and Curtis 2003, R198–R206). Consequently, different coating techniques (see Figure 13.28) have been developed to protect the magnetic nanoparticles and to meet the requirements already mentioned.

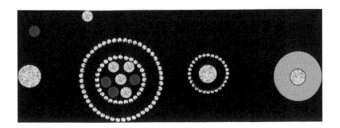

FIGURE 13.28 Schematic representation of differently coated magnetic nanoparticles.

13.2.3.2.2.1 Polymer Coatings   Polymer coatings meet the requirements above mentioned (a steric barrier prevents nanoparticle agglomeration and opsonization) (Sun et al. 2008, 1252–1265). Other questions such as hydrophilicity or biodegradation depend more specifically on the nature of the polymer coating. There are block copolymers that can change the hydrophobic surface into a hydrophilic one (Qin et al. 2007, 1874–1878). Another question is the possibility of losing the coated polymers under harsh in vivo conditions, resulting in aggregation. (Moghimi et al. 2001, 283–318; Zhang et al. 2002, 1553–1561). To avoid aggregation thermal cross-linking is an adequate choice (Lee et al. 2006a, 7383–7389; Park et al. 2009, 1553–1566).

Let us start with the most extended polymer, which is dextran. Its combination with iron oxide nanoparticles has been named CLIO (cross-linked iron oxide nanoparticles) (Josephson et al. 2002, 554–560). This method has been widely used for MR/optical imaging by coupling to fluorophores such as rhodamine (Banerjee and Chen 2009), AlexaFluor (Maxwell et al. 2008, 517–524), oligothiophene 71 (Quarta et al. 2008, 10545–10555) or Cy5.5 (Pittet et al. 2006, 73–79). Among these fluorophores the combination of CLIO with Cy5.5 permits to achieve good biocompatibility, small hydrodynamic size, and stable hydrophilic polymer coating. Moreover, Cy5.5 is an adequate fluorophore because it is excited at 675 nm and emits at 694 nm, which is not harmful for in-vivo applications (the wavelength is very close to the window of low absorbance between 700 and 1000 nm) and permits a deeper penetration in the tissue. That is why many applications, mainly targeted, have been developed based on this technique.

One of the problems with CLIO-Cy5.5 based probes is the quenching of the fluorescence. It seems that it has to do with the coupling between CLIO and Cy5.5 (Louie 2010, 3146–3195). In conjugations of Cy5.5 with other compounds the same quenching does not occur. Anyway the quenching is a typical problem with fluorescent materials and there are strategies to solve this problem. The dye molecules can be incorporated in a protective matrix, which protects the dye from quenching. In Burns et al. (2006a, 1028–1042), the protective matrix is made of silica. This adds biocompatibility, water solubility, and the possibility of modifying the surface with a wide range of functional groups, which permits conjugation of targeting molecules. Based on these advantages, iron oxide nanoparticles have been assembled with fluorescent silica to make multimodal imaging probes.

Another widely used polymer is poly(ethlylene glycol) (PEG) (Zhang et al. 2004, 33–40). As an example, in Veiseh et al. (2005, 1003–1008) iron oxide nanoparticles have been conjugated to Cy5.5 by first coating the iron oxides with a layer of PEG. An interesting property of this polymer is that it resists the protein adsorption and thus avoids the particle recognition by macrophage cells. This facilitates the nanoparticle uptake to specific cancer cells for cancer therapy and diagnosis (Zhang et al. 2002, 1553–1561).

Polymer coatings have also been used to introduce optical probes in the shell around iron oxide cores. For instance NIR dyes have been introduced in a polymeric coating of poly(acrylic acid) (PAA) (Santra et al. 2009, 1862–1868) used as a shell for iron oxide nanoparticles. NIR dyes own the good properties of biocompatible excitation wavelength and deep penetration.

Apart from the selection of the polymer, much research is being dedicated to the development of new techniques that allow for the development of additional properties. For instance, copolymerization is a technique used for the development of iron oxide nanoparticles and coating in a single step (Lee et al. 2006a, 7383–7389). In this way, the synthesis method is simplified and good properties are obtained such as reduced agglomeration, at a cost of a less accurate control of the size structure and morphology of the crystal (Sun et al. 2008, 1252–1265).

13.2.3.2.2.2 Liposomes and Micelles   Liposomes are spherical self-closed structures formed by one or several concentric lipid bilayers with an aqueous phase in the core and between the lipid bilayers (Mulder et al. 2007, 307–324). There are two main approaches. In the first one the nanoparticles are set in the core of the liposome, as it is the case in Figure 13.28 (Mamot et al. 2004, 1–9). In the second approach, the nanoparticles are incorporated into the lipid bilayer (Mulder et al. 2006b, 142–164).

Regarding micelles, they consist of a single layer and they are good carriers of poorly soluble materials (Mulder et al. 2007, 307–324). As an example, in Nasongkla et al. (2006, 2427–2430) mixtures of hydrophobic nanomaterials are included in the core of micelles. In Dubertret et al. (2002, 1759–1762), the nanoparticles have been included in the core of micelles formed from PEGylated lipids. This permits the probes to simultaneously provide efficient fluorescence, a great reduction in photobleaching, colloidal stability in a variety of bioenvironments, and low nonspecific adsorption. Another approach was done in Nitin et al. (2004, 706–712), which consisted of PEG-modified, phospholipid micelle coating. This allowed for both water solubility and conjugation of TAT peptides and a fluorescent label to the distal end of PEG chains of the phospholipids to coat the iron oxide particles. In this way both fluorescent and MRI monitorization was possible of the targeting of cellules.

13.2.3.2.2.3 Core/Shell Iron oxide nanoparticles with a dextran coating could be considered as a core/shell configuration (Louie 2010, 3146–3195). Similarly, liposomes encapsulating IO nanoparticles described in the previous paragraph could be also considered core-shell. So the condition for considering core/shell configuration is often unclear and depends on the perception of the author. The key factor which determines the consideration of a core/shell structure is the organic or inorganic nature of the coating. The previous cases studied (polymers, liposomes, and micelles) were organic, whereas for core-shell inorganic coatings are considered (Sun et al. 2008, 1252–1265).

The most typical inorganic shell is silica. As it was mentioned above, silica presents good properties that make it adequate for biological applications: water soluble, biocompatible, easy modification of the surface for conjugation, porosity, and transparency. So different fluorophores such as rhodamine (Yoon et al. 2005, 1068–1071) (it was used also for conjugation), FITC (Lu et al. 2007a, 149–154; Wu et al. 2008, 53–57), or pyrene (Nagao et al. 2008, 9804–9808), and other materials such as quantum dots (Salgueiriño-Maceira et al. 2009, 3684–3688) are included in the silica shell that coats the IO core. The methods used by most of them are modifications of Stöber method (Stöber et al. 1968, 62–69), which solve the problem of being only adequate for a specific range of colloidal particles (Graf et al. 2003, 6693–6700). The second main group of methods is Water-oil (W/O) reverse microemulsion and will be explained in section dedicated to silica afterwards.

An example of modification of Stöber method can be found in Ma et al. (2009, 1368–1371), where the core-shell structure was fabricated by a synthesis process consisting of preparation of citric acid-modified Fe_3O_4 nanoparticles by a coprecipitation method, and a subsequent coating with a SiO_2 shell formed by the copolymerization of tetraethylorthosilicate (TEOS) and a silane precursor containing an organic chromophore (p-aminobenzoic acid, PABA) and a chelate (diethyl enetriaminepentaaceticacid DTPA), which chelates the Tb^{3+} ions with the rest of the structure. See Figure 13.29.

In a second example, where the core is FeCo instead of iron oxide (Yoon et al. 2005, 1068–1071), the poly-(vinylpyrrolidone) PVP method is applied to particles having ionic surface charges to generate a sol–gel silica coating. Its thickness can be altered by varying the amount of tetraethoxysilane (TEOS) loaded (Graf et al. 2003, 6693–6700). In Figure 13.30 it can be observed the influence of TEOS on the thickness of the silica shell.

Similarly cobalt ferrite ($CoFe_2O_4$) nanoparticles were coated with dye doped silica shell, whose surface was functionalized with PEG and conjugated with an antibody to target cancer cells (Yoon et al. 2005, 1068–1071). In this case, the dye is coupled to silica before de polymerization, but it can be also coupled afterwards in order to avoiding quenching in the core-shell structure (Lu et al. 2007c, 1222–1244). For instance, in Salgueiriño-Maceira et al. (2006, 509–514), silica nanoparticles with a magnetite core were deposited with an LbL dye coating.

Another widely extended shell is gold (Choi et al. 2007a, 861–867). Gold offers the possibility of photoacustic imaging, reflectance, photothermal imaging therapy (Louie 2010, 3146–3195), and it also permits to form self-assembly monolayers on its surface by using alkanethiols (Bain et al. 1989, 321–335). There is a variety of methods that permit to synthesize gold coated IO nanoparticles (Lu et al. 2007c, 1222–1244). The main problems are the low solubility and high size (250 nm) (Louie 2010, 3146–3195).

A less explored possibility is graphite coating good biocompatibility and in combination with FeCo core it presents both magnetic properties and high absorbance in the near infrared domain. (Seo et al. 2006, 971–976)

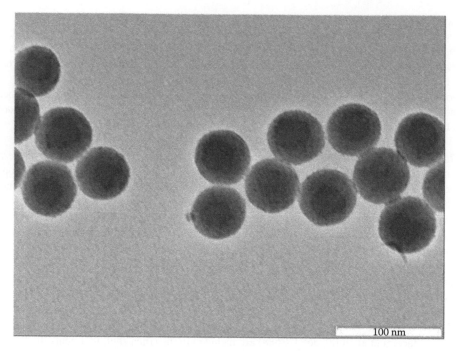

FIGURE 13.29 TEM image of multifunctional magnetic silica nanocomposites. (Reprinted from *J. Magn. Magn. Mater.*, 321, Ma, Z., Dosev, D., Nichkova, M., Dumas, R.K., Gee, S.J., Hammock, B.D., Liu, K., and Kennedy, I.M., Synthesis and characterization of multifunctional silica core-shell nanocomposites with magnetic and fluorescent functionalities, 1368–1371, 2009, Copyright 2009, with permission from Elsevier.)

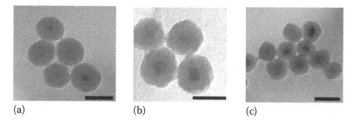

      (a)   (b)   (c)

FIGURE 13.30 TEM images of Co ferrite–silica (core–shell) MNPs with controlled shell thicknesses. (a) TEOS/MNP = 0.12 mg/4 mg, scale bar = 100 nm, (b) TEOS/MNP = 0.06 mg/4 mg, scale bar = 50 nm, (c) TEOS/MNP = 0.03 mg/4 mg, scale bar = 50 nm. As the ratio of TEOS/MNP (w/w) decreases, the shell thickness decreases. (From Yoon, T.-J., Kim, J.S., Kim, B.G., Yu, K.N., Cho, M.-H., and Lee, J.-K.: Multifunctional nanoparticles possessing a "magnetic motor effect" for drug or gene delivery. *Angew. Chem. Int. Ed.* 2005. 44. 1068–1071. Copyright Wiley-VCH Verlag GmbH & Co. KGaA. Reprinted with permission.)

13.2.3.2.2.4 Other Structures   Instead of including the structures presented in the following section in the core-shell group, it is better to consider them apart in view of their complexity. In Figure 13.31, one interesting configuration consists of core-satellite structured hybrid nanoparticles composed of dye-doped silica as the core and multiple iron oxide nanoparticles as the satellite (Lee et al. 2006b, 8160–8162).

    Another approach is to coat the iron oxide with a QD shell. This approach was used to add CdSe shells to iron oxide cores (Selvan et al. 2007, 2448–2452; Lai et al. 2008, 218–224). However, the synthesis produces not uniform coating of QDs over the IO shell, which could be considered as heterodymers. The QD was responsible for optical emission, and the wavelength could be controlled with the reaction time in the growing step.

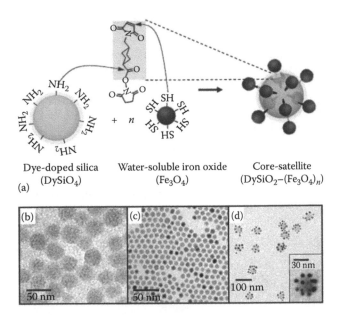

(a) Dye-doped silica (DySiO$_4$)   Water-soluble iron oxide (Fe$_3$O$_4$)   Core-satellite (DySiO$_2$–(Fe$_3$O$_4$)$_n$)

(b)   (c)   (d)

50 nm   50 nm   100 nm   30 nm

FIGURE 13.31 (a) Schematic diagram for the synthesis of core–satelliteDySiO$_2$–(Fe$_3$O$_4$)$_n$ nanoparticles, (b–d) TEM images of (b) rhodaminedoped silica (DySiO$_2$), (c) ferrite (Fe$_3$O$_4$), and (d) core–satellite DySiO$_2$–(Fe$_3$O$_4$)$_n$ nanoparticles. (From Lee, J.-H., Jun, Y.-W., Yeon, S.-I., Shin, J.-S., and Cheon, J.: Dual-mode nanoparticle probes for high-performance magnetic resonance and fluorescence imaging of neuroblastoma. *Angew. Chem. Int. Ed.* 2006b. 45. 8160–8162. Copyright Wiley-VCH Verlag GmbH & Co. KGaA. Reprinted with permission.)

13.2.3.3 Paramagnetic Nanoparticles

Gadolinium Gd^{3+} complexes own some properties that make them adequate for MRI. The inherent toxicity of these compounds makes it necessary to combine them with ions to form less toxic metal complexes. The most extended compound is Gd-DTPA (Kim et al. 2009, 372–390). However, it was mentioned in Section 13.2.3.1 the recent controversy regarding toxicity of gadolinium based agents observed in certain subsets of patients with pre-existing kidney disease (Idée et al. 2008, 77–88; Shellock and Spinazzi 2008, 1129–1139), which implies to be careful in in-vivo applications.

13.2.3.3.1 Synthesis

In order to obtain good T1 contrast agents, the best strategy is to introduce in the probe as many Gd compounds as possible. Gd is incorporated in quantum dots, dendrimers, silica nanoparticles, and mesoporous silica. Moreover, it is even possible to obtain a combination of IO and paramagnetic nanoparticles. In Dosev et al. (2007, 055102) cobalt- and neodymium-doped IO are encapsulated in shells of europium-doped gadiolinium oxide.

In this sense, it is important to obtain a high surface area that permits to include many Gd compounds. Hence the spherical geometry is the best choice. As an example, in Mulder et al. (2006a, 1–6) quantum dots were coated with a water-soluble and paramagnetic micellular coating. The quantum dots preserve their optical properties and have a very high relaxivity, r$_1$, which is typical of a good T1 contrast agent (the coating was composed of a pegylated phospholipid, PEG-DSPE and a paramagnetic lipid, Gd-DTPA-BSA).

However, even though the spherical geometry is adequate for a maximum number of Gd compounds, there are structures such as dendrimers that are adequate for maximum relaxivity because they own many groups that can conjugate with Gd (Mulder et al. 2007, 307–324). In Talanov et al. (2006, 1459–1463) (PAMAM) dendrimers own 256 surface amino groups conjugated with Gd3$^+$-chelating ligands and NIR fluorescent Cy5.5.

It is important to note that the relaxivity can be increased also by increasing the payload of Gd^{3+} ions (Kim et al. 2009, 372–390). The accessibility of the Gd centers to water molecules is the key to design highly efficient T1 MR contrast agents in nanoparticulate form. In Rieter et al. (2007, 3680–3682) silica nanoparticles containing $Ru(bpy)_3$ where synthesized via the reverse microemulsion method and the surface of the silica with Gd^{3+}-chelating ligands was functionalized. The result was a higher amount of Gd^{3+} ions (see Figure 13.32a). A second strategy (see Figure 13.32b) is to apply layer-by-layer (LbL) method (Kim et al. 2007b, 8962–+). As the number of layers was increased, the relaxivities per particle were increased proportionally. This indicates that water molecules could readily access the Gd centers thanks to the hydrophilic nature of the layer composed of Gd-DOTA and PSS.

Obviously mesoporous silica nanoparticles, which are capable enclosing various nanoparticles (see Section 13.2.5), constitute another way increasing the relaxivity (Kim et al. 2009, 372–390).

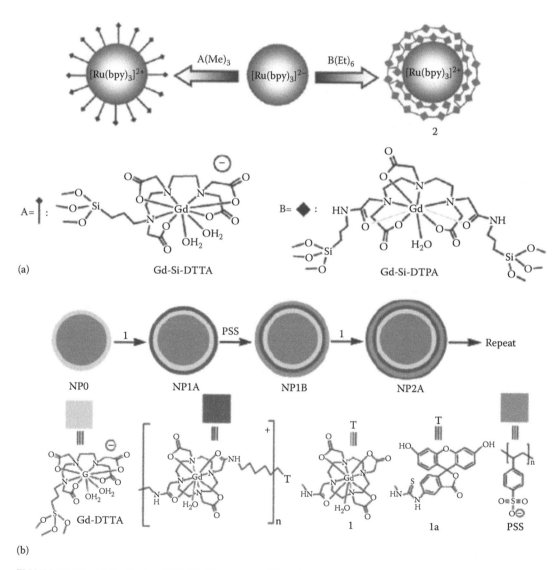

FIGURE 13.32 (a) Synthesis of hybrid silica nanoparticles. (b) LbL self-assembly strategy for magnetic fluorescent nanoparticles. (From Kim, J., Piao, Y., and Hyeon, T., Multifunctional nanostructured materials for multimodal imaging and simultaneous imaging and therapy, *Chem. Soc. Rev.*, 38(2), 372–390, 2009, Reprinted with permission of The Royal Society of Chemistry.)

The relaxivities per particle of the mesoporous silica nanoparticles loaded with Gd^{3+} (Taylor et al. 2008, 2154–+), were even higher than those of non-porous silica nanoparticles loaded with Gd^{3+} via the LbL method.

13.2.3.4 Biological Applications

During the last years there has been a growing interest in the development of biological applications based on the fine properties of superparamagnetic nanoparticles, which allow for multifunctionality. In the near future, the new generation of multifunctional magnetic nanoparticles combined with therapeutic drugs, targeting moieties, and MR contrast agents will allow for the investigation of diseases across a number of platforms, and accumulation of a vast amount of information in clinics (Park et al. 2009, 1553–1566). There are two main disciplines: imaging and detection, and therapy. There are many techniques developed but, according to the title of this chapter, focus is stressed on optical ones, such as the fluorescence technique used in combination with MRI in Huh et al. (2005, 12387–12391).

As a general rule, the in-vitro application requirements are less subtle than the in-vivo ones. Three key questions must be satisfied for in-vivo applications: biocompatibility (the probes must be soluble in water and biologically harmless), optical wavelength not located in the UV domain, which permits its utilization in-vivo, and finally adequate size that avoids on the one hand splenic filtration (particles smaller than 5.5 nm are rapidly removed by the renal system) (Chen and Weiss 1973, 529–537), and detection by phagocytotic cells of the spleen (particles higher than 200 nm) (Soo Choi et al. 2007, 1165–1170). Moreover, as it was mentioned before, opsonization of the particles must be avoided by adequate selection of the nanoparticle probe coating. Otherwise it will be detected and eliminated by the RES system.

Let us consider an example. In Kang et al. (1996, 2209–2211) IOs were coated with Y:Er:NaYF4. Since the shell coating reaction occurs in water under normal atmospheric conditions, some of the particles oxidize to γ-Fe_2O_3. Their particles had iron oxide cores from 5 to 15 nm with 20–30 nm thick shells and overall diameter in TEM is quite large; the particles seem to be 100–200 nm diameter overall. Excitation at 980 nm resulted in up-conversion emission at 539 and 658 nm. So the conditions of biocompatibility and optical wavelength are satisfied, but the size of the particles is two high and they will be cleared fast by the RES.

13.2.3.4.1 Multimodal Imaging and Detection

As it was mentioned in Section 13.2.3.1, MRI and fluorescent imaging complement each other. MRI owns a higher spatial resolution than fluorescent imaging, but fluorescent imaging permits real time monitorization (Kim et al. 2009, 372–390). As an example, dual mode imaging permits to monitorize lymph nodes in Talanov et al. (2006, 1459–1463). There are a wide range of experiments around multimodal imaging. Aminated CLIO (see Section 13.2.3.2.2) was conjugated with Cy5.5, which allowed for detection of brain tumor in rats (Kircher et al. 2003, 8122–8125). In the MRI images, the tumor region exhibited a reduction of the signal intensity, indicating that the brain tumor could be diagnosed. Also protease-sensing probes have been developed based on coupling of Cy5.5-labeled peptides to CLIO through disulfide or thioeteher (Josephson et al. 2002, 554–560). CLIO-Cy5.5 has also been conjugated with vascular cell adhesion molecule-1 imaging (it identifies inflammatory activation of cells in atherosclerosis (Nahrendorf et al. 2006, 1504–1511). In addition to this, chlorotoxin, a targeting peptide for glioma tumors, was also conjugated with the CLIO. In this way, primary brain tumor could be detected (Kohler et al. 2004, 7206–7211). CLIO-Cy5.5 has also been conjugated to peptides targeted to VCAM1 to target endothelium (Kelly et al. 2005, 327–336).

As an alternative to CLIO, thermally cross-linked superparamagnetic iron oxide nanoparticles (TCL-SPION) were prepared for their application to the dual imaging of cancer in vivo (Lee et al. 2007, 12739–12745).

In Park et al. (2008, 7284–7288) micellar hybrid nanoparticles that contain magnetic nanoparticles, QDs, and the anticancer drug DOX within a single PEG-modified phospholipid micelle have been prepared. The strong interaction of the hydrophobic chains of the PEG–phospholipids with hydrophobic chains attached to the magnetic nanoparticles and QDs leads to high dispersibility and stability for in vitro and in vivo applications. Dual mode imaging (optic and MRI) could be obtained with this approach.

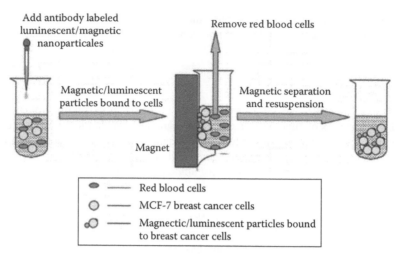

FIGURE 13.33 Magnetic separation of MCF-7 breast cancer cells. (Reprinted with permission from Wang, D., He, J., Rosenzweig, N., and Rosenzweig, Z., Superparamagnetic Fe_2O_3 beads-CdSe/ZnS quantum dots core-shell nanocomposite particles for cell separation, *Nano Lett.*, 4(3), 409–413. Copyright 2004 American Chemical Society.)

In Wang et al. (2004, 409–413) a coupling between QDs and γ-Fe_2O_3 SPIOs is obtained. The SPIOs were coated with a polymer and its surface was thiol-modified to allow the conjugation with QDs. By using a permanent magnet, the nanoparticles can be easily separated from solution. This can be used for bioanalytical assays. As an example, anticycline E antibodies are immobilized on the surface and then, the antibody coated particles permit to separate MCF-7 breast cancer cells from serum solutions (see Figure 13.33). In this case multimodal imaging was not used but it could be so. Simply by analyzing optically by fluorescence microscopy it was possible to analyze the presence of cancer cells.

Similarly, in Law et al. (2008, 7972–7977) solution-phase synthesis of multifunctional nanoprobes for imaging live cancer cells is reported. More specifically, quantum dots (QDs) and magnetite (Fe_3O_4) nanoparticles were coencapsulated within organically modified silica (ORMOSIL) nanoparticles. These optically and magnetically doped ORMOSIL nanoparticles were systematically characterized, and used for magnetically guided in vitro delivery to a human cancer cell line. Two-photon imaging was used to confirm the uptake of nanoparticle-doped ORMOSIL nanoparticles into the cancer cells.

In Mulder et al. (2006a, 1–6) targeting ligands can be coupled to paramagnetic QDs. They were functionalized by conjugating them with cyclic RGD peptides and were successfully targeted to human endothelial cells in vitro, which is a good multimodal probe for the detection of angiogenesis.

There are also commercial clinical products such as Cetuximab. It consists of conjugated fluorescent magnetic nanohybrids (FMNH) for detection of cancer using MR and optical imaging. The nanohybrids are composed of $MnFe_2O_4$ magnetic nanocrystals encapsulated in pyrene-labeled surfactant prepared by a nano-emulsion method. Cetuximab is used for MR and fluorescence optical imaging of cancer cell lines (Yang et al. 2008b, 2548–2555).

Regarding paramagnetic applications, liposomes have also been developed. As an example, a pegylated liposomal contrast agent for dual-mode imaging was introduced with both MRI and fluorescence techniques (Mulder et al. 2004, 799–806). Another paramagnetic applications can be found in Dosev et al. (2007, 055102), where antibodies were physisorbed to the nanoparticle surface of shells of europium-doped gadolinium oxide to demonstrate the ability to use them for magnetic immunoassays.

Finally, Gd^{3+} loaded mesoporous silica nanoparticles (in spherical or nanorod shape [Taylor et al. 2008, 2154–+; Tsai et al. 2008, 186–191]), in view of the improvement in relaxivity they induce, are adequate probes for multimodal imaging. Moreover, the inherent toxicity of Gd is suppressed thanks to the internalization of Gd in the pores (Hsiao et al. 2008, 1445–1452). In Lin et al. (2006, 5170–5172) and Taylor et al. (2008, 2154–+) the nanoparticles were also labeled with a fluorophore in order to allow multimodal imaging. A study on long-term stability is done in Hsiao et al. (2008, 1445–1452), where the labeled cells can be visualized in an MRI system in vivo for at least 14 days.

13.2.3.4.2 Therapy

Compared to the administration of small molecules, the nanoparticle delivery systems can inject a higher amount of drugs (10- to 100-fold) to the vicinity of the tumors (Kaul and Amiji 2002, 1061–1067), thus improving therapeutic efficacy and reducing harmful non-specific side effects. The use of nanoparticles as drug-delivery systems for anticancer therapeutics has great potential to revolutionize the future of cancer therapy (Park et al. 2009, 1553–1566). Moreover, the development of multimode imaging techniques presented above, allows a parallel observation of the evolution of the therapy, which is not possible with administration of small molecules.

The same combination of CLIO-Cy5.5 used for detection of cancer in Kircher et al. (2003, 8122–8125), was attached to synthetic EPPT peptides on a dextran coating to target the underglycosylated mucin-1 (uMUC-1) antigen of various cancer cells (Moore et al. 2004, 1821–1827), with RGD peptides to target integrins (Montet et al. 2006, 214–222) and with annexin V to target apoptotic cells (Schellenberger et al. 2004, 1062–1067). The disadvantage of fluorescence quenching can be converted into an application (Josephson et al. 2002, 554–560). If the fluorophore is conjugated to CLIO particles in such a way that it can be cleaved by an enzyme, the fluorescence is restored in the presence of the enzyme, which will break the bindings between CLIO and fluorphores. This is what is called activatable probes that react against the presence of a specific enzyme.

Thermal crosslinking, as indicated in Section 13.2.3.2.2, is a good alternative to polymer coating in case it is desired to avoid aggregation under harsh conditions (Lee et al. 2006a, 7383–7389). Recently, it has been possible to develop thermally cross linked SPIOs including doxorubicin (Dox), and hence performing dual functions of drug carrier and MRI agent (Yu et al. 2008, 5362–5365). By controlling the pH the Dox was either released or not. In this way, systemically administered Dox@TCL-SPIONs displayed a much lower toxicity in organs than free Dox, but exerted therapeutic response to cancer cells.

Liposomes have received a lot of attention during the last decades as pharmaceutical carriers. More recently, many new developments have been seen in the area of liposomal drugs, from clinically approved products to new experimental applications, with gene delivery and cancer therapy still being the principal areas of interest (Torchilin 2005, 145–160). Fluorescent dyes were incorporated in liposome coated IO nanoparticles (Soenen et al. 2009, 257–267). It was proved that magnetoliposiomes are not actively exocytosed and are very stable over time. Moreover, their retention times greatly exceeded those of the more commonly used SPIOs. They can be detected for up to at least one month (equivalent to approximately 30 cell doublings).

The combination of QDs and magnetic nanoparticles is also widely extended. In Selvan et al. (2007, 2448–2452), this combination is used for marking cells, which permits cell tracking. The emission characteristics can be controlled during the fabrication process. Following the same combination, in Zebli et al. (2005, 4262–4265) the magnetic functionalized and luminescent labeled particles were specifically concentrated in a region by application of a permanent magnet on a specific region. The uptake of capsules by cells could be conveniently monitored with a fluorescence microscope by the luminescence of QDs composed of CdTe nanocrystals that had been embedded. The application was adhesion to pathogenic parts of a tissue. In Kim et al. (2008, 478–+), QDs, magnetic nanoparticles and doxorubicin (DOXO) were embedded in a probe, which was coated with biodegradable poly(D,L-lactic-co-glycolic acid) (PLGA) polymer nanoparticles. The probes could provide effective targeting to folate receptors, detection by MRI and optical imaging, and cell growth inhibition in KB cancer cells. Another good work related to cancer cells showed PAA coated IO nanoparticles containing both an anticancer drug and a near-infrared dye (Santra et al. 2009, 1862–1868). In this work it is studied how the functionalization of the nanoparticles dramatically impacts on the cell intake processes, and therefore in the imaging and tumor cell killing capability of the magnetic nanoparticles. For example it was found that the PAA functional coating is critical for the release time of the anticancer drug. As an example, it has been analyzed in the same work that at pH 4 the release time is much faster than at pH 7 (Santra et al. 2009, 1862–1868).

Regarding mesoporous silica nanoparticles, they were successfully embedded with both semiconductor quantum dots (QD) and ferrite (Fe_3O_4) nanocrystals for both optical encoding and magnetic separation (Sathe et al. 2006, 5627–5632). The short term biocompatibility of this type of nanoparticles has been analyzed in Tsai et al. (2008, 186–191). Finally, a good work that proves the ability of MSNs containing iron oxide nanocrystals and quantum dots to adsorb and release drugs was done in Kim et al.

(2006c, 688–689). The probes adsorbed ibuprofen, and the release rate of ibuprofen was controlled by the surface properties the mesoporous silica nanospheres. Monitorization was also possible thanks to luminescence detection and separation of particles by applying a magnetic field.

13.2.3.4.2.1 Photodynamic Therapy (PDT)   It is a form of light activated chemotherapy using light-sensitive drugs, that is, photosensitizers (Park et al. 2009, 1553–1566). The molecules with the photosensitizer are uptaken by the tumor cells. Once excited by light of the appropriate wavelength, cytotoxic reactive oxygen species are generated, which induces destruction of diseased cells and tissues. Photofrin is one of the most efficient photosensitizers approved for PDT of cancer. However, there are some drawbacks in its utilization: Photofrin can cause prolonged skin photosensitization (patients are required to avoid direct exposure to sunlight for a period of 4–6 weeks) and it is required to wait for one day between the injection of the PDT and the application of irradiation. In Reddy et al. (2006, 6677–6686), polymeric polyacrylamide (PAA) nanoparticles were embedded with iron oxide nanoparticles and Photofrin and the probes were targeted with vascular homing peptide F3. In this way, a more selective uptake was obtained which avoided the necessity to wait for one day for irradiation. Moreover, the efficacy of the treatment was improved compared to experiments developed with untargeted probes. In the experiments, a 630 nm laser source was used, which avoids the UV region.

13.2.3.4.2.2 NIR and RF Photothermal Therapy   The basic procedure consists of applying a radiation that does not damage the tissues and which is absorbed by adequately designed nanoparticles. Typical wavelength domains are near infrared irradiation (NIR) and radiofrequency (RF). The result is that there is a thermal increase in those cells targeted by nanoparticles, which permits to reduce tumors. There are studies such as that of Seo et al. (2006, 971–976), where FeCo/graphitic shells were injected in a rabbit. These probes are adequate contrast agents and it was checked that there is a maximum absorbance in the NIR domain (808 nm). So they are able to convert the near infrared photon energy in thermal energy. Examples in the RF domain can be found in Xu et al. (2010, 167–176).

13.2.3.4.2.3 Gene Delivery   A final group of applications are those aimed for gene delivery. Nanoparticles have the potential to deliver specific molecules into cells for therapeutic purposes. The potential of combining magnetic nanoparticles with fluorophores has been used to monitorize the delivery of molecules in cells.

Antisense oligonucleotides have been applied for effective antisense therapy, but there are some issues that must be solved (Park et al. 2009, 1553–1566). In order to avoid these problems, polyamidoamine (PAMAM) dendrimer-conjugated magnetic nanoparticles have been used as a gene transfection vector (Pan et al. 2007, 8156–8163). Moreover, dendrimers own good properties such as easy production, simplicity and safety.

The well-known CLIO system is used for simultaneous noninvasive imaging and siRNA delivery to tumors (Deng et al. 2003, 1729–+; Medarova et al. 2007, 372–377; 2008, 1170–1177). The siRNA molecules can act as mediators of RNA interference (RNAi) within the cytoplasm of cells. Therapeutic application of siRNAs requires the effective delivery of siRNAs in to the target cells. However, the simple delivery of siRNAs does not permit them to enter into the cells. Again CLIO system is conjugated with NIR fluorescent Cy5.5 dyes and siRNA molecules are set on the surface of the dextran coating. There are other methods for delivery such as lipid based agents, antibody-protein fusion proteins, and liposomes (Park et al. 2009, 1553–1566).

13.2.4 Lanthanides

13.2.4.1 Synthesis and Sensing Techniques

Lanthanides series comprises the following elements: Lanthanum (La), Cerium (Ce), Praseodymium (Pr), Neodymium (Nd), Promethium (Pm), Samarium (Sm), Europium (Eu), Gadolinium (Gd), Terbium (Tb), Dysprosium (Dy), Holmium (Ho), Erbium (Er), Thulium (Tm), Ytterbium (Yb), and Lutetium (Lu).

Gd based applications were included in Section 13.2.3 because of its magnetic properties. Consequently, only applications of the rest of elements will be presented in this section.

Lanthanide f-f electronic transitions are Laporte-forbidden, and lanthanide ions are generally considered to be photophysically inert (Sabbatini et al. 1993, 201–228). However, in the presence of a proximal fluorophore, indirect excitation can be achieved through energy transfer, which is called in some works sensitization (Fu et al. 2005, 747–750; Viguier and Hulme 2006, 11370–11371). Lanthanide chelates present good properties: long fluorescence lifetime, large Stoke shift, and sharp line-like emission bands (very sharp compared to organic fluorophores, typically with a full width at half maximum of less than 10 nm [Pandya et al. 2006, 2757–2766]). For these reasons lanthanide-based luminescent sensors have gained a great deal of attention in recent years and they offer considerable advantages over the use of standard fluorescent dyes for detection in vivo, especially when there is significant autofluorescence (Ai et al. 2009, 304–308).

It was indicated in section dedicated to magnetic nanoparticles in this chapter that fluorescence imaging permits to monitorize biological experiments in real time, which is vital for in-vivo applications. Conventional fluorescence is a phenomenon where the excitation energy is higher than the emission energy. Hence, the excitation wavelength is lower than the emission wavelength. If the emission wavelength is located in the visible range, it is necessary to excite at the UV wavelength range, which is negative for biological tissues (Louie 2010, 3146–3195). Consequently, two photon excitation, where the emitted wavelength is due to the contribution of two low energy emission states, is a better choice. This phenomenon is called upconversion (Chatterjee et al. 2008, 937–943). In this way, the excitation occurs in the NIR domain, which is safe for biological tissues and which permits a deeper penetration of light.

Another interest technique which is also based on lanthanides is Lanthanide Resonance Energy Transfer (LRET) (Kupcho et al. 2007, 13372–13373). This technique is similar to Förster resonance energy transfer (FRET), where a donor fluorophore transfers its excitation to an acceptor one (Förster 1948, 55) (see Section 13.2.2.2). The particularity this time is that the donor is a lanthanide. In Kupcho et al. (2007, 13372–13373) the distance between chelated terbium and two spectrally distinct acceptor fluorophores (fluorescein and Alexa Fluor 633) is calculated. Ligand-dependent association and dissociation of differentially labeled peptides was monitored simultaneously, with no mathematical deconvolution required to separate the signals from the distinct acceptor fluorophores. These results demonstrate the utility of terbium chelates for the analysis of complex multicomponent binding events by LRET.

Lanthanides present two major advantages in comparison with quantum dots. First, they present a lower toxicity than other particles, such as semiconductor. As an example, experiments in-vitro have proved that cadmium is more toxic than cerium, lanthanum, or neodymium (Palmer et al. 1987, 142–156). However, there is still a concern about the toxicity of lanthanides (Louie 2010, 3146–3195). In fact, it has been observed liver induced toxicity (Sarkander and Brade 1976, 1–17). Second, as an example of the efficiency of lanthanides, colloidal Yb/Er and Yb/Tm co-doped $NaYF_4$ nanoparticles have been reported as the most efficient infrared to visible upconversion fluorescent material (Heer et al. 2004, 2102–2105). The fluorescence obtained is seven orders of magnitude higher than CdSe-ZnS QDs (Chatterjee et al. 2008, 937–943). Apart from $NaYF_4$ codoped nanoparticles, there are other structures such as LaF_3, Y_2O_3, $NaGdF_4$ or $LaPO_4$ codoped ones (Stouwdam et al. 2003, 4604–4616; Sivakumar et al. 2006, 5878–5884; Vetrone et al. 2009, 2924–2929; Louie 2010, 3146–3195).

Regarding the synthesis there are several methods to develop lanthanide co-doped nanocrystals. Most of them are based on hydrothermal (Yi et al. 2002, 2910–2914) or solvothermal method (Zeng et al. 2005, 2119–2123). The addition of chelating agents such as ethylenediamine tetraacetic acid (EDTA) (Zeng et al. 2005, 2119–2123), or polyvinylpyrrolidone (PVP) (Li and Zhang 2006, 7732–7735), permits to control the size and stability of the nanoparticles (see Figure 13.34).

In view of the well-known biocompatibility of silica nanospheres, it is a good choice, like in the case of magnetic nanoparticles, to coat them with silica (Van De Rijke et al. 2001, 273–276). However, there is a difficulty in the coating of hydrophobic nanoparticles. Consequently, there is a need for additional processes that add hydrophilicity. For example, $NaYF_4$ was coated with polyethyleneimine (PEI) for this purpose (Wang et al. 2006a, 5786–5791).

Europium and Terbium are the most common lanthanides used because they emit in the visible spectrum, in the red and green regions respectively (Bottrill et al. 2006, 557–571). In addition to

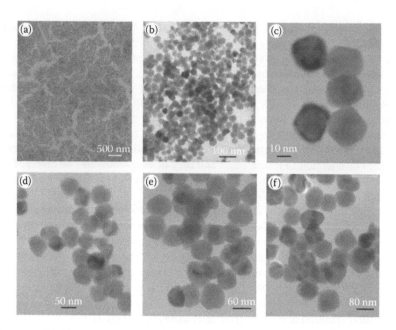

FIGURE 13.34 (a–c) TEM images with different resolutions of PVP/NaYF4: Yb,Er codoped nanocrystals with a size of 30 nm. (d–f) TEM images of nanocrystals with different sizes: (d) 48; (e) 65; (f) 87 nm. (From Li, Z. and Zhang, Y.: Monodisperse silica-coated polyvinyl-pyrrolidone/NaYF4 nanocrystals with multicolor upconversion fluorescence emission. *Angew. Chem. Int. Ed.* 2006. 45. 7732–7735. Copyright Wiley-VCH Verlag GmbH & Co. KGaA. Reprinted with permission.)

this, they possess long luminescent lifetimes, which is advantageous in applications where there is a short-lived background fluorescence (Zheng et al. 2005, 16178–16188). Its utilization is discussed in other works such as (Viguier and Hulme 2006, 11370–11371; Ai et al. 2009, 304–308). Traditionally Tb^{3+} has been considered as a better option in sensors due to the general acceptance that excited states of Eu^{3+} can couple more efficiently with the high-frequency OH oscillators of water molecules than Tb^{3+}, resulting in enhancement of nonradiative quenching of the Eu^{3+} emission and hence lower detection sensitivity. However, Europium^{3+} offers several advantages over Tb^{3+} in terms of larger Stoke shift, red emission, and exclusion of second-order scattering interference in the maximum fluorescence intensity (Ai et al. 2009, 304–308). These advantages apply particularly to the case of a nanoparticle sensor. More importantly, false-positive results arising from the nonselective binding of aromatic compounds to Tb^{3+} can be decreased with the europium-based sensor (Ai et al. 2009, 304–308). As an example, in it has been proved that europium-based nanoparticles are two orders of magnitude more sensitive than terbium ones, and can discriminate anthrax from a host of interfering compounds more effectively than terbium (Chun 2008). In Viguier and Hulme (2006, 11370–11371) it is explained the sensitization of europium; by Huisgen 1,3-dipolar cycloaddition reaction catalyzed by the GS$^-$-Cu(I) complex (where GS is the anion of glutathione). Moreover, in Fu et al. (2005, 747–750) efficient two-photon sensitization of Eu^{3+} luminescence in [Eu(tta)$_3$dpbt] was demonstrated showing a high-purity red emission, which is a good range for its applicability in biological assays.

The behavior of Europium is improved in Dasary et al. (2008, 187–190) where surface enhanced fluorescence (SEF) is used. SEF is the modification of fluorescence in a surface enhanced Raman scattering (SERS) (Jeanmaire and Van Duyne 1977, 1–20) type environment. Fluorescence from Eu^{3+} ions that are bound within the electromagnetic field of gold nanoparticles exhibit a strong enhancement. This technique, which is based on the near field coupling between the emitter and surface modes, leads to an overall improvement in the fluorescence detection efficiency through modification and control of the local electromagnetic environment of the emitter (the near field coupling between the emitter and the surfaces modes) (Fort and Grésillon 2008).

Another form of increasing the fluorescence is also based on silica shell. In this case it is a combination of Ag and SiO_2. Good results are obtained for fluorescent materials; among them Eu-TDPA, [Tris(dibenzoylmethane) mono(5-amino phenanthroline)] (Aslan et al. 2007, 1524–1525).

Another good lanthanide is Yb (Feng et al. 2010, 3596–3600). After ligand-mediated excitation, Yb(DBM)$_3$phen-MMS (phen-MMS means a framework of magnetic mesoporous silica with a chelate ligand), emits in the NIR region (1000 nm). At this wavelength both biological tissues and fluids (e.g., blood) are relatively transparent. Therefore, it is suggested that mesoporous silica nanoparticles with magnetite embedded and covalently binded with Yb have a great potential in drug delivery or optical imaging (Feng et al. 2010, 3596–3600). As an example of this, in Figure 13.35 images of the probes are shown at different scales.

For lanthanide doped nanoparticles there are also techniques for improving the fluorescence such as that described in Sivakumar et al. (2006, 5878–5884), where lanthanide ion (Ln^{3+}) doped LaF3 nanoparticles coated with a silica nanoshell were synthesized and bioconjugated with FITC–avidin. A wide range of emission lines by up- and down conversion processes have been achieved by doping with different lanthanide ions. The surface modification of the silica-coated nanoparticles with 3-APS, followed by biotin–avidin binding, resulted in a 25-fold increase in the FITC signal relative to nonbiotin-functionalized silica coated nanoparticles.

Similarly in Vetrone et al. (2009, 2924–2929), core/shell structure has also permitted to increase the intensity of NIR-to-visible upconversion in Ln^{3+}-doped nanoparticles by employing an active-core/active-shell architecture. NaGdF$_4$:Er^{3+} 2%, Yb^{3+} 20% active-core/NaGdF$_4$:Yb^{3+} 20% active-shell nanoparticles were synthetized, and results were better than those obtained with an inert shell and better than those obtained without shell (core-only nanoparticles). The active-shell serves two purposes: it protects the luminescing Er3+ ions from the non-radiative decay, and it transfers NIR absorbed radiation to the luminescing core.

Similarly in Stouwdam et al. (2003, 4604–4616), it has been explored the factors that permit to increase the luminescence lifetime in surface-coated nanoparticles of LaF$_3$ and LaPO$_4$ doped with the lanthanide ions Eu^{3+}, Nd^{3+}, Er^{3+}, Pr^{3+}, Ho^{3+}, and Yb^{3+}. These ions emit in the visible and in the near-infrared part of the electromagnetic spectrum. The ions Nd^{3+}, Er^{3+}, Pr^{3+}, and Ho^{3+} are the main

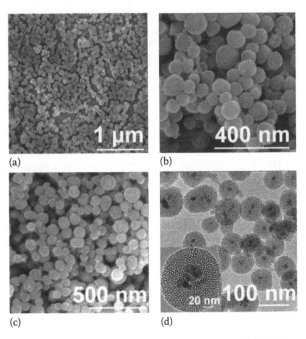

(a)   (b)

(c)   (d)

FIGURE 13.35 SEM images of phen-MMS: (a) Nd(DBM)3phen-MMS, (b) Yb(DBM)3phen-MMS, (c) nanospheres, (d) TEM image of Yb(DBM)3phen-MMS nanospheres. (Reprinted with permission from Feng, J., Song, S.-Y., Deng, R.-P., Fan, W.-Q., and Zhang, H.-J., Novel multifunctional nanocomposites: Magnetic mesoporous silica nanospheres covalently bonded with near-infrared luminescent lanthanide complexes, *Langmuir*, 26(5), 3596–3600. Copyright 2010 American Chemical Society.)

focus in this research because they show emissions in telecommunication windows. However, the interest in biological applications is more focused on Yb^{3+} and Eu^{3+} ions, which are more adequate for biological applications (they emit in wavelengths where the tissues are transparent). The increase in the luminescence lifetime is indicative of an effective shielding of the lanthanide ions from non-radiative decay of the excited state by the high-energy vibrations of the solvents and the coordinated organic ligands.

13.2.4.2 Applications

Though the applications of lanthanide nanoparticles in the biological and the chemical sensors domain has not reached a level of development similar to magnetic nanoparticles or quantum dots, there is a group of interesting publications, which indicate the potential of this research field. The advantages of lanthanides over other elements were stressed in this section, and that is why they have been successful in the applications that will be explained now. In addition to this, the photophysical pathway defining sensitized lanthanide luminescence offers three ways for sensing, as either the two short-lived excited states of the sensitizer or the lanthanide excited state may be perturbed (Pandya et al 2006, 2757–2766).

One of the most important domains of application is obviously time-resolved cellular imaging. Lanthanide based nanoparticles present a luminescence with special characteristics, which is exploited in some works (Pandya et al. 2006, 2757–2766). However, for in-vivo applications, in view of the region where skin is transparent (700–900 nm), NIR emitting luminophores are required. This include cyanine dyes and Nd^{3+}/Yb^{3+} complexes and can be extended by utilization of two-photon excitation protocols to Eu^{3+}/Tb^{3+} luminescent probes. Independently of the lanthanide ion used, the idea is to label the cellules and there are two different types of probes. The first group is based on stable luminescent complexes that serve as "tags" in bioconjugates for use in FRET or 'tracking' assays (in [Hanaoka et al. 2007, 13502–13509] the long-time luminescence of europium based complexes permitted to track living cellules). The second group is composed of responsive systems wherein the spectral emission profile, lifetime or circular polarization varies as a function of the local concentration of a target intracellular analyte (in [Hanaoka et al. 2007, 13502–13509] variations of intracellular Zn^{2+} was detected) (Pandya et al. 2006, 2757–2766). In any case the conjugate must be prepared against interferants present in the assay, which may induce a suppression of a part of the luminescence. Some techniques for avoiding this question were already explained in Section 13.2.4.1.

Multimodal imaging is also possible in the domain of cell tracking. In Zheng et al. (2005, 16178–16188) lipophilic chelates for MRI and fluorescence imaging are designed. They are based on a ligand that is the combination of two known metal ligands (diethylenetriamine pentaacetate (DTPA) and 2,6-pyridin-edimethaneamine (PDA)) in the form Ln/DTPA-PDA-C_n. As an example, in Figure 13.36 a ligand of the form Tb/DTPA-PDA-C_{10} is used. The samples were monitorized with a diffusion enhanced fluorescence resonance energy transfer DEFRET technique. Tb/DTPA-PDA-C_{10} was the donor and calcein was chosen as the energy acceptor. Tb/DTPA-PDA-C_{10} was excited at 262 nm and an emission peak was generated at 490 nm, which is exactly the excitation wavelength of calcein. Finally, calcein emits between 500 and 530 nm (Figure 13.36a). In Figures 13.36b and c two different cases of DEFRET between calcein outside and inside cells with takeup Tb/DTPA-PDA-C_{10} are presented. In both the cases the emission band of calcein between 500 and 530 nm is observed. The idea behind creating a lipophilic chelate was to tag the cell membrane with the contrast agent.

In Dosev et al. (2007, 055102) paramagnetic, superparamagnetic, and luminescent detection is possible with cobalt- and neodymium-doped IO encapsulated in shells of europium-doped gadiolinium oxide. Like in the case where the nanoparticles are only luminescent, in cases where multimodal imaging is desired (typically a combination of MRI and optical imaging) it is also required for biological in-vivo applications to excite the complexes with low energy photons. As an example, upconversion was done in IO nanoparticles coated with $NaYF_4$ codoped with Y and Er shells (excitation 980 nm and emission at 539 and 658 nm) (Lu et al. 2004b, 1336–1341). However, in this case the size of the probes was too high (80–150 nm) and aggregation is observed, which would lead to a fast elimination by the RES system in case they were used in-vivo.

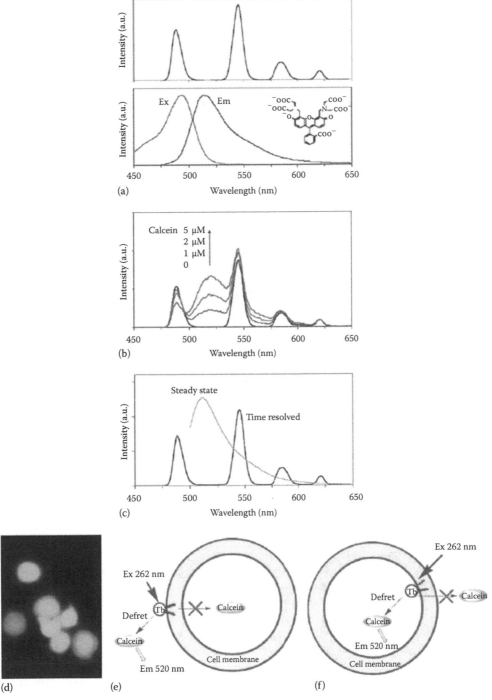

FIGURE 13.36 Probe cellular localization of Tb/DTPA-PDA-C$_{10}$ by DEFRET. (a) The spectral overlap of Tb/DTPA-PDA-C$_{10}$ emission (upper trace) and calcein excitation (lower trace, calcein structure also shown). (b) DEFRET between extracellular calcein and Tb/DTPA-PDA-C$_{10}$ taken up by cells. (c) DEFRET between intracellular calcein and Tb/DTPA-PDA-C$_{10}$ taken up by cells. (d) A fluorescence image (Ex 490 ± 10 nm, Em 530 ± 20 nm) of Jurkat cells loaded with calcein/AM. (e, f) Illustrations of DEFRET between calcein and Tb/DTPA-PDA-C$_{10}$ with respect to their cellular localizations. (Reprinted with permission from Zheng, Q., Dai, H., Merritt, M.E., Malloy, C., Pan, C.Y., and Li, W.-H., A new class of macrocyclic lanthanide complexes for cell labeling and magnetic resonance imaging applications, *J. Am. Chem. Soc.*, 127(46), 16178–16188. Copyright 2005 American Chemical Society.)

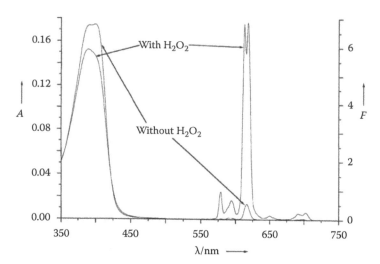

FIGURE 13.37 Absorption and emission spectra of the Eu^{3+}-tetracycline complex in pH 6.9 buffer, in the presence and absence of $0.8\,mmol \cdot L^{-1}$ of hydrogen peroxide. Fluorescence excitation 405 nm. (From Wolfbeis, O.S., Dürkop, A., Wu, M., and Lin, Z.: A europium-ion-based luminescent sensing probe for hydrogen peroxide. *Angew. Chem. Int. Ed.* 2002. 41. 4495–4498. Copyright Wiley-VCH Verlag GmbH & Co. KGaA. Reprinted with permission.)

Another multimodal imaging example is doubly luminescent core/shell structure nanoparticles (Louis et al. 2005, 1673–1682). The core is composed of gadolinium oxide (Gd_2O_3) doped with the luminescent Tb^{3+} ions. The water sensitivity of these particles, which is detrimental for the Tb ion's luminescence, was overcome by embedding the oxide core in a functionalized polysiloxane shell prepared by hydrolysis condensation of a mixture of APTES and TEOS. Due to the presence of amino groups, organic dyes, and biotargeting groups were covalently linked to the polysiloxane network. Consequently, not only an imaging system was developed, but it was also possible to detect biomolecules whose presence is revealed by the high fluorescence of organic dyes and/or the photostable Tb^{3+} ion's luminescence.

Regarding molecular detection applications, in Wolfbeis et al. (2002, 4495–4498) a hydrogen sensor was developed based on an Eu^{3+}-tetracycline complex. On addition of H_2O_2 a 15-fold increase in luminescence intensity at 616 nm occurs (see Figure 13.37).

Immunoassays are also a domain where lanthanide based nanoparticles can be used. In Hai et al. (2004, 245–246) an europium based immunoassy has been developed, whose aim was to detect hepatitis B surface antigen. In Ai et al. (2009, 304–308) the realization of a europium-based nanoparticle sensor for the rapid and ultrasensitive detection of B. anthracis spores in aqueous solution. In comparison with reported terbium-based sensors, the sensitivity of the europium-based sensor toward CaDPA, an anthrax biomarker, was higher and has a remarkable selectivity over aromatic ligands in aqueous. This is possible with a design strategy based on combination of ethylenediamine tetraacetic acid dianhydride (EDTAD) with $EuCl_3$. Under the presence of CaDPA, [Eu(EDTA)-(DPA)] complex is formed, which minimizes the non-radiative quenching of Eu^{3+} emission. In Figure 13.38 the EDTAD with $EuCl_3$ based nanoparticle sensor (sensor 1) presents a much higher luminescence than a free $EuCl_3$ sensor agent (sensor 2), under the presence of $1\,\mu m$ CaDPA.

Other immunoassays can be found in Tan et al. (2004, 2896–2901), where fluorescent europium nanoparticles are used for biolabeling of streptavidin and used in sandwich-type time-resolved fluoroimmunoassay (TR-FIA) of carcinoembryonic antigens (CEA) and hepatitis B surface antigens (HBsAg) in human sera. The same labeling of strepavidin was done in Ye et al. (2004, 513–518), with silica-coated terbium^{3+} chelate fluorescent nanoparticles.

Another interesting application is detection of organophosphorus agents (OPA), which represent a serious concern to public safety as nerve agents and pesticides (Dasary et al. 2008, 187–190). A gold nanoparticle based SEF probe is used. Fluorescent from Eu^{3+} ions that are bound within the electromagnetic field of gold nanoparticles exhibit a strong enhancement (Dasary et al. 2008, 187–190). Eu^{3+} owns

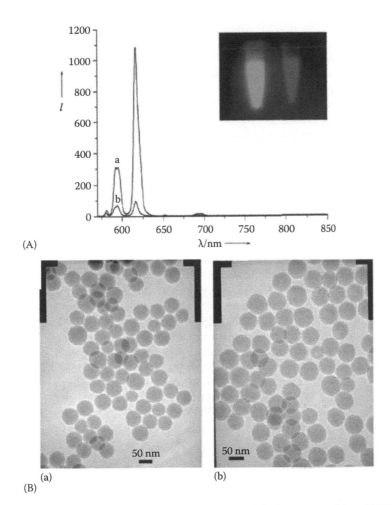

(A)

(B)

(a)   (b)

FIGURE 13.38 (A) Fluorescence spectra of (a) Sensor 1 and (b) Sensor 2 in the presence of 1 μm CaDPA at pH 6.5. The inset shows visual fluorescence color of Sensor 1 (left) and Sensor 2 (right) in the presence of 10 μm CaDPA (filtration with a CB565 filter). (B) TEM images of FITC doped silica nanoparticles: (a) with combination of EDTAD and $EuCl_3$, (b) with free $EuCl_3$. (From Ai, K., Zhang, B., and Lu, L.: Europium-based fluorescence nanoparticle sensor for rapid and ultrasensitive detection of an anthrax biomarker. *Angew. Chem. Int. Ed.* 2009. 48. 304–308. Copyright Wiley-VCH Verlag GmbH & Co. KGaA. Reprinted with permission.)

several fluorescence emission bands. In the presence of gold, the band at 615 is increased, and this band is sensitive to OPAs up to the μM level.

Vascular imaging is also possible with europium based chelates. In this case a Eu^{3+} ligand was used to determine the binding affinities with human serum albumin (HSA) from luminescent enhancement of the bound species (Hamblin et al. 2005, 657–659).

Lanthanide probes based on dual-lanthanide-chelated silica nanoparticles were generated with precisely controlled ratios of Eu^{3+}:Tb^{3+} (Zhang et al. 2007, 5875–5881). In this way, specific ratios of luminescence intensity, at two different emission wavelengths, under a single excitation wavelength can be obtained. As an application example, human hepatitis B surface antigen could be detected. The results showed much increased sensitivity and quantification ranges than the enzyme-linked immunosorbent assay, demonstrating the suitability and advantages of these nanoparticles for bioassays (Zhang et al. 2007, 5875–5881).

Finally, lanthanide probes can be used for detection of ions. In Bruce et al. (2000, 9674–9684) reversible oxy-anion binding in aqueous media at chiral Eu^{3+} and Tb^{3+} centers has been characterized by nuclear magnetic resonance (NMR) and by changes in the emission intensity. In Parker (2000, 109–130), sensors are presented for detection pH, oxygen and selected anions. Further efforts have been done towards combination of lanthanides with inocuous nanoparticles. In Ipe et al. (2006, 1907–1913) a dramatic decrease in the luminescence of Eu^{3+}/Tb^{3+} ions and bipyridines, functionalized on the surface of Au nanoparticles was observed upon addition of alkaline earth metal ions (Ca^{2+}, Mg^{2+}) and transition metal ions (Cu^{2+}, Zn^{2+}, Ni^{2+}).

As indicated in Gunnlaugsson and Leonard (2005, 3114–3131), even though many applications based on lanthanide complexes have been developed, there are some drawbacks that must be solved: they have low quantum yields and due to potential toxicity they have to be formed as kinetically and thermodynamically stable complexes. So in general there is a lot to do in the development of luminescent lanthanide probes that guarantee can be used for in-vivo biological tissues (Louie 2010, 3146–3195). Despite the great promise shown by lanthanide-doped nanoparticles in biological labeling, their synthesis and biological applications are still in the early stages (Knopp et al. 2009, 14–30).

13.2.5 Silica

The integration of multiple nanomaterials into a single nanosystem or a "lab-on-a-particle" (see Figure 13.39), is one of the main objectives of scientifics nowadays (Burns et al. 2006a, 1028–1042; 2006b, 723–726; Piao et al. 2008, 3745–3758). This requires the careful control of many factors. Of the various platforms explored, integrated nanoparticle systems based on silica nanoparticles have attracted a great deal of interest (Piao et al. 2008, 3745–3758). For the sake of simplicity they will be divided into two groups: silica nanoparticle based configurations and mesoporous silica.

The explanation for the success of silica particles can be explained by its inherent properties. Silica is "generally recognized as safe" (GRAS) by the US Food and Drug Administration (FDA) (Piao et al. 2008, 3745–3758) and, in view of the low-toxicity they present (Piao et al. 2008, 3745–3758), they are considered as biocompatible (Burns et al. 2006a, 1028–1042). In addition to this, they are water soluble, uniform in size, chemically stable, and it is possible to modify its surface with a wide range of functional groups, which permits conjugation of targeting molecules (Burns

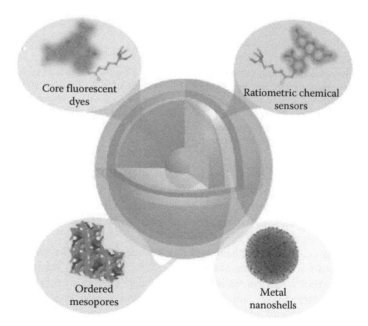

FIGURE 13.39 Lab on a particle concept, which incorporates probe/beacon capabilities with metal nanoshells and/or mesoporous materials for therapeutic delivery to develop materials for nanomedicine applications. (From Piao, Y., Burns, A., Kim, J., Wiesner, U., and Hyeon, T.: Designed fabrication of silica-based nanostructured particle systems for nanomedicine applications. *Adv. Funct. Mater.* 2008. 18. 3745–3758. Copyright Wiley-VCH Verlag GmbH & Co. KGaA. Reprinted with permission.)

et al. 2006a, 1028–1042). Moreover, they are optically transparent and resistant to swelling (Louie 2010, 3146–3195). Finally, the preparation is also cost effective with simple methods that will be discussed now.

However, there are drawbacks. Even though it has been mentioned that silica particles present a good level of biocompatibility, there is still a level of toxicity that must be completely avoided if applications such as in-vivo cellular uptake are to be developed. Liposomes or micelles, which are composed of phospholipids, are examples that overcome silica particles (Van Schooneveld et al. 2008, 2517–2525). Another problem is particle aggregation, an effect which is enhanced when silica particles are administered intravenously due to the absorption of opsonins (markers which enhance endocytosis). The result is a fast suppression of these particles by the RES (Barbé et al. 2004, 1959–1966). As an example, it has been checked that the tendency for aggregation in blood is higher than it is for those nanoparticles coated in dextran or carbodextran (Weissleder et al. 1989a, 835–839). In this sense, great efforts are being dedicated, as we will show in the applications section.

13.2.5.1 Silica Nanoparticle–Based Configurations

13.2.5.1.1 Synthesis of Silica Nanoparticles

According to many authors (Koole et al. 2008, 2471–2479; Piao et al. 2008, 3745–3758; Van Schooneveld et al. 2008, 2517–2525; Knopp et al. 2009, 14–30; Louie 2010, 3146–3195), there are two principal methods for synthesizing silica nanoparticles. The first one, Stöber sol-gel method (Stöber et al. 1968, 62–69), was introduced in the section dedicated to magnetic nanoparticles. Stöber method is primarily used for the preparation of hydrophobic dye-doped nanoparticles (Knopp et al. 2009, 14–30). Recent papers use a modified layer-by-layer approach to incorporate mixtures of probes in silica (Louie 2010, 3146–3195).

Water-oil (W/O) reverse microemulsion is the second method, where the micelles are used to confine the seed particles and control the deposition of silica within the micelle (Piao et al. 2008, 3745–3758). It requires more steps that Stöber method (it mixes surfactants, oil and water to form nanoreactors for the synthesis of silica nanoparticles) (Louie 2010, 3146–3195). Hydrophilic dyes tend to be encapsulated using the microemulsion method (Knopp et al. 2009, 14–30). Another important difference between both methods refers to the size of the particles. The Stöber synthesis generates 50–1000 nm particles, whereas the microemulsion synthesis produces <50 nm particles (Van Schooneveld et al. 2008, 2517–2525).

13.2.5.1.2 Configurations

According to Piao et al. (2008, 3745–3758), there are three different types of structures based on silica nanoparticles. There are more complex classifications, but the simplicity of Figure 13.40 is preferred for the purpose of the explanation in this section.

The first one consists of employing the silica matrix as a core. Once the silica nanoparticle is grown, several techniques can be used to attach the material such as LbL (Osseo-Asare and Arriagada 1990, 321–339) or chemical adsorption of nanoparticles onto functionalized silica nanoparticles (Westcott et al. 1998, 5396–5401). As an example of the versatility of this configuration, core–satellite structured hybrid nanoparticles composed of dye-doped silica as the core and multiple iron oxide nanoparticles as the satellite is explained in Lee et al. (2006b, 8160–8162). However, the simple dye-doped silica nanoparticle configuration permits also to develop interesting applications (Santra et al. 2001, 4988–4993).

The second strategy is to set the silica as a shell. The methods used to this purpose are explained in the synthesis of silica section. An additional method can be found in Garcia et al. (2003, 13310–13311), which is based on synthesis of nanoparticles from reverse block copolymer mesophases. This permits to control both the size distributions of the nanoparticles and their shape, including forms such as nanospheres, nanocylinders, or nanoplates.

The shell may contain combinations of dyes and magnetic materials, such as $Ru(bpy):Gd^{3+}/SiO_2$ (Santra et al. 2005, 593–602). Both the W/O (Santra et al. 2005, 593–602) or the Stöber method can be applied leading to different results. The perfection of development methods is crucial and the efficiency and the stability of the fluorescence has improved the results obtained with simple dye doped

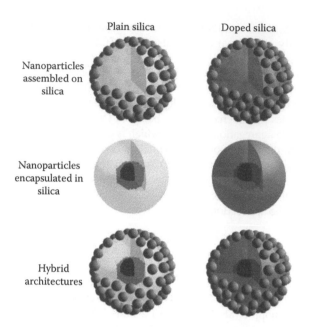

FIGURE 13.40 Schematic illustration of several basic architectures for the development of multifunctional silica nanoparticles. (From Piao, Y., Burns, A., Kim, J., Wiesner, U., and Hyeon, T.: Designed fabrication of silica-based nano-structured particle systems for nanomedicine applications. *Adv. Funct. Mater.* 2008. 18. 3745–3758. Copyright Wiley-VCH Verlag GmbH & Co. KGaA. Reprinted with permission.)

silica nanoparticles. In this sense it is important to highlight the C-dots, which are cited in some works (Fuller et al. 2008, 1526–1532; Piao et al. 2008, 3745–3758). These particles allow for high fluorescence, because multiple fluorophores can be entrapped in the core of the nanoparticle (Santra et al. 2001, 4988–4993; Ow et al. 2005, 113–117; Choi et al. 2007; Knopp et al. 2009, 14–30).

In Figure 13.41 the brightness of C-dots is compared with its constituent fluorophore (showing a 20 fold increase (Ow et al. 2005, 113–117)) and with quantum dots (the difference with quantum dots is only

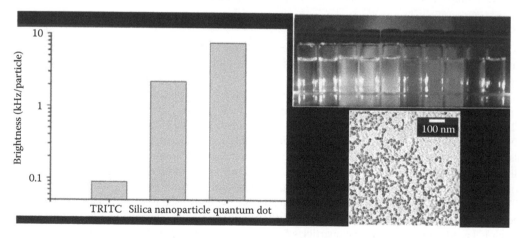

FIGURE 13.41 Brightness comparison for TRITC, silica nanoparticles and quantum dots. The brightness values are obtained from FCS measurements of the count rate/particle at intensities below which fluorescence saturation and photo-bleaching occur. Excitation: 860 nm. Power: 0.4 mW. The resultant 30 nm particles as characterized by transmission electron microscopy (TEM). Because of the slow water evaporation process used to prepare the TEM grids, the particles are aggregated on bottom right image. Different organic dyes incorporated in the nanoparticles are presented on top right image. (Reprinted with permission from Ow, H., Larson, D.R., Srivastava, M., Baird, B.A., Webb, W.W., and Wiesner, U., Bright and stable core-shell fluorescent silica nanoparticles, *Nano Lett.*, 5(1), 113–117. Copyright 2005 American Chemical Society.)

a factor of two to three and may be overcome with improved synthetic methods for the silica nanoparticles). Moreover, in Knopp et al. (2009, 14–30) it is indicated that the silica nanoparticles can contain hundreds to thousands of dye molecules and, therefore, an intense fluorescence signal is obtained. However, a careful control must be done of the amount of dye stored in the core of the particle in order to avoid self quenching (Knopp et al. 2009, 14–30).

Magnetite-doped silica nanoparticles are also preferable to other carriers, such as membranes and microcapsules, since they provide a regular spherical surface that can allow an even distribution of immobilized proteins, reduce the diffusion limitations on both reactants and products, and facilitate fast biochemical reactions between specific combinations of bioactive molecules and their counterparts (Knopp et al. 2009, 14–30). As an example, in Deng et al. (2005, 5548–5550) magnetic nanoparticles are coated with silica in a core-shell configuration. An interesting application was presented in Section, 13.2.3.4, where the utilization of a permanent magnet permitted to separate cancer cells in an immunoassay (Wang et al. 2004, 409–413)

Lanthanide doped nanoparticles are also a good candidate for core-shell structures. In Section 13.2.4 it was presented the 25 fold increase of the FITC signal obtain by adequate modification of the silica surface (Sivakumar et al. 2006, 5878–5884). In the same section, the phenomenon of Surface Enhanced Raman Scattering (SERS) was mentioned, which for the case of fluorescence is called SEF. Nanoprobes containing silica based on this phenomenon have been developed, and they consist of core-shell nanoparticles with a metallic core for optical enhancement, a reporter molecule for spectroscopic signature, and an encapsulating silica shell for protection and conjugation (Doering and Nie 2003, 6171–6176).

The applications of quantum dots can even be widened by coating with silica, which allows for biocompatibility. As an example, in Tan et al. (2007, 3112–3117) a simple reverse microemulsion method is used for the fabrication of silica-coated PbSe QDs that were both water soluble and non-cytotoxic.

The core can also be a combination of two types of particles. For instance, it was seen in Section 13.2.3.2 that dimmers composed of magnetite and QD shell can be developed (Selvan et al. 2007, 2448–2452). In order to confer this multiple purpose nanoparticle better properties, a silica shell is added. However, it is typical that a quenching is induced in the dye by the magnetic nanoparticle.

That is why a core-shell-shell architecture is often a good solution. In Li et al. (2008, 114–117), magnetite nanoparticles are embedded in a silica shell, and QDs are embedded in a second silica shell. The aim is to reduce the quenching induced by magnetite nanoparticles in QDs. Similarly, in Ren et al. (2009, 640–645) a two layer configuration was developed (see Figure 13.42). First, iron oxide nanoparticles

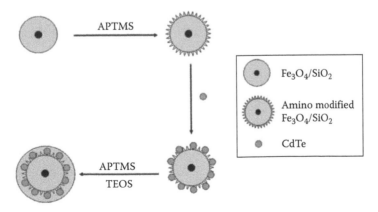

FIGURE 13.42 Preparation procedure of multifunctional nanoparticles composed of iron oxide nanoparticles and CdTe nanocrystals. (Reprinted with permission from Ren, C., Sun, J., Li, J., Chen, X., Hu, Z., and Xue, D., Bi-functional silica nanoparticles doped with iron oxide and cdte prepared by a facile method, *Nanoscale Res. Lett.*, 4(7), 640–645. Copyright 2009 Springer Science+Business Media.)

were coated with silica. Then the surface was modified with amino group. After that, CdTe nanocrystals were assembled on the particle surfaces, and a silica shell was deposited.

Multilayer configurations based on coating with several polymer layers is also possible (Kim et al. 2007b, 8962–+).

Finally, hollow silica nanoparticles are a possibility for drug upload and delivery (Yang et al. 2008a, 3417–3421). Initially the nanoparticles were composed of ferrite nanoparticles core and a silica shell. After treatment with hydrochloric acid and calcination at high temperature, the core was removed.

The last configuration, the hybrid design, is the combination of the previous two methods as well as the incorporation of materials into the silica matrix such as molecular and atomic species. As an example, in Rieter et al. (2007, 3680–3682) the reverse microemulsion method is used for silica nanoparticles containing a luminescent $[Ru(bpy)_3]Cl_2$ core (bpy = 2,2'-bypyridine) and a paramagnetic monolayer coating of a silylated Gd complex.

13.2.5.1.3 Applications

Here applications of the configurations explained in the previous section will be presented.

Biological imaging is one of the main fields where silica nanoparticles are applied. In this sense, the good properties of silica make it a good candidate for optical probes. For example the good biocompatibility of C-dots was tested and they induced no toxicity and they were erased by RES after some days (Ow et al. 2005, 113–117), which permits to use them for in-vivo imaging (Choi et al. 2007b). Moreover, multimodal imaging is possible by combining fluorophores and magnetic nanoparticles in the core of a silica shell (Santra et al. 2005, 593–602; Rieter et al. 2007, 3680–3682), as seen in Section 13.2.3.

Controlling the toxicity and the time when nanoparticles are erased has also been explored in other works. One possibility is based on the covalent conjugation of silanols on the silica surface with silanols conjugated with a short functional group, which results in free amine, thiol, carboxyl, carboxylate, or phosphonate groups on the particle exterior. A variation of this technique is the chemical attachment of silanols conjugated with large molecules, that is, block copolymers, such as PEG. However, short functional groups are easily subject to opsonization and not always stable against aggregation, whereas the chemical attachment of large molecules is likely to result in low-density surface coatings (Van Schooneveld et al. 2008, 2517–2525). To solve these issues in Koole et al. (2008, 2471–2479) a strategy based on a monolayer lipid is presented (see Figure 13.43).

The results showed that monolayer lipid coating increased the blood circulation halflife time of silica particles by a factor of 10 (165 vs. 15 min for lipid coated or bare silica particles, respectively). This relatively long circulation half-life time as compared to earlier reported values in the order of a few minutes, suffices for most targeting applications of silica-based diagnostics and therapeutics. Furthermore, the lipid coating of silica particles results in a favorable tissue distribution profile as compared to the bare silica particles (Van Schooneveld et al. 2008, 2517–2525). In this work the results in living mice are presented, and it is shown that bare silica nanoparticles are immediately uptaken by the liver, whereas lipid coated silica nanoparticles circulate for a much longer time and are progressively uptaken.

Application in-vitro for cellular labeling was found in Koole et al. (2008, 2471–2479), where $\alpha_v\beta_3$-specific nanoparticles were uptaken by human umbilical vein derived endothelial cells. The particles had both fluorescent and paramagnetic properties, which permitted to monitorize $\alpha_v\beta_3$ integrin expression on the cells.

However, the harsh environmental conditions into which nanoparticles are placed, often causes inactivation of the sensitive biological targets. To deal with these challenges various surface modifications and immobilization procedures such as physisorption, affinity interaction, covalent conjugation, and entrapment in sol–gel matrices, have been explored and developed (Knopp et al. 2009, 14–30).

Regarding the question of cell tracking, in order to be able to enter the cytoplasm and nucleus of the cellules, the nanoparticles can be coated electrostatically with cationic polymers, changing their surface charge and enabling them to escape from endosomes and enter the cytoplasm and nucleus (Fuller et al. 2008, 1526–1532). Moreover, they can be complexed with DNA, and mediate and trace DNA delivery and gene expression. DNA hybridization can also be analyzed (Wang et al. 2006b, 646–654).

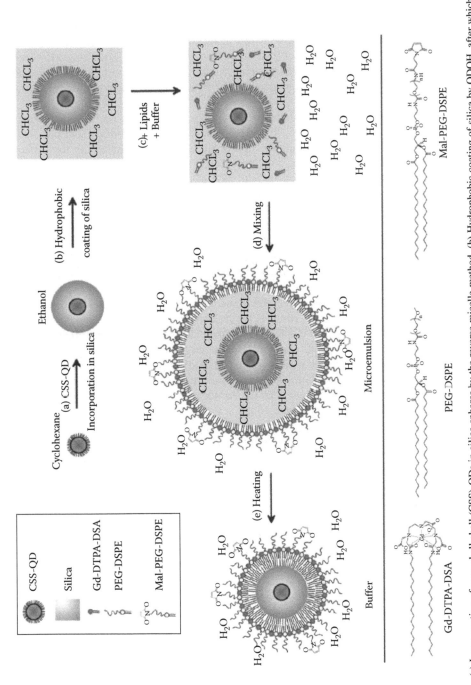

FIGURE 13.43 (a) Incorporation of core-shell-shell (CSS)-QDs in silica spheres by the reverse micelle method. (b) Hydrophobic coating of silica by ODOH, after which they can be dispersed in chloroform. (c) Addition of the different lipids to the QD/silica particles in chloroform, which is subsequently added to a HEPES buffer. (d) Vigorous stirring results in an emulsion, with the chloroform and nanoparticles enclosed in a lipid monolayer. (e) Chloroform is evaporated by heating the mixture, resulting in water-soluble Q-SiPaLCs (QD containing silica particles with a paramagnetic lipid coating). (Reprinted with permission from Koole, R., Van Schooneveld, M.M., Hilhorst, J., Castermans, K., Cormode, D.P., Strijkers, G.J., Donegá, C.D.M. et al., Paramagnetic lipid-coated silica nanoparticles with a fluorescent quantum dot core: A new contrast agent platform for multimodality imaging, *Bioconjugate Chem.*, 19(12), 2471–2479. Copyright 2008 American Chemical Society.)

An alternative method for cell labeling can be found in Zhelev et al. (2006, 6324–6325), where The silica-shelled QD micelles can be highly concentrated into the cells. This improves the quality of the images. The easy penetration of these nanoparticles into cells and the possibility for conjugation with other chemical species makes this probe adequate for drug delivery. This last application can also be developed with hollow silica nanoparticles (HSNP) (Yang et al. 2008a, 3417–3421). In order to permit a controlled release of the drug, a surface modification was necessary. The best choice was to use PEG on the surface due to the pore narrowing effect induced by this molecules.

Biological applications such as the detection of glucose by trapping QDs with silica have also been developed. In Cavaliere-Jaricot et al. (2008, 375–383), an indirect method is presented which is based on the quenching of the fluorescence induced by the presence of the H_2O_2 reaction product produced during glucose oxidation.

Another widely explored field is the improvement of specificity. To this purpose, targeting agents can be added. For instance, in Wu et al. (2009, 1600–1607), horseradish peroxidise (HRP), a well-known enzyme used in many biological applications, is immobilized with an antibody on the surface of silica nanoparticles. Similarly, in Wei et al. (2008, 3687–3693) [Ru(bpy)$_3$]$^{2+}$-doped silica nanoparticles were immobilized on a glassy carbon electrode with biomolecules such as BSA, lysozyme, and ctDNA for biosensing. Regarding biomedical applications, neuroblastoma model cells expressing PSAs, were detected with probes based on conjugated DySiO$_2$–(Fe$_3$O$_4$)$_n$ hybrid nanoparticles with HmenB1 antibodies through sulfo-SMCC conjugation. HmenB1 antibody has been known to specifically target cells with PSAs (Lee et al. 2006b, 8160–8162). In Santra et al. (2001, 4988–4993) identification of leukemia cells is possible with antibody dye doped silica nanoparticles. In Lian et al. (2004, 135–144) particles are coupled with antibodies for sensitive detection of antigens using various formats. In Tapec et al. (2002, 405–409), the immobilization of glutamate dehydrogenase on the surface of organic-dye-doped silica nanoparticles, enables its utilization for glutamate determination. In Yang et al. (2004, 163–169) a fluoroimmunoassay is developed for AFP and hepatitis B surface antigens (HBsAg) based on fluorescein-doped silica nanoparticle with immobilized antibodies on its surface. In Zhao et al. (2004, 15027–15032), monoclonal antibodies (mAbs) against *E. coli* O157 were covalently immobilized onto the surface of fluorescent-bioconjugated silica nanoparticles (RuBpy-doped silica). The mAb was highly selective for *E. coli* O157:H7 in the immunoassay because the antibody-conjugated nanoparticles specifically associated with *E. coli* O157:H7 cell surfaces (Figure 13.44a) but not with *E. coli* DH5α, which lacks the surface O157:H7 antigen (Figure 13.44b). In Figure 13.44c a fluorescence image is shown of a single *E. coli* O157 cell. The probe contains thousands of mAbs conjugated with fluorescent dye molecules, which amplifies the fluorescence. Response times were less than 20 min.

In Kim et al. (2006b, 6967–6973) SERS tagging of cancer cells is achieved. SERS dots composed of silver nanoparticle-embedded silica spheres and organic Raman are used as labels for cellular cancer targeting in living cells. SERS dots showed linear dependency of Raman signatures on their different amounts, allowing their possibility for the quantification of targets. In addition, the antibody-conjugated SERS dots were successfully applied to the targeting on cellular membranes of proteins HER2 and CD10, which are overexpressed in some cancers, and exhibited good specificity. Another application of SERS tagging nanoparticles is the detection of DNA sequence related to HIV. To this purpose Ag/SiO$_2$ core-shell nanoparticle-based Raman tags and the amino group modified silica-coated magnetic nanoparticles as immobilization matrix and separation tool are used. The hybridization is performed between Raman tags functionalized with 3′-amino-labeled oligonucleotides as detection probes and the amino group modified silica-coated magnetic nanoparticles functionalized with 5′-amino-labeled oligonucleotides as capture probes. The detection is done with Raman spectroscopy (Liang et al. 2007, 443–449). A more complex design is done in Gong et al. (2007, 1501–1507), where a sandwich-type immunoassay was performed between polyclonal antibody functionalized Ag/SiO$_2$ nanoparticle-based Raman tags and monoclonal antibody modified silica-coated magnetic nanoparticles. The presence of the analyte and the reaction between the antigen and antibody can be monitored by the Raman spectra of the Ag/SiO$_2$ tags. The main advantages of this strategy are the high stability of Raman tags derived from the silica shell-coated silver core-shell nanostructure and the use of silica-coated magnetic nanoparticles as immobilization matrix and separation tool, which permits to avoid complicated pretreatment and washing steps.

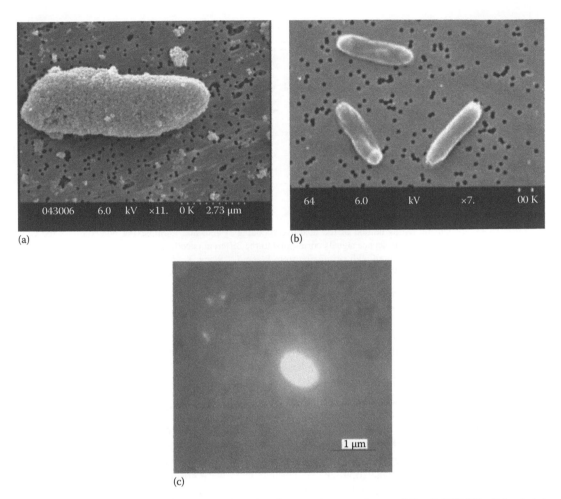

(a)

(b)

(c)

FIGURE 13.44 Images of bacterial cells. (a) Scanning electron microscope image of *E. coli* O157:H7 cell incubated with antibody-conjugated nanoparticles. (b) Scanning electron microscope image of *E. coli* DH5α cell (negative control) incubated with nanoparticles conjugated with antibody for *E. coli* O157:H7. (c) Fluorescence image of *E. coli* O157:H7 after incubation with antibody-conjugated nanoparticles. The fluorescence intensity is strong, enabling single-bacterium cell identification in aqueous solution. (Reprinted with permission from Zhao, X., Hilliard, L.R., Mechery, S.J., Wang, Y., Bagwe, R.P., Jin, S., and Tan, W., A rapid bioassay for single bacterial cell quantization using bioconjugated nanoparticles, *Proc. Natl. Acad. Sci. USA*, 101(42), 15027–15032. Copyright 2004 National Academy of Sciences, U.S.A.)

An interesting application is found in Wang et al. (2006b, 646–654), where dye doped silica nanoparticles are used as novel substrates for multiplexed optical signaling. If the type and concentration of the doped dyes are changed, different barcoding nanoparticles can be prepared that exhibit unique optical signatures. These nanoparticles can then subsequently identify multiple biomolecules on the target surface (see Figure 13.45).

However, the previous method lacks a reference. By doping the silica shell with a dye a pH sensor can be developed where the core acts as a reference dye (Burns et al. 2006b, 723–726). In this way, the probes can be used for quantitatively measurements where the dye core acts as a reference and the shell allows maximum surface of interaction for the dye used for sensing. (Burns et al. 2006a, 1028–1042; 2006b, 723–726). Otherwise the results are qualitative: the vast majority of these molecular sensors can only provide qualitative data, as the measured fluorescence intensity is dependent not only on the analyte concentration, but also on the concentration of the sensor, and the free dye molecules are limited in brightness due to photobleaching, cellular toxicity, and solvatochromic shifts (Burns et al. 2006a, 1028–1042). This idea has been named nanosized photonic explorers for

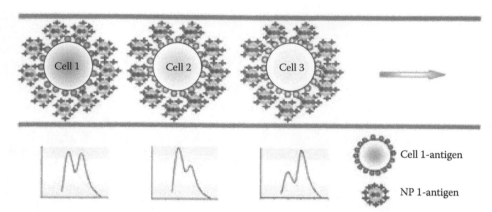

FIGURE 13.45   Schematic representation of the dual-dye-encoding system and multiplexing potential. As the cells pass through the channel, the fluorescence signals correspond to the different encoding nanoparticles (NPs) specifically attached to the target cells. (Reprinted with permission from Wang, L., Wang, K., Santra, S., Zhao, X., Hilliard, L.R., Smith, J.E., Wu, Y., and Tan, W., Watching silica nanoparticles glow in the biological world, *Anal. Chem.*, 78(3), 646–654. Copyright 2006 American Chemical Society.)

bioanalysis with biologically localized embedding (PEBBLEs) and it can be used for intracellular monitorization of small analytes such as H^+, Ca^{2+}, Mg^{2+}, Zn^{2+}, O_2, K^+, Na^+, Cl^-, OH, and glucose (Buck et al. 2004, 540–546). The probes are based on the inclusion of fluorescent analyte-sensitive indicator dyes and analyte-insensitive reference dyes in different matrix containing structures, silica being one of them.

An example of a commercialized product can be found in Corpuscular Inc.

Finally, some applications of therapy are also based on silica nanoparticles. It was experimentally proved that gold coated silica nanoparticles can be used in photothermal therapy (Gobin et al. 2007, 1929–1934), whereas photodynamic therapy could be applied with silica nanoparticle (up to 30 nm) based probes (Kim et al. 2007a, 2669–2675). The particles were co-encapsulated with a photosensitizing anticancer drug, HPPH, and fluorescent aggregates of a two-photon absorbing dye, BDSA. Two-photon spectroscopic measurements showed that BDSA aggregates could efficiently up-convert the energy of near-IR light and transfer it to the HPPH molecules through FRET. This permitted to increase the generation of singlet oxygen in water, which is used for cell killing. A multimodal imaging alternative with the inclusion of magnetic nanoparticles can be found in McCarthy et al. (2006, 983–987). However, the interaction of magnetic nanoparticles induces a quenching in the fluorescence. This can be avoided with the design of Lai et al. (2008, 218–224), where Fe_3O_4/SiO_2 core/shell nanoparticles are functionalized with phosphorescent iridium complexes (Ir), which avoids self-quenching.

13.2.5.2 Mesoporous Silica Nanoparticle

13.2.5.2.1 Characteristics

Mesoporous silica nanoparticles (MSNs) appear in the nineties as a variant of silica nanoparticles. The first and the most extended type is Mobil crystalline materials (MCM-41) (Beck et al. 1992, 10834–10843). However, there are others such as SBA-15 (Dai et al. 2008, 1070–1076), or TUD-1. The good properties of silica nanoparticles were presented at the beginning of this section. MSNs widen these properties thanks to the presence of pores. The modification the size and morphology of pores and the whole particles lead to important changes in the sensing character of probes based on MSNs (Slowing et al. 2007, 1225–1236). A variety of shapes and sizes ranging from 20 to 500 nm, and with pore sizes ranging from 2 to 6 nm, has been obtained (see Figure 13.46) (Lai et al. 2003, 4451–4459; Slowing et al. 2007, 1225–1236).

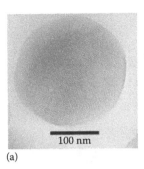
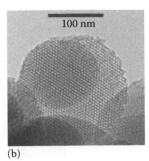
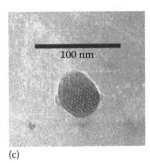

(a)　　　　　　　　　　　　(b)　　　　　　　　　　　　(c)

FIGURE 13.46 Transmission electron microscopy images of three spherical MSNs with different particle and pore sizes: (a) Particle size ca. 250 nm; pore diameter ca. 2.3 nm. (b) Particle size ca. 200 nm; pore diameter ca. 6.0 nm. (c) Particle size ca. 50 nm; pore diameter ca. 2.7 nm. (From Slowing, I.I., Trewyn, B.G., Giri, S., and Lin, V.S.-Y.: Mesoporous silica nanoparticles for drug delivery and biosensing applications. *Adv. Funct. Mater.* 2007. 17. 1225–1236. Copyright Wiley-VCH Verlag GmbH & Co. KGaA. Reprinted with permission.)

The structure of MSNs allows its utilization as a shell capable of enclosing various nanoparticles. This is used for carrying drugs, as it will be seen in the applications. Moreover, the porosity of these probes constitutes another way of increasing the relaxivity of the contrast agent (i.e., the sensitivity) in case it contains magnetic nanoparticles, as it was explained in Section 13.2.3.

13.2.5.2.2 Applications

In view of the fact that MSNs are made of the same material as silica, it is logical that they will be used in similar applications. However, the porosity of MSNs permit them to improve the capacity of silica nanoparticles in domains such as biomedical imaging, drug delivery, gene therapy, and biosensing (Slowing et al. 2007, 1225–1236; Trewyn et al. 2007, 3236–3245).

Regarding monitorization, MSNs are used for imaging by uploading the pores with contrast agents. As an example, in Lee et al. (2009b, 215–222) indocyanine green (ICG), an NIR imaging moiety, was used for NIR fluorescence imaging. To this purpose trimethylammonium groups modified MSN (MSN-TA) particles were loaded with ICG molecule. In consideration of the ICG loading only decreased 5% of the surface area on MSN, the availability of remaining ample space could be further modified for carry of therapeutic drugs. Moreover, since the ICG molecules are widespread in the surface, this reduces both self-quenching and aggregation, and the inclusion in the holes protects them from degradation. The NIR wavelength based system is adequate for application in human tissues. Regarding the combined utilization of magnetic nanoparticles and fluorophores for multimodal imaging an explanation was given in Section 13.2.3.4 (Lin et al. 2006, 5170–5172; Taylor et al. 2008, 2154–+; Kim et al. 2009, 372–390).

Another important field is drug delivery. The good properties of MSNs were mentioned at the beginning of this section. That is why, by surface functionalization, these structures become a good candidate for drug release. An important first question is cellular uptake. In this sense, it has been studied the efficiency of uptake by nonphagocytic eukaryotic cells as a function of the particle size, showing that particles of size above 500 nm were not uptaken, and that the best size was 200 nm (Rejman et al. 2004, 159–169). One of the key parameters for drug release is the delivery time. According to Trewyn et al. (2004, 2139–2143) the key parameter that determines the release and upload delivery times is the shape of the holes. As an example a 1000-fold increase is obtained with parallel hexagonal channels compared to disordered wormhole pores (Trewyn et al. 2004, 2139–2143).

The second question is to control when the drug delivery is done. As an example, if a toxic drug against cancerous cells is carried by an MSN, it is critical to deliver it exactly when it targets the cancerous cell, and not before. In this sense there are several ways of doing this. The first one is chemical. As an example, with disulfide-reduction-based gating, the holes are filled with the drug and with CdS nanoparticles via a chemically cleavable disulfide linkage to the MSN surface (Lai et al. 2003, 4451–4459). The controlled-release mechanism of the system is based on chemical reduction of the disulfide linkage between the CdS caps and the MSN hosts. Another example is found in Hernandez et al. (2004, 3370–3371), where a redox control is done (see Figure 13.47).

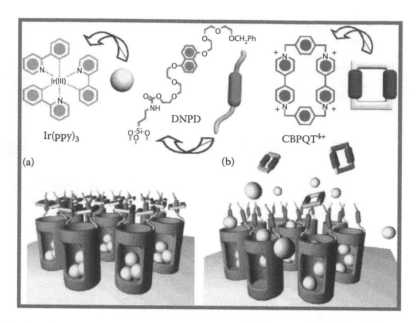

FIGURE 13.47 Graphical representations of operation of nanovalves. (a) The orifices of the nanopores (diameter 2 nm) are covered with pseudorotaxanes (formed between DNPD and CBPQT4+) which trap the luminescent Ir- (ppy)3 molecules inside the nanopores. (b) Upon their reduction, the CBPQT2+ bisradical dications are released and so allow the Ir(ppy)3 to escape. (Reprinted with permission from Hernandez, R., Tseng, H.-R., Wong, J.W., Stoddart, J.F., and Zink, J.I., An operational supramolecular nanovalve, *J. Am. Chem. Soc.*, 126(11), 3370–3371. Copyright 2004 American Chemical Society.)

Pseudorotaxane, a molecule used to cap the probes, is broken by chemical reduction, releasing the Ir(ppy)$_3$; a fluorophore that emits at 506 nm. A variation of the chemical method consists of including a magnetic material in the capping material (magnet-MSNs) (Giri et al. 2005, 5038–5044). Guest molecules smaller than 3 nm, such as fluorescein, were encapsulated and released from the magnet-MSN delivery system by using cell-produced antioxidants (e.g., dihydrolipoic acid) as triggers in the presence of an external magnetic field (Giri et al. 2005, 5038–5044). The probes were biocompatible and efficiently uptaken by human cervical cancer cells, and they offer a promising potential in utilization for human treatment. Other works related to the utilization of magnetic nanoparticles and fluorophores in MSNs for drug release can be found in Section 13.2.3.4 (Kim et al. 2006c, 688–689; Sathe et al. 2006, 5627–5632; Wu et al. 2008, 53–57). In the same manner MSNs containing magnetic nanoparticles and lanthanide based luminescent are also explained in the same section (Feng et al. 2010, 3596–3600).

A second method is optical. In Mal et al. (2003, 350–353), coumarin-functionalized mesoporous silica particles were developed. Successful functionalization requires uncalcined MSNs still filled with the template molecules that directed the formation of its pores, to ensure that coumarin derivatives attach preferentially to the pore outlets, rather than their inside walls. MSNs are filled with the drug and the pores are closed by photodimerization of coumarin (irradiation of sample with UV light with wavelengths longer than 310 nm). The release of the drug is done by impinging 250 nm UV light, which leads to photocleavage of the coumarin dimmers. These wavelengths prevent its utilization for in-vivo tissues applications.

Finally, dendrimers can be covalently attached to the surface of an MSN material. This dendrimers used as capping material can also be used to complex with a plasmid DNA and in this way they will be used as gene transfection agents (see Figure 13.48). The probes also can be previously uploaded with a fluorescent material for the sake of monitorization (Radu et al. 2004, 13216–13217).

An interesting work developed for cancer therapy based on MSNs as drug carriers can be found in Lu et al. (2007b, 1341–1346). The hydrophobic anticancer drug camptothecin is incorporated into fluorescent MSNs and delivered to various cancer cells to induce cell death. Camptothecin is insoluble in water, but its encapsulation in the MSN probe avoids this problem and permits to enter the cells. This method solves the problem of other drugs that are also insoluble in water.

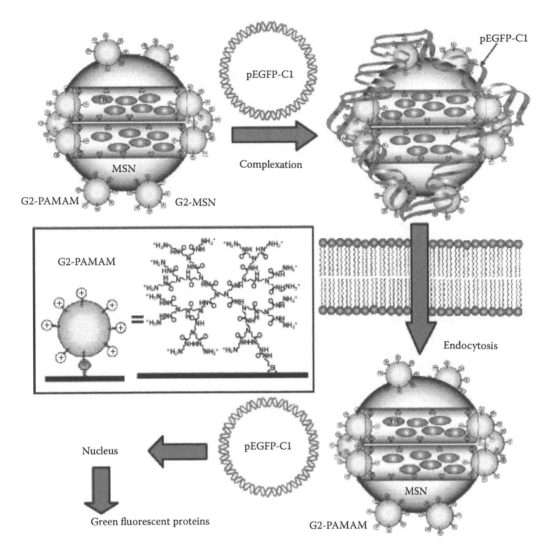

FIGURE 13.48 Schematic representation of a nonviral gene transfection system based on a Texas Red (TR)-loaded, G2-PAMAM dendrimer-capped MSN material complexed with an enhanced green fluorescence protein (Aequorea Victoria) plasmid DNA (pEGFP-C1). (Reprinted with permission from Radu, D.R., Lai, C.-Y., Jeftinija, K., Rowe, E.W., Jeftinija, S., and Lin, V.S.-Y., A polyamidoamine dendrimer-capped mesoporous silica nanosphere-based gene transfection reagent, *J. Am. Chem. Soc.*, 126(41), 13216–13217. Copyright 2004 American Chemical Society.)

Sensing of different molecular species is another field where MSNs have been successfully developed. Initial works were based on microporous MSNs developed with sol-gel method (Coradin et al. 2006, 99–108; Dave et al. 1994). However, in view of the already mentioned influence of pore shape on the release time of drugs, it is obvious that this also affects the quality of biosensing. That is why MSNs are preferred rather than microporous silica nanoparticles. Good reviews of microporous silica nanoparticles for biosensing can be found in Coradin et al. (2006, 99–108) and Sakai-Kato et al. (2006, 273–294). MSNs have much larger surface areas and pore diameters, which allows for the detection of larger analyte molecules and the incorporation of a great amount of receptors/sensors into the porous matrix to achieve a better detection limit, and permit a faster diffusion of the analytes through the mesoscale pores of these materials (Slowing et al. 2007, 1225–1236).

There are two different ways of developing biosensors based on MSNs. The first one is to use molecular receptors such as enzymes or proteins. In Wei et al. (2002, 802–808) and Dai et al. (2008, 1070–1076) glucose sensors are developed. However, these receptors do not present long-term stability and are expensive, which implies the utilization of other strategies based on materials with long-term stability (Slowing et al. 2007, 1225–1236). As an example functionalization of MSNs with o-phthalic hemithioacetal permitted to distinguish between dopamine and glucosamine (Lin et al. 2001, 11510–11511).

13.3 Nanolayers

Most of the research dedicated to nanolayer based sensors is developed in the domain of optical fiber. For this reason we refer to Chapter 7 regarding that topic. The rest of applications are explained here.

There are many techniques used for the deposition of nanolayers on different substrates. The two most extended ones are Langmuir Blodgett and Layer-by-Layer, and they are also explained in Chapter 7. They own good properties such as reproducibility (Decher 1997, 1232–1237; Mitsuishi et al. 2003, 2875–2879), which makes them adequate for sensing purposes. However, there are others such as spin-coating (Bai et al. 2008, 777–782), sputtering, or electrospun nanofibrous membranes (Wang et al. 2002a, 1273–1275), that are also used. Sol-gel is a technique that does not permit such an accurate control of the nanolayer thickness and only applications where nanometer scale sensing films are deposited will be included here. However, it presents some important properties such as its ability to be fabricated at low temperatures and physiological pHs, its optical transparency from the UV to the NIR, the possibility to modify the index of refraction (waveguiding depends on the refractive index of the thin-film), its permeation selectivity (small ions and molecules can enter the sol-gel matrix, whereas large molecules such as proteins remain outside), its chemical and mechanical stability and the possibility to be fabricated in many geometries similarly to LB and LbL techniques (Reisfeld 2001, 1–7). Sol-gel technique can be combined with spin-coating for a higher accuracy of the nanolayer (Botzung-Appert et al. 2004, 1609–1611).

The substrates where nanolayers are deposited are typically glass and gold. However, glass presents better properties than gold (it is transparent to light). That is why it has been frequently used for fluorescent biosassays and biological studies: protein, DNA, microschips, and so forth (Basabe-Desmonts et al. 2007, 993–1017).

Teflon amorphous fluoropolymer (TEFLON®-AF) coated devices induced by pressure is another method for depositing controlled thin-film coatings. They are less time consuming that spin-coating, but the coating thickness is in the micrometer regime (Cho et al. 2009, 1057–1059).

Another question is the detection method, which includes optical absorbance, fluorometry, interferometry and Surface Plasmon Resonance (SPR) as the most typical ones. SPR has been explained in chapter "Nano Surface Plasmon Resonance Sensors," so focus will be centered on the other three. A final section is dedicated to magneto-optic based nanolayer sensors as a promising technique that may improve in the future the performance of SPR based sensors.

In the development of sensors several questions must be solved: suppression of non-specific binding, securely attaching and stability of the film and good sensitivity (Mukundan et al. 2009, 5783–5809). Modified biopolymers (Gauglitz 2005, 141–155), physically adsorbed phospholipid bilayers or polymers (Mukundan et al. 2009, 5783–5809) are used to this purpose. However, they do not achieve a sufficient low non-specific binding. Consequently, two main strategies are followed. The first one is silanization of the substrates with subsequent covalent binding of biopolymers. In contrast to physical techniques such as dissolution, adsorption and entrapment in a porous network, covalent immobilization of the luminescent indicator improves the long-term stability of the sensitive system (Basabe-Desmonts et al. 2007, 993–1017). The first sensing system using covalently bonded dyes to glass was used for pH sensing (Harper 1975, 348–351). With silanization the durability of the film increases from days to months (Gauglitz 2005, 141–155). In addition to this, it is possible to combine poly(ethylene glycol) of different chain length to the silanized surface to produce a kind of a polymer brush. These layers resist non-specific binding but have a reduced number of interaction sites, because they are restricted to the surface and not to the volume (Piehler et al. 2000, 473–481). In this case the monitorization is based on the modification of the film properties (thickness or refractive index). In other cases where a fluorophore is deposited, the fluorescence quenching during the presence of the analyte will be the method of detection. It is also important to avoid cross-sensitivity due to swelling, which is avoided with porous glass materials (Basabe-Desmonts et al. 2007, 993–1017).

It is also interesting to allow sensors for multianalyte sensing (Nagl and Wolfbeis 2007, 507–511), which permits to detect various parameters simultaneously. The simplest case is compensating sensors, meaning that apart from the analyte itself, the sample is also evaluated for a second species which interferes with the result. This interference is then corrected for. As an example, in Hradil et al. (2002, 1552–1557) a fluorescence based sensor for pressure is compensated for temperature cross sensitivity.

In Pearton et al. (2010, 1–59) it is shown the recent interest in the use of surface functionalized nanolayer and nanowire wide bandgap semiconductors, principally GaN, In N, ZnO and SiC, for different purposes. However, for sensing of gases, heavy metals, UV photons and biological molecules only the domain of nanowire wide bandgap semiconductors has been explored.

13.3.1 Absorbance-Based Sensors

Absorbance is based on the attenuation of light due the characteristics of the material light is guided through. The selection of materials that modify its absorbance in the presence of a specific parameter permits to develop absorbance based sensors. In this section some representative works of optical sensors based on this phenomenon are presented. The detection of pH is a parameter that has concentrated much attention due to the fact that many materials present sensitivity to pH changes once immersed in a solution. The well known redox indicator Prussian Blue serves for this purpose. In Koncki and Wolfbeis (1998, 355–358) it is analyzed the influence of pH on the optical absorbance of thin-films of Prussian Blue immersed in a polypyrrole matrix.

The detection of ions is another research field, not only in absorbance based sensors, but also for fluorescence based ones. One of the first works is a sensor for potassium where the nanolayer is deposited with LB technique (Wolfbeis and Schaffar 1987, 1–12). Since that moment other ions have been detected such as Na^+, Li^+, and Ca^{2+} (Nagl and Wolfbeis 2007, 507–511).

Gas sensing is another interesting field. It is well-known that Pd can be used for H_2 sensing. That is why Pd has been successfully deposited on optical waveguides for detection of this gas (Alam et al. 2007). Ammonia is another gas used in industry which can be detected with a planar optical waveguide made of a thin film of polymer polyimide doped with indicator dye bromocresol purple (Sarkisov et al. 2004, 33–44), or with WO_3 sputtered thin films on an optical waveguide (Lazcano-Hernndez et al. 2008).

Phthalocyanine combined with metals or other chemical species can also be used for gas sensing. They can be deposited by LB technique in different substrates such as interdigital gold electrodes, and in this way it is possible to measure NO_2 or H_2S (Valli 2005, 13–44). In addition to this, bisphthalocyanines, a subfamily phthalocyanines, can be used for detection of volatile organic compounds and gases such NO_2, and they improve the results obtains with the rest of phthalocyanines (De Saja and Rodríguez-Méndez 2005, 1–11). Phthalocyanines can be used combined with other materials. As an example, the combination with porphyrin permits to obtain UV-Visible variations caused by the exposure of the sensing layers to alcohols, amines, ketones, alkanes, and pyridines with application in food quality control (Spadavecchia et al. 2004, 2083–2090).

Conjugated polydiacetylenes based sensors are reviewed in Yoon et al. (2009, 1958–1968), and they can be used for monitorization of ambient parameters both with fluorescence and absorbance detection modalities.

Refractive index of a liquid is another parameter that can be detected with absorbance based sensors. In Ramsden (1993, 439–442), the deposition LB lipid films permits to enhance the sensitivity to refractive index changes of the waveguide without the LB lipid film. The sensing mechanism is based on the phase velocities of the guide waves in the optical waveguide. The utilization of an optical waveguide evanescent field refractometer with a deposited titanium dioxide (TiO_2) thin film (Kwon et al. 2010, 431–435), permits to improve the results of the other work, by adequate selection of the coating thickness.

Devices for water quality control are also reviewed in Mizaikoff (2003, 35–42), where the chemical modification of the surface of waveguides is analyzed, which leads to enhanced analyte recognition. Another example can be found in Han et al. (1999, 381–389), where surface-modified sol-gel-coated Si attenuated total reflectance (ATR) mid-infrared sensors were used to detect amounts of isopropanol and acetone in water, improving for both cases in several orders the sensitivity compared to non coated devices.

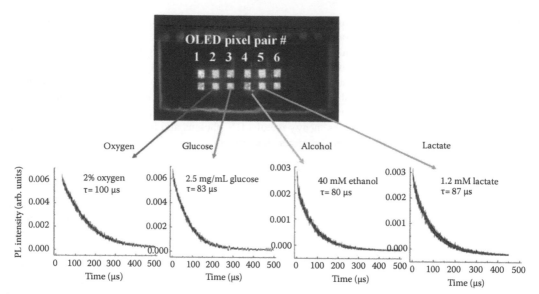

FIGURE 13.49 Structurally integrated OLED-based photoluminescent multianalyte sensor for sequential monitoring of oxygen, glucose, alcohol, and lactate. All of the sensing elements were based on a PS:PtOEP film, positioned above OLED pixel pairs 2–5. The analytes were monitored via the photoluminescence lifetime of the PtOEP. The figure shows the intensity as a function of time. Measurements were conducted in air at 23°C. (Reprinted from *Sens. Actuat. B Chem.*, 134, Cai, Y., Shinar, R., Zhou, Z., and Shinar, J., Multianalyte sensor array based on an organic light emitting diode platform, 727–735, Copyright 2008, with permission from Elsevier.)

The quality of air is also been explored in Korposh et al. (2008, 473–480), where the influence of water vapor and ammonia on the optical absorption bacteriorhodopsin based thin films deposited on glass substrates is analyzed.

Finally, multiparameter sensing is possible with different structures. In Ruano et al. (2002, 175–184) a 16-element array device fabricated with flame hydrolysis deposition permitted the monitorization of signals origined from a 24 mM fluorophore concentration. In addition to this, several liquids could be detected by using a sensor array, where each sensor was excited by pulsed light emitting diode pixels and the signal detected with photodiodes (Cai et al. 2008, 727–735). Glucose, lactate, ethanol, and oxygen (see Figure 13.49) could be detected based on monitoring the concentration of dissolved oxygen (DO) at the completion of the enzymatic oxidation reactions of these analytes in sealed cells.

13.3.2 Fluorometry-Based Sensors

There are many fluorescence based sensors in the literature. Most of them are based on direct detection of a fluorescence signal induced at the surface of a waveguide by the evanescent wave (Mukundan et al. 2009, 5783–5809). Among them, the most explored molecular species is the detection of oxygen, which is also the most successful in commercial terms (Nagl and Wolfbeis 2007, 507–511). In Grant and McShane (2003, 139–146) LbL films, containing Ru(bpy)2(mcbpy) conjugated to PAH as a model indicator, were shown to retain oxygen sensitivity and did not exhibit significant self quenching. The films where deposited on quarz slides, fiber optic and polymer microspheres and a fluorescence quenching was observed in the presence of oxygen. LB technique was also used for the same purpose in Chu and Yam (2006, 7437–7443), where again ruthenium indicator was included in the sensing film. The signal of oxygen sensor can be amplified by using antenna dyes with higher brightness, which absorb light and transfer the energy to an indicator dye. This strategy is also used for ammonia sensing. A more complex device capable of detecting both oxygen in air and water was developed in Nock et al. (2008, 1300–1307). Sensor patterns were integrated into a PDMS microfluidic device by plasma bonding and detection was done by fluorescence microscopy. In Schröder et al. (2007, 60–70) a sensor for both pH and oxygen detection is presented.

Ruthenium was used previously for temperature sensing, but in this case the quenching in the presence of oxygen had to be avoided. Sol-gel glass and polyacrylonitrile were identified as being the most suitable candidates for embedding it in a non-quenchable form (Liebsch et al. 1999, 1296–1299).

Like absorbance based sensors, pH detection it is possible with fluorescent sol-gels of about 300 nm thickness covalently immobilized on glass substrates with good long-term stability (Lobnik et al. 1998, 159–165) or with a fluorescent molecule assembled with a polycation in an LbL deposited structure (Lee et al. 2000, 10482–10489).

Different substances immersed in liquids can also be detected with fluorescence based nanolayers. In Mulder et al. (2006a, 1–6) rhodamine dye doped LB films are deposited on ITO substrate and a fluorescence intensity variation is experimented as a function of NaCl, KCl and $MnSO_4$.

The detection of metal ions such as (Fe^{3+} and Hg^{2+}) can be done either by a fluorescent molecule assembled with a polycation in an LbL structure (Lee et al. 2000, 10482–10489), or by electrospinning nanofibrous membranes including a fluorescent polymer (Wang et al. 2002a, 1273–1275). Moreover, organic and toxic compound, 2,4-Dinitrotoluene (DNT) can be detected with the same techniques already mentioned for ion detection (Lee et al. 2000, 10482–10489; Wang et al. 2002a, 1273–1275). In another work (Zimmerman et al. 2005, 2772–2777), cations in water can be detected by fluorescence spectroscopy. The sensor consists of a glass substrate where fluorescent self-assembled monolayers are deposited.

Organic Cu^{2+} salts can be detected by detection of fluorescence quenching in films based on immobilization of anthracene on a glass plate surface (Lü et al. 2007, 4123–4131) or by immobilization of pyrene (Lü et al. 2006, 841–845). Moreover, the devices can be at the same time insensitive to inorganic Cu^{2+} salts (Lü et al. 2006, 841–845; 2007, 4123–4131).

Detection of liquids is another research area that was presented in absorbance based sensors. In Ionov et al. (2006, 1453–1457), different liquids ethanol and toluene are detected with a device based on fluorescence interference contrast of semiconductor nanocrystals near a reflecting silicon surface. The fluorescence intensity of the hydrophobic CdSeS nanocrystals adsorbed onto a polymer brush layer was studied (see Figure 13.50), and the fluorescence intensity depended on the liquid the sensor is immersed in.

Nitroaromatic explosives are detected by using two different strategies. In the first one (Bai et al. 2008, 777–782), conjugated oligopyrene with high fluorescence quantum yield was synthesized and deposited by spin-coating technique. A quenching of the fluorescence was observed in the presence of vapors such as trinitrotoluene. Alternatively, the chemical assembly of oligo(diphenylsilane)s on a glass plate surface was used for the same purpose (He et al. 2009, 1494–1499).

In Rhee et al. (2008, 784–785) a fluorescent chemosensor composed of pyrene and bis(Zn^{2+}-dipicolylamine) is used for the detection of (p)ppGpp, a bacterial and plant alarmone.

VOCs can also be detected with a metal-organic $Zn_3(BTC)_2$ film. Among other VOCs the sensor was highly sensitive to dimethylamine dissolved in ethanol.

Another interesting approach can be found in Dickert et al. (1999, 4559–4563), where a sensor for polycyclic aromatic hydrocarbons in water is developed by depositing a molecular imprinted sensor layer. A highly cross-linked polymer is synthesized around a template molecule. When the template molecule is removed, a geometrically adapted polymer skeleton with fitting cavities and diffusion pathways for analyte inclusion is left behind. Depending on the size of the template molecule, different analytes can be detected with high selectivity.

Detection of DNA is possible in several publications. In Duveneck et al. (1997, 88–95) small amounts of fluorescently labeled DNA are detected based on luminescence generation in the evanescent field of high-refractive-index single-mode planar waveguides, whereas in Lee et al. (2009c, 3317–3325) a molecular beacon probe is designed based on conjugated poly(oxadiazole) derivative exhibiting amine and thiol functional groups (POX-SH), which is covalently immobilized onto a maleimido-functionalized glass slide by means of its thiol group. Selective hybridization of the molecular beacon probes with the target DNA sequence opens up the molecular beacon probes and affects the FRET between POX-SH and the dye or quencher, producing label-free detection method.

Biochips are another widely explored field where nanolayer based sensors can be used. LB films of cinnamoylbutylether-cellulose were transferred onto planar waveguides (Furch et al. 1996, 220–226)

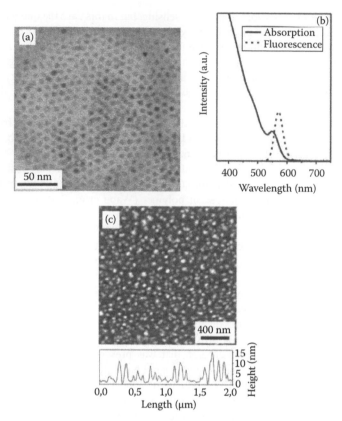

FIGURE 13.50 Characterization of the semiconductor nanocrystals. (a) Transmission electron microscopy (TEM) image of the nanocrystals. (b) Absorption and emission spectra of the nanocrystals. (c) AFM image of the brush layer after adsorption of the nanocrystals. (From Ionov, L., Sapra, S., Synytska, A., Rogach, A.L., Stamm, M., and Diez, S.: Fast and spatially resolved environmental probing using stimuli-responsive polymer layers and fluorescent nanocrystals. *Adv. Mater.* 2006. 18. 1453–1457. Copyright Wiley-VCH Verlag GmbH & Co. KGaA. Reprinted with permission.)

and these films served as matrices for the immobilization of biotinylated oligonucleotides via streptavidin. The generated streptavidin layers offered a good stability which permitted to use the device for detection of Salmonella by DNA hybridization. In Agnarsson et al. (2010, 56–61) the waveguides consist of a polymer layer on top of a fluoropolymer (CytopTM) cladding. The construction steps are shown in Figure 13.51. The waveguide is composed of a silicon wafer and a PMMA layer surrounded by two cladding regions fluoropolymer (CytopTM), where the upper cladding owns a cavity where the sample in aqueous solution can be applied. The monitorization process is explained on the right side of Figure 13.51. A single-mode optical fiber couples light into the planar waveguide. The fluoropolymer is closely index-matched to water sample, providing a symmetric cladding environment which simplifies optical excitation and provides tunability in penetration depth. The fluorescence signal emitted is imaged by the microscope objective. With this technique it was possible to monitorize cultured cells.

In addition to this, in Hofmann et al. (2002, 5243–5250) a microchip-based flow confinement method for rapid delivery of small sample volumes to sensor surfaces is described. The makeup flow confines the sample into a thin layer above the sensing area and increases its velocity, with application in efficiency of DNA hybridization or heterogeneous immunoassays performance. As an example, rabbit IgG was immobilized onto a silicon nitride waveguide and Cy5-labeled anti-rabbit IgG was hydrodynamically pumped over the immobilized zone. Evanescent field-based fluorescence detection enabled monitoring of the binding event.

More complex devices that permit to detect various elements can be developed (Touahir et al. 2010, 17–25). The probe molecules are anchored to a substrate in the form of small spots. The biomolecules in

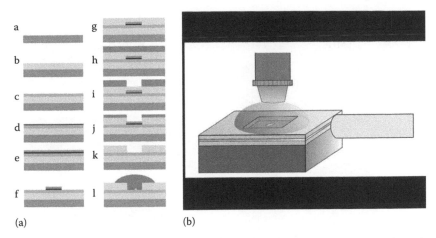

FIGURE 13.51 (a) Main steps in the chip fabrication process of a biochip. (b) Experimental configuration. Excitation light is coupled from a single-mode optical fiber into the planar waveguide. A sample in aqueous solution is excited by the evanescent tail of the bound mode, emitting a fluorescence signal, which is imaged by the microscope objective, typically through a cover glass. (Reprinted from *Microelectron. Eng.*, 87, Agnarsson, B., Halldorsson, J., Arnfinnsdottir, N., Ingthorsson, S., Gudjonsson, T., and Leosson, K., Fabrication of planar polymer waveguides for evanescent-wave sensing in aqueous environments, 56–61, Copyright 2010, with permission from Elsevier.)

the solution to be analyzed are labeled with a fluorophore. If target molecules matching one of the spotted probes are present in the solution, that spot will become fluorescent upon the molecular recognition event. By adequate selection of a substrate such as silicon in thin-film form on a conventional glass substrate, the fluorescence loss is reduced compared with a glass substrate. Similarly in Basabe-Desmonts et al. (2004, 7293–7299) sensing systems can be fabricated using microcontact printing, a soft lithography technique that permits to easily make controlled size features down to 100 nm that will change its fluorescence as a function of the analyte to detect. A different approach based on the simultaneous intensity measurements at four different thicknesses of SiO_2 deposited on a silicon substrate (see Figure 13.52) (Ionov et al. 2006, 1453–1457).

Multiparameter sensing is also used for immunosensors. In Plowman et al. (1999, 4344–4352) silicon oxynitride integrated optical waveguide was used to evanescently excite fluorescence from a multianalyte sensor surface in a rapid, sandwich immunoassay format. The system was proved both for polyclonal and monoclonal antibodies, and the results were compared with those obtained with a single analyte immunoassay.

In Kim et al. (2006a, 1313–1323), a biosensor array is developed where the coating is composed of T-cells and B-cells. B-cells present the antigen B cells capture target pathogens and proteolyze them

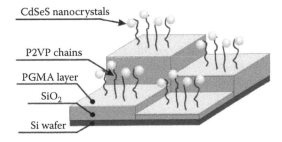

FIGURE 13.52 Quantification of the thickness of the polymer layer by fluorescence intensity contrast: schematic of the simultaneous intensity measurements at four different thicknesses of SiO_2. (From Ionov, L., Sapra, S., Synytska, A., Rogach, A.L., Stamm, M., and Diez, S.: Fast and spatially resolved environmental probing using stimuli-responsive polymer layers and fluorescent nanocrystals. *Adv. Mater.* 2006. 18. 1453–1457. Copyright Wiley-VCH Verlag GmbH & Co. KGaA. Reprinted with permission.)

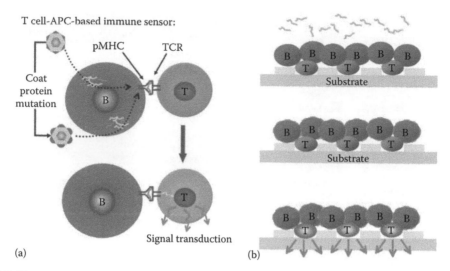

T cell-APC-based immune sensor:

FIGURE 13.53 (a) Conceptual view of a live T-cell/APC (antigen presenting cell) immunosensor. (b) Schematic description of living T-cell/B-cell-based immune sensor. (From Kim, H., Cohen, R.E., Hammond, P.T., and Irvine, D.J.: Live lymphocyte arrays for biosensing. *Adv. Funct. Mater.* 2006a. 16. 1313–1323. Copyright Wiley-VCH Verlag GmbH & Co. KGaA. Reprinted with permission.)

into fragments for presentation as peptide-MHC (pMHC) complexes on the cell surface, where they are recognized by TCRs on the responding T lymphocyte (see Figure 13.53).

Finally, the signal separation in multiparameter sensing is an important task. This is typically done by fluorescent lifetime imaging. However, in some works the setup of color CCD and CMOS cameras is used (Schröder et al. 2007, 60–70). The principle of operation is based on the application of three different types of pixels, each of them sensitive towards different wavelength ranges. These detectors can be used to record and separate multiple sensor signals. However, the sensitive dyes must emit at different wavelengths in order to discriminate them.

13.3.3 Interferometry-Based Sensors

In interferometry based sensors the evanescent electric field strength and overlap with bound biological species must be optimized to maximize the impact on the effective refractive index as seen by the guided optical wave (Mukundan et al. 2009, 5783–5809). There is a variety of different sensing structures and the most relevant ones are presented here.

Similarly to tapered optical fiber, tapered optical waveguides are also developed. In Qi et al. (2000, 1106–1110), the waveguide consists of a single-mode potassium ion-exchanged planar waveguide overlaid with a high-index thin film that has two tapered ends and supports only the TE0 mode. The TE0 and TM0 modes coexisting in the potassium ion-exchanged layer were separated in the thin film region of the waveguide: the TE0 mode was coupled into the thin film while the TM0 mode was confined in the potassium ion-exchanged layer. Interference occurs between TE and TM-polarized output components when a single output beam is passed through a 45-polarized analyzer. The phase difference between both orthogonal output components used for sensing refractive index changes of up to 3.71×10^{-6}.

Reflectometric interference spectroscopy is another well known technique that is used as chemical sensor. In Reichl et al. (2000, 583–586), this technique is used for monitoring organic solvent vapors, such as toluene, in air. The mechanism of detection is based on the wavelength shift of the interference pattern caused by a change of the optical thickness in the sensing film. Instead of the entire spectrum only four wavelengths are used.

Interferometry also permits to improve the sensitivity of nanolayer based sensors. As an example, in Busse et al. (2002, 704–710) a comparison of surface plasmon spectroscopy, waveguide mode

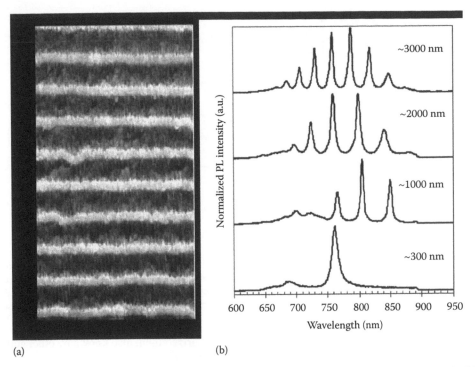

(a)   (b)

FIGURE 13.54 (a) Cross-sectional SEM micrograph of a 10 period p+ oxidized porous silicon multilayer stack with overall thickness of 2.4 μm. The thickness of the low porosity layers (43%) is 80 nm and that of the high porosity layers (62%) is 160 nm. (b) Bands in the electromagnetic spectrum as a function of the thickness of the layer inserted between the two Bragg gratings. (Reprinted from *Mater. Sci. Eng. C*, 15, Chan, S., Li, Y., Rothberg, L.J., Miller, B.L., and Fauchet, P.M., Nanoscale silicon microcavities for biosensing, 277–282, Copyright 2001, with permission from Elsevier.)

spectroscopy and an integrated optical Mach/Zehnder-interferometer are compared and the last one offers an improved sensitivity of an order of magnitude.

Bragg gratings is another interferometric technique widely explored in the domain of optical fiber which is successfully used in planar waveguides. In Chan et al. (2001, 277–282), a luminescent porous silicon layer is inserted between two Bragg reflectors. Porous silicon presents a visible luminescence at room temperature. The presence of the Bragg reflectors permits to narrow the band in the electromagnetic spectrum, which is interesting for detecting small changes in the refractive index, caused by particles that fit into the porous matrix, such as DNA. In Figure 13.54a one the two 10 layer Bragg mirrors used in the experiment is shown, whereas in Figure 13.54b the electromagnetic spectra for different central silicon layer thicknesses is plotted.

Array biosensors also include interferometric based sensors. In Schneider et al. (2000, 597–604), an optical chip based on the Hartman interferometer is developed. It uses a single planar lightbeam to address multiple interferometers, each comprising a signal-reference pair of sensing regions developed for detection antigens coupled to antibodies deposited in the sensing region.

13.3.4 Magnetooptic-Based Sensors

Surface Plasmon resonance (SPR) and all applications related to this topic are explained in the chapter "Nano Surface Plasmon Resonance Sensors" in this book. However, the magneto-optic SPR (MOSPR) effect is a less explored phenomenon which is worth explaining. The magneto-optic effect is based on the modification of the light polarization induced by a magnetic field. In combination with SPR, it improves the limit of detection of SPR based sensors, which is 10^{-5} refractive index units (RIU) (Homola et al. 1999, 3–15), and the sensitivity limit, located between 10^3 and 10^4 nm per RIU (Lee et al. 2009a, 209–221). In Sepúlveda et al. (2006, 1085–1087) it is explained that the magnetooptic effect

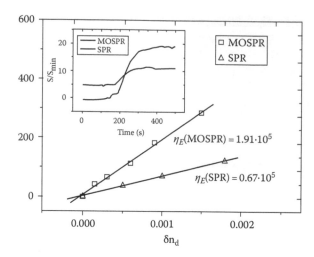

FIGURE 13.55 Comparison of the experimental normalized signals of the MOSPR and the SPR sensors that are due to refractive-index changes, and evaluation of their experimental sensitivities. Inset: normalized signal of the detection of the physical adsorption of bovine serum albumin proteins. (Reprinted from Sepúlveda, B., Calle, A., Lechuga, L.M., and Armelles, G., Highly sensitive detection of biomolecules with the magneto-optic surface-plasmon-resonance sensor, *Opt. Lett.*, 31(8), 1085–1087, 2006. With permission of Optical Society of America.)

produces a relative change of the reflectivity R_{pp} of the p-polarized light, also known by other authors as TM-polarized light (see the explanation of SPR in Section 13.2.1):

$$\frac{\Delta R_{pp}}{R_{pp}} = \frac{R_{pp}(M) - R_{pp}(0)}{R_{pp}(0)} \qquad (13.2)$$

where $R_{pp}(M)$ and $R_{pp}(0)$ represent the reflectivities with and without magnetization, respectively. When the surface plasmon is excited, R_{pp} is largely reduced while the magneto-optic components (ΔRpp) are maintained or even increased, producing a sharp enhancement of the magneto-optic effect in the reflected light. The result is an enhancement of the sensitivity to variations in the thickness or refractive index of the thin-film or the dielectric medium surrounding the thin-film. In Figure 13.55 it is represented the normalized signals of SPR and MOSPR based sensors for the same variations in the refractive index. The sensitivity is improved by a factor of 3.

Instead of combining both SPR and magneto-optic effect, it is also possible to develop magneto-optic effect based nanosensors. As an example, in Zayat et al. (2009, 254–259), a sol-gel nanocomposite containing γ-Fe$_2$O$_3$ is used in NANOSTAT (the first nanosatellite). The material used for the sensor meets the requirements of using SPIO nanoparticles, transparency, high Verdet constant and good Good mechanical properties, to allow its integration in the magneto-optical device (Guerrero et al. 1997, 2698–2700).

13.4 Final Remarks

In the recent decades nanoparticles have revolutionized the world of optical biosensing. This field has been traditionally limited to applications based on organic dyes, selective stains, and organic fluorophores, typically using technologies such as colorimetric or spectroscopic techniques and fluorescence microscopy.

Nevertheless, the apparition of new materials with outstanding new properties has made possible to obtain new applications or, in other cases, to improve significantly the existing ones. In the present chapter the most relevant and useful contributions of nanomaterials applied to optical biosensing have been reviewed, including metallic nanoparticles, QDs, magnetic nanoparticles, lanthanides, silica nanoparticles, and finally coatings and nanostructures.

The use of metallic nanoparticles and their plasmon resonant coupling of light has allowed the development of LSPR-based biosensors capable to detect very precisely DNA strands or even to be integrated into lab-on-a-chip device. SERS technique also has become a very valuable tool for bio-identification, allowing the tracking of the Raman fingerprint from simple biomolecules such as glucose, or even living organisms like bacteria.

In the case of semiconductor nanocrystals (QDs) this new technology has been successfully integrated into the existing assays and techniques. In most of the applications QDs are used as high performance and highly photostable fluorophores. In this sense, they have overcome what conventional organic fluorophores could do and they are widely used in research fields such as in vitro assays or in vivo assisted imaging of target tissues.

Looking to magnetic nanoparticles they offer the possibility of selective binding using a proper surface biochemical funtionalization, and the possibility of mechanically separate the NPs from the sample, or even it makes possible to remove selectively undesired tissues using techniques as the photodynamic therapy.

Silica nanoparticles is another important area were important advances have been obtained such as drug delivery, preserving at the same time biocompatibility, and permitting to monitorize the evolution in the tissues. Many configurations are possible, including mesoporous silica, a variation of traditional silica nanoparticles, with outstanding properties that permit to improve the applicability of silica.

In summary, this field of nanoparticles has significantly improved the analysis, diagnosis, therapeutics, and other biosensing applications, permitting the scientists to build complex structures with targeted behavior. Some of the approaches and innovations shown in this chapter will doubtlessly change bioassays, advanced medical optical diagnosis tools, and enhanced therapies in the near future.

REFERENCES

Agnarsson, B., J. Halldorsson, N. Arnfinnsdottir, S. Ingthorsson, T. Gudjonsson, and K. Leosson. 2010. Fabrication of planar polymer waveguides for evanescent-wave sensing in aqueous environments. *Microelectronic Engineering* 87 (1): 56–61.

Ai, K., B. Zhang, and L. Lu. 2009. Europium-based fluorescence nanoparticle sensor for rapid and ultrasensitive detection of an anthrax biomarker. *Angewandte Chemie—International Edition* 48 (2): 304–308.

Alam, M. Z., J. Moreno, J. S. Aitchison, and M. Mojahedi. 2007. An integrated optic hydrogen sensor for fast detection of hydrogen, *Proceedings of SPIE—the International Society for Optical Engineering* 6758: 67580D.

Ali, E. M., Y. Zheng, H.-H. Yu, and J. Y. Ying. 2007. Ultrasensitive Pb2+ detection by glutathione-capped quantum dots. *Analytical Chemistry* 79 (24): 9452–9458.

Alivisatos, A. P., K. P. Johnsson, X. Peng, T. E. Wilson, C. J. Loweth, M. P. Bruchez Jr., and P. G. Schultz. 1996. Organization of 'nanocrystal molecules' using DNA. *Nature* 382 (6592): 609–611.

Aslan, K., M. Wu, J. R. Lakowicz, and C. D. Geddes. 2007. Fluorescent core-shell Ag@SiO2 nanocomposites for metal-enhanced fluorescence and single nanoparticle sensing platforms. *Journal of the American Chemical Society* 129 (6): 1524–1525.

Bai, H., C. Li, and G. Shi. 2008. Rapid nitroaromatic compounds sensing based on oligopyrene. *Sensors and Actuators, B: Chemical* 130 (2): 777–782.

Bain, C. D., E. B. Troughton, Y.-T. Tao, J. Evall, G. M. Whitesides, and R. G. Nuzzo. 1989. Formation of monolayer films by the spontaneous assembly of organic thiols from solution onto gold. *Journal of the American Chemical Society* 111 (1): 321–335.

Banerjee, S. S., and D.-H. Chen. 2009. A multifunctional magnetic nanocarrier bearing fluorescent dye for targeted drug delivery by enhanced two-photon triggered release. *Nanotechnology* 20 (18): 185103 *doi:10.1088/0957-4484/20/18/185103*.

Barbé, C., J. Bartlett, L. Kong, K. Finnie, H. Q. Lin, M. Larkin, S. Calleja, A. Bush, and G. Calleja. 2004. Silica particles: A novel drug-delivery system. *Advanced Materials* 16 (21): 1959–1966.

Basabe-Desmonts, L., J. Beld, R. S. Zimmerman, J. Hernando, P. Mela, M. F. García Parajó, N. F. Van Hulst, A. Van Den Berg, D. N. Reinhoudt, and M. Crego-Calama. 2004. A simple approach to sensor discovery and fabrication on self-assembled monolayers on glass. *Journal of the American Chemical Society* 126 (23): 7293–7299.

Basabe-Desmonts, L., D. N. Reinhoudt, and M. Crego-Calama. 2007. Design of fluorescent materials for chemical sensing. *Chemical Society Reviews* 36 (6): 993–1017.

Beck, J. S., J. C. Vartuli, W. J. Roth, M. E. Leonowicz, C. T. Kresge, K. D. Schmitt, C. T.-W. Chu, et al. 1992. A new family of mesoporous molecular sieves prepared with liquid crystal templates. *Journal of the American Chemical Society* 114 (27): 10834–10843.

Berry, C. C., and A. S. G. Curtis. 2003. Functionalisation of magnetic nanoparticles for applications in biomedicine. *Journal of Physics D-Applied Physics* 36 (13) (July 7): R198–R206.

ProteOn™ series from Bio-Rad Inc. More information available at www.bio-rad.com. Date of access: 16/07/2012.

Blanco Gomis, D., D. Muro Tamayo, and J. Mangas Alonso. 2001. Determination of monosaccharides in cider by reversed-phase liquid chromatography. *Analytica Chimica Acta* 436 (1): 173–180.

Bottrill, M., L. Kwok, and N. J. Long. 2006. Lanthanides in magnetic resonance imaging. *Chemical Society Reviews* 35 (6): 557–571.

Botzung-Appert, E., V. Monnier, T. H. Duong, R. Pansu, and A. Ibanez. 2004. Polyaromatic luminescent nanocrystals for chemical and biological sensors. *Chemistry of Materials* 16 (9): 1609–1611.

Bruce, J. I., R. S. Dickins, L. J. Govenlock, T. Gunnlaugsson, S. Lopinski, M. P. Lowe, D. Parker, et al. 2000. The selectivity of reversible oxy-anion binding in aqueous solution at a chiral europium and terbium center: Signaling of carbonate chelation by changes in the form and circular polarization of luminescence emission. *Journal of the American Chemical Society* 122 (40): 9674–9684.

Bruchez Jr., M., M. Moronne, P. Gin, S. Weiss, and A. P. Alivisatos. 1998. Semiconductor nanocrystals as fluorescent biological labels. *Science* 281 (5385): 2013–2016.

Buck, S. M., Y.-E L. Koo, E. Park, H. Xu, M. A. Philbert, M. A. Brasuel, and R. Kopelman. 2004. Optochemical nanosensor PEBBLEs: Photonic explorers for bioanalysis with biologically localized embedding. *Current Opinion in Chemical Biology* 8 (5): 540–546.

Bulte, J. W. M. and D. L. Kraitchman. 2004. Iron oxide MR contrast agents for molecular and cellular imaging. *Nmr in Biomedicine* 17 (7) (November): 484–499.

Burns, A., H. Ow, and U. Wiesner. 2006a. Fluorescent core-shell silica nanoparticles: Towards "lab on a particle" architectures for nanobiotechnology. *Chemical Society Reviews* 35 (11): 1028–1042.

Burns, A., P. Sengupta, T. Zedayko, B. Baird, and U. Wiesner. 2006b. Core/shell fluorescent silica nanoparticles for chemical sensing: Towards single-particle laboratories. *Small* 2 (6): 723–726.

Busse, S., V. Scheumann, B. Menges, and S. Mittler. 2002. Sensitivity studies for specific binding reactions using the biotin/streptavidin system by evanescent optical methods. *Biosensors and Bioelectronics* 17 (8): 704–710.

Cai, Y., R. Shinar, Z. Zhou, and J. Shinar. 2008. Multianalyte sensor array based on an organic light emitting diode platform. *Sensors and Actuators, B: Chemical* 134 (2): 727–735.

Caravan, P. 2006. Strategies for increasing the sensitivity of gadolinium based MRI contrast agents. *Chemical Society Reviews* 35 (6): 512–523.

Cavaliere-Jaricot, S., M. Darbandi, E. Kuçur, and T. Nann. 2008. Silica coated quantum dots: A new tool for electrochemical and optical glucose detection. *Microchimica Acta* 160 (3): 375–383.

Chan, S., Y. Li, L. J. Rothberg, B. L. Miller, and P. M. Fauchet. 2001. Nanoscale silicon microcavities for biosensing. *Materials Science and Engineering C* 15 (1–2): 277–282.

Chan, W. C. W. and S. Nie. 1998. Quantum dot bioconjugates for ultrasensitive nonisotopic detection. *Science* 281 (5385): 2016–2018.

Chatterjee, D. K., A. J. Rufaihah, and Y. Zhang. 2008. Upconversion fluorescence imaging of cells and small animals using lanthanide doped nanocrystals. *Biomaterials* 29 (7): 937–9343.

Chen, L. T. and L. Weiss. 1973. The role of the sinus wall in the passage of erythrocytes through the spleen. *Blood* 41 (4): 529–537.

Chen, J.-L. and C.-Q. Zhu. 2005. Functionalized cadmium sulfide quantum dots as fluorescence probe for silver ion determination. *Analytica Chimica Acta* 546 (2): 147–153.

Cho, S. H., J. Godin, and Y.-H. Lo. 2009. Optofluidic waveguides in teflon AF-coated PDMS microfluidic channels. *IEEE Photonics Technology Letters* 21 (15): 1057–1059.

Choi, J., A. A. Burns, R. M. Williams, Z. Zhou, A. Flesken-Nikitin, W. R. Zipfel, U. Wiesner, and A. Y. Nikitin. 2007a. Core-shell silica nanoparticles as fluorescent labels for nanomedicine. *Journal of Biomedical Optics* 12 (6): 064007.

Choi, J. H., F. T. Nguyen, P. W. Barone, D. A. Heller, A. E. Moll, D. Patel, S. A. Boppart, and M. S. Strano. 2007b. Multimodal biomedical imaging with asymmetric single-walled carbon nanotube/iron oxide nanoparticle complexes. *Nano Letters* 7 (4) (Apr): 861–867.

Chu, B. W.-K. and V. W.-W. Yam. 2006. Sensitive single-layered oxygen-sensing systems: Polypyridyl ruthenium(II) complexes covalently attached or deposited as langmuir-blodgett monolayer on glass surfaces. *Langmuir* 22 (17): 7437–7443.

Chun, A. L. 2008. Lanthanide nanoparticles: Sensing danger. *Nature Nanotechnology.*

Cole Parmer Inc. Available from http://www.coleparmer.com. Equipment reference: EW-75955-71. Access date: 16/07/2012.

Coradin, T., M. Boissière, and J. Livage. 2006. Sol-gel chemistry in medicinal science. *Current Medicinal Chemistry* 13 (1): 99–108.

Corpuscular Inc. Fluorescent SiO2 silica nanospheres and microspheres. Available from http://www.micro-spheres-nanospheres.com. Access date 16/07/2012.

Corot, C., K. G. Petry, R. Trivedi, A. Saleh, C. Jonkmanns, J.-F Le Bas, E. Blezer et al. 2004. Macrophage imaging in central nervous system and in carotid atherosclerotic plaque using ultrasmall superparamagnetic iron oxide in magnetic resonance imaging. *Investigative Radiology* 39 (10): 619–625.

Costa-Fernández, J. M., R. Pereiro, and A. Sanz-Medel. 2006. The use of luminescent quantum dots for optical sensing. *TrAC—Trends in Analytical Chemistry* 25 (3): 207–218.

Courty, S., C. Luccardini, Y. Bellaiche, G. Cappello, and M. Dahan. 2006. Tracking individual kinesin motors in living cells using single quantum-dot imaging. *Nano Letters* 6 (7): 1491–1495.

Crouch, D. J., P. O'Brien, M. A. Malik, P. J. Skabara, and S. P. Wright. 2003. A one-step synthesis of cadmium selenide quantum dots from a novel single source precursor. *Chemical Communications* 9 (12): 1454–1455.

Dai, Z., J. Bao, X. Yang, and H. Ju. 2008. A bienzyme channeling glucose sensor with a wide concentration range based on co-entrapment of enzymes in SBA-15 mesopores. *Biosensors and Bioelectronics* 23 (7): 1070–1076.

Danek, M., K. F. Jensen, C. B. Murray, and M. G. Bawendi. 1994. Electrospray organometallic chemical vapor deposition—A novel technique for preparation of II-VI quantum dot composites. *Applied Physics Letters* 65 (22): 2795–2797.

Dasary, S. S. R., U. S. Rai, H. Yu, Y. Anjaneyulu, M. Dubey, and P. C. Ray. 2008. Gold nanoparticle based surface enhanced fluorescence for detection of organophosphorus agents. *Chemical Physics Letters* 460 (1–3): 187–190.

Dave, B. C., B. Dunn, J. S. Valentine, and J. L. Zink. 1994. Sol-gel encapsulation methods for biosensors. *Analytical Chemistry* 66 (22): 1120A–1127A.

De Saja, J. A. and M. L. Rodríguez-Méndez. 2005. Sensors based on double-decker rare earth phthalocyanines. *Advances in Colloid and Interface Science* 116 (1–3): 1–11.

Decher, G. 1997. Fuzzy nanoassemblies: Toward layered polymeric multicomposites. *Science* 277 (5330): 1232–1237.

Deng, Y., C. Deng, D. Yang, C. Wang, S. Fu, and X. Zhang. 2005. Preparation, characterization and application of magnetic silica nanoparticle functionalized multi-walled carbon nanotubes. *Chemical Communications*(44): 5548–5550.

Deng, Z., F. L. Lie, S. Shen, I. Ghosh, M. Mansuripur, and A. J. Muscat. 2009. Water-based route to ligand-selective synthesis of ZnSe and cd-doped ZnSe quantum dots with tunable ultraviolet A to blue photoluminescence. *Langmuir* 25 (1): 434–442.

Deng, Y. H., W. L. Yang, C. C. Wang, and S. K. Fu. 2003. A novel approach for preparation of thermoresponsive polymer magnetic microspheres with core-shell structure. *Advanced Materials* 15 (20) (Oct 16): 1729+.

Derfus, A. M., W. C. W. Chan, and S. N. Bhatia. 2004. Intracellular delivery of quantum dots for live cell labeling and organelle tracking. *Advanced Materials* 16 (12): 961–966.

Dick, L. A., A. D. McFarland, C. L. Haynes, and R. P. Van Duyne. 2002. Metal film over nanosphere (MFON) electrodes for surface-enhanced raman spectroscopy (SERS): Improvements in surface nanostructure stability and suppression of irreversible loss. *Journal of Physical Chemistry B* 106 (4): 853–860.

Dickert, F. L., M. Tortschanoff, W. E. Bulst, and G. Fischerauer. 1999. Molecularly imprinted sensor layers for the detection of polycyclic aromatic hydrocarbons in water. *Analytical Chemistry* 71 (20): 4559–4563.

Doering, W. E. and S. Nie. 2003. Spectroscopic tags using dye-embedded nanoparticles and surface-enhanced raman scattering. *Analytical Chemistry* 75 (22): 6171–6176.

Dosev, D., M. Nichkova, R. K. Dumas, S. J. Gee, B. D. Hammock, K. Liu, and I. M. Kennedy. 2007. *Nanotechnology* 18: 055102.

Dubertret, B., P. Skourides, D. J. Norris, V. Noireaux, A. H. Brivanlou, and A. Libchaber. 2002. In vivo imaging of quantum dots encapsulated in phospholipid micelles. *Science* 298 (5599) (November 29): 1759–17562.

Duveneck, G. L., M. Pawlak, D. Neuschäfer, E. Bär, W. Budach, U. Pieles, and M. Ehrat. 1997. Novel bioaffinity sensors for trace analysis based on luminescence excitation by planar waveguides. *Sensors and Actuators, B: Chemical* 38 (1–3): 88–95.

El-Deab, M. S., T. Sotomura, and T. Ohsaka. 2006. Oxygen reduction at au nanoparticles electrodeposited on different carbon substrates. *Electrochimica Acta* 52 (4): 1792–1798.

Elghanian, R., J. J. Storhoff, R. C. Mucic, R. L. Letsinger, and C. A. Mirkin. 1997. Selective colorimetric detection of polynucleotides based on the distance-dependent optical properties of gold nanoparticles. *Science* 277 (5329): 1078–1081.

Feng, J., S.-Y. Song, R.-P. Deng, W.-Q. Fan, and H.-J. Zhang. 2010. Novel multifunctional nanocomposites: Magnetic mesoporous silica nanospheres covalently bonded with near-infrared luminescent lanthanide complexes. *Langmuir* 26 (5): 3596–3600.

Fernández-Argüelles, M. T., J. J. Wei, J. M. Costa-Fernández, R. Pereiro, and A. Sanz-Medel. 2005. Surface-modified CdSe quantum dots for the sensitive and selective determination of cu(II) in aqueous solutions by luminescent measurements. *Analytica Chimica Acta* 549 (1–2): 20–25.

Ferrari, M. 2005. Cancer nanotechnology: Opportunities and challenges. *Nature Reviews Cancer* 5 (3): 161–171.

Förster, Th. 1948. Zwischenmolekulare energiewanderung und fluoreszenz. *Ann. Physik.* 437: 55.

Fort, E. and S. Grésillon. 2008. Surface enhanced fluorescence. *Journal of Physics D: Applied Physics* 41 (1): 013001 *doi:10.1088/0022-3727/41/1/013001.*

Fu, L.-M. X.-F. Wen, X.-C. Ai, Y. Sun, Y.-S. Wu, J.-P. Zhang, and Y. Wang. 2005. Efficient two-photon-sensitized luminescence of a europium(III) complex. *Angewandte Chemie—International Edition* 44 (5): 747–750.

Fujiwara, K., H. Watarai, H. Itoh, E. Nakahama, and N. Ogawa. 2006. Measurement of antibody binding to protein immobilized on gold nanoparticles by localized surface plasmon spectroscopy. *Analytical and Bioanalytical Chemistry* 386 (3): 639–644.

Fuller, J. E., G. T. Zugates, L. S. Ferreira, H. S. Ow, N. N. Nguyen, U. B. Wiesner, and R. S. Langer. 2008. Intracellular delivery of core-shell fluorescent silica nanoparticles. *Biomaterials* 29 (10): 1526–1532.

Furch, M., J. Ueberfeld, A. Hartmann, D. Bock, and S. Seeger. 1996. Ultrathin oligonucleotide layers for fluorescence-based DNA sensors. *Proceedings of SPIE—the International Society for Optical Engineering* 2928: 220–226.

Gao, X., Y. Cui, R. M. Levenson, L. W. K. Chung, and S. Nie. 2004. In vivo cancer targeting and imaging with semiconductor quantum dots. *Nature Biotechnology* 22 (8): 969–976.

Gao, J., B. Zhang, Y. Gao, Y. Pan, X. Zhang, and B. Xu. 2007. Fluorescent magnetic nanocrystals by sequential addition of reagents in a one-pot reaction: A simple preparation for multifunctional nanostructures. *Journal of the American Chemical Society* 129 (39): 11928–11935.

Garcia, C. B. W., Y. Zhang, S. Mahajan, F. DiSalvo, and U. Wiesner. 2003. Self-assembly approach toward magnetic silica-type nanoparticles of different shapes from reverse block copolymer mesophases. *Journal of the American Chemical Society* 125 (44): 13310–13311.

Gauglitz, G. 2005. Direct optical sensors: Principles and selected applications. *Analytical and Bioanalytical Chemistry* 381 (1): 141–155.

Germano, G., H. Kiat, P. B. Kavanagh, M. Moriel, M. Mazzanti, H. T. Su, K. F. Vantrain, and D. S. Berman. 1995. Automatic quantification of ejection fraction from gated myocardial perfusion spect. *Journal of Nuclear Medicine* 36 (11) (November): 2138–2147.

Ghadiali, J. E., S. B. Lowe, and M. M. Stevens. 2011. Quantum-dot-based FRET detection of histone acetyltransferase activity. *Angewandte Chemie - International Edition* 50 (15): 3417–3420.

Giri, S., B. G. Trewyn, M. P. Stellmaker, and V. S.-Y. Lin. 2005. Stimuli-responsive controlled-release delivery system based on mesoporous silica nanorods capped with magnetic nanoparticles. *Angewandte Chemie—International Edition* 44 (32): 5038–5044.

Gobin, A. M., M. H. Lee, N. J. Halas, W. D. James, R. A. Drezek, and J. L. West. 2007. Near-infrared resonant nanoshells for combined optical imaging and photothermal cancer therapy. *Nano Letters* 7 (7): 1929–1934.

Goicoechea, J., F. J. Arregui, and I. R. Matias. 2008. Quantum dots for sensing. In ed. F. J. Arregui. New York: Springer.

Goldman, E. R., E. D. Balighian, M. K. Kuno, S. Labrenz, G. P. Anderson, J. M. Mauro, and H. Mattoussi. 2002a. Luminescent quantum dot-adaptor protein-antibody conjugates for use in fluoroimmunoassays. *Physica Status Solidi (B) Basic Research* 229 (1): 407–414.

Goldman, E. R., E. D. Balighian, H. Mattoussi, M. K. Kuno, J. M. Mauro, P. T. Tran, and G. P. Andersont. 2002b. Avidin: A natural bridge for quantum dot-antibody conjugates. *Journal of the American Chemical Society* 124 (22): 6378–6382.

Goldman, E. R., A. R. Clapp, G. P. Anderson, H. T. Uyeda, J. M. Mauro, I. L. Medintz, and H. Mattoussi. 2004. Multiplexed toxin analysis using four colors of quantum dot fluororeagents. *Analytical Chemistry* 76 (3): 684–688.

Goldman, E. R., I. L. Medintz, J. L. Whitley, A. Hayhurst, A. R. Clapp, H. T. Uyeda, J. R. Deschamps, M. E. Lassman, and H. Mattoussi. 2005. A hybrid quantum dot—antibody fragment fluorescence resonance energy transfer-based TNT sensor. *Journal of the American Chemical Society* 127 (18): 6744–6751.

Gong, J.-L. Y. Liang, Y. Huang, J.-W. Chen, J.-H. Jiang, G.-L. Shen, and R.-Q. Yu. 2007. Ag/SiO$_2$ core-shell nanoparticle-based surface-enhanced Raman probes for immunoassay of cancer marker using silica-coated magnetic nanoparticles as separation tools. *Biosensors and Bioelectronics* 22 (7): 1501–1507.

Graf, C., D. L. J. Vossen, A. Imhof, and A. Van Blaaderen. 2003. A general method to coat colloidal particles with silica. *Langmuir* 19 (17): 6693–6700.

Grant, P. S. and M. J. McShane. 2003. Development of multilayer fluorescent thin film chemical sensors using electrostatic self-assembly. *IEEE Sensors Journal* 3 (2): 139–146.

Guerrero, H., G. Rosa, M. P. Morales, F. Del Monte, E. M. Moreno, D. Levy, R. Pérez Del Real, T. Belenguer, and C. J. Serna. 1997. Faraday rotation in magnetic γ-Fe2O3/SiO2 nanocomposites. *Applied Physics Letters* 71 (18): 2698–2700.

Gunnlaugsson, T. and J. P. Leonard. 2005. Responsive lanthanide luminescent cyclen complexes: From switching/sensing to supramolecular architectures. *Chemical Communications* (25): 3114–3131.

Haacke, E. M., R. W. Brown, M. R. Thompson, and R. Venkatesan. 1999. *Magnetic Resonance Imaging: Physical Principles and Sequence Design.* In ed. E. Haacke. John Wiley & Sons Inc.

Haes, A. J. and R. P. Van Duyne. 2002. A nanoscale optical biosensor: Sensitivity and selectivity of an approach based on the localized surface plasmon resonance spectroscopy of triangular silver nanoparticles. *Journal of the American Chemical Society* 124 (35): 10596–10604.

Hai, X., M. Tan, G. Wang, Z. Ye, J. Yuan, and K. Matsumoto. 2004. Preparation and a time-resolved fluoroimmunoassay application of new europium fluorescent nanoparticles. *Analytical Sciences* 20 (2): 245–246.

Hamblin, J., N. Abboyi, and M. P. Lowe. 2005. A binaphthyl-containing eu(III) complex and its interaction with human scrum albumin: A luminescence study. *Chemical Communications* (5): 657–659.

Han, M., X. Gao, J. Z. Su, and S. Nie. 2001. Quantum-dot-tagged microbeads for multiplexed optical coding of biomolecules. *Nature Biotechnology* 19 (7): 631–635.

Han, L., T. M. Niemczyk, D. M. Haaland, and G. P. Lopez. 1999. Enhancing IR detection limits for trace polar organics in aqueous solutions with surface-modified sol-gel-coated ATR sensors. *Applied Spectroscopy* 53 (4): 381–389.

Hanaoka, K., K. Kikuchi, S. Kobayashi, and T. Nagano. 2007. Time-resolved long-lived luminescence imaging method employing luminescent lanthanide probes with a new microscopy system. *Journal of the American Chemical Society* 129 (44): 13502–13509.

Harper, G. B. 1975. Reusable glass-bound pH indicators. *Analytical Chemistry* 47 (2): 348–351.

Haynes, C. L. and R. P. Van Duyne. 2003. Plasmon-sampled surface-enhanced raman excitation spectroscopy. *Journal of Physical Chemistry B* 107 (30): 7426–7433.

Haynes, C. L., C. R. Yonzon, X. Zhang, and R. P. Van Duyne. 2005. Surface-enhanced raman sensors: Early history and the development of sensors for quantitative biowarfare agent and glucose detection. *Journal of Raman Spectroscopy* 36 (6–7): 471–484.

He, G., G. Zhang, F. Lü, and Y. Fang. 2009. Fluorescent film sensor for vapor-phase nitroaromatic explosives via monolayer assembly of oligo(diphenylsilane) on glass plate surfaces. *Chemistry of Materials* 21 (8): 1494–1499.

Heer, S., K. Kömpe, H.-U. Güdel, and M. Haase. 2004. Highly efficient multicolour upconversion emission in transparent colloids of lanthanide-doped NaYF$_4$ nanocrystals. *Advanced Materials* 16 (23–24): 2102–2105.

Hernandez, R., H.-R. Tseng, J. W. Wong, J. F. Stoddart, and J. I. Zink. 2004. An operational supramolecular nanovalve. *Journal of the American Chemical Society* 126 (11): 3370–3371.

Ho, H. P., Y. H. Huang, S. Y. Wu, and S. K. Kong. 2012. Detecting phase shifts in surface plasmon resonance: A review. *Advances in Optical Technologies*, art. no. 471957.

Hofmann, O., G. Voirin, P. Niedermann, and A. Manz. 2002. Three-dimensional microfluidic confinement for efficient sample delivery to biosensor surfaces. application to immunoassays on planar optical wave-guides. *Analytical Chemistry* 74 (20): 5243–5250.

Homola, J., S. Y. Sinclair, and G. Gauglitz. 1999. Surface plasmon resonance sensors: Review. *Sensors and Actuators B: Chemical* 54 (1–2) (1/25): 3–15.

Hone, D. C., A. H. Haines, and D. A. Russell. 2003. Rapid, quantitative colorimetric detection of a lectin using mannose-stabilized gold nanoparticles. *Langmuir* 19 (17): 7141–7144.

Hradil, J., C. Davis, K. Mongey, C. McDonagh, and B. D. MacCraith. 2002. Temperature-corrected pressure-sensitive paint measurements using a single camera and a dual-lifetime approach. *Measurement Science and Technology* 13 (10): 1552–1557.

Hsiao, J.-K. C.-P. Tsai, T.-H. Chung, Y. Hung, M. Yao, H.-M. Liu, C.-Y. Mou, C.-S. Yang, Y.-C. Chen, and D.-M. Huang. 2008. Mesoporous silica nanoparticles as a delivery system of gadolinium for effective human stem cell tracking. *Small* 4 (9): 1445–1452.

Hu, M., J. Tian, H.-T. Lu, L.-X. Weng, and L.-H. Wang. 2010. H_2O_2-sensitive quantum dots for the label-free detection of glucose. *Talanta* 82 (3): 997–1002.

Huang, C., K. Bonroy, G. Reekman, K. Verstreken, L. Lagae, and G. Borghs. 2009. An on-chip localized surface plasmon resonance-based biosensor for label-free monitoring of antigen-antibody reaction. *Microelectronic Engineering* 86 (12): 2437–2441.

Huang, H., J. Liu, B. Han, and S. Xu. 2011. Aqueous synthesis of highly luminescent CdTe/CdS/ZnS core-shell-shell quantum dots with biocompatibility. *Advanced Materials Research* 287–290, 348–351.

Huh, Y. M., Y. W. Jun, H. T. Song, S. Kim, J. S. Choi, J. H. Lee, S. Yoon, et al. 2005. In vivo magnetic resonance detection of cancer by using multifunctional magnetic nanocrystals. *Journal of the American Chemical Society* 127 (35) (September 7): 12387–12391.

Hutter, E. and J. H. Fendler. 2004. Exploitation of localized surface plasmon resonance. *Advanced Materials* 16 (19): 1685–16706.

Idée, J.-M. M. Port, C. Medina, E. Lancelot, E. Fayoux, S. Ballet, and C. Corot. 2008. Possible involvement of gadolinium chelates in the pathophysiology of nephrogenic systemic fibrosis: A critical review. *Toxicology* 248 (2–3): 77–88.

Ionov, L., S. Sapra, A. Synytska, A. L. Rogach, M. Stamm, and S. Diez. 2006. Fast and spatially resolved environmental probing using stimuli-responsive polymer layers and fluorescent nanocrystals. *Advanced Materials* 18 (11): 1453–1457.

Ipe, B. I., K. Yoosaf, and K. G. Thomas. 2006. Functionalized gold nanoparticles as phosphorescent nanomaterials and sensors. *Journal of the American Chemical Society* 128 (6): 1907–1913.

Jarvis, R. M. and R. Goodacre. 2003. Discrimination of bacteria using surface-enhanced raman spectroscopy. *Analytical Chemistry* 76 (1) (11/19; 2011): 40–47.

Jeanmaire, D. L. and R. P. Van Duyne. 1977. Surface raman spectroelectrochemistry part I. heterocyclic, aromatic, and aliphatic amines adsorbed on the anodized silver electrode. *Journal of Electroanalytical Chemistry* 84 (1): 1–20.

Jiang, W., A. Singhal, B. Y. S. Kim, J. Zheng, J. T. Rutka, C. Wang, and W. C. W. Chan. 2008. Assessing near-infrared quantum dots for deep tissue, organ, and animal imaging applications. *JALA—Journal of the Association for Laboratory Automation* 13 (1): 6–12.

Josephson, L., M. F. Kircher, U. Mahmood, Y. Tang, and R. Weissleder. 2002. Near-infrared fluorescent nanoparticles as combined MR/optical imaging probes. *Bioconjugate Chemistry* 13 (3) (May-Jun): 554–560.

Jun, Y.-W., J.-H. Lee, and J. Cheon. 2008. Chemical design of nanoparticle probes for high-performance magnetic resonance imaging. *Angewandte Chemie—International Edition* 47 (28): 5122–5135.

Kang, Y. S., S. Risbud, J. F. Rabolt, and P. Stroeve. 1996. Synthesis and characterization of nanometer-size Fe_3O_4 and γ-Fe_2O_3 particles. *Chemistry of Materials* 8 (9): 2209–2211.

Kaul, G. and M. Amiji. 2002. Long-circulating poly(ethylene glycol)-modified gelatin nanoparticles for intracellular delivery. *Pharmaceutical Research* 19 (7): 1061–1067.

Kelly, K. A., J. R. Allport, A. Tsourkas, V. R. Shinde-Patil, L. Josephson, and R. Weissleder. 2005. Detection of vascular adhesion molecule-1 expression using a novel multimodal nanoparticle. *Circulation Research* 96 (3): 327–336.

Khatei, J. and K. S. R. Koteswara Rao. 2011. Hydrothermal synthesis of CdTe QDs: Their luminescence quenching in the presence of bio-molecules and observation of bistable memory effect in CdTe QD/PEDOT:PSS heterostructure. *Materials Chemistry and Physics* 130 (1–2): 159–164.

Kim, H., R. E. Cohen, P. T. Hammond, and D. J. Irvine. 2006a. Live lymphocyte arrays for biosensing. *Advanced Functional Materials* 16 (10): 1313–1323.

Kim, J.-H. J.-S. Kim, H. Choi, S.-M. Lee, B.-H. Jun, K.-N. Yu, E. Kuk, et al. 2006b. Nanoparticle probes with surface enhanced raman spectroscopic tags for cellular cancer targeting. *Analytical Chemistry* 78 (19): 6967–6973.

Kim, J., J. E. Lee, J. Lee, J. H. Yu, B. C. Kim, K. An, Y. Hwang, et al. 2006c. Magnetic fluorescent delivery vehicle using uniform mesoporous silica spheres embedded with monodisperse magnetic and semiconductor nanocrystals. *Journal of the American Chemical Society* 128 (3) (January 25): 688–689.

Kim, J., J. E. Lee, S. H. Lee, J. H. Yu, J. H. Lee, T. G. Park, and T. Hyeon. 2008. Designed fabrication of a multifunctional polymer nanomedical platform for simultaneous cancer-targeted imaging and magnetically guided drug delivery. *Advanced Materials* 20 (3) (February 4): 478,+.

Kim, S., Y. T. Lim, E. G. Soltesz, A. M. De Grand, J. Lee, A. Nakayama, J. A. Parker, et al. 2004a. Near-infrared fluorescent type II quantum dots for sentinel lymph node mapping. *Nature Biotechnology* 22 (1): 93–97.

Kim, J. H., D. Morikis, and M. Ozkan. 2004b. Adaptation of inorganic quantum dots for stable molecular beacons. *Sensors and Actuators, B: Chemical* 102 (2): 315–319.

Kim, J., Y. Piao, and T. Hyeon. 2009. Multifunctional nanostructured materials for multimodal imaging, and simultaneous imaging and therapy. *Chemical Society Reviews* 38 (2): 372–390.

Kim, J. S., W. J. Rieter, K. M. L. Taylor, H. An, W. L. Lin, and W. B. Lin. 2007b. Self-assembled hybrid nanoparticles for cancer-specific multimodal imaging. *Journal of the American Chemical Society* 129 (29) (July 25): 8962,+.

Kim, S., T. Y. Ohulchanskyy, H. E. Pudavar, R. K. Pandey, and P. N. Prasad. 2007a. Organically modified silica nanoparticles co-encapsulating photosensitizing drug and aggregation-enhanced two-photon absorbing fluorescent dye aggregates for two-photon photodynamic therapy. *Journal of the American Chemical Society* 129 (9): 2669–2675.

Kim, S.-W., J. P. Zimmer, S. Ohnishi, J. B. Tracy, J. V. Frangioni, and M. G. Bawendi. 2005. Engineering $InAs_xP_{1-x}/InP/ZnSe$ III-V alloyed core/shell quantum dots for the near-infrared. *Journal of the American Chemical Society* 127 (30): 10526–10532.

Kircher, M. F., U. Mahmood, R. S. King, R. Weissleder, and L. Josephson. 2003. A multimodal nanoparticle for preoperative magnetic resonance imaging and intraoperative optical brain tumor delineation. *Cancer Research* 63 (23) (December 1): 8122–8125.

Kneipp, K., Y. Wang, H. Kneipp, L. T. Perelman, I. Itzkan, R. R. Dasari, and M. S. Feld. 1997. Single molecule detection using surface-enhanced raman scattering (SERS). *Physical Review Letters* 78 (9): 1667–1670.

Knopp, D., D. Tang, and R. Niessner. 2009. Review: Bioanalytical applications of biomolecule-functionalized nanometer-sized doped silica particles. *Analytica Chimica Acta* 647 (1): 14–30.

Kohler, N., G. E. Fryxell, and M. Zhang. 2004. A bifunctional poly(ethylene glycol) silane immobilized on metallic oxide-based nanoparticles for conjugation with cell targeting agents. *Journal of the American Chemical Society* 126 (23): 7206–7211.

Koncki, R. and O. S. Wolfbeis. 1998. Optical chemical sensing based on thin films of prussian blue. *Sensors and Actuators, B: Chemical* 51 (1–3): 355–358.

Koole, R., M. M. Van Schooneveld, J. Hilhorst, K. Castermans, D. P. Cormode, G. J. Strijkers, C. D. M. Donegá, et al. 2008. Paramagnetic lipid-coated silica nanoparticles with a fluorescent quantum dot core: A new contrast agent platform for multimodality imaging. *Bioconjugate Chemistry* 19 (12): 2471–2479.

Korposh, S. O., Y. P. Sharkan, and J. J. Ramsden. 2008. Response of bacteriorhodopsin thin films to ammonia. *Sensors and Actuators, B: Chemical* 129 (1): 473–480.

Koyama, T., S. Ohtsuka, H. Nagata, and S. Tanaka. 1992. Fabrication of microcrystallites of II-IV compound semiconductors by laser ablation method. *Journal of Crystal Growth* 117 (1–4): 156–160.

Kupcho, K. R., D. K. Stafslien, T. DeRosier, T. M. Hallis, M. S. Ozers, and K. W. Vogel. 2007. Simultaneous monitoring of discrete binding events using dual-acceptor terbium-based LRET. *Journal of the American Chemical Society* 129 (44): 13372–13373.

Kwon, S. W., W. S. Yang, H. M. Lee, W. K. Kim, H.-Y. Lee, G. S. Son, W. J. Jeong, and D. H. Yoon. 2010. Response properties of waveguide-optic evanescent field refractive index sensors according to titanium dioxide thin film conditions. *Sensor Letters* 8 (3): 431–435.

Lacoste, T. D., X. Michalet, F. Pinaud, D. S. Chemla, A. P. Alivisatos, and S. Weiss. 2000. Ultrahigh-resolution multicolor colocalization of single fluorescent probes. *Proceedings of the National Academy of Sciences of the United States of America* 97 (17): 9461–9466.

Lai, C.-Y., B. G. Trewyn, D. M. Jeftinija, K. Jeftinija, S. Xu, S. Jeftinija, and V. S.-Y. Lin. 2003. A mesoporous silica nanosphere-based carrier system with chemically removable CdS nanoparticle caps for stimuli-responsive controlled release of neurotransmitters and drug molecules. *Journal of the American Chemical Society* 125 (15): 4451–4459.

Lai, C. W., Y. H. Wang, C. H. Lai, M. J. Yang, C. Y. Chen, P. T. Chou, C. S. Chan, Y. Chi, Y. C. Chen, and J. K. Hsiao. 2008. Iridium-complex-functionalized Fe_3O_4/SiO_2 core/shell nanoparticles: A facile three-in-one system in magnetic resonance imaging, luminescence imaging, and photodynamic therapy. *Small* 4 (2) (Feb): 218–224.

Landin, L., M. S. Miller, M.-E. Pistol, C. E. Pryor, and L. Samuelson. 1998. Optical studies of individual In As quantum dots in GaAs: Few-particle effects. *Science* 280 (5361): 262–264.

Larrió.n, B., M. Hernáez, F. J. Arregui, J. Goicoechea, J. Bravo, and I. R. Matías. 2009. Photonic crystal fiber temperature sensor based on quantum dot nanocoatings. *Journal of Sensors* 2009: 932471 doi:10.1155/2009/932471.

Law, W. C., K. T. Yong, I. Roy, G. Xu, H. Ding, E. J. Bergey, H. Zeng, and P. N. Prasad. 2008. Optically and magnetically doped organically modified silica nanoparticles as efficient magnetically guided biomarkers for two-photon imaging of live cancer cells. *Journal of Physical Chemistry C* 112 (21) (May 29): 7972–7977.

Lazcano-Hernndez, H. E., C. Snchez-Pérez, and A. García-Valenzuela. 2008. An optically integrated NH3 sensor using WO_3 thin films as sensitive material. *Journal of Optics A: Pure and Applied Optics* 10 (10): 104016 doi:10.1088/1464-4258/10/10/104016.

Lee, C.-H. S.-H. Cheng, Y.-J. Wang, Y.-C. Chen, N.-T. Chen, J. Souris, C.-T. Chen, C.-Y. Mou, C.-S. Yang, and L.-W. Lo. 2009b. Near-infrared mesoporous silica nanoparticles for optical imaging: Characterization and in vivo biodistribution. *Advanced Functional Materials* 19 (2): 215–222.

Lee, J.-H., Y.-W. Jun, S.-I. Yeon, J.-S. Shin, and J. Cheon. 2006b. Dual-mode nanoparticle probes for high-performance magnetic resonance and fluorescence imaging of neuroblastoma. *Angewandte Chemie—International Edition* 45 (48): 8160–8162.

Lee, S.-H., J. Kumar, and S. K. Tripathy. 2000. Thin film optical sensors employing polyelectrolyte assembly. *Langmuir* 16 (26): 10482–10489.

Lee, H., E. Lee, D. K. Kim, N. K. Jang, Y. Y. Jeong, and S. Jon. 2006a. Antibiofouling polymer-coated superparamagnetic iron oxide nanoparticles as potential magnetic resonance contrast agents for in vivo cancer imaging. *Journal of the American Chemical Society* 128 (22): 7383–7389.

Lee, H., K. Y. Mi, S. Park, S. Moon, J. M. Jung, Y. J. Yong, H.-W. Kang, and S. Jon. 2007. Thermally cross-linked superparamagnetic iron oxide nanoparticles: Synthesis and application as a dual imaging probe for cancer in vivo. *Journal of the American Chemical Society* 129 (42): 12739–12745.

Lee, B., S. Roh, and J. Park. 2009a. Current status of micro- and nano-structured optical fiber sensors. *Optical Fiber Technology* 15 (3) (6): 209–221.

Lee, K., J.-M. Rouillard, B.-G. Kim, E. Gulari, and J. Kim. 2009c. Conjugated polymers combined with a molecular beacon for label-free and self-signal-amplifying DNA microarrays. *Advanced Functional Materials* 19 (20): 3317–3325.

Lenz, J. and A. S. Edelstein. 2006. Magnetic sensors and their applications. *IEEE Sensors Journal* 6 (3) (June): 631–649.

Li, L., E. S. G. Choo, Z. Liu, J. Ding, and J. Xue. 2008. Double-layer silica core-shell nanospheres with superparamagnetic and fluorescent functionalities. *Chemical Physics Letters* 461 (1–3): 114–117.

Li, Z. and Y. Zhang. 2006. Monodisperse silica-coated polyvinyl-pyrrolidone/NaYF4 nanocrystals with multicolor upconversion fluorescence emission. *Angewandte Chemie—International Edition* 45 (46): 7732–7735.

Lian, W., S. A. Litherland, H. Badrane, W. Tan, D. Wu, H. V. Baker, P. A. Gulig, D. V. Lim, and S. Jin. 2004. Ultrasensitive detection of biomolecules with fluorescent dye-doped nanoparticles. *Analytical Biochemistry* 334 (1): 135–144.

Liang, Y., J.-L. Gong, Y. Huang, Y. Zheng, J.-H. Jiang, G.-L. Shen, and R.-Q. Yu. 2007. Biocompatible core-shell nanoparticle-based surface-enhanced raman scattering probes for detection of DNA related to HIV gene using silica-coated magnetic nanoparticles as separation tools. *Talanta* 72 (2): 443–449.

Liao, H. and J. H. Hafner. 2005. Gold nanorod bioconjugates. *Chemistry of Materials* 17 (18): 4636–4641.

Liebsch, G., I. Klimant, and O. S. Wolfbeis. 1999. Luminescence lifetime temperature sensing based on sol-gels and poly(acrylonitrile)s dyed with ruthenium metal-ligand complexes. *Advanced Materials* 11 (15): 1296–1299.

Lin, V. S.-Y., C.-Y. Lai, J. Huang, S.-A. Song, and S. Xu. 2001. Molecular recognition inside of multifunction-alized mesoporous silicas: Toward selective fluorescence detection of dopamine and glucosamine [8]. *Journal of the American Chemical Society* 123 (46): 11510–11511.

Lin, Y. J. and A. P. Koretsky. 1997. Manganese ion enhances T-1-weighted MRI during brain activation: An approach to direct imaging of brain function. *Magnetic Resonance in Medicine* 38 (3) (September): 378–388.

Lin, Y. S., S. H. Wu, Y. Hung, Y. H. Chou, C. Chang, M. L. Lin, C. P. Tsai, and C. Y. Mou. 2006. Multifunctional composite nanoparticles: Magnetic, luminescent, and mesoporous. *Chemistry of Materials* 18 (22) (October 31): 5170–5172.

Liz-Marzán, L. M. 2004. Nanometals: Formation and color. *Materials Today* 7 (2): 26–31.

Lobnik, A., I. Oehme, I. Murkovic, and O. S. Wolfbeis. 1998. pH optical sensors based on sol-gels: Chemical doping versus covalent immobilization. *Analytica Chimica Acta* 367 (1–3): 159–165.

Louie, A. 2010. Multimodality imaging probes: Design and challenges. *Chemical Reviews* 110 (5): 3146–3195.

Louis, C., R. Bazzi, C. A. Marquette, J.-L. Bridot, S. Roux, G. Ledoux, B. Mercier, L. Blum, P. Perriat, and O. Tillement. 2005. Nanosized hybrid particles with double luminescence for biological labeling. *Chemistry of Materials* 17 (7): 1673–1682.

Lu, A. H., E. L. Salabas, and F. Schuth. 2007c. Magnetic nanoparticles: Synthesis, protection, functionaliza-tion, and application. *Angewandte Chemie-International Edition* 46 (8): 1222–1244.

Lü, F., L. Gao, L. Ding, L. Jiang, and Y. Fang. 2006. Spacer layer screening effect: A novel fluorescent film sensor for organic copper(II) salts. *Langmuir* 22 (2): 841–845.

Lü, F., L. Gao, H. Li, L. Ding, and Y. Fang. 2007. Molecular engineered silica surfaces with an assembled anthracene monolayer as a fluorescent sensor for organic copper(II) salts. *Applied Surface Science* 253 (9): 4123–4131.

Lu, C. W., Y. Hung, J. K. Hsiao, M. Yao, T. H. Chung, Y. S. Lin, S. H. Wu, et al. 2007a. Bifunctional magnetic silica nanoparticles for highly efficient human stem cell labeling. *Nano Letters* 7 (1) (January): 149–154.

Lu, J., M. Liong, J. I. Zink, and F. Tamanoi. 2007b. Mesoporous silica nanoparticles as a delivery system for hydrophobic anticancer drugs. *Small* 3 (8): 1341–1346.

Lu, H., G. Yi, S. Zhao, D. Chen, L.-H. Guo, and J. Cheng. 2004. Synthesis and characterization of multi-functional nanoparticles possessing magnetic, up-conversion fluorescence and bio-affinity properties. *Journal of Materials Chemistry* 14 (8): 1336–1341.

Ludolph, B., M. A. Malik, P. O'Brien, and N. Revaprasadu. 1998. Novel single molecule precursor routes for the direct synthesis of highly monodispersed quantum dots of cadmium or zinc sulfide or selenide. *Chemical Communications* (17): 1849–1850.

Ma, Z., D. Dosev, M. Nichkova, R. K. Dumas, S. J. Gee, B. D. Hammock, K. Liu, and I. M. Kennedy. 2009. Synthesis and characterization of multifunctional silica core-shell nanocomposites with magnetic and fluorescent functionalities. *Journal of Magnetism and Magnetic Materials* 321 (10): 1368–1371.

Mafuné, F., J.-Y. Kohno, Y. Takeda, T. Kondow, and H. Sawabe. 2001. Formation of gold nanoparticles by laser ablation in aqueous solution of surfactant. *Journal of Physical Chemistry B* 105 (22): 5114–5120.

Mal, N. K., M. Fujiwara, and Y. Tanaka. 2003. Photocontrolled reversible release of guest molecules from coumarin-modified mesoporous silica. *Nature* 421 (6921): 350–353.

Mamot, C., J. B. Nguyen, M. Pourdehnad, P. Hadaczek, R. Saito, J. R. Bringas, D. C. Drummond, et al. 2004. Extensive distribution of liposomes in rodent brains and brain tumors following convection-enhanced delivery. *Journal of Neuro-Oncology* 68 (1): 1–9.

Marqués-Hueso, J., R. Abargues, J. Canet-Ferrer, S. Agouram, J. L. Valdés, and J. P. Martínez-Pastor. 2010. Au-pva nanocomposite negative resist for one-step three-dimensional e-beam lithography. *Langmuir* 26 (4): 2825–2830.

Maxwell, D. J., J. Bonde, D. A. Hess, S. A. Hohm, R. Lahey, P. Zhou, M. H. Creer, D. Piwnica-Worms, and J. A. Nolta. 2008. Fluorophore-conjugated iron oxide nanoparticle labeling and analysis of engrafting human hematopoietic stem cells. *Stem Cells* 26 (2): 517–524.

McCarthy, J. R., F. A. Jaffer, and R. Weissleder. 2006. A macrophage-targeted theranostic nanoparticle for biomedical applications. *Small* 2 (8–9): 983–987.

Medarova, Z., M. Kumar, S.-W. Ng, J. Yang, N. Barteneva, N. V. Evgenov, V. Petkova, and A. Moore. 2008. Multifunctional magnetic nanocarriers for image-tagged SiRNA delivery to intact pancreatic islets. *Transplantation* 86 (9): 1170–1177.

Medarova, Z., W. Pham, C. Farrar, V. Petkova, and A. Moore. 2007. In vivo imaging of siRNA delivery and silencing in tumors. *Nature Medicine* 13 (3) (March): 372–377.

Medintz, I. L., H. T. Uyeda, E. R. Goldman, and H. Mattoussi. 2005. Quantum dot bioconjugates for imaging, labelling and sensing. *Nature Materials* 4 (6): 435–446.

Mirkin, C. A., R. L. Letsinger, R. C. Mucic, and J. J. Storhoff. 1996. A DNA-based method for rationally assembling nanoparticles into macroscopic materials. *Nature* 382 (6592): 607–609.

Mitchell, J. 2010. Small molecule immunosensing using surface plasmon resonance. *Sensors* 10 (8): 7323–7346.

Mitsuishi, M., S. Kikuchi, T. Miyashita, and Y. Amao. 2003. Characterization of an ultrathin polymer optode and its application to temperature sensors based on luminescent europium complexes. *Journal of Materials Chemistry* 13 (12): 2875–2879.

Mizaikoff, B. 2003. Infrared optical sensors for water quality monitoring. *Water Science and Technology* 47 (2), 35–42.

Moghimi, S. M., A. C. Hunter, and J. C. Murray. 2001. Long-circulating and target-specific nanoparticles: Theory to practice. *Pharmacological Reviews* 53 (2): 283–318.

Montet, X., K. Montet-Abou, F. Reynolds, R. Weissleder, and L. Josephson. 2006. Nanoparticle imaging of integrins on tumor cells. *Neoplasia* 8 (3): 214–222.

Moore, A., Z. Medarova, A. Potthast, and G. Dai. 2004. in vivo targeting of underglycosylated MUC-1 tumor antigen using a multimodal imaging probe. *Cancer Research* 64 (5): 1821–1827.

Mukundan, H., A. S. Anderson, W. K. Grace, K. M. Grace, N. Hartman, J. S. Martinez, and B. I. Swanson. 2009. Waveguide-based biosensors for pathogen detection. *Sensors* 9 (7): 5783–5809.

Mulder, W. J. M., A. W. Griffioen, G. J. Strijkers, D. P. Cormode, K. Nicolay, and Z. A. Fayad. 2007. Magnetic and fluorescent nanoparticles for multimodality imaging. *Nanomedicine* 2 (3): 307–324.

Mulder, W. J. M., R. Koole, R. J. Brandwijk, G. Storm, P. T. K. Chin, G. J. Strijkers, C. D. Donega, K. Nicolay, and A. W. Griffioen. 2006a. Quantum dots with a paramagnetic coating as a bimodal molecular imaging probe. *Nano Letters* 6 (1) (January): 1–6.

Mulder, W. J. M., G. J. Strijkers, A. W. Griffioen, L. Van Bloois, G. Molema, G. Storm, G. A. Koning, and K. Nicolay. 2004. A liposomal system for contrast-enhanced magnetic resonance imaging of molecular targets. *Bioconjugate Chemistry* 15 (4): 799–806.

Mulder, W. J. M., G. J. Strijkers, G. A. F. van Tilborg, A. W. Griffioen, and K. Nicolay. 2006b. Lipid-based nanoparticles for contrast-enhanced MRI and molecular imaging. *NMR in Biomedicine* 19 (1): 142–164.

Murray, C. B., D. J. Norris, and M. G. Bawendi. 1993. Synthesis and characterization of nearly monodisperse CdE (E = S, se, te) semiconductor nanocrystallites. *Journal of the American Chemical Society* 115 (19): 8706–8715.

Nagao, D., M. Yokoyama, N. Yamauchi, H. Matsumoto, Y. Kobayashi, and M. Konno. 2008. Synthesis of highly monodisperse particles composed of a magnetic core and fluorescent shell. *Langmuir* 24 (17): 9804–9808.

Nagl, S. and O. S. Wolfbeis. 2007. Optical multiple chemical sensing: Status and current challenges. *Analyst* 132 (6): 507–511.

Nahrendorf, M., F. A. Jaffer, K. A. Kelly, D. E. Sosnovik, E. Aikawa, P. Libby, and R. Weissleder. 2006. Noninvasive vascular cell adhesion molecule-1 imaging identifies inflammatory activation of cells in atherosclerosis. *Circulation* 114 (14): 1504–1511.

Nahrendorf, M., H. W. Zhang, S. Hembrador, P. Panizzi, D. E. Sosnovik, E. Aikawa, P. Libby, F. K. Swirski, and R. Weissleder. 2008. Nanoparticle PET-CT imaging of macrophages in inflammatory atherosclerosis. *Circulation* 117 (3) (January 22): 379–387.

Nasongkla, N., E. Bey, J. M. Ren, H. Ai, C. Khemtong, J. S. Guthi, S. F. Chin, A. D. Sherry, D. A. Boothman, and J. M. Gao. 2006. Multifunctional polymeric micelles as cancer-targeted, MRI-ultrasensitive drug delivery systems. *Nano Letters* 6 (11) (November 8): 2427–2430.

Nath, N. and A. Chilkoti. 2002. A colorimetric gold nanoparticle sensor to interrogate biomolecular interactions in real time on a surface. *Analytical Chemistry* 74 (3): 504–509.

Neuberger, T., B. Schöpf, H. Hofmann, M. Hofmann, and B. Von Rechenberg. 2005. Superparamagnetic nanoparticles for biomedical applications: Possibilities and limitations of a new drug delivery system. *Journal of Magnetism and Magnetic Materials* 293 (1): 483–496.

Nie, S. and S. R. Emory. 1997. Probing single molecules and single nanoparticles by surface-enhanced raman scattering. *Science* 275 (5303): 1102–1106.

Nikiforov, T. T. and J. M. Beechem. 2006. Development of homogeneous binding assays based on fluorescence resonance energy transfer between quantum dots and alexa fluor fluorophores. *Analytical Biochemistry* 357 (1): 68–76.

Nitin, N., L. E. W. LaConte, O. Zurkiya, X. Hu, and G. Bao. 2004. Functionalization and peptide-based delivery of magnetic nanoparticles as an intracellular MRI contrast agent. *Journal of Biological Inorganic Chemistry* 9 (6): 706–712.

Nock, V., R. J. Blaikie, and T. David. 2008. Patterning, integration and characterisation of polymer optical oxygen sensors for microfluidic devices. *Lab on a Chip—Miniaturisation for Chemistry and Biology* 8 (8): 1300–1307.

Ordal, M. A., L. L. Long, R. J. Bell, S. E. Bell, R. W. Alexander Jr., C. A. Ward, and R. R. Bell. 1983. Optical properties of the metals al, co, cu, au, fe, pb, ni, pd, pt, ag, ti, and W in the infrared and far infrared. *Applied Optics* 22 (7): 1099–1119.

Osseo-Asare, K. and F. J. Arriagada. 1990. Preparation of SiO_2 nanoparticles in a non-ionic reverse micellar system. *Colloids and Surfaces* 50 (C): 321–339.

Ow, H., D. R. Larson, M. Srivastava, B. A. Baird, W. W. Webb, and U. Wiesnert. 2005. Bright and stable core-shell fluorescent silica nanoparticles. *Nano Letters* 5 (1): 113–117.

Palmer, R. J., J. L. Butenhoff, and J. B. Stevens. 1987. Cytotoxicity of the rare earth metals cerium, lanthanum, and neodymium in vitro: Comparisons with cadmium in a pulmonary macrophage primary culture system. *Environmental Research* 43 (1): 142–156.

Pan, B., D. Cui, Y. Sheng, C. Ozkan, F. Gao, R. He, Q. Li, P. Xu, and T. Huang. 2007. Dendrimer-modified magnetic nanoparticles enhance efficiency of gene delivery system. *Cancer Research* 67 (17): 8156–8163.

Pandya, S., J. Yu, and D. Parker. 2006. Engineering emissive europium and terbium complexes for molecular imaging and sensing. *Dalton Transactions* (23): 2757–2766.

Pankhurst, Q. A., J. Connolly, S. K. Jones, and J. Dobson. 2003. Applications of magnetic nanoparticles in biomedicine. *Journal of Physics D: Applied Physics* 36 (13): R167 *doi:10.1088/0022-3727/36/13/201.*

Park, K., S. Lee, E. Kang, K. Kim, K. Choi, and I. C. Kwon. 2009. New generation of multifunctional nanoparticles for cancer imaging and therapy. *Advanced Functional Materials* 19 (10): 1553–1566.

Park, J. H., G. von Maltzahn, E. Ruoslahti, S. N. Bhatia, and M. J. Sailor. 2008. Micellar hybrid nanoparticles for simultaneous magnetofluorescent imaging and drug delivery. *Angewandte Chemie-International Edition* 47 (38): 7284–7288.

Parker, D. 2000. Luminescent lanthanide sensors for pH, p O_2 and selected anions. *Coordination Chemistry Reviews* 205 (1): 109–130.

Pearton, S. J., F. Ren, Y.-L. Wang, B. H. Chu, K. H. Chen, C. Y. Chang, W. Lim, J. Lin, and D. P. Norton. 2010. Recent advances in wide bandgap semiconductor biological and gas sensors. *Progress in Materials Science* 55 (1): 1–59.

Peng, Z. A. and X. Peng. 2001. Formation of high-quality CdTe, CdSe, and CdS nanocrystals using CdO as precursor [6]. *Journal of the American Chemical Society* 123 (1): 183–184.

Pérez-Juste, J., I. Pastoriza-Santos, L. M. Liz-Marzán, and P. Mulvaney. 2005. Gold nanorods: Synthesis, characterization and applications. *Coordination Chemistry Reviews* 249 (17–18 SPEC. ISS.): 1870–1901.

Phelps, M. E. 2000. PET: The merging of biology and imaging into molecular imaging. *Journal of Nuclear Medicine* 41 (4) (April): 661–681.

Piao, Y., A. Burns, J. Kim, U. Wiesner, and T. Hyeon. 2008. Designed fabrication of silica-based nanostructured particle systems for nanomedicine applications. *Advanced Functional Materials* 18 (23): 3745–3758.

Piehler, J., A. Brecht, R. Valiokas, B. Liedberg, and G. Gauglitz. 2000. A high-density poly(ethylene glycol) polymer brush for immobilization on glass-type surfaces. *Biosensors and Bioelectronics* 15 (9–10): 473–481.

Pittet, M. J., F. K. Swirski, F. Reynolds, L. Josephson, and R. Weissleder. 2006. Labeling of immune cells for in vivo imaging using magnetofluorescent nanoparticles. *Nature Protocols* 1 (1): 73–79.

Plowman, T. E., J. D. Durstchi, H. K. Wang, D. A. Christensen, J. N. Herron, and W. M. Reichert. 1999. Multiple-analyte fluoroimmunoassay using an integrated optical waveguide sensor. *Analytical Chemistry* 71 (19): 4344–4352.

Qi, Z. M., K. Itoh, M. Murabayashi, and H. Yanagi. 2000. Composite optical waveguide-based polarimetric interferometer for chemical and biological sensing applications. *Journal of Lightwave Technology* 18 (8): 1106–1110.

Qin, J., S. Laurent, Y. S. Jo, A. Roch, M. Mikhaylova, Z. M. Bhujwalla, R. N. Müller, and M. Muhammed. 2007. A high-performance magnetic resonance imaging T2 contrast agent. *Advanced Materials* 19 (14): 1874–1878.

Quarta, A., R. Di Corato, L. Manna, S. Argentiere, R. Cingolani, G. Barbarella, and T. Pellegrino. 2008. Multifunctional nanostructures based on inorganic nanoparticles and oligothiophenes and their exploitation for cellular studies. *Journal of the American Chemical Society* 130 (32): 10545–10555.

Quarta, A., R. Di Corato, L. Manna, A. Ragusa, and T. Pellegrino. 2007. Fluorescent-magnetic hybrid nano-structures: Preparation, properties, and applications in biology. *Ieee Transactions on Nanobioscience* 6 (4) (December): 298–308.

Radu, D. R., C.-Y. Lai, K. Jeftinija, E. W. Rowe, S. Jeftinija, and V. S.-Y. Lin. 2004. A polyamidoamine den-drimer-capped mesoporous silica nanosphere-based gene transfection reagent. *Journal of the American Chemical Society* 126 (41): 13216–13217.

Ramsden, J. J. 1993. Sensitivity enhancement of integrated optic sensors using langmuir-blodgett lipid films. *Sensors and Actuators: B.Chemical* 16 (1–3): 439–442.

Reddy, G. R., M. S. Bhojani, P. McConville, J. Moody, B. A. Moffat, D. E. Hall, G. Kim, et al. 2006. Vascular targeted nanoparticles for imaging and treatment of brain tumors. *Clinical Cancer Research* 12 (22) (November 15): 6677–6686.

Reichl, D., R. Krage, C. Krummel, and G. Gauglitz. 2000. Sensing of volatile organic compounds using a sim-plified reflectometric interference spectroscopy setup. *Applied Spectroscopy* 54 (4): 583–586.

Reisfeld, R. 2001. Prospects of sol-gel technology towards luminescent materials. *Optical Materials* 16 (1–2): 1–7.

Rejman, J., V. Oberle, I. S. Zuhorn, and D. Hoekstra. 2004. Size-dependent internalization of particles via the pathways of clathrin-and caveolae-mediated endocytosis. *Biochemical Journal* 377 (1): 159–169.

Ren, T., P. K. Mandal, W. Erker, Z. Liu, Y. Aviasevich, L. Puhl, K. Mullen, and T. Basché. 2008. A simple and versatile route to stable quantum dot-dye hybrids in nonaqueous and aqueous solutions. *Journal of the American Chemical Society* 130 (51): 17242–17243.

Ren, C., J. Sun, J. Li, X. Chen, Z. Hu, and D. Xue. 2009. Bi-functional silica nanoparticles doped with iron oxide and cdte prepared by a facile method. *Nanoscale Research Letters* 4 (7): 640–645.

Rhee, H.-W., C.-R. Lee, S.-H. Cho, M.-R. Song, M. Cashel, H. E. Choy, Y.-J. Seok, and J.-I. Hong. 2008. Selective fluorescent chemosensor for the bacterial alarmone (p)ppGpp. *Journal of the American Chemical Society* 130 (3): 784–785.

Rieter, W. J., J. S. Kim, K. M. L. Taylor, H. Y. An, W. L. Lin, T. Tarrant, and W. B. Lin. 2007. Hybrid silica nanoparticles for multimodal imaging. *Angewandte Chemie-International Edition* 46 (20): 3680–3682.

Rivero, P. J., A. Urrutia, J. Goicoechea, C. R. Zamarreño, F. J. Arregui, and I. R. Matías. 2011. An antibacterial coating based on a polymer/sol- gel hybrid matrix loaded with silver nanoparticles. *Nanoscale Research Letters* 6 (1): 1–7.

Ruan, G., A. Agrawal, A. I. Marcus, and S. Nie. 2007. Imaging and tracking of tat peptide-conjugated quantum dots in living cells: New insights into nanoparticle uptake, intracellular transport, and vesicle shedding. *Journal of the American Chemical Society* 129 (47): 14759–14766.

Ruano, J. M., A. Glidle, A. Cleary, A. Walmsley, J. S. Aitchison, and J. M. Cooper. 2002. Design and fabrication of a silica on silicon integrated optical biochip as a fluorescence microarray platform. *Biosensors and Bioelectronics* 18 (2–3): 175–184.

Russ A. W. and U. J. Krull. 2011. Characterization of the adsorption of oligonucleotides on mercaptopropionic acid-coated CdSe/ZnS quantum dots using fluorescence resonance energy transfer. *Journal of Colloid and Interface Science* 359 (1): 148–154.

Sabbatini, N., M. Guardigli, and J.-M. Lehn. 1993. Luminescent lanthanide complexes as photochemical supra-molecular devices. *Coordination Chemistry Reviews* 123 (1–2): 201–228.

Sakai-Kato, K., Kato, M., Utsunomiya-Tate, N., and Toyo'oka, T. 2006. Encapsulated biomolecules using sol-gel reaction for high-throughput screening. *Frontiers in Drug Design and Discovery* 2 (1), 273–294.

Salgueiriño-Maceira, V., M. A. Correa-Duarte, M. A. López-Quintela, and J. Rivas. 2009. Advanced hybrid nanoparticles. *Journal of Nanoscience and Nanotechnology* 9 (6): 3684–3688.

Salgueiriño-Maceira, V., M. A. Correa-Duarte, M. Spasova, L. M. Liz-Marzán, and M. Farle. 2006. Composite sil-ica spheres with magnetic and luminescent functionalities. *Advanced Functional Materials* 16 (4): 509–514.

Santra, S., D. Dutta, G. A. Walter, and B. M. Moudgil. 2005. Fluorescent nanoparticle probes for cancer imag-ing. *Technology in Cancer Research and Treatment* 4 (6): 593–602.

Santra, S., C. Kaittanis, J. Grimm, and J. M. Perez. 2009. Drug/dye-loaded, multifunctional iron oxide nanopar-ticles for combined targeted cancer therapy and dual optical/magnetic resonance imaging. *Small* 5 (16): 1862–1868.

Santra, S., P. Zhang, K. Wang, R. Tapec, and W. Tan. 2001. Conjugation of biomolecules with luminophore-doped silica nanoparticles for photostable biomarkers. *Analytical Chemistry* 73 (20): 4988–4993.

Sarkander, H. I. and W. P. Brade. 1976. On the mechanism of lanthanide induced liver toxicity. *Archives of Toxicology* 36 (1): 1–17.

Sarkar, S. K., N. Chandrasekharan, S. Gorer, and G. Hodes. 2002. Reversible adsorption-enhanced quantum confinement in semiconductor quantum dots. *Applied Physics Letters* 81 (26): 5045–5047.

Sarkisov, S. S., M. J. Curley, C. Boykin, D. E. Diggs, J. Grote, and F. Hopkins. 2004. Planar optical waveguide sensor of ammonia. *Proceedings of SPIE - The International Society for Optical Engineering* 5586, art. no. 05, 33–44.

Sathe, T. R., A. Agrawal, and S. M. Nie. 2006. Mesoporous silica beads embedded with semiconductor quantum dots and iron oxide nanocrystals: Dual-function microcarriers for optical encoding and magnetic separation. *Analytical Chemistry* 78 (16) (August 15): 5627–5632.

Sato, Y., S. Nishizawa, and N. Teramae. 2011. Label-free molecular beacon system based on DNAs containing abasic sites and fluorescent ligands that bind abasic sites. *Chemistry—A European Journal* 17 (41): 11650–11656.

Schellenberger, E. A., D. Sosnovik, R. Weissleder, and L. Josephson. 2004. Magneto/optical annexin V, a multimodal protein. *Bioconjugate Chemistry* 15 (5): 1062–1067.

Schneider, B. H., E. L. Dickinson, M. D. Vach, J. V. Hoijer, and L. V. Howard. 2000. Optical chip immunoassay for hCG in human whole blood. *Biosensors and Bioelectronics* 15 (11–12): 597–604.

Schröder, C. R., L. Polerecky, and I. Klimant. 2007. Time-resolved pH/pO2 mapping with luminescent hybrid sensors. *Analytical Chemistry* 79 (1): 60–70.

Selvan, S. T., P. K. Patra, C. Y. Ang, and J. Y. Ying. 2007. Synthesis of silica-coated semiconductor and magnetic quantum dots and their use in the imaging of live cells. *Angewandte Chemie—International Edition* 46 (14): 2448–2452.

Seo, W. S., J. H. Lee, X. Sun, Y. Suzuki, D. Mann, Z. Liu, M. Terashima, et al. 2006. FeCo/graphitic-shell nanocrystals as advanced magnetic-resonance-imaging and near-infrared agents. *Nature Materials* 5 (12): 971–976.

Sepúlveda, B., P. C. Angelomé, L. M. Lechuga, and L. M. Liz-Marzán. 2009. LSPR-based nanobiosensors. *Nano Today* 4 (3): 244–251.

Sepúlveda, B., A. Calle, L. M. Lechuga, and G. Armelles. 2006. Highly sensitive detection of biomolecules with the magneto-optic surface-plasmon-resonance sensor. *Optics Letters* 31 (8): 1085–1087.

Shellock, F. G. and A. Spinazzi. 2008. MRI safety update 2008: Part 1, MRI contrast agents and nephrogenic systemic fibrosis. *American Journal of Roentgenology* 191 (4): 1129–1139.

Sivakumar, S., P. R. Diamente, and F. C. J. M. Van Veggel. 2006. Silica-coated Ln3+-doped LaF3 nanoparticles as robust down- and upconverting biolabels. *Chemistry - A European Journal* 12 (22): 5878–5884.

Slowing, I. I., B. G. Trewyn, S. Giri, and V. S.-Y. Lin. 2007. Mesoporous silica nanoparticles for drug delivery and biosensing applications. *Advanced Functional Materials* 17 (8): 1225–1236.

Smith, D. K. and B. A. Korgel. 2008. The importance of the CTAB surfactant on the colloidal seed-mediated synthesis of gold nanorods. *Langmuir* 24 (3): 644–649.

Soenen, S. J. H., D. Vercauteren, K. Braekmans, W. Noppe, S. De Smedt, and M. De Cuyper. 2009. Stable long-term intracellular labelling with fluorescently tagged cationic magnetoliposomes. *ChemBioChem* 10 (2): 257–267.

Soo Choi, H., W. Liu, P. Misra, E. Tanaka, J. P. Zimmer, B. Itty Ipe, M. G. Bawendi, and J. V. Frangioni. 2007. Renal clearance of quantum dots. *Nature Biotechnology* 25 (10): 1165–1170.

Spadavecchia, J., G. Ciccarella, T. Stomeo, R. Rella, S. Capone, and P. Siciliano. 2004. Variation in the optical sensing responses toward vapors of a porphyrin/phthalocyanine hybrid thin film. *Chemistry of Materials* 16 (11): 2083–2090.

Stöber, W., A. Fink, and E. Bohn. 1968. Controlled growth of monodisperse silica spheres in the micron size range. *Journal of Colloid and Interface Science* 26 (1): 62–69.

Storhoff, J. J., R. Elghanian, R. C. Mucic, C. A. Mirkin, and R. L. Letsinger. 1998. One-pot colorimetric differentiation of polynucleotides with single base imperfections using gold nanoparticle probes. *Journal of the American Chemical Society* 120 (9): 1959–1964.

Storhoff, J. J., A. A. Lazarides, R. C. Mucic, C. A. Mirkin, R. L. Letsinger, and G. C. Schatz. 2000. What controls the optical properties of DNA-linked gold nanoparticle assemblies? *Journal of the American Chemical Society* 122 (19): 4640–4650.

Storhoff, J. J., A. D. Lucas, V. Garimella, Y. P. Bao, and U. R. Müller. 2004. Homogeneous detection of unamplified genomic DNA sequences based on colorimetric scatter of gold nanoparticle probes. *Nature Biotechnology* 22 (7): 883–887.

Stouwdam, J. W., G. A. Hebbink, J. Huskens, and F. C. J. M. Van Veggel. 2003. Lanthanide-doped nanoparticles with excellent luminescent properties in organic media. *Chemistry of Materials* 15 (24): 4604–4616.

Sudeep, P. K., S. T. S. Joseph, and K. G. Thomas. 2005. Selective detection of cysteine and glutathione using gold nanorods. *Journal of the American Chemical Society* 127 (18): 6516–6517.

Sun, C., J. S. H. Lee, and M. Zhang. 2008. Magnetic nanoparticles in MR imaging and drug delivery. *Advanced Drug Delivery Reviews* 60 (11): 1252–1265.

Sweatlock, L. A., S. A. Maier, H. A. Atwater, J. J. Penninkhof, and A. Polman. 2005. Highly confined electromagnetic fields in arrays of strongly coupled ag nanoparticles. *Physical Review B—Condensed Matter and Materials Physics* 71 (23): 1–7.

Talanov, V. S., C. A. S. Regino, H. Kobayashi, M. Bernardo, P. L. Choyke, and M. W. Brechbiel. 2006. Dendrimer-based nanoprobe for dual modality magnetic resonance and fluorescence imaging. *Nano Letters* 6 (7) (July 12): 1459–1463.

Tan, T. T., S. T. Selvan, L. Zhao, S. Gao, and J. Y. Ying. 2007. Size control, shape evolution, and silica coating of near-infrared-emitting PbSe quantum dots. *Chemistry of Materials* 19 (13): 3112–3117.

Tan, M., G. Wang, X. Hai, Z. Ye, and J. Yuan. 2004. Development of functionalized fluorescent europium nanoparticles for biolabeling and time-resolved fluorometric applications. *Journal of Materials Chemistry* 14 (19): 2896–2901.

Tapec, R., X. J. Zhao, and W. Tan. 2002. Development of organic dye-doped silica nanoparticles for bioanalysis and biosensors. *Journal of Nanoscience and Nanotechnology* 2 (3–4): 405–409.

Tartaj, P., T. González-Carreño, and C. J. Serna. 2002. Synthesis of nanomagnets dispersed in colloidal silica cages with applications in chemical separation. *Langmuir* 18 (12): 4556–4558.

Taylor, K. M. L., J. S. Kim, W. J. Rieter, H. An, W. L. Lin, and W. B. Lin. 2008. Mesoporous silica nanospheres as highly efficient MRI contrast agents. *Journal of the American Chemical Society* 130 (7) (February 20): 2154,+.

Tokumasu, F. and J. Dvorak. 2003. Development and application of quantum dots for immunocytochemistry of human erythrocytes. *Journal of Microscopy* 211 (3): 256–261.

Torchilin, V. P. 2005. Recent advances with liposomes as pharmaceutical carriers. *Nature Reviews Drug Discovery* 4 (2): 145–160.

Torchilin, V. P. 2006. Multifunctional nanocarriers. *Advanced Drug Delivery Reviews* 58 (14): 1532–1555.

Touahir, L., P. Allongue, D. Aureau, R. Boukherroub, J.-N. Chazalviel, E. Galopin, A. C. Gouget-Laemmel, et al. 2010. Molecular monolayers on silicon as substrates for biosensors. *Bioelectrochemistry* 80 (1): 17–25.

Trewyn, B. G., S. Giri, I. I. Slowing, and V. S.-Y. Lin. 2007. Mesoporous silica nanoparticle based controlled release, drug delivery, and biosensor systems. *Chemical Communications* (31): 3236–3245.

Trewyn, B. G., C. M. Whitman, and V. S.-Y. Lin. 2004. Morphological control of room-temperature ionic liquid templated mesoporous silica nanoparticles for controlled release of antibacterial agents. *Nano Letters* 4 (11): 2139–2143.

Tsai, C. P., Y. Hung, Y. H. Chou, D. M. Huang, J. K. Hsiao, C. Chang, Y. C. Chen, and C. Y. Mou. 2008. High-contrast paramagnetic fluorescent mesoporous silica nanorods as a multifunctional cell-imaging probe. *Small* 4 (2) (February): 186–191.

Valli, L. 2005. Phthalocyanine-based langmuir-blodgett films as chemical sensors. *Advances in Colloid and Interface Science* 116 (1–3): 13–44.

Van De Rijke, F., H. Zijlmans, S. Li, T. Vail, A. K. Raap, R. S. Niedbala, and H. J. Tanke. 2001. Up-converting phosphor reporters for nucleic acid microarrays. *Nature Biotechnology* 19 (3): 273–276.

Van Schooneveld, M. M., E. Vucic, R. Koole, Y. Zhou, J. Stocks, D. P. Cormode, C. Y. Tang, et al. 2008. Improved biocompatibility and pharmacokinetics of silica nanoparticles by means of a lipid coating: A multimodality investigation. *Nano Letters* 8 (8): 2517–2525.

Vazquez, M., A. Asenjo, M. del Puerto, K. Roberto, G. Baldini, and M. Hernandez. 2008. Nanostructrured magnetic sensors. In ed. F. J. Arregui. New York: Springer.

Veiseh, O., C. Sun, J. Gunn, N. Kohler, P. Gabikian, D. Lee, N. Bhattarai, et al. 2005. Optical and MRI multifunctional nanoprobe for targeting gliomas. *Nano Letters* 5 (6) (June): 1003–1008.

Vetrone, F., R. Naccache, V. Mahalingam, C. G. Morgan, and J. A. Capobianco. 2009. The active-core/active-shell approach: A strategy to enhance the upconversion luminescence in lanthanide-doped nanoparticles. *Advanced Functional Materials* 19 (18): 2924–2929.

Viguier, R. F. H. and A. N. Hulme. 2006. A sensitized europium complex generated by micromolar concentrations of copper(I): Toward the detection of copper(I) in biology. *Journal of the American Chemical Society* 128 (35): 11370–11371.

Wang, F., D. K. Chatterjee, Z. Li, Y. Zhang, X. Fan, and M. Wang. 2006a. Synthesis of polyethylenimine/NaYF4 nanoparticles with upconversion fluorescence. *Nanotechnology* 17 (23): 5786–5791.

Wang, D., J. He, N. Rosenzweig, and Z. Rosenzweig. 2004. Superparamagnetic Fe_2O_3 beads-CdSe/ZnS quantum dots core-shell nanocomposite particles for cell separation. *Nano Letters* 4 (3): 409–413.

Wang, X., C. Drew, S.-H. Lee, K. J. Senecal, J. Kumar, and L. A. Samuelson. 2002a. Electrospun nanofibrous membranes for highly sensitive optical sensors. *Nano Letters* 2 (11): 1273–1275.

Wang, T. C., M. F. Rubner, and R. E. Cohen. 2002b. Polyelectrolyte multilayer nanoreactors for preparing silver nanoparticle composites: Controlling metal concentration and nanoparticle size. *Langmuir* 18 (8): 3370–3375.

Wang, L., K. Wang, S. Santra, X. Zhao, L. R. Hilliard, J. E. Smith, Y. Wu, and W. Tan. 2006b. Watching silica nanoparticles glow in the biological world. *Analytical Chemistry* 78 (3): 646–654.

Wei, Y., H. Dong, J. Xu, and Q. Feng. 2002. Simultaneous immobilization of horseradish peroxidase and glucose oxidase in mesoporous sol-gel host materials. *ChemPhysChem* 3 (9): 802–808.

Wei, H., J. Liu, L. Zhou, J. Li, X. Jiang, J. Kang, X. Yang, S. Dong, and E. Wang. 2008. $[Ru(bpy)_3]^{2+}$-doped silica nanoparticles within layer-by-layer biomolecular coatings and their application as a biocompatible electrochemiluminescent tag material. *Chemistry—A European Journal* 14 (12): 3687–3693.

Weissleder, R., G. Elizondo, L. Josephson, C. C. Compton, C. J. Fretz, D. D. Stark, and J. T. Ferrucci. 1989a. Experimental lymph node metastases: Enhanced detection with MR lymphography. *Radiology* 171 (3): 835–839.

Weissleder, R., D. D. Stark, B. L. Engelstad, B. R. Bacon, C. C. Compton, D. L. White, P. Jacobs, and J. Lewis. 1989b. Superparamagnetic iron oxide: Pharmacokinetics and toxicity. *American Journal of Roentgenology* 152 (1): 167–173.

Westcott, S. L., S. J. Oldenburg, T. R. Lee, and N. J. Halas. 1998. Formation and adsorption of clusters of gold nanoparticles onto functionalized silica nanoparticle surfaces. *Langmuir* 14 (19): 5396–5401.

Willard, M. A., L. K. Kurihara, E. E. Carpenter, S. Calvin, and V. G. Harris. 2004. Chemically prepared magnetic nanoparticles. *International Materials Reviews* 49 (3–4): 125–170.

Wolfbeis, O. S., A. Dürkop, M. Wu, and Z. Lin. 2002. A europium-ion-based luminescent sensing probe for hydrogen peroxide. *Angewandte Chemie—International Edition* 41 (23): 4495–4498.

Wolfbeis, O. S. and B. P. H. Schaffar. 1987. Optical sensors: An ion-selective optrode for potassium. *Analytica Chimica Acta* 198 (C): 1–12.

Wu, Y., C. Chen, and S. Liu. 2009. Enzyme-functionalized silica nanoparticles as sensitive labels in biosensing. *Analytical Chemistry* 81 (4): 1600–1607.

Wu, S.-H., Y.-S. Lin, Y. Hung, Y.-H. Chou, Y.-H. Hsu, C. Chang, and C.-Y. Mou. 2008. Multifunctional mesoporous silica nanoparticles for intracellular labeling and animal magnetic resonance imaging studies. *ChemBioChem* 9 (1): 53–57.

Xiang, Y., X. Wu, D. Liu, Z. Li, W. Chu, L. Feng, K. Zhang, W. Zhou, and S. Xie. 2008. Gold nanorod-seeded growth of silver nanostructures: From homogeneous coating to anisotropic coating. *Langmuir* 24 (7): 3465–3470.

Xu, Y., M. Mahmood, A. Fejleh, Z. Li, F. Watanabe, S. Trigwell, R. B. Little et al. 2010. Carbon-covered magnetic nanomaterials and their application for the thermolysis of cancer cells. *International Journal of Nanomedicine* 5 (1): 167–176.

Yang, J., J. Lee, J. Kang, K. Lee, J.-S. Suh, H.-G. Yoon, Y.-M. Huh, and S. Haam. 2008a. Hollow silica nanocontainers as drug delivery vehicles. *Langmuir* 24 (7): 3417–3421.

Yang, J., E. K. Lim, H. J. Lee, J. Park, S. C. Lee, K. Lee, H. G. Yoon, J. S. Suh, Y. M. Huh, and S. Haam. 2008b. Fluorescent magnetic nanohybrids as multimodal imaging agents for human epithelial cancer detection. *Biomaterials* 29 (16) (June): 2548–2555.

Yang, W., C. G. Zhang, H. Y. Qu, H. H. Yang, and J. G. Xu. 2004. Novel fluorescent silica nanoparticle probe for ultrasensitive immunoassays. *Analytica Chimica Acta* 503 (2): 163–169.

Ye, Z., M. Tan, G. Wang, and J. Yuan. 2004. Preparation, characterization, and time-resolved fluorometric application of silica-coated terbium(III) fluorescent nanoparticles. *Analytical Chemistry* 76 (3): 513–518.

Yeh, H.-C., J. Sharma, J. J. Han, J. S. Martinez, and J. H. Werner. 2011. NanoCluster beacon—A new molecular probe for homogeneous detection of nucleic acid targets. *Paper presented at NEMS 2011—6th IEEE International Conference on Nano/Micro Engineered and Molecular Systems.*

Yi, G., B. Sun, F. Yang, D. Chen, Y. Zhou, and J. Cheng. 2002. Synthesis and characterization of high-efficiency nanocrystal up-conversion phosphors: Ytterbium and erbium codoped lanthanum molybdate. *Chemistry of Materials* 14 (7): 2910–2914.

Yonzon, C. R., C. L. Haynes, X. Zhang, J. T. Walsh Jr., and R. P. Van Duyne. 2004. A glucose biosensor based on surface-enhanced raman scattering: Improved partition layer, temporal stability, reversibility, and resistance to serum protein interference. *Analytical Chemistry* 76 (1): 78–85.

Yonzon, C. R., D. A. Stuart, X. Zhang, A. D. McFarland, C. L. Haynes, and R. P. Van Duyne. 2005. Towards advanced chemical and biological nanosensors—an overview. *Talanta* 67 (3): 438–448.

Yoon, T.-J., J. S. Kim, B. G. Kim, K. N. Yu, M.-H. Cho, and J.-K. Lee. 2005. Multifunctional nanoparticles possessing a "magnetic motor effect" for drug or gene delivery. *Angewandte Chemie—International Edition* 44 (7): 1068–1071.

Yoon, B., S. Lee, and J.-M. Kim. 2009. Recent conceptual and technological advances in polydiacetylene-based supramolecular chemosensors. *Chemical Society Reviews* 38 (7): 1958–1968.

Yu, H.-W., A. Jang, L. H. Kim, S.-J. Kim, and I. S. Kim. 2011. Bead-based competitive fluorescence immunoassay for sensitive and rapid diagnosis of cyanotoxin risk in drinking water. *Environmental Science and Technology* 45 (18): 7804–7811.

Yu, M. K., Y. Y. Jeong, J. Park, S. Park, J. W. Kim, J. J. Min, K. Kim, and S. Jon. 2008. Drug-loaded superparamagnetic iron oxide nanoparticles for combined cancer imaging and therapy in vivo. *Angewandte Chemie—International Edition* 47 (29): 5362–5365.

Zayat, M., R. Pardo, G. Rosa, R. P. Del Real, M. Diaz-Michelena, I. Arruego, H. Guerrero, and D. Levy. 2009. A sol-gel based magneto-optical device for the NANOSAT space mission. *Journal of Sol-Gel Science and Technology* 50 (2): 254–259.

Zebli, B., A. S. Susha, G. B. Sukhorukov, A. L. Rogach, and W. J. Parak. 2005. Magnetic targeting and cellular uptake of polymer microcapsules simultaneously functionalized with magnetic and luminescent nanocrystals. *Langmuir* 21 (10) (May 10): 4262–4265.

Zeng, J.-H., J. Su, Z.-H. Li, R.-X. Yan, and Y.-D. Li. 2005. Synthesis and upconversion luminescence of hexagonal-phase NaYF 4:Yb, Er3+ phosphors of controlled size and morphology. *Advanced Materials* 17 (17): 2119–2123.

Zhang, Y., N. Kohler, and M. Zhang. 2002. Surface modification of superparamagnetic magnetite nanoparticles and their intracellular uptake. *Biomaterials* 23 (7) (4): 1553–15561.

Zhang, Y., C. Sun, N. Kohler, and M. Zhang. 2004. Self-assembled coatings on individual monodisperse magnetite nanoparticles for efficient intracellular uptake. *Biomedical Microdevices* 6 (1): 33–40.

Zhang, H., Y. Xu, W. Yang, and Q. Li. 2007. Dual-lanthanide-chelated silica nanoparticles as labels for highly sensitive time-resolved fluorometry. *Chemistry of Materials* 19 (24): 5875–5881.

Zhang, X., M. A. Young, O. Lyandres, and R. P. Van Duyne. 2005. Rapid detection of an anthrax biomarker by surface-enhanced raman spectroscopy. *Journal of the American Chemical Society* 127 (12) (03/01; 2011): 4484–4489.

Zhao, X., L. R. Hilliard, S. J. Mechery, Y. Wang, R. P. Bagwe, S. Jin, and W. Tan. 2004. A rapid bioassay for single bacterial cell quantitation using bioconjugated nanoparticles. *Proceedings of the National Academy of Sciences of the United States of America* 101 (42): 15027–15032.

Zhelev, Z., H. Ohba, and R. Bakalova. 2006. Single quantum dot-micelles coated with silica shell as potentially non-cytotoxic fluorescent cell tracers. *Journal of the American Chemical Society* 128 (19): 6324–6325.

Zheng, Q., H. Dai, M. E. Merritt, C. Malloy, C. Y. Pan, and W.-H. Li. 2005. A new class of macrocyclic lanthanide complexes for cell labeling and magnetic resonance imaging applications. *Journal of the American Chemical Society* 127 (46): 16178–16188.

Zhong, D. L., F. L. Yuan, L. Lian, and Z. H. Cheng. 2009. A localized surface plasmon resonance light-scattering assay of mercury (II) on the basis of Hg2+-DNA complex induced aggregation of gold nanoparticles. *Environmental Science and Technology* 43 (13): 5022–5027.

Zimmerman, R., L. Basabe-Desmonts, F. Van Der Baan, D. N. Reinhoudt, and M. Crego-Calama. 2005. A combinatorial approach to surface-confined cation sensors in water. *Journal of Materials Chemistry* 15 (27–28): 2772–2777.

Zou, W.-S., J.-Q. Qiao, X. Hu, X. Ge, and H.-Z. Lian. 2011. Synthesis in aqueous solution and characterisation of a new cobalt-doped ZnS quantum dot as a hybrid ratiometric chemosensor. *Analytica Chimica Acta* 708 (1–2): 134–140.

14

Linear and Nonlinear Spectroscopy at Nano Scale

Anna Chiara De Luca, Giuseppe Pesce, Giulia Rusciano, and Antonio Sasso

CONTENTS

14.1 Basic Theory of SERS

14.1.1 Nanoplasmonic

Many modern optical techniques, such as surface plasmon resonance spectroscopy, surface-enhanced fluorescence, or surface-enhanced Raman scattering, are based on the unique and fascinating optical properties of nanosized metallic structures. The peculiar properties of such nanostructures, when properly illuminated by laser radiation, are at the basis of the current interest in many different fields, such as biomedicine or environment control [1,2].

Plasmonics, or better nanoplasmonics, is the new field in nanoscience that studies the electromagnetic response of metal nanostructures to an incident electro-magnetic (EM) field. As the name suggests, this response is mainly related to the oscillation of the so-called plasmons. A plasmon is a wavepacket induced by collective oscillations of conduction electrons. If the size of the metallic structure is small compared to the wavelength of the incident light, the plasmon wavepacket will remain confined and forms a surface plasmon (SP). If the incident EM field has a frequency close to the SP oscillation frequency, the field extinction (absorption + scattering) reaches a maximum. It is not difficult to image that the SP oscillation frequency depends on both the metal dielectric function and the nanoparticle shape. This is the reason for which, although the gold bulk plasma frequency is in the UV region, the SP frequency for a gold nanoparticle can be found in the green region (near 520 nm).

This and all the other properties of noble metal nanoparticles have attracted considerable interest since historical times when metallic nanoparticles were used as decorative pigments in stained glasses and artworks. However, the phenomenon of SP was firstly investigated by Gustav Mie in 1908 [3], which first pointed out that the interaction of light with metal nanoparticles can result in the collective oscillation of the metal-free electrons with respect to the nanoparticle lattice in resonance with the light field (localized surface plasmon resonance, LSPR).

To get an insight into the origin of SPR in metallic nanostructures, we have no other options but to actually solve Maxwell's equations with appropriate boundary conditions. This can be, in general, a rather difficult task. One useful approximation scheme, widely used in the literature, is the electrostatic approximation. In this case, the electric field of the light is considered to be constant over distances comparable to the object size so that any phase delay of the field is neglected. It is obvious that this approximation works well when the object size is much smaller than the wavelength. In practice, with wavelengths in the visible region, it means that the electrostatic approximation will be mostly valid for objects of typical sizes up to ~10 nm. Full analytical solutions of Maxwell equations exist in a few selected simple geometries (Mie theory). The sphere is one of these cases [4]. However, some essential aspects can be observed in the much simplest electrostatic approximation. It turns out that the electrostatic boundary conditions on the sphere can be fully satisfied by considering the superposition of an induced dipole at the origin (p) with the external applied field (E). The induced dipole is given by

$$p = \alpha E \tag{14.1}$$

being α the dipole polarizability, given by the Clausius–Mossotti relation:

$$\alpha = 3\varepsilon_m V \frac{\varepsilon - \varepsilon_m}{\varepsilon + 2\varepsilon_m} \tag{14.2}$$

where

V is the particle volume
$\varepsilon(\omega) = \varepsilon_r(\omega) + i\varepsilon_{im}(\omega)$ is the complex frequency-dependent dielectric function of the metal
ε_m is the dielectric constant of the surrounding medium

The most important feature of Equation 14.2 is the presence of the denominator $\varepsilon + 2\varepsilon_m$. In fact, assuming $\varepsilon_{im}(\omega) \sim 0$ (low metal absorption), a strong resonance occurs roughly at the EM frequency ω where $\varepsilon = -2\varepsilon_m$. This resonance defines the LSPR of the sphere.

For gold, silver, and copper, the resonance condition can be fulfilled at visible frequencies, where they exhibit a negative dielectric function and low absorption. It is worth noticing that the resonance condition is purely induced by geometrical aspects, deriving from the need to satisfy the boundary conditions of the geometry under consideration. It should be also noticed that the condition $\varepsilon \sim -2\varepsilon_m$ introduces a small dependence of the resonance wavelength on the embedding medium. As a result, the SPR is red-shifted in media with a larger ε_m (for example, in water compared to air).

It should be noticed that, in the electrostatic approximation, the LSPR does not depend on the actual size of the sphere. Nevertheless, the conditions under which the electrostatic approximation will represent the real solution of the problem (the solution of Maxwell's equations) is when the sphere size is less than a few tens of nanometers for an incident radiation in the visible region. For larger particles, the size effect on the LSPR can be observed mostly on the numerical solution of Maxwell's equations.

As a general rule, it is possible to say that as the size increases, LSPR (i) shifts to the red, (ii) is strongly damped and spectrally broadened, and (iii) new resonances appear, which are typically related to the activation of multipolar resonances (such as quadrupolar resonance) [5,6].

Objects with different shapes will have different resonances (sometimes more than one), and also this property makes metals so interesting for nanooptics. The fact that metals have negative dielectric constant spanning a wide range in magnitude gives place to different resonance conditions. For example, a nanorod has both transverse and longitudinal modes, as shown in Figure 14.1. In transverse modes, the directions of the collective oscillation of conductive electrons are perpendicular to the longer axis of the rod, whereas in the longitudinal modes electronic oscillation is parallel to the axis. The transverse modes give the resonance peak at ~520 nm while longitudinal modes have wavelengths which become higher by increasing the rod aspect ratio [4].

An additional level of complexity of the plasmon resonances arises from the existence of coupled plasmon resonances for two or more closely spaced objects. As two spherical nanoparticles approach each other, the fields produced by their respective induced dipoles start to interact. This interaction can reinforce or weaken the field in certain regions of space. Additionally, resonances coming from higher-order multipoles can be activated by the interaction. Considering, for instance, the very simple case of

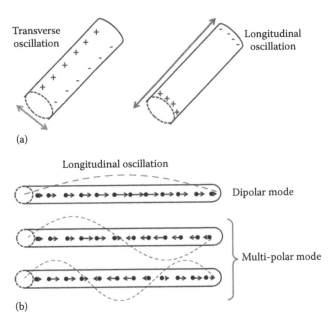

FIGURE 14.1 (a) Charge distribution associated with transverse and longitudinal plasmon modes in a nanowire. (b) Dipolar and multipolar (quadrupole and hexapole) longitudinal plasmon oscillation in a nanowire.

two metallic nanobeads, it is possible to demonstrate that resonance coupling produces a red shift of the plasmon resonance, with its intensity mainly concentrated in the middle of the two beads [7].

14.1.2 Surface Enhanced Raman Scattering

Before discussing the SERS, we briefly recall the spontaneous Raman effect, from which SERS derives. In the photon picture, the Raman effect can be seen as an inelastic scattering process between a photon and a molecule in which the molecule undergoes an excitation to a virtual state followed by a nearly simultaneous deexcitation toward a vibrational energy level different from the initial state. The scattering event occurs in 10^{-14} s or less. The description of this process is schematically shown in Figure 14.2.

Since the virtual state are not real states but are created only when photons interact with electrons, the frequency of these states is determined by the frequency of the used light source. From the virtual state, photons may follow three different paths. Most of them return to the initial state through the emission of a photon, in a process named Rayleigh scattering. This process does not involve any energy change of the molecule, so that the frequency itself of the remitted photons matches the frequency of the incoming photon. However, it is also possible that the scattering process leads to an increase of energy of the molecule, which, after excitation, reaches a higher vibrational level. This is the Stokes scattering. The energy difference between the initial and final vibrational levels (Raman shift) expressed in wavenumbers (cm^{-1}), is given through the relation:

$$\hat{v} = \frac{1}{\lambda_{inc}} - \frac{1}{\lambda_{scat}} \tag{14.3}$$

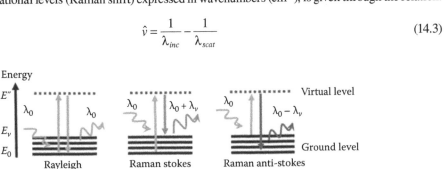

FIGURE 14.2 Energy level diagram for Rayleigh scattering, Stokes Raman scattering, and anti-Stokes Raman scattering.

in which λ_{inc} and λ_{scat} are the wavelengths (in cm^{-1}) of the incident and Raman scattered photons, respectively. Finally, a third process can take place. Indeed, due to thermal energy, some molecules may lie in an excited vibrational level. Scattering from this state to the ground level is called anti-Stokes scattering, and involves an energy transfer from the molecule to the photon. Due to the quite small population of excited levels at room temperature (ruled by Boltzmann statistic), anti-Stokes scattering is much less probable than the Stokes scattering (typically by a factor around 1000). For this reason, Raman analysis is usually limited to the observation of the Stokes scattering. Therefore, a Raman spectrum is obtained by dispersing the scattered photons and it is characterized by several peaks which correspond to different vibrational modes of the studied molecules. Hence, it is possible to collect spectra which contain the same information of infrared absorption spectra, with an intrinsic "fingerprint" character. The greatest disadvantage of Raman spectroscopy relies on the extremely low cross section of the Raman process, which is 12–14 orders of magnitude below fluorescence cross sections. During 1970s, an unexpectedly high Raman signal from pyridine on a rough silver electrode attracted considerable attention. Numerous experiments performed in different laboratories crucially demonstrated that this enormously strong Raman signal was caused by a true enhancement of the Raman scattering efficiency and not by an anomalous concentration of scattering molecules on the electrodes [8]. Within the next few years, strongly enhanced Raman signals were verified for many different molecules which had been attached to various rough metal surfaces, and the effect was named SERS. For an overview see Ref. [9].

In the first SERS experiments, Van Duyne and coworkers estimated enhancement factors of the order of 10^5–10^6 for pyridine on rough silver electrodes. The magnitude of the enhancement was found to be dependent on electrode roughness, which suggests a strong electromagnetic field enhancement. On the other hand, the experimental observation of the dependence of the enhancement factor on the electrode potential is an indication that chemical enhancement must be operative as well.

EM SERS enhancement mechanism can be easily understood by looking Figure 14.3. The metallic nanostructure is a small sphere with complex dielectric constant $\varepsilon(\nu)$, in a surrounding medium with dielectric constant ε_m. The diameter of the sphere $(2r)$ is small compared with the wavelength of light (electrostatic regime). A molecule in proximity of the sphere (at distance d) is exposed to a field E_{tot}, which is the superposition of the incoming field E_0 and the field of a dipole, E_{dip}, induced in the metal sphere. Taking into account Equations 14.1 and 14.2, this field results as follows:

$$E_{tot} = E_0 + E_{dip} = E_0 + r^3 \frac{\varepsilon - \varepsilon_m}{\varepsilon + 2\varepsilon_m} E_0 \frac{1}{(r+d)^3} \qquad (14.4)$$

It is possible to define an enhancement factor as ratio of the total field (E_{tot}) at the molecule position and the incoming field (E_0):

$$A(\nu) = \frac{E_{tot}}{E_0} \sim \frac{\varepsilon(\nu) - \varepsilon_m}{\varepsilon(\nu) + 2\varepsilon_m} E_0 \left(\frac{r}{r+d} \right)^3 \qquad (14.5)$$

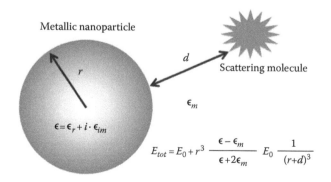

FIGURE 14.3   Cartoon of the electromagnetic enhancement effect in SERS spectroscopy.

The enhancement factor $A(v_L)$, at the incoming beam frequency v_L, is particularly strong when $\varepsilon(v_L)$ is equal to $-2\varepsilon_m$. Additionally, for a strong electromagnetic enhancement, the imaginary part of the dielectric constant needs to be small. These conditions, as shown in the former paragraph, describe the resonant excitation of surface plasmons of the metal sphere. In an analogous fashion, the Stokes and anti-Stokes fields will be enhanced too, if they are in resonance with surface plasmons of the sphere. So far, taking into account both effects, the electromagnetic enhancement factor for the Stokes signal is [10]

$$G_{em}(v_S) = |A(v_L)|^2 |A(v_S)|^2 \sim E_0^4 \left| \frac{\varepsilon(v_L) - \varepsilon_m}{\varepsilon(v_L) + 2\varepsilon_m} \right|^2 \left| \frac{\varepsilon(v_S) - \varepsilon_m}{\varepsilon(v_S) + 2\varepsilon_m} \right|^2 \left(\frac{r}{r+d} \right)^{12} \qquad (14.6)$$

being v_S the frequency of the inelastically scattered photon. This formula shows us that the EM enhancement factor scales as the forth power of the local field and reaches its maximum when both excitation and Stokes field are in resonance with the plasmons. Moreover, EM enhancement is effective when the analyte is very close to the nanostructure, decreasing as $(1/d)^{12}$.

SERS effect has also a strong molecular selectivity and a clear dependence on the chemical nature of the Raman molecule. This suggests the existence of an additional chemical SERS enhancement mechanism. Among the different mechanisms proposed to explain this chemical effect, the charge transfer mechanism is the most reliable. It involves four steps:

1. An incident photon induces an intraband transition in the metallic nanoparticle.
2. This new "excited" electron transfers from the metal to the lowest unoccupied molecular orbital (LUMO) of the analyte.
3. The electron with slightly altered energy, due to interactions with internal molecular vibrations in the molecule, transfers from the LUMO of the adsorbate back to the metal.
4. The electron returns in its initial state causing a photoemission at Stokes shifted frequency.

The SERS enhancement caused by the chemical mechanism is generally limited to those analyte molecules in direct contact with the metal surface and it is therefore a shorter range enhancement process. In general, the chemical SERS enhancement mechanisms contribute to the enhancement by a factor of $10–10^3$.

14.2 SERS Substrate

The first step in the design of SERS experiments is the choice and preparation of the SERS substrate. Numerous protocols for the synthesis of metallic nanoparticles are available. Metals commonly employed for making SERS substrates are Au or Ag, although Cu [11] and Al [12] also reveal plasmon resonances in visible/near-infrared spectral window. Aluminum presents complementary plasmon resonances in UV region and therefore it can extend the spectral range of applicability of standard Ag/Au SERS devices. Unfortunately, Al and Cu can rapidly form oxides that sharply affect the SERS efficiency [13].

The most popular substrates used in SERS studies are colloidal nanoparticles and roughened metal surfaces. Silver and gold colloidal nanoparticles can be produced by the Lee and Meisel [14] citrate reduction method while roughened metal surfaces are usually prepared by performing several oxidation–reduction cycles [15]. Although high efficiencies have been observed from these substrates, technical difficulties, as low reproducibility and tunability, have limited their practical application.

Production of metallic substrates with definite shapes and sizes is crucial to achieve stable surface plasmon resonance frequencies and reproducible spectral features by SERS measurements. Therefore, there is an increasing interest in developing novel SERS substrates, as nanoshells, nanoarrays, nanoholes, or sharp tips.

Nanoshells are spherical nanoparticles consisting of a dielectric core (generally silica nanoparticles) and a thin metallic shell. These structures are characterized by tunable plasmon resonances in a broad spectral region (visible/near-infrared) according to the relative dimensions of the core and shell layer.

SERS enhancements observed with nanoshells are quite strong and highly reproducible [16]: 60 nm sized Au–Ag nanoshells show an enhanced SERS signal eight times larger than Au-nanoparticles of the same size [17].

As previously underlined, surface enhanced spectroscopies crucially require reproducible signals, which are an important characteristic of well-ordered arrays of nanostructures. Recently, lithographic techniques, such as e-beam lithography, photolithography, and nanosphere lithography, have been employed for the fabrication of periodic arrays with specific shapes at the nanoscale [18–20]. Zhang et al. in a recent review summarized the most advances in the fabrication of size-tunable nanoparticles using lithographic techniques [21]. By modifying the dimensions and the geometry of nanostructures pattern it is possible to tune the optical properties and, consequently, the sensitivity of the substrates can be modulated to different wavelengths.

Another class of reproducible SERS substrates is based on the combination of the unique physical properties of photonic crystal devices with highly reliable semiconductor manufacturing techniques [22,23]. By coating photonic crystal surfaces with metallic nanoparticles it is possible to realize devices, which concentrate and localize the optical field as sort of antennae, and, at the same time, can efficiently couple laser light with absorbed molecule.

14.3 TERS

Tip-enhanced Raman spectroscopy (TERS) relies on the combination of scanning probe microscopy and Raman spectroscopy. The basic idea is the use of a metaled tip as a sort of optical nanoantenna, which gives place to SERS effect close to the tip end. Since the tip behaves as a nano-antenna for SERS effect, now an intrinsic high spatial resolution (below the Abbe limit) is achieved [24]. TERS allows obtaining simultaneously topographic and chemical information of the analyzed surface region, which constitute an important step for surface analysis in its broadest sense. The first TERS study dates back to 1999; it was presented by Kawata and coworkers [25] shortly followed by Zenobi's group [26]. These pioneer studies have elucidated the main TERS features, such as sensitivity, near-field enhancement, and optical resolution. Nowadays, TERS has reached a high level of maturity, and is now applied to different systems, mainly in physical and biological sciences.

For the last years, several companies have been working on commercializing TERS instruments, by integrating scanning probe microscopy (SPM) and micro-Raman system. The market attention constitutes a clear indication of the potential impact of this novel analytical tool. However, these first systems are not (yet) completely stand-alone machines, although huge steps have been done recently in this direction.

Different kinds of SPM techniques can be used for TERS systems, including scanning tunneling microscopy and shear-force feedback. However, atomic force microscopy (AFM) with a metaled tip attached to the cantilever is the most commonly used SPM method.

Depending on the sample transparency, different illumination geometries can be chosen. If thick, nontransparent bulk material is under investigation, a top-or a side-illumination geometry (by using a long working distance objective) can be employed, as sketched in Figure 14.4a and b. Although the first geometry is easier aligned, it can be demonstrated that plasmon excitation at the tip is more efficient by using side illumination, and by using the focusing objective also to collect the Raman photons, the optimum angle at which the optical axis of the objective is set with respect to the surface normal is 55°. If transparent samples are investigated, in-line illumination becomes possible (Figure 14.4c), where the excitation beam is focused onto the probe apex through the sample. In this case, high NA objectives can be used at short working distances, which maximize the collection efficiency.

Vapor deposition is the most employed technique to metalize AFM tips [27]. It should be clearly said that the production of tips useful for TERS analysis still represents a crucial challenge. As a matter of fact, heating of the cantilever during fast vapor deposition rates will distort it and make it unsuitable for scanning. Another key issue is to produce TERS tips that are not only highly enhancing, but also robust enough to be reused for several scans. For etched metal tips, slight contact between them and the sample can be sufficient to damage the tip apex, although the use of tips based on metallic alloys could contribute to fix this problem. Silver tips also undergo oxidation, which reduces their usefulness within a day.

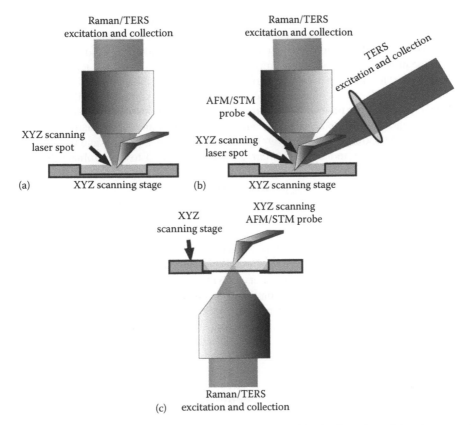

FIGURE 14.4 Typical optical configurations for a TERS system. (a) Up illumination using a right microscope, (b) side illumination with a long working distance objective, and (c) bottom illumination using an inverted microscope.

14.4 SERS Applications

High sensitivity provided by the plasmonic enhancement combined with the specificity of Raman spectroscopy have led to a wide variety of applications including sensitive detection of analytes with biochemical significance, such as proteins and DNA/RNA detection. This section provides an overview of the latest SERS applications in biosensing and single biomolecule detection.

14.4.1 SERS for Proteins Detection

There are different methods generally used for protein detection and identification. The Western blot combined with immunoassay is a widely used analytical technique, which uses gel electrophoresis to separate native or denatured proteins by the length of the polypeptide. The proteins are then transferred to a membrane where they are probed using antibody–antigen interactions. This method allows the detection of only one specific protein and significant technical problems arise with quantitative analysis [28]. Mass spectroscopy is general used to characterize protein structure and sequences. The cost of instrumentation and its destructive nature are the major drawbacks of this technique [29]. Fluorescence is another common protocol used to explore proteins structure and dynamics [30], but typical fluorescence spectra of molecular fluorophores are broad and photobleaching-dependent. These issues make multiplexing detection very difficult.

To overcome all these limitations and characterize different proteins with high sensitivity and photostability, several SERS-based sensors have been developed. These protocols can be divided in two main types: label-free and Raman dye-labeled method [31]. The label-free method is based on direct absorption of the proteins on the SERS substrate and the analysis of vibrational features of the proteins themselves [32–34]. This analysis is not straightforward as for small molecules or DNA, etc. In fact, the band assignment of SERS spectra of large molecules as proteins, enzymes, and antibodies is often based on the detection of subcompartment of the macromolecules rather than the localized vibrational modes. Typical spectral features used to characterize and detect proteins are the aromatic amino acid bands for C–COO⁻ stretch at 950 cm⁻¹ and for COO⁻ symmetric stretch at 1400 cm⁻¹ and the amide (CO–NH) I and III at 1600–1700 cm⁻¹ and 1200–1350 cm⁻¹, respectively, used to discriminate the primary and secondary structure.

Ozaki et al. using several different approaches, reported the most extensive evaluations of protein detection with SERS. They used the described specific Raman bands to characterize, for example, two different types of proteins: myoglobin (Mb) and bovine serum albumin (BSA) [35]. More precisely, they combined the SERS and surface enhanced resonant Raman scattering (SERRS) spectroscopies to the traditional Western blot analysis for label-free detection of Mb and BSA. Initially, they separated the protein mixture by gel electrophoresis then the proteins were electroblotted onto a sheet of nitrocellulose membrane where they were immobilized. After applying Ag nanoparticles, they acquired label-free SERS spectra of different proteins directly on the nitrocellulose membrane without digestion, extraction, and other pretreatments. In another recent work, Ozaki et al. detected SERS spectra of several label free proteins (lysozyme, ribonulease B, avidin, catalase, and hemoglobin) for the first time in aqueous solutions [36]. They demonstrated that it is possible to detect low concentrations of proteins (down to 50 ng/mL) in solutions even without resonance effect. Anyway, much work must still be done to determine sensor detection limits as well as sensing in presence of interfering species. Intrinsic SERS analysis has been also used to identify an important part of proteins: enzymes. Bjerneld et al. analyzed enzymatic activity of horseradish peroxidase by measuring SERS signals from isolated and immobilized protein-nanoparticle aggregates [37]. SERS based biosensors can be also used for measurement in vivo. Cullum and coworkers, in their paper, described the development and optimization of a novel class of SERS-based immuno-nanosensors for detection of specific proteins in complex environments as cell culture matrices and intracellular environments [38]. In particular, the SERS substrate was realized by depositing multiple layers of silver on silica nanospheres, followed by binding of the antibody of interest to the silver surface via a short rigid crosslinker. The nanosensors were evaluated by monitoring their response to various antigens (i.e., proteins), as human insulin or interleukin II, in complex environments. While the protein spectra detection can be considered a promising tool to characterize the difference between various proteins, the signal reproducibility and multiplexing capability of SERS measurements remains a big challenge. To overcome these problems, there is an increasing interest in using Raman dye-labeled based SERS protocols for protein and antibodies biosensing. This extrinsic SERS method is based on the indirect detection of the SERS signal from Raman dye linked to the target molecules [39,40].

Two different schemes proposed for SERS-based immunoassay are shown in Figure 14.5. In the proposed cases, antibodies are chemically immobilized on the coverslip then exposed to antigens. This system successively interacts with gold/silver nanoparticles conjugated to antibodies as well as Raman dyes. In the case of Figure 14.5a, the SERS probe is a gold-coated Ag-nanoparticle directly linked to both Raman dyes and antibodies. Otherwise, in the configuration of Figure 14.5b, a gold nanoparticle is firstly linked to the Raman dyes and then to the antibodies. Finally, SERS signals can be measured for the immobilized immunocomplexes.

In the recent literature, there are many examples where SERS has been applied for proteins detection and characterization, with various metal nanoparticles (e.g., Au, Ag) and dyes (e.g., Cy3, Cy5, MBA), revealing important advantages in terms of sensitivity and selectivity. Cao and coworkers have designed Raman dye-functionalized nanoparticle and used them to perform multiplexed screening of protein-protein interactions [41]. Recently, Cui et al. presented two immunoassay methods for analysis of proteins, based on the sandwich structure concept capture antibody substrate/antigen/Raman reporter-labeled immuno-nanoparticle" [42].

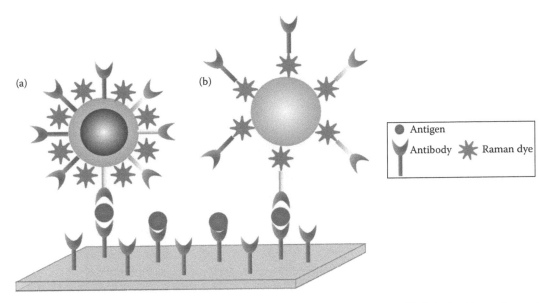

FIGURE 14.5 Two schemes for SERS-based immunoassay: (a) gold-coated Ag-nanoparticle linked to Raman dyes and antibodies; (b) gold nanoparticle linked to the antibodies through the Raman dyes.

In conclusion, the main limitation of direct SERS detection is due to the spectral similarities of various proteins. On the contrary, indirect SERS detection through immunoassays is quite promising tool, although it requires a more complex sample preparation procedure. Anyway, in terms of detection limits and multiplexing, SERS-based immunoassay could replace the more widespread enzyme-based immunoassays for clinical and diagnostic use.

14.4.2 DNA/Nucleic Acids SERS Detection

A general method that allows the detection of multiple DNAs in one assay is the fluorescence [43], which requires complex labeling and sample preparation.

The goal of direct optical detection of DNAs, oligonucleotides, and aptamers remain a key analytical challenge. Recent developments of SERS substrate in terms of detection limits and reproducibility make SERS a tool for single or double-stranded DNA (ssDNA and dsDNA, respectively) analysis. For example, Bell et al. reported SERS spectra of all the DNA/RNA mononucleotides [44]. By using citrate-reduced silver colloids aggregated with $MgSO_4$ the SERS spectra of adenine, guanine, thymine, cytosine, and uracil were recorded. Recently, Barhoumi and coworkers used gold nanoshells as SERS substrate to characterize single and double-stranded thiolated DNA oligomers [45]. More precisely, they demonstrated that SERS spectra of DNA were dominated by the Stokes modes of adenine at $729\,cm^{-1}$. A spectral correlation function analysis useful for assessing signals reproducibility was introduced. Our group has recently demonstrated the SERS capability, in terms of sensitivity and reproducibility, to characterize complex DNA structures, such as G-quadruplex [46]. TERS has been also proposed to detect DNA and RNA nucleotides. Bailo and Deckert, for example, demonstrated the TERS capabilities to identify and sequence the nucleobases in a single RNA strand, with controlled movement of the TERS probe from base to base [47].

An alternative approach for indirect DNA and aptamers SERS detection is shown in Figure 14.6. The SERS probe consist of a functionalized gold nanoparticle linked to a single stranded DNA trough Raman dye-labeled oligonucleotides. After the hybridization with a complementary strand of DNA, the SERS signal of the reporter molecule is acquired. Numerous groups have employed such detection scheme

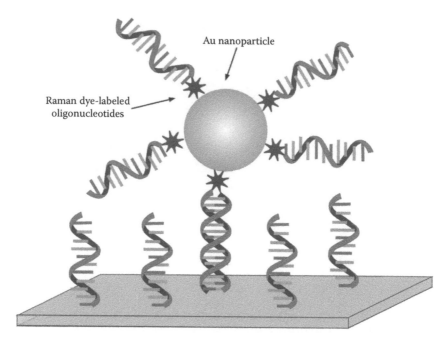

Au nanoparticle

Raman dye-labeled
oligonucleotides

FIGURE 14.6 SERS-based assay for DNA detection: gold nanoparticle modified with Raman dye-labeled oligonucle-otide strands used as probes to monitor the presence of specific target DNA strands.

for different applications. Cao et al. for the first time, showed that nanoparticles functionalized with oligonucleotides and Raman labels (Cy3), coupled with SERS spectroscopy, can be used as a spectroscopic fingerprint to monitor the presence of a specific target oligonucleotide strand with high sensitivity and specificity [48]. Fabris and coworkers presented an identification method for single-stranded DNA based on the ssDNA/PNA hybridization [49]. In such example, there was no need to prefabricate the metallic SERS substrate. The enhancement was obtained by electrostatic deposition of metallic clusters. Banholzer et al. used functionalized silver nanowires, produced by on-wire lithography (OWL), for the detection of DNA binding [50].

In principle, the results obtained with SERS-based techniques offer the possibility for quick diagnosis of specific diseases using relatively simple sample preparation and avoiding amplification of the DNA.

14.4.3 In Vivo SERS Biosensing

SERS biosensing of DNA, proteins, and other biological molecules has also been extended to in vivo systems. In 2006 Stuart et al. [51] presented the first in vivo application of SERS. This sensitive spectroscopy was used to obtain quantitative glucose measurements from a mouse. At this purpose, functionalized silver film over nanosphere surfaces were subcutaneously implanted under the mouse skin. The SERS signal of the interstitial fluid in contact with the SERS substrate was measured through a window placed along the midline of the mouse back. The obtained results of the glucose concentration were in good agreement with the data from a commercial glucometer.

One year later X. Qian and coworkers described, for the first time, the use of nanoparticles and SERS for in vivo tumor targeting and detection [52]. More precisely, thiol-PEG coated gold nanoparticles have been used as SERS substrates with good reproducibility and signal enhancement under very harsh conditions including strong acids or bases, concentrated salts, and organic solvents. For tumor detection, the SERS nanoparticles were conjugated to tumor-targeting ligands. The conjugated nanoparticles were able to target tumor biomarkers such as epidermal growth factor receptors which are sovra expressed in many human cancer cells. Functionalized particles allowed in vivo spectroscopic detection of small tumors ($0.03\,cm^3$) in a mouse at a penetration depth of 1–2 cm, as shown in Figure 14.7.

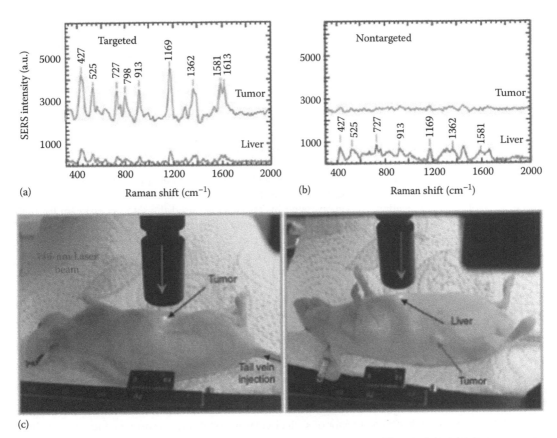

FIGURE 14.7 SERS detection of cancer in a mouse in vivo by using antibody-conjugated gold nanoparticles. (a) SERS spectra obtained from the cancer and non-cancer locations by using targeted (b) and nontargeted nanoparticles. (c) Photographs showing a laser beam focusing on the tumor site or on the anatomical location of liver. In vivo SERS spectra were obtained with 2-s signal integration and at 785 nm excitation. (Figure Reproduced by permission from Macmillan Publishers Ltd., *Nat. Biotechnol.*, Qian, X., Peng, X., Ansari, D., Yin-Goen, Q., Chen, G., Shin, D., Yang, L., Young, A., Wang, M., and Nie, S., In vivo tumor targeting and spectroscopic detection with surface-enhanced Raman nanoparticle tags, 26, 83–90. Copyright 2007.)

14.4.4 SERS-Based Microfluidic Device

Microfluidic chip devices for sensitive chemical and biological analyses have attracted significant attention over the past years. The miniaturization of reaction systems associated to micro fluidic devices offers practical advantages over conventional bench-top systems, including minimal sample requirement and reduced reaction times. Because of the extremely small volume in a micro fluidic channel, a high sensitivity is intrinsically required for the detection of analytes. Due to its potentially "single-molecule" sensitivity, SERS is an excellent candidate to be integrated in microfluidic chips.

As a matter of fact, many groups have reported on SERS-based microfluidic devices [53,54]. These experiments can be schematically divided into two broad categories. In the first, microfluidic systems have been used to manipulate metallic nanoparticles within liquids, while in the second, SERS-active sites metal nanostructure have been embedded in microfluidic systems. The former solution has been more intensively used, due to the easy synthesis of silver and gold nanoparticles.

In order to generate reproducible SERS signals, it is important to achieve a uniform and controlled mixing of metallic colloids with the analyte. To date, a wide range of passive and active microfluidic mixing devices have been developed to improve the diffusive mixing efficiencies in laminar flow regimes. Special geometries in the mixing channel that generate chaotic advection have been also used. For example, Yea and coworkers used a teeth-shaped channel formed in polydimethylsiloxane (PDMS) to allow efficient mixing of cyanide anions and silver colloids within short times [55].

A problem associated with operation within continuous flow regimes is the adhesion of both, colloidal nanoparticles as well as analyte molecules, to the channel walls, which can deeply affect the sensitivity as well as the reproducibility of data, due to the so-called memory effect. For these problems, two possible solutions are feasible. Microfluidic glass-based chips can be regenerated through rinsing with piranha solution, which dissolves the sedimented aggregates. Otherwise, low-cost disposable poly (dimethylsiloxane) chips can be used. An original solution to address the memory effect was introduced by Popp and coworkers. In particular, they developed a system in which the adhesion of nanoparticles to glass walls of the cell is prevented, as the analyte is contained in an aqueous droplet which is conducted in a stream of lipophilic tetradecane. Within such droplets, sample volumes down to a few nanoliters can be processed in a reliable and efficient way [56].

To control the aggregation of plasmonic particles, a number of different approaches involving external stimuli have been reported. For example, optical tweezers can be used to trap colloidal particles and to control their aggregation, as demonstrated by Tong, which used the same laser beam for both Raman excitation and optical tweezing [57]. Electrokinetic effects have been also used to enhance solution-phase mixing between analytes and metallic colloids and to physically concentrate product into microwells embedded in microfluidic channel. In fact, by controlling the electric potential between the upper and lower electrodes, species could be either locally attracted into the microwells or repulsed from them [58].

The use of metallic nanostructures directly incorporated into microfluidic devices can alleviate some of the issues associated with nanoparticle-based SERS detection. Liu and Lee have reported on the fabrication of nanowell-based silver SERS substrates in microfluidic devices by soft lithography [20]. More recently, Kho et al. demonstrated that a periodic array of subwavelength apertures (nanoholes) is a promising substrate for SERS analyses due to the strong EM field in the holes proximity which enhances inherently weak Raman signals and, at the same time, restricts the analysis to a defined region of space [59]. The authors suggest that microfluidic systems containing immobilized nanostructures should be more stable in a biofluid analysis than colloid-based systems due to variations in the ionic strength of the biofluid. However, the enhancement factor of such circular nanohole arrays was found to be not as high as that of normal SERS substrates.

14.4.5 SERS-Active Optical Fibers

As for spontaneous Raman scattering spectroscopy, attention is currently paid to the development of SERS-active optical fibers for signal detection in remote locations. Optical fiber SERS sensors represent a potentially robust means of extending SERS into biomedical applications, such as in-vivo sensing. Several types of optical fiber SERS sensors have been developed to achieve sensitive detection, together with the convenience of optical fiber coupling [60].

The most used geometry is the so-called optrode [61]. The basic optrode design is illustrated in Figure 14.8. The SERS substrate is built up on the tip of the sensor fiber, thereby ensuring automatic overlap between the excitation and collection fields. The optrode is interfaced with a conventional Raman microscope or a conventional remote Raman probe. The optrode configuration minimizes the required number of optical elements, avoids free-space light propagation, and also reduces the scattering or absorption by suspended particulate matter. With this technique, Schmidt et al. have measured a number of different polynuclear aromatic hydrocarbons in seawater [62]. Effective SERS-active optrodes have been produced using several approaches. For instance, a silver nanoparticle film can deposited on fiber end faces by photochemical modification [63] or by slow evaporation of metal island films and vacuum deposition of metal films [64]. The compact size of the optrode makes it well suited for in-situ and in-vivo applications. The main drawback of SERS optrodes is the background (fluorescence and Raman scattering) that is generated by the laser excitation within the fiber core.

Generally, in optrodes, the excitation of the localized surface plasmon resonance is induced by propagating EM fields. An alternative approach is to induce evanescent fields by total internal reflection at the boundary between the optical fiber core and cladding (cladding-coupled sensors). In such a geometry, the waveguide structure plays a key role in determining the strength of the interaction with the analyte. Early examples of cladding-coupled SERS sensors were provided by Stokes and coworkers [65].

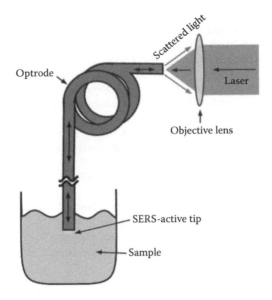

FIGURE 14.8 Basic design for the SERS optrode.

Recently, hollow-core photonic crystal fibers (PCF) have attracted attention as platforms for SERS sensing. In a PCF, the light is confined by a photonic bandgap, a phenomenon similar to Bragg reflection from a periodic cladding structure. In the earliest reported demonstration, gold nanoparticles were deposited on the inner surface of the air holes to create a SERS substrate with an interaction length of a few cm at the tip of a fiber [23]. The backscatter geometry used in this measurement led to some interference due to background from the silica scaffold. In a further work, a hollow-core PCF was modified to fill only the core with analyte molecules in solution with silver nanoparticles. The technique was demonstrated for detection of rhodamine 6G, human insulin, and tryptophan [66].

REFERENCES

1. Novotny, L. and Hecht, B. (2006) *Principles of Nano-Optics.* Cambridge University Press, Cambridge, U.K.
2. Ozbay, E. (2006) Plasmonics: Merging photonics and electronics at nanoscale dimensions. *Science*, 311, 189–193.
3. Wriedt, T. (2008) Mie theory 1908–2008. Introduction to the conference. W. Hergert, T. Wriedt (Eds.), Springer Series in Optical Sciences, Springer-Verlag, Berlin Heidelberg.
4. Noguez, C. (2007) Surface plasmons on metal nanoparticles: The influence of shape and physical environment. *J. Phys. Chem. C*, 111, 3806–3819.
5. Jain, P. K., Huang, X., El-Sayed, I. H., and El-Sayed, M. A. (2007) Review of some interesting surface plasmon resonance-enhanced properties of noble metal nanoparticles and their applications to biosystems. *Plasmonics*, 2, 107–118.
6. Okamoto, H. and Imura, K. (2011) Imaging of optical field distributions and plasmon wavefunctions in metal nanoparticles. *Proc. SPIE*, 6642, 66420A-1.
7. Swanglap, P., Slaughter, L., Chang, W.-S., Willingham, B., Khanal, B., Zubarev, E., and Link, S. (2011) Seeing double: Coupling between substrate image charges and collective plasmon modes in self-assembled nanoparticle superstructures. *ACS Nano*, 5, 4892–4901.
8. Fleischmann, M., Hendra, P., and Mcquillan, A. (1974) Raman spectra of pyridine adsorbed at a silver electrode. *Chem. Phys. Lett.*, 26, 163–166.
9. Kneipp, K., Kneipp, H., Itzkan, I., Dasari, R., and Feld, M. (1999) Ultrasensitive chemical analysis by Raman spectroscopy. *Chem. Rev.*, 99, 2957–2976.
10. Kneipp, K., Kneipp, H., Itzkan, I., Dasari, R., and Feld, M. (2002) Surface-enhanced Raman scattering and biophysics. *J. Phys. Condens. Matter.*, 14, R597–R624.

11. Chan, G., Zhao, J., Hicks, E., Schatz, G., and Duyne, R. V. (2007) Plasmonic properties of copper nanoparticles fabricated by nanosphere lithography. *Nano Lett.*, 7, 1947–1952.
12. Langhammer, C., Schwind, M., Kasemo, B., and Zoric, I. (2008) Localized surface plasmon resonances in aluminum nanodisks. *Nano Lett.*, 8, 1461–1471.
13. Chan, G. H., Zhao, J., Schatz, G. C., and Duyne, R. P. V. (2008) Localized surface plasmon resonance spectroscopy of triangular aluminum nanoparticles. *J. Phys. Chem. C*, 112, 13958–13963.
14. Lee, P. and Meisel, D. (1982) Adsorption and surface-enhanced Raman of dyes on silver and gold sols. *J. Phys. Chem.*, 86, 3391–3395.
15. Weitz, D., Gramila, T., Genack, A., and Gersten, J. (1980) Anomalous low-frequency Raman scattering from rough metal surfaces and the origin of surface-enhanced Raman scattering. *Phys. Rev. Lett.*, 45, 355–358.
16. Wang, H., Brandl, D., Nordlander, P., and Halas, N. (2007) Plasmonic nanostructures: Artificial molecules. *Acc. Chem. Res.*, 40, 53–62.
17. Schlucker, S. (2009) SERS microscopy: nanoparticle probes and biomedical applications. *Chem. Phys. Chem.*, 10, 1344–1354.
18. Shanmukh, S., Jones, L., Driskell, J., Zhao, Y., Dluhy, R., and Tripp, R. (2006) Rapid and sensitive detection of respiratory virus molecular signatures using a silver nanorod array SERS substrate. *Nano Lett.*, 6, 2630–2636.
19. Baia, L., Baia, M., Popp, J., and Astilean, S. (2006) Gold films deposited over regular arrays of polystyrene nanospheres as highly effective SERS substrates from visible to NIR. *J. Phys. Chem. B*, 110, 23982–23986.
20. Liu, G. L. and Lee, L. P. (2005) Nanowell surface enhanced Raman scattering arrays fabricated by soft-lithography for label-free biomolecular detections in integrated microfluidics. *Appl. Phys. Lett.*, 87, 074101–074104.
21. Zhang, X., Whitney, A., Zhao, J., Hicks, E., and Duyne, R. V. (2006) Advances in contemporary nanosphere lithographic techniques. *J. Nanosci. Nanotechnol.*, 6, 1920–1934.
22. Xie, Z., Lu, Y., Wei, H., Yan, J., Wang, P., and Ming, H. (2009) Broad spectral photonic crystal fiber surface enhanced Raman scattering probe. *Appl. Phys. B*, 95, 751–755.
23. Yan, H., Gu, C., Yang, C., Liu, J., Jin, G., Zhang, J., Hou, L., and Yao, Y. (2006) Hollow core photonic crystal fiber surface-enhanced Raman probe. *Appl. Phys. Lett.*, 89, 204101–204104.
24. Domke, K. F. and Pettinger, B. (2010) Studying surface chemistry beyond the diffraction limit: 10 years of TERS. *Chem. Phys. Chem.*, 11, 1365–1373.
25. Inouye, Y., Hayazawa, N., Hayashi, K., Sekkat, Z., and Kawata, S. (2003) Near-field scanning optical microscope using a metallized cantilever tip for nanospectroscopy. *Proc. SPIE*, 3791, 40.
26. Stockle, R., Suh, Y., Deckert, V., and Zenobi, R. (2000) Nanoscale chemical analysis by tip enhanced Raman spectroscopy. *Chem. Phys. Lett.*, 318, 131–136.
27. Yeo, B., Stadler, J., Schmid, T., Zenobi, R., and Zhang, W. (2009) Tip-enhanced Raman spectroscopy-its status, challenges and future directions. *Chem. Phys. Lett.*, 472, 1–13.
28. Murphy, G., Elgamal, A., Su, S., Bostwick, D., and Holmes, E. (1998) Current evaluation of the tissue localization and diagnostic utility of prostate specific membrane antigen. *Cancer*, 83, 2259–2269.
29. Li, J., Zhang, Z., Rosenzweig, J., Wang, Y., and Chan, D. (2002) Proteomics and bioinformatics approaches for identification of serum biomarkers to detect breast cancer. *Clin Chem.*, 48, 1296–1304.
30. Nienhaus, G. U. (2006) Exploring protein structure and dynamics under denaturing conditions by single-molecule fret analysis. *Macromol. Biosci.*, 6, 907–922.
31. Han, X. X., Zhao, B., and Ozaki, Y. (2009) Surface-enhanced Raman scattering for protein detection. *Anal. Bioanal. Chem.*, 394, 1719–1727.
32. Lecomte, S., Wackerbarth, H., Soulimane, T., Buse, G., and Hildebrandt, P. (1998) Time-resolved surface-enhanced resonance Raman spectroscopy for studying electron-transfer dynamics of heme proteins. *J. Am. Chem. Soc.*, 120, 7381–7382.
33. Kahraman, M., Sur, I., and Culha, M. (2010) Label-free detection of proteins from self-assembled protein-silver nanoparticle structures using surface-enhanced Raman scattering. *Anal. Chem.*, 82, 7596–7602.
34. Abdali, S., Johannessen, C., Nygaard, J., and Nørbygaard, T. (2007) Resonance surface enhanced Raman optical activity of myoglobin as a result of optimized resonance surface enhanced Raman scattering conditions. *J. Phys.: Condens. Matter.*, 19, 285205–285208.

35. Han, X., Jia, H., Wang, Y., Lu, Z., Wang, C., Xu, W., Zhao, B., and Ozaki, Y. (2008) Analytical technique for label-free multi-protein detection based on western blot and surface-enhanced Raman scattering. *Anal. Chem.*, 80, 2799–2804.

36. Han, X., Huang, G., Zhao, B., and Ozaki, Y. (2009) Label-free highly sensitive detection of proteins in aqueous solutions using surface-enhanced Raman scattering. *Anal. Chem.*, 81, 3329–3333.

37. Bjerneld, E., Foldes-Papp, Z., Kall, M., and Rigler, R. (2002) Single-molecule surface-enhanced Raman and fluorescence correlation spectroscopy of horseradish peroxidase. *J. Phys. Chem. B*, 106, 1213–1218.

38. Li, H., Sun, J., and Cullum, B. (2006) Label-free detection of proteins using sers-based immunonanosensors. *Nanobiotechnology*, 2, 17–28.

39. Xu, S., Ji, X., Xu, W., Li, X., Wang, L., Bai, Y., Zhao, B., and Ozaki, Y. (2004) Immunoassay using probe-labelling immunogold nanoparticles with silver staining enhancement via surface-enhanced Raman scattering. *Analyst*, 129, 63–68.

40. Grubisha, D., Lipert, R., Park, H., Driskell, J., and Porter, M. (2003) Femtomolar detection of prostate-specific antigen: An immunoassay based on surface-enhanced Raman scattering and immunogold labels. *Anal. Chem.*, 75, 5936–5943.

41. Cao, Y., Jin, R., Nam, J., Thaxton, C., and Mirkin, C. (2003) Raman dye-labeled nanoparticle probes for proteins. *J. Am. Chem. Soc.*, 125, 14676–14677.

42. Cui, Y., Ren, B., Yao, J.-L., Gu, R.-A., and Tian, Z.-Q. (2007) Multianalyte immunoassay based on surface-enhanced Raman spectroscopy. *J. Raman Spectrosc.*, 38, 896–902.

43. Li, Y., Cu, Y. T. H., and Luo, D. (2005) Multiplexed detection of pathogen DNA with DNA-based fluorescence nanobarcodes. *Nat. Biotechnol.*, 23, 885–889.

44. Bell, S. and Sirimuthu, N. (2006) Surface-enhanced Raman spectroscopy (SERS) for sub-micromolar detection of DNA/RNA mononucleotides. *J. Am. Chem. Soc.*, 128, 15580–15581.

45. Barhoumi, A., Zhang, D., Tam, F., and Halas, N. (2008) Surface-enhanced Raman spectroscopy of dna. *J. Am. Chem. Soc.*, 130, 5523–5529.

46. Rusciano, G. et al. (2011) Label-free probing of G-quadruplex formation by surface-enhanced Raman scattering. *Anal. Chem.*, 83, 6849–6855.

47. Bailo, E. and Deckert, V. (2008) Tip-enhanced Raman spectroscopy of single RNA strands: Towards a novel direct-sequencing method. *Angew. Chem. Int. Ed.*, 47, 1658–1661.

48. Cao, Y. C. (2002) Nanoparticles with Raman spectroscopic fingerprints for DNA and RNA detection. *Science*, 297, 1536–1540.

49. Fabris, L., Dante, M., Braun, G., Lee, S., Reich, N., Moskovits, M., Nguyen, T., and Bazan, G. (2007) A heterogeneous PNA-based SERS method for DNA detection. *J. Am. Chem. Soc.*, 129, 6086–6087.

50. Banholzer, M. J., Qin, L., Millstone, J. E., Osberg, K. D., and Mirkin, C. A. (2009) On-wire lithography: Synthesis, encoding and biological applications. *Nat. Protocol.*, 4, 838–848.

51. Stuart, D., Yuen, J., Shah, N., Lyandres, O., Yonzon, C., Glucksberg, M., Walsh, J., and Duyne, R. V. (2006) In vivo glucose measurement by surface-enhanced Raman spectroscopy. *Anal. Chem.*, 78, 7211–7215.

52. Qian, X., Peng, X., Ansari, D., Yin-Goen, Q., Chen, G., Shin, D., Yang, L., Young, A., Wang, M., and Nie, S. (2007) In vivo tumor targeting and spectroscopic detection with surface-enhanced Raman nanoparticle tags. *Nat. Biotechnol.*, 26, 83–90.

53. Chen, L. and Choo, J. (2008) Recent advances in surface enhanced Raman scattering detection technology for microfluidic chips. *Electrophoresis*, 29, 1815–1828.

54. Lim, C., Hong, J., Chung, B., and Choo, J. (2010) Optofluidic platforms based on surface-enhanced Raman scattering. *Analyst*, 135, 837–844.

55. Yea, K., Lee, S., Kyong, J., Choo, J., Lee, E., Joo, S., and Lee, S. (2005) Ultra-sensitive trace analysis of cyanide water pollutant in a PDMS microfluidic channel using surface-enhanced Raman spectroscopy. *Analyst*, 130, 1009–1011.

56. Strehle, K., Cialla, D., Rosch, P., Henkel, T., Kohler, M., and Popp, J. (2007) A reproducible surface-enhanced Raman spectroscopy approach. Online SERS measurements in a segmented microfluidic system. *Anal. Chem.*, 79, 1542–1547.

57. Tong, L., Righini, M., Gonzalez, M., Quidant, R., and Kall, M. (2008) Optical aggregation of metal nanoparticles in a microfluidic channel for surface-enhanced Raman scattering analysis. *Lab. Chip*, 9, 193–195.

58. Huh, Y. S., Lowe, A. J., Strickland, A. D., Batt, C. A., and Erickson, D. (2009) Surface-enhanced Raman scattering based ligase detection reaction. *J. Am. Chem. Soc.*, 131, 2208–2213.

59. Kho, K. W., Qing, K. Z. M., Shen, Z. X., Ahmad, I. B., Lim, S. S. C., Mhaisalkar, S., White, T. J., Watt, F., Soo, K. C., and Olivo, M. (2011) Polymer-based microfluidics with surface-enhanced Raman-spectroscopy-active periodic metal nanostructures for biofluid analysis. *J. Biomed. Opt.*, 13, 054026–054033.

60. Stoddart, P. and White, D. (2009) Optical fibre SERS sensors. *Anal. Bioanal. Chem.*, 394, 1761–1774.

61. Huy, N. Q., Jouan, M., and Dao, N. Q. (1993) Use of a mono-fiber optrode in remote and in situ measurements by the Raman/laser/fiber optics (RLFO) method. *Appl. Spectrosc.*, 47, 2013–2016.

62. Schmidt, H., Ha, N. B., Pfannkuche, J., Amann, H., Kronfeldt, H., and Kowalewska, G. (2004) Detection of pahs in seawater using surface-enhanced Raman scattering (SERS). *Mar. Pollut. Bull.*, 49, 229–234.

63. Polwart, E., Keir, R. L., Davidson, C. M., Smith, W. E., and Sadler, D. A. (2000) Novel SERS-active optical fibers prepared by the immobilization of silver colloidal particles. *Appl. Spectrosc.*, 54, 522–527.

64. Viets, C. and Hill, W. (1998) Comparison of fibre-optic SERS sensors with differently prepared tips. *Sens. Actuat. B*, 51, 92–99.

65. Stokes, D. and Vo-Dinh, T. (2000) Development of an integrated single-fiber SERS sensor. *Sens. Actuat. B*, 69, 28–36.

66. Zhang, Y., Shi, C., Gu, C., Seballos, L., and Zhang, J. (2007) Liquid core photonic crystal fiber sensor based on surface enhanced Raman scattering. *Appl. Phys. Lett.*, 90, 1935041–1935044.

15

Plasmonic Nanostructures and Nano-Antennas for Sensing

Anuj Dhawan and Tuan Vo-Dinh

CONTENTS

15.1 Introduction

Surface plasmons are collective oscillations of the conduction band electrons that are excited in metallic nanostructures (or nano-scale thin films) by propagating light when the size of the nanostructures is smaller than the wavelength of the incident light. In the case of metallic films, coupling of propagating light to plasmons requires either prism coupling or the use of a diffraction grating structure to satisfy the requirements for wave-vectors—of the surface plasmon waves and the component of the incident radiation in the direction of the surface plasmon waves—to be equal. Upon excitation of the surface plasmons into the metallic thin films or metallic thin films with nanohole arrays, one can observe resonances—called surface plasmon resonances (SPRs)—at certain wavelengths or angles of the incident radiation depending on the mechanism employed for SPR coupling [1–16]. Plasmon resonances in metallic nanoparticles or nanostructures—called localized surface plasmon resonance or LSPR—are excited when light is incident on metallic nanostructures that have dimensions smaller than the wavelength of the incident light. At specific wavelengths, resonant multipolar modes are excited in the nanostructures and nanoparticles [17–31], leading to enhancement in absorbed and scattered light and a strong increase in the electromagnetic (EM) fields in the vicinity of the particles. Noble metal (e.g., gold, silver, and copper) nanostructures and nanoparticles resonantly scatter and absorb light in the visible and near-infrared spectra and lead to the excitation of localized surface plasmon resonances [25–28].

Localized surface plasmons can be detected as resonance peaks (or dips) in the reflection or transmission spectra from the individual or array of metallic nanoparticles or nanostructures [17–31]. The shape and magnitude of the peaks or dips in the reflection or transmission spectra depend on the size and geometry of nanostructures as well as the polarization and angle of incidence of light. Excitation of localized surface plasmons leads to an increase in the localized EM fields in the vicinity of the metallic nanoparticles. The increase in the EM fields are dependent on the shape and size of the nanoparticles, higher fields being observed in metallic nanoparticles having nonspherical geometries such as triangular or ellipsoidal nanoparticles as well as in the spacing between the metallic nanoparticles.

The enhancement of EM fields in the vicinity of the plasmonic nanostructures critical for developing SPR and LSPR based sensors and other devices [30,31]. The high EM fields around the nanostructures also lead to higher SERS [32–98] signals obtained from the region containing these nanostructures as compared to the SERS signals obtained without the structures and are important for the development of SERS-based sensors. Intensities of the SERS peaks are dependent on the nanostructure geometry and the spacing between the nanostructures. Recently, the development of plasmonic "nano-antenna" structures [31–36] has been reported, such that the EM fields in the nano-scale spacings between or around nano-antenna structures could be maximized. Theoretical and experimental studies on the development of nano-antennas have been reported and their applications in sensing have been discussed in recent literature.

In this chapter, the design, modeling, and fabrication of individual or arrays of nano-sized structures, apertures, and antennas from plasmonic thin films are described, such that these plasmonic nanostructures and their arrays enable coupling of the propagating light to surface plasmons. This chapter also describes how the plasmonic nanostructures and antennas are employed for sensing and imaging chemical and biological molecules using SPR or LSPR, or SERS techniques.

15.2 Plasmonic Nano-Antennas

In the past few years, significant amount of research has been carried out in the careful design of EM "hot spots," which are regions where extremely high concentration and enhancement of the EM fields can be achieved [99–110]. The design of such regions as well as high density of such regions has involved engineering the sizes and geometries of plasmonic nanoparticles in solution and nanostructures on substrates (forming one- or two-dimensional arrays of the nanostructures), as well as the gaps between individual nanostructures or nanoparticles. Moreover, plasmonic nanoparticles and nanostructures can be arranged in a certain manner so as to form "nano-antennas," which can achieve directional concentration of the optical radiation scattered by the plasmonic nano-antennas [99–109]. In order to make resonant metallic "nano-antennas," the geometries and arrangements of metallic nanoparticles and nanostructures are designed such that the plasmon resonance frequency of the overall arrangement of the nanoparticles or nanostructures can be tuned to the wavelength of the radiation incident on these structures.

Recent research work has focused on designing and developing plasmonic nano-antennas to achieve concentrations and enhancement of EM fields in certain regions in the vicinity of the nano-antennas [99–109]. Greffet [99] as well as Alu and Engheta et al. [100] have described analogies between radio antennas (RF and microwave frequency antennas) and optical antennas [99]. Greffet [99] describes that a coaxial wire can act as a source of radio waves, whose direction of propagation can be modified and intensity amplified by connecting the coaxial wire to an antenna such as a dipole antenna (see Figure 15.1). Similarly, light emission from a light source can be modified by placing the light source inside an optical analogue of a dipole antenna which consists of two nanorod-shaped pillars separated by a gap. Moreover, Greffet [99] described that photons are emitted in a direction perpendicular to the antenna. Alu and Engheta [100] have described that optical nano-antennas such as dipole nano-antennas consist of a pair of plasmonic nanostructures forming the plasmonic dipole, such that this antenna is loaded with combinations of nanoparticles acting as nanocircuit elements. This is analogous to a regular radio dipole antenna that (consisting of two arms of a highly conductive metal separated by a gap) is

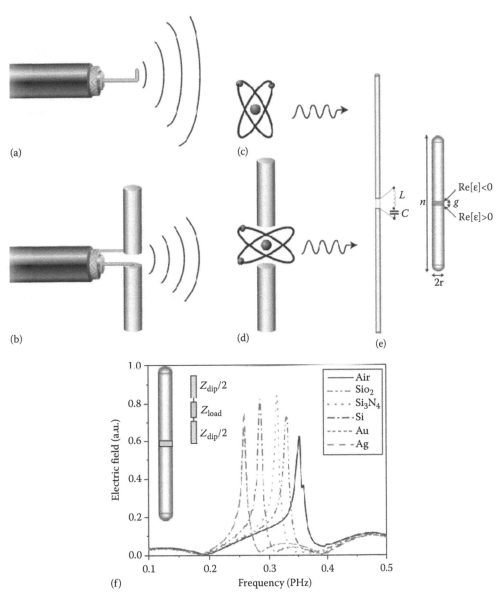

FIGURE 15.1 (a–d) Radio and optical antennas. The end of a coaxial wire (a) is a source of radio waves. Connecting the wire to an antenna (b) amplifies the radio emission and modifies its direction. Light emission can be modified in a similar way by placing a light source such as an atom (c) between two rods (d). In (c), the photon is emitted in almost any direction, whereas in (d), the emission direction is concentrated in directions perpendicular to the antenna. (Reprinted from Greffet, J.-J., *Science*, 308, 1561, 2005. With permission.) (e) Analogy between loading of a regular RF dipole and an optical nano-dipole antenna. a, A regular RF dipole antenna made of highly conductive metal is loaded with lumped circuit elements, as is commonly carried out at RF and microwave frequencies. b, A plasmonic optical nanodipole antenna loaded with combinations of nanoparticles acting as nanocircuit elements. (Reprinted from Alu, A. and Engheta N., *Nat. Photonics*, 2, 307, 2008. With permission.) (f) Far-field scattered electric-field amplitude numerically evaluated on the side of the nanodipole silver nanodipole (with $h = 110$ nm, $r = 5$ nm, and $g = 3$ nm) indicating a shift in the scattering resonance frequency as material of a single nanodisk filling the gap was varied. Inset: circuit model for nanodipole scattering. (Reprinted from Alu, A. and Engheta N., *Nat. Photonics*, 2, 307, 2008. With permission.)

loaded with lumped circuit elements. Alu and Engheta [100] describe that the resonant frequency of the far-field electric field shifts as the nano-scale material (forming nano-disks inside the nano-antenna gap) inside the gap of the plasmonic dipole antenna is varied (see Figure 15.1f).

Different kinds of plasmonic nano-antennas have been designed or developed including dipole nano-antennas (see Figure 15.2)—a single plasmonic nanorod [101] or two plasmonic nanorods separated by a feed gap [102,106], bow-tie nano-antennas [104,105], Yagi-Uda nano-antenna [103], or patch nano-antennas [108]. In traditional long-wavelength dipole antennas (such as RF and microwave antennas), the dipolar resonance wavelength occurs at a wavelength of the incident radiation "λ" when the length "L" of the antenna is "$\lambda/2$." Bryant et al. [101] describe single plasmonic nanorods—not having a feed gap—as optical dipole antennas such that the dipolar plasmon resonance wavelength of these nano-antennas depends on the length of the nanorods (see Figure 15.2a and b). Bryant et al. [101] discuss that in the case of single nanorod antenna, the length "L" at which the dipolar plasmon resonance occurs is less than "$\lambda/2$," where "λ" is the wavelength of the incident light on the nanorod dipole antenna. Moreover, they describe that although the plasmon resonance wavelength depends significantly on the nanorod length "L" and radius "R," but it does not scale with the aspect ratio of the nanorods (see Figure 15.2b).

Cubukcu et al. [102] and Muhlschlegel et al. [106] have described the development of plasmonic dipole antennas based on two metallic nanorod-shaped pillars that are separated by a gap. Based on numerical EM field calculations, they observed very high concentration of the electric field in the gap between the two nanorods when the dipole nano-antenna is resonant (see Figure 15.2c and 15.3a). Muhlschlegel et al. [106] also found from Green's tensor calculations that the near-field enhancement is maximum only for certain overall antenna lengths (lengths of two nanorods and the feed gap length), which are less than the one-half the wavelength of the incident light. Maximum enhancement of the electric field in the gap between the nanorods forming a resonant dipole nano-antenna occurs when the incident light is polarized in the line connecting the nanorods. Cubukcu et al. [102] employed near-field scanning optical microscopy (NSOM) to experimentally determine the electric field in the vicinity of a resonant plasmonic dipole antenna developed on a facet of a commercial diode laser (see Figure 15.3b) and confirmed the concentration and maximum electric field enhancement in the nano-scale gap between the two gold nanorods forming the dipolar nano-antenna.

Kinkhabwala et al. [104] and Schuck et al. [105] have described the development of gold "bow-tie" antennas that increase the concentration and enhancement of electric field intensity (see Figure 15.2e and f) in the gap between the tips of the two metallic nanotriangles facing each other. Significant enhancement of electric field (E-field) intensity (~1500) is obtained in a resonant bow-tie antenna (higher enhancement than a nanorod-based dipolar antenna having the same rod lengths and gap) due to not only the presence of sharp metal tips but also due to coupled plasmon resonant pairs of nanostructures [104,105,109]. Schuck et al. [105] have described through finite difference time domain calculations and through measurement of two-photon excited photoluminescence (TPPL) that the enhancement of the E-field intensity in the gap between the nanotriangles forming the "bow-tie" nano-antenna increases with a decrease in the bow-tie gap.

Li et al. [110] employed multipole spectral expansion method to design nao-antennas or nano-scale plasmonic lens structures that could focus light at certain regions in the assembly of the metallic nanoparticles. These nano-scale lenses consisted of a self-similar linear chain of several silver nanospheres with progressively decreasing sizes and separations of the nanospheres. When excited by optical radiation, the nano-scale lens structure focuses light onto a nano-scale region in between the smallest nanospheres (see Figure 15.4a and b) where the localized electric fields are focused and substantially enhanced. This occurs due to the cascade effect of the geometry of the nano-scale lenses with progressively decreasing sizes and separations of the nanospheres and high Q-factor associated with surface plasmon resonance excited in these nano-scale lens structures. Careful design and development of such nano-scale lenses from a linear arrays of plasmonic nanospheres can lead to very high EM field enhancements, which can be very useful in areas such as surface enhanced Raman scattering (SERS), two-photon excited photoluminescence (TPPL), and plasmon enhanced fluorescenece (PEF). Novontny et al. [103] have described the design and development of gold tri-mer nano-antennas (see Figure 15.4c), based on nanoparticles of decreasing sizes (180, 90, and 50 nm) supported by a dielectric tip, and

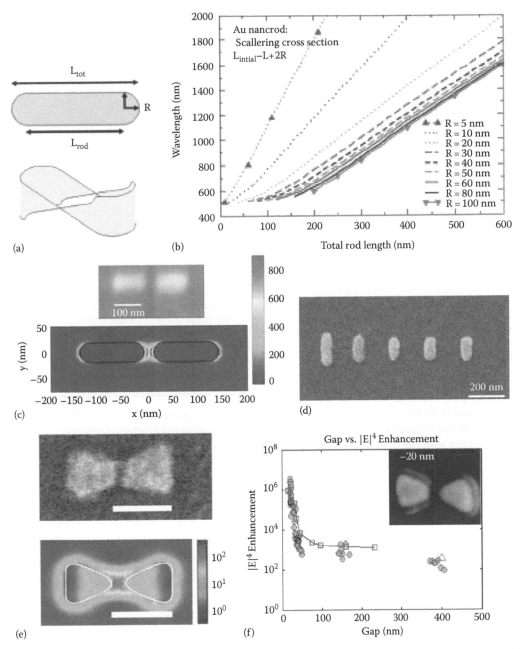

FIGURE 15.2 **(See color insert.)** (a) Schematic of a cylindrical nanorod nano-antenna with hemispherical caps and the surface charge density typical for a dipolar resonance. (b) The dependence of the dipole resonance wavelength on L tot for different R. The resonance wavelength extracted from the far-field scattering is shown. (Reprinted from Bryant, G.W. et al., *Nano Lett.*, 8(2), 631, 2008. With permission.) (c) Dipole nano-antenna, consisting of two rod-shaped nanopillars separated by a small gap: (above) SEM micrograph of a resonant optical dipole antenna; (below) numerical simulation of the total electric field intensity enhancement with respect to the incident intensity. (Reprinted from Cubukcu, E. et al., *Appl. Phys. Lett.*, 89, 093120-1, 2006. With permission.) (d) Yagi-Uda nano-antenna, fabricated by employing electron beam lithography and lift-off. (Reprinted from Novotny, L. and Van Hulst N., *Nat. Photon.*, 5, 83, 2011. With permission.) (e) Bow-tie nano-antenna, consisting of two triangular-shaped nanopillars separated by a small gap between them: (above) SEM micrograph of a gold bow-tie nano-antenna. Scale bar is 100 nm, (below) Finite-difference time-domain calculation of local intensity enhancement. (Reprinted from Kinkhabwala, A. et al., *Nat. Photon.*, 3, 654, 2009. With permission.) (f) Figure showing the effect of spacing between the triangles forming the bow-tie nano-antenna on the electric field enhancement in the region between the nano-triangles. (Reprinted from Schuck, P.J. et al., *Phys. Rev. Lett.*, 94, 017402-1, 2005. With permission.)

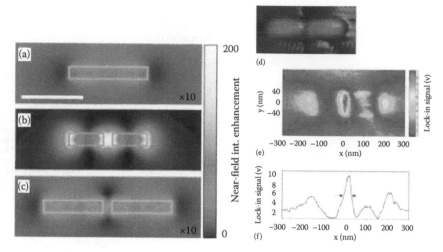

FIGURE 15.3 Near-field intensity (Electric field^2) enhancement factor computed 10 nm above (a) A gold stripe (250 × 40 nm^2), (b) resonant gold dipole nano-antenna (250 × 40 nm^2), and (c) an off-resonant gold dipole nano-antenna (410 × 40 nm^2), the nano-antenna feed gap being 30 nm. Enhancement factor refers to electric field of an evanescent field in the absence of the antenna. Scaling factor in (a) and (c), 10×. Scale bar in (a–c) is 200 nm. (Reprinted from Muhlschlegel, P. et al., *Science*, 308, 1607, 2005. With permission.) (d) AFM topography and (e) a-NSOM image of resonant optical dipole antenna fabricated on one of the facets of a commercial diode laser operating at 830 nm wavelength. (f) Line scan of the near-field distribution along the antenna axis. (Reprinted from Cubukcu, E. et al., *Appl. Phy. Lett.*, 89, 093120-1, 2006. With permission.)

showed a significant increase in the fluorescence signals from a metallofullerene as the distance from the tri-mer nano-antenna was decreased.

The development of different kinds of gold and silver nano-antenna structures [111,112] was carried out by employing focused ion beam (FIB) milling. These structures have included individual gold bow-tie nanopillars with different angle of the triangles or a periodic array of gold bow-tie nano-antennas developed on a silica substrate (see Figure 15.5a through c) A periodic array of gold dimer nanopillars with a 20 nm gap between the nanopillars forming the dimer was also developed using FIB milling. FIB development of linear arrays of elliptical gold nanopillars—separated by only a 15 nm gap by first milling out lines of the plasmonic metals followed by another FIB milling step to cut these lines and obtain a one-dimensional array of nanorods (see Figure 15.5e)—was reported [111,112]. Nano-antenna structures such as gold nanostar pillars (see Figure 15.5f) and diamond-shaped gold nanostructures (see Figure 15.5g) were also developed using FIB milling, such that adjacent nanostructures had their tips facing each other.

Recently, diamond-shaped nanowire structures, based on lateral epitaxial growth of silicon germanium on certain facets of silicon nanowires, were developed [113]. The silicon germanium nanowires were overcoated with plasmonic metals such as gold and silver to develop plasmonic nanowire structures with precisely controlled nano-scale gaps and geometrical structures that would allow EM fields to be concentrated in certain regions between the diamond-shaped nanowires (see Figure 15.6a). Between two parallel silicon germanium diamond-shaped nanowires, one can observe small triangular sections (see Figure 15.6b) due to growth silicon germanium on the bottom silicon region between the lithographically formed silicon nanowires with sub-10 nm gaps—and therefore, high EM field enhancement—between the gold-coated nanostructures present in the diamond-shaped nanowires. When an SOI wafer is used as the substrate to develop the nanowires, the triangular sections in between the diamond-shaped nanowires are not present. The diamond-shaped nanowires shown in Figure 15.6c have a sharp diamond-shaped geometry that can enable concentration and enhancement of EM fields between the tips of the metal-coated diamond nanostructures similar to dipole nano-antenna structures when the length of the nano-antenna is tuned to maximize the EM fields in the feed gap of the nano-antenna. In order to carefully design the plasmonic nanostructures and nano-antennas, we employed the following analytical and numerical modeling procedures.

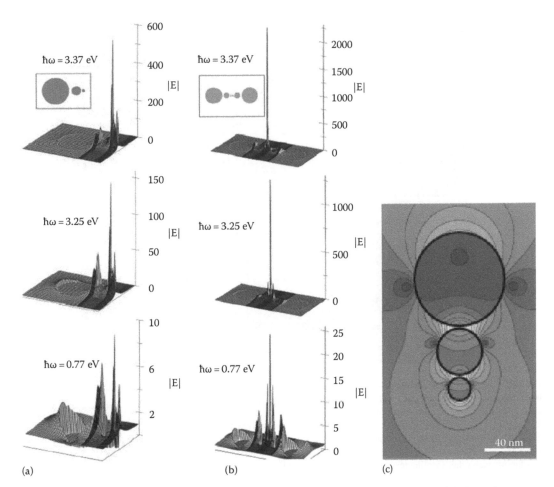

(a)   (b)   (c)

FIGURE 15.4 (a) Local fields (absolute value relative to that of the excitation field) in the equatorial plane of symmetry for the linear self-similar chain of three silver nanospheres. The ratio of the consecutive radii is $R_{i+1}/R_i = 1/3$; the distance between the surfaces of the consecutive nanospheres $d_{i,i+1} = 0.6R_{i+1}$. Inset: the geometry of the system in the cross section through the equatorial plane of symmetry. (b) Local fields (absolute value relative to that of the excitation field) in the equatorial plane of symmetry for the linear self-similar chain of six silver nanospheres with two chains of three silver nanospheres of decreasing sizes facing each other. (a and b: Reprinted from Li, J. et al., *Phys. Rev. B*, 76, 245403-1, 2007. With permission.) (c) Computed intensity near a gold trimer nano-antenna—having gold nanospheres of decreasing size, irradiated at a wavelength of 650 nm. Adjacent contour lines differ by a factor of two in intensity. (Reprinted from Novotny, L. and Van Hulst, N., *Nat. Photon.*, 5, 83, 2011. With permission.)

15.3 Design and Modeling of Plasmonic Nanostructures and Nano-Antennas

15.3.1 Analytical Modeling

EM fields in the near-field or far-field region of individual metallic nanoparticles having spherical geometries can be accurately obtained by employing Mie theory. But the analytical calculations employing Mie theory become very difficult these are carried out for an array of nanoparticles and nanostructures. In this chapter, we describe recent research [114,115] that involves determination of EM fields around plasmonics-active nanoparticle and nanoshell dimers using the multipole expansion [116]. The multipole expansion, also called the superposition method for multiple nanostructures [114], refers to total scattered field being computed as a superposition of the fields scattered from the individual spheres. In the multipole expansion method to determine the EM fields around gold and silver dimer pairs [114], EM fields for each nanosphere forming the dimer were represented as a multipole expansion with respect to

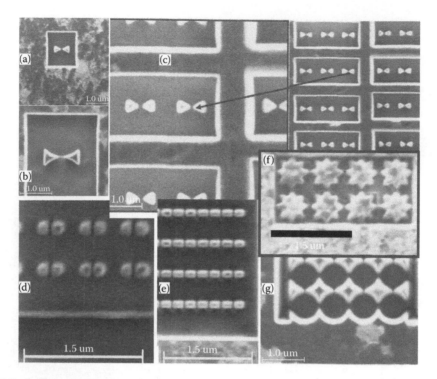

FIGURE 15.5 Scanning ion microscope images showing FIB fabricated gold nano-antenna structures of different kinds by Dhawan, Vo-Dinh et al: (a) and (b) Individual gold bow-tie nanopillars with different angle of the triangles. (c) Periodic array of gold bow-tie nano-antennas developed on a silica substrate. (d) A periodic array of gold dimer nanopillars with a 20 nm gap between the nanopillars forming the dimer. (Reprinted from Dhawan, A. et al., *J. Vac. Sci. Technol. B*, 26(6), 2168, 2008. With permission.) (e) Linear arrays of elliptical gold nanopillars separated by a 15 nm gap. (Reprinted from Dhawan, A. et al., *Nanobiotechnology*, 3, 1, 2007. With permission.) (f) A periodic array of gold nanostar pillars with the tips facing each other forming resonant nano-antennas. (g) Periodic arrays of diamond-shaped gold nanostructures.

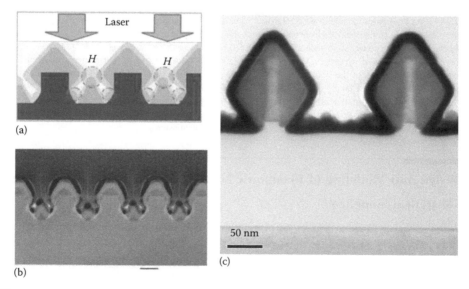

FIGURE 15.6 Gold-coated diamond-shaped silicon germanium nanowires that can act as resonant nano-antennas to concentrate and focus light in the nanoscale gaps in between the nano-antennas as shown in the schematic in (a). (b) TEM cross section of the gold-coated diamond-shaped nanowires with small nano-triangles in between. (c) Diamond-shaped silicon germanium nanowires coated conformally with a thin layer of platinum. (Reprinted from Dhawan, A. et al., *Small*, 7, 727, 2011. With permission.)

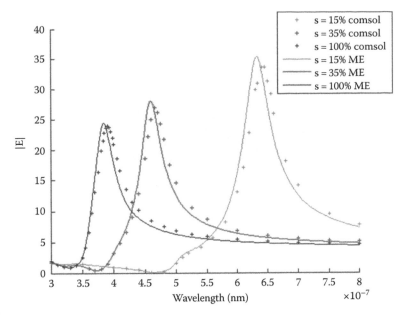

FIGURE 15.7 Electric-field magnitude in the particle gap versus wavelength. The solid curves are the calculations using the ME method and the symbols indicate the FEM calculations. The three cases considered are dimers whose particles are silver nanoshells with a shell thickness of 15% and 35% of the outer shell radius, and a dimer whose particles are solid silver spheres (labeled as 100%). (Reprinted from Vo-Dinh, T. et al., *J. Phys. Chem. C*, 114, 7480, 2010. With permission.)

a coordinate system centered on that sphere. The coefficients in each multipole expansion were obtained by employing translational formulas based on the spherical harmonic addition theorem [117]. The multipole expansion (ME) method provides accurate results as it is a semi-analytical method and the convergence of the expansion can be established.

Using the ME method, quasi-static approximation was employed along with the assumption for axial symmetry in their analytical calculations [114]. As the gold nanoparticles forming the dimer were much smaller than the wavelength of the incident radiation, the employed the quasi-static approximation was valid and we do not need to use vector wave functions due to this assumption. Employing quasi-static approximation [114], Laplace's equation was solved instead of Helmholtz equation resulting in calculation times significantly shorter than the time required for full-wave algorithms. EM field enhancement values between the multipole expansion (ME) method and the numerical simulations for nanosphere dimers (using finite difference time domain modeling, employing Fullwave 6.0) and nanoshell dimers (using finite element modeling, employing COMSOL 3.5a) were compared [114–116] and it was found that these methods (the ME and the numerical methods) were in good agreement (see Figure 15.7).

15.3.2 Finite-Difference Time-Domain (FDTD) Modeling

In the case of nanostructures having more complex geometries or for an assembly of multiple nanostructures, determination of EM fields using the analytical methods becomes very difficult and numerical methods such as the finite-difference time-domain method (FDTD method) are employed. FDTD algorithms [118–121] involve solving the differential form of coupled Maxwell's equations in order to obtain the EM fields (electric and magnetic field distributions, energy and power distributions, etc.) in the near-field or far-field region of the different types of structures being analyzed. Some of these structures include optical waveguides, metallic nanoparticles and their arrays, plasmonic waveguides, arrays of nanoapertures, etc. In FDTD modeling, E and H fields are calculated at different positions and time-steps by discretizing the Maxwell's equations in both time and the space domains.

FDTD modeling has been extensively employed for calculating EM fields around plasmonic nanoparticles of different kinds—nanospheres [119], nanoshells [120], nanostars [121], etc. FDTD analysis was employed [114] to determine electric and magnetic field distributions in the nano-scale gaps between two

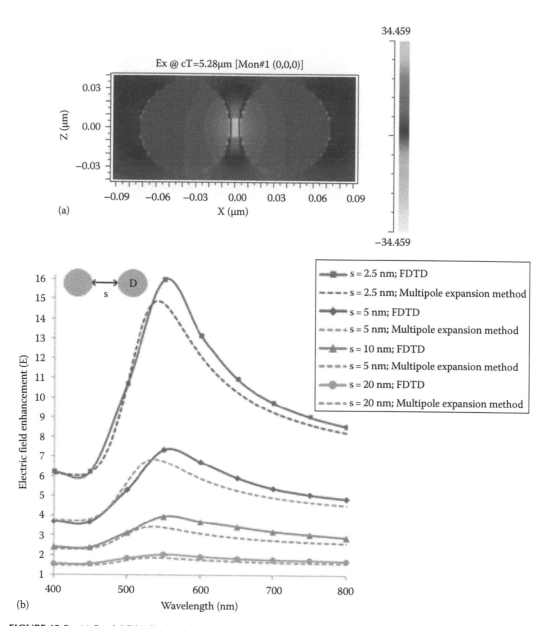

FIGURE 15.8 (a) Spatial Distribution of the E-field as a function of the incident field (i.e., E-field enhancement), polarized along the axis of two 70 nm gold spheres forming a dimer. At the "hot spot," the spacing between the two nanospheres was 5 nm. (b) Effect of the spacing s between two adjacent gold nanospheres forming a dimer, on the magnitude of the electric field as a ratio of the incident electric field, i.e., the electric field enhancement (E), as a function of wavelength of the incident field. Evaluations were carried out using FDTD simulations and analytical calculations using the multipole expansion method. Diameter D of the nanospheres was 20 nm. (Reprinted from Dhawan, A. et al., *Opt. Express*, 17, 9688, 2009. With permission.)

metallic spheres forming a dimer (see Figure 15.8). Fullwave 6.0 software was employed for the FDTD analysis and the effects of dispersion relations of the dielectric constants of the metallic media were included. An extended Debye model was employed for modeling the dielectric constant of gold. The time steps employed in the FDTD simulations were selected to be small enough such that the Courant stability criteria was satisfied for the different grid sizes employed in the simulations. The EM field enhancement and the SERS EM enhancement (calculated using the FDTD method) as a function of the gap between

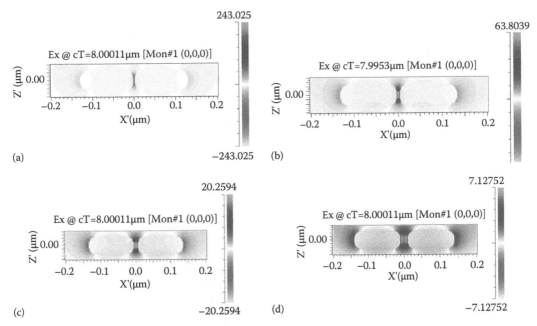

FIGURE 15.9 Three-dimensional FDTD calculations, of electromagnetic fields in the spacing between two spheroidal nanorods, using Fullwave 6.0. E-field, as a ratio of the incident E-Field, i.e., (E_x/E_0), in the direction of the long axis of the nanorods when the spacing between the nanorods was (a) 2 nm, (b) 5 nm, (c) 10 nm, and (d) 20 nm. (Reprinted from Dhawan, A. et al., *Nanobiotechnology*, 3, 1, 2007. With permission.)

gold and silver dimers are shown in Figure 15.8. It can be observed that as the spacing—between two nanospheres of metallic (gold or silver) dimers—is decreased, the EM field and the EM enhancement of SERS are substantially increased. Figure 15.8 also shows the differences between the ME and FDTD methods that occur in the different nanoparticle size regimes as well as different spacings between the nanoparticles. It was seen that a very good correlation—for the amplitude of the E-field enhancement in the middle of the nanosphere dimers—was obtained between the ME and FDTD methods. The maximum difference between the two methods was 5%–10% for gold nanosphere dimers and 20% for silver nanosphere dimers.

Three-dimensional FDTD simulations were also carried out for calculating EM fields around gold nano-antennas, single or multiple gold nanorods, cylinders, and elliptical nanopillars [112]. The EM fields in the vicinity of the nanorods were calculated for a pair of elliptical nanopillars separated by a gaps varying between 5 and 20 nm (see Figure 15.9). The effect of varying the wavelength on the enhancement of the EM fields in the middle of the nano-gap between the spheroidal nanorods was also studied. In the FDTD calculations for two elliptical nanopillars as well as for the two spheroidal nanorods, the incident field was taken to be in the direction of the long axis of the nanopillars structures and the spheroidal nanorods.

Recently, FDTD calculations were carried out to determine EM field enhancement in the nanoscale gaps in the gold-coated diamond-shaped nanowire substrates [113]. One can observe from Figure 15.10 that the E-field enhancement as a function of the incident field increases from ~ 9.5 to 25.6 (for $\lambda_{incident\ radiation}$ = 785 nm) when the minimum spacing between the gold-coated diamond-shaped silicon nanowires and the triangular nanowire structures in between them is reduced from 20 to 4 nm. Similarly, the E-field enhancement increases from ~7.6 to 18.5 when the minimum spacing between the nanowires is reduced from 20 to 4 nm (for $\lambda_{incident\ radiation}$ = 633 nm). It has reported in past literature [32–34] that the EM SERS enhancement is approximately equal to the fourth power of the E-field enhancement. Hence, the increase in E-field on decreasing the spacing between the diamond-shaped and trinagilar nanowires can lead to a very large increase in the EM SERS enhancement from molecules present in these nano-scale gaps between the nanowires.

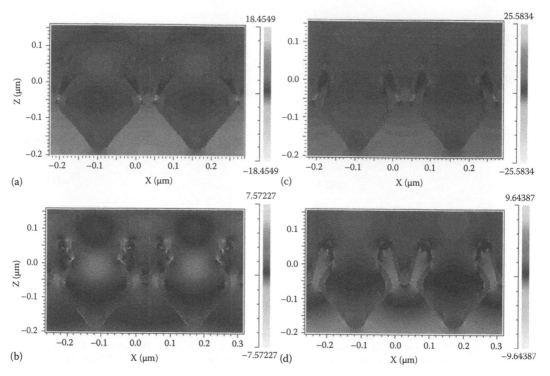

FIGURE 15.10 Simulations show enhancement of the electric field in the spacings between the diamond-shaped nanowires and the triangular nanowires. (a–d) Enhancement of electric field when the nearest spacing "*d*" between adjacent tips of diamond-shaped and triangular nanowires was (a) 4 nm, and (b) 20 nm, when 785 nm radiation is normally incident on the nanowires, and (d) 4 nm, and (e) 20 nm when 633 nm radiation is incident on the DNW structures. (Reprinted from Dhawan, A. et al., *Small*, 7, 727, 2011. With permission.)

15.3.3 Rigorous Coupled Wave Analysis (RCWA)

RCWA algorithms allow a full vectorial solution of the Maxwell's equations in the Fourier domain, wherein periodic permittivity functions are represented using Fourier harmonics and the EM fields are represented as summations over coupled waves. The RCWA calculations require significantly less computational time and memory as compared to the FDTD method and are therefore an excellent choice for numerical calculations on periodic structures. 2-D and 3-D Rigorous Coupled Wave Analysis (RCWA) calculations [122,123] have been described to determine reflectance from periodic metallic nanostructures of different kinds—nanohole arrays in gold and silver films, deep groove metallic gratings, metallic gratings employed for surface plasmon resonance imaging, etc. DiffractMOD 3.1 software was employed to carry out the RCWA calculations. It provides very accurate calculations of the EM field distribution, as well as diffraction efficiency, for different kinds of periodic photonic structures and devices.

Dhawan, Vo-Dinh et al. [122] carried out RCWA calculations to determine reflectance values form gold and silver deep groove nano-gratings when TM polarized plane waves were incident normally on the nano-gratings. They varied the nano-grating dimensions (line width, height, periodicity, and filling factor) to study the effect of these variables on the plasmon resonances associated with the deep groove nano-gratings. This allows tuning of the plasmon resonance associated with these nano-gratings over several spectral regimes and matching these plasmon resonance wavelengths with the conventional wavelengths associated with the LEDs or Lasers. The RCWA calculations carried out by Dhawan, Canva et al. [122] determined changes in reflectance values when the bulk refractive index as well as localized refractive index around the nano-gratings is modified. The localized changes in refractive index could arise from a biomolecule target binding to a capture probe that is functionalized of the surface of a nano-grating. The RCWA calculations demonstrated substantially higher differential reflectance signals (difference in the reflectance signals after and before the introduction of a localized change of refractive index around the nano-gratings)—when changes of refractive index occur in the localized region around the narrow groove plasmonic gratings—as compared to those obtained from SPR-based sensing systems based on the conventional Kretschmann configuration. These simulations enabled us to calculate the narrow-groove nano-grating dimensions (height, periodicity, line width, filling factor) that would provide the highest sensitivity to localized changes of refractive index.

15.4 Fabrication of Plasmonic Nanostructures and Nano-Antennas

15.4.1 Focused Ion Beam Milling

Focused ion beam (FIB) milling has is an important technique for the development of cross-sectional cuts of micro- and nano-scale materials such that their TEM or SEM cross-sections could be obtained. More recently, FIB milling has evolved into a nanofabrication tool and has been employed for the development of a wide array of nano- and microstructures such nano-aperture arrays, nanopillar arrays, photonic crystals, etc. [15,16,111,112].

FIB milling has been employed for developing gold nanopillar arrays on tips of optical fibers and planar substrates for developing LSPR sensors [111]. They described the development of individual and arrays of plasmonic nanostructures of different geometries–triangular, elliptical, circular, and square, etc. and having sub-20 nm spacing between the nanostructures—on planar substrates and cleaved optical fiber tips. Dhawan et al. [111] described the development of plasmonic nanostructures by carving out nanostructures of different geometries from 40 to 100 nm thick layers of gold and silver thin films deposited—on planar substrates such as glass slides, quartz slides, mica sheets, and optical fibers—using electron beam deposition. Focused ion beam instrument, with a liquid gallium ion source, was used to mill the metallic nanoparticles, nanorods, and nanopillars with different sizes, shapes, and separations in the gold layer (see Figure 15.11). The desired nanostructures were milled at very high magnifications (between 6,000× and 18,000×) by rastering the ion beam and employing a beam blanker that turns the beam on and off according to an image file. It was observed that the milling of the nanopillars was considerably time intensive as the bulk of the metallic material was removed to carve out the plasmonic nanopillars on the substrates. A multistep FIB milling process was developed such that nanopillars of larger sizes and of more simpler geometries (such as square-shaped nanopillars) were developed in the initial FIB milling step, which were further milled to modify their size and geometry in the subsequent milling steps. Use of these multistep milling techniques can achieve a minimum separation between the plasmonic nanopillar structures of ~15 nm.

Recently, SERS substrates were developed by employing a focused ion beam milling [98,112]. In order to develop the SERS substrates, arrays of plasmonics-active (gold and silver) nano-pillars arranged in a diagonal geometry, linear arrays of nanorod-shaped pillars, arrays of nanostars-shaped pillars, and FIB roughened metallic thin films were developed. We employed focused ion beams at very high beam magnifications (between 10,000× and 20,000×) and employed a beam blanker, such that the nanostructures of desired geometries and sizes could be developed. Figure 15.12 shows a scanning ion microscope image of an array of square-shaped gold nanopillars (pillar width being 200 nm), such that opposing diagonal tips face each other and gaps of ~20 nm are present between the diagonal tips of the pillars. It was observed that fabrication of this geometry required less amount of milling time, while accurate nanostructure dimensions and spacings were achieved. Moreover, FIB milling was employed for roughening of a 150 nm thick gold film deposited on a silica substrate by exposing the gold film to a focused ion beam thereby leading to creation of nano-scale roughness on the surface of the film. In order to develop SERS substrates, we also developed an array of star-shaped nanopillars—the star-shaped nanopillar containing eight tips such that the number of points where the EM fields would be concentrated was increased—with ~25 nm spacing between neighboring nanostructures [98]. FIB milling was also employed for patterning out linear arrays of metallic nanorods as well as an array of nanopillar pairs to develop nano-antenna arrays that could be employed as SERS substrates [112]. In order to develop linear arrays of nanorod-shaped pillars, arrays of metallic nanolines were first milled out followed by a further milling step to cut the sub-20 nm gaps between nanorod-shaped pillar of desired lengths. It was

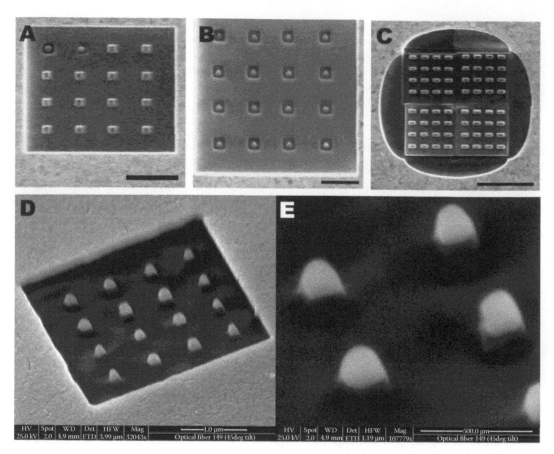

FIGURE 15.11   (A) SIM micrograph of 4 × 4 square nanopillar array, developed by FIB, at tip of a 50 nm gold-coated multimode optical fiber. The first and second elements of the nanopillar array show how particles of smaller size and the desired shape can be formed by employing a first step involving formation of a nanopillar array consisting of particles of larger size (200 nm by 200 nm) and then employing a second step to form nanopillars of the desired size and shape. Scale bar is 2 μm. (B) Triangular-shaped nanopillars formed by the above method. Scale bar is 1 μm. (C) 8 × 8 array of nanorods formed on the core of a 4-mode fiber coated with a 200 nm thick Au layer. The height of the nanorods can be thinned down to the desired height (80 nm) after the first stage of nanorods formation as is shown in the top 4 rows of the nanorods array. Scale bar is 2.5 μm. (D) SEM micrograph, taken at a 30° tilt, showing an array of gold nanopillars formed by FIB. Scale bar is 1 μm. (E) Higher magnification SEM image of the nanopillars shown in (D). Scale bar is 500 nm. (Reprinted from Dhawan, A. et al., *J. Vac. Sci. Technol. (JVST) B*, 26(6), 2168, 2008. With permission.)

observed that this two-step milling procedure led to fabrication of very small gaps between adjacent nanorod-shaped pillars in the linear array. Moreover, the procedure also led to reduced milling time in the development of cuts between the adjacent nanorods in the linear array.

The development of nanohole arrays on tips of optical fibers was reported using FIB milling [15–16], such that these optical fibers could be employed for chemical and biological sensing applications. These optical fiber sensors were fabricated by first coating the prepared tips of the optical fibers—step-index and graded-index optical fibers—with an optically thick layer of gold (using electron beam deposition), followed by FIB milling to fabricate arrays of sub-wavelength apertures. These periodic nano-aperture arrays in optically thick metallic films (100–250 nm films of gold and silver) were fabricated reproducibly with relative ease using the FIB system. Moreover, Ag and Au thin films were also deposited on tapered step-index multimode fibers formed by employing a fusion splicer. During the FIB milling of the nano-aperture arrays, magnification was varied between 6,000 and 10,000 depending on the desired minimum feature size. In Figure 15.13, an array of nanoholes can be observed such that the hole diameter and the center-to-center distance between neighboring holes are 200 and 600 nm. One can observe from Figure 15.13 that the different nanoholes in the array are nearly uniform in terms of their size and geometry.

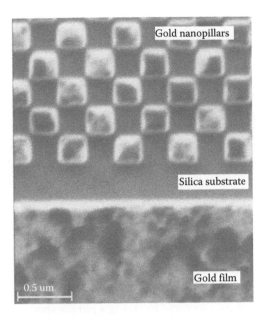

FIGURE 15.12 SIM image of an array of 250 nm wide square-shaped nanopillar arranged in a manner that there is ~18 nm gap between the diagonal tips of the square pillars. These nanopillars were developed on planar silica substrates by FIB milling. (Reprinted from Dhawan, A. et al., *IEEE Sens. J.*, 10, 608, 2010. With permission.)

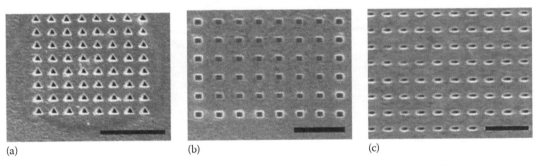

FIGURE 15.13 (a) SIM micrograph showing triangular-shaped nanoaperture arrays formed on a gold-coated tip, cleaved end-face of a 4-mode optical fiber (on the core region), created by FIB milling. Scale bar is 2.5 μm. (b) Rectangular FIB mills created this nanohole array. Scale bar is 1.5 μm. (c) Elliptical nanoholes in an array. Scale bar is 2 μm. (Reprinted from Dhawan, A. et al., *J. Vac. Sci. Technol. (JVST) B*, 26(6), 2168, 2008. With permission.)

The development of fiber sensors [16] with different nano-aperture periodicities, aperture shapes, and gold film thicknesses was described (see Figure 15.13). Figure 15.13a through c shows periodic nanohole arrays having triangular, rectangular, and elliptical shapes, such that the individual apertures in the array had the same cross-sectional area. It was observed that developing nanohole arrays by employing FIB took substantially less time as compared to nanopillar arrays, thereby establishing FIB milling as a method of choice for the development of periodic arrays of nanoholes.

Moreover, FIB milling can be employed for developing nanostructures on cleaved tips of optical fibers (Figure 15.14) as well as on nonplanar substrates (such as tapered optical fibers), which cannot be done by nanolithography processes such as E-Beam Lithography and Deep UV Lithography.

15.4.2 Deep UV Lithography

Totzeck et al. [124] have described that while a 1.35-NA immersion deep UV lithography scanner operating at a wavelength of 193 nm is capable of printing down to a 36.5 nm half-pitch at full scan speed under laboratory conditions using polarized illumination, further reductions in the dimensions of

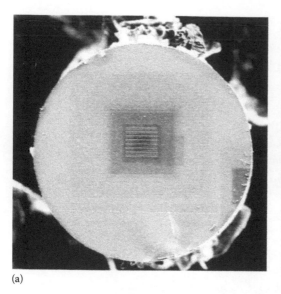 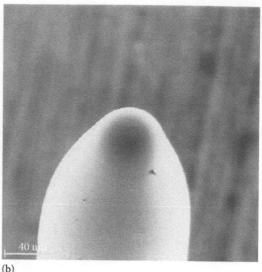

(a)   (b)

FIGURE 15.14   SIM images showing a 180 nm gold film having a periodic square array of nanoholes on the tip of an optical fiber tip: (a) a cleaved end-face of a step-index multimode fiber with a thick layer of gold and a periodic array of nanoholes, (b) a tapered multimode fiber (top fiber view) having the nanohole array. (Reprinted from Dhawan, A. and Muth, J.F., *Mater. Sci. Eng. B Solid-State Mater. Adv. Technol.*, 149(3), 237, 2008. With permission.)

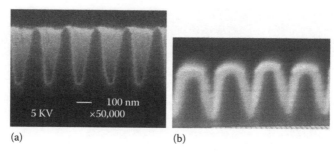

(a)   (b)

FIGURE 15.15   (a) SEM cross section of silicon nanowires formed by using deep-UV lithography and etching of the silicon wafer using TMAH chemical etch. (b) SEM cross section of triangular silicon nanowires overcoated (using electron beam evaporation) with a gold film (in white) to develop the SERS substrates. (Reprinted from Dhawan, A. et al., *IEEE Sens. J.*, 10, 608, 2010. With permission.)

nanostructures could be achieved by employing extreme-UV lithography (which operates at a far shorter wavelength of 13 nm).

Dhawan, Vo-Dinh et al. [98] described the development of SERS substrates, such as gold-coated nanowire reflection gratings, on a wafer-scale by employing deep UV lithography. The fabrication of the gold-coated nanowires firstly involved the development of silicon nanowires on a 6 in. wafer by employing 193 nm deep UV lithography (ASML 5500/950B Scanner employing 193 nm radiation) and wet etching using Tetramethylammonium Hydroxide (TMAH). One-dimensional arrays of triangle-shaped silicon nanowires (see Figure 15.15), having periodicities that ranged from 24 to 20 μm, were developed in different regions on each chip. The development of SERS substrates was carried out by depositing a layer of gold film (20–60 nm thick) on the underlying silicon nanowires by employing electron beam deposition. It was observed that sub-20 nm gaps between the gold film layers on neighboring triangle-shaped nanowires were formed after the electron beam deposition step. Other methods for depositing the metallic thin films were also employed such as thermal evaporation and sputter deposition.

Surface plasmon resonance imaging (SPRI) substrates, containing gold micro- and nano- lines, were fabricated on 6 in. borosilicate glass wafers [123] by employing deep UV lithography and the lift-off

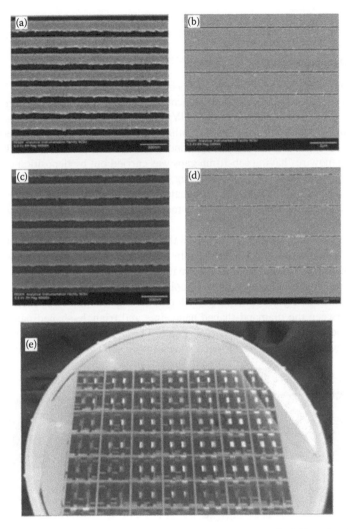

FIGURE 15.16 SEM micrographs showing examples of gold micro- and nano-sized lines developed using deep-UV lithography with different gap sizes/structure periods: (a) 103 ± 10 nm/300 ± 9 nm, (b) 56 ± 5 nm/1655 ± 10 nm, (c) 127 ± 9 nm/402 ± 10 nm, (d) 58 ± 8 nm/2207 ± 15 nm. (e) Picture showing a 6 in. wafer containing several gold micro- and nano-line chips, with borosilicate glass substrate. (Reprinted from Dhawan, A. et al., *Nanotechnology*, 22, 165301, 2011. With permission.)

process (see Figure 15.16). Gold nano- and micro-lines were fabricated by firstly exposing a positive photoresist-coated (with an antireflective film layer underneath the photoresist layer) 6 in. wafer to 193 nm deep UV radiation. A reticle (mask), containing nanolines structures of different periodicities and sizes on the same chip (varying between 100 nm and 10 μm), was employed for the deep UV lithography. Different parameters—such as deep UV exposure dose and depth of focus—were varied during the deep UV lithography process in order to develop nano- and micro-lines of different dimensions. Before carrying out the metal deposition on the developed photoresist, reactive ion etching (RIE) was employed for removal of the antireflective coating layer. Thin films of titanium (~5 nm) and gold (~50 nm thick) were deposited by employing electron beam deposition. In order to carry out the lift-off process, the wafer was placed in an ultrasonic bath containing NMP (N- methyl Pyrrolidone) solution at 80°C. Employing the liftoff process in conjunction with deep UV lithography, nanolines having widths of ~120 nm and a minimum gap between neighboring lines of ~25 nm were developed.

In order to achieve nanostructures having sizes and dimensions less than 40 nm, Totzeck et al. [124] describe the use of a process called double patterning. Double patterning using 193 nm deep UV lithography can be employed for developing very small plasmonic nanostructures and gaps between plasmonic nanostructures for SERS, SPR, and SPRI applications.

15.4.3 Hybrid Approaches

Recently, a hybrid top-down and bottom-up approach for developing sub-10 nm gaps between plasmonic nanostructures was described [113]. The development of diamond-shaped silicon germanium nanowires was carried out by using a combination of deep UV lithography (a top-down approach) for developing silicon nanostructures, with epitaxial growth of silicon germanium on certain facets of silicon nano-structures (bottom-up approach) to controllably reduce gap between nanostructures. The development of plasmonics-active nanowires having sub-10 nm gaps was carried out by deposition of plasmonic metals on the silicon germanium nanowires (by sputtering or electron beam evaporation). These gaps lead to very high enhancement of EM fields and SERS signals from molecules lying in between these nano-scale gaps. Both periodic one-dimensional arrays of diamond-shaped nanowires as well as two-dimensional arrays [113] were fabricated (see Figure 15.17).

The development of silicon nanowires on 6 in. wafer silicon and silicon-on-insulator wafers was carried out using a combination of deep UV lithography (193-nm UV lithography) and RIE. To develop the diamond-shaped nanowire structures, silicon germanium epitaxial films were grown on certain facets of silicon nanowires using the ultrahigh vacuum rapid thermal chemical vapor deposition process. This bottom-up approach involving the epitaxial growth of silicon germanium nanowires led to decrease of nano-scale gaps between the diamond-shaped nanowires and triangle-shaped nanostructures lying in between the diamond-shaped nanostructures (see Figure 15.17). This was followed by overcoating of the silicon germanium wires with a plasmonics-active metal layer such as silver or gold.

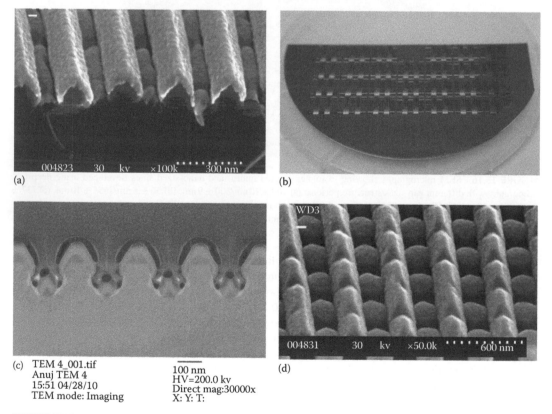

FIGURE 15.17   (a) SEM image of a one-dimensional array of gold-coated $Si_{1-x}Ge_x$ nanowires. (b) A 6 in. wafer containing nanowire arrays developed on the entire wafer. (c) TEM cross-section image showing small triangular sections formed in between the diamond NWs. (d) SEM image of two-dimensional gold-coated $Si_{1-x}Ge_x$ nanowires. (Reprinted from Dhawan, A. et al., *Small*, 7, 727, 2011. With permission.)

Another bottom-up fabrication process called atomic layer deposition (ALD) that can be employed in conjunction with top-down nanofabrication processes for developing very small gaps between nano-structures [113]. As ALD leads to the development of highly conformal films, it can be employed for controlling the nano-scale dimensions of gaps between plasmonic nanostructures developed by other nanolithography processes such as FIB milling, electron beam lithography, and deep UV lithography before coating with plasmonic-active metals. State-of-the-art ALD technology does not allow atomic layer deposition of plasmonics-active metals such as gold, silver, or copper. Future improvements in ALD technology could lead to both conformal ALD deposition of plasmonics-active metals as well as the reduction of nano-scale gaps between adjacent plasmonic nanostructures.

15.4.4 Electron Beam Induced Deposition

Electron beam induced deposition (EBID) [125] involves electron beam irradiation of a substrate in certain regions leading to the generation of secondary electrons from the substrate. This causes chemical decomposition of organometallic gas precursors present above the substrate, which in turn leads to the fabrication of metallic nanostructures regions on the substrates that are irradiated with an electron beam. In order to develop gold nanostructures using EBID [126], precursor Dimethyl Au (III) Actylacetonate was employed as the ionization energy required for decomposition of this precursor matched that of secondary electrons (between 5 and 50 eV). One-dimensional and two-dimensional gold nanopillar arrays were developed by rastering the electron beam in a predefined pattern (see Figure 15.18). For the EBID process, conductive substrates (coated with thin continuous films of ITO or titanium) were employed to avoid accumulation of charge from the electron beam. As accumulation of charge leads to deflection of the electron beam, development of gold nanostructures on uncoated silicon dioxide surfaces was not successful. The development of gold nanopillar arrays using EBID was carried out using a dual beam (FIB and Scanning Electron Microscope) instrument. Localized injection of the precursor molecules using a gas injection needle system ensured the adsorption of the precursor molecules to the sample surface. Although chemical composition evaluation of the EBID

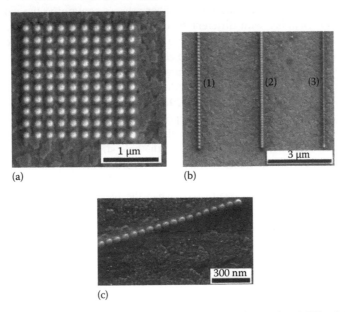

(a)   (b)

(c)

FIGURE 15.18 (a) SEM micrograph of a two-dimensional array of metallic nanodots (~100 nm) deposited on an ITO-coated glass substrate. This nanoparticle array was fabricated using EBID methodology at the CNMS at ORNL. (b) SEM micrographs showing gold nanopillars developed on ITO-coated glass slides using EBID (for developing plasmonic wave-guides). The gold nanopillars shown in (1), (2), and (3) had different sizes and spacings between the nanopillars forming the one-dimensional arrays. (c) EBID of gold nanopillars on Ti-coated glass slides. (Reprinted from Dhawan, A. et al., *Scanning*, 31, 1, 2009. With permission.)

nanostructures was not carried out, the chemical composition of larger (4 mm by 4 mm by 700 nm) gold structures—that were fabricated using the same deposition conditions as the gold nanostructures—was found to be ~10–19 atomic wt% Au. Two-dimensional arrays of plasmonic nanopillars were developed (see Figure 15.18a) such that each nanopillar had a width of ~100 nm. Moreover, one-dimensional arrays of metallic nanopillars—having dimensions varying from 50 to 100 nm and gaps between adjacent nanopillars varying from 17 to 50 nm—were also developed using the EBID process such that these one-dimensional arrays of nanopillars could be employed as plasmonic waveguides (see Figure 15.18b and c).

15.5 Plasmonic Nanostructures for SPR Sensing and Imaging

15.5.1 SPR Sensing Employing Nanohole Arrays

The development of nanohole-array-based surface plasmon resonance chemical and biological fiber-optic sensors has recently been described [15,16]. One observes extraordinary transmission [7] of light—through a periodic two-dimensional array of nanoholes in a metallic film (of plasmonics-active metals such as gold and silver)—at certain wavelengths. The sensing mechanism in these sensors was based on shifts in plasmon resonance peaks in the transmission spectra upon changing the refractive index of the medium next to the metallic film (of plasmonics-active metal) containing the nanohole arrays. In order to develop the fiber-optic sensors, nanohole arrays were reproducibly milled—in the metallic films deposited on tips of optical fibers—using the focused ion beam milling system.

The advantages of employing nanohole-array-based optical fiber sensors are that they allow direct coupling of the incident radiation into surface plasmons and do not require prism coupling using the Kretschmann or Otto configuration. Therefore, they can be employed as fiber-optic probes for in vivo biomolecular detection. Moreover, these sensors are highly sensitive to changes in the bulk refractive index of the media surrounding the metallic film containing the nanohole arrays.

In order to study the effect of changing the refractive index of the medium surrounding the nanohole array fiber sensors, tips of the gold-coated optical fiber sensors containing the nanohole arrays were placed inside a sensor chamber with different fluids (such as methanol, ethanol, isopropyl alcohol, and water) inserted into the chamber. Light from a tungsten halogen lamp was coupled into the end of the fiber-optic sensor that was present outside the sensor chamber. Optical transmission through the nanohole arrays on fiber-optic tips was measured by a collector fiber aligned to the sensor fiber. Changes in the transmission spectrum—as the medium surrounding the tip of a nanohole array sensor was changed from air (refractive index, $n \sim 1.00029$) to methanol ($n \sim 1.329$), water ($n \sim 1.33$), ethanol ($n \sim 1.36$), isopropyl alcohol ($n \sim 1.376$)—was reported. The sensor fiber contained a 24 by 24 array of nanoholes in a 230 nm gold film deposited on the tip of a multimode fiber such that the spacing between the nanoholes in this array was 600 nm and the hole size was 200 nm. It is observed that the peak corresponding to the (1, 0) diffraction order of the nanohole array grating shifts on changing the media surrounding the optical fiber tips (see Figure 15.19). A shift of the (1, 0) diffraction order peak of ~16 nm—for a refractive index change of 0.03 when the medium surrounding the optical fiber is changed from methanol and ethanol—was observed. This gives a sensitivity value of ~533 nm/RIU of these sensors, which is higher than the values reported for LSPR sensors. Moreover, the nanohole-array-based optical fiber sensors were employed to detect the presence of biomolecules. Changes in the transmission spectra of the nanohole fiber sensors were monitored upon binding of biotin-HPDP to the gold surface (the gold film containing the nanohole array on the tip of the optical fiber) and streptavidin binding to the biotin molecules.

15.5.2 Plasmonic Nano- and Microstructures for SPR Imaging (SPRI)

Surface plasmon resonance (SPR) sensors and localized surface plasmon resonance (LSPR) sensors are extremely sensitive to changes in refractive index in the vicinity of the metallic surfaces employed for the development of these sensors. Conventional SPR sensors employ continuous metallic (plasmonic metals

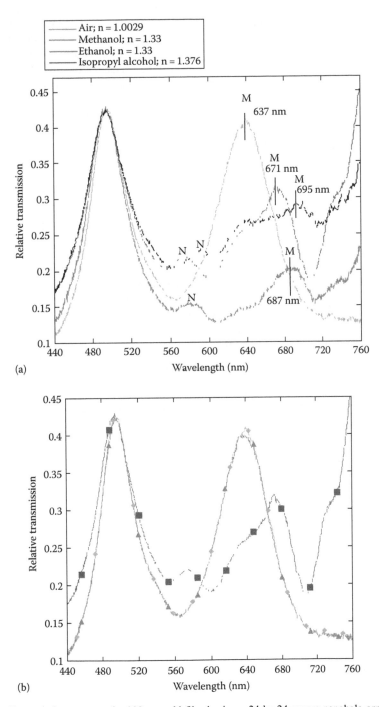

FIGURE 15.19 Transmission spectra of a 230 nm gold film having a 24 by 24 square nanohole array on the tip of a multimode fiber. The spacing between the holes was 600 nm. The transmission spectrum was evaluated in the wavelength range 440–760 nm upon changing the medium surrounding the fiber tip: (a) From air to methanol, ethanol, and isopropyl alcohol. (b) From air (red color) to methanol (blue color), to air again after drying the methanol (green color). (Reprinted from Dhawan, A. et al., *IEEE Sens. J.*, 8, 942, 2008. With permission.)

such as gold and silver) thin films and are very effective in detecting changes in the bulk refractive index around the sensors as the plasmonic field at the metal–dielectric interface has a significant overlap with the media surrounding the metallic film. Surface plasmon resonance is employed for both sensing and imaging (SPRI) to monitor real-time changes of refractive indices. On the other hand, the LSPR sensors employ metallic nanoparticles and nanopillar arrays and are extremely sensitive to detecting changes in the refractive index that occur very close to ($<$ ~10 nm) the surface of the metallic nanostructures, as the EM field enhancement in the near-field of the metallic nanoparticles or nanopillars decreases within ~10 nm of the metallic surface.

It has been reported in recent research that nano- and microstructured plasmonic chips employed for surface plasmon resonance sensing have higher sensitivity as compared to the chips employing continuous metallic thin films [127]. Kim et al. [128] and Alleyne et al. [129] carried out theoretical calculations to demonstrate that nanostructuration of the gold film surface leads to an increase in the sensitivity of the SPR sensors. Recently, Malic et al. [127] developed two-dimensional arrays of metallic nanopillars on the surface of metallic films and demonstrated an increase in sensitivity of biomolecular detection. Masson et al. [130] developed hexagonal arrays of nanoholes in gold films and observed an increase in sensitivity to localized refractive index changes. Recently, deep UV Lithography followed by a lift-off process was employed for developing surface plasmon resonance imaging (SPRI) chips having gold nano- and microstructures on borosilicate glass substrates [123]. An angulo-spectral SPRI measurement setup was employed for scanning of the wavelength and coupling angle and capturing full-field surface plasmon resonance images of the biochip. The plasmon dispersion relation was obtained by capturing a series of SPR images at each angle of incidence ranging from 68.2° to 73.1° (in 0.16° steps) and each wavelength, ranging from 600 to 850 nm (in 5 nm steps). In order to reduce the noise, we took an average of 4 SPR images. Figure 15.20a shows the SPR image of the nano- and microstructured chip with different areas of the chip visible on the image as vertical and horizontal rectangles. Once can observe that each rectangular region's gray level intensity provides information about the reflectivity of the chip as well as the coupling efficiency of the incident light and SP modes. Region 1 on the image corresponds to a nanostructure with a 150 nm gap and a 300 nm periodicity while region 2 corresponds to a nanostructure with a 150 nm gap and a 1650 nm periodicity. We varied the dose of deep UV radiation to achieve gaps between the gold nano- and microstructures that are smaller than the limit of the ASML deep UV lithography equipment. By applying a high dose, positive photoresist under the deep UV mask was exposed which leads to larger widths of the micro- and nanolines and smaller gaps between the lines.

15.5.3 LSPR Sensing Employing Nanostructure Arrays

Collective oscillations of conduction band electrons are excited in metallic nanoparticles by the incident optical radiation due to absorption and scattering of the incident radiation. This leads lead to resonances (called localized surface plasmon resonances or LSPRs) in the optical extinction (absorption and scattering) spectra of these particles such that these resonances are dependent on the geometry, size, spacing, and the refractive index of the media in the vicinity of the nanoparticles. LSPR excitation leads to large enhancements in electromagnetic fields in the vicinity of the nanoparticles. Unlike surface plasmon resonances (SPRs) excited in planar metallic films that have a field decay length of ~100 nm, LSPRs have a field decay length of ~10 nm. Hence, LSPR based sensors are suitable for detecting localized changes (~10 nm above the nanoparticle) in refractive indices around the metallic nanoparticles such as those caused by binding of molecular targets (antigens, DNAs, etc.) to probe molecules immobilized on the nanoparticles. LSPR [17–31] sensors are developed by either characterizing the metallic (of plasmonics active metals such as gold, silver, copper, etc.) nanoparticles in solution, by immobilization of the metallic nanoparticles on planar substrates, as well as by fabricating metallic nanopillars on planar substrates. In order to develop LSPR sensors, different geometries of nanoparticles or nanopillars have been immobilized or fabricated on planar substrates such as gold nanospheres [25], nanorods [21], nanoprisms [31], nanorings [26], etc. The sensitivity ($\delta l_{max}/\delta n$) of an LSPR sensor is measured in terms of the shift of the peak wavelength (l_{max})—associated with the plasmon resonances of metallic nanostructures—for a given change in either the bulk refractive index or the localized refractive index around

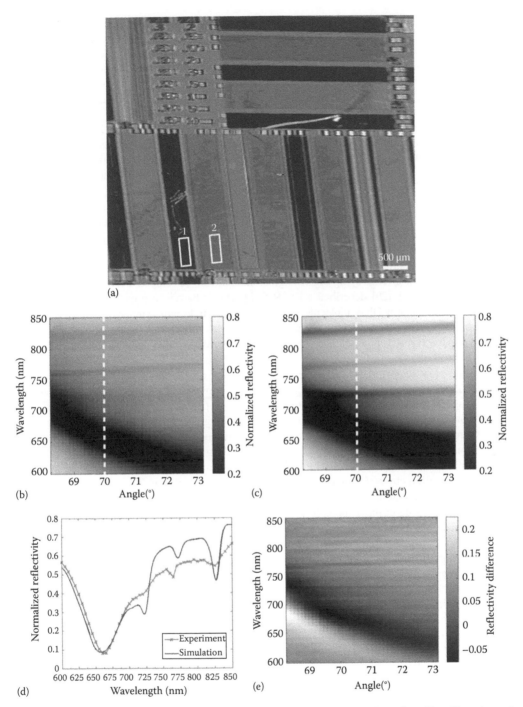

FIGURE 15.20 (a) Surface plasmon resonance image showing the gold micro- and nanoline chip with regions where the SPRI measurement were taken, marked as 1 ($g_s = 103 \pm 10$ nm, $p_s = 300 \pm 9$ nm) and 2 ($g_s = 56 \pm 5$ nm, $p_s = 1655 \pm 10$ nm). The image was taken for p polarization at 800 nm and 70°. The plasmon propagation follows the west–east direction of the image. (b) Experimental angulo-spectral reflectivity map of region 2 (measurement steps 0.25° and 5 nm). (c) Corresponding RCWA simulation (computation steps: 0.1° and 1 nm, truncation order $N = 30$). (d) Spectral reflectivity curve at 70° comparing the experiment (with line and crosses) and the simulation (only line). (e) Experimental SPRI data showing the differential reflectance map when the media around the SPRI substrate is changed from water to water containing 2.5% glycerol (v/v). The micro- and nanolines had spacing/size of 56 ± 5 nm/1655 ± 10 nm (region 2 in the SPRI image of Figure 15.4). (Reprinted from Dhawan, A. et al., *Nanotechnology*, 22, 165301, 2011. With permission.)

the nanoparticles or nanostructures. This localized change can occur due to a change in the length of the molecules on the surface of nanoparticles upon binding of the target molecules. The sensitivity of the LSPR sensors is denoted by a figure-of-merit term ($\{\delta l_{max}/\delta n\}$/FWHM) that also takes the full width half maxima (FWHM) of the peak in the extinction spectra into consideration [31]. McFarland et al. [25] have shown that the bulk refractive index sensitivity of LSPR sensors made from chemically synthesized spherical, triangular, and rodlike nanoparticles are 161 nm/RIU, 197 nm/RIU, 235 nm/RIU, respectively (see Figure 15.21a through d).

While Marinakos et al. [21] had reported a bulk refractive index sensitivity of 252 nm/RIU employing gold nanorods immobilized on a planar substrate, Larsson et al. [26] employed gold nanorings for developing LSPR sensors with bulk refractive index sensitivity of ~880 nm. Recently Verellen et al. [30] described employing nanocrosses as LSPR sensors with bulk refractive index sensitivity of ~1000 for a subradiant Bonding Dipole-Dipole or BDD mode (see Figure 15.22).

Gish et al. [28] developed an LSPR biosensor by developing an array of nanoparticles using glancing angle deposition and reported a low pM limit of detection of anti-rabbit immunoglobulin G (anti-rIgG) concentration when rabbit immunoglobulin G (rIgG) was functionalized as receptor molecule on the silver nanoparticle film (see Figure 15.23).

Willets et al. [27] reported detecting concanavalin A and amyloid-beta-derived diffusible ligand (see Figure 15.24a through c) employing Ag triangular nanopillars developed by employing nanosphere lithography. Anker et al. [29] described a LSPR DNA sensor based on a molecular plasmonic ruler, where the attachment of a DNA molecule to a gold nanoparticle pair connected with a single DNA molecule strand leads to separation of the gold nanoparticles (due to formation of a double stranded DNA), and a spectral shift (of the scattering cross section) associated with the gold nanoparticle pair (see Figure 15.24d through f).

FIB milling was employed [111] for the development of arrays of nanopillars and nanorods on planar substrates and optical fiber tips, such that these nanostructure arrays could be employed for LSPR sensing. In order to develop the nanopillar arrays on tips of cleaved optical fibers, gold films (~50–100 nm thick) were first deposited on the fiber tips using electron beam deposition. Figure 15.14 shows a scanning ion microscope (SIM) image of an array of nanopillars (24 × 24 array) that were developed on a cleaved multimode optical fiber tip by milling the gold film from the core region such that the nanopillars were formed. A FIB milling system (Hitachi FB2100) was employed for fabricating the gold nanopillars by rastering the gallium ion beam and employing beam blanking. Developing nanostructures on the core region of an optical fiber can ensure maximum interaction of the optical fiber modes propagating in the 4-mode fiber to interact with the plasmonic nanostructures and excite localized surface plasmon resonance. They also developed multiple 4 by 4 nanopillars regions on the cleaved tip of a multimode optical fiber (100 µm diameter core region) coated with a 200 nm thick layer of gold film in such that gold film blocked most of the light propagating through the gold film and the light propagating through the milled regions could interact effectively with the gold nanopillar arrays. Figure 15.11 shows SIM micrographs of gold nanopillar arrays of square, triangular, elliptical, and circular geometries, formed on tips of optical fibers. Nanostructures below 60–80 nm side dimensions (and nanostructures of different geometries) were fabricated by employing a two-step fabrication process that involved initial milling of larger nanostructures (~200 nm by 200 nm) followed by a second FIB milling step to reduce the nanostructure dimensions. Figure 15.11 (see the first and second nanopillars in the 4 by 4 array of gold nanopillars developed on the tip of a multimode optical fiber) shows a SIM micrograph demonstrating the development of nanopillars of desired size and shape by employing the two step process. The multistep FIB milling process can also be employed for developing nanopillars of varying heights to be present on the tip of the optical fiber. In order to remove the gold material that re-deposits on the regions where gold film is milled to develop the nanopillar arrays, an extra FIB milling step was employed to ensure that all the gold material around the nanopillars were removed. The nanopillar fabrication process takes significantly more time as compared with the development of nanohole arrays as it involves removal of a majority of the bulk gold film material to leave behind the array of nanopillars. In order to increase the viability of FIB milling technology for develop LSPR sensors on a large scale, employing reactive gases could be carried out such that the FIB etch rate is enhanced.

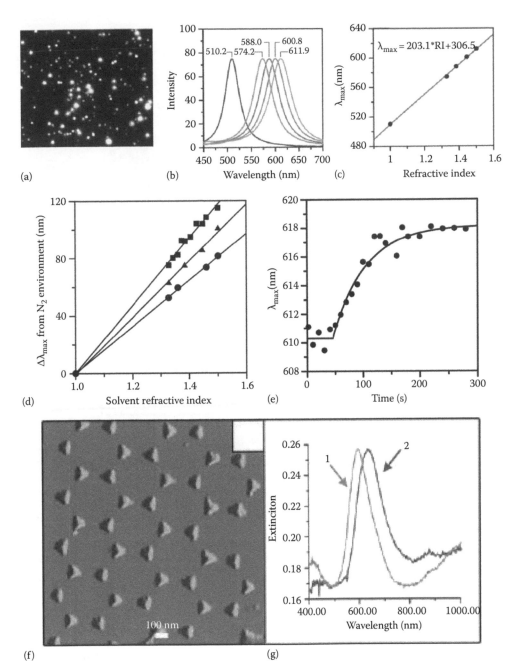

FIGURE 15.21 (a) Dark-field optical image of Ag nanoparticles. (b) Single Ag nanoparticle resonant scattering spectrum in various solvent environments (left to right): nitrogen, methanol, 1-propanol, chloroform, and benzene. (c) Plot depicting the linear relationship between the solvent refractive index and the LSPR λ_{max}. (d) Comparison of refractive index sensitivity for Ag nanoparticles with different geometries. The spherical nanoparticle has a sensitivity of 161 nm/RIU (solid circles in the plot), the triangular nanoparticle has a sensitivity of 197 nm/RIU (solid triangles), and the rodlike nanoparticle (solid squares) has a sensitivity of 235 nm/RIU. (e) Real-time LSPR response of a single Ag nanoparticle as 1.0 mM 1-octanethiol was injected into the flow cell. The circles represent the experimental data, and the line is a first-order response profile with a rate constant of 0.0167 s⁻¹. (a–e: Reprinted from Mcfarland, A.D. and Van Duyne, R.P., *Nano Lett.*, 3, 1057, 2003. With permission.) (f) Tapping-mode AFM image of the Ag nanoparticles developed on a mica substrate by employing nanosphere lithography. (g) LSPR spectra of triangular nanoparticles on a mica substrate (with a 100 nm width and 50.0 nm height) before chemical modification (1) and after modification (2) with 1 mM hexadecanethiol. (f and g: Reprinted from Haes, A.J. and Van Duyne, R.P., *Anal. Bioanal. Chem.*, 379, 920, 10.1007/s00216-004-2708-9, 2004. With permission.)

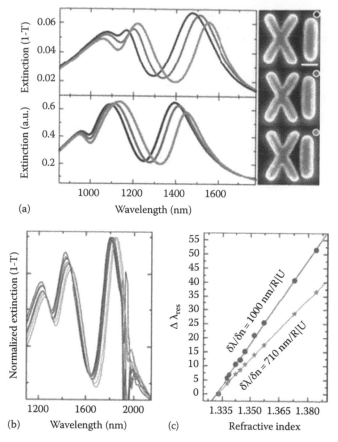

FIGURE 15.22 (a) Experimental extinction spectra for XI cavities with decreasing cross-bar spacing *G*: black, *G* ~ 40 nm; dark grey, *G* ~ 30 nm; light grey, *G* ~ 20 nm. Insets: SEM images of the corresponding fabricated XI cavities (scale bar, 100 nm). (b) Refractive index sensing results for the XI cavity with a cross-bar spacing *G* ~ 20 nm: Extinction spectra for 0, 4, 5, 8, 10, 12, 16, 20, 32, and 40% (concentrations increasing from left curve to right curve) glycerol solutions normalized to the subradiant BDD mode (spectral resolution 32 cm^{-1}). (c) Spectral shift $\Delta \lambda_{res}$ relative to water (n = 1.333) versus refractive index of the surrounding liquid. Dots are the experimental data points and lines are linear fits to the data giving a RI sensitivity of 1000 nm/RIU and 710 nm/RIU for the subradiant Bonding dipole–dipole or BDD mode (line with circular dots) and Bonding Quadruple-Dipole mode or BQD Fano resonance (line with star dots), respectively. (Reprinted from Verellen, N. et al., *Nano Lett.*, 11, 391, 2011. With permission.)

15.6 Plasmonic Nanostructures for SERS-Based Sensing

Raman scattering is an inelastic light scattering process which involves one photon being absorbed and one being emitted either at a lower frequency than the incident light frequency (called Stokes scattering) or at a higher frequency (called Anti-Stokes scattering). The cross-sections associated with Raman scattering are extremely small (10^{-30} to 10^{-25} cm^2 per molecule) and therefore places a restriction on the lowest concentration of molecules that can be detected using this process. The Raman scattering cross section is substantially increased when the surface enhanced Raman scattering (SERS) [32–37] effect is employed wherein the Raman signals from molecules lying in the vicinity of a plasmonics-active surface are greatly enhanced. The enhancement of the SERS signals can occur due to either the EM field enhancement or chemical enhancement [32]. The EM enhancement occurs firstly due to excitation of surface plasmons polaritons on the surface of the plasmonics-active SERS substrates. When surface plasmons are excited in the near-field of the plasmonic nanostructures, there is an increase in the Raman emission intensity proportional to the square of the optical electric field incident on the

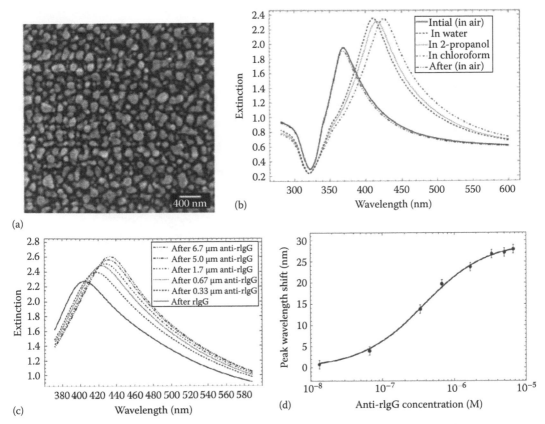

FIGURE 15.23 SEM images of (a) the top surface of the Ag nanoparticle film developed by glancing angle deposition on a fused-silica substrate. (b) Extinction spectra of the Ag nanoparticle film in air and in solvents of varying index of refraction, and in air again after the solvent measurements. The refractive index of each solvent is 1.33, 1.38, and 1.45 for water, 2-propanol, and chloroform, respectively. (c) Extinction spectrum of a rabbit immunoglobulin G (rIgG) functionalized sample and a selection of the extinction spectra after exposure to solutions of varying concentrations of anti-rIgG. (d) Peak LSPR extinction wavelength shift as a function of anti-rIgG concentration. (Reprinted from Gish, D.A. et al., *Anal. Chem.*, 79, 4228, 2007. With permission.)

molecule. The plasmonic EM enhancement decreases substantially as the distance from the surface of the plasmonic substrate is increased [59]. Moreover, the Raman signals emanating from the molecules are also enhanced by the surface plasmons. This leads to the EM SERS enhancement being proportional to the fourth power of the localized electric field at the surface of the SERS substrate, when the Stokes frequency of the emitted Raman signal is spectrally close to the incident light frequency. On the other hand, the chemical enhancement is related to electronic coupling between the metallic substrate and the molecules thereby resulting in a change in the Raman cross section of the coupled molecule–substrate complex. One of the biggest advantages of SERS is that the SERS spectra exhibit narrow spectral features characteristic of vibrational states of the detected molecular species. This allows specific detection of a target analyte molecule in the presence of other molecules [59].

SERS has evolved into a powerful spectroscopic tool for sensitive and specific detection of chemical, biological, and biomedical analytes since the discovery of the SERS effect in the 1970s. Due to recent advances in nanofabrication techniques, reproducible development of plasmonic nanostructures has become possible. SERS-based analytical techniques are either based on metallic nanoparticles of different geometries, dimers, trimers, and aggregations of these metallic nanoparticles in solution, or on nanoparticles and nanostructures deposited or fabricated on a solid surface. Since the mid-1980s, Vo-Dinh laboratory has previously carried out extensive research work in the development of SERS substrates [38,40–42,44–49,51,59], both nanoparticles in solution as well as nanostructures on solid substrates (see Figure 15.25). SERS substrates based on metal-coated half nanoshells were developed

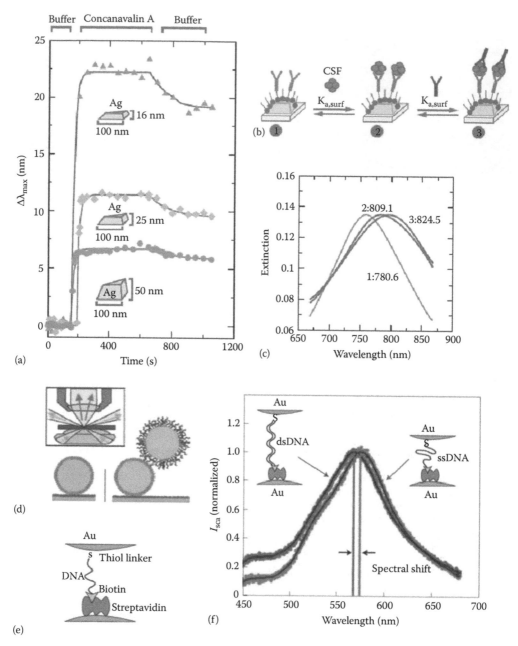

FIGURE 15.24 (a) Real-time response of mannose-functionalized Ag nanosensor (Ag triangular nanopillars developed by employing nanosphere lithography) of different out-of-plane heights as 19-μM concanavalin A is injected in the cell following buffer injection. (b) Surface chemistry for the possible amyloid-beta-derived diffusible ligand (ADDL) detection in human cerebrospinal fluid (CSF) samples using the antibody sandwich assay. The three steps include (1) functionalization of silver triangular nanopillars with anti-ADDL, (2) the introduction of human CSF, and (3) the introduction of the second capping antibody. (c) LSPR spectra for each step of the assay for an Alzheimer's patient. (a–c: Reproduced from Willets, K.A. and Van Duyne, R.P., *Annu. Rev. Phys. Chem.*, 58, 267, 2007. With permission.) (d) A molecular plasmonic ruler. a, schematic illustration of nanoparticle functionalization and immobilization. Inset: principle of transmission dark-field microscopy. (e) A schematic showing the attachment of a gold nanoparticle—connected to a single DNA molecule strand via a thiol linker and having a Biotin molecule on the other end of the DNA molecule—to another gold nanoparticle covered with a layer of streptavidin molecule and attached to a planar surface. (f) Spectral shift between a gold nanoparticle pair connected with single-stranded dNA (red) and double-stranded dNA (blue). (d–f: Reprinted from Anker, J.N. et al., *Nat. Mater.*, 7, 442, 2008. With permission.)

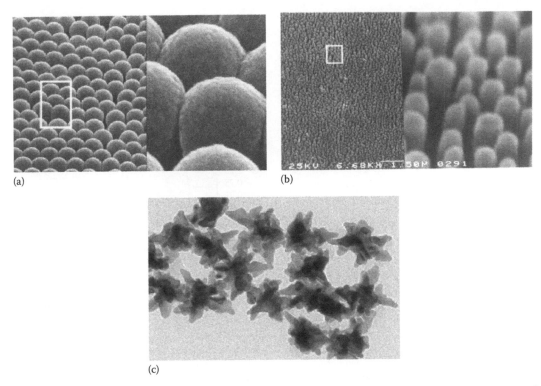

(a)

(b)

(c)

FIGURE 15.25 Plasmonics platforms developed for SERS applications: (a) Substrates based on nanosphere arrays coated with silver. (b) Nanorod arrays fabricated using submicron lithography and plasma etching. (c) Chemically developed gold nanostars. (Reprinted from Vo-Dinh, T., *J. Phys. Chem. C*, 114, 7480, 2010. With permission.)

by self-assembly of metallic and dielectric nanoparticles on planar substrates, followed by overcoating the particles with a thin layer (~50 nm) of a plasmonic metal. Other SERS substrates developed included the development of islanded silver films and metallic microplates on planar substrates, as well as optical fibers on which silver-coated dielectric nanoparticles were deposited. Moreover, nanoparticles of different geometries such as spherical nanoparticles, nanorods, nanostars, etc. have been chemically developed as probes for SERS detection in solution.

This has tremendously increased research interests in the development of SERS substrates and applying these substrates to fields such as defense and security, medicine, industrial process monitoring, chemical and biological sensing, as well as biomedical diagnostics. Vo-Dinh laboratory was the first to employ SERS for detecting trace organic molecules such as homocyclic and heterocyclic polyaromatic compounds as well as organophosphorus compounds [45–47]. Moreover, a wide array of novel analytical methods for detecting chemical and biological molecules, as well as biomedical species have been developed [48,49]. This includes the application of the SERS effect for detection of DNA targets [40–42] and evolution of this method as a medical diagnostics tool. Moreover, the application of DNA nanoprobes called molecular sentinels for detecting DNA targets such as such as HIV and breast cancer genes has been reported [131,132]. In recent work, cells were incubated with Raman dye labeled nanoparticles and Raman imaging was carried out [133].

In order to carry out extremely sensitive SERS-based sensing, such as that required for single molecule detection [54–58], it is extremely important to develop SERS substrates that achieve the highest value of SERS enhancement factor. It is been reported that the extremely large enhancement of Raman signals is obtained from "hot spots," or regions where two or more plasmonic nanoparticles are close to each other. The high-intensity SERS is due to mutual enhancement of localized surface plasmon EM fields of different nanoparticles forming the "hot spots" that determine the dipole moment of molecules trapped in the small gaps between the plasmonic nanoparticles [32,59]. SERS "hot spots" are engineered by controlling

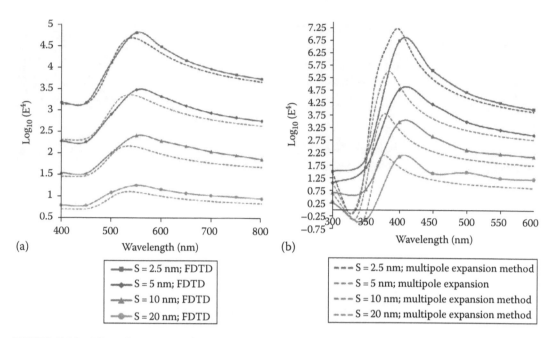

FIGURE 15.26 Effect of spacing s between two metallic nanospheres forming a dimer on the magnitude of the SERS EM enhancement factor plotted on a logarithmic scale as a function of wavelength of the incident field, i.e., $\text{Log}_{10}(E^4)$: (a) evaluations were carried out for gold nanosphere dimers using FDTD simulations and analytical calculations using the multipole expansion method and (b) evaluations were carried out for silver nanosphere dimers using FDTD simulations and the multipole expansion method. Diameter D of the gold and silver nanospheres was 20 nm. (Reprinted from Dhawan, A. et al. *Opt. Express*, 17, 9688, 2009. With permission.)

the size, geometry, and spacing between the plasmonic nanostructures such that the plasmon resonance wavelength of the overall arrangement of the nanostructures can be tuned to the wavelength employed for SERS measurements, as well as by developing nano-antenna structures (discussed in Section 15.2). Detection of SERS signals from individual nano-antenna structures is important for single molecule SERS and can be realized by employing confocal or near field scanning optical microscope in conjunction with SERS spectroscopy.

Characterizing a SERS substrate conventionally involves measurement of SERS signals that are an average value obtained from the surface area of the metallic nanostructures on which there is adsorption of the molecules generating the SERS. This includes the SERS signals obtained from high EM field regions—such as nano-scale gaps (See Figure 15.26) between plasmonic nanostructures or SERS "hot spots"—as well as those from regions on top of the nanostructures or from planar metallic regions in between the arrays of nanostructures. The maximum enhancement factor of SERS [32–37]—obtained from the SERS "hotspots"—can be greater than the average value of SERS from the SERS substrates by several orders of magnitude.

The development of SERS substrates using different methodologies—such as annealing of islanded gold films, focused ion beam milling of gold films, and gold-coating of silicon nanowires developed by employing deep-UV lithography—was recently reported [98]. Figure 15.27a shows SERS signals from multiple chemical molecules (cresyl fast violet and p-mercaptobenzoic acid) that are detected by employing gold-coated silicon nanowires chips. As the different SERS peaks corresponding to the two molecules are clearly distinguishable, this demonstrates the feasibility of employing SERS as an analytical tool for the identification of multiple molecules present in the mixture. SERS signals from molecules—dipicolinic acid (a marker for spores of bacteria such as Anthrax), pMBA (see Figure 15.276b), and a dye-labeled breast cancer gene sequence (Figure 15.27c)—were also obtained by employing gold-coated silicon nanowires and annealed gold island as SERS substrates [98,115].

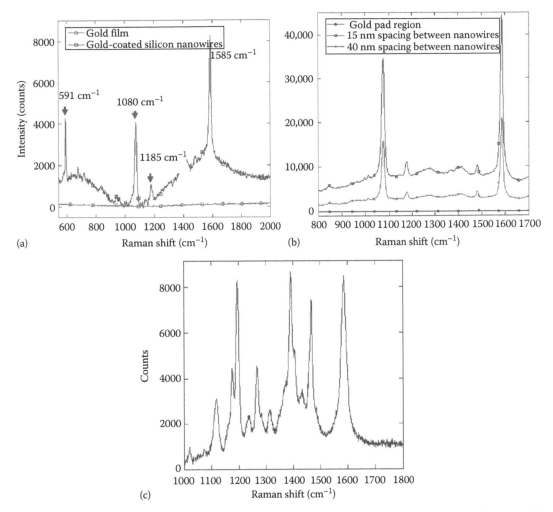

FIGURE 15.27 SERS spectra of a mixture of two chemicals—5 μM CFV and 2 μM p-mercaptobenzoic acid (p-MBA)—together deposited on a linear array of gold-coated nanowires having 40 nm between the adjacent nanowires; 10% laser power was employed. The SERS signal from the gold nanowires (line with squares) is compared with that (line with circles) obtained from a gold pad, i.e., a film deposited on a planar substrate region. The peaks at 591 cm⁻¹ and 1185 cm⁻¹ are the CFV SERS peaks and the peaks at 1080 cm⁻¹ and 1585 cm⁻¹ are the p-MBA SERS peaks (a: Reprinted from Dhawan, A., et al., *IEEE Sens. J.*, 10, 608, 2010. With permission.) (b) SERS signals from pMBA molecules on gold-coated silicon nanowire substrates, for different spacing between neighboring nanowires in the one-dimensional array of nanowires. The SERS substrate chips were coated with pMBA molecules by dipping the chips in a 1 mM ethanol solution of pMBA. (c) SERS signal from Cy3 dye-labeled breast cancer gene sequence (ERBB2) molecules on gold-coated silicon nanowire substrates. (b and c: Reprinted from Vo-Dinh, T. et al., *J. Phys. Chem. C*, 114, 7480, 2010. With permission.)

A hybrid approach consisting of both top-down and bottom-up processes was employed for reproducible development of diamond-shaped plasmonic nanowire substrates on a wafer scale. SERS detection of molecules such as p-mercaptobenzoic acid (pMBA) molecules lying in the SERS hotspots—such as sub-10 nm gaps regions between the plasmonic nanowires—leads to significant enhancement of SERS signals from molecules lying in these regions (see Figure 15.28). Intensity of SERS peaks are substantially higher for the regions containing the diamond shaped nanowires—and having small triangular nanostructures between the diamond nanowires with less than 10 nm gap between neighboring metal-coated nanostructures—as compared with continuous metallic thin film regions (see Figure 15.28c). The gold-coated diamond-shaped nanowires were employed for detection of labeled breast cancer DNA (ERBB2) sequence having the Cy3 SERS-active dye attached to the DNA molecules.

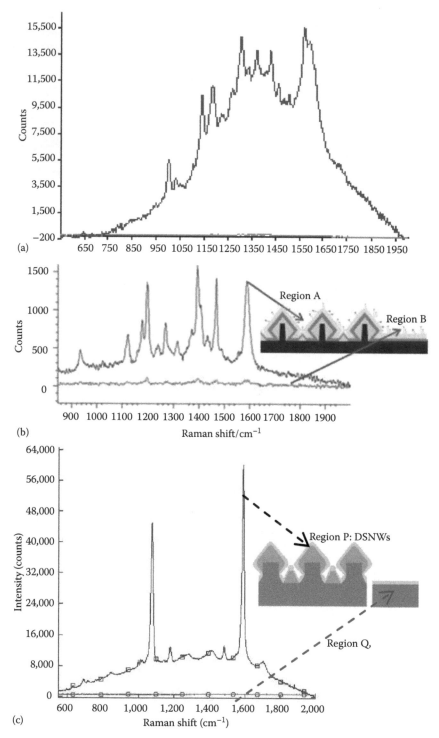

FIGURE 15.28 (a) SERS signals from gold-coated diamond-shaped nanowires (DNW) substrates showing detection of 80 ppm DPA, (b) SERS signals from gold coated DNW substrates showing detection of 1 μM Cy3 dye-labeled ERBB2 DNA segment solution (region A), as compared with SERS signal from gold region having no nanowires (region B). (c) SERS signals from 1 mM pMBA (line with squares) on gold-coated DNW substrates as compared to that from a gold film (line with circles). (Reprinted from Dhawan, A. et al., *Small*, 7, 727, 2011. With permission.)

15.7 Conclusion

Different EM modeling methodologies employed in the design of plasmonic nanostructures and nano-antennas are described in this chapter. The EM modeling methods discussed in this chapter involves numerical methods (FDTD and RCWA) as well as an analytical method (multipole expansion method). Moreover, we discuss in detail the different techniques that are involved in the development of plasmonics-active nanostructures and nano-antennas, comparing the merits and demerits of the different processes. Finally, this chapter describes current research work in several areas such as surface-enhanced Raman scattering (SERS), localized surface plasmon resonance (LSPR), as well as surface plasmon resonance (SPR) sensing and imaging. The applications of these processes for detecting chemical, biological, and biomedical species are discussed.

REFERENCES

1. H. Raether, *Surface Plasmons on Smooth and Rough Surfaces and on Gratings*, Springer Verlag, Berlin, Germany, 1988.
2. J. Homola, On the sensitivity of surface-plasmon resonance sensors with spectral interrogation, *Sens. Actuat. B*, 41, 207–211, 1997.
3. R. C. Jorgenson, and S. S. Yee, A fiber-optic chemical sensor based on surface plasmon resonance, *Sens. Actuat. B*, 12, 213–220, 1993.
4. R. Slavik, J. Homola, and J. Ctyroky, Single-mode optical fiber surface plasmon resonance sensor, *Sens. Actuat. B*, 54, 74–79, 1999.
5. R. Slavik, J. Homola, J. Ctyroky, and E. Brynda, Novel spectral fiber optic sensor based on surface plasmon resonance, *Sens. Actuat. B*, 74, 106–111, 2001.
6. H. A. Bethe, Theory of diffraction by small holes, *Phys. Rev.*, 66, 163–182, 1944.
7. W. L. Barnes, A. Dereux, and T. W. Ebbesen, Surface plasmon subwavelength optics, *Nature*, 424, 824–830, 2003.
8. H. Cao and A. Nahata, Resonantly enhanced transmission of terahertz radiation through a periodic array of subwavelength apertures, *Opt. Express*, 12, 1000–1004, 2004.
9. J. H. Kim and P. J. Moyer, Thickness effects on the optical transmission characteristics of small hole arrays on thin gold films, *Opt. Express*, 14, 6595–6602, 2006.
10. T. J. Kim, T. Thio, T. W. Ebbesen, D. E. Grupp, and H. J. Lezec, Control of optical transmission through metals perforated with subwavelength hole arrays, *Opt. Lett.*, 24, 256–258, 1999.
11. A. Degiron and T. W. Ebbesen, The role of localized surface plasmon modes in enhanced transmission of periodic subwavelength apertures, *J. Opt. A Pure Appl. Opt.*, 7, 590–596, 2005.
12. J. Dintinger, S. Klein, and T. W. Ebbesen, Molecule-surface plasmon interactions in hole arrays: Enhanced absorption, refractive index changes, and all-optical switching, *Adv. Mater.*, 18, 1267–1270, 2006.
13. H. J. Lezec and T. Thio, Diffracted evanescent wave model for enhanced and suppressed optical transmission through subwavelength hole arrays, *Opt. Express*, 12, 3629–3651, 2004.
14. A. Krishnan, T. Thio, T. J. Kim, H. J. Lezec, T. W. Ebbesen, P. A. Wolff, J. Pendry, L. Martin-Moreno, and F. J. Garcia-Vidal, Evanescently coupled resonance in surface plasmon enhanced transmission, *Opt. Commun.*, 200, 1–7, 2001.
15. A. Dhawan and J. F. Muth, Engineering surface plasmon based fiber optic sensors, *Mater. Sci. Eng. B Solid-State Mater. Adv. Technol.*, 149(3), 237–241, 2008.
16. A. Dhawan, M. D. Gerhold, and J. F. Muth, Plasmonic structures based on sub-wavelength apertures for chemical and biological sensing applications, *IEEE Sens. J.*, 8, 942–950, 2008.
17. K. Fujiwara, H. Watarai, H. Itoh, E. Nakahama, and N. Ogawa, Measurement of antibody binding, to protein immobilized on gold nanoparticles by localized surface plasmon spectroscopy, *Anal. Bioanal. Chem.*, 386, 639–644, 2006.
18. A. J. Haes and R. P. Van Duyne, A unified view of propagating and localized surface plasmon resonance biosensors, *Anal. Bioanal. Chem.*, 379, 920–930, 10.1007/s00216-004-2708-9, 2004.
19. T. Okamoto, Near-field spectral analysis of metallic beads, in *Near-Field Optics and Surface Plasmon Polaritons*, ed. Kawata, S., Springer-Verlag, Berlin, Germany, 2001.

20. J. J. Mock, D. R. Smith, and S. Schultz, Local refractive index dependence of plasmon resonance spectra from individual nanoparticles, *Nano Lett.*, 3, 485–491, 2003.

21. S. M. Marinakos, S. H. Chen, and A. Chilkoti, Plasmonic detection of a model analyte in serum by a gold nanorod sensor, *Anal. Chem.*, 79(14), 5278–5283, 2007.

22. M. Futamata, Y. Maruyama, and M. Ishikawa, Local electric field and scattering cross section of Ag nanoparticles under surface plasmon resonance by finite difference time domain method, *J. Phys. Chem. B*, 107(31), 7607–7617, 2003.

23. A. Dhawan and J. F. Muth, In-line optical fiber sensors for environmental sensing applications, *Opt. Lett.*, 31, 1391–1393, 2006.

24. A. Dhawan and J. F. Muth, Plasmon resonances of gold nanoparticles incorporated inside an optical fibre matrix, *Nanotechnology*, 17, 2504–2511, 2006.

25. A. D. Mcfarland and R. P. Van Duyne, Single silver nanoparticles as real-time optical sensors with zeptomole sensitivity, *Nano Lett.*, 3, 1057–1062, 2003.

26. E. M. Larsson, J. Alegret, M. Käll, and D. S. Sutherland, Sensing characteristics of NIR localized surface plasmon resonances in gold nanorings for application as ultrasensitive biosensors, *Nano Lett.*, 7, 1256–1263, 2007.

27. K. A. Willets and R. P. Van Duyne, Localized surface plasmon resonance spectroscopy and sensing, *Annu. Rev. Phys. Chem.*, 58, 267–297, 2007.

28. D. A. Gish, F. Nsiah, M. T. McDermott, and M. J. Brett, Localized surface plasmon resonance biosensor using silver nanostructures fabricated by glancing angle deposition, *Anal. Chem.*, 79, 4228–4232, 2007.

29. J. N. Anker, W. Paige Hall, O. Lyandres, N. C. Shah, J. Zhao, and R. P. Van Duyne, Biosensing with plasmonic nanosensors, *Nature Mater.*, 7, 442–453, 2008.

30. N. Verellen, P. Van Dorpe, C. Huang, K. Lodewijks, G. A. E. Vandenbosch, L. Lagae, and V. V. Moshchalkov, Plasmon line shaping using nanocrosses for high sensitivity localized surface plasmon resonance sensing, *Nano Lett.*, 11, 391–397, 2011.

31. L. J. Sherry, S.-H. Chang, G. C. Schatz, R. P. V. Duyne, B. J. Wiley, and Y. Xia, Localized surface plasmon resonance spectroscopy of single silver triangular nanoprisms, *Nano Lett.*, 5, 2034–2038, 2005.

32. A. Otto, I. Mrozek, H. Grabhorn, and W. Akemann, Surface-enhanced Raman scattering, *J. Phys. Condens. Matter*, 1992, 4, 1143–1212.

33. M. Moskovits, Surface-enhanced spectroscopy, *Rev. Mod. Phys.*, 57, 783–826, 1985.

34. R. K. Chang and T. E. Furtak, eds. *Surface-Enhanced Raman Scattering*, Plenum, New York, 1982.

35. M. Fleischmann, P. J. Hendra, and A. J. McQuillan, Raman spectra of pyridine adsorbed at a silver electrode, *Chem. Phys. Lett.*, 26, 163, 1974.

36. D. L. Jeanmaire and R. P. Van Duyne, Surface Raman spectroelectrochemistry. 1. Heterocyclic, aromatic, and aliphatic-amines adsorbed on anodized silver electrode, *J. Electroanal. Chem.*, 84, 1–20, 1977.

37. M. G. Albrecht and J. A. Creighton, Anomalously intense Raman spectra of pyridine at a silver electrode, *J. Am. Soc.*, 99, 5215–5217, 1977.

38. T. Vo-Dinh, M. Y. K. Hiromoto, G. M. Begun, and R. L. Moody, Surface enhanced Raman spectrometry for trace organic-analysis, *Anal. Chem.*, 56, 1667–1670, 1984.

39. Y. C. Cao, J. Rongchao, and C. A. Mirkin, Nanoparticles with Raman spectroscopic fingerprints for DNA and RNA detection, *Science*, 297, 1536, 2002.

40. T. Vo-Dinh, K. Houck, and D. L. Stokes, Surface-enhanced Raman gene probes, *Anal. Chem.*, 66, 3379–3383, 1994.

41. N. R. Isola, D. L. Stokes, and T. Vo-Dinh, Surface-enhanced Raman gene probes for HIV detection, *Anal. Chem.*, 70, 1352, 1998.

42. T. Vo-Dinh, L. R. Allain, and D. L. Stokes, Cancer gene detection using surface-enhanced Raman scattering (SERS), *J. Raman Spectrosc.*, 33, 511–516, 2002.

43. J. Kneipp, H. Kneipp, and K. Kneipp, SERS—A single molecule and nanoscale tool for bioanalytics, *Chem. Soc. Rev.*, 37, 1052–1060, 2008.

44. M. B. Wabuyele, F. Yan, G. D. Griffin, and T. Vo-Dinh, Hyperspectral surface-enhanced Raman imaging of labeled silver nanoparticles in single cells, *Rev. Sci. Instr.*, 76, 063710, 2005.

45. N. Taranenko, J. P. Alarie, D. L. Stokes, and T. Vo-Dinh, Surface-enhanced Raman detection of nerve agents simulants (DMMP and DIMP) vapor on electrochemically prepared silver oxide substrates, *J. Raman Spectrosc.*, 27, 379, 1996.

46. A. M. Alak, and T. Vo-Dinh, Surface-enhanced Raman spectrometry of chlorinated pesticides, *Anal. Chim. Acta*, 206, 333–337, 1988.
47. A. M. Alak, and T. Vo-Dinh, Surface-enhanced Raman spectrometry of organophosphorous chemical agents, *Anal. Chem.*, 59, 2149–2153, 1997.
48. F. Yan, M. B. Wabuyele, G. D. Griffin, A. A. Vass, and T. Vo-Dinh, Surface-enhanced Raman scattering detection of chemical and biological agent simulants, *IEEE Sens. J.*, 5, 665–670, 2005.
49. T. Vo-Dinh, SERS chemical sensors and biosensors: New tools for environmental and biological analysis, *Sens. Actuat.*, 29, 183–189, 1995.
50. X. Zhang, M. A. Young, O. Lyandres, and R. P. Van Duyne, Rapid detection of an anthrax biomarker by surface-enhanced Raman spectroscopy, *J. Am. Chem. Soc.*, 127, 4484–4489, 2005.
51. D. Zeisel, V. Deckert, R. Zenobi, and T. Vo-Dinh, Near-field surface-enhanced Raman spectroscopy of dye molecules adsorbed on silver island films, *Chem. Phys. Lett.*, 283, 381, 1998.
52. C. R. Yonzon, C. L. Haynes, X. Zhang, J. T. Walsh, and R. P. A VanDuyne, Glucose biosensor based on surface-enhanced Raman scattering: Improved partition layer, temporal stability, reversibility, and resistance to serum protein interference, *Anal. Chem.*, 76, 78–85, 2004.
53. J. D. Ingle and S. R. Crouch, eds., *Spectrochemical Analysis*, Prentice Hall, Englewood Cliffs, NJ, 1988.
54. S. Nie and S. R. Emory, Probing single molecules and single nanoparticles by surface-enhanced Raman scattering, *Science*, 275, 1102–1106, 1997.
55. A. Otto, A. Bruckbauer, Y. X. Chen, On the chloride activation in SERS and single molecule SERS, *J. Mol. Struct.*, 661–662, 501–514, 2003.
56. K. Kneipp, Y. Wang, H. Keipp, L. T. Perelman, I. Itzkan, R. R. Dasari, and M. S. Feld, Single molecule detection using surface-enhanced Raman scattering (SERS), *Phys. Rev. Lett.*, 78, 1667–1670, 1997.
57. W. E. Doering, and S. Nie, Single-molecule and single-nanoparticles SERS: Examining the roles of surface active sites and chemical enhancement, *J. Phys. Chem. B*, 106, 311–317, 2002.
58. H. Xu, E. J. Bjerneld, M. Kall, and L. Borjesson, Spectroscopy of single hemoglobin molecules by surface enhanced Raman scattering, *Phys. Rev. Lett.*, 83, 4357–4360, 1999.
59. T. Vo-Dinh, Surface-enhanced Raman spectroscopy using metallic nanostructures, *Trends Anal. Chem.*, 17, 557–582, 1998.
60. K. Arya and R. Zeyher, *Light Scattering in Solids*, Vol. 4, eds. M. Cardona and G. Guntherodt, chapter 7, Springer, Berlin, Germany, 1984.
61. L. J. Radziemski, R. W. Solarz, and J. A. Paisner, *Laser Spectroscopy and Its Applications, Optical Engineering*, Vol. 11, Marcel Decker, Inc., New York, 1987.
62. E. C.Le Ru, P. G. Etchegoin, and M. Meyer, Enhancement factor distribution around a single surface-enhanced Raman scattering hot spot and its relation to single molecule detection, *J. Chem. Phys.*, 125, 204701, 2006.
63. H. Xu, J. Aizpurua, M. Käll, P. Appell, Electromagnetic contributions to single-molecule sensitivity in surface-enhanced Raman scattering, *Phys. Rev. E: Stat. Phys. Plasmas Fluids Relat. Interdiscipl. Top.*, 4318, 4318–4324, 2000.
64. G. C. Schatz, Theoretical-studies of surface enhanced Raman-scattering, *Acc. Chem. Res.*, 17, 370–376, 1984.
65. M. Kerker, Electromagnetic model for surface-enhanced Raman scattering (SERS) on metal colloids, *Acc. Chem. Res.*, 17, 271–277, 1984.
66. P. K. Aravind and H. Metiu, The enhancement of Raman and fluorescent intensity by small surface-roughness—Changes in dipole emission, *Chem. Phys. Lett.*, 74, 301–305, 1980
67. F. J. Garcia-Vidal and J. B. Pendry, Collective theory on surface enhanced Raman scattering, *Phys. Rev. Lett.*, 77, 1163–1166, 1996.
68. Otto, A., The 'chemical' (electronic) contribution to surface-enhanced Raman scattering, *J. Raman Spectrosc.*, 36, 497–509, 2005.
69. A. Campion, J. E. Ivanecky, III, C. M. Child, M. Foster, On the mechanism of chemical enhancement in surface-enhanced Raman scattering, *J. Am. Chem. Soc.*, 117, 11807–11808, 1995.
70. J. R. Lombardi, R. L. Birke, T. Lu, and J. Xu, Charge-transfer theory of surface enhanced Raman spectroscopy: Herzberg-Teller contributions, *J. Chem. Phys.*, 84, 4174–4180, 1986.
71. J. R. Lombardi and R. L. Birke, Time-dependent picture of the charge-transfer contributions to surface enhanced Raman spectroscopy, *J. Chem. Phys.*, 126, 244709, 2007.

72. S. ILecomte, P. Matejka, and M. H. Baron, Correlation between surface enhanced Raman scattering and absorbance changes in silver colloids. Evidence for the chemical enhancement mechanism, *Langmuir*, 14, 4373–4377, 1998.

73. B. N. J. Persson, K. Zhao, and Z. Zhang, Chemical contribution to surface-enhanced Raman scattering, *Phys. Rev. Lett.*, 96, 207401, 2006.

74. A. Campion and P. Kambhampati, Surface-enhanced Raman scattering, *Chem. Soc. Rev.*, 27, 241–250, 1998.

75. L. Jensen, L. L. Zhao, J. Autschbach, and G. C. Schatz, Theory and method for calculating resonance Raman scattering from resonance polarizability derivatives, *J. Chem. Phys.*, 123, 174110, 2005.

76. J. Neugebauer and B. A. Hess, Resonance Raman spectra of uracil based on Kramers–Kronig relations using time-dependent density functional calculations and multireference perturbation theory, *J. Chem. Phys.*, 120, 11564–11577, 2004.

77. L. Jensen, J. Autschbach, M. Krykunov, and G. C. Schatz, Resonance vibrational Raman optical activity: A time-dependent density functional theory approach, *J. Chem. Phys.*, 127, 134101, 2007.

78. L. Jensen and G. C. Schatz, Resonance Raman scattering of Rhodamine 6G as calculated using time-dependent density functional theory, *J. Phys. Chem. A*, 110, 5973–5977, 2006.

79. L. H. Quan, A. Inoue, and M. W. Chen, Large surface enhanced Raman scattering enhancements from fracture surfaces of nanoporous gold, *Appl. Phys. Lett.*, 92, 093113, 2008.

80. D. J. Maxwell S. R. Emory, and S. Nie, Nanostructured thin-film materials with surface-enhanced optical properties, *Chem. Mater.*, 13, 1082–1088, 2001.

81. M. Schierhorn, S. J. Lee, S. W. Boettcher, G. D. Stucky, and M. Moskovits, Metal-silica hybrid nanostructures for surface-enhanced Raman spectroscopy, *Adv. Mater.*, 18, 2829–2832, 2006.

82. L. L. Zhao, L. Jensen, and G. C. Schatz, Surface-enhanced Raman scattering of pyrazine at the junction between two Ag nanoclusters, *Nano Lett.*, 6, 1229–1234, 2006.

83. L. Qin, S. Zou, C. Xue, A. Atkinson, G. C. Schatz, and C. A. Mirkin, Designing, fabricating, and imaging Raman hot spots, *Proc. Nat. Acad. Sci. USA*, 104, 13300–13303, 2006.

84. B. Cui, L. Clime, K. Li, and T. Veres, Fabrication of large area nanoprism arrays and their application for surface enhanced Raman spectroscopy, *Nanotechnology*, 19, 145302, 2008.

85. K. Li, L. Clime, B. Cui, and T. Veres, Surface enhanced Raman scattering on long-range ordered nobel-metal nanocrescent arrays, *Nanotechnology*, 19, 145305, 2008.

86. N. Felidj, J. Aubard, G. Levi, J. R. Krenn, A. Hohenau, G. Schider, A. Leitner, and F. R. Aussenegg, Optimized surface-enhanced Raman scattering on gold nanoparticle arrays, *Appl. Phys. Lett.*, 82, 3095–3097, 2003.

87. G. L. Liu and L. P. Lee, Nanowell surface enhanced Raman scattering arrays fabricated by soft-lithography for label-free biomolecular detections in integrated microfluidics, *Appl. Phys. Lett.*, 87, 074101, 2005.

88. P. F. Liao, J. G. Bergman, D. S. Chemla, A. Wokaun, J. Melngailis, A. M. Hawryluk, and N. P. Economou, Surface-enhanced Raman scattering from microlithographic silver particle surfaces, *Chem. Phys. Lett.*, 82, 355, 1981.

89. T. A. Alexander, Development of methodology based on commercialized SERS-active substrates for rapid discrimination of poxviridae virions, *Anal. Chem.*, 80 (8), 2817–2825, 2008.

90. T. A. Alexander and D. M. Le, Characterization of a commercialized SER-active substrate and its applications to the identification of intact Bacillus endospores, *Appl. Opt.*, 46, 3878–3890, 2007.

91. P. Sharma, C. Y. Liu, C. F. Hsu, N. W. Liu, and Y. L. Wang, Ordered arrays of Ag nanoparticles grown by constrained self-organization, *Appl. Phys. Lett.*, 89, 163110, 2006.

92. H. H. Wang, C. Y. Liu, S. B. Wu, N. W. Liu, C. Y. Peng, T. H. Chan, C. F. Hsu, and J. K. Wang, Highly Raman-enhancing substrates based on silver nanoparticle arrays with tunable sub-10nm gaps, *Adv. Mater.*, 18, 491–495, 2006.

93. J. A. Dieringer, A. D. McFarland, N. C. Shah, D. A. Stuart, A. V. Whitney, C. R. Yonzon, M. A. Young, X. Zhang, and R. P. Van Duyne, Surface enhanced Raman spectroscopy: New materials, concepts, characterization tools, and applications, *Faraday Discuss*, 132, 3–8, 2006.

94. S. J. Lee, A. R. Morill, and M. Moskovits, Hot spots in silver nanowire bundles for surface-enhanced Raman spectroscopy, *J. Am. Chem. Soc.*, 128, 2200–2201, 2006.

95. L. Billot, M. Lamy de la Chapelle, A.-S. Grimault, A. Vial, D. Barchiesi, J.-L. Bijeon, P.-M. Adam, and P. Royer, Surface enhanced Raman scattering on gold nanowire arrays: Evidence of string multipolar surface plasmon resonance enhancement, *Chem. Phys. Lett.*, 422, 303–307, 2006.

96. T. Qiu, X. L. Wu, J. C. Shen, P. C. T. Ha, and P. K. Chu, Surface-enhanced Raman characteristics of Ag cap aggregates on silicon nanowire arrays, *Nanotechnology*, 17, 5769–5772, 2006.
97. M. A. Khan, T. P. Hogan, and B. Shanker, Surface-enhanced Raman scattering from gold-coated germanium oxide nanowires, *J. Raman Spectrosc.*, 39(7), 893–900, 2008.
98. A. Dhawan, Y. Du, F. Yan, M. D. Gerhold, V. Misra, and T. Vo-Dinh, Methodologies for developing surface-enhanced Raman scattering (SERS) substrates for detection of chemical and biological molecules, *IEEE Sens. J.*, 10, 608–616, 2010.
99. J.-J. Greffet, Nanoantennas for light emission, *Science*, 308, 1561–1562, 2005.
100. A. Alu and N. Engheta, Tuning the scattering response of optical nanoantennas with nanocircuit loads, *Nat. Photonics*, 2, 307–310, 2008.
101. G. W. Bryant, F. Javier Garcia de Abajo, and J. Aizpurua, Mapping the plasmon resonances of metallic nanoantennas, *Nano Lett.*, 8 (2), 631–636, 2008.
102. E. Cubukcu, E. A. Kort, K. B. Crozier, and F. Capasso, Plasmonic laser antenna, *Appl. Phys. Lett.*, 89, 093120-1–093120-1, 2006.
103. L. Novotny and N. Van Hulst, Antennas for light, *Nat. Photonics*, 5, 83–90, 2011.
104. A. Kinkhabwala, Z. Y. Fan, Y. Avlasevich, K. Mullen, and W. E. Moerner, Large single-molecule fluorescence enhancements produced by a bowtie nanoantenna, *Nat. Photon.*, 3, 654–657, 2009.
105. P. J. Schuck, D. P. Fromm, A. Sundaramurthy, G. S. Kino, and W. E. Moerner, Improving the mismatch between light and nanoscale objects with gold bowtie nanoantennas, *Phys. Rev. Lett.*, 94, 017402-1–017402-4, 2005.
106. P. Muhlschlegel, H.-J. Eisler, O. J. F. Martin, B. Hecht, and D. W. Pohl, Resonant optical antennas, *Science*, 308, 1607, 2005.
107. J. Li, A. Salandrino, and N. Engheta, Shaping light beams in the nanometer scale: A Yagi-Uda nanoantenna in the optical domain, *Phys. Rev. B*, 76, 245403-1–245403-4, 2007.
108. R. Esteban, T. V. Teperik, and J. J. Greffet, Optical patch antennas for single photon emission using surface plasmon resonances, *Phys. Rev. Lett.*, 104, 026802-1–026802-4, 2010.
109. D. P. Fromm, A. Sundaramurthy, P. J. Schuck, G. Kino, and W. E. Moerner, Gap-dependent optical coupling of single "Bowtie" nanoantennas resonant in the visible, *Nano Lett.*, 4, 957–961, 2004.
110. K. Li, M. I. Stockman, and D. J. Bergman, Self-similar chain of metal nanospheres as an efficient Nanolens, *Phys. Rev. Lett.*, 91(22), 227402-1–227402-4, 2003.
111. A. Dhawan, J. F. Muth, D. N. Leonard, M. D. Gerhold, J. Gleeson, T. Vo-Dinh, and P. E. Russell, FIB fabrication of metallic nanostructures on end-faces of cleaved optical fibers for chemical sensing applications, *J. Vac. Sci. Technol. B*, 26 (6), 2168–2173, 2008.
112. A. Dhawan, M. D. Gerhold, and T. Vo-Dinh, Theoretical simulation and focused ion beam fabrication of gold nanostructures for surface-enhanced Raman scattering (SERS), *Nanobiotechnology*, 3, 1–8, 2007.
113. A. Dhawan, Y. Du, H. Wang, D. Leonard, V. Misra, M. Ozturk, M. Gerhold, and T. Vo-Dinh, Hybrid top-down and bottom-up fabrication approach for wafer-scale plasmonic nano-platforms, *Small*, 7, 727–731, 2011.
114. T. Vo-Dinh, A. Dhawan, C. Khoury, S. Norton, H. Wang, V. Misra, and M. Gerhold, Plasmonic nanoparticles and nanowires: Design, fabrication and application in sensing, *J. Phys. Chem. C*, 114, 7480–7488, 2010.
115. C. G. Khoury, S. J. Norton, and T. Vo-Dinh, Plasmonics of 3-D nanoshell dimers using multipole expansion and finite element method, *ACS Nano*, 3, 2776–2788, 2009.
116. A. Dhawan, S. J. Norton, M. D. Gerhold, and T. Vo-Dinh, Comparison of FDTD numerical computations and analytical multipole expansion method for plasmonics-active nanosphere dimers, *Opt. Express*, 17, 9688–9703, 2009.
117. S. J. Norton and T. Vo-Dinh, Optical response of linear chains of metal nanospheres and nanospheroids, *J. Opt. Soc. Am. A*, 25, 2767–2775, 2008.
118. A. Taflove and S. C. Hagness, *Computational Electrodynamics: The Finite-Difference Time Domain Method*, 2nd edn., Artech, Boston, MA, 2000.
119. P. Nordlander, C. Oubre, E. Prodan, K. Li, and M. I. Stockman, Plasmon hybridization in nanoparticle dimers, *Nano Lett.*, 4, 899–903, 2005.
120. F. Hao, C. L. Nehl, J. H. Hafner, and P. Nordlander, Plasmon resonances of a gold nanostar, *Nano Lett.*, 7, 729–732, 2007.
121. C. Oubre and P. Nordlander, Finite-difference time-domain studies of the optical properties of nanoshell dimers, *J. Phys. Chem. B*, 109, 10042–10051, 2005.

122. A. Dhawan, M. Canva, and T. Vo-Dinh, Narrow groove plasmonic nano-gratings for surface plasmon resonance sensing, *Opt. Express*, 19, 787–813, 2011.

123. A. Dhawan, A. Duval, M. Nakkach, G. Barbillon, J. Moreau, M. Canva, and T. Vo-Dinh, Development of surface plasmon resonance imaging (SPRI) sensor chips based on gold nano- and micro- structures, *Nanotechnology*, 22, 165301, 2011.

124. M. Totzeck, W. Ulrich, A. Göhnermeier, and W. Kaiser, Semiconductor fabrication: Pushing deep ultra-violet lithography to its limits, *Nat. Photon.*, 1, 629–631, 2007.

125. S. Graells, R. Alcubilla, G. Badenes, and R. Quidant, Growth of plasmonic gold nanostructures by electron beam induced deposition, *Appl. Phys. Lett.*, 91, 121112, 2007.

126. A. Dhawan, M. D. Gerhold, A. Madison, J. Fowlkes, P. E. Russell, T. Vo-Dinh, and D. N. Leonard, Fabrication of metallic nanodot structures using focused ion beam (FIB) milling and electron beam-induced deposition for plasmonic waveguides, *Scanning*, 31, 1–8, 2009.

127. L. Malic, B. Cui, T. Veres, and M. Tabrizian, Enhanced surface plasmon resonance imaging detection of DNA hybridization on periodic gold nanoposts, *Opt. Lett.*, 32(21), 3092–3094, 2007.

128. K. M. Byun, M. L. Shuler, S. J. Kim, S. J. Yoon, and D. Kim, Sensitivity enhancement of surface plasmon resonance imaging using periodic metallic nanowires, *J. Lightwave Technol.*, 26, 1472–1478, 2008.

129. C. J. Alleyne, A. G. Kirk, R. C. McPhedran, N-A. P. Nicorovici, and D. Maystre, Enhanced SPR sensitivity using periodic metallic structures, *Opt. Express*, 15(13) 8163–8169, 2007.

130. L. S. Live and J.-F. Masson, High sensitivity of plasmonic microstructures near the transition from short-range to propagating surface plasmon, *J. Phys. Chem. C*, 113, 10052–10060, 2009.

131. H.-N. Wang and T. Vo-Dinh, Multiplex detection of breast cancer biomarkers using plasmonic molecular sentinel nanoprobes, *Nanotechnology*, 20, 065101-1–065101-6, 2009.

132. M. Wabuyele and T. Vo-Dinh, Detection of HIV Type 1 DNA sequence using plasmonics nanoprobes, *Anal. Chem.*, 77, 7810–7815, 2005.

133. M. K. Gregas, F. Yan, J. Scaffidi, H.-N. Wang, T. Vo-Dinh, Characterization of nanoprobe uptake in single cells: Spatial and temporal tracking via SERS labeling and modulation of surface charge, *Nanomed. Nanotechnol. Biol. Med.*, 7, 115–122, 2011.

16

Overcoming Mass-Transport Limitations with Optofluidic Plasmonic Biosensors and Particle Trapping

Hatice Altug, Ahmet Ali Yanik, Alp Artar, Arif Engin Cetin, Min Huang, and John H. Connor

CONTENTS

16.1 Introduction

Miniaturized label-free biosensors, providing an alternate route for diagnostics of infectious diseases and cancer markers, offer tremendous opportunities for point-of-care diagnostics and global health [1–3]. These platforms do not require labeling (unlike fluorescence and radioactive techniques) and circumvent the need for enzymatic detection. In general, they provide simple, compact, and low-cost sensing capabilities. Apart from some general similarities, label-free biosensors are based on a variety of signal transduction mechanisms utilizing optical [1,3,4], electrical [5,6], and mechanical [7,8] detection techniques. Among these platforms, optical biosensors are particularly promising, since they allow remote detection without any need for an actual physical contact to the sensing volume [9,10]. As a result, they are compatible with physiological solutions and insensitive to the variations in ionic strengths of the samples [11,12]. However, an important limiting factor in the utilization of optical biosensors is the need for a precise coupling of the excitation light source to the sensing volume. As a result, these systems are not readily compatible with point-of-care applications. Nanoplasmonic biosensors, on the other hand, overcome such stringent alignment requirements by enabling direct coupling of a perpendicularly incident light to the sensing volume while allowing massive multiplexing capabilities enabled by miniaturization [1,3,13].

Despite the recent breakthroughs, significant challenges remain in the transition of label-fee biosensing technologies from research laboratories to clinical and on-field settings. One of the fundamental limitations, inherent to many biochemical reactions (catalytic, enzymatic, protein–protein, etc.), is the random diffusion of reactants to each other. This limitation, known as the mass-transport problem, is of particular concern when surface-sensitive nanosensors are considered [14–16]. Recent studies have demonstrated that the randomized nature of biomolecular transport could lead to impractically long detection

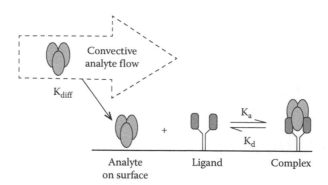

FIGURE 16.1 Accumulation of target molecules/bioparticles on a sensor surface is not only controlled by the affinities of these targets to the surface immobilized ligands but also by the availability of targets to the surface capturing agents. Accordingly, accumulation of target molecules/bioparticles on the sensing surface is limited by the diffusion process at low concentrations of analytes.

times from days to weeks when samples with medically relevant low concentrations of target molecules are tested [16,17]. Accordingly, performances of surface biosensors are often restricted by the inefficient biomolecular diffusive transport instead of intrinsic detection limits of the sensors.

As illustrated in Figure 16.1, the biomolecular bindings or accumulation of analytes on the surface-immobilized ligands are controlled not only by the affinities of the biomolecules but also by the availability of the analytes to the sensing surface. For medically low concentration of analytes, diffusion rates can take control over the rate of binding processes or analyte accumulations. Within the last decade, extensive effort has been made to overcome mass-transport limitations for the direct detection of biomolecules as quickly as the chemical reaction kinetics allow. Mainly microfluidic approaches are proposed, offering only moderate improvements [18,19]. One of the major conceptual constraints in previous studies was that the microfluidics and biosensing are always considered to be different parts of a sensor platform completing each other, but not a fully merged single modality. In this chapter, we focus on nanoplasmonic biosensing platforms enabling targeted delivery and trapping of bioparticles into the desired optically active detection volumes to overcome mass-transport limitations [20–22]. In the following, we show that these approaches can lead to dramatically improved sensor response. Our platforms are based on periodically nanopatterned surfaces bringing complementary characteristics of localized and extended surface plasmons, electromagnetic waves trapped at the metal/dielectric interfaces [23]. Our discussion on overcoming mass-transport limitations is organized into three different approaches: (i) targeted nanofluidic, (ii) structural, and (iii) optical trapping.

16.2 Nanoplasmonics and Label-Free Biosensing

In this section, we briefly discuss surface plasmons and working principles of nanoplasmonic biosensing devices. An illustration of a plasmonic biosensor for label-free detection of intact viruses is also presented.

16.2.1 Localized and Extended Surface Plasmons

Surface plasmon resonances are strongly sensitive to the refractive index changes within their near vicinity [24]. Unlike techniques based on external labeling, such effective refractive index-dependent resonance behavior operates as a reporter of biomolecular binding phenomena in a label-free fashion. Surface plasmon excitations can be grouped into (i) *surface plasmon polaritons* (SPPs), (ii) *localized surface plasmons* (LSPs), or (iii) a mixture of both [23].

SPPs are surface electromagnetic waves associated with electron oscillations confined to the surface of a continuous metal film. SPPs can freely propagate at the metal/dielectric interface albeit with losses [25]. However, direct excitation of the SPPs on a continuous metal film surface is not a straightforward process.

Due to the momentum–energy mismatch between the electromagnetic waves in the free-space and the surface-confined SPPs, photonic excitations cannot be converted to surface plasmon excitation in a readily manner [25]. Different excitation techniques utilizing *prism* and *grating coupling* mechanisms can be utilized to overcome this limitation. In *prism coupling* configuration, a thin layer of metallic film is employed where the light coupling is achieved by using a high-refractive-index prism. When the incident light is impinged at a non-perpendicular angle from the prism-coupled metallic surface, SPPs can be created at the metallic interface with lower refractive index by evanescent coupling of light through the thin metallic film. These systems provide very sensitive label-free biosensors that operate in real-time and have been the workhorse for the biomedical and pharmacological research for decades [4,24]. However, strict alignment requirements related to the angular incidence and angular collection of the reflected light impose serious limitations on practical applications of this scheme in point-of-care settings. Similarly, these platforms offer limited multiplexing capabilities due to the difficulty of integrating high numerical aperture (NA) imaging systems to the angular coupling configuration [24]. On the other hand, techniques based on *grating coupling* of incident light to SPPs enable direct excitation with normally incident light, and provide a much more practical platform suitable for point-of-care diagnostics. In this configuration, grating can consist of a periodic pattern of metallic or dielectric surface corrugations on a metallic film. Here, the periodicity of the structure provides the necessary in-plane momentum enabling conversion of incident photons directly to the SPPs. Assuming a two-dimensional (2-D) grating with same periodicities for both directions (a_0), the resonance condition for the spectral positions where the external light coupling to SPPs is given [25,26] by

$$\lambda_{res} \approx \frac{a_0}{\sqrt{i^2 + j^2}} \sqrt{\frac{\varepsilon_m \varepsilon_d}{\varepsilon_m + \varepsilon_d}} \qquad (16.1)$$

Here, (i, j) denotes the grating order and ε_m, ε_d are the dielectric constants for the metal film and the dielectric layer, respectively.

Unlike SPPs, LSPs are localized surface plasmon excitation on metallic nanoparticles of a finite nanoscale volume [23,27]. These excitations are associated to the coherent oscillation of the charge density of the nanoparticle with the incident electromagnetic wave. LSPs can be directly created with the incident electromagnetic field. The resonance spectra of these excitations are controlled by the structural properties of the metallic particle (i.e., shape, material, dielectric environment, etc.).

Depending on the spectral and spatial overlap of the SPPs and LSPs, hybrid excitations consisting of a mixture of both can also be created. Furthermore, SPPs and LSPs can be utilized to excite each other and enable reciprocal conversion from one form of excitation to another [28,29]. Hybrid nature of plasmonic excitations is observed in many nanoplasmonic phenomena. For example, when the periodic corrugations on an optically thick metal film are in the form of sub-wavelength nano-apertures, a novel phenomenon occurs. As it was shown in the pioneering work by Ebbesen et al., the transmission of light through nano-apertures can be highly efficient at resonances corresponding to excitation of the SPPs even for wavelengths longer than the dimensions of the nanohole [26]. This unexpected behavior is contrary to Bethe's predictions that the light transmission should decay with the third power of the ratio of aperture radius to wavelength of the incident light [30]. This phenomenon is associated to the excitation of the SPPs and their reciprocal coupling to LSPs. As SPPs propagate along the incidence surface, they couple to the out-coupling surface through the nanohole openings [29]. Since the wavelengths of surface plasmons are much shorter than the wavelength of same energy photons, this coupling could be very efficient. Surface plasmons are then converted back to photons on the out-coupling surface and reradiate in the same direction as the incident beam. In fact, orders of magnitude stronger transmissions than the predictions based on classical diffraction theory are experimentally demonstrated [22,26,31,32]. This phenomenon named as extraordinary optical transmission (EOT) is highly advantageous for label-free biosensing applications.

16.2.2 Nanoplasmonic Biosensors for Label-Free Sensing

Exponential decay of the plasmonic excitations from the surface of the metallic surface results in sub-wavelength confinement of the electromagnetic field to the metal/dielectric interface [4]. This decaying

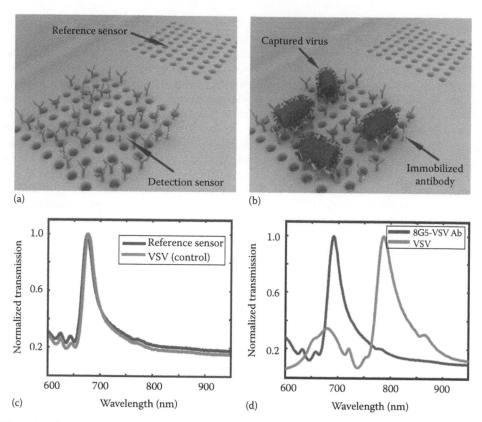

FIGURE 16.2 (See color insert.) Renderings (not drawn to scale) illustrate (a) before and (b) after incubation of viruses on a reference and a detection sensor with antibodies functionalized on it. (c) Minimal red shifting in resonance frequencies is observed for reference sensors after the VSV incubation, and washing. (d) As a result of accumulation of the VSV, increasing effective refractive index results in a strong red-shifting of the plasmonic resonances (~100 nm). (From Yanik, A.A. et al., *Nano Lett.*, 10(12), pp. 4962–4969, 2010.)

behavior has two important consequences for label-free biosensing applications: (i) Plasmonic resonances are highly sensitive to the refractive index changes in the near vicinity of the metal dielectric interface such as accumulation processes due to the binding of the biomolecules and pathogens. (ii) The effects of the uncontrollable variations in bulk medium (such as temperature fluctuations) are minimal, enabling high signal-to-noise measurements. An application of a plasmonic biosensor for the selective detection of vesicular stomatitis virus (VSV) at a concentration of 10^9 PFU/mL by using virus-specific antibodies immobilized on metallic surface [33] is illustrated in Figure 16.2. Here, the transmission light spectra are collected from a sub-wavelength nanohole array of 90 μm × 90 μm, with a periodicity of 600 nm and an aperture diameter of 220 nm in a set of experimental end-point measurements. The extraordinary light transmission resonances are observed at specific wavelengths corresponding to the excitation of the surface plasmons at a given periodicity by Equation 16.1. As intact viruses bind to the antibody-functionalized surface, the effective refractive index at close vicinity of the metal surface increases and red-shifting of the resonances occurs. The dependence of the resonant wavelength to the immediate environment of the sensor system can be directly observed from Equation 16.1 where the binding of the pathogens causes medium dielectric constant ε_d to increase. In Figure 16.2c and d, spectra are presented for before (blue curve) and after (red curve) the incubation of the virus-containing sample on the plasmonic nanohole sensors [33]. The strong transmission resonance observed at 690 nm (blue curve) with 25 nm full width at half-maximum is a result of the extraordinary light transmission through the optically thick gold film. This transmission resonance (blue curve) corresponds to the excitation of the (1,0) grating order SPP mode at the metal/dielectric interface of the antibody-immobilized detection sensor. After the incubation process (enabling the diffusive delivery of analytes) with the virus-containing sample, a strong red-shifting

(~100 nm) of the plasmonic resonance peak (red curve) is observed as a result of the accumulated biomass on the functionalized sensing surface. As shown in Figure 16.2c, for the unfunctionalized control sensors, a negligible red-shifting (~1 nm) of the resonances is observed (blue vs. red curves), possibly due to the nonspecific binding events. These proof-of-concept experimental measurements clearly demonstrate that the plasmonic surface biosensors are highly promising candidates for the specific detection of intact viruses in a label-free fashion. Circumventing the requirements for angular excitation of SPPs is also a crucial step toward achieving point-of-care diagnostics in a multiplexing format.

16.3 Overcoming Mass-Transport Limitations

As we pointed out in the introduction, performances of surface biosensors are often limited in a fluidic environment by the inefficient mass transport instead of the intrinsic detection capabilities of the biosensors. In this section, we are going to discuss different approaches to overcome these mass-transport limitations.

16.3.1 Nanofluidic Targeted Delivery of Analytes

Nanohole plasmonic devices, resonantly transmitting light through the EOT effect, are intrinsically suitable for targeted delivery of analytes. When the metallic nanohole pattern is defined through a suspended dielectric film with matching nanohole openings, these apertures can be also used as nanofluidic channels by connecting two separate chambers on the each side of the nanohole arrays [20]. This configuration uniquely offers an extra degree of freedom in microfluidic design by connecting separate layers of microfluidic circuits through biosensors. In the following, we show 14-fold improvement in mass-transport constant that appear in the exponential terms. To fabricate these structures, we introduced a lift-off free plasmonic device fabrication approach that eliminates the need for ion milling and lift-off processes and allows fabrication of high optical-quality plasmonic devices on suspended membranes.

Lift-off free fabrication scheme, summarized in Figure 16.3a through f, consists of three stages: (i) fabrication of the free-standing membrane, (ii) patterning on the membrane, and (iii) direct deposition of metallic plasmonic devices [20,34].

Free-standing membranes: An important consideration here is the mechanical strength of the membranes. As a result, highly robust low pressure chemical vapor deposition (LPCVD) SiN$_x$ films are used. Starting with a silicon wafer coated with LPCVD grown SiN$_x$ on double sides, we define etching windows on the back surface SiN$_x$ layer using photolithography and a reactive ion etching (RIE) process. Later, the chips are immersed in KOH solution to create free-standing membranes as the etching stops at the SiN$_x$ layer.

Patterning of the membrane: E-beam lithography is performed to define nanohole openings on the top surface after a positive e-beam resist spin coating on the suspended structure (Figure 16.3a). Nanohole pattern (with hole diameters of 220 nm and a periodicity of 600 nm) is transferred to the suspended SiN$_x$ film through a dry etching process (Figure 16.3c). E-beam resist is later removed with an oxygen plasma cleaning process, leaving only a patterned SiN$_x$ film with air on both sides (Figure 16.3c).

Direct deposition of plasmonic nanoholes: A directional e-beam evaporator is used to deposit Ti (5 nm) and Au (125 nm) metal layers, defining the suspended plasmonic sensors with nanohole openings (Figure 16.3d). This deposition process is extremely reliable and enables fabrication of suspended metallic nanohole arrays over large areas with high yield. Minimal clogging of the hole is observed, resulting in a small shrinking in nanohole diameter (<4%) after gold deposition due to a slight coverage of the nanohole sidewalls (Figure 16.3e and f).

These suspended nanoholes are then mounted in a custom-designed multilayered microfluidic channel system based on poly-(dimethylsiloxane) [20]. The plasmonic sensor layer divides the microfluidic system into two chambers (upper/lower chamber as illustrated in Figure 16.4).

There are multiple inlets and outlets to the fluidic system, enabling active manipulation of the analyte flow. Convective flow corresponding to conventional microfluidic approaches is realized by running

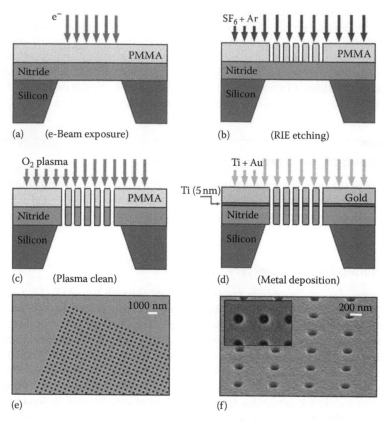

FIGURE 16.3 Lift-off free fabrication of suspended plasmonic nanohole arrays is illustrated (a–d). Scanning electron microscopy images are shown with high-quality structures (e, f). (From Yanik, A.A. et al., *Nano Lett.*, 10(12), pp. 4962–4969, 2010.)

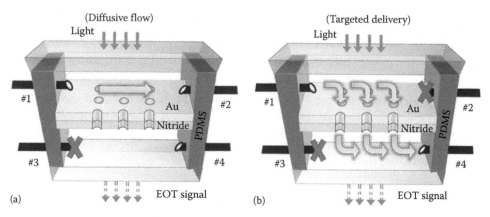

FIGURE 16.4 3-D control of fluidic platform is achieved by using multiple inlet and outlets. (a) For the conventional fluidic scheme, flow is maintained in a single chamber by using inlets and outlets on the same side of the suspended device. (b) Targeted delivery of analytes is achieved by allowing flow only through one of the inlet/outlets of the top/bottom channels. (From Yanik, A.A. et al., *Appl. Phys. Lett.*, 96(2), 021101, 2010.)

the analyte solutions in between input and output lines defined in the same chamber on either side of the nanohole sensors (Figure 16.4a). The actively controlled (targeted) delivery scheme is realized by steering the convective flow perpendicularly to the plasmonic sensing surface by allowing the flow only through one inlet/outlet on opposite sides of the suspended sensor system (Figure 16.4b).

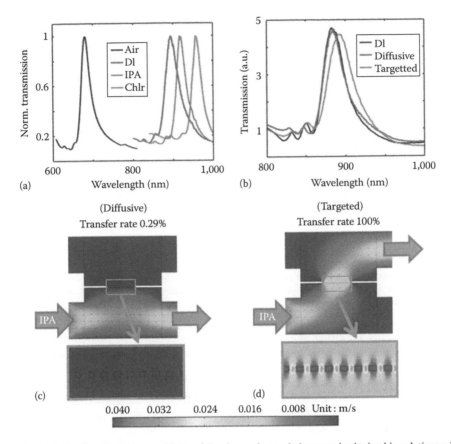

FIGURE 16.5 (a) Bulk refractive index sensitivity of the plasmonic nanohole arrays is obtained in solutions with changing refractive index. (b) Resonance shifts for the passively and actively controlled mass-transport schemes are compared after running analyte solution for 10 min at 20 μL/min flow rate. Microfluidic simulations demonstrate that (c) transfer rates for the passive transport scheme due to the weaker perpendicular flow of the analytes, while (d) much more efficient mass transport toward the surface is observed for the targeted delivery scheme. (From Yanik, A.A. et al., *Appl. Phys. Lett.*, 96(2), 021101, 2010.)

Targeted delivery of analytes to the sensing surface is shown in Figure 16.5a. Initially, a low refractive index liquid (DI water with $n_{DI} = 1.333$) is filled in both chambers. This causes a spectral shift of the EOT resonance from $\lambda_{air} = 679$ nm to $\lambda_{DI} = 889$ nm, which corresponds to a bulk refractive index sensitivity of $\Delta\lambda/\Delta n = 630$ nm/RIU. To quantify the analyte transport efficiency of the both delivery schemes, a lower viscosity analyte solution (IPA) with higher refractive index is introduced from the bottom inlet. The plasmonic sensor measures only the refractive index change due to the perpendicularly diffused or actively delivered IPA solution, depending on the implemented flow scheme. This configuration, reflecting the perfect collection case, provides a good way of quantifying the transport limit by separating binding kinetics and mass-transport phenomena. For the conventional fluidic scheme (diffusive transport), IPA solution is pumped into the bottom channel (20 μL/min) and collected from the bottom side while the top outlet is kept open. A small red-shifting of the plasmonic resonance (1.2 nm, blue curve in Figure 16.5b) is observed after running the IPA solution for 10 min.

This small resonance shift demonstrates that the analyte accumulation due to the diffusive delivery of the IPA is very weak. These observations are also validated using microfluidic simulations (Figure 16.5c and d) based on finite-element method and incompressible Navier–Stokes equations. To quantify the performance of the delivery scheme, *transfer rate* is defined as the ratio of the perpendicular flow rate to the inlet flow rate. For the targeted delivery of the convective current to the surface, analyte solution is directed from down-to-top direction by flowing the analyte solution

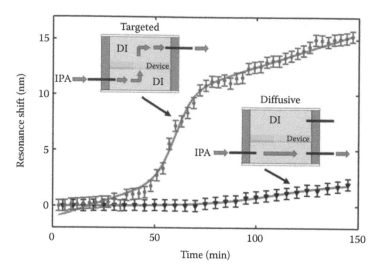

FIGURE 16.6   Real-time measurements are performed to compare the efficiencies of the passive (triangles) and targeted (squares) delivery of the analytes. Fourteen-fold improvement in mass-transport rate constant is shown using the targeted delivery scheme. (From Yanik, A.A. et al., *Appl. Phys. Lett.*, 96(2), 021101, 2010.)

through an inlet/outlet on either side of the suspended structure. In this configuration, a larger red-shifting of the plasmonic resonance is observed (10 nm), corresponding to much more efficient analyte delivery.

Time-dependent spectral measurements are also performed to characterize the efficiency of the analyte delivery to the sensing surface for both diffusive and targeted delivery methods. Experimentally observed resonance shifts are (least squares) fitted to a sigmoid function of form $A_b + (A_t - A_b)/(1 + e^{-k(t-t_0)})$ as shown in Figure 16.6.

The mass-transport rate constants appearing in the exponents are obtained as $k_{diff} = 0.0158\,min^{-1}$ and $k_{targ} = 0.2193\,min^{-1}$ for the diffusive and targeted transport schemes, respectively. This corresponds to more than 14-fold improvement in rate constants, which is crucial for enhancing the performance in immunoassay-based applications. After more than 10 h of operation at compatible flow rates for bio-sensing applications (4 µL/min), we observed that the membranes and the plasmonic devices are intact. Similarly, no structural deformations are observed in the spectral measurements, showing the reliability of the devices.

16.3.2 Optical Trapping in Plasmonic Nanopillar Antennas

Instead of relying on the geometrical properties of the structure for trapping biological particles, an active trapping scheme can be employed. Optical trapping allows accurate manipulation of nanoscale particles through controlling the mechanical force exerted on them by an incident electromagnetic field [35]. In conventional optical tweezers, high laser powers are needed for a sufficient control over the position of a small particle preventing Brownian motion. However, low-power optical tweezers with minute trapping volumes well below the diffraction-limited volume of light can be realized with nanostructured plasmonic systems [36,37].

A periodic array of nanopillars that are formed on a continuous metal sheet allows enhanced device performances for biosensing, nanospectroscopy, and optical trapping applications simultaneously. High-refractive-index sensitivities (~675 nm/RIU) can be achieved with such a structure. In this configuration, plasmonic hot spots, created at the tips of the nanopillar structures, are easily accessible to the analytes. This latter feature is important for optical trapping of nanoparticles with low-power excitation sources. Polarization-dependent optical force gradient enabled by the nanopillar arrays also provides the ability to control the direction of the trapping force.

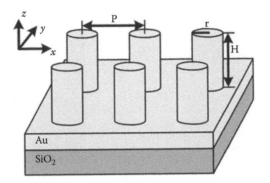

FIGURE 16.7 An illustration of the nanopillar arrays is shown where, H and r denote the height and the radius of the nanopillars, respectively. P is the periodicity of the nanopillar array and is symmetric in x and y directions. (From Çetin, A.E. et al., *Appl. Phys. Lett.*, 98, 111110, 2011.)

Nanopillar array structure is illustrated in Figure 16.7.

The gold nanopillars stand on a gold metal sheet with a thickness of 150 nm, which is placed on a SiO_2 substrate and the whole structure is immersed in DI water. Figure 16.8a shows the strong role of the localized and the propagating SPP couplings in periodic nanopillar array response. In periodic arrays,

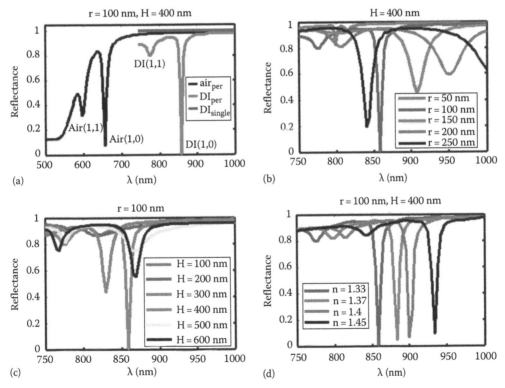

FIGURE 16.8 (a) Reflectance from free-standing nanopillar structures ($r = 100$ nm, $H = 400$ nm, and $P = 600$ nm) consisting of single and periodic nanopillar systems is compared in DI water. The SPP modes corresponding to different grating orders for the periodic structures are indicated in the figure. (b) Spectrum is shown for nanopillar arrays for varying r values at a fixed H (=400 nm) and P (=600 nm) in DI water. (c) Spectrum is shown for nanopillar arrays with different H values at a fixed r (=100 nm) and P (=600 nm) in DI water. (d) Resonance shift is observed for nanopillar arrays (with device parameter $r = 100$ nm, $H = 400$ nm, and $P = 600$ nm) with changing refractive index of the bulk medium. (From Çetin, A.E. et al., *Appl. Phys. Lett.*, 98, 111110, 2011.)

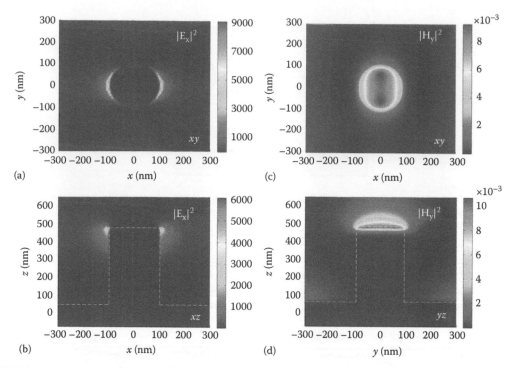

FIGURE 16.9 Electric and magnetic field intensities of a nanopillar structure in a periodic array are shown for the x-component of E-field (a-b) and the y-component of H-field (c-d) for different cross sections. Nanopillars are assumed to be immersed in water. The device parameters are r = 100 nm, H = 400 nm, and P = 600 nm. (xy) cross section for the electric field is obtained at the top surface of the nanopillar. The position of the rod is indicated by white dashed line for (xz) and (yz) cross sections. (From Çetin, A.E. et al., *Appl. Phys. Lett.*, 98, 111110, 2011.)

two resonance dips in the reflection spectra are observed, corresponding to the excitation of the SPPs by different grating orders (Equation 16.1). Relatively sharper and stronger resonance dip observed at longer wavelengths is due to the excitation of the LSP modes of the nanopillar structure, driven by the first grating order in the Au/DI interface. The weaker resonance dip is due to the excitation of the higher grating order mode (1,1). For a single nanopillar structure, resonant modes are not observable due to the lack of a grating coupling mechanism. This observation proves the contribution of SPPs in the excitation of strongly localized plasmon resonances in the structure. Figure 16.8b shows the radius dependence of the resonant mode for a nanopillar array of fixed height (H = 400 nm) in DI water (n = 1.333). Small radii results in a weak resonant behavior because of lower polarizabilities of the nanopillars. As the radius is increased, the larger polarizability of the structure enables a stronger coupling of the grating-coupled SPPs to the LSPs. An optimum resonant behavior is obtained with a sharp spectral feature for r = 100 nm.

Increasing the height of the structure is another way of optimizing the coupling between the SPPs and the LSPs as shown in Figure 16.8c.

An optimum height for a structure with r = 100 nm is H = 400 nm. Figure 16.8d shows spectral shifting of the resonant modes as the bulk refractive index changes. Refractive index sensitivity of 675 nm/RIU in good agreement with finite-difference time-domain (FDTD) simulations is obtained.

Figure 16.9 shows the near-field intensity distribution of the E_x and H_y field components for localized resonances of the optimized array excited at the fundamental grating order.

Plasmonic excitations lead to dipolar E_x distribution around the rims of the nanopillar structure at the top surface, with intensity enhancement factors close to four order of magnitude (Figure 16.9a and b). Strong localization of the electromagnetic field is demonstrated in the xz-cross-sectional field profile (Figure 16.9b). As shown in Figures 16.9c and d, near-field profile of the magnetic field component also

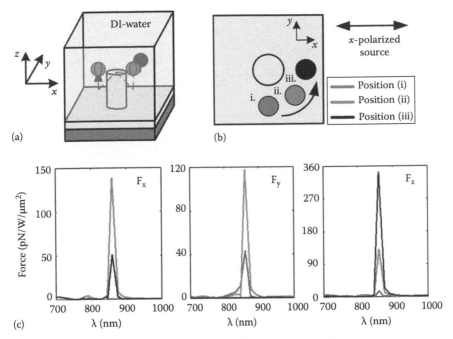

FIGURE 16.10 (a) 3-D-schematic view of the nanopillar structure with three different bead locations. The medium is assumed to be DI-water. (b) 2-D cross section of the bead locations are shown. (c) x-, y-, and z-components of the optical gradient force for three beads are calculated using Maxwell stress tensor. The device parameters for the nanopillar array are r = 100 nm, H = 400 nm, and P = 600 nm. (From Çetin, A.E. et al., *Appl. Phys. Lett.*, 98, 111110, 2011.)

demonstrates strong near-field enhancements on the top surface of the pillar structure extending into the dielectric medium above. Such near-field intensity enhancements strongly overlapping with the dielectric medium are ideal for optical trapping applications.

Optical forces can be calculated numerically by evaluating the Maxwell stress tensor over a surface surrounding the nanoparticle.

Three-dimensional (3-D) and 2-D illustrations are presented in Figure 16.10a and b for three different locations of a dielectric spherical bead of radius 50 nm, where the optical forces are calculated in the presence of the near-field gradient of the nanopillar structure. The bead is positioned 200 nm above the top surface pillar structure. In Figure 16.10c, different components of the optical forces acting on the bead are compared for different locations of the bead. Due to the dipolar mode characteristics of the LSPs of the nanopillar, the near-field gradients are symmetric in the x-direction when illuminated with an x-polarized light, hence F_x drops to zero for bead at *position (i)* (Figure 16.10b). Similarly F_y drops to zero for bead at *position (iii)*. For the diagonal *position (ii)*, both F_x and F_y exist and the stronger F_x component pulls the bead toward the hotspots where the highest near-field gradient is expected. The strongest optical force (F_z = 350 pN/W/μm²) is obtained once the bead is located at *position (iii)*. The ability to control the locations of the hotspots with the polarization of the incident light source also allows tunability of the optical gradient force. The observed strong optical forces with polarization control can be utilized for optical trapping applications. Since the system can be simultaneously used for sensing applications at the same time, it's possible to enhance the sensing properties by utilizing the optical trapping feature.

16.3.3 Structural Trapping of Pathogens

An important advantage of using patterned plasmonic structures for biosensing systems is their potential for structural trapping of certain biological particles, such as pathogens. For structural trapping of bioparticles, geometries that extend to the third dimension are necessary.

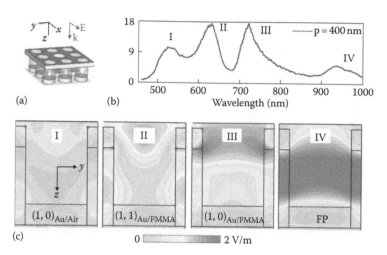

FIGURE 16.11 (a) Schematics of the multilayered plasmonic structure are shown. Here the blue arrows indicate the incident light polarization and direction. (b) The absolute transmission spectra of the structures for crystals with diameter (d) 300 nm and periodicity (p) 400 are obtained experimentally. (c) Cross section electric field patterns for the modes are calculated using FDTD simulations. (From Artar, A. et al., *Appl. Phys. Lett.*, 95(5), 051105, 2009.)

A multilayered system that is formed by the coupling of two physically separated metal nano-apertures and an array of nanoparticle layers can be used (Figure 16.11a) [22]. These structures have nanoscale openings due to the nano-apertures that extend down to the nanoparticles and provide a natural structural trapping geometry. Zero-order transmission spectra obtained at normal incidence exhibit strong wavelength dependence as a result of complex interplay of diffraction, interference, and plasmonic resonances. Figure 16.11b shows the transmission spectra of an array with 300 nm aperture diameter and 400 nm of periodicity. Absolute transmissions are calculated by normalizing the transmitted light intensity with the incident light. Signal strengths of the observed peaks are higher than the predicted values by Bethe's theory, confirming the extraordinary nature of the light transmission phenomena. The spectral positions of certain transmission resonances depend on the lattice constant as in Equation 16.1, corresponding to excitation of SPPs.

One of the observed resonances in the transmission spectra (peak IV in Figure 16.11b) cannot be explained with any of the SPP excitations due to different grating orders. According to our simulation results (Figure 16.11c), the upper surface of the gold nanoparticles in the bottom interface and the bottom surfaces of the nano-apertures on the top interface create a resonant Fabry–Perot (FP) cavity with strong field confinement. The field profile of the FP mode is coupled in the lateral direction and well confined in the z-direction with strong field concentration in the dielectric region.

As shown in Figure 16.12, FP nanocavity resonance is both controlled by the physical separation between the nanostructured layers and by the dimension and the periodicity of the apertures. This behavior can be explained using a simple cavity model, where the governing formula for the spectral locations of the FP resonance is given by $\lambda_{FP} = 2dn_{eff}$. Here n_{eff} is the effective refractive index of the nanocavity and d is the separation between the nano-apertures and the nanoparticles. The field is confined approximately equally in the air and the dielectric layer (mode IV in Figure 16.11c). As a result, the effective refractive index parameter can be incorporated as a weighted average of the refractive indices of the air and the dielectric in proportion with their volumes in a unit cell. Transmission spectra shown in Figure 16.12 support this model.

As it is discussed before, this 3-D plasmonic structure provides a natural trapping mechanism with its nanoscale opening. The FP mode is shown to have a huge field overlap with this opening and can be utilized as a probe for sensing the trapping of a biological particle inside that region. On the other hand, cross-sectional field distributions at EOT resonances show less field overlap inside the opening region (Figure 16.11c, modes I–II–III). Figure 16.12d shows the spectral shifts in the FP (red line) and the EOT (blue line) resonances as a function of the background refractive index of the medium. According to our FDTD simulations, FP resonance is nearly three times more sensitive to the refractive index changes

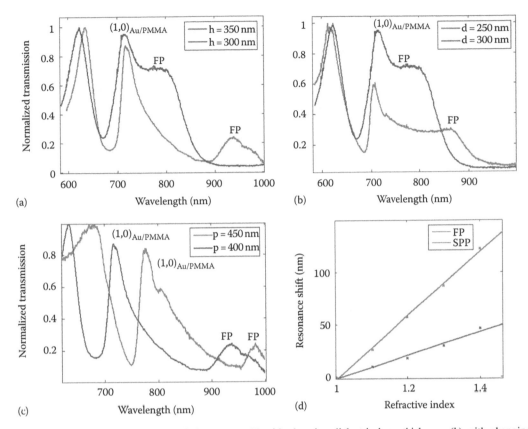

FIGURE 16.12 Comparison of transmission spectra (a) with changing dielectric layer thickness, (b) with changing diameter, and (c) with changing periodicity. (d) Shifts in spectral position with changing RI for FP peak (red) and EOT peak I (blue). (From Artar, A. et al., *Appl. Phys. Lett.*, 95(5), 051105, 2009.)

than the EOT resonances. Field confinement in the nanowell region is also highly advantageous for sensing applications since the nanowells can also serve for structural trapping of pathogens.

16.4 Conclusion

In conclusion, we have shown that patterned nanoplasmonic biosensors offer unique opportunities for highly sensitive label-free and real-time detection of biomolecules and pathogens by overcoming mass-transport limitations. With the utilization of periodic patterns of nanostructures, direct coupling of incident light to surface excitations can be achieved, which circumvents the need for precise alignment of the system components and enables massive multiplexing capabilities in a readily manner with small detection volumes. For point-of-care applications, we show that plasmonic sensing platforms can be merged with the nanofluidics, resulting in huge improvements in the analyte delivery rates. Also, optical and structural trapping of bioparticles is offered in periodically patterned plasmonic structures, taking advantage of complementary characteristics of localized and extended surface plasmons.

Acknowledgments

We gratefully acknowledge support from NSF CAREER Award (ECCS-0954790), ONR Young Investigator Award, Massachusetts Life Science Center New Investigator Award, NSF Engineering Research Center on Smart Lighting (EEC-0812056), Boston University Photonics Center, and Army Research Laboratory.

REFERENCES

1. Anker, J.N. et al., Biosensing with plasmonic nanosensors. *Nature Materials*, 2008. **7**(6): 442–453.
2. Homola, J., Present and future of surface plasmon resonance biosensors. *Analytical and Bioanalytical Chemistry*, 2003. **377**(3): 528–539.
3. Lal, S., S. Link, and N.J. Halas, Nano-optics from sensing to waveguiding. *Nature Photonics*, 2007. **1**(11): 641–648.
4. Homola, J., Surface plasmon resonance sensors for detection of chemical and biological species. *Chemical Reviews*, 2008. **108**(2): 462–493.
5. Cui, Y. et al., Nanowire nanosensors for highly sensitive and selective detection of biological and chemical species. *Science*, 2001. **293**(5533): 1289–1292.
6. Bunimovich, Y.L. et al., Quantitative real-time measurements of DNA hybridization with alkylated non-oxidized silicon nanowires in electrolyte solution. *Journal of the American Chemical Society*, 2006. **128**(50): 16323–16331.
7. Savran, C.A. et al., Micromechanical detection of proteins using aptamer-based receptor molecules. *Analytical Chemistry*, 2004. **76**(11): 3194–3198.
8. Lee, J. et al., Real-time detection of airborne viruses on a mass-sensitive device. *Applied Physics Letters*, 2008. **93**: 013901.
9. Pineda, M.F. et al., Rapid specific and label-free detection of porcine rotavirus using photonic crystal biosensors. *Sensors Journal, IEEE*, 2009. **9**(4): 470–477.
10. Lee, J. et al., Diffractometric detection of proteins using microbead-based rolling circle amplification. *Analytical Chemistry*, 2009. **82**(1): 197–202.
11. Gupta, A.K. et al., Anomalous resonance in a nanomechanical biosensor. *Proceedings of the National Academy of Sciences*, 2006. **103**(36): 13362–13367.
12. Stern, E. et al., Importance of the Debye screening length on nanowire field effect transistor sensors. *Nano Letters*, 2007. **7**(11): 3405–3409.
13. Bishnoi, S.W. et al., All-optical nanoscale pH meter. *Nano Letters*, 2006. **6**(8): 1687–1692.
14. Squires, T.M., R.J. Messinger, and S.R. Manalis, Making it stick: Convection, reaction and diffusion in surface-based biosensors. *Nature Biotechnology*, 2008. **26**(4): 417–426.
15. Sheehan, P.E. and L.J. Whitman, Detection limits for nanoscale biosensors. *Nano Letters*, 2005. **5**(4): 803–807.
16. Nair, P.R. and M.A. Alam, Screening-limited response of nanobiosensors. *Nano Letters*, 2008. **8**(5): 1281–1285.
17. Nair, P.R. and M.A. Alam, Theoretical detection limits of magnetic biobarcode sensors and the phase space of nanobiosensing. *Analyst*, 2010. **135**(11): 2798–2801.
18. Yoon, S.K., G.W. Fichtl, and P.J.A. Kenis, Active control of the depletion boundary layers in microfluidic electrochemical reactors. *Lab on a Chip*, 2006. **6**(12): 1516–1524.
19. Vijayendran, R.A. et al., Evaluation of a three-dimensional micromixer in a surface-based biosensor. *Langmuir*, 2002. **19**(5): 1824–1828.
20. Yanik, A.A. et al., Integrated nanoplasmonic-nanofluidic biosensors with targeted delivery of analytes. *Applied Physics Letters*, 2010. **96**(2): 021101–021103.
21. Çetin, A.E. et al., Monopole antenna arrays for optical trapping, spectroscopy, and sensing. *Applied Physics Letters*, 2011. **98**: 111110.
22. Artar, A., A.A. Yanik, and H. Altug, Fabry—P[e-acute]rot nanocavities in multilayered plasmonic crystals for enhanced biosensing. *Applied Physics Letters*, 2009. **95**(5): 051105–051103.
23. Maier, S., *Plasmonics—Fundamentals and Applications*. 2007, New York: Springer, p. 245.
24. Phillips, K. and J. Homola (Eds.), Surface plasmon resonance-based sensors. *Analytical and Bioanalytical Chemistry*, 2008. **390**(5): 1221–1222.
25. Raether, H., *Surface Plasmons on Smooth and Rough Surfaces and on Gratings*. 1988, Berlin, Germany: Springer.
26. Ebbesen, T.W. et al., Extraordinary optical transmission through sub-wavelength hole arrays. *Nature*, 1998. **391**(6668): 667–669.
27. Link, S. and M.A. El-Sayed, Shape and size dependence of radiative, non-radiative and photothermal properties of gold nanocrystals. *International Reviews in Physical Chemistry*, 2000. **19**(3): 409–453.

28. Lalanne, P. et al., Surface plasmons of metallic surfaces perforated by nanohole arrays. *Journal of Optics A: Pure and Applied Optics*, 2005. **7**(8): 422.
29. Liu, H. and P. Lalanne, Microscopic theory of the extraordinary optical transmission. *Nature*, 2008. **452**(7188): 728–731.
30. Bethe, H.A., Theory of diffraction by small holes. *Physical Review*, 1944. **66**(7–8): 163.
31. Yanik, A.A. et al., Extraordinary midinfrared transmission of rectangular coaxial nanoaperture arrays. *Applied Physics Letters*, 2008. **93**(8): 081104–081103.
32. Lezec, H.J. et al., Beaming light from a subwavelength aperture. *Science*, 2002. **297**(5582): 820–822.
33. Yanik, A.A. et al., An optofluidic nanoplasmonic biosensor for direct detection of live viruses from biological media. *Nano Letters*, 2010. **10**: 4962–4969.
34. Yanik, A.A. et al., Seeing protein monolayers with naked eye through plasmonic Fano resonances. *Proceedings of the National Academy of Sciences*, 2011.**108**: 11784.
35. Ashkin, A. and J. Dziedzic, Optical trapping and manipulation of viruses and bacteria. *Science*, 1987. **235**(4795): 1517–1520.
36. Righini, M. et al., Surface plasmon optical tweezers: Tunable optical manipulation in the femtonewton range. *Physical Review Letters*, 2008. **100**(18): 186804.
37. Volpe, G. et al., Surface plasmon radiation forces. *Physical Review Letters*, 2006. **96**(23): 238101.

17

Optical Micro-Ring Resonators for Chemical Vapor Sensing

Yuze Sun and Xudong Fan

CONTENTS

Real-time chemical vapor detection and analysis have broad applications in environmental monitoring, homeland security, military surveillance, and biomedical diagnosis. As an emerging new sensing technology, optical micro-ring resonator has unique advantages in developing low-cost and portable vapor sensors. This chapter introduces ring resonator vapor sensing principles, describes representative micro-ring resonator vapor sensor designs and performances, and then focuses on one special micro-ring resonator design—the optofluidic ring resonator (OFRR), which inherently integrates the ring resonator vapor sensor and gas fluidics. The OFRR sensor has fast response to a wide spectrum of vapor analytes, with a detection limit of around 100 pg. Moreover, the OFRR is a versatile sensing platform that can be integrated with micro-gas chromatography (μGC) for rapid and in situ identification of organic chemical compounds (VOCs) in complex interfering vapor mixtures. Combination with a μGC system greatly enhances the vapor detection specificity, which is one of the major limitations in optical vapor sensors. Applications of the OFRR-μGC in detection of explosives and rapid tandem-column separation of VOCs are also presented. Future research/development directions in the ring resonator vapor sensor are discussed in the end.

17.1 Introduction

The optical ring resonator is an emerging sensing technology that has been under intense study in the past decades for applications in healthcare, biochemistry, homeland security, and environmental protection and monitoring. The optical ring resonator sensor was initially developed for detection of analytes in liquid solution. However, in the recent 5 years, ring resonators in gas detection have gradually drawn research attention, which represents a new frontier for the ring resonator sensing technology that has yet to be fully explored.

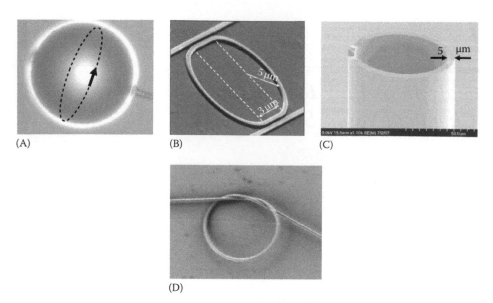

(A)   (B)   (C)

(D)

FIGURE 17.1 Various ring resonator configurations that can be used for gas sensing. (A) Dielectric microsphere. (B) Planar ring resonator. (C) Optofluidic ring resonator (OFRR) fabricated by the drawing method. (D) Micro/nano-fiber coil based ring resonator. (Reprinted from Yebo, N.A. et al., *Opt. Express*, 18, 11859, 2010; Tong, L. et al., *Nature*, 426, 816, 2003. With permission.)

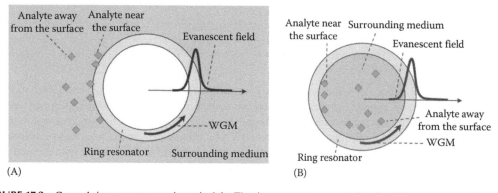

(A)   (B)

FIGURE 17.2 General ring resonator sensing principle. The ring resonator supports the circulating resonant mode called the whispering gallery mode (WGM), whose evanescent field interacts with the analyte near the sensor exterior (A) or interior (B) surface to produce the sensing transduction signal.

While implementations may be different for the ring resonator liquid and gas sensor, they share the similar ring resonator configurations such as dielectric microspheres/cylinders [1–3], ring-shaped wave-guides on substrate [4–8], thin-walled capillaries or OFRRs [9–15], and nano-fiber knots [16–18], as shown in Figure 17.1. Figure 17.2 illustrates the general sensing principle for the optical ring resonator sensor (the details of gas sensing mechanism will be described in Section 17.2). In a ring resonator, the circulating optical resonant mode, usually called the whispering gallery mode (WGM), forms as a result of total internal reflection of light at the curved boundary. The resonant wavelength, λ, is determined by [19]:

$$\lambda = \frac{2\pi r n_{eff}}{m} \tag{17.1}$$

where

r is the resonator radius

n_{eff} is the effective refractive index (RI) experienced by the WGM

m is an integer number

The resonant light travels along the ring resonator circumference and has the evanescent field of several hundred nanometers into the surrounding medium and interacts repetitively with the analytes near the resonator surface to provide quantitative and temporal information of the presence of the analytes.

As shown earlier, the optical ring resonator relies on the light–analyte interaction to produce the sensing transduction signal. In contrast to linear optical waveguide-based sensor where the light–analyte interaction length is essentially the physical length of the sensor, the circulating nature of the WGM leads to extremely long effective interaction length determined by [19,20]:

$$L_{eff} = \frac{Q\lambda}{(2\pi n_{eff})}, \tag{17.2}$$

where Q is the resonator quality factor (Q-factor), representing the number of round trips that the resonant light can circulate along the ring resonator. Depending on the ring resonator configurations, the Q-factor usually ranges from 10^4 to 10^8. The circulating nature of the resonant mode offers the ring resonator–based sensor an excellent detection limit while having miniaturized footprint.

17.2 Sensing Principles

There are a number of optical properties that can be employed to generate the sensing transduction signal, such as RI, optical absorption, fluorescence, and Raman scattering. RI detection is the most popular sensing mechanism used in gas detection, followed by optical absorption. We will discuss these two sensing principles as follows.

17.2.1 RI-Based Gas Detection

The generic setup for the ring resonator RI detection is illustrated in Figure 17.3A. The WGM of the ring resonator is excited by a tunable diode laser via an optical waveguide or fiber taper coupled to the ring resonator. The laser periodically scans while the detector at the output of the waveguide or fiber taper measures the optical intensity. As shown in Figure 17.3B, when the laser wavelength satisfies the WGM resonant condition, the light couples into the ring resonator and causes the measured transmission intensity to drop, leaving a spectral dip at the detector. The WGM shifts in response to the RI change near the ring resonator surface due to the presence of analyte, as will be discussed in detail later. The sensorgram, as shown in Figure 17.3C, can be obtained if we monitor the WGM spectral shift in real time, which provides the quantitative and temporal information of the analyte within the sensing region.

When used for gas sensing, most of the ring resonators are coated with a layer of vapor sensitive polymer [4,8,9] or even are made of the polymers [5,6]. To better understand the ring resonator gas sensor principle, we use an OFRR-based gas sensor as a model system to study its sensing performance. The OFRR is a piece of thin-walled glass capillary with a diameter in the range of 50–200 µm and a wall thickness of only 2–4 µm [10–13]. Gas samples are flowed inside of the capillary while detection can be carried out at any location along the capillary via a tapered optical fiber, where a ring resonator sensor forms.

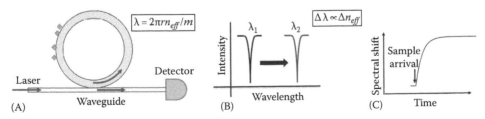

FIGURE 17.3 (A) Experimental setup for ring resonator RI detection. (B) The WGM resonance wavelength shifts in response to the presence of the analyte in the evanescent field of the WGM. (C) The sensorgram is obtained by monitoring the WGM spectral shift as a function of time, which provides both quantitative and temporal information about the analytes near the ring resonator surface.

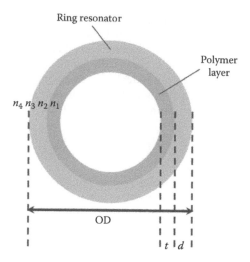

FIGURE 17.4 Cross-sectional view of the OFRR coated with a layer of vapor-sensitive polymer on its interior surface. OD: ring resonator outer diameter. t: polymer thickness. d: ring resonator wall thickness. n_1, n_2, n_3, and n_4 are the refractive indices for the medium inside (usually, air), polymer, ring resonator, and the medium outside (usually, air), respectively.

Due to this unique and neat fluidics and optics integration, the OFRR achieves excellent sensing performance such as μL sample volume, ultra-fast sensor response, and complete sensor regeneration. Figure 17.4 shows a cross-sectional view of the OFRR gas sensor. The polymer layer is coated on the interior surface of the ring resonator. The polymer coating has a few functions. (1) It captures and concentrates vapor molecules through absorption/adsorption. As compared to the vapor density outside the polymer (vapor density in the capillary), the analyte density inside the polymer is increased by a factor of K, i.e.,

$$\rho = K\rho_0, \tag{17.4}$$

where ρ_0 and ρ are the analyte density outside and inside the polymer, respectively. K is the partition coefficient, which ranges from hundreds to hundreds of thousands [21,22]. (2) The polymer RI and thickness change in response to the presence of the analyte, which, in turn, causes the WGM spectral shift, as described by

$$S = \frac{d\lambda}{d\rho} = \frac{\partial\lambda}{\partial t} \cdot \frac{\partial t}{\partial\rho} + \frac{\partial\lambda}{\partial n_2} \cdot \frac{\partial n_2}{\partial t} \cdot \frac{\partial t}{\partial\rho} + \frac{\partial\lambda}{\partial n_2} \cdot \frac{\partial n_2}{\partial\rho}, \tag{17.5}$$

where S is the ring resonator sensitivity. $\partial\lambda/\partial t$ and $\partial\lambda/\partial n_2$ refer to the WGM thickness sensitivity (S_t) and RI sensitivity (S_{RI}), respectively, which are the intrinsic properties associated with the optical modes of the coated ring resonator. $\partial t/\partial\rho$ and $\partial n_2/\partial\rho$ are the polymer swelling/shrinkage and the RI change due to the vapor molecule absorption, which depend on the polymer-analyte interaction, thus providing a certain degree of selectivity for different analytes. RI change can be caused by either the polymer volume change induced by vapor molecules or by the doping effect due to the presence of the vapor molecules in the polymer matrix [23], as described respectively by the second and the third term on the right-hand-side of Equation 17.5. The RI change due to the doping effect can be modeled by the Lorentz–Lorenz equation [23]:

$$\delta n_2 = \frac{\left(n_2^2 + 2\right)^2}{6n_2} \frac{1}{3\varepsilon_0}(\delta\rho)\alpha, \tag{17.6}$$

where α is the vapor molecule polarizability. The RI sensitivity for the polymer RI change, S_{RI}, is related to the fraction of light in the polymer, η, by [24–26]

$$S_{RI} = \frac{\partial \lambda}{\partial n_2} = \frac{\lambda}{n_{eff}} \eta. \tag{17.7}$$

The rest parameters may need to be determined experimentally. (3) The polymer coating on the interior surface of the OFRR is essentially a stationary phase, which can be used to separate vapor chemical compounds carried by the mobile phase gas flow according to their volatilities and polarities. We will come back to this function later when we discuss the ring resonator–based gas chromatography (GC).

The WGM spectral position and the corresponding field distribution can be analytically calculated based on the Mie theory using the four-layer model [11] described in Figure 17.4. The corresponding electric field radial distribution, $E_{m,l}(r)$, is given by

$$E_{m,l}(r) = \begin{cases} AJ_m(kn_1 r) & (r \leq OD/2 - d - t) \\ BJ_m(kn_2 r) + CH_m^{(1)}(kn_2 r) & (OD/2 - d - t \leq r \leq OD/2 - d) \\ DJ_m(kn_3 r) + EH_m^{(1)}(kn_3 r) & (OD/2 - d \leq r \leq OD/2) \\ FH_m^{(1)}(kn_4 r) & (r \geq OD/2) \end{cases}, \tag{17.8}$$

where J_m and $H_m^{(1)}$ are the mth Bessel function and the mth Hankel function of the first kind, respectively. $k = 2\pi/\lambda$, where λ is the WGM wavelength in vacuum. This model provides a foundation for the ring resonator vapor sensor based on the RI detection, and allows us to predict the sensor performance and optimize the sensor design. One example of the electric field radial distribution of OFRR gas sensor with various polymer thicknesses is shown in Figure 17.5.

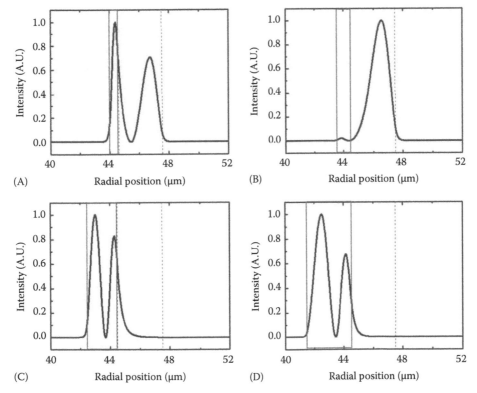

FIGURE 17.5 Normalized electric field radial distribution of the second order WGM based on Equation 17.8 and Figure 17.4 for various polymer thicknesses. (A)–(D) The polymer thickness, $t = 0.5$, 1.0, 2.0, and 3.0 μm, respectively. The relevant parameters are OD = 95 μm. $d = 3$ μm. $n_1 = 1$. $n_2 = 1.7$. $n_3 = 1.45$. $n_4 = 1$. $m = 257$. Vertical dash lines indicate the boundaries of the ring resonator surface. The boxed areas to the left indicate the polymer regions. (Reprinted from Sun, Y. and Fan, X., *Opt. Express*, 16(14), 10254, 2008. With permission.)

The ring resonator can also be used to directly detect the RI change induced by the presence of the analyte [7]. The corresponding WGM spectral shift is given by

$$\frac{\Delta\lambda}{\lambda} = \eta\frac{\Delta n_{gas}}{n_{eff}},$$ (17.9)

where η is the fraction of light interacting with the analyte, which satisfies $\Delta n_{eff} = \eta\Delta n_{gas}$. Δn_{gas} is the RI change due to the presence of the analyte.

17.2.2 Absorption-Based Gas Detection

Optical ring resonators have also been implemented for absorption detection. Due to the enhanced interaction length resulting from the high-Q cavity, the absorption is significantly increased for sensitive vapor measurement. Rosenberger et al. recently proposed to use the depth of the WGM resonance dip (see Figure 17.3B for illustration) as the sensing signal [3,27]. Referring to Figure 17.3A, if r and t are the transmission coefficient and the coupling coefficient of the electric field, respectively, at the contact region between the waveguide and the ring resonator, then the normalized resonant light intensity at the distal end of the waveguide is

$$R = \left|\frac{r - \exp(-\alpha L/2)}{1 - r\exp(-\alpha L/2)}\right|^2,$$ (17.10)

where αL is the ring resonator round-trip loss. L is the circumference of the ring resonator. When $\alpha L \ll 1$, the resonance dip depth at the detector is given by

$$M = \frac{4x}{(1+x)^2},$$ (17.11)

where $x = t^2/\alpha L$ is the ratio between the out-coupling loss and the ring resonator round-trip loss. When $x = 1$, the critical coupling is reached, for which $M = 0$. When the analyte is attached to the ring resonator surface, the depth change due to the additional absorption from the analyte is given by

$$\frac{1}{M}\frac{\partial M}{\partial \alpha} = \frac{x-1}{x+1}\frac{\eta}{\alpha}.$$ (17.12)

Using the resonance dip depth as the sensing signal is advantageous in that the sensitivity is determined only by the intrinsic loss (α) of the cavity, whereas in other types of absorption measurement techniques such as Q-degradation (or linewidth broadening) and cavity ring-down spectroscopy, the sensitivity is determined by the total loss.

In addition to measure the absorption by monitoring the resonance dip depth change, Hu proposed highly sensitive vapor detection based on photo-thermal spectroscopy (PTS) carried out in conjunction with a high-Q ring resonator [28]. In PTS, the WGM spectral position shifts in response to the heat generated by the analyte near the ring surface when the pump light is tuned into the analyte's absorption band. According to Hu's calculation, it is suggested that the enhancement of the sensing signal is proportional to the ring resonator Q-factor. With a moderate Q-factor of 10^5, detection of vapor molecules down to the parts-per-trillion (ppt) level can potentially be achieved.

17.3 Examples

17.3.1 RI-Based Detection

17.3.1.1 Static Gas Sensing

As discussed previously, the ring resonator gas sensor is typically coated with a layer of vapor-sensitive materials such as polymer or inorganics [4,8,10] or the ring resonator itself is made of these materials [5,6]. For the static gas measurement, the sensor is exposed to the analyte vapor continuously.

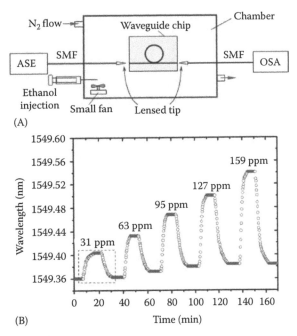

FIGURE 17.6 (A) Typical setup for ring resonator gas sensor in static exposure of vapor analyte. ASE: amplified spontaneous emission. SMF: single mode fiber. OSA: optical spectral analyzer. (B) The WGM spectral shift in response to different concentrations of ethanol vapor. (Reprinted from Pang, F. et al., *Sens. Actuat. B*, 120, 610, 2007. With permission.)

Figure 17.6A illustrates a typical setup for the static vapor sensing. The analyte is injected in liquid and vaporized by a small fan or saturated headspace vapor is diluted with inert gas such as nitrogen or argon to achieve desired concentration in the chamber. Using the setup in Figure 17.6A, Pang et al. placed the ring resonator made of sol–gel into the chamber and detected ethanol vapor (see Figure 17.6B). The rise and fall of the WGM spectral position were due to the presence of ethanol vapor and the purge by pure nitrogen flow. A detection limit on the order of 10 parts per million (ppm) was achieved. In addition to ethanol, detection of other vapors such as isopropanol [4] and ammonia [29] has been investigated. In particular, using the ring resonator made of chromophore-doped polymer, Chen et al. were able to detect trace amount of explosive simulant, 2,4-dinitrotoluene, down to the parts per billion (ppb) level [6].

One of the major issues related to the ring resonator vapor sensing described earlier is the slow response time or detection speed and hence long sensor dead-time, which is unacceptable in many applications that require rapid vapor detection and real-time monitoring. As exemplified in Figure 17.6B, it takes tens of minutes after injection for the sensor (i.e., sensing polymer) to reach equilibrium and subsequently similar amount of time for purging. Such slow response is due mainly to the ineffective fluidic designs, for example, the 4.8 L gas chamber used to obtain the results shown in Figure 17.6B. As mentioned previously, the capillary-based OFRR gas sensor can perfectly solve this problem. As shown in Figure 17.4A, the OFRR is a thin-walled micron-sized glass capillary whose interior surface is coated with polymer. The electric field of the WGM, as described by Equation 17.8, is present in the polymer layer and interacts with the vapor molecules flowing through the capillary [10]. The OFRR enables rapid detection with small sample volumes. Figure 17.7A depicts a typical experimental setup for the OFRR vapor sensing. In this experiment, the OFRR has a diameter of 75 µm and a wall thickness of approximately 4 µm [14]. Its interior surface can be coated with different kinds of polymer, for example, moderately polar methyl phenol polysiloxane (OV-17) or highly polar polyethylene glycol (PEG). The coating thickness is approximately 200 nm and can be adjusted by varying the coating solution concentration. The OFRR response to ethanol and hexane vapors is shown in Figure 17.7B and C. The detection limit was about 200 ppm for ethanol. In addition, thorough purge can be achieved, as evidenced by the WGM returns to the baseline completely. Sensorgrams plotted in the insets of Figure 17.7B and C show the response time of only tens of seconds, which is limited

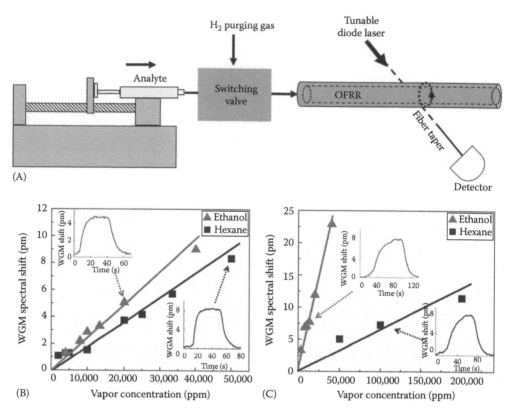

FIGURE 17.7   (A) Setup for the OFRR vapor sensor. A syringe pump is used to inject the vapor analyte into the OFRR capillary. A switching valve is used to quickly switch between the injection mode and the purge mode. (B) and (C) OFRR responses to various concentrations of ethanol and hexane vapors. The OFRR is coated with a 200 nm thick OV-17 (B) and PEG-400 (B). Insets are the sensorgrams taken by monitoring the WGM shift in real time. (Reprinted from Sun, Y. et al., *Opt. Lett.*, 33, 788, 2008. With permission.)

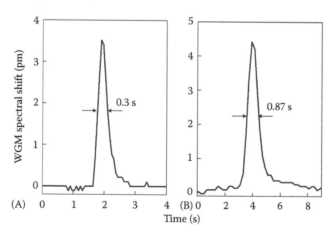

FIGURE 17.8   Rapid detection of (A) ethanol vapor and (B) hexane vapor when the analyte is injected in a pulsed manner. (Reprinted from Sun, Y. et al., *Opt. Lett.*, 33, 788, 2008. With permission.)

by the injection and switching speed. To truly test the OFRR response, sub-second pulses were injected into the OFRR. As shown in Figure 17.8, the sub-second OFRR response time is demonstrated for ethanol and hexane vapors. This peak width is limited by the laser scanning rate used in the experiment. An even sharper peak can be obtained with a higher scanning rate. Such high detection speed makes the OFRR an excellent GC detector that usually requires sub-second response time, as discussed later.

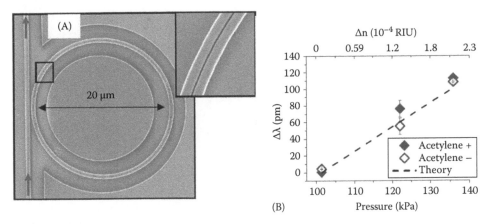

FIGURE 17.9 (A) SEM image of a silicon slotted waveguide. (B) Detection of acetylene. (Reprinted from Robinson, J.T. et al., *Opt. Express*, 16(6), 4296, 2008. With permission.)

It is noted that although it is not specific, polymers may exhibit different response to different analytes, as shown in Figure 17.7B and C. The magnitude of the response depends on the strength of polymer–analyte interaction, often related to analyte volatility and polarity, which allows us to choose the optimal polymer to detect the analytes of interest. In addition, an array of polymers can be employed in vapor sensing and their collective response pattern can be used to enhance the detection specificity.

Direct detection of the acetylene vapor induced RI change without the mediation of the polymer has recently been carried out by Robinson et al. using a chip-based slotted ring resonator [7] (see Figure 17.9). The slot waveguide geometry significantly enhances the electric field in the nano-sized gap where the vapor analyte resides, leading to stronger light–analyte interaction and hence enhanced sensitivity. A sensitivity of 490 nm/RIU (RIU = refractive index units) was achieved, 20 times higher than the same device without a slot. Such sensitivity resulted in a detection limit of 10^{-4} RIU, corresponding to acetylene of 100 kPa at room temperature.

Although the ring resonators feature high sensitivities and miniaturized footprints, the most significant drawback with the RI detection is the lack of detection specificity, as the sensing polymer responds to nearly all vapors. While the collective response from a sensor array coated with different polymers may mitigate the specificity problem, recent studies have shown that such detection still lacks the specificity to identify more than three analytes from one another [30–33], particularly when dealing with tens of unknown analytes.

In the following section, we will discuss the implementation of the ring resonator vapor sensor with GC technology to improve the detection specificity.

17.3.1.2 Ring Resonator Vapor Sensors in Conjunction with Micro-Gas Chromatography (μGC)

GC in conjunction with mass spectrometry (GC-MS) is commonly regarded as the standard technology in vapor analysis. In a GC system, as illustrated in Figure 17.10A and B, a mixture of vapor analytes is introduced into the gas injector and then released as a pulse carried by the carrier gas that travels along the GC column coated with a layer of polymeric stationary phase. Since different vapor molecules have different interactions with the polymer, they arrive at the end-column detector (e.g., MS) with unique elution times (or retention times) and can thus be identified individually on the chromatogram. Despite its popularity, GC-MS is extremely bulky (>50 kg) and power intensive, and requires dedicated personnel. Therefore, it is not poised for field deployment to perform real-time in situ gas detection and monitoring.

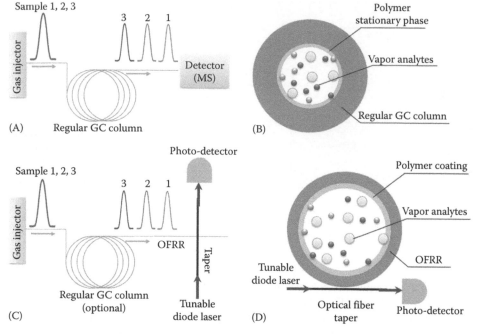

FIGURE 17.10 (A) Conceptual illustration of a regular GC setup. (B) Cross-sectional view of a regular GC column. (C) Conceptual illustration of the OFRR-based μGC setup. (D) Cross-sectional view of the OFRR.

μGC is a particularly attractive technology for in situ VOC analysis due to its small size, low power consumption, and the capability for analyzing vapor mixtures of arbitrary composition. In the past several decades, μGC has undergone significant breakthroughs, especially with respect to miniaturization and rapid detection [34–38]. There are multiple configurations to implement the μGC, among which the OFRR-based μGC, developed by Fan and co-workers, innovatively integrate the capillary ring resonator sensing technology with the gas separation technique [9,11–13]. As shown in Figure 17.10C and D, the OFRR is very similar to the regular GC column except that it has a thin wall, which allows the electric field of the WGM to penetrate into the polymer layer (stationary phase) to quantitatively respond to the interaction of vapor analytes and the polymer in real time. The OFRR has a dual-function of a GC separation column and an on-column optical detector that is capable of detecting the vapor analytes at any locations along the OFRR capillary. Compared to conventional end-column GC detection, on-column detection not only provides greater flexibility in terms of sample detection but could also provide additional separation information about co-eluted analytes, which are impossible to be identified by an end-column detector. Furthermore, the OFRR-based μGC has a few additional distinct advantages. First, as an optical μGC detector, it has ultra-fast intrinsic response time and ultra-low intrinsic noise level, both of which are key points for a μGC system to achieve rapid separation and detection of trace level of analytes. Second, efficient capillary microfluidics and in situ analyte separation and detection within the same column reduce the fluidic dead volume to a minimum, and thus reduce the band broadening effect along the column. Third, the OFRR's circular shape makes it highly compatible with a regular GC column (see Figure 17.10C) without introducing any additional dead volumes. Therefore, the OFRR can either be used independently as a separation column and on-column detector or can be used in conjunction with a regular GC column. In this case, polar and nonpolar stationary phase coated tandem-column system can be developed, which can improve the GC separation capability and increase the sensitivity and selectivity for detection of particular classes of analytes. Fourth, well-established GC column-coating techniques can be used without any significant modifications to provide uniform stationary phase coating on the OFRR. Fifth, the unique on-column analysis enables multi-point detection at any predetermined locations along the OFRR without interfering with the gas flow inside, which is not feasible with conventional GC detectors that are separated from the column.

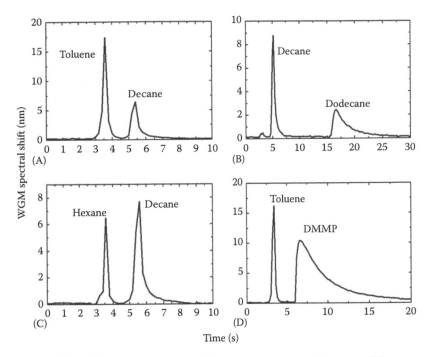

FIGURE 17.11 Separation and detection of different analytes with a PEG-400 coated OFRR μGC.

To demonstrate the proof-of-concept of the OFRR's separation and on-column detection, several vapor analytes were used as a model system. The OFRR was coated with 100 nm PEG using the static coating method. The detection was achieved by optical taper coupling of WGM at a position of 3 cm downstream from the OFRR capillary inlet. As shown in Figure 17.11, all the analytes could be well separated at room temperature. Toluene, hexane, and decane vapors have a rapid response, with a peak width of sub second. The rise time for DMMP was 1.25 seconds and it had strong tailing effect due to the strong interaction of DMMP and polar PEG coating. Dodecane had a relatively low response time of 2 s because of the low volatility at room temperature. In Figure 17.12, the capability of OFRR-based μGC in explosive detection was explored, in which 2,4-dinitrotoluene (DNT) was chosen as a model system, due to its chemical structure similarity to trinitrotoluene (TNT). DNT is also an additive or contaminant in many

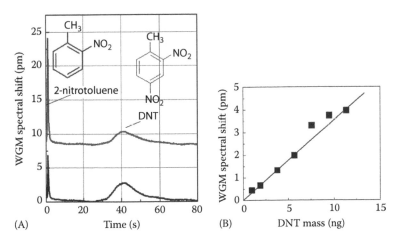

FIGURE 17.12 (A) Highly repeatable separation and detection of DNT from 2-nitrotoluene and DNT mixtures. Different heights in two runs are due to the different mass injected. Curves are vertically shifted for clarity. (B) WGM spectral shift as a function of the injected DNT mass. (Reprinted from Sun, Y. et al., *Analyst*, 134, 1386, 2009. With permission.)

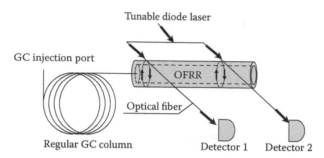

FIGURE 17.13 Conceptual illustration of tandem-column separation based on OFRR μGC. A nonpolar phase regular GC column is connected with a relatively short polar phase coated OFRR column through a press-tight universal connector. A tunable diode laser is coupled into two fibers in contact with the OFRR. The first fiber is placed at the inlet of the OFRR and the second one a few centimeters downstream along the OFRR, defining two detection locations, respectively. (Reprinted from Sun, Y. et al., *Analyst*, 135, 165, 2010. With permission.)

explosives and thus can be used as an indicator to predict the existence of TNT. Figure 17.12A shows the separation between DNT vapor and 2-nitrotoluene vapor. Although 2-nitrotoluene and DNT only have one nitro group difference in chemical structure, OFRR vapor sensors achieved very efficient separation between them. In addition, the signal amplitude is proportional to the DNA mass injected as shown in Figure 17.12B. The detection limit was estimated to be on the order of 100 pg.

As demonstrated earlier, μGC achieves rapid separation and detection of simple vapor analytes. However, the improvement in separation speed is obtained at the cost of the most essential separation capability for GC, owing to insufficient interaction of the short μGC columns with analytes. Oftentimes, multiple analytes co-elute out of the column and become unidentifiable. This inherently low chromatographic resolution is one of the most severe issues in μGC, making it difficult to identify analytes embedded in co-eluted peaks and significantly limiting the complexity of the mixtures that μGC can effectively analyze. In order to maintain fast separation while improving the resolution, tandem-column μGC systems have been investigated [36,37,39]. A tandem-column μGC typically consists of two connected columns coated with different stationary phases. A nonpolar or slightly polar phase column is usually used as the first separation column, followed by a short polar phase column. Analytes are basically separated according to their volatility on the first column and then further separated according to their polarity on the second column, resulting potentially in an increased separation power. However, the tandem-column configuration may still suffer from the same co-elution problem as in a short single-column μGC, because analytes separated by the first column may still co-elute after passing through the second column. Since current GC detectors are placed at the end of the second column, the co-eluted analytes cannot be differentiated by this configuration and therefore need sophisticated modulator or programmed switch after the first column [36,40,41].

In contrast, the OFRR, due to its on-column, multi-point detection capability, is particularly attractive to tackle this co-elution problem in the tandem-column μGC system. As illustrated in Figure 17.13, the OFRR (~10 cm long) coated with polar stationary phase (such as PEG) was connected with a regular GC column (~100–200 cm long) coated with nonpolar or low-polarity polymer (such as Rtx-1 and Rtx-5). Two detection locations were chosen, one at the inlet of the OFRR, right after the first column, and the other one a few centimeters away from the first detection point, to monitor the separation in the first column and the second column, respectively. Since OFRR on-column detectors record the retention time independently and provide complementary chromatograms for each chemical compound, co-elution at the terminal end of columns is no longer an issue for the OFRR-based tandem-column μGC. By monitoring retention times at multiple locations along the OFRR column, analytes can be well separated on at least one detection location. Figure 17.14 demonstrates that 12 analytes with different volatilities and polarities could be separated by the OFRR-based tandem-column μGC system within 4 minutes. In particular, after the first column separation Analytes #3 and #8, and Analyte #4 and #9 co-eluted, but were all resolved after the second column. Certainly, the opposite cases exist, where the co-elution occurs after the second column. However, those co-eluted analytes may be well resolved after the first column separation, as demonstrated in Figure 17.15.

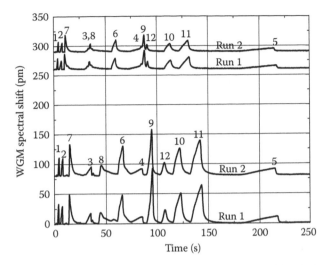

FIGURE 17.14 Chromatograms of 12 analytes with various volatilities and polarities obtained by the two detection channels. The whole analysis was completed within 4 min. Two groups of co-elution (#3/#8 and #4/#9) from the first GC column (upper traces) were well resolved after the second 6 cm OFRR column separation (lower traces). The experiments were performed twice (Run 1 and Run 2) to show the separation repeatability. Curves are vertically shifted for clarity. 1: Heptane. 2: Octane. 3: Decane. 4: Undecane. 5: Dodecane. 6: 1-octanol. 7: Dimethyl methyl phosphonate (DMMP). 8: Diethyl methyl phosphonate (DEMP). 9: 2-nitrotoluene. 10: 3-nitrotoluene. 11: 4-nitrotoluene. 12: Dimethyl dinitro butane (DMNB). (Reprinted from Sun, Y. et al., *Analyst*, 135, 165, 2010. With permission.)

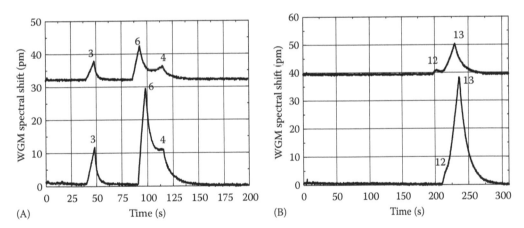

FIGURE 17.15 (A) Analytes #4 and #6 partially co-eluted after the second column (bottom curve), but were resolved after the first column (upper curve). (B) Analytes #12 and #13 co-eluted after the second column (bottom curve), but were resolved after the first column (upper curve). Curves are vertically shifted for clarity.13: Methyl salicylate. Other analytes' names are given in Figure 17.14. (Reprinted from Sun, Y. et al., *Analyst*, 135, 165, 2010. With permission.)

On-column detection capability also enables the monitoring of the vapor separation process inside the OFRR to further enhance the GC resolution. Figure 17.16 demonstrates the monitoring of the separation process of a target analyte (Analyte #6: 1-octanol) in a vapor mixture as it traveled along the OFRR. This was accomplished by placing the second channel at two different locations along the OFRR column. As shown in Chromatogram #1 in Figure 17.16, Analyte #6 came out of the first column earlier than Analyte #4, but got delayed more in the OFRR column, resulting in the merging of these two analytes. Such delay became more apparent if the second detection point was moved 4.5 cm downstream, as shown in Chromatogram #2 in Figure 17.16.

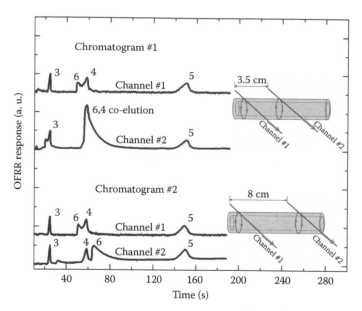

FIGURE 17.16 Monitoring of the separation process of the target analyte (Analyte #6: 1-octanol) by acquiring three chromatograms at three different locations along the OFRR column. The upper traces in Chromatograms #1 and #2 were recorded from the tapered fiber (Channel #1) located at the OFRR column inlet. The lower traces in Chromatograms #1 and #2 were recorded from another tapered fiber (Channel #2) located 3.5 and 8 cm downstream from Channel #1, respectively. Curves are vertically shifted for clarity. The analytes' names are given in Figure 17.14. (Reprinted from Sun, Y. et al., *Analyst*, 135, 165, 2010. With permission.)

17.3.2 Absorption Based Detection

Ring resonator enhanced laser absorption spectroscopy was recently carried out by Farca et al. using the experimental setup illustrated in Figure 17.17A [3]. An optical fiber forms a ring resonator with a Q-factor in excess of 10^6. The ring resonator was sandwiched in a PZT clamp, which adjusted the pressure exerted on the ring resonator to lock the WGM resonance to an external laser. Three analytes, methane, methyl chloride, and ethylene, were chosen, as they all have the absorption band within the laser wavelength tuning range. Figure 17.17B shows the corresponding WGM dip depth change, as described by Equation 17.12. For comparison, the transmission spectra from a 16-cm long chamber are also plotted and virtually no absorption is observed.

17.4 Outlook

We have overviewed the state-of-the-art of ring resonators in chemical vapor sensing applications. Various ring resonator structures are described and different detection mechanisms are introduced. To date, RI-based detection is still the most widely used technique in vapor sensing. Although RI detection is universal in that it does not require any particular detection wavelength, it lacks the detection specificity (which is different from biosensing in which bio-recognition molecules are available to provide the specificity), even with the aid of the sensor arrays that generate response patterns. In contrast, absorption detection is analyte-specific, but it requires lasers that cover the absorption band of the analytes of interest, a requirement that is difficult to meet when dealing with multiple analytes. In addition, the detection limit using absorption technique is usually not as good as the RI detection.

   In this chapter, we demonstrated the powerful analysis capability when we combine the ring resonator technology with μGC technology, which adds the detection specificity by separating each vapor analyte in the sample mixture. This synergy may represent one of important research directions where both photonic sensing technology and separation technology can be leveraged. How to improve the GC detection

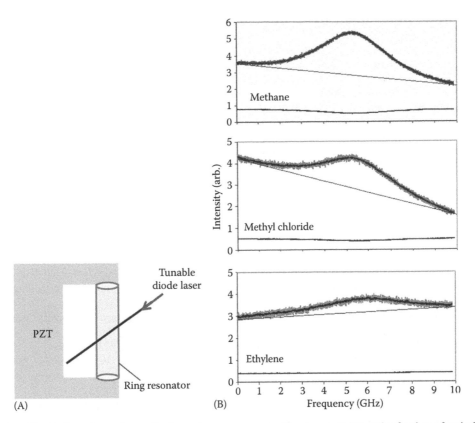

FIGURE 17.17 (A) Experimental setup for absorption measurement with a ring resonator made of a piece of optical fiber. A PZT was used to lock the WGM resonance to a tunable laser. (B) The measured absorption profiles of methane, methyl chloride, and ethylene. In each case, the top curve shows the variation in dip depth and the bottom curve is the transmission profile of the gas in a 16 cm absorption cell for comparison. (Reprinted from Farca, G. et al., *Opt. Express*, 15(25), 17443, 2007. With permission.)

sensitivity, how to better solve co-elution problems in μGC while maintaining the detection speed, and how to integrate the ring resonator sensor with the μGC system are imminent issues.

In addition to RI and absorption detection, the fluorescence, Raman, and PTS-based detection in the ring resonator may be explored for sensitive and specific vapor detection. One example is conjugated polymers that respond quite specifically to the presence of explosives through fluorescence quenching [42,43]. Research on fluorescence, Raman spectroscopy, and PTS in the ring resonator will certainly lead to new detection paradigms that may provide unique features complementary to RI and absorption detection.

REFERENCES

1. Vollmer F et al. (2002) Protein detection by optical shift of a resonant microcavity. *Appl. Phys. Lett.* 80:4057–4059.
2. Hanumegowda NM, Stica CJ, Patel BC, White I, and Fan X (2005) Refractometric sensors based on microsphere resonators. (Translated from English) *Appl. Phys. Lett.* 87(20):201107 (in English).
3. Farca G, Shopova SI, and Rosenberger AT (2007) Cavity-enhanced laser absorption spectroscopy using microresonator whispering-gallery modes. *Opt. Express* 15(25):17443–17448.
4. Ksendzov A, Homer ML, and Manfreda AM (2004) Integrated optics ring-resonator chemical sensor with polymer transduction layer. *Electron. Lett.* 40:63–65.
5. Pang F et al. (2007) Sensitivity to alcohols of a planar waveguide ring resonator fabricated by a sol-gel method. *Sens. Actuat. B* 120:610–614.
6. Chen A et al. (2008) Chromophore-containing polymers for trace explosive sensors. *J. Phys. Chem. C* 112:8072–8078.

7. Robinson JT, Chen L, and Lipson M (2008) On-chip gas detection in silicon optical microcavities. *Opt. Express* 16(6):4296–4301.

8. Yebo NA, Lommens P, Hens Z, and Baets R (2010) An integrated optic ethanol vapor sensor based on a silicon-on-insulator microring resonator coated with a porous ZnO film. *Opt. Express* 18:11859–11866.

9. Shopova SI et al. (2008) On-column micro gas chromatography detection with capillary-based optical ring resonators. *Anal. Chem.* 80:2232–2238.

10. Sun Y, Shopova SI, Frye-Mason G, and Fan X (2008) Rapid chemical-vapor sensing using optofluidic ring resonators. *Opt. Lett.* 33:788–790.

11. Sun Y and Fan X (2008) Analysis of ring resonators for chemical vapor sensor development. *Opt. Express* 16(14):10254–10268.

12. Sun Y et al. (2009) Optofluidic ring resonator sensors for rapid DNT vapor detection. *Analyst* 134:1386–1391.

13. Sun Y et al. (2010) Rapid tandem-column micro-gas chromatography based on optofluidic ring resonators with multi-point on column detection. *Analyst* 135:165–171.

14. Zamora V, Diez A, Andres MV, and Gimeno B (2007) Refractometric sensor based on whispering-gallery modes of thin capillaries. *Opt. Express* 15:12011–12016.

15. Bernardi A et al. (2008) On-chip Si/SiO$_x$ microtube refractometer. *Appl. Phys. Lett.* 93:094106.

16. Tong L et al. (2003) Subwavelength-diameter silica wires for low-loss optical wave guiding. *Nature* 426:816–819.

17. Xu F, Horak P, and Brambilla G (2007) Optical microfiber coil resonator refractometric sensor. *Opt. Express* 15(12):7888–7893.

18. Sumetsky M (2004) Optical fiber microcoil resonator. *Opt. Express* 12(10):2303–2316.

19. Chang RK and Campillo AJ eds. (1996) *Optical Processes in Microcavities* (World Scientific, Singapore).

20. Gorodetsky ML, Savchenkov AA, and Ilchenko VS (1996) Ultimate Q of optical microsphere resonators. *Opt. Lett.* 21:453–455.

21. Liron Z, Kaushansky N, Frishman G, Kaplan D, and Greenblatt J (1997) The polymer-coated SAW sensor as a gravimetric sensor. *Anal. Chem.* 69:2848–2854.

22. Potyrailo RA and Sivavec TM (2004) Boosting sensitivity of organic vapor detection with silicone block polyimide polymers. *Anal. Chem.* 76:7023–7027.

23. Podgorsek RP and Franke H (2002) Selective optical detection of aromatic vapors. *Appl. Opt.* 41:601–608.

24. Zhu H, White IM, Suter JD, Dale PS, and Fan X (2007) Analysis of biomolecule detection with optofluidic ring resonator sensors. (Translated from English) *Opt. Express* 15(15):9139–9146 (in English).

25. White IM and Fan X (2008) On the performance quantification of resonant refractive index sensors. *Opt. Express* 16:1020–1028.

26. Mortensen NA, Xiao S, and Pedersen J (2008) Liquid-infiltrated photonic crystals: Enhanced light-matter interactions for lab-on-a-chip applications. *Microfluid. Nanofluid.* 4:117–127.

27. Rosenberger AT (2007) Analysis of whispering-gallery microcavityenhanced chemical absorption sensors. *Opt. Express* 15:12959–12964.

28. Hu J (2010) Ultra-sensitive chemical vapor detection using micro-cavity photothermal spectroscopy. *Opt. Express* 18:22174–22186.

29. Passaro VMN, Dell'Olio F, and De Leonardis F (2007) Ammonia optical sensing by microring resonators. *Sensors* 7:2741–2749.

30. Park J, Groves WA, and Zellers ET (1999) Vapor recognition with small arrays of polymer-coated microsensors. A comprehensive analysis. *Anal. Chem.* 71:3877–3886.

31. Hsieh M-D and Zellers ET (2004) Limits of recognition for simple vapor mixtures determined with a microsensor array. *Anal. Chem.* 76:1885–1896.

32. Jin C and Zellers ET (2008) Limits of recognition for binary and ternary vapor mixtures determined with multitransducer arrays. *Anal. Chem.* 80:7283–7293.

33. Jin C, Kurzawski P, Hierlemann A, and Zellers ET (2008) Evaluation of multitransducer arrays for the determination of organic vapor mixtures. *Anal. Chem.* 80:227–236.

34. Veriotti T and Sacks R (2001) A tandem column ensemble with an atmospheric pressure junction-point vent for high-speed GC with selective control of peak-pair separation. *Anal. Chem.* 73(4):813–819.

35. Libardoni M, McGuigan M, Yoo YJ, and Sacks R (2005) Band acceleration device for enhanced selectivity with tandem-column gas chromatography. *J. Chromatogr. A* 1086(1–2):151–159.

36. Lu C-J, Whiting J, Sacks RD, and Zellers ET (2003) Portable gas chromatograph with tunable retention and sensor array detection for determination of complex vapor mixtures. *Anal. Chem.* 75(6):1400–1409.
37. Rowe MP, Steinecker WH, and Zellers ET (2007) Chamber evaluation of a portable GC with tunable retention and microsensor-array detection for indoor air quality monitoring. *J. Environ. Monit.* 8(2):270–278.
38. Kim S-J et al. (2010) Microfabricated thermal modulator for comprehensive two-dimensional micro gas chromatography: Design, thermal modeling, and preliminary testing *Lab Chip* 10:1647–1654.
39. Deans DR and Scott I (2002) Gas chromatographic columns with adjustable separation characteristics. *Anal. Chem.* 45(7):1137–1141.
40. Akard M and Sacks R (2002) Pressure-tunable selectivity for high-speed gas chromatography. *Anal. Chem.* 66(19):3036–3041.
41. Lambertus G and Sacks R (2005) Stop-flow programmable selectivity with a dual-column ensemble of microfabricated etched silicon columns and air as carrier gas. *Anal. Chem.* 77(7):2078–2084.
42. Yang J-S and Swager TM (1998) Fluorescent porous polymer films as TNT chemosensors: Electronic and structural effects. *J. Am. Chem. Soc.* 120:11864–11873.
43. McQuade DT, Pullen AE, and Swager TM (2000) Conjugated polymer-based chemical sensors. *Chem. Rev.* 100:2537–2574.

Part III

Applications

18

Nano-Optical Sensors for the Detection of Bioterrorist Threats

Vinod Kumar Khanna

CONTENTS

18.1 Introduction

The aftermath of the catastrophic terrorist events of September 11, 2001, and subsequent anthrax perpetration via the U.S. postal system (as powders in mailed packages), was the worldwide awakening and recognition of the critical need of reliable, unambiguous, and early detection systems to trace biological weapons of mass destruction. Such systems can enable early-stage detection so that the patient can be treated before becoming contagious.

A bioterrorist attack uses disease and incapacitation-inducing agents as arsenal to cause epidemics and death. It involves deliberate dissemination of living organisms or replicating entities such as protozoa, fungi, bacteria, protists, viruses, or toxins in air, water, and food of the opponent's territory by overt and covert means, for action on humans, animals, and plants for military or criminal purposes. Prominent bioterrorism agents include anthrax (*Bacillus anthracis*), botulism (*Clostridium botulinum* toxin), smallpox (Variola major), tularemia (*Francisella tularensis*), and hemorrhagic fever, due to Ebola or

Marburg virus. Human exposure to pathogens occurs through inhalation, skin exposure, or ingestion of contaminated food or water.

Biological pathogens are broadly classified into three categories. Category *A* includes the highest priority agents posing a risk to national security. It comprises agents that are easily disseminated from one person to another, have high mortality rates, and cause public panic or social disruption. Category *B* agents rank the second highest priority being moderately easy to disseminate and having modest morbidity rates. Category *C* agents are placed at the third highest priority and consist of emerging pathogens and infectious disease threats.

A closely allied field to biological or germ warfare is *chemical warfare*. Another related branch is *toxic industrial chemicals* (TICs), which sometimes leak through manufacturing plants harming nearby civilian population. Explosives hidden by terrorists at public places constitute another threat. All these agents constitute somewhat similar threats to mankind and are summarized in Tables 18.1 and 18.2 (Webber et al. 2005).

Biosecurity has become an increasingly important element in the battle against terrorist acts. Biological warfare fixing installation requires low detection limits, high specificity, portability, strength, cheapness, and the ability to simultaneously detect several broad-range potential threat agents. In recent decades, a diversity of optochemical nanosensors have been developed for detecting various spores, viruses, bacteria, and toxins with high sensitivity and selectivity. This chapter will address mainly the nanosensors useful for combating bioterrorism, but some aspects of associated fields will also be touched upon because these can also be sometimes used in wars. It is earnestly aspired that all the nanotools built by mankind are used for peaceful purposes only for ensuring a better quality of life on planet Earth and providing happiness, health, and comfort to its inhabitant Homo sapiens, flora, and fauna.

TABLE 18.1

Biological and Chemical Warfare Agents (BWAs and CWAs), Toxic Industrial Chemicals (TICs), and Explosives

BWAs	CWAs	TICs	Explosives
Bacteria (inhalational anthrax, pneumonic plague, cholera, tularemia, brucellosis, Q fever); *Viruses* (smallpox, monkeypox, viral encephalitides, viral hemorrhagic fevers); *Toxins* (Staphylococcal enterotoxin B, ricin, Botulinum toxin, Mycotoxins)	Nerve agents [tabun (GA), sarin (GB), soman (GD), cyclohexylsarin (GF), and VX ($C_{11}H_{26}NO_2PS$)]; vesicating or blistering agents (mustards, lewisite); choking agents or lung toxicants (chlorine, phosgene, diphosgene)	Ammonia, arsine, boron trichloride, ethylene oxide, nitric acid, HBr, HCl, HF	Trinitrotoluene (TNT), Pentaerythritol tetranitrate (PETN)

TABLE 18.2

Categorywise BWAs

Category *A*	Category *B*	Category *C*
Organisms that can be easily transmitted from person to person resulting in high mortality rates: anthrax, botulism, plague, smallpox, tularemia, viral hemorrhagic fevers	Agents that are moderately easy to disseminate resulting in medium morbidity rates: Brucellosis, Epsilon toxin of *Clostridium perfringens, Salmonella* species, *Escherichia coli* O157:H7, *Shigella,* Glanders, Melioidosis, Psittacosis, Q fever, Ricin toxin, Staphylococcal enterotoxin B, Typhus fever, *Vibrio cholerae, Cryptosporidium parvum*	Emerging infectious diseases are pathogens that could be engineered for mass dissemination in the future such as Nipah virus and hantavirus

18.2 Hybrid Silica Nanoparticle-Based Dipicolinic Acid Luminescence Sensor

Dipicolinic acid (DPA, 6-pyridinedicarboxylic acid: $C_7H_5NO_4$) is a unique, universal, and specific component (5%–14% of their dry weight) of endospores of *Bacillus* and *Clostridium* bacteria. These genera can actively metabolize, decomposing plant and animal matter and various inorganic nutrients. Endospores are dormant alternate life forms produced by them. They are metabolically inert and capable of withstanding a variety of harsh conditions, e.g., radiation, abrasion, extreme heat and cold, and starvation due to lack of nutrients and water. This ability of survival forbearing deleterious conditions makes DPA detection a reliable indicator of bacteria spore content, implying that DPA found in a sample invariably indicates spore presence.

Lanthanide luminescence is a familiar method for the detection of DPA (Taylor and Lin 2009). The visible emitting lanthanides (Sm, Eu, Tb, and Dy) each display a characteristic luminescence color. The luminescence detection is based on the coordination of DPA to Tb(III) forming the complex $[Tb(DPA)]^+$, resulting in luminescence exhibited by the complex at the absorbance maximum of DPA after ultraviolet excitation (Figure 18.1). The origin of the luminescence increase lies in energy transference.

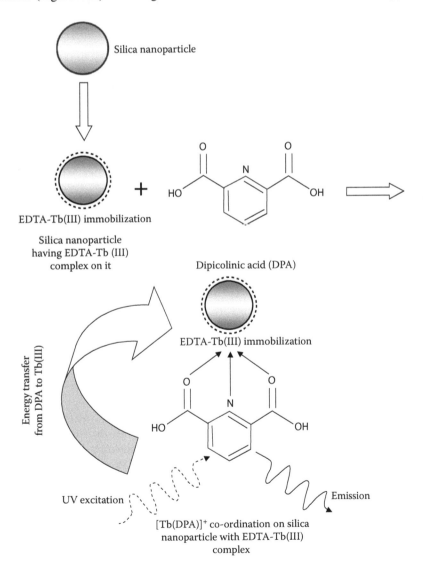

FIGURE 18.1 DPA binds to Tb^{3+} with luminescence emission. DPA content is linked to bacterial spore concentration (number of spores per mL).

Energy transfer from the ligand (DPA) to the terbium excited states is responsible for the enhanced luminescence shown when compared to aquated Tb(III) ions alone.

Thus, unbound Tb^{3+} ion by itself has a low absorption cross-section ($<1 M^{-1} cm^{-1}$) and consequently has low luminescence intensity. Binding of the light-harvesting DPA (absorption cross-section $>10^4 M^{-1} cm^{-1}$), originating from endospores, gives rise to intense Tb luminescence via an absorption, energy transfer, emission mechanism.

Ethylenediaminetetraacetic acid (EDTA)-Tb(III) complexes are immobilized on silica nanoparticles. EDTA ($C_{10}H_{16}N_2O_8$) is a strong chelating agent whose molecules are able to form several bonds to a single metal ion. It must be emphasized that a nanoparticle-based sensor has the advantageous feature of allowing the incorporation of an internal reference into the nanoparticle for providing ratiometric detection, thereby circumventing the need for instrument-specific calibration curves, and also alleviating the corrections necessary for dilution.

Ultraviolet excitation of a suspension of the nanoparticle sensors does not cause any emission. However, upon the addition of the DPA disodium salt, characteristic Tb emission is observed with peaks appearing at 489, 544, and 584 nm. Titration of a suspension of the nanoparticle sensors with DPA linearly increases the emission intensity. This persists until all of the Tb complexes have become coordinatively saturated.

18.3 Photoluminescence Detection of Bacillus cereus Spores by Nanosensor with Dipicolinic Acid Imprinted Nanoshell

Molecular imprinting is a technique for the preparation of tailor-made fashioned synthetic polymers with specific binding sites for target molecules and therefore capable of molecular recognition for the given molecules. Molecularly imprinted polymers (MIPs) are utilized as artificial recognition elements for aimed chemical analytes of interest. The imprinting process involves arranging polymerizable functional monomers around a template, followed by polymerization and template removal. A template molecule or ionic species associates with one or more functional monomers forming a complex. This complex is polymerized with a matrix-forming monomer (cross-linker) to produce a resin. Upon removal of the template species, cavities are used to selectively rebind the template from a mixture of chemical species.

In recent years, the combination of nanoparticles and MIPs has been applied in selective recognition and sensing of spores. For *Bacillus cereus* spores recognition, Gültekin et al. (2010) proposed a thiol ligand-exchange (LE) method using gold–silver (Au–Ag) nanoclusters with a surface layer of methacryloylamidocysteine (MAC) and reconstructing the surface shell by synthetic host polymers based on molecular imprinting method. Methacryloylamidoantipyrine–terbium [$(MAAP)_2$–Tb(III)] was used as a metal-chelating monomer via metal coordination–chelation interactions. Considering the ability of DPA to chelate to Tb (III) ion of MAAP monomer to create reminiscent LE-assembled binding sites for *Bacillus spores* identification, they combined nanoscale materials (Au–Ag nanoclusters) with MIPs. Figure 18.2 illustrates the process sequence for realization of nanoshell based on DPA template reconstruction on Au–Ag/nanoclusters.

Nanoshell sensors with templates serve a dual purpose: (i) They offer a cavity that is selective for DPA. (ii) The DPA simultaneously chelates to Tb(III) metal ion and fits into the shape-selective cavity. The selective binding ability and detection of DPA imprinted Au–Ag nanocluster sensor were examined with fluorescence spectroscopy. The DPA imprinted Au–Ag nanoclusters showed a large separation between the excitation and emission wavelengths, simplifying fluorescence measurements of the recorded photoluminescence spectra using spectrofluorometer. DPA addition caused significant decreases in fluorescence intensity because it induced photoluminescence emission from Au–Ag nanoclusters through the specific binding to the recognition sites of the cross-linked nanoshell polymer matrix. The fluorescence intensity of the DPA imprinted Au–Ag nanoclusters was quenched by DPA. The quenching fluorescence intensity was proportional to DPA concentration. From the linear relationship between spore concentration (CFU mL^{-1}) and dipicolinate content, it was found that 556 $\mu mol L^{-1}$ dipicolinate corresponded to 5.79×10^7 CFU mL^{-1}. The results showed that the change in fluorescence could be attributed to the high

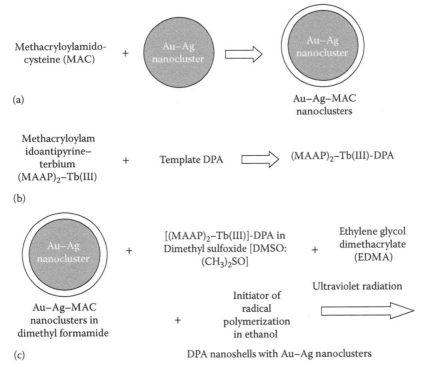

FIGURE 18.2 Schematic representation of the synthesis of nonimprinted (NIP) Au–Ag nanocluster sensor: (a) Preparation of Au–Ag–MAC nanoclusters. (b) Preorganization of (MAAP)$_2$–Tb(III)–DPA using (MAAP)$_2$–Tb(III) and the template, DPA. (c) DPA recognized nanoshell/polymer Au–Ag nanocluster formation by using 1 mol% of the initiator 2, 2 dimethoxy-2-phenylacetophenone with magnetic stirring in a glass polymerization tube and under UV irradiation.

complexation geometric shape affinity (or DPA memory) between DPA molecules and DPA cavities, which occurred on the Au–Ag nanoshells.

The DPA imprinted (MAAP)$_2$–Tb(III)–EDMA copolymer of Au–Ag nanoshell binds DPA and its analogs for *B. cereus* spores sensing. This method is useful to determine spore concentrations in other *Bacillus* species also.

DPA imprinted nanoshell sensor has earned widespread appreciation as a sensor for DPA because the imprinting methods generate a nanoenvironment based on shape of cavity memorial, size, and positions of functional groups that recognize the imprinted molecule, DPA, based on ligand-exchange imprinting methods.

18.4 SERS-Based Anthrax Biomarker Detection

Unenhanced or unaided Raman spectroscopy has an extremely small scattering cross-section. The small cross-section has the effect of restricting its application as a low-level bioanalytical sensor. SERS is a variation of Raman spectroscopy in which the incoming laser beam interacts with electrons in plasmon oscillations in metallic nanostructures to augment, by several orders of magnitude, the vibrational spectra of molecules adsorbed to the surface. SERS is based on the million-fold or higher Raman signal improvement accomplished when a molecule interacts with surface plasmon modes of metal nanoparticles (Moskovits 2005). Success of SERS also relies on the unique set of Raman spectral peaks associated with the molecular vibrational modes of each molecule. Since SERS is a valuable tool for determining molecular structural information, and because SERS provides ultrasensitive detection limits including single molecule sensitivity, it has been utilized to detect bacteria and viruses using direct spectroscopic characterization. In SERS analysis, capture layers provide a means of bringing the target analyte to the nanoparticle surface without obscuring the desired Raman spectrum.

Haynes et al. (2005), Bell et al. (2005), Zhang et al. (2005), and Willets and Van Duyne (2007) elucidated the application of SERS for the detection of *Bacillus subtilis* spores, which are benign simulants for *Bacillus anthracis*. Structurally, a *Bacillus* spore consists of protective layers and a core cell. Calcium dipicolinate exists in these protective layers and is used as the *spore biomarker* because other potentially interfering species do not contain this chemical compound. Calcium dipicolinate was extracted from spores by sonicating in HNO_3 solution. A 3.1×10^{-13} M spore suspension was deposited onto an AgFON (silver film on nanosphere) substrate for the SERS measurement.

The AgFON substrate is a special substrate fabricated by a process called *nanosphere lithography* (NSL) for SERS measurements (Yonzon et al. 2005). Figure 18.3 illustrates the process of NSL. Nanospheres are drop coated onto a surface and allowed to self-assemble into a hexagonally close-packed array (steps 3a–c). The nanosphere assembly is followed by metal deposition, typically 15–100 nm, (step 3d). Two situations are of interest: (i) In the first case (step 3e), the nanosphere mask is washed away. After washing away the mask, the triangular nanoparticle array is left behind. (ii) In the second case, a thicker layer of metal (~200 nm) is deposited over the nanospheres, producing a metal film over nanosphere (FON). This substrate is particularly effective for SERS applications due to the stability of the local surface roughness on this curved substrate (step 3f).

A SERS spectrum was registered with a high signal-to-noise ratio in a 1-min data acquisition period. The spectrum is subjugated by bands associated with CaDPA. Previous Raman studies on Bacillus spores (Carmona 1980) also found such bands. In order to determine the saturation binding capacity of the AgFON surface as an internal standard to reduce the effect of intensity variations arising from laser power fluctuations, the SERS signal from extracted CaDPA was measured over the spore concentration range 10^{-14}–10^{-12} M. The data demonstrated that the limit of detection (LOD) by SERS is far lower than the anthrax infectious dose of 10^4 spores.

Using a commercially available transportable Raman instrument, a SERS spectrum of 10^4 *B. subtilis* spores dosed onto a 1 month old AgFON substrate was readily acquired. This was an achievement in the pursuit toward the goal of performing SERS detection of biowarfare agents in the field.

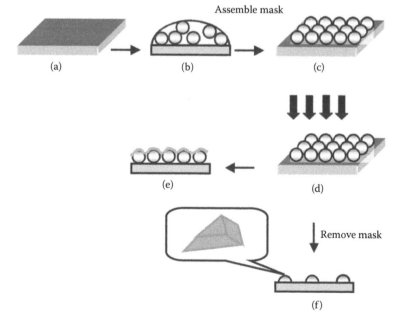

FIGURE 18.3 Fabrication of nanoparticle arrays and film over nanosphere (FON) surfaces for SERS by nanosphere lithography. (a) Subtrate cleaning, (b) Droplet coating, (c) Drying, (d) Metal deposition, (e) Film over nanosphere (FON), and (f) Triangular nanopartilcles.

A high signal-to-noise ratio spectrum was recorded within a data acquisition period of 5 s. Thus a range of possibilities were opened in the detection of harmful chemicals or bioagent weapons in real-world situations by exploiting the portability and ease of use of this type of device in conjunction with the molecular specificity and spectral sensitivity inherent to SERS.

It is worthwhile mentioning here that Pestov et al. (2008) considered a complementary approach, coherent *anti-Stokes Raman scattering (CARS) spectroscopy.* They showed the acquisition of a CARS spectrum from ~10^4 *Bacillus subtilis* spores in a single, femtosecond laser shot. Their scheme merged broadband Raman excitation by a pair of ultrashort laser pulses and a time-delayed narrowband probing of the consequential coherent molecular vibrations. The broadband preparation allows the monitoring of manifold Raman lines in a single measurement, thereby yielding specificity. The narrowband probing permits frequency-resolved acquisition. Recording of the whole CARS spectrum straight away without delay, and without tuning the laser wavelengths, makes the technique relatively protected from fluctuations. Their results suggested the efficacy of the technique and its promise for "on-the-fly" detection of biohazards, such as *Bacillus anthracis.*

18.5 Distinguishing Viral Strains by Silver Nanorod Arrays Using SERS

Strains are a group of organisms of the same species, having divergent characteristics but not usually considered a separate breed or variety. Influenza A virus strains are classified on the basis of initials of two virus surface proteins, followed by a number, e.g., in H1N1, H stands for hemagglutinin, an antigenic glycoprotein permitting the attachment of the virus to the cell that is being infected, and N denotes neuraminidase, an enzyme allowing the release of the virus from the host cell. All influenza A viruses enclose both the proteins, but the structure of these proteins is different from strain to strain, due to rapid mutation in the genome of the virus. An H number and an N number are allocated to each virus strain, based on the subtype of these two proteins found in the strain. In human beings, the only common subtypes found are H 1, 2, and 3, and N 1 and 2.

The state-of-the-art viral diagnostic methods involve isolation and cultivation of viruses (Dluhy et al. 2008). Two popular methods are as follows: (i) enzyme-linked immunosorbant assay (ELISA), which uses antibodies linked to an enzyme whose activity is used for quantitative determination of the antigen with which it reacts, and (ii) the polymerase chain reaction (PCR), which entails amplification of fragments of genetic material so that they become detectable. These diagnostic methods are cumbersome and time-consuming.

Zhao et al. (2006) showed that a silver nanorod array fabricated by oblique-angle deposition (OAD) acts as an extremely sensitive SERS substrate. Enhancement factors greater than 10^8 were within the realms of possibility (Figure 18.4). For the OAD method, they slanted the substrate in such a manner that the vapor arrived at nearly the grazing angle. This artifice caused preferential growth of nanorods on the substrate in the direction of deposition, through a shadowing effect.

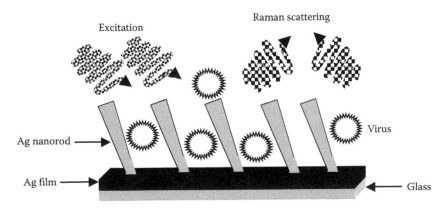

FIGURE 18.4 Spectroscopic assay using silver nanorod array SERS substrate.

The SERS substrates produced by OAD offer the advantages of large area, uniformity, and reproducibility (Driskell et al. 2008, Shanmukh et al. 2008). These substrates enable the speedy and economic development of SERS-based biosensors suitable for the detection of exceedingly low levels of viruses. The aforementioned researchers used the substrate to reveal that SERS can distinguish among different RNA viruses, namely, adenovirus, rhinovirus, and human immunodeficiency virus (HIV). The spectra were recorded within 1 min span at a virus volume <5 μL. These observations amply supported the assertion that SERS is able to establish reproducible molecular fingerprints of several crucial human respiratory viruses, as well as HIV.

Since SERS could detect and differentiate between different viral pathogens, it was vital to determine if different viral strains from a single pathogen could be demarcated (Shanmukh et al. 2006). To address this prospect, attention was focused on the SERS analysis of influenza virus (flu) A strain viruses A/HKx31, A/WSN/33, and A/PR/8/34. The analysis inferred that the spectra of the three viruses are identical. Nevertheless, dissimilarities exist in the relative intensities of the peaks in the spectra. This was attributable to a difference in the nature of binding of the surface proteins of this particular virus on the SERS substrate surface. Thus, SERS detection of flu strain spectra is satisfactorily different to allow identification of individual strains in an intricate mixture. This effect is unequivocally evident in the spectral regions between 900 and 700 cm^{-1}.

18.6 Plasmonic Sensing of Concavalin A and Optofluidic Nanoplasmonic Biosensor for Detection of Live Viruses

Surface plasmon resonance (SPR), a coherent oscillation of the surface conduction electrons excited by electromagnetic (EM) radiation, is supported by materials that possess a negative real and small positive imaginary dielectric constant. *Plasmonics* is the study of these particular light–matter interactions, which are finding a vast array of applications in biological and chemical sensing (Stuart et al. 2005, Vo-Dinh 2008).

The propagation constant of the surface plasma wave propagating at the interface between a semi-infinite dielectric and metal is expressed by the equation (Homola et al. 1999)

$$\beta = k\sqrt{\frac{\varepsilon_m n_s^2}{\varepsilon_m + n_s^2}} \tag{18.1}$$

where

k denotes the free space wave number
ε_m the dielectric constant of the metal ($\varepsilon_m = \varepsilon_{mr} + i\varepsilon_{mi}$)
n_s is the refractive index (RI) of the dielectric

As irresistibly clear from Equation 18.1, the surface plasma wave may be sustained by the structure provided that the condition $\varepsilon_{mr} < -n_s^2$ is satisfied. At optical wavelengths, several metals comply with this condition, of which gold and silver are the most commonly used.

Hwang et al. (2008) performed plasmonic sensing of Conconavalin A (Con A) using a nanohole array detector etched into a thin gold film as a transmission-based SPR sensor. In order to capture specific pathogens, the Au film was coated with carbohydrate receptor molecules. Real-time interaction between an infectious agent simulant, Con A carbohydrate-binding lectin, an infectious agent simulant, and mannos was measured on the glycoprotein ovomucoid. Taken as a whole, a paired polarization-sensitive detector enabled a detection resolution of 6.6×10^{-5} refractive index units (RIU).

Plasmonic nanohole arrays (PNAs) comprise arrays of nanoscale apertures (holes, with diameters of 250–350 nm). The apertures are defined periodically with pitch ~500–800 nm, on optically thick noble metal films such as gold films ~100 nm thickness. At particular wavelengths, these nanohole arrays are capable of transmitting light much more strongly than predicted by the classical aperture theory. This important phenomenon is called extraordinary optical transmission (EOT) effect. EOT signals are a consequence of the involvement of surface plasmon-polariton resonances (SPR).

The resonance wavelength of EOT signal is impressively correlated with the effective dielectric constant of the medium surrounding the plasmonic sensor. As in conventional nanohole arrays, the position of certain transmission resonances depends on the lattice constant obeying the following formula (Artar et al. 2009):

$$\lambda_{res} = \frac{p}{\sqrt{i^2 + j^2}} \sqrt{\frac{\varepsilon_d \varepsilon_m}{\varepsilon_d + \varepsilon_m}} \tag{18.2}$$

where

λ_{res} is the resonant frequency
ε_d is the dielectric constant of the medium, and as before
ε_m is the dielectric constant of metal
p is the periodicity of nanoholes
(i, j) are the grating orders

In comparison to traditional SPR-based detection, nanohole array-based detection is appropriate for device-level miniaturization/integration due to two factors, viz., the small footprint of the nanohole arrays and the collinear optical arrangement afforded by transmission mode operation. The integration of nanohole arrays in a fluidic chip environment has been a focus of several works. Yanik et al. (2010) demonstrated that the astonishing light transmission phenomena on PNA is adaptable to pathogen detection without being confounded by the flanking biological media by convincingly showing that the optofluidic nanoplasmonic sensors enable direct detection of uninjured viruses from biologically relevant media in a label-free fashion with trivial sample preparation.

The fabrication scheme is summarized in Figure 18.5. At the outset, free-standing SiN$_x$ membranes are produced employing a series of photolithographic and chemical wet etching (KOH) process steps

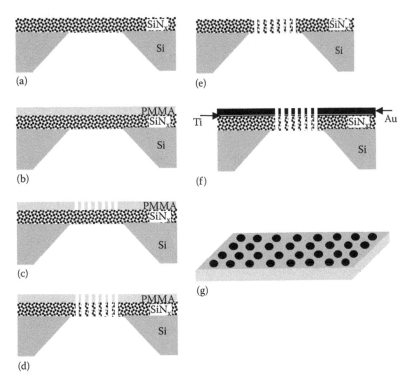

FIGURE 18.5 Process sequence for the fabrication of nanohole array starting from silicon nitride membrane: (a) silicon nitride membrane resting on silicon supports; (b) PMMA coating; (c) electron beam photolithography; (d) reactive ion etching (RIE) of silicon nitride; (e) removal of photoresist; (f) Ti–Au metallization; and (g) top view of nanohole array.

(not shown in the diagram). The membranes (Figure 18.5a) are then coated with positive electron-beam resist poly(methyl methacrylate) (PMMA) (Figure 18.5b). In the next stage, electron-beam lithography is performed to define the nanohole pattern in the resist (Figure 18.5c). This pattern is delineated in the SiN$_x$ membrane through a reactive ion etching (RIE) process (Figure 18.5d). An oxygen plasma etching process removes the resist (Figure 18.5e). Thus, a photonic crystal-like free-standing SiN$_x$ membrane is defined. Sequential deposition of the metal layers (5 nm Ti, 100 nm Au) results in free-standing plasmonic nanoholes transmitting light at resonance (Figure 18.5f). Figure 18.5g shows the top surface of the membrane.

The fabricated device consists of a suspended nanohole array grating. This grating couples the perpendicularly incident light to surface plasmons, electromagnetic waves trapped at metal/dielectric interface in coherence with collective electron oscillations (Figure 18.6). As biomolecules/pathogens fasten to the metal surface or to the ligands immobilized on the metal surface, the effective RI of the medium rises, and the red shifting of the plasmonic resonance occurs, i.e., the resonance phenomenon is displaced toward lower frequencies or larger wavelengths. Selectivity of sensing is secured by the surface immobilized highly specific antiviral immunoglobulins showing strong affinity to the viral membrane proteins.

Subwavelength confinement of the electromagnetic field to the metal/dielectric interface is made possible by the exponential decay of the extent of the plasmonic excitation. As a consequence, the sensitivity of the biosensor to the RI changes decreases severely with the increasing distance from the surface, thereby abating the effects of RI variations caused by temperature fluctuations in the bulk medium.

They used a genetically derived vesicular stomatitis virus (VSV)-pseudotyped Ebola (PT-Ebola), where the Ebola glycoproteins are expressed on the virus membrane instead of the VSV's own glycoprotein. Three sensors were reserved for reference measurements. Antibodies against the Ebola glycoprotein were immobilized on the remaining 9 out of the 12 sensors on a single chip. Successful functionalization of the protein antigens and the antibodies was corroborated by spectral measurements. After the immobilization of the antibodies, PT-Ebola in a PBS buffer solution was added onto the chips and incubation was done. Transmission spectra were collected after the washing process. Unfailing red shifting of the plasmonic resonances was noticed on antibody-coated spots indicating PT-Ebola detection ($\geq$14 nm red shift). In comparison, the reference sensors displayed no spectral shift. This occurred with high repeatability (9 out of 9 sensors, i.e., 100%) and excellent signal-to-noise ratios. Similarly, they tested the platform for the detection of enveloped DNA poxviruses. For this purpose, they utilized Vaccinia virus, a poxvirus that is frequently used as a prototype for more pathogenic viruses such as smallpox and monkeypox. An analogous approach (A33L Vaccinia antibody and immobilized on 9 of 12 sensors,

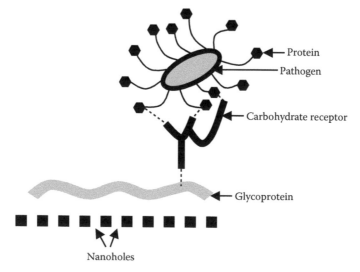

FIGURE 18.6 Mechanism of capturing pathogen through carbohydrate receptor and protein binding on a nanohole array.

incubation with undamaged vaccinia virus at the same concentration of 10^8 PFU/mL) yielded comparable affirmative results to those seen with Ebo-VSV. All the reference sensors indicated nominal binding. Contrarily, all the 9 sensors detected the virus.

18.7 Multiplexed Quantitation of Four Toxins Using Quantum Dots

Quantum dots (QDs) are semiconductor nanocrystals composed of elements from groups II–VI or III–V in the Periodic Table (Liang et al. 2005), typically chalcogenides (selenides or sulfides) of metals like cadmium or zinc (CdSe or ZnS), which range from 2 to 10 nm in diameter (about the width of 50 atoms). They are artificial atoms that confine electrons to a small space. The confinement of electrons and phonons is made possible by semiconductor nanocrystals, when the particle size is of the order of magnitude of the Bohr radius of the exciton.

QDs have several distinctive advantages compared to traditional organic fluorophores. Mention may be made of their narrow emission peaks, excellent photostability, and a broad range of fluorescence emission wavelengths extending from blue to infrared, depending on their physical size. A noteworthy fact is that when the same types of QDs with varying sizes are excited by single-wavelength radiation, the QDs emit at different wavelengths. These different but narrow emission bands make QDs very useful as photoluminescent labels. The wavelength of these photon emissions depends not on the material from which the quantum dot is made, but its size. The smaller the dot, the closer it is to the blue end of the spectrum, and the larger the dot, the closer to the red end. Dots can even be tuned beyond visible light, into the infrared or into the ultraviolet. Such broad emission of quantum dot spectra provides the unique opportunity of simultaneous excitation of different particle sizes at a single wavelength with emission at multiple wavelengths. Hence, they are used for the detection of multiple targets at the same time (Sapsford et al. 2006).

Goldman et al. (2004, 2006) coupled antitoxin antibodies to QDs with emission maximums at 510, 555, 590, and 610 nm (Figure 18.7). Antibody-conjugated QDs were prepared through a *mixed*

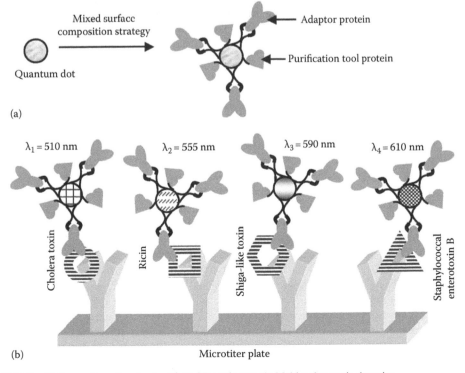

FIGURE 18.7 (a) Preparation of a mixed-surface QD conjugate. (b) Multianalyte toxin detection.

surface strategy. In this scheme, both an adaptor protein designed to bind antibodies through their Fc domain (PG-zb) and a purification tool protein (MBP-zb) were conjugated to each QD. The responsibilities assigned to these proteins are as follows: *adaptor protein* serves to bridge the QD and antibody while the *purification tool protein* is used to separate the QD fluoroimmunoreagent from any uncoupled antibody using an amylose affinity resin. Multiplexed sandwich fluoroimmunoassays were performed using antibody-conjugated QDs. These assays were aimed at the detection of cholera toxin, ricin, shiga-like toxin 1, and staphylococcal enterotoxin B simultaneously with all ingredients in single well of a microtiter plate. For each QD, individual spectra were measured. Subsequently, single and multianalyte assay results were deconvoluted by assuming a superposition of independent QD spectra. QD-based multianalyte assays show promise to be expanded to assay six or more toxins simultaneously.

18.8 Fluorescent Man-Au Nanodots (NDs) for the Detection of ConA and *E. coli*

A long-standing concern in the application of QDs for the detection in biological samples has been their possible cytotoxicity due to their heavy-metal components. Toxicity is a major drawback of QDs. To overcome this shortcoming, there is a demand for the development of nontoxic fluorescent nanoparticles (NPs) to substitute potentially toxic QDs in bioassays.

Huang et al. (2009) devised a simple and convenient method for the preparation of water-soluble biofunctional Au nanodots (NDs) for the detection of lectin (LOD = 75 pM), Figure 18.8. When the concentration of Con A was >25 nM, they observed by the naked eye that after centrifugation, green pellets of a Con A-Man-Au ND complex could be seen under ultraviolet lamp illumination, supporting the notion of Con A-induced aggregation of the Man-Au NDs. As a result of aggregation of the Man-Au NDs, the fluorescence intensities of the supernatants (the clear liquid overlying material deposited by settling, precipitation, or centrifugation) decreased upon increasing the concentration of Con A. They also developed a new method for fluorescence detection of bacteria using these water-soluble Man-Au NDs. Incubation with *E. coli* revealed that the Man-Au NDs bind to the bacteria, yielding brightly fluorescent cell clusters. This aggregation is due to the multivalent interactions between the mannosylated Au NDs and mannose receptors located on the bacterial pili. The pili are used for attachment of bacteria to soil, rocks, teeth, etc.

18.9 QD-Based Detection of *E. coli* O157:H7 and Salmonella Typhimurium

Pathogenic bacteria represent the foremost concern in food industries and water treatment facilities because of their rapid proliferation and lethal effects on human health (Hahn et al. 2005). Although standard microbiological methods of cell culture and plating are confirmative to identify bacterial strains, it commonly takes long duration, approximately several days to complete the processes. In addition, most of the conventional methods require complicated instrumentation and cannot be used on-site. Thus, both private and government sectors immediately need biosensors that have detection capability of pathogens in a fast and accurate manner (Heo and Hua 2009). Extremely sensitive immunosensors are suitable for fast screening in terms of response time and sensitivity.

Yang and Li (2006) used semiconductor QDs as fluorescence labels in immunoassays for detecting *Escherichia coli* O157:H7 and *Salmonella* Typhimurium, two of the most widespread foodborne pathogenic bacteria. Streptavidin-coated QDs 525 and streptavidin-coated QDs 705 were conjugated with biotinylated anti-*E. coli* antibody and biotinylated anti-Salmonella antibody, respectively (Figure 18.9). For quantitative detection of *E. coli* O157:H7 and *S.* Typhimurium simultaneously, the intensities of fluorescence emission peaks of the final complexes were measured at 525 and 705 nm. The fluorescence intensity as a function of cell number was determined for both the analytes. Then regression models were obtained for these analytes. The detection limit of this method was 10^4 cfu mL^{-1}.

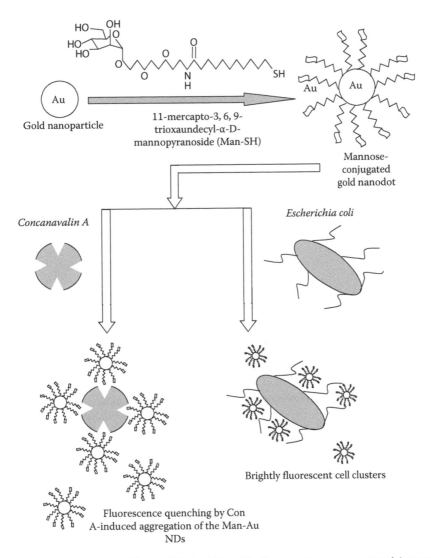

FIGURE 18.8 Detection of Concanavalin A and *Escherichia coli* by fluorescent mannose-protected Au nanodots.

18.10 Liposome-Caged AChE-Based Fluorescent Nanobiosensor for Pesticides Analysis

Liposomes are artificially prepared vesicles made of lipid bilayer. They are used to convey vaccines, drugs, enzymes, or other substances to target cells or organs. Vamvakaki and Chaniotakis (2007) developed nanobiosensors for the direct detection of two widely used organophosphorus pesticides, dichlorvos and paraoxon, and the determination of total toxicity in drinking water specimens. The inherently unstable enzyme acetylcholinesterase (AchE) is encapsulated in the internal nano-environment of liposomes. This method of enzyme entrapment has been established to greatly improve its stabilization against unfolding, denaturation and dilution effects. For the optical transduction of the enzymatic activity, the pH-sensitive fluorescent indicator pyranine was also immobilized within the liposomes. Increasing amounts of pesticides led to the decrease of the enzymatic activity for the hydrolysis of the acetylcholine and thus to a decline in the fluorescent signal of the pH indicator. It was found that the decrease of the liposome biosensor signal is dependent on the concentration of dichlorvos and paraoxon, downwards to 10^{-10} M levels.

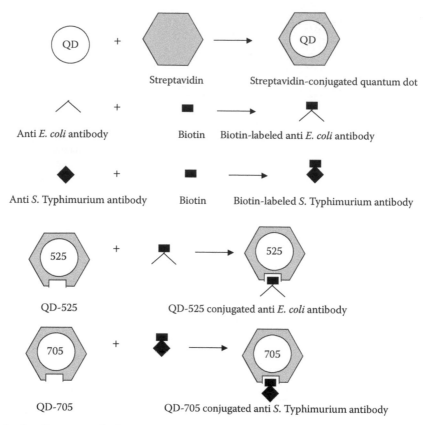

FIGURE 18.9 Coupling strategy for forming QD–antibody conjugates.

18.11 SERS Analysis of Fonofos Pesticide, Adsorbed on Ag and Au Nanoparticles

Vongsvivut et al. (2010) recorded the SERS spectra of fonofos pesticide using citrate-reduced silver (Ag) and gold (Au) colloidal nanoparticles in the form of dried films. Fonofos concentrations down to ~10 ppm could be detected with the Ag colloids, signifying a potential for the technique in the analysis of fonofos residues. Using the prepared Ag colloidal solution, an experiment was conducted to detect fonofos in a carbonated lemonade soft drink starting at ~10 ppm concentration. The storage vessel of the soft drink was left open overnight before the experiment or until the carbon dioxide dissipated, as evidenced by an absence of bubbles and a clear flat top surface. Furthermore, prior to adding fonofos and mixing an aliquot of Ag colloidal solution, the drink was brought to a neutral pH from the original pH of 3.5, using 0.1 M NaOH.

The SERS spectra collected from a dried film of the soft drink, treated as mentioned earlier, was compared to the Raman spectrum of an untreated soft drink. Interestingly, the SERS spectra presented clearly the fonofos signature without any noticeable interference from other additives contained in the drink (such as sugar, caffeine, flavorings, colors, preservatives). This was clearly discernible even though the signal achieved appeared to be substantially lower than that observed in SERS of pure fonofos at the same concentration and under the same sufficiently strong affinity to replace the citrate residue and to overcome other components in the same matrix to achieve a strong binding onto the Ag colloidal surfaces. It should however be remarked that the intensity of the signals varied substantially from one aggregate to another.

18.12 Layer-by-Layer (LBL) Biosensor Assembly Incorporating Functionalized Quantum Dots for Paraxon Detection

Constantine et al. (2003) showed that TGA-capped CdSe QDs can be sandwiched into the LbL film, improving the photoluminescence of the QDs. Consequent upon the improved optical photoluminescence, the system was used as a sensor to detect the presence of paraoxon at different concentrations.

Chitosan (CS) is a poly glucosamine which adsorbs strongly onto negatively charged surfaces and the adsorbed CS layer adopts a flat conformation providing a homogeneous film on which the thioglycolic acid (TGA)-capped CdSe QDs are adsorbed. Five bilayers of CS and QDs were prepared to produce a stable supramolecular film and to ensure surface charge uniformity.

Organophosphorus hydrolase (OPH) is an enzyme that catalyzes the hydrolysis of a large variety of organophosphorus compounds by producing risk-free products such as p-nitrophenol (PNP) and diethyl phosphate.

Multilayers of OPH/TGA-capped CdSe QDs were then integrated into the LbL system, so that on exposure to aqueous paraoxon solution, the hydrolytic product, pnitrophenol (PNP), is liberated and identified using UV-vis spectroscopy. Paraoxon was also perceived using the change in the photoluminescence intensity of the QDs.

18.13 TNT Detection Based on Molecularly Imprinted Polymers (MIPs) and SERS

2,4,6-Trinitrotoluene (TNT) is a powerful explosive for which detection techniques on a person's body or in one's baggage are considered important for assuring safety of airports and air travel. Holthoff et al. (2011) reported a nanosensor strategy that integrates molecularly imprinted sol–gel derived xerogel polymers using a noncovalent imprinting approach in combination with SERS. To detect the explosive, TNT, xerogels were molecularly imprinted for TNT using noncovalent interactions with the polymer matrix. Binding of the TNT within the polymer matrix results in unique SERS bands, which permit recognition of the molecule in the molecularly imprinted polymer.

18.14 Portable Surface Plasmon Resonance Biosensor System for Real-time Toxin Detection

The fundamental Kretschmann structure forms the basis of the SPR biosensors. In Kretschmann design, the intensity of transverse magnetic (TM) polarized light reflected of a thin gold layer (~50nm) on the surface of a prism shows a dependence on the angle of incidence or wavelength of the incident light (Figure 18.10). The SPR curve or profile is a graph of the intensity of reflection versus the angle of reflection. Harmonization of the frequency and momentum of the incident light with that of the surface plasmon is accompanied by the formation of a valley or dip in the SPR curve. The RI of the medium in contact with the outer surface of the gold layer determines the angle or wavelength at which the minimum of reflection takes place.

A condition for monitoring the presence of specific targets in real time is furnished by the attachment of specific recognition elements on the gold surface (usually antibodies), and passivation of the gold surface to nonspecific binding. Because the RI of protein ($n = 1.45$) is greater than that of usual aqueous buffers ($n \sim 1.334$), when an analyte of RI greater than that of water/buffer and of sufficient size is bound at the surface, the change in RI is substantial enough to alter the position of the minimum of the SPR curve. Through the instrumentation software, the alteration in SPR minima is transformed into RI as a function of time. This facilitates the analysis of the binding event in real time.

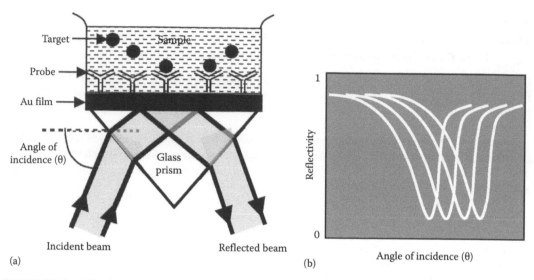

(a)   (b)

FIGURE 18.10 (a) Surface plasmon resonance (SPR) sensor surface with attached antibodies that bind specific analytes. (b) Diagram showing that the minima of SPR curves are displaced toward the right as the analyte is bound.

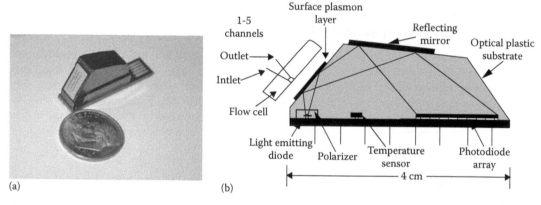

(a)   (b)

FIGURE 18.11 Spreeta sensor element around which the portable sensor system is constructed: (a) Small-size three-channel sensor element with a U.S. dime (18 mm) placed close by for dimensional comparison, and (b) components of Spreeta sensor element. (Reprinted with permission from Springer Science+Business Media: *J. Ind. Microbiol. Biotechnol.* A portable surface plasmon resonance sensor system for real-time monitoring of small to large analytes, 32, 669–674, Soelberg, S. D., Geiss T. C. G., Spinelli C. B., Near R.S. S., Kauffman P., and Furlong, S. Y. C. E., 2005, DOI 10.1007/s10295-005-0044-5. Copyright 2005 Society for Industrial Microbiology.)

Advancements in miniaturization of SPR technology have rendered possible the development of portable systems that are adaptable to rapid identification of possible agents of biological or chemical warfare. Soelberg et al. (2005) described a portable SPR sensor system (Figure 18.11). This miniature SPR sensing chip (Figure 18.10) is a fully integrated SPR sensor element whose chief components are an LED as a light source, a gold SPR surface, a reflecting mirror for directing the reflected light to a photodiode array, and a temperature sensor. The output signal derived from the sensor element is processed by a digital signal processing chip and fed to a computer, which generates and displays both the SPR curves and a plot of RI with respect to real time.

When groping for the presence of toxic agents, it is essential to steer clear of false positives. Once an expected detect signal has been noticed, a second antibody with specificity for a different target epitope than the one used to capture the analyte is made to flow through the sensor system.

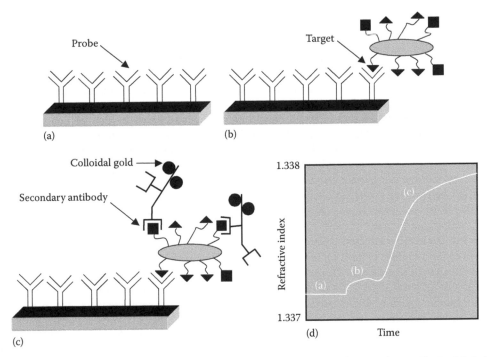

FIGURE 18.12 (a) Open condition. (b) Signal detection. (c) Signal amplification with secondary antibodies labeled with colloidal gold particles. (d) Refractive index time graph.

Specific binding of this second antibody both authenticates that the target analyte was bound with the added advantage of amplification of the original signal. As an example, a general antibody for Salmonella strains is used as the capture antibody on the sensor surface. Upon detecting a signal, antibodies specific for different strains are used to identify the specific serological strain of organisms bound on the sensor surface. The step also amplifies the original signal if colloidal gold is attached to the secondary antibody (Figure 18.12).

18.15 Portable Raman Integrated Tunable Sensor for SERS Detection of Biological and Chemical Agents

SERS originates from molecules in close vicinity of nanostructured metal surfaces, primarily silver and gold, that can support plasmon resonances in the visible spectral region (Hudson and Chumanov 2009). Because of the aggressive development of SERS substrates and application to a wide range of chemicals, the potential of SERS as a routine analytical technique has long been recognized. Yan and Vo-Dinh (2007) reported the construction of a fully integrated portable field-deployable Raman sensor instrument. It consists of the following components: (i) an 830-nm diode laser for excitation, (ii) holographic notch filters for Rayleigh rejection, (iii) an acousto-optic tunable filter (AOTF) for wavelength discrimination, and (iv) an avalanche photodiode (APD) for detection. It provides direct identification of chemical and biological samples in a few seconds under field conditions.

They discussed its applications for the detection of methyl parathion and DPA. Chemical warfare simulants such as dimethyl methylphosphonate (DMMP), pinacolyl methylphosphonate (PMP), diethyl phosphoramidate (DEPA), and 2-chloroethyl ethylsulfide (CEES) were cited as also biological warfare simulants, e.g., *Bacillus globigii* (BG), *Erwinia herbicola* (EH), and *Bacillus thuringiensis* (BT).

18.16 Discussion

Various types of optochemical nanosensors warning about the bioterrorist agents were described. Numerous nanosensing approaches and techniques for the detection of BWAs and CWAs were delved into. Each technique has its own advantages and disadvantages. The most critical parameters of these nanosensors include (i) sensitivity, (ii) specificity, (iii) short detection time, (iv) simplicity, (v) low cost, and (vi) wide concentration range. More consideration should be given to a simple, cheap, sensitive, selective, and field-deployable detector (Khanna 2008, 2009, Ahmed et al. 2011).

Mainly fluorescent, plasmonic and SERS techniques have been applied in these nanosensors (Abraham et al. 2008). QDs promise more sensitive, multiplexed sensors. For the sake of comparison, a few nanosensors for chemical terrorist agents were also briefly mentioned, indicating the universal applicability of aforesaid methods to the chemical field. General approaches for realization of all these nanosensors are essentially the same. Therefore, nanosensors for bioterrorism form a subset of the bigger warfare agent group of nanosensors.

It is evident that in many of these endeavors a suitable bio- or chemical marker of the toxic agent is chosen and attention is directed toward its identification, e.g., DPA and calcium dipicolinate (CaDPA) were selected as markers of anthrax spores. Then by building nanosensors for these marker species, the eventual analyte of interest is easily recognized. Moreover, for safety considerations, harmless simulants of BWAs/CWAs are used for laboratory experiments before trying the nanosensors in field application.

Furthermore, it must be reiterated that these nanosensors are also useful for facing the threats of epidemics during peace time, e.g., the pathogen nanosensors of the biological family are utilized for detecting the onslaught of infectious diseases, whereas pesticide nanosensors of chemical family are used for environmental and ecological surveillance, so that the research and efforts made for developing these sophisticated devices are useful for improving the living and health conditions of human beings under normal conditions at large. Table 18.3 provides a glimpse of some of the milestones achieved by different researchers. SERS has been attractive to several researchers because it is an extremely sensitive and selective technique that involves enhancements in the Raman scattering intensities of analytes adsorbed on a roughened metal surface (typically, gold or silver) (Dluhy et al. 2008, Krause et al. 2008). These enhancements (up to 14–15 orders of magnitude as compared to spontaneous Raman) are due to chemical and electromagnetic enhancement, which results when the incident light in the experiment strikes the metal surface and excites localized surface plasmons. The detection capabilities of SERS make it an excellent transduction method for selective, full compound identification. SERS provides ultrasensitive detection limits, even approaching single molecule sensitivity.

Needless to say that these optical nanobiosensors share all the advantages of optical nanosensors, notably, optical biosensors allow remote transduction of the biomolecular binding response from the sensing region without any physical connection between the excitation source and the detection channel. Unlike mechanical and electrical sensors, they are also compatible with physiological solutions and are not sensitive to the changes in the ionic strengths of the solutions. Further, the sensing elements and propagating light do not in any way interfere with the surrounding fluid medium and the biochemistry, making them extremely biocompatible.

18.17 Conclusions and Outlook

Significant progress has been achieved in the battle against bioterrorism. Success of these nanosensors outside the laboratory in adverse field conditions prevalent during a war will be assured by creating suitable nanorobots that could tap the full potential of these nanosensors in tough situations (Cavalcanti et al. 2008a,b). Nanosensors serve as the cornerstone of accurate and timely measurements. The increased sophistication of nanosensors and nanosystems calls for improved signal processing, data logging, and performance measurements. Research continues toward perfection of technologies because in public settings there is a need for the early detection of BWAs and CWAs. This is essential in order that parts of buildings can be rapidly isolated or evacuated. Also, the probability of false positives is avoided keeping in view the adverse economic impact caused by hoax alarms, leading to unnecessary evacuations.

TABLE 18.3

Anti-Bioterrorism/Chemical Bioterrorism Optochemical Nanosensors at a Glance

Sl. No.	BWA/CWA	Indicator of BWA/CWA	Construction of the Nanosensor	Optochemical Principle Used
1.	Anthrax	Dipicolinic acid	(EDTA)-Tb(III) complexes immobilized on silica nanoparticles	Luminescence signal accompanying the coordination of DPA to Tb^{3+}
		Dipicolinic acid	Dipicolinic acid imprinted Au-Ag nanoshell clusters	Change in fluorescence due to complexation between DPA molecules and DPA cavities on Au–Ag nanoshells
		Calcium dipicolinate	Spore suspension deposition on an AgFON substrate	SERS
2.	RNA viruses	Adenovirus, rhinovirus, and human immunodeficiency virus (HIV)	SERS substrate	SERS
3.	Infectious diseases	Concavalin A	Nanohole array detector	Interaction between Con A and mannose on a transmission-based SPR sensor
4.	Hemorrhagic fever, small pox, monkey pox	PT-Ebola, Vaccinia	Optofluidic nanoplasmonic sensors	Red shifting of the plasmonic resonances on antibody coated spots
5.	Toxins	Cholera toxin, ricin, shiga-like toxin, and staphylococcal enterotoxin B	Multiplexed QD operation	Sandwich fluoroimmunoassay
6.	Lectin and bacteria	Con A and *E. coli*	Mannose-conjugated Au nanodots	Man-Au aggregation induced fluorescent intensity changes
7.	Foodborne bacteria	*Escherichia coli* O157:H7 and *Salmonella* Typhimurium	Anti-*E. coli* antibody and anti-Salmonella antibody coated QDs	The intensities of fluorescence emission peaks of QDs
8.	Organophosphorus pesticides	Dichlorvos and paraoxon	Liposome-caged AChE-based fluorescent nanobiosensor	Fluorescent signal changes with pesticide concentration
		Fonofos	Ag and Au colloidal nanoparticles as dried films	SERS spectra analysis
		Paraxon	OPH/TGA-capped CdSe QDs integrated into the LbL system	Change in the photoluminescence intensity of QD
9.	TNT	TNT analog	QD-anti-TNT antibody bioconjugate	Fluorescence immunoassay
10.	Miscellaneous	Viruses, microbes, and spores	Both the SPR curves and a plot of RI versus real time	SPR
		Methyl parathion, dipicolinic acid, DMMP, PMP, DEPA, CEES, BG, EH, BT	Raman integrated tunable sensor	SERS

Acknowledgment

The author thanks the Director, CSIR-CEERI, Pilani for encouragement and guidance during the course of this work.

REFERENCES

Abraham A. M., R. Kannangai, and G. Sridharan. 2008. Nanotechnology: A new frontier in virus detection in clinical practice. *Ind. J. Med. Microbiol.* 26(4): 297–301.

Ahmed S. R., S. C. Hong, and J. Lee. 2011. Optical and electrical nano eco-sensors using alternative deposition of charged layer. *Front. Mater. Sci.* 5(1): 40–49. DOI: 10.1007/s11706-011-0117-5.

Artar A., A. A Yanik, and H. Altug. 2009. Fabry–Pérot nanocavities in multilayered plasmonic crystals for enhanced biosensing. *Appl. Phys. Lett.* 95(5): 051105.

Bell S. E. J., N. Joseph, J. N. Mackle, and N. M. S. Sirimuthu. 2005. Quantitative surface-enhanced Raman spectroscopy of dipicolinic acid—Towards rapid anthrax endosporedetection. *Analyst* 130: 545–549.

Carmona P. 1980. Vibrational spectra and structure of crystalline dipicolinic acid and calcium dipicolinate trihydrate. *Spectrochim. Acta A Molec. Spectrosc.* 36(7): 705–712.

Cavalcanti A., B. Shirinzadeh, R. A. Freitas Jr., and T. Hogg. 2008a. Nanorobot architecture for medical target identification. *Nanotechnology*, 19(1): 015103(15pp). DOI: 10.1088/0957-4484/19/01/015103.

Cavalcanti A., B. Shirinzadeh, M. Zhang, and L. C. Kretly. 2008b. Nanorobot hardware architecture for medical defense. *Sensors* 8: 2932–2958. DOI: 10.3390/s8052932.

Constantine C. A., K. M. Gatta's-Asfura, S. V. Mello, G. Crespo, V. Rastogi, T.-C. Cheng, J. J. DeFrank, and R. M. Leblanc. 2003. Layer-by-layer biosensor assembly incorporating functionalized quantum dots. *Langmuir* 19: 9863–9867.

Dluhy R. A., R. A. Tripp, Y. Zhao, and J. Driskell. 2008. Surface enhanced Raman spectroscopy (SERS) systems for the detection of viruses and methods of use thereof, United States Patent Application 20090086201, Filing Date: June 09, 2008, Publication Date: April 02, 2009.

Driskell J. D., S. Shanmukh, Y. Liu, S. B. Chaney, X.-J. Tang, Y.-P. Zhao, and R. A. Dluhy. 2008. The use of aligned silver nanorod arrays prepared by oblique angle deposition as surface enhanced Raman scattering substrates. *J. Phys. Chem. C* 112(4): 895–901. DOI: 10.1021/jp075288u.

Goldman E. R., A. R. Clapp, G. P. Anderson, H. T. Uyeda, J. M. Mauro, I. L. Medintz, and H. Mattoussi. 2004. Multiplexed toxin analysis using four colors of quantum dot fluororeagents. *Anal. Chem.* 76: 684–688.

Goldman E. R., I. L. Medintz, and H. Mattoussi. 2006. Luminescent quantum dots in immunoassays. *Anal. Bioanal. Chem.* 384: 560–563, DOI 10.1007/s00216-005-0212-5.

Gültekin A., A. Ersöz, N. Y. Sarıözlü, A. Denizli, and R. Say. 2010. Nanosensors having dipicolinic acid imprinted nanoshell for Bacillus cereus spores detection. *J. Nanopart. Res.* 12: 2069–2079.

Hahn M. A., J. S. Tabb, and T. D. Krauss. 2005. Detection of single bacterial pathogens with semiconductor quantum dots. *Anal. Chem.* 77: 4861–4869.

Haynes C. L., C. R. Yonzon, X. Zhang, and R. P. Van Duyne. 2005. Surface-enhanced Raman sensors: Early history and the development of sensors for quantitative biowarfare agent and glucose detection. *J. Raman Spectrosc.* 36: 471–484.

Heo J. and S. Z. Hua. 2009. An overview of recent strategies in pathogen sensing. *Sensors* 9: 4483–4502. DOI:10.3390/s90604483.

Holthoff E. L., D. N. Stratis-Cullum, and M. E. Hankus. 2011. A nanosensor for TNT detection based on molecularly imprinted polymers and surface enhanced Raman scattering. *Sensors* 11: 2700–2714. DOI:10.3390/s110302700.

Homola J., S. S. Yee, and G. Gauglitz. 1999. Surface plasmon resonance sensors: Review. *Sensor Actuat B Chem.* 54(1–2): 3–15.

Huang C.-C., C.-T. Chen, Y.-C. Shiang, Z.-H. Lin, and H.-T. Chang. 2009. Synthesis of fluorescent carbohydrate-protected Au nanodots for detection of concanavalin A and escherichiacoli. *Anal. Chem.* 81: 875–882.

Hudson S. D. and G. Chumanov. 2009. Bioanalytical applications of SERS (surface-enhanced Raman spectroscopy). *Anal. Bioanal. Chem.* 394: 679–686. DOI: 10.1007/s00216-009-2756-2.

Hwang G. M., L. Pang, E. H. Mullen, and Y. Fainman. 2008. Plasmonic sensing of biological analytes through nanoholes. *IEEE Sens. J.* 8(12): 2074–2079.

Khanna V. K. 2008. Nanoparticle-based sensors. *Defence Sci. J.* 58(5): 608–616.

Khanna V. K. 2009. Frontiers of nanosensor technology. *Sensor. Transduc. J.* 103(4): 1–16.

Krause D. C., S. M. L. Hennigan, R. A. Dluhy, J. Driskell, Y. Zhao, and R. A Tripp. 2008. Surface enhanced Raman spectroscopy (SERS) systems for the detection of bacteria and methods of use thereof, United States Patent 7889334, Filing Date: June 16, 2008, Publication Date: February 15, 2011.

Liang S., D. T. Pierce, C. Amiot, and X. Zhao. 2005. Photoactive nanomaterials for sensing trace analytes in biological samples. *Syn. React. Inorg Met-Org. Nano-Met. Chem.* 35: 661–668. DOI: 10.1080/15533170500299859.

Moskovits M. 2005. Surface-enhanced Raman spectroscopy: A brief retrospective. *J. Raman Spectrosc.* 36(6–7): 485–496. DOI:10.1002/jrs.1362.

Pestov D., X. Wang, G. O. Ariunbold, R. K. Murawski, V. A. Sautenkov, A. Dogariu, A. V. Sokolov, and M. O. Scully. 2008. Single-shot detection of bacterial endospores via coherent Raman spectroscopy. *PNAS* 105(2): 422–427.

Sapsford K. E., T. Pons, I. L. Medintz, and H. Mattoussi. 2006. Biosensing with luminescent semiconductor quantum dots. *Sensors* 6: 925–953.

Shanmukh S., L. Jones, Y.-P. Zhao, R. A. Dluhy, and R. A. Tripp. 2006. Rapid and sensitive detection of respiratory virus molecular signatures using a silver nanorod array SERS substrate. *Nano Lett.* 6(11): 2630–2636. DOI:10.1021/nl061666f.

Shanmukh S., L. Jones, Y.-P. Zhao, J. D. Driskell, R. A. Tripp, and R. A. Dluhy. 2008. Identification and classification of respiratory syncytial virus (RSV) strains by surface-enhanced Raman spectroscopy and multivariate statistical techniques. *Anal. Bioanal. Chem.* 390(6): 1551–1555. DOI: 10.1007/s00216-008-1851-0.

Soelberg S. D., T. C. G. Geiss, C. B. Spinelli, R. S. S. Near, P. Kauffman, and S. Y. C. E. Furlong. 2005. A portable surface plasmon resonance sensor system for real-time monitoring of small to large analytes. *J. Ind. Microbiol. Biotechnol.* 32: 669–674.

Stuart D. A., A. J. Haes, C. R. Yonzon, E. M. Hicks, and R. P. Van Duyne. 2005. Biological applications of localised surface plasmonic phenomenae. *IEE Proc. Nanobiotechnol.* 152(1): 13–32.

Taylor K. M. L. and W. Lin. 2009. Hybrid silica nanoparticles for luminescent spore detection. *J. Mater. Chem.* 19: 6418–6422.

Vamvakaki V. and N. A. Chaniotakis. 2007. Pesticide detection with a liposome-based nano-biosensor. *Biosens. Bioelectron.* 22: 2848–2853.

Vo-Dinh T. 2008. Nanobiosensing using plasmonic nanoprobes. *IEEE J. Sel. Top. Quantum Electron.* 14(1): 198–205.

Vongsvivut J., E. G. Robertson, and D. McNaughton. 2010. Surface-enhanced Raman spectroscopic analysis of fonofos pesticide adsorbed on silver and gold nanoparticles. *J. Raman Spectrosc.* 41: 1137–1148.

Webber M. E., M. Pushkarsky, C. Kumar, and N. Patel. 2005. Optical detection of chemical warfare agents and toxic industrial chemicals: Simulation. *J. Appl. Phys.* 97: 113101-1 to 113101-11.

Willets K. A. and R. P. Van Duyne. 2007. Localized surface plasmon resonance spectroscopy and sensing. *Annu. Rev. Phys. Chem.* 58: 267–297.

Yan F. and T. Vo-Dinh. 2007. Surface-enhanced Raman scattering detection of chemical and biological agents using a portable Raman integrated tunable sensor. *Sensor Actuat. B Chem.* 121: 61–66.

Yang L. and Y. Li. 2006. Simultaneous detection of *Escherichia coli* O157:H7 and *Salmonella* Typhimurium using quantum dots as fluorescence labels. *Analyst* 131: 394–401.

Yanik A. A., M. Huang, O. Kamohara, A. Artar, T. W. Geisbert, J. H. Connor, and H. Altug. 2010. An opto-fluidic nanoplasmonic biosensor for direct detection of live viruses from biological media. *Nano Lett.* 10: 4962–4969.

Yonzon C. R., D. A. Stuart, X. Zhang, A. D. McFarland, C. L. Haynes, and R. P. Van Duyne. 2005. Towards advanced chemical and biological nanosensors—An overview. *Talanta* 67: 438–448.

Zhang X., M. A. Young, O. Lyandres, and R. P. Van Duyne. 2005. Rapid detection of an anthrax biomarker by surface-enhanced Raman spectroscopy. *J. Am. Chem. Soc.* 127: 4484–4489.

Zhao Y., S. Shanmukh, Y. Liu, L. Jones, R. A. Dluhy, and R. A. Tripp. 2006. Silver nanorod arrays can distinguish virus strains. *SPIE Int. Soc. Opt. Eng.* 10.1117/2.1200610.0438: 1–3.

19

Nano-Optical Sensors for Food Safety and Security

Euiwon Bae and Arun K. Bhunia

CONTENTS

19.1 Introduction

19.1.1 Overview

Recent incidence of foodborne outbreaks due to pathogen contamination demands a microbial surveillance strategy that could routinely monitor food systems for the presence of harmful microorganisms administered naturally or intentionally. Continued outbreaks from food commodities remind us the vulnerability of the food supply chain to microbial contamination that may occur at any step of the food production chain: during harvesting of raw materials, processing, packaging, transport, or retail distribution. Foodborne infections may be of bacterial, viral, mold, or parasitic origin. Centralized food manufacturing practices and globalized raw material sourcing and processed food distribution made food safety a global issue that requires concerted efforts to improve food safety and food defense, and management practices. Globally food-related outbreaks are widespread; however, foodborne infection statistics are available only for a few developed countries [1]. The Centers for Disease Control and Prevention (CDC) in the United States estimates 48 million cases of foodborne illness occur each year, among which 3000 are fatal. The major foodborne pathogens include *Listeria monocytogenes*, *Salmonella* (non-typhoidal serotypes), *Escherichia coli* O157:H7, *E. coli* non-O157 STEC (Shiga-toxin–producing *E. coli*), *Campylobacter*, Norovirus, and *Toxoplasma gondii* and are responsible for a majority of outbreaks [2]. The economic losses to the food industry account in billions of dollars ($152 billion/year) [3].

Traditional technique for detecting pathogenic bacteria from food involves culturing and plating. Even though this method is regarded as a gold standard [4], it involves many labor-intensive steps; sample preparation, rinsing or homogenization, growing, plating, and subculturing, and takes up to 3–10 days to provide results. To expedite detection, many researchers have used antibody or nucleic acid–based methods [5,6]. The actual assay time is reasonably fast, requiring a few minutes to hours, but a prolonged sample preparation and enrichment step (18–48 h) made many of these tests rather lengthy [7]. In recent years, nucleic acid–based technologies such as multiplex-PCR [8], PCR coupled with mass spectrometry [9], or high-density genome sequencing or microarray [10,11] are shown to be promising novel technologies for food safety and food defense applications. Alternatively, biosensor-based methods including electrical, electrochemical, mass-based, and optical sensors have been developed to facilitate rapid detection and diagnosis of infective agents in food or clinical samples [12–19].

19.1.2 Optical Methods for Nanometer Scale Detection

Application of optical diagnostics tools in pathogen detection offers many advantages: methods are generally non-destructive, i.e., they maintain original sample integrity, deliver results quickly, and the system can be engineered to make portable. Depending on the technology platform and the probes or reagents used, they can be made highly sensitive or specific. Some optical sensor platform works independent of any probes or labeling reagents and the inherent natural biomolecular differences provide specificity for a given pathogen. The signals generated from a nano-scale object are roughly proportional to their volumes. Researchers have also found a number of ways to amplify sensor signals by (i) increasing the detector sensitivity, (ii) increasing the transducer signal by filtering the unwanted signal or noise, and (iii) amplifying the signal generated from the sample itself. Detector sensitivity can be improved by selecting ultra-sensitive devices such as photomultiplier tube (PMT), where a single photon is amplified about 1000 times, or cooled charge coupled device (CCD), where thermal and electronic noise levels are very low. Selection of appropriate target-specific labels can also increase the sensitivity. Signal-to-noise ratio (SNR) can be enhanced by spectrally blocking unwanted signals and by accepting only desired ones. Finally, sensitivity can be enhanced by increasing the amount of the target analyte deposited on the sensor platform. For example, PCR allows amplification of the target DNA from a pathogen, which can be readily detected by an optical sensor. Furthermore, amplification can be achieved by allowing microbial cells to grow to a level that are within the threshold detection limit for the sensor. Here, we reviewed the five major optical sensors that are currently being developed as nano-optical sensors for food safety and biosecurity applications.

19.2 Elastic Light Scattering

19.2.1 Introduction

Elastic light scattering (ELS) is defined as an optical measurement technique that utilizes the characteristics of the spatial distribution of the scattered light. ELS signal strength is very high compared to other spectroscopic and inelastic scattering techniques. By analyzing the ELS signal, it is possible to solve an inverse-scattering problem without any specific labeling reagent such as nucleic acid (DNA or RNA) or antibody probes, fluorophore molecules, or enzymes. Due to its unparallel performance, ELS has been used in diverse science and engineering fields such as astronomy, semiconductor industry, and biology. Furthermore, ELS method is non-destructive, i.e., it maintains the sample integrity during interrogation and the signal measurement is instantaneous.

19.2.2 Measurement Principle and Instrument

In ELS, the scattered light emanating from a sample is captured by a photo detector, PMT or CCD. PMT works well as a point detector since it has high gain amplification capacity while CCD is better suited for detecting 2-D spatial distribution of the scattered light.

As shown in Figure 19.1, when an incident beam with wavelength (λ) impinges on a mixture of particles with diameter (D), the elastically scattered light (I_s) spreads out to the 4π spherical direction with different polar and azimuthal angles (θ, Φ). The scattered light can be measured and analyzed in three different ways: (i) integration of the amount of all the scattered light called total scattering cross-section (TSC). This is proportional to the size of the individual scatterer, i.e., sample, and it can estimate the diameter of the sample. (ii) Measurement of the point intensity using PMT at a scattering angle. This determines the differential scattering cross-section (DSC) for the designated scattering angle to provide angle-resolved scattering (ARS) patterns. Typically, this type of measurement is performed in a goniometric setup to rotate the point detector in both azimuthal and polar directions. (iii) Finally, generation of a TSC versus wavelength, which is defined as an elastic scattering spectroscopy (ESS). To achieve this, a broadband light source and series of filters are required to provide a band-limited incident beam that sweeps across the different wavelengths. Typical ELS measurements have strong signals and are label-free in nature but they suffer from high background noise when coherent light is used. This worsens as the size of the scatterer (sample size) becomes smaller since their scattering intensity decrease with the sixth power of the diameter. Therefore, to improve detection at the nano-scale, researchers proposed using high refractive index materials such as gold or silver nano-particle as a scattering contrast agent since metals have larger scattering cross-section per volume [20,21].

Figure 19.2 shows typical ELS instruments for both single cell and bacterial colony detection. Bacterial growth on solid surface results in the formation of colonies. Even though optical sensors are also developed to detect and identify bacterial colonies at micrometer range, approaches to detect and identify colonies at the submicron level are often difficult since individual colonies of different species or serovars may have biomaterial signatures that may not be adequate to provide distinguishing features. Prolonged growth, resulting in higher cell mass, may provide differential signals that can be detected and identified rapidly. Figure 19.2a shows the principles of fluorescence activated cell sorter (FACS), which is also called flow cytometer. Even though fluorescence tag is generally used, the system still measures the forward and right angle scatter to differentiate the cell size. Figure 19.2b displays the BActerial Rapid

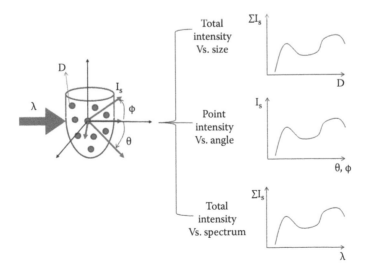

FIGURE 19.1 Schematic diagram depicting principles of elastic light scattering sensor (ELS). When the incident light with wavelength (λ) impinges samples consists of particle of diameter (D), scattered light (I_s) will spread out in hemispherical direction (θ and Φ). Scattered light intensity can be measured and plotted against D, or θ and Φ, or λ to retrieve sample information.

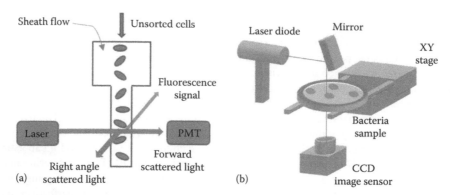

FIGURE 19.2 Examples of elastic light scattering sensor (ELS): (a) fluorescence activated cell sorter (FACS) and (b) BActerial Rapid Detection using Optical scattering Technology (BARDOT) instrument, which consists of a diode laser of 635 nm and a CCD detector to capture 2D scattering patterns from bacterial colonies.

Detection using Optical scattering Technology (BARDOT) instrument which consists of a diode laser of 635 nm and a CCD detector to capture 2D scattering patterns from bacterial colonies [22,23]. Once the scattering patterns are captured, it is stored in the database to be used as a fingerprint library for future detection and identification of pathogens or non-pathogenic bacteria using advanced classification algorithm [22,24].

19.2.3 Application of ELS

The first reported attempt to differentiate bacterial culture by using light scattering was by Wyatt [25,26], who used ARS to differentiate different microbial species suspended in liquid. Since then, many researchers [27–30] manipulated the incident light and used polarized light to determine the size and diameter distribution [27–31], sporulation and differentiation [31], structure and changes in the morphology [29,32], and metal toxicity on bacterial cells [32]. Furthermore, the light scattering pattern of *E. coli* cells was investigated using a scanning flow cytometer (SFC) where the scattering polar angle was set to 5–100 [33]. Various cell types suspended in liquid were differentiated by measuring the hologram from a lens-less holographic imaging device [34–36]. To identify bacterial colony growth on solid agar plate using 2-D spatial scattering pattern, BARDOT device was introduced [22,23,37]. Later, this system was used to measure bacterial colony growth and differentiation [23,38,39]. A quantitative image processing algorithm was developed using Zernike polynomial and a support vector machine [24]. The BARDOT system was automated to include a secondary camera to map colony locations in the Petri dish, and colony centering and traveling salesman algorithm to generate scatter images of colonies of interest [39]. This system was successful in detecting and identifying pathogenic bacteria such as *L. monocytogenes, Staphylococcus, Salmonella, Vibrio*, and *E. coli* O157:H7 from inoculated meat, vegetable, and seafood samples [37,40].

19.3 Fourier Transform Infrared Spectroscopy

19.3.1 Introduction

Fourier transform infrared spectroscopy (FTIR) is one of the non-destructive chemical imaging techniques which measures the wavelength dependent light absorption characteristics to identify samples based on the chemical composition and unique molecular vibration [41,42]. Typical wavelength range is from near-IR (714 nm–2.5 μm), mid-IR (2.5–25 μm), to far-IR (25–1000 μm). Since molecules and their structures respond differently to the incoming IR light, it is possible to provide a unique chemical fingerprint signature of molecules by understanding their fundamental and rotational vibrations and recording their spectral signatures (Figure 19.3).

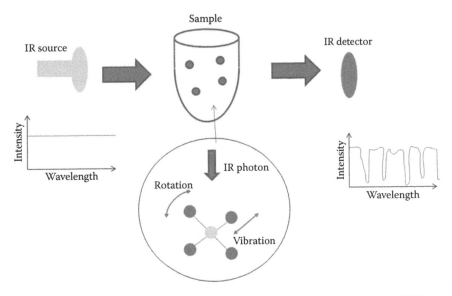

FIGURE 19.3 Schematic diagram depicting principles of Fourier transform infrared spectroscopy (FTIR).

19.3.2 Measurement Principle and Instrument

There are two major variations in the instrument structure of FTIR: dispersive and Fourier-transform. The dispersive-type spectrometer uses typical diffraction gratings to spread the incoming broadband IR source and filter narrow bandwidths to record the spectrum by scanning the dispersed light. In contrast, the Fourier-transform-type measures the time-domain data and converts them into frequency domain yielding a spectrum via Fourier transform technique. Since none of the existing detectors respond to the optical frequencies (~10^{14} Hz), interferometric setup is used to convert this time-domain signal into measurable and visible fringe patterns. As shown in Figure 19.4, infrared photons from light sources

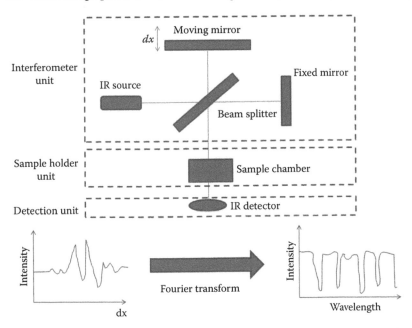

FIGURE 19.4 Interoferometric setup of FTIR. When infrared photons from light sources enter the Michelson interferometer unit, which is then split and directed toward fixed and moving mirror, they are recombined to the sample chamber in the sample holder unit.

enter the Michelson interferometer unit which is split and directed towards fixed and moving mirror and recombined to the sample chamber (in sample holder unit). The outgoing spectrum is then detected by infrared detectors (either thermal or pyroelectric types). This process is repeated at different moving mirror positions (labeled dx) and the resulting raw interferograms were transformed into spatial (dx) and intensity relationship. Finally, once the data are transformed using Fourier transform, the absorption spectra are obtained.

19.3.3 Application in Food Safety

Naumann and his colleagues [41] were the first to apply FTIR for the identification of different bacterial species. Since then, it has been extensively studied because of its unique capability of correlating the chemical signature of molecules unique to bacterial genera or species. FTIR coupled with various chemo-metrics such as HCA (hierarchical cluster analysis), PCA (principle component analysis), and ANN (artificial neural network) was used to identify and differentiate bacteria, yeast, and other microorganisms [43,44]. Specifically, FTIR was successfully applied in the identification of *E. coli*, *Bacillus*, *Pseudomonas*, *Listeria*, and *Staphylococcus* [44]. A variation of FTIR called photoacoustic spectroscopy (PAS) that utilizes the photoacoustic effect (modulated incident photon energy is converted to pressure oscillation and measured by microphone or piezotransducers) was able to demonstrate bacterial contamination on the surface of produce [45,46]. In food safety and food defense applications, FTIR coupled with ANN was used to analyze the presence of foodborne pathogens or bio-threat agents [12,47,48]. Analytical procedure for direct application of FTIR for food testing was also reported [49–51]. Since FTIR does not require any labeling reagent for discrimination, it is highly attractive for rapid and label-free detection of pathogenic bacteria: *Listeria monocytogenes* [52], *Staphylococcus aureus* [53], *Salmonella* [54], and *E. coli* O157:H7 [50]. Even though the reported detection limit for FTIR to be in the range of 10^3 CFU/mL [41,50], most experiments were conducted using bacterial concentrations ranging from 10^8 to 10^9 CFU/mL to ensure good signal-to-noise ratio [44,47,50,51].

19.4 Hyperspectral Imaging

19.4.1 Introduction

Hyperspectral imaging technology is often referred to as imaging spectrometry, and it was originally developed to perform satellite imaging for geophysics and remote sensing applications. Hyperspectral imaging system collects the spatial intensity information across many electromagnetic spectra and generates complete spatio-spectral map of the terrain or the object. Multispectral measurement has also been used similarly but the measurement technique typically uses spectral band consisting of 10–20 discrete wavelengths. In contrast, in hyperspectral measurement, the spectrum can be a sweep from visible to infrared ranges with very small bandwidth (several nm). Hyperspectral imaging technique has been used widely in crop production and disease monitoring in agriculture, mineralogy, physics, and land surveillance. Recently, it has drawn significant interests in biological sensor development because it has the capability to generate unique fingerprints in both spatial and spectral domain without the use of any labeling reagents or biological probes [55–57]. Spatial information is typically recorded as intensity images while the spectral information consists of reflection intensity and generates three-dimensional data set called hyperspectral data cube (Figure 19.5).

19.4.2 Measurement Principle and Instrument

Typical hyperspectral measurement system used in food industry consists of three core components: broadband light source, sample stages, and collection optics equipped with a detector. Broadband light source, typically tungsten light, provides the wide spectrum of electromagnetic energy to impinge the sample under investigation. For 2-D spatial measurement (Figure 19.6), sample stages require a 1-D

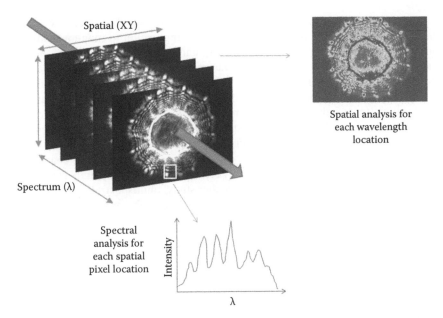

FIGURE 19.5 **(See color insert.)** Schematic diagram showing principles of hyperspectral or multispectral imaging setup. Spatial information is typically recorded as intensity images while the spectral information consists of reflection intensity and generates three-dimensional data set called hyperspectral data cube.

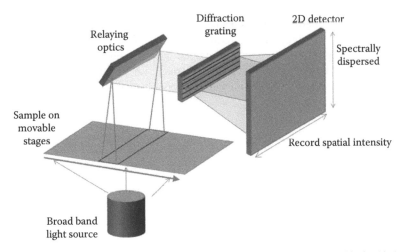

FIGURE 19.6 Schematic diagram showing typical setup for hyperspectral imaging setup used in food industry. It consists of three core components: broadband light source, sample stages, and collection optics equipped with a detector.

translation movement when the image from the reflection or transmission is collected in line-scan mode. The reflected or transmitted light from this narrow 1-D strip is collected via relaying optics and travels through dispersive component such as prism or grating in such a way that different electromagnetic energy is decomposed and projected onto different spatial location. Continuing scanning of the sample thus generates a three-dimensional data cube as shown in Figure 19.5.

19.4.3 Application in Food Safety

One of the earliest reports on application of hyperspectral imaging was by Park et al. [57], who examined microbial contamination on poultry carcasses. They applied multispectral measurement coupled with linear discriminate model to predict contamination with 83%–97% accuracy. Since then most application in food industry was directed toward online application to monitor quality of products such as tomatoes [58], apples [59,60], cucumbers [61], and poultry [56,62]. One of the most interesting applications of hyperspectral measurement was proposed by Kim and his colleagues [63,64] to inspect apples for possible contamination with fecal matters online in real time during sorting and quality assessment. They used both visible and near-infrared spectra. Using fluorescence signal emitted from the fecal matter as indicator, the system was able to accurately (100%) detect the contamination but the accuracy was slightly compromised (99.5%) when two band classification methods were used. Recently, Yoon et al. [65,66] used hyperspectral imaging system to classify *Campylobacter* colonies from non-*Campylobacter* organisms based on spectral absorption characteristics using 400–900 nm wavelengths. The results were highly reproducible with very high accuracy (97%–99%) and data were obtained in 24 h after plating. In another study, multispectral imaging using three narrow bands of 743, 458, and 541 nm was employed to detect and differentiate toxigenic fungi with 97% of accuracy [67].

19.5 Fiber-Optic Biosensor

19.5.1 Introduction

Fundamental properties of optical fiber are its ability to deliver the incoming light signal and to act as a transducer of outgoing light [68]. Due to the mass production and lower cost of optical fibers, fiber-optic detection platform has been a popular detection system that has found wide usage in various fields. It had been used to measure pressure [69], displacement [70], temperature [71], acceleration [72], and liquid levels [73]. In addition, it is one of the earliest biosensing platforms developed for the detection of biological and chemical agents primarily in the U.S. Naval Research Laboratory [74–76]. As shown in the top part of Figure 19.7, overall measurement is performed by comparing the net change from interrogating light source to the detected light through the fiber. As the incident light travels through the fiber, the biological recognition element, i.e., the analyte and fluorophore-labeled receptor interaction, generates a disturbance in the fiber signal, which is then measured by an output detector.

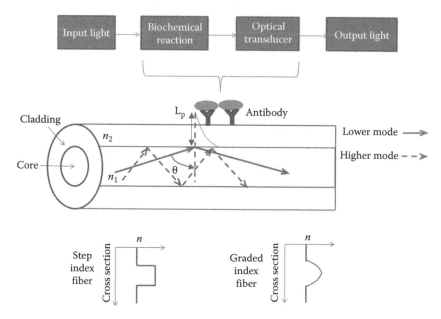

FIGURE 19.7   Schematic diagram showing light propagation and detection of analyte on fiber-optic sensor.

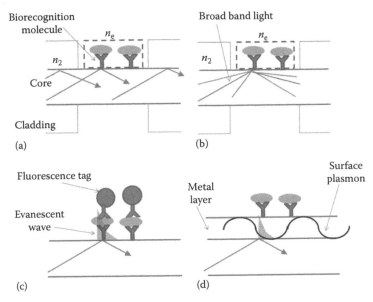

FIGURE 19.8 Fundamental measurement principle of fiber-optic sensor. Signal intensity depends on coupling between biochemical reaction and optical transducer that is achieved in four different ways: (a) intensity-based fiber-optic sensor uses a section that is devoid of cladding to immobilize biorecognition molecule that captures target analyte; (b) displays the similar concept but it uses the broadband light; (c) displays the evanescent wave fiber-optic sensor with sandwich configuration where the immobilized biorecognition molecule first binds to the analyte, which is then detected by fluorophore-labeled antibody or biomolecule; (d) shows the layout for surface plasmon resonance (SPR)-fiber-optic sensor, which combines the principle of SPR and the fiber-optic signal transduction setup.

19.5.2 Measurement Principle and Instrument

Fundamental measurement principle of fiber-optic sensor relies on coupling between biochemical reaction and optical transducer, which is achieved in four different ways (Figure 19.8). Figure 19.8a shows the intensity-based fiber-optic sensor, which uses a section that is devoid of cladding to immobilize biorecognition molecule to capture target analyte. Since refractive indices of cladding (n_2) are different from the effective refractive index (n_e) of recognition and target analyte complex, the reflection coefficient of each beam interacting in this area will be affected, which will result in intensity variation at the measurement end. Figure 19.8b displays the similar concept but it uses the broadband light. Since refractive indices are function of the wavelengths used, broadband source will provide spectroscopic means of differentiating the molecule–molecule interactions rather than pure intensity measurement. Figure 19.8c displays the evanescent wave fiber-optic sensor with sandwich configuration where the immobilized biorecognition molecule first binds to the analyte, which is then detected by fluorophore-labeled antibody or biomolecule. In this setup, cladding is removed from the wave-guide. The launched incident light is generally larger than the critical angle. The evanescent wave penetrating 100–200 nm into the surrounding core acts as the probe beam and fluorescence tag serves as a signal transducer. Figure 19.8d shows the layout for SPR-fiber-optic sensor, which combines the principle of SPR and the fiber-optic signal transduction setup. Metal layer with negative permittivity is deposited outside the dielectric fiber. When the incident beam inside the core is larger than the critical angle and matches the SPR condition, energy is transferred from the evanescent wave to the surface plasmon on the metal layer to disturb the reflected light from the recognition molecule and target analyte complex.

19.5.3 Application in Food Safety and Food Defense

The first report on fiber-optic sensor for food safety application was made by Lim and his colleagues [77], who used this sensor for detection of *E. coli* O157:H7 from a complex food matrix. Later, Ferreira et al. [78] reported intensity-based measurement of *E. coli* O157:H7 from a set number of bacterial cells to calculate sensitivity of 0.016 dB/h per bacterium. Analyte 2000 is the core fiber-optic instrument developed by Research International (Monroe, WA, USA). It is widely used for the detection of various pathogens including *E. coli* O157:H7 [77,79], *L. monocytogenes* [80–83], *Salmonella enteritidis* [84,85], staphylococcal enterotoxin B [86], and *Vaccinia* virus [87] in a sandwich configuration using fluorophore-labeled antibody. Later, a portable and semi-automated system was developed called RAPTOR (Research International, Monroe, WA) to report the detection of foodborne pathogens or biothreat agents [88] such as *S. typhimurium* [89], *Bacillus anthracis* and *Francisella tularensis* [90], and staphylococcal enterotoxins [91]. Using the evanescent wave principle, planner waveguide technology (NRL Array biosensor) was developed by Ligler and her colleagues [92] at the Naval Research Laboratory (USA) to detect multiple biothreat agents including bacterial pathogens, mycotoxins, and microbial and non-microbial toxins in an array format using antibodies.

19.6 Surface Plasmon Resonance

19.6.1 Introduction

Surface plasmon resonance (SPR) sensor utilizes the coherent electron oscillation that is generated at the metal–dielectric interface during coupling of excited light of specific wavelength. This sensor generates surface wave that travels parallel to the metal–dielectric interface and the field strength shows an evanescent characteristics, which diminishes quickly from the surface, typically up to several hundred nanometers. When the biorecognition molecule, i.e., antibody, enzymes, receptors, or nucleotides, captures target analyte, it results in a change in the reflection intensity, which is translated into the refractive index variation. In general, SPR is considered a label-free detection system since it does not require any differentiating reagents or fluorophores at the sample preparation stage. However, it requires antibody or biorecognition molecules to be pre-deposited onto the SPR surface to provide specificity for target analyte.

19.6.2 Measurement Principle and Instrument

The measurement principle of SPR is illustrated in Figure 19.9. Depending on SPR configuration, the analyte containing samples are either deposited or flown through fluidic channels to the SPR surface, which is pre-coated with biorecognition molecules. An incident p-polarized light beam striking the prism surface at a particular angle greater than the total internal reflection (TIR) angle generates surface

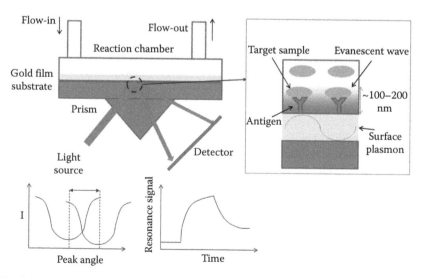

FIGURE 19.9 Schematic diagram depicting analyte detection using surface plasmon resonance sensor (SPR). The biorecognition molecule that lies within the penetration depth of the evanescent waves captures the target analyte. The detection is typically achieved by measuring the shift in the reflection intensity profile or recording the time-resolved resonance signal.

plasmons at the surface of metal layer (gold or silver). Surface plasmons are special modes of the electromagnetic field comprised of transverse magnetic (TM) polarized waves, i.e., light polarization occurs parallel to the plane of incidence where the height of the evanescent wave is approximately one-fourth of the wavelength of the incoming light that strikes the prism surface. As binding events unfold at the sensor surface, the refractive index of the medium near the dielectric–metal interface changes and as a result there is a shift in the reflected light, which is detected via photodiode array (PDA) or CCD. For surface plasmons to occur at the interface of metal–dielectric, the real part of the permittivity of the two media should be of opposite charges [93]. Metals such as gold and silver fulfill such conditions when interrogated by light in the IR-visible spectrum. Gold is widely used in most of the commercially available SPR systems as it provides the added advantage of enabling well-established surface chemistry for immobilization of bioreceptors. As shown in Figure 19.9, the target molecules are captured by the biorecognition molecule that lies within the penetration depth of the evanescent waves. The detection of target molecules by SPR is typically achieved by measuring the shift in the reflection intensity profile or recording the time-resolved resonance signal. The first case provides the thickness measurement by observing the reflectivity variation and matching with the Fresnel multilayer reflection as

$$I_R = |R|I_0 \qquad (19.1)$$

where
I_R is the reflected intensity measured by the detector
R is the total reflection coefficient
I_0 is the incident intensity

If we define r_{12} as reflection coefficient between prism and metal layer, r_{23} as between metal and air, the total reflection coefficient and a phase delay, δ, from the multilayer is expressed as

$$R = \frac{r_{12} + r_{23}\exp(2i\delta)}{1 + r_{12}r_{23}\exp(2i\delta)} \qquad \delta = \frac{2\pi n_2 D\cos(\theta_2)}{\lambda} \qquad (19.2)$$

where
n_2 is the refractive index of the medium
D is the thickness
θ_2 is the refracted angle

Secondly, SPR can also provide real-time measurement of resonance signal, which is directly related to the binding kinetics of the molecule. This type of time-dependent dynamic measurement monitors the real-time changes in the reflectivity close to the resonance point and reports association and dissociation of molecules.

19.6.3 Application in Food Safety and Food Defense

Among the optical sensors, SPR sensors marketed by several manufacturers had been tested with many pathogens and toxins from wide varieties of food matrices. SPR has been used to detect heat-killed or live cells of *L. monocytogenes* at 10^6 cell/mL using mouse or rabbit antibodies or phage-displayed single chain fragment variable (scFv) antibodies that were covalently immobilized on sensor platform [94–97]. Oh et al. [98] detected *S. enteric* serovar Typhimurium at a concentration of 10^2 CFU/mL using protein G to immobilize anti-*Salmonella* antibody on gold surface. SPR biosensors were also used effectively for the detection of *Salmonella* in milk at 1.25×10^5 CFU/mL [99], *E. coli* O157:H7 [100–103], short DNA sequences of *Brucella*, *E. coli* and *S. aureus* [104], enterotoxins from *Staphylococcus* [105,106], and mycotoxins [107].

19.7 Future Trends

Currently developed detection technologies are capable of providing results within 12–24 h and are proven useful to address food safety and biosecurity needs. With the introduction of novel nano-optical sensors, the detection time, however, can be substantially shortened to 4–8 h. Most of the nano-optical sensors, such as SPR and fiber-optic, however, require specific biorecognition molecules such as antibodies, bacteriophage proteins, bioreceptors, nucleic acids, or other capture molecules for specificity. The future scope of these types of biosensors lies on their ability to detect multiple analyte or wide varieties of biothreat agents. Strategies to achieve such goals would be to use biorecognition molecule(s) that can interact with wide varieties of pathogens/toxins. In contrast, label-free methods such as ELS and FTIR do not require any labeling reagents or probes and can provide identity of the agent instantly provided the database contain the spectral information. The future challenge is to develop algorithm that would allow detection/identification of previously unknown or unclassified organisms. Rajwa et al. [40] in a recent study reported the use of Bayesian approach to learning with nonexhaustive training dataset for the classification/identification of previously unknown bacterial classes. Another bottleneck for these label-free methods is the time required for signal amplification since the minute changes in nanometer scales have to be amplified to be macroscopically measurable. For example, BARDOT requires the incubation time of 12–24 h for a colony of most pathogens to reach to a certain size for detection. To reduce detection time to less than 8 h, a generally practiced food industry work shift, a new concept of micro-colony detection method was introduced [108,109]. The first experimental results [110] show both promise and limitation on differentiating bacterial colonies in the size regime. Pathogens belonging to three different genera (*Salmonella enterica* serovar Montevideo, *Listeria monocytogenes* F4244, and *E. coli* DH5α) were tested when the colony size (diameter) was around 100–200 μm range after about 7–11 h of growth. Even though the scattering pattern displayed differences in the number of rings and circularity, the differential characteristics among genera were lowered compared to the fully-grown 1-mm-diameter colonies. Further investigation on detecting micron-sized colonies with a more sensitive optical method such as Mueller Matrix formalism [111] may aid in early detection. Since each measurement modality has its own strengths and weaknesses, there is a great interest in using hybrid methodology such as combining SPR and fiber-optics sensors or enhancing ELS sensor signals by using high refractive index materials such as gold or silver nano-particle as a scattering contrast agent have been proposed [20,21].

Acknowledgments

The USDA (1935-42000-035) and NIH (1R56AI089511-01) provided the funds to conduct biosensor-related research in the author's laboratories.

REFERENCES

1. Flint JA et al. (2005) Estimating the burden of acute gastroenteritis, foodborne disease, and pathogens commonly transmitted by food: An international review. *Clin. Infect. Dis.* 41 (5):698–704.
2. Scallan E et al. (2011) Foodborne illness acquired in the United States—Major pathogens. *Emerg. Infect. Dis.* 17 (1):7–15.
3. Scharff R. (2010) Health-related costs from foodborne illness in the United States. The Produce Safety Project at Georgetown University. Available from www.producesafetyproject.org
4. Gracias KS and McKillip JL. (2004) A review of conventional detection and enumeration methods for pathogenic bacteria in food. *Can. J. Microbiol.* 50 (11):883–890.
5. Swaminathan B and Feng P. (1994) Rapid detection of foodborne pathogenic bacteria. *Annu. Rev. Microbiol.* 48:401–426.
6. Banada PP and Bhunia AK. 2008. Antibodies and immunoassays for detection of bacterial pathogens. In *Principles of Bacterial Detection: Biosensors, Recognition Receptors and Microsystems*, eds. M. Zourob, S. Elwary, and A. Turner. Manchester, U.K.: Cambridge University.
7. Brehm-Stecher B, Young C, Jaykus L-A, and Tortorello ML. (2009) Sample preparation: The forgotten beginning *J. Food Protect.* 72:1774–1789.
8. Fukushima H et al. (2010) Simultaneous screening of 24 target genes of foodborne pathogens in 35 foodborne outbreaks using multiplex real-time SYBR green PCR analysis. *Int. J. Microbiol.* (Article ID 864817):18.
9. Ecker DJ et al. (2008) Innovation—Ibis T5000: A universal biosensor approach for microbiology. *Nat. Rev. Microbiol.* 6 (7):553–558.

10. Fang H et al. (2010) An FDA bioinformatics tool for microbial genomics research on molecular characterization of bacterial foodborne pathogens using microarrays. *BMC Bioinformatics* 11 (Suppl 6):S4.

11. Leski TA et al. (2009) Testing and validation of high density resequencing microarray for broad range biothreat agents detection. *PLoS ONE* 4 (8):e6569.

12. Yu C and Irudayaraj J. (2006) Identification of pathogenic bacteria in mixed cultures by FTIR spectroscopy. *Trans. ASABE* 49 (5):1623–1632.

13. Bhunia AK. (2008) Biosensors and bio-based methods for the separation and detection of foodborne pathogens. In *Advances in Food and Nutrition Research*, ed. S. Taylor. Amsterdam, the Netherlands: Elsevier, Inc.

14. Privett BJ, Shin JH, and Schoenfisch MH. (2010) Electrochemical sensors. *Anal. Chem.* 82 (12):4723–4741.

15. Velusamy V, Arshak K, Korostynska O, Oliwa K, and Adley C. (2010) An overview of foodborne pathogen detection: In the perspective of biosensors. *Biotechnol. Adv.* 28 (2):232–254.

16. Yang L and Bashir R. (2008) Electrical/electrochemical impedance for rapid detection of foodborne pathogenic bacteria. *Biotechnol. Adv.* 26 (2):135–150.

17. Fischer NO, Tarasow TM, and Tok JBH. (2007) Heightened sense for sensing: Recent advances in pathogen immunoassay sensing platforms. *Analyst* 132 (3):187–191.

18. Bhunia AK, Nanduri V, Bae E, and Hirleman ED. 2010. Biosensors, foodborne pathogen detection. In *Encyclopedia of Industrial Biotechnology*, ed. M. C. Flickinger. Hoboken, NJ: John Wiley & Sons, Inc.

19. Gehring AG and Tu SI. (2011) High-throughput biosensors for multiplexed food-borne pathogen detection. *Ann. Rev. Anal. Chem.* 4:151–172.

20. Loo C et al. (2004) Nanoshell-enabled photonics-based imaging and therapy of cancer. *Technol. Cancer Res. Treat.* 3 (1):33–40.

21. Chen K, Liu Y, Ameer G, and Backman V. (2005) Optimal design of structured nanospheres for ultrasharp light-scattering resonances as molecular imaging multilabels. *J. Biomed. Opt.* 10 (2):024005.

22. Banada PP et al. (2007) Optical forward-scattering for detection of *Listeria monocytogenes* and other *Listeria* species. *Biosens. Bioelectron.* 22 (8):1664–1671.

23. Bae E, Banada PP, Huff K, Bhunia AK, Robinson JP, and Hirleman ED. (2007) Biophysical modeling of forward scattering from bacterial colonies using scalar diffraction theory. *Appl. Opt.* 46 (17):3639–3648.

24. Bayraktar B, Banada PP, Hirleman ED, Bhunia AK, Robinson JP, and Rajwa B. (2006) Feature extraction from light-scatter patterns of *Listeria* colonies for identification and classification. *J. Biomed. Opt.* 11 (3):034006.

25. Wyatt PJ. (1969) Identification of bacteria by differential light scattering. *Nature* 221 (5187):1257–1258.

26. Wyatt PJ. (1968) Differential light scattering—A physical method for identifying living bacterial cells. *Appl. Opt.* 7 (10):1879–1896.

27. VandeMerwe WP, Li ZZ, Bronk BV, and Czege J. (1997) Polarized light scattering for rapid observation of bacterial size changes. *Biophys. J.* 73 (1):500–506.

28. Bronk BV, Druger SD, Czege J, and Vandemerwe WP. (1995) Measuring diameters of rod-shaped bacteria in vivo with polarized light scattering. *Biophys. J.* 69 (3):1170–1177.

29. Bickel WS, Davidson JF, Huffman DR, and Kilkson R. (1976) Application of polarization effects in light scattering—New biophysical tool. *Proc. Natl. Acad. Sci. USA* 73 (2):486–490.

30. Bohren CF and Huffman DR. (1983) *Absorption and Scattering of Light by Small Particles*. New York: Wiley.

31. Diaspro A, Radicchi G, and Nicolini C. (1995) Polarized light scattering—A biophysical method for studying bacterial cells. *IEEE Trans. Biomed. Eng.* 42 (10):1038–1043.

32. Bronk BV, Li ZZ, and Czege J. (2001) Polarized light scattering as a rapid and sensitive assay for metal toxicity to bacteria. *J. Appl. Toxicol.* 21 (2):107–113.

33. Shvalov AN et al. (2000) Individual *Escherichia coli* cells studied from light scattering with the scanning flow cytometer. *Cytometry* 41 (1):41–45.

34. Seo S, Su TW, Tseng DK, Erlinger A, and Ozcan A. (2009) Lensfree holographic imaging for on-chip cytometry and diagnostics. *Lab Chip* 9 (6):777–787.

35. Ozcan A and Demirci U. (2008) Ultra wide-field lens-free monitoring of cells on-chip. *Lab Chip* 8 (1):98–106.

36. Mudanyali O et al. (2010) Compact, light-weight and cost-effective microscope based on lensless incoherent holography for telemedicine applications. *Lab Chip* 10 (11):1417–1428.

37. Banada PP et al. (2009) Label-free detection of multiple bacterial pathogens using light-scattering sensor. *Biosens. Bioelectron.* 24 (6):1685–1692.

38. Bae E, Banada PP, Huff K, Bhunia AK, Robinson JP, and Hirleman ED. (2008) Analysis of time-resolved scattering from macroscale bacterial colonies. *J. Biomed. Opt.* 13 (1):014010.

39. Bae E, Aroonnual A, Bhunia AK, Robinson JP, and Hirleman ED. (2009) System automation for a bacterial colony detection and identification instrument via forward scattering. *Meas. Sci. Technol.* 20 (1):015802.

40. Rajwa B et al. (2010) Discovering the unknown: Detection of emerging pathogens using a label-free light-scattering system. *Cytom. Part A* 77A (12):1103–1112.

41. Naumann D, Helm D, and Labischinski H. (1991) Microbiological characterizations by FT-IR spectroscopy. *Nature* 351 (6321):81–82.

42. Berthomieu C and Hienerwadel R. (2009) Fourier transform infrared (FTIR) spectroscopy. *Photosynth. Res.* 101 (2):157–170.

43. Mariey L, Signolle JP, Amiel C, and Travert J. (2001) Discrimination, classification, identification of microorganisms using FTIR spectroscopy and chemometrics. *Vib. Spectrosc.* 26 (2):151–159.

44. Helm D and Naumann D. (1995) Identification of some bacterial cell components by FT-IR spectroscopy. *FEMS Microbiol. Lett.* 126 (1):75–79.

45. Sivakesava S, Irudayaraj J, and DebRoy C. (2004) Differentiation of microorganisms by FTIR-ATR and NIR spectroscopy. *Trans. ASAE* 47 (3):951–957.

46. Foster NS, Valentine NB, Thompson SE, Johnson TJ, and Amonette JE. (2004) FTIR transmission and photoacoustic spectroscopy for the statistical identification of bacteria. *SPIE Proc.* 5269:172–182.

47. Gupta MJ, Irudayaraj JM, Debroy C, Schmilovitch Z, and Mizrach A. (2005) Differentiation of food pathogens using FTIR and artificial neural networks. *Trans. ASAE* 48 (5):1889–1892.

48. Al-Holy MA, Lin MS, Al-Qadiri H, Cavinato AG, and Rasco BA. (2006) Classification of foodborne pathogens by Fourier transform infrared spectroscopy and pattern recognition techniques. *J. Rapid Methods Automat. Microbiol.* 14 (2):189–200.

49. Puzey KA, Gardner PJ, Petrova VK, Donnelly CW, and Petrucci GA. (2008) Automated species and strain identification of bacteria in complex matrices using FTIR spectroscopy—Art. no. 695412. *SPIE Proc.* 6954:95412–95412.

50. Burgula Y et al. (2006) Detection of *Escherichia coli* O157:H7 and *Salmonella typhimurium* using filtration followed by Fourier-transform infrared spectroscopy. *J. Food Prot.* 69 (8):1777–1784.

51. Rebuffo CA, Schmitt J, Wenning M, von Stetten F, and Scherer S. (2006) Reliable and rapid identification of *Listeria monocytogenes* and *Listeria* species by artificial neural network-based Fourier transform infrared spectroscopy. *Appl. Environ. Microbiol.* 72 (2):994–1000.

52. Rebuffo-Scheer CA, Schmitt J, and Scherer S. (2007) Differentiation of *Listeria monocytogenes* serovars by using artificial neural network analysis of Fourier-transformed infrared spectra. *Appl. Environ. Microbiol.* 73 (3):1036–1040.

53. Lamprell H, Mazerolles G, Kodjo A, Chamba JF, Noel Y, and Beuvier E. (2006) Discrimination of *Staphylococcus aureus* strains from different species of *Staphylococcus* using Fourier transform infrared (FTIR) spectroscopy. *Int. J. Food Microbiol.* 108 (1):125–129.

54. Baldauf NA, Rodriguez-Romo LA, Männig A, Yousef AE, and Rodriguez-Saona LE. (2007) Effect of selective growth media on the differentiation of *Salmonella enterica* serovars by Fourier-transform mid-Infrared spectroscopy. *J. Microbiol. Methods* 68 (1):106–114.

55. Gray PC et al. (1998) Distinguishability of biological material by use of ultraviolet multispectral fluorescence. *Appl. Opt.* 37 (25):6037–6041.

56. Kong SG, Chen YR, Kim I, and Kim MS. (2004) Analysis of hyperspectral fluorescence images for poultry skin tumor inspection. *Appl. Opt.* 43 (4):824–833.

57. Park B, Chao YR, and Chao KL. (1999) Multispectral imaging for detecting contamination in poultry carcasses. *SPIE Proc.* 3544:156–165.

58. Polder G, van der Heijden G, and Young IT. (2002) Spectral image analysis for measuring ripeness of tomatoes. *Trans. ASAE* 45 (4):1155–1161.

59. Mehl PM, Chen YR, Kim MS, and Chan DE. (2004) Development of hyperspectral imaging technique for the detection of apple surface defects and contaminations. *J. Food Eng.* 61 (1):67–81.

60. Lu R. (2003) Detection of bruises on apples using near-infrared hyperspectral imaging. *Trans. ASAE* 46 (2):523–530.

61. Ariana D and Lu R. (2008) Quality evaluation of pickling cucumbers using hyperspectral reflectance and transmittance imaging: Part I. Development of a prototype. *Sens. Instrum. Food Quality Saf.* 2 (3):144–151.

62. Nakariyakul S and Casasent DP. (2008) Hyperspectral waveband selection for contaminant detection on poultry carcasses. *Opt. Eng.* 47 (8):087202.

63. Kim M et al. (2007) Hyperspectral reflectance and fluorescence line-scan imaging for online defect and fecal contamination inspection of apples. *Sens. Instrum. Food Qual. Saf.* 1 (3):151–159.

64. Kim MS, Cho BK, Lefcourt AM, Chen YR, and Kang S. (2008) Multispectral fluorescence lifetime imaging of feces-contaminated apples by time-resolved laser-induced fluorescence imaging system with tunable excitation wavelengths. *Appl. Opt.* 47 (10):1608–1616.

65. Yoon S et al. (2010) Detection of *Campylobacter* colonies using hyperspectral imaging. *Sens. Instrum. Food Qual. Saf.* 4 (1):35–49.

66. Yoon SC, Lawrence KC, Siragusa GR, Line JE, Park B, and Feldner PW. (2009) Hyperspectral reflectance imaging for detecting a foodborne pathogen: Campylobacter. *Trans. ASABE* 52 (2):651–662.

67. Yao H, Hruska Z, Kincaid R, Brown R, and Cleveland T. (2008) Differentiation of toxigenic fungi using hyperspectral imagery. *Sens. Instrum. Food Qual. Saf.* 2 (3):215–224.

68. Leung A, Shankar PM, and Mutharasan R. (2007) A review of fiber-optic biosensors. *Sens. Actuat. B Chem.* 125 (2):688–703.

69. Pahler RH and Roberts AS. (1977) Design of a fiber optic pressure transducer. *J. Eng. Ind. Trans. ASME* 99 (1):274–280.

70. Lagakos N, Litovitz T, Macedo P, Mohr R, and Meister R. (1981) Multimode optical fiber displacement sensor. *Appl. Opt.* 20 (2):167–168.

71. Lee CE, Atkins RA, and Taylor HF. (1988) Performance of a fiber optic temperature sensor from 200°C to1050°C. *Opt. Lett.* 13 (11):1038–1040.

72. Jonsson L and Hok B. (1984) Multimode fiber optic accelerometers. *SPIE Proc.* 514:191–194.

73. Ilev IK and Waynant RW. (1999) All-fiber-optic sensor for liquid level measurement. *Rev. Sci. Instrum.* 70 (5):2551–2554.

74. Anderson GP, Golden JP, and Ligler FS. (1993) A fiber optic biosensor—Combination tapered fibers designed for improved signal acquisition. *Biosens. Bioelectron.* 8 (5):249–256.

75. Anderson GP, Wijesuriya DC, Ogert RA, Shriverlake LC, Golden JP, and Ligler FS. (1993) Fiber optic-based biosensor—Applications for environmental monitoring. *Biophys. J.* 64 (2):A219–A219.

76. Anderson GP, Golden JP, Shriverlake LC, and Ligler FS. (1992) Optimization of sensor probe taper angle for a fiber optic biosensor. *FASEB J.* 6 (1):A169–A169.

77. DeMarco DR, Saaski EW, McCrae DA, and Lim DV. (1999) Rapid detection of *Escherichia coli* O157:H7 in ground beef using a fiber-optic biosensor. *J. Food Protect.* 62 (7):711–716.

78. Ferreira AP, Werneck MM, and Ribeiro RM. (2001) Development of an evanescent-field fibre optic sensor for *Escherichia coli* O157:H7. *Biosens. Bioelectron.* 16 (6):399–408.

79. Geng T, Uknalis J, Tu SI, and Bhunia AK. (2006) Fiber-optic biosensor employing Alexa-Fluor conjugated antibody for detection of *Escherichia coli* O157:H7 from ground beef in four hours. *Sensors* 6:796–807.

80. Tims TB, Dickey SS, Demarco DR, and Lim DV. (2001) Detection of low levels of *Listeria monocytogenes* within 20 hours using an evanescent wave biosensor. *Am. Clin. Lab.* 20 (8):28–29.

81. Nanduri V et al. (2006) Antibody immobilization on waveguides using a flow-through system shows improved *Listeria monocytogenes* detection in an automated fiber optic biosensor: RAPTOR (TM). *Sensors* 6 (8):808–822.

82. Geng T, Morgan MT, and Bhunia AK. (2004) Detection of low levels of *Listeria monocytogenes* cells by using a fiber-optic immunosensor. *Appl. Environ. Microbiol.* 70 (10):6138–6146.

83. Ohk SH, Koo OK, Sen T, Yamamoto CM, and Bhunia AK. (2010) Antibody-aptamer functionalized fibre-optic biosensor for specific detection of *Listeria monocytogenes* from food. *J. Appl. Microbiol.* 109 (3):808–817.

84. Bhunia AK, Geng T, Lathrop AA, Valadez A, and Morgan MT. (2004) Optical immunosensors for detection of *Listeria monocytogenes* and *Salmonella enteritidis* from food. *Proc. SPIE* 5271:1–6.

85. Valadez A, Lana C, Tu S-I, Morgan M, and Bhunia A. (2009) Evanescent wave fiber optic biosensor for *Salmonella* detection in food. *Sensors* 9 (7):5810–5824.

86. Tempelman LA, King KD, Anderson GP, and Ligler FS. (1996) Quantitating staphylococcal enterotoxin B in diverse media using a portable fiber-optic biosensor. *Anal. Biochem.* 233 (1):50–57.

87. Donaldson KA, Kramer MF, and Lim DV. (2004) A rapid detection method for Vaccinia virus, the surrogate for smallpox virus. *Biosens. Bioelectron.* 20 (2):322–327.
88. Bhunia AK, Banada PP, Banerjee P, Valadez A, and Hirleman ED. (2007) Light scattering, fiber optic-and cell-based sensors for sensitive detection of foodborne pathogens. *J. Rapid Methods Automat. Microbiol.* 15:121–145.
89. Kramer MF and Lim DV. (2004) A rapid and automated fiber optic-based biosensor assay for the detection of salmonella in spent irrigation water used in the sprouting of sprout seeds. *J. Food Protect.* 67 (1):46–52.
90. Jung CC, Saaski EW, McCrae DA, Lingerfelt BM, and Anderson GP. (2003) RAPTOR: A fluoroimmunoassay-based fiber optic sensor for detection of biological threats. *IEEE Sens. J.* 3 (4):352–360.
91. Anderson GP and Nerurkar NL. (2002) Improved fluoroimmunoassays using the dye Alexa Fluor 647 with the RAPTOR, a fiber optic biosensor. *J. Immunol. Methods* 271 (1–2):17–24.
92. Ligler FS et al. (2007) The array biosensor: Portable, automated systems. *Anal. Sci.* 23 (1):5–10.
93. Homola J. (2006) Electromagnetic theory of surface plasmons. *Springer Ser. Chem. Sens. Biosens.* 4:3–44.
94. Leonard P, Hearty S, Quinn J, and O'Kennedy R. (2004) A generic approach for the detection of whole *Listeria monocytogenes* cells in contaminated samples using surface plasmon resonance. *Biosens. Bioelectron.* 19 (10):1331–1335.
95. Koubova V et al. (2001) Detection of foodborne pathogens using surface plasmon resonance biosensors. *Sens Actuat. B Chem.* 74 (1–3):100–105.
96. Nanduri V, Bhunia AK, Tu SI, Paoli GC, and Brewster JD. (2007) SPR biosensor for the detection of *L. monocytogenes* using phage-displayed antibody. *Biosens. Bioelectron.* 23:248–252.
97. Lathrop AA, Jaradat ZW, Haley T, and Bhunia AK. (2003) Characterization and application of a *Listeria monocytogenes* reactive monoclonal antibody C11E9 in a resonant mirror biosensor. *J. Immunol. Methods* 281 (1–2):119–128.
98. Oh BK, Kim YK, Park KW, Lee WH, and Choi JW. (2004) Surface plasmon resonance immunosensor for the detection of *Salmonella typhimurium. Biosens. Bioelectron.* 19 (11):1497–1504.
99. Mazumdar SD, Hartmann M, Kampfer P, and Keusgen M. (2007) Rapid method for detection of *Salmonella* in milk by surface plasmon resonance (SPR). *Biosens. Bioelectron.* 22 (9–10):2040–2046.
100. Fratamico PM, Strobaugh TP, Medina MB, and Gehring AG. (1998) Detection of *Escherichia coli* O157: H7 using a surface plasmon resonance biosensor. *Biotechnol. Tech.* 12 (7):571–576.
101. Oh BK, Kim YK, Bae YM, Lee WH, and Choi JW. (2002) Detection of *Escherichia coli* O157: H7 using immunosensor based on surface plasmon resonance. *J. Microbiol. Biotechnol.* 12 (5):780–786.
102. Subramanian A, Irudayaraj J, and Ryan T. (2006) A mixed self-assembled monolayer-based surface plasmon immunosensor for detection of *E. coli* O157:H7. *Biosens. Bioelectron.* 21 (7):998–1006.
103. Meeusen CA, Alocilja EC, and Osburn WN. (2005) Detection of *E. coli* O157:H7 using a miniaturized surface plasmon resonance biosensor. *Trans. ASAE* 48 (6):2409–2416.
104. Piliarik M, Párová L, and Homola J. (2009) High-throughput SPR sensor for food safety. *Biosens. Bioelectron.* 24 (5):1399–1404.
105. Homola J, Dostalek J, Chen SF, Rasooly A, Jiang SY, and Yee SS. (2002) Spectral surface plasmon resonance biosensor for detection of staphylococcal enterotoxin B in milk. *Int. J. Food Microbiol.* 75 (1–2):61–69.
106. Rasooly A. (2001) Surface plasmon resonance analysis of staphylococcal enterotoxin B in food. *J. Food Prot.* 64 (1):37–43.
107. Schnerr H, Vogel RF, and Niessen L. (2002) A biosensor-based immunoassay for rapid screening of deoxynivalenol contamination in wheat. *Food Agri. Immunol.* 14 (4):313–321.
108. Goodwin JR, Hafner LM, and Fredericks PM. (2006) Raman spectroscopic study of the heterogeneity of microcolonies of a pigmented bacterium. *J. Raman Spectrosc.* 37 (9):932–936.
109. Choo-Smith LP et al. (2001) Investigating microbial (micro)colony heterogeneity by vibrational spectroscopy. *Appl. Environ. Microbiol.* 67 (4):1461–1469.
110. Bae E, Bai N, Aroonnual A, Bhunia AK, and Hirleman ED. (2011) Label-free identification of bacterial microcolonies via elastic scattering. *Biotechnol. Bioeng.* 108 (3):637–644.
111. Bohren CF and Huffman DR. (1998) *Absorption and Scattering of Light by Small Particles.* New York: Wiley-Interscience.

20

Multifunctional Fiber-Optic Nanosensors for Environmental Monitoring

Alessio Crescitelli, Marco Consales, Antonello Cutolo,
Michele Giordano, and Andrea Cusano

CONTENTS

20.1 Introduction

In order to preserve the environment from toxic contaminants and pathogens that can be released into a variety of media (such as air, soil, and water), the detection of biological and chemical species in atmosphere, process gases, water, and soil is of a great concern. In particular, the ability to quickly and efficiently detect the presence or absence of specific chemicals can be a matter of life or death. Leaks of toxic gases, monitoring of glucose or anesthetics in the bloodstream, and testing for harmful compounds in foods all require selective, sensitive, fast, and reliable devices.

In particular, air pollutants include sulfur dioxide, carbon monoxide, nitrogen dioxide, and VOCs, which originate from sources such as vehicle emissions, power plants, refineries, and industrial and laboratory processes. Soil and water contaminants can be classified as microbiological, radioactive, inorganic, synthetic organic, and VOCs. Pesticides and herbicides are applied directly to plants and soils, and incidental releases of other noxious waste can derive from spills, leaking pipes, underground storage tanks, etc. Some of these contaminants can endure for many years and migrate through large regions of soil until they reach water resources, where they may be an ecological or human-health threat. The projected increase in global energy usage and unwanted release of pollutants has led to a serious focus on advanced monitoring technologies for environmental protection, remediation, and restoration. For this reason, a need exists for accurate, inexpensive, continuous, and long-term monitoring of environmental contaminants using sensors that can be operated on site.

In this scenario, nanomaterials and nanotechnologies are being extensively applied in the design of sensors system for environmental monitoring. Indeed, the last decade has witnessed an exponential growth in the number of publications dedicated to nanosensors able to detect pollutants in real time with good sensing performance.

Nanoscale science implies basic research into properties (physical, chemical, and biological) of materials on the nanometer scale, e.g., of objects with 1–100 nm for at least one of the three dimensions (Gusev 2005, Kobayashi 2005, Pitkethly 2004, Roco et al. 2000). Thus, we may visualize the nanomaterials as structures produced by reducing one, two, or three dimensions of a bulk material, thereby resulting in 2D nanolayers, 1D nanowires, or 0D nanoclusters. Such length scales are close to atomic sizes and at the nanoscale, the physical, chemical, and biological properties of materials differ in fundamental and valuable ways from the properties of individual atoms and molecules or bulk matter (Lieber 1998). As a matter of fact, a change from macro- and micro- to nanoscale qualitatively modifies the physicochemical properties (electric conductivity, magnetism, light absorption and emission, optical refraction, thermal stability, and strength) and causes the resulting compounds and materials to display catalytic activity or reactivity and properties lacking in macro- and microscopic objects of the same chemical nature. Indeed, at the nanoscale, most of the atoms are surface atoms, so there is an increase in the actual number of sites available for reactions. Increase in surface area to volume ratio with the size reduction is very important for the sensing capability. Enhancing the interactions that occur at the nanoscale allows the realization of devices that may offer significant advantages versus conventional sensors, in terms of greater sensitivity and selectivity, as well as lower production costs. As proof of these considerations, a fairly broad spectrum of nanomaterials is used in chemical and biological sensors as suggested by several publications (Costa-Fernandez et al. 2006, Gouma 2010, Huang and Choi 2007, Huang et al. 2005, James and Tatam 2006, Jeronimo et al. 2007, Ju et al. 2011, Riu et al. 2006, Shi et al. 2004, Tansil and Gao 2006, Vaseashta and Dimova-Malinovska 2005).

Chemical and biological sensors have become an indispensable part of our technology-driven society and can be found in chemical process, pharmaceutical, food, biomedical, environmental, security, industrial safety, and indoor monitoring applications (Johnson et al. 2007, Li et al. 2007, Pejcic et al. 2007, Stetter et al. 2003, Wang et al. 2008). Like many fields in science, chemical sensors have benefited from the growing power of computers, integrated electronics, new materials, novel designs, and processing tools.

In particular, a chemical sensor consists of a recognition element that is sensitive to stimuli produced by various chemical compounds and a transduction part that generates a signal whose magnitude is functionally related to the concentration of the analyte. They are generally classified into gas, liquid, and solid particulate sensors based on the phases of the analyte. Another categorization of the chemical sensors is based on the operating principle of the transducer: optical, electrochemical, thermometric, and gravimetric (mass sensitive) (Janata 2009).

The sensitive element is the heart of the chemical sensor since it is the interface between the transducer and the external environment so that the nature, the selectivity, and sensitivity of the sensor depend upon these interactive materials. High-quality materials to use as sensing part should optimize specific interactions with a target analyte or narrow class of analytes that should supply a fast and reversible diffusion of the penetrants and small recovery times, and also they should preserve the physical state as well as their geometry over numerous cycles of use to avoid hysteretic effects, and thus to ensure the reproducibility (Grate and Abraham 1991). The most exploited materials for chemical sensors include polymers; organic monolayers; ceramics; metals semiconductors; nanostructured materials, such as carbon nanotubes, nanobelts, nanowires, and quantum dots; biomolecules; and combination thereof.

Once the most suitable material to detect a specific analyte has been selected, the natural step following concerns the choice of an appropriate transducer and of an opportune technique to read the physical or chemical changes occurring at the sensing part. Transducing approaches can include mechanical, electrochemical, optical, thermal, and electronic types. Each type has strengths and weaknesses relative to the particular application. The most exploited transduction principles in chemical sensing are the mass change and the resistivity/conductivity change of the sensitive part occurring on exposure to and consequent sorption of the molecules of target environmental analytes (Barsan and Weimar 2003, Consales et al. 2006a,

Fine et al. 2010, Grate 2000, Hajjam et al. 2010, Hartmann et al. 1994, Huang et al. 2010, James et al. 2005, Kepley et al. 1992, Korotcenkov et al. 2007, Korposh et al. 2010, Kukla et al. 2009, Penza et al. 2005a, 2006a, Rodriguez-Pardo et al. 2005, Shen et al. 2008, Zhang et al. 2004).

However, during the last two decades, a remarkable interest has also been focused on optical transduction principles for the measurement of chemical and biological quantities (Wolfbeis et al. 2006). Since the development of the first optical sensors for the measurement of CO_2 and O_2 concentrations (Lübbers and Opitz 1975), a large variety of devices based upon optical methods have been utilized in chemical sensors and biosensors including ellipsometry, spectroscopy (luminescence, phosphorescence, fluorescence, Raman), interferometry (white light interferometry, modal interferometry in optical waveguide structures), spectroscopy of guided modes in optical waveguide structures (grating coupler, resonant mirror), and surface plasmon resonance (Arregui et al. 2003, Byun et al. 2007, Daghestani and Day 2010, Fan et al. 2008, Homola et al. 1999, Lee et al. 2007, Leopold et al. 2009, Mignani et al. 2005a,b, Orellana 2004, Shankaran et al. 2007, Steinberg et al. 2003, Wolfbeis 1991, Zudans et al. 2004). In these sensors, pollutant detection is performed through the measurement of refractive index, absorbance, and fluorescence properties of the analyte molecules or of a chemo-optical transducing medium interacting with them.

Among the optical sensors, optical fiber sensor development has matured to the point where the impact of this new technology is now evident. They are able to offer a number of advantages, such as small size, light weight, increased sensitivity over existing techniques, geometric versatility, dielectric construction (so that they can be used in high-voltage, electrically noisy, high-temperature, corrosive, or other stressing environments), and inherent compatibility with optical fiber telemetry technology. For these unique characteristics, optical fiber sensors are also very attractive in chemical sensing applications, enabling the transmission of light over large distances with low losses, thus allowing a sensor head to be remotely located from the interrogation unit. Furthermore, optical fibers have the capability of carrying a huge amount of information, much greater than that carried by electrical wires.

Fiber-optic sensing is very versatile, since the intensity, wavelength, phase, and polarization of light can all be exploited as measurement parameters, and several wavelengths launched in the same fiber in either direction form independent signals. This gives the possibility of monitoring several chemicals with the same fiber sensor or even simultaneously monitoring changes in unwanted environment parameters which could drastically affect the chemical measurements, such as temperature and humidity.

In what follows, we provide an overview of the development of high-performance optochemical nanosensors based on the integration of the optical fiber technology with CNT (carbon nanotube)-based sensitive overlays. In particular, the excellent sensing capabilities of the SWCNT (single-walled carbon nanotube)-based nanosensors, demonstrated in the recent years against VOCs and other pollutants in different environments (air and water) and operating conditions (room temperature and cryogenic temperatures), are reported and discussed.

20.2 CNTs as Advanced Materials for Environmental Monitoring

Carbon-based nanostructures exhibit unique properties and morphological flexibility, which renders them inherently multifunctional and compatible with organic and inorganic systems. In particular, CNTs were first discovered by Sumio Iijima in the early 1990s (Iijima 1991), and since then, they have become a prominent material for an amazing breadth of scientific and technological disciplines ranging from structural and material science to chemistry, biology, and electronics (Rivas et al. 2009).

CNTs are structures from the fullerene family consisting of a honeycomb sheet of sp2 bonded carbon atoms rolled seamlessly into itself to form a cylinder with diameter of few nanometers and length ranging from 1 to 100 μm. They have been widely used as sensitive materials for the construction of chemical and biological sensors thanks to their unique electronic, chemical, structural, optical, mechanical, and thermal properties depending on their specific hollow nanostructure (Rivas et al. 2009, Yellampalli 2011). Indeed, due to their unique morphology, CNTs possess an excellent ability to reversibly absorb molecules of environmental pollutants undergoing a modulation of their electrical, geometric, and optical properties, such as conductivity, thickness, and refractive index, thus resulting particularly suitable to be employed as sensitive materials in chemical and biological sensing (Ye and Sheu 2007).

CNTs can be distinguished in SWCNTs or multiwalled carbon nanotubes (MWCNTs) depending on whether only one layer or many layers of graphite are concentrically rolled up together, and can behave either as metallic or semiconducting, depending upon their diameter and chirality (Terrones 2003). A further distinction is usually made between closed-end or opened-end CNTs depending upon whether the ends of the tubes are capped or uncapped (Dresselhaus et al. 2006).

The first CNT-based chemical sensor was designed in 1997 (Tans et al. 1977), after the problem of electric contact between the nanotubes and electrode was solved. Initially, CNTs were mainly applied in field effect transistor (FET) (Chen et al. 2003, Collins et al. 2000, Kong et al. 2000, Qi et al. 2003, Varghese et al. 2001, Zahab et al. 2000) and amperometric-based sensors for the detection of various gases (such as NO_2, NH_3, O_2, CO, and CO_2) and H_2O vapors (Luque et al. 2007, Manso et al. 2007, Tkac et al. 2007, Yan et al. 2007, Zare and Nasirizadeh 2007).

In particular, the operation of semiconductor CNT-based FETs is underlain by extremely high sensitivity of their electric properties to the adsorbed substances and the charge transfer effect.

Wei et al. (2003) demonstrated a gas sensor depositing CNT bundles onto a piezoelectric quartz crystal, which detected CO, NO_2, N_2, and hydrogen (H_2) by detecting changes in oscillation frequency and was more effective at higher temperatures (200°C). Recently, Marzari et al. studied sensing mechanisms for CNT-based NH_3 sensors (Peng et al. 2009). They found that at higher temperatures (150°C or above), charge transfer process contributes to the sensing signal and NH_3 adsorption is to be facilitated by environmental oxygen.

Bao et al. fabricated thin-film transistor (TFT) sensors consisting of aligned, sorted nanotube networks (Roberts et al. 2009). These SWCNT-TFTs were used to detect trace concentrations (down to 2 ppb) of dimethyl methylphosphonate and trinitrotoluene in aqueous solutions.

Penza et al. (2004a) developed surface acoustic wave and quartz crystal microbalance sensors coated with SWCNTs and MWCNTs and used them to detect VOCs such as ethanol, ethylacetate, and toluene by measuring the downshift in the resonance frequency of the acoustic transducers.

Another possibility, exploited by several researchers, is to integrate CNTs in chemical sensors modifying the carbon nanotube surface with polymer films enhancing the selectivity of sensors (Du et al. 2010b, Firdoz et al. 2010, Qi et al. 2003, Zhang et al. 2008). For example, sensors with a polyethylenimine coating are capable of selectively measuring very low NO_2 concentrations (about 100 ppt) against background of many other gases, and sensors with a Nafion film coating can detect NH_3 against the NO_2 background (Qi et al. 2003).

On the same principle, nanoparticles made from transition metals such as gold, platinum, palladium (Pd), copper, silver, and nickel (well known for their high catalytic activity) have been widely utilized to enhance the performance of carbonaceous materials, increasing their sensitivity toward a specific analyte (Day et al. 2005, Du et al. 2010a, Penza et al. 2010).

CNTs have been successfully employed as sensitive materials also in chemo-optical sensors as demonstrated by Barone et al. (2005), who developed a device for β-D-glucose sensing in solution phase.

In the last years, great effort to develop optical biochemical sensors based on CNTs has been given by two groups (Massachusetts Institute of Technology and Rice University) who solved single-molecule adsorption on the SWCNT sidewall through the quenching of excitons (Cognet et al. 2007, Heller et al. 2009, Jin et al. 2008, 2010, Siitonen et al. 2010a,b), enabling a new generation of optical sensors capable of the ultimate detection limit (e.g., single molecules). The recorded fluorescence modulation, supplemented with proper calibration, provides a means for evaluating the concentration of the quencher molecule, even at low concentrations (Jin et al. 2008, 2010, Kim et al. 2009).

Moreover, in the last year, Heller et al. used a peptide structure to modulate the SWCNT fluorescence to realize a chaperone sensor for nitroaromatics (Heller et al. 2011). They described indirect detection mechanism, as well as an additional exciton quenching-based optical nitroaromatic detection method, illustrating that functionalization of CNT surface can result in completely unique sites for recognition, resolvable at the single-molecule level.

Finally, great attention was recently focused on the capability of SWCNT overlays, deposited on the facet of standard single-mode optical fibers, to undergo changes in their geometrical properties as a consequence of the adsorption of target analyte molecules, enabling to exploit such materials for the development of fiber nanosensors for a wide range of strategic environmental applications, such as chemical detection in gaseous and liquid phase at room temperature (Consales et al. 2006a, 2007a, 2009a, Cusano et al. 2006a,b, Penza et al. 2005b) as well as for hydrogen detection at cryogenic temperatures.

In particular, the possibility to integrate such materials with the optical fiber technology has enabled the development of advanced nanosensors capable of air and water quality monitoring, characterized by part per million (ppm) and sub-ppm resolutions, good recovery features, and fast responses, as will be seen in the following sections.

20.3 Fiber-Optic Nanosensor Configuration

In this section, the attention is focused on the description and operating principle of the optoelectronic sensing configuration adopted for the realization of the CNT-based fiber-optic nanosensors (e.g., the reflectometric configuration) as well as the optoelectronic scheme exploited for their robust, continuous, and real-time interrogation.

As schematically represented in Figure 20.1, the reflectometric configuration is essentially based on a low finesse and extrinsic FP interferometer created by the deposition of a thin sensitive layer on the facet of standard monomode silica optical fibers (SOF). The thin film acts as an optical cavity where the fiber-sensitive layer interface and the sensitive layer/external medium interface represent respectively the first and the second mirror of the interferometer.

In this configuration, the light is partially reflected each time it reaches the second surface, resulting in multiple offset beams which can interfere one with each other. The amount of light reflected at the first interface can be calculated as the sum of the multiple reflected beams and is strongly influenced even by very small changes of the distance between the two surfaces (the sensitive layer thickness) or its optical properties (the sensitive layer refractive index) (Dakin and Culshow 1998). This explains the massive use of such configuration in fiber-optic-based sensing in the last two decades, especially for the detection and measurements of various physical, chemical, and biomedical parameters (Chan et al. 1994, Jackson 1994). All these characteristics, combined with the possibility of integrating a number of sensitive materials with the optical fibers by means of very simple, low-cost, and versatile deposition techniques, make the FP one of the most attractive and useful configurations for practical applications.

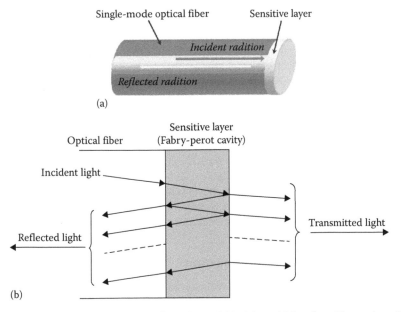

(a)

(b)

FIGURE 20.1 (a) Schematics of the FP-based configuration and (b) of the multiple reflected beams in an FP cavity.

The operating principle of a fiber-optic nanosensor based on the reflectometric FP configuration thus relies on the modulation of the intensity of light reflected at the fiber-sensitive layer interface induced by the changes in the layer thickness (d_{film}) and/or its complex refractive index ($\tilde{n}_{film}$).

Indeed, the fiber-film reflectance can be expressed as follows (Macleod 2001):

$$R = \left| \frac{r_{12} + r_{23} \cdot e^{-i \cdot \tilde{K}_{film}}}{1 + r_{12} \cdot r_{23} \cdot e^{-i \cdot \tilde{K}_{film}}} \right|^2 \tag{20.1}$$

with

$$r_{12} = \frac{n_f - \tilde{n}_{film}}{n_f + \tilde{n}_{film}}; \quad r_{23} = \frac{\tilde{n}_{film} - n_{ext}}{\tilde{n}_{film} + n_{ext}} \tag{20.2}$$

$$\tilde{K}_{film} = \frac{2\pi \cdot \left(2\tilde{n}_{film} \cdot d_{film}\right)}{\lambda} = \frac{4\pi \cdot n \cdot d_{film}}{\lambda} - i \frac{4\pi \cdot k \cdot d_{film}}{\lambda} = \beta_{film} - i\alpha \cdot d_{film}$$

where

$\tilde{n}_{film} = n - i \cdot k$; $\alpha = (4\pi \cdot k)/\lambda$ is the absorption coefficient of the sensitive coating

n_f and n_{ext} are the optical fiber and external medium refractive index

λ is the optical wavelength

In particular, the interaction between sensitive overlay and target analyte molecules (eventually present in the environment under test) is able to produce a modulation of both $\tilde{n}_{film}$ and d_{film}, thus leading to a change in the reflectance R (ΔR), which can be expressed as follows:

$$\Delta R = \left(\frac{\delta R}{\delta n}\right) \cdot \Delta n + \left(\frac{\delta R}{\delta \alpha}\right) \cdot \Delta \alpha + \left(\frac{\delta R}{\delta d_{film}}\right) \cdot \Delta d_{film} = S_n \cdot \Delta n + S_\alpha \cdot \Delta \alpha + S_{d_{film}} \cdot \Delta d_{film} \tag{20.3}$$

In Equation 20.3, S_n, S_α, and S_{dfilm} represent the sensitivities against the variations in the refractive index, absorption coefficient, and thickness of the overlay, respectively. They strongly depend upon the geometrical and electro-optical properties of the sensitive nanocoatings and upon the environmental condition (e.g., vapor or liquid phase), and for this reason they have to be properly considered case by case.

Several effects are involved to promote a reflectance change as a consequence of the analyte molecule adsorption within the sensitive overlay: first of all, swelling of the SWCNT-based nanoscale overlay that leads to a consequent increase of the film thickness; also, refractive index variations are expected due to the film density variation as expressed by the Lorentz-Lorentz law (Kingery et al. 1976). In addition, according to the plasma optic effect (Heinrich 1990, Soref and Bennet 1987, Wooten 1972), a change either in the real part of the refractive index or in the absorption coefficient is possible as a consequence of the free carrier concentration change induced by charge transfer mechanisms during analyte sorption. Modifications of film reflectance could be also possible due to optical absorption modifications induced by the chemical interaction with the target analyte.

In addition, it is noteworthy that when the concentration of the analyte in the environment under investigation is very low (as in the cases reported in this chapter), it can be safely assumed that analyte molecule adsorption occurs at constant overlay thickness ($\Delta d_{film} = 0$ in Equation 20.3).

20.3.1 Fiber-Optic Nanosensor Interrogation

An important issue to address when dealing with sensors is the design and development of a proper demodulation unit able to provide a continuous interrogation of single or multiple sensor probes by minimizing size and complexity and increasing the cost effectiveness. So far, a variety of schemes have

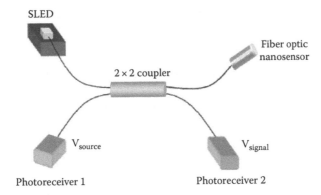

FIGURE 20.2 Schematics representation of the typical optoelectronic scheme adopted for the interrogation of a fiber nanosensor based on the FP reflectometric configuration. (From *Sens. Actuators B*, 118(1–2), Consales, M., Campopiano, S., Cutolo, A., Penza, M., Aversa, P., Cassano, G., Giordano, M., and Cusano, A., Carbon nanotubes thin films fiber optic and acoustic VOCs sensors: Performances analysis, 232–242. Copyright 2006a, from Elsevier.)

been proposed for the interrogation of a fiber-optic nanosensor based on the FP configurations, the most used ones relying on spectrum-modulating approach and single wavelength reflectometry (Kersey and Dandrich 1990). Our attention has always been focused on this last technique, which is simple to implement and requires just few widespread commercial and low-cost optoelectronic components while preserving excellent performance. In addition, it enables the fabrication of cost-effective, reliable, robust, and portable equipment, which are factors of crucial importance for in situ and long-term monitoring applications and for the desired technology transfer to the market.

The exploited interrogation system, schematically represented in Figure 20.2 and described more in detail in Consales et al. (2007b), basically involves a superluminescent light emitting diode (with central wavelength $\lambda = 1310\,nm$ and a bandwidth of approx. 40 nm), an optical isolator, a 2×2 coupler, and two photodetectors.

It provides an output signal I that is proportional to the fiber-film interface reflectance R and that is insensitive to eventual fluctuations of the optical power levels along the whole measurement chain. In what follows, the relative change of the sensor output $\Delta I/I_0$ has always been considered (where I_0 is the output signal in the reference condition, e.g., in the absence of analyte molecules in the environment under analysis), which in turn corresponds to the relative reflectance change occurring at the fiber-sensitive layer interface ($\Delta R/R_0$).

To increase the signal-to-noise ratio, and consequently to enhance the system performance, a synchronous detection approach has been typically used by amplitude modulating the light source at 500 Hz and retrieving the photodetector voltages by using a dual-channel lock-in amplifier. The minimum $\Delta R/R_0$ that can be detected by means of the described interrogation system, calculated by considering the maximum variation on the sensor response in a steady-state level for a time interval of at least 10 min, is typically in the range $1-6 \times 10^{-4}$. In addition, the use of a time division multiplexing approach enabled to perform the quasi-simultaneous interrogation of more than one fiber-optic nanosensor by means of a multichannel fiber-optic switch.

20.4 Integration of CNTs with Optical Fibers

The realization of homogeneous thin films of CNTs with a controllable thickness is an important basis for the development of their scientific understanding and technological applications. As a matter of fact, since CNTs are nanosized materials and tend to aggregate in mats, their precise and controlled handling represents one of the most challenging issues to deal with for their integration into devices.

Proper manipulation techniques are required to apply thin films of CNTs onto small and unconventional substrates (such as the optical fiber one) in such a way to control their morphological and geometrical characteristics as well as to fully benefit of their properties and to mathematically schematize the final device for a rational design of its performance.

At the same time, simple and low-cost fabrication procedures and equipment are required for a fast and cost-effective transition of the CNT-based devices from the laboratories to the market.

So far, several techniques have been implemented to integrate CNTs with optical fibers for the development of in-fiber devices (either for sensing and telecommunications applications). In what follows, we describe the reported approaches, focusing the attention on the Langmuir-Blodgett (LB) deposition method, which is the technique we used for the realization of the fiber-optic nanosensor in reflectometric configuration.

20.4.1 Spraying Method

The CNT spraying is a very simple method proposed in 2004 by Set et al. to coat the terminal face of an optical fiber with a thin layer of CNTs (Kashiwagi and Yamashita 2010, Set et al. 2004). It mainly consists of two steps: (1) first a dispersed solution of CNTs is sprayed on the fiber facet (Kashiwagi and Yamashita 2010, Set et al. 2004) and (2) successively the residual solvent is evaporated by heat treatments.

Since CNTs tend to be entangled with one another, not many types of solvents can be used to disperse them with high uniformity and fewer entanglements. Set et al. used dimethylformamide, which is one of the most commonly employed solvents to disperse CNTs. The main advantage of this method relies on its simplicity (it is possible to realize optical devices using a simple setup); however, it has a low control of the morphological and geometrical features of the sprayed layer and suffers from a poor efficiency of CNT use (the sprayed solution typically spread around the chosen target position).

20.4.2 Direct Synthesis

A further integration technique was proposed by Yamashita et al. (2004) in the same year and relies on the direct growth of CNTs onto the facet of standard optical fibers. With this technique, the cleaved end of an optical fiber is first submerged into bimetal acetate solution of Mo and Co and then is placed in a furnace maintained at 400°C in air to decompose acetates or any other organic residues to form an oxide of bimetallic Mo-Co catalyst. Successively the realized device is placed on a quartz boat, which in turn is positioned in a quartz tube inside an electric furnace. Ar/H$_2$ gas is supplied during the heat-up, stopped at the desired temperature, and evacuated. Then, ethanol vapor is supplied. After the reaction, the electric furnace is turned off and cooled down to room temperature with Ar/H$_2$ flow. Figure 20.3 shows the field emission scanning electron microscope (FE-SEM) images of the synthesized SWCNTs, where a tangled network of SWCNT is clearly appreciable on the cleaved face of the fiber. This direct growth approach has the advantage of direct integration with the fibers; however, there is no method to remove impurities in as-synthesized CNT, and consequently it requires the ability to manufacture high-purity nanotubes (Yamashita et al. 2004).

20.4.3 Transferring Using Hot Water

A new method able to transfer films of vertically aligned single-walled carbon nanotube (VA-SWCNT) onto a D-shaped fiber using hot water was also introduced (Song et al. 2007).

The CNT film is first grown on a quartz substrate by the alcohol catalytic chemical vapor deposition method. To achieve this, catalyst particles are loaded onto a quartz substrate by dip-coating into an ethanol solution incorporating Mo-Co bimetallic acetates, and then the catalyst-coated substrate is inserted into a tube furnace and heated under 40 kPa of Ar/H$_2$ gas (3% H$_2$). Ethanol is supplied as a carbon source, forming into SWCNTs on the surface of the metal catalyst particles. As the CNTs grow, the high nanotube density causes the initially randomized growth direction to align perpendicular to the substrate, resulting in vertically aligned growth. Successively, the VA-SWCNT film is peeled from the

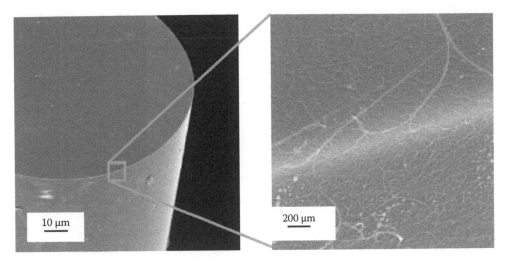

FIGURE 20.3 FE-SEM images of SWCNTs directly synthesized onto cleaved end of a single-mode fiber. (From Yamashita, S. et al., *Opt. Lett.*, 29, 1581, 2004. With permission.)

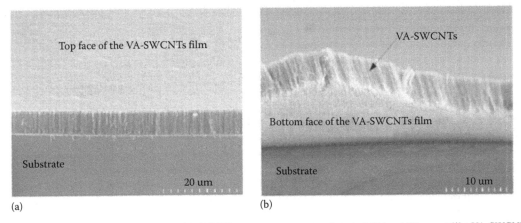

(a) (b)

FIGURE 20.4 (a) SEM image of the VA-SWCNT film grown on a quartz substrate (×2 k) and (b) carpet-like VA-SWCNT film peeled away from the substrate (×3 k). (From Song, Y.-W. et al., *Opt. Lett.* 32(11), 1399, 2007. With permission.)

substrate by submersion into hot water (60°C) and attached onto the flat face of the D-fiber. The realized device is finally dried at 80°C for 2 h. The optical absorption properties of the film remain the same before and after the film transfer. In Figure 20.4, the SEM images of the VA-SWCNT film grown on a quartz substrate (Figure 20.4a) and the VA-SWCNT film peeled away from the substrate (Figure 20.4b) are reported.

This new preparation procedure can guarantee almost 100% yield by the safe and easy process to form CNT devices and enables to maximize the CNT interaction with the field of propagating light.

20.4.4 Optically Assisted Deposition

Recently, a novel technique was also proposed that exploits the optical radiation propagating through an optical fiber as a means to integrate CNTs on its facet (Kashiwagi et al. 2007, Nicholson 2007, Nicholson et al. 2007). With reference to the schematic representation of Figure 20.5, the method requires only a light source and enables the deposition of CNTs preferentially onto the core region of the fiber by injecting a light beam from the fiber end into a CNT-dispersed dimethylformamide (Kashiwagi et al. 2007)

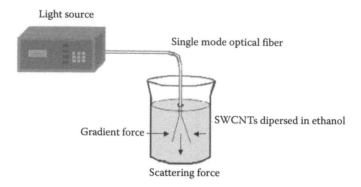

FIGURE 20.5 Setup for depositing carbon nanotubes on the facet of cleaved optical fibers using optical radiation.

or ethanol (Nicholson 2007, Nicholson et al. 2007) solution. This allows for optimal interaction with propagating radiation, while at the same time minimizing the waste of the nanotubes during device preparation.

The basic principle of the optically assisted CNT deposition is not yet confirmed; however, the authors presume that one possible mechanism could be the optical tweezer effect, which is caused by the optical intensity diversion of light in a solution. Another possible mechanism could be the flow of solution due to the injected light. Indeed, the light might thermally induce convection and swirl nearby the core, enabling the attachment of entangled CNTs.

This technique possesses many advantages in terms of efficiency of CNT use and simplicity of the device fabrication process compared to alternative fabrication techniques. However, it requires very precise control of the light injection power to deposit uniform and less scattering CNT layer. In particular, smaller CNT entanglements require higher injection power, whereas high-power injection makes the CNT layer to form around the core, not on the core. The upper limit of optical intensity depends on the flow speed caused by the injected light. Additional techniques are therefore needed to optimize the injection power for each solution.

On this line of argument, Kashiwagi et al. in 2009 proposed the use of optical reflectometry as a means to provide an in situ monitoring of CNT layer deposition process and to precisely control the optical power injection Kashiwagi et al. (2009). With this method the thickness can be controlled by changing the light injection period after the increase of reflectivity due to the first entanglement deposition. In Figure 20.6, the Fe-SEM images of thin and thick CNTs layers deposited by this technique on an area of about 15 μm around the fiber core are reported.

20.4.5 Langmuir-Blodgett Layer-by-Layer Deposition

The LB technique is a well-known method that allows the manipulation of material at the molecular level for depositing defect-free, ultrathin organic films with an accurate control over the architecture of the films at the nanoscale level (Petty 1996). The technique, schematically represented in Figure 20.7, is based on the production of organic monolayer films, first oriented on a subphase and subsequently transferred, layer by layer, onto a solid surface at room temperature and molecule-specific surface pressure.

Usually, the LB film is constituted by amphiphilic molecules (they have hydrophobic and hydrophilic tail or head groups) (Peterson 1990, Petty 1996). A monolayer of the material is at first formed by dispersing the molecules onto the surface of a subphase in the form of a solution. The solvent evaporates, leaving the molecules dispersed across the water surface and oriented with the hydrophobic part upward and the hydrophilic part in water, creating a suspended monolayer. Reducing the surface area by means of moving barriers, the molecules start to repel one another, modifying the surface pressure (Peterson 199). At the first compression, the molecules are randomly oriented, and experience weak interaction with each other. Upon further compression, they are pressed

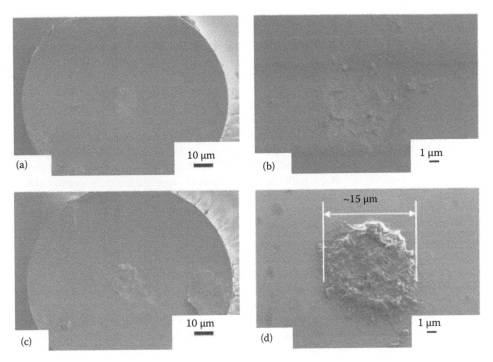

FIGURE 20.6 FE-SEM images of the fiber end with the thin layer ((a) whole fiber end (b) magnified around the core region) and with the thick layer ((c) whole fiber end (d) magnified around the core region). (From Kashiwagi, K. et al., *Opt. Exp.*, 17(7), 5711, 2009. With permission.)

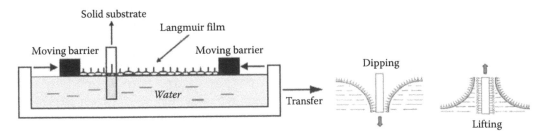

FIGURE 20.7 Schematics of the LB deposition procedure.

closer together with the hydrophobic tails of the molecules that initiate to lift from the surface. The molecules undergo a transition from the gas phase to a liquid-condensed phase, accompanied by a steep rise in surface pressure. Additional compression decreases the surface area occupied by each molecule close to its cross-sectional area, producing a solid phase in which the molecules are packed in an ordered array. From this phase, the molecules may be transferred to a properly cleaned and prepared solid substrate, dipping it through the condensed Langmuir layer to realize the LB film (Peterson 1990, Petty 1996). Repeated dipping of the same substrate is also possible, resulting in the deposition of a thin film one monolayer at a time.

This procedure provides a high-resolution control and uniformity of film thickness (about 1–3 nm per layer) and is thus ideal for the optical interference principle used in the reflectometric configuration to convert dielectric constant changes into optical intensity variations. An additional advantage of the LB technique is that monolayers can be transferred on almost any kind of solid substrate. However, these advantages have to be traded with the low speed of the deposition procedure as well as the limited number of materials suitable for this technique.

20.4.5.1 Fiber-Optic Nanosensors Fabrication

Due to the precise control and uniformity of film thickness in the submicron range allowed by the LB deposition, we chose this technique as a way to transfer nanometer-scale layers of either SWCNTs or cadmium arachidate (CdA)/SWCNT-based nanocomposite upon the distal end of properly cleaved and cleaned single-mode optical fibers. The possibility to integrate SWCNTs inside a host-matrix of a foreign material has been explored to obtain a better adhesion of the CNTs to the fiber end while preserving their excellent sensing properties, as will be shown in the section on experimental results. CdA has been chosen to incorporate the SWCNTs in the nanocomposite due to its peculiar amphiphilic molecular structure suitable for the LB deposition process. Moreover, multilayers of CdA were at the beginning used as buffer materials for enhancing the adhesion of the carbon tubes upon the fiber end (Penza et al. 2004b, 2005b). However, the use of this buffer layer unfortunately resulted in a deterioration of the sensor performance, in terms of low sensitivity and resolution.

Before the deposition procedure, the standard SOFs are previously accurately polished from the acrylic protection and cleaved with a precision cleaver in order to obtain a smooth and plane surface. Then, they are washed in chloroform and dried with gaseous nitrogen.

For the direct SWCNT deposition, a solution (0.2 mg/mL) of SWCNT pristine material (purchased from Carbon Nanotechnologies Inc., Houston, United States) in chloroform is spread onto a subphase constituted by deionized water (18 MΩ) with 10^{-4} M of $CdCl_2$. The subphase pH and temperature are kept constant at 6.0°C and 23°C, respectively. The monolayer is compressed with a barrier rate of 15 mm/min up to a surface pressure of 45 mN/m. The single layer is deposited with a dipping rate of 3 mm/min, and the transfer ratio of the monolayer from the subphase to the substrate surface is typically in the range 0.5–0.7.

Instead, for the deposition of nanocomposite overlays, two separate solutions of arachidic acid in chloroform and SWCNTs in chloroform are mixed to obtain a final solution of chloroform with arachidic acid (0.25 mg/mL) and SWCNTs (0.19 mg/mL). Different concentrations of arachidic acid and SWCNTs in the final solution can be obtained for the preparation of composites with different percentages of CNTs inside the CdA matrix. This mixed solution is accurately dispersed and stirred in an ultrasonic bath for 1 h. Then, only 160 μL of the mixed solution are spread onto a subphase constituted by acetate buffer with $CdCl_2$ 10^{-4} M. The subphase pH and temperature were kept constant at 6.0°C and 20°C. The monolayer of the nanocomposite is compressed with a barrier rate of 15 mm/min up to a surface pressure of 27 mN/m. The single composite layer is deposited upon the optical fiber with a dipping rate of 14 mm/min. Also in this case, the transfer ratio of the monolayer from the subphase to the surface of the sensor is typically in the range of 0.6–0.7.

The thickness of the realized SWCNT-based overlays can range from several nanometers to tens of nanometers, depending upon the number of SWCNT-based monolayers that is transferred upon the distal end of the optical fiber, and can be tailored by a proper choice of the process parameters (Penza et al. 2005b).

The deposited SWCNT LB films have been characterized from a structural and morphological point of view by means of high-resolution transmission electron microscopy (HRTEM) and SEM observations (Consales et al. 2009a, Cusano et al. 2009, Penza et al. 2006b).

In particular, from the SEM photogram reported in Figure 20.8a, the peculiarity of CNTs to adhere with one another forming bundles or ropes with mean diameter in the range of about 4–40 nm and length of about 1–15 μm can be observed (Consales et al. 2006a). As already mentioned, this represents one of the typical limitations of their integration with the optical fiber technology. In fact, this causes the repeatability of the deposition process to be a critical aspect to be addressed, especially with concern to the distribution of the tubes over the fiber substrate and, as a consequence, of the optical properties of the sensitive overlay itself. The alignment of the carbon tubes onto the sensor substrate (Valentini et al. 2004) is seen as possible way to overcome this drawback and to enhance the reproducibility of the deposition process.

The nanometric structure of the carbon tubes is confirmed by the HRTEM image reported in Figure 20.8b, where either single carbon tubes or agglomerates of tubes can be distinguished.

The presence of Fe metal particles can also be revealed between them. This is another typical issue to address, as generally the SWCNT powders contain not only carbon tubes but also carbonaceous

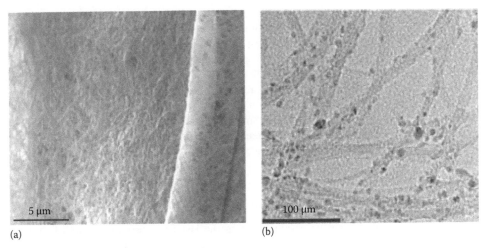

(a)   (b)

FIGURE 20.8 (a) SEM photogram of an LB SWCNT layer deposited on the fiber facet. (From *Sens. Actuators B*, 118(1–2), Consales, M., Campopiano, S., Cutolo, A., Penza, M., Aversa, P., Cassano, G., Giordano, M., and Cusano, A., Carbon nanotubes thin films fiber optic and acoustic VOCs sensors: Performances analysis, 232–242. Copyright 2006a, from Elsevier.); (b) HR-TEM image of SWCNT powder. (From Consales, M. et al., *J. Sens.*, 2008, 29, 2008.)

particles, such as amorphous carbon and fullerenes, and transition metals introduced as catalysts during the synthesis (Journet et al. 1997). These impurities sometimes hold back the fine characteristics of CNTs and limit the performance of SWCNT-based devices. For this reason, effective purifications for the removal of the by-products (amorphous carbon, fullerenes, and catalyst metal particles) of SWCNT synthesis are typically needed in order to fully exploit their outstanding properties.

A detailed morphological and structural characterization of the CNT-based nanocomposites has also been carried out by Penza et al. (2006b).

Here, we report (in Figure 20.9a) the typical SEM images of a CdA/SWCNT nanocomposite overlay, where the nanoscale dimensions as well as the typical bundle disposition of the carbon tubes can be seen, together with their good CdA-assisted adhesion to the fiber surface can also be revealed. Raman spectroscopy analyses have also been carried out to characterize the fabricated probes (Consales et al. 2009a). The results are shown in Figure 20.9b, where typical Raman spectra of fiber-optic probes coated with SWCNT and SWCNT-based composite layers are reported. The characteristic multi-peak feature

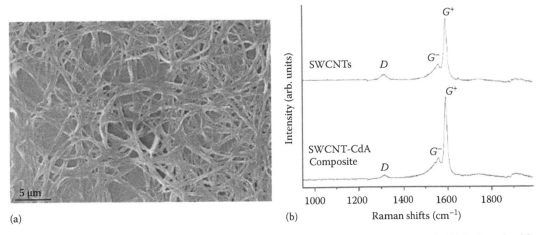

(a)   (b)

FIGURE 20.9 (a) SEM image of a CdA/SWCNT nanocomposite overlay. (From *Sens. Actuators B,* 138(1), Consales, M., Crescitelli, A., Penza, M. et al., SWCNT nano-composite optical sensors for VOC and gas trace detection, 351–361. Copyright 2009a, from Elsevier.); (b) Raman spectra of LB SWCNT and CdA/SWCNT layers directly deposited on the optical fiber facet.

"G-band" at about $1580\,cm^{-1}$, corresponding to carbon atom vibration tangentially with respect to the nanotube walls (Zhang et al. 2004), together with the less remarkable disorder-induced "D-band" peak typically in the range $1300–1400\,cm^{-1}$, representing the degree of defects or dangling bonds (Zhang et al. 2004), can be easily revealed for both samples. In particular, the observation of the two most intense G peaks (labeled G^+ and G^-) confirm the single-walled nature of the carbon tubes while their predominant semiconducting behavior can be derived by the Lorentzian lineshape of the G^- feature which, on the contrary, is broadened for metallic SWCNTs (Zhang et al. 2004). In addition, the large ratio of G to D peaks observed here gives us an indication of an ordered structure of the deposited SWCNT overlay. Between the two recorded Raman spectra, no significant differences can be observed, thus revealing that no structural degradation of the carbon nanotubes occurred as a consequence of their inclusion within the CdA matrix.

20.5 Environmental Monitoring Applications: Results

In this section, we review the results obtained during the last few years of research focused on the development of SWCNT-based fiber-optic nanosensors to exploit for a variety of strategic environmental monitoring applications. In particular, experimental results pertaining to VOC detection in air, chemical trace monitoring in water, and H_2 detection at cryogenic temperature are presented.

The charge transfer effects on the optical properties of CNT-based overlays as a consequence of their interaction with electron-donating (hydrocarbons and alcohols) and electron-accepting analytes (NO_2) are also discussed.

Finally, we show how the use of standard pattern recognition method, such as the principal component analysis (PCA), applied on both static and dynamic responses of the single elements of a fiber-optic nanosensor array, enables to enhance the typically low discrimination ability of SWCNT-based transducers.

20.5.1 Room Temperature Detection of Environmental Pollutants in Air

The capability of SWCNT-based fiber-optic nanosensors to reveal the presence of pollutants in the environment through reflectivity changes, caused by the adsorption of target analyte molecules within the sensitive coating, was demonstrated for the first time in 2004 (Penza et al. 2004b). In that case, LB films consisting of tangled bundles of SWCNTs, transferred onto the fiber tip by using a linker-buffer CdA multilayer (pre-deposited on the sensor surface to promote their adhesion), were used as highly sensitive coatings for the detection of several VOCs, such as isopropanol, methanol, ethanol, toluene, and xylene. In 2006, multilayer of SWCNTs with different thickness was successfully deposited directly on the fiber facet by a modification of the LB process (Consales et al. 2006b), resulting in an improvement of the sensing performance of the unbuffered configurations with respect to the buffered cases, both in terms of sensitivity and response times. In particular, both devices demonstrated to be able to detect traces of the chemical under investigation at a few tens of ppm levels, with a fast response, complete recovery of the reference output signal (which grants the reuse of the sensor after a given measurement) as well as a marked dependence of the response times on the analyte concentration (Consales et al. 2006b). However, optical probes directly coated by SWCNT layer turned out to be one order of magnitude more sensitive than those based on the CdA-buffered configuration. As an example, Figure 20.10 shows the highest relative reflectance changes $\Delta R/R_0$ of the unbuffered probe when exposed to 30 min decreasing concentration pulses of xylene vapors.

In light of these results, the attention was then focused on the investigation of the sensing capabilities of SWCNT-based nanosensors in the unbuffered configuration, whose sensing performance toward different chemicals in air (and at room temperature) was characterized in-depth during an intense experimental campaign (Consales et al. 2006a). There, the amazing sensing features of such devices were widely confirmed. In particular, they demonstrated to be able to provide a linear response toward the investigated pollutants in the concentration range from a few ppm to a few hundreds of ppm (see, e.g., Figure 20.11, which reports the calibration curves of a SWCNT-based nanosensor, exposed at room temperature to toluene and xylene vapors), high repeatability, and very high sensitivities and detection limits of the order of few hundreds of part per billion (ppb).

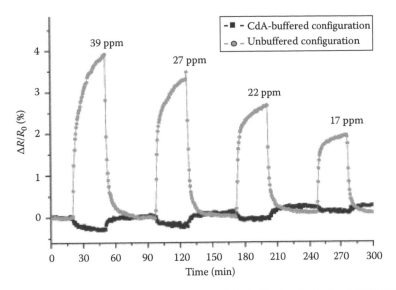

FIGURE 20.10 $\Delta R/R_0$ occurred upon xylene vapor exposure to CdA-buffered and unbuffered SWCNT-based nanosensors, at room temperature. (From Consales, M. et al., *J. Sensor.*, 2008, 29, 2008. With permission.)

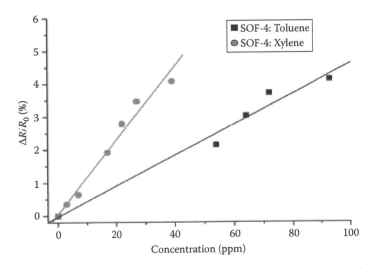

FIGURE 20.11 Calibration curves of a SWCNT-based nanosensor, exposed to toluene and xylene vapors at room temperature. (From *Sens. Actuators B*, 118(1–2), Consales, M., Campopiano, S., Cutolo, A., Penza, M., Aversa, P., Cassano, G., Giordano, M., and Cusano, A., Carbon nanotubes thin films fiber optic and acoustic VOCs sensors: Performances analysis, 232–242. Copyright 2006a, from Elsevier.)

Moreover, it was found that the fiber-optic sensor sensitivity can be tailored by a proper choice of the number of deposited SWCNT monolayer (and consequently the amount of SWCNTs). This choice, however, has to be made by taking into account the trade-off existing between sensor sensitivity and response time (Consales et al. 2006a). Indeed, it turned out that an increase of the SWCNTs on the fiber facet results in an increase in sensor sensitivity, but at the same time it drastically increases its response time.

As a matter of fact, in Figure 20.12, a bar plot shows the mean response and recovery times of two fiber nanosensors, coated by 2 and 4 monolayers of SWCNTS (namely, SOF-2 and SOF-4, respectively), obtained for a 17 ppm xylene vapor exposure. As observed, even if a decrease of the sensor response time (from 9 to 6 min) as well as of the mean recovery time (from 5 to 2 min) can be achieved by reducing the SWCNT layer thickness, this results in a strong diminution of the sensor sensitivity, from 1.1×10^{-3} to 3×10^{-4} ppm^{-1}.

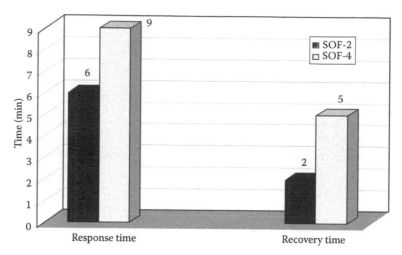

FIGURE 20.12 Bar plots showing the mean response and recovery times of two fiber nanosensors coated by 2 and 4 monolayers of SWCNT$_S$ (namely, SOF-2 and SOF-4), respectively. (From Consales, M. et al., *J. Sens.*, 29, 2008. With permission.)

It is worth noting that fiber-optic nanosensors based on nanoscale layer of SWCNTs were also found to be characterized by a slight cross sensitivity toward temperature and relative humidity (RH) changes (Consales et al. 2006b). This means that a constant and continuous monitoring of these physical parameters, aimed at compensating their effects on the sensor responses, is strictly required during standard operation in order to avoid a reduction of the system performance, especially when high accuracy is necessary. However, temperature and RH monitoring can be implemented by the use of fiber Bragg grating sensors, which could be separately inserted in the test ambient or even integrated in the same fiber at the end of which the SWCNT nanosensor is realized (close to the coated facet) (Berruti et al. 2011, Cusano et al. 2004, Yeo et al. 2008).

One of the most critical issues to address pertaining to the exploitation of the described fiber nanosensors in practical applications relies on the not very strong adhesion of CNTs to the fiber end-face. As mentioned earlier, even though the use of buffered configurations addresses this issue, it severely limits the sensing performance of the final devices.

In light of these considerations, we recently proposed to exploit novel nanoscale fiber coatings composed of SWCNTs embedded in CdA matrixes (Consales et al. 2009a). Indeed, the use of CdA matrix ensures a better adhesion of the carbon tubes to the fiber-optic surface, thus improving the sensor robustness and reliability.

In what follows, we focus the attention on the sensing performance in air of fiber-optic nanosensors coated by nanoscale overlays of CdA/SWCNTs and compare it to that obtained with the same transducers coated by standard SWCNT overlays. In particular, we report the trace detection capability of a sensor coated by 10 monolayers of nanocomposite (with a SWCNT filler weight percentage of approximately 25 wt.%) toward several VOCs (such as toluene, xylene, ethanol, and isopropanol) and one gas (NO_2).

In Figure 20.13, the typical relative reflectance changes $\Delta R/R_0$ occurring as a consequence of the exposure to toluene and ethanol vapors are reported. It shows that the analyte adsorption within the SWCNT-based nanocomposite overlay was able to induce a significant increase of the fiber-film reflectance as a consequence of the complex refractive index changes in the FP sensing cavity. Experimental data clearly reveal the capability to detect very low concentrations of the tested pollutants at ppm levels, as well as its quite good attitude to recover the initial baseline signal upon the complete analyte molecule desorption. As already mentioned, this feature is of great importance for environmental monitoring applications, since it enables the sensor to be easily and quickly reused after a given measurement, avoiding ad hoc cleaning procedures, which are costly and time-consuming.

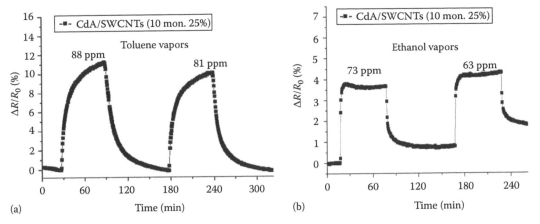

FIGURE 20.13 Relative reflectance changes of the fiber nanosensor coated by 10 monolayers of CdA/SWCNT nano-composite (25 wt.%), exposed to different concentrations of (a) toluene and (b) ethanol vapors, at room temperature. (From *Sens. Actuators B*, 138(1), Consales, M., Crescitelli, A., Penza, M., Aversa, P., Veneri, P.D., Giordano, M., and Cusano, A., SWCNT nano-composite optical sensors for VOC and gas trace detection, 351–361. Copyright 2009a, from Elsevier.)

In particular, the sensor exhibited a complete and fast reversibility of the response in case of toluene exposure; however, this was not the case of ethanol exposure for which the reference signal corresponding to the condition of uncontaminated ambient is not fully recovered. This behavior was also observed on isopropanol exposure and can be attributed to the higher polarity of this kind of chemical species (e.g., alcohols).

Different toluene and ethanol sensitivities (approx. $3.2 \times 10^{-3}\,ppm^{-1}$ and approx. $5 \times 10^{-4}\,ppm^{-1}$, respectively), as well as different response dynamics, were observed for the CdA/SWCNT-based fiber probe, the latest being mainly attributable to the different molecule diffusivities within the nanocomposite overlay. Figure 20.14 reports the typical response of the same CdA/SWCNT-based sensor to five NO_2 pulses with concentrations in the range 1–10 ppm.

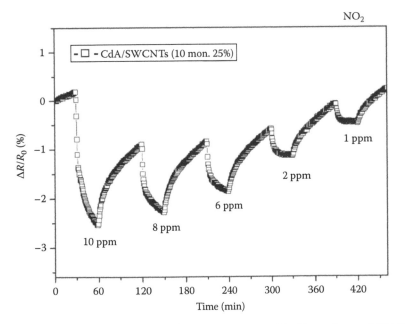

FIGURE 20.14 Relative reflectance changes of the fiber nanosensor coated by 10 monolayers of CdA/SWCNT nano-composite (25 wt.%), exposed to different concentrations of gaseous NO_2, at room temperature. (From *Sens. Actuators B*, 138(1), Consales, M., Crescitelli, A., Penza, M., Aversa, P., Veneri, P.D., Giordano, M., and Cusano, A., SWCNT nano-composite optical sensors for VOC and gas trace detection, 351–361. Copyright 2009a, from Elsevier.)

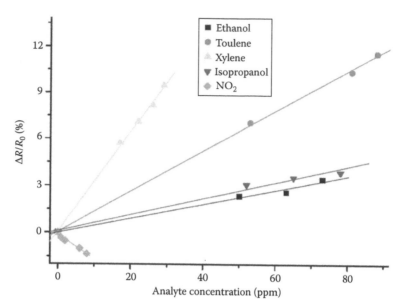

FIGURE 20.15 Comparison between the sensor characteristic curves obtained for the five tested chemicals (From *Sens. Actuators B*, 138(1), Consales, M., Crescitelli, A., Penza, M., Aversa, P., Veneri, P.D., Giordano, M., and Cusano, A., SWCNT nano-composite optical sensors for VOC and gas trace detection, 351–361. Copyright 2009a, from Elsevier.)

Also in this case, the sensor confirmed its capability of detecting very small traces of the gas under test, and similarly to what was observed for alcohol detection, a baseline shift occurred upon multiple exposures of NO_2. However, in this case, the drift is thought to be mainly caused by the fact that the time extent between two successive NO_2 exposures is not sufficiently high to let the sensor completely recover its initial baseline. This is demonstrated by the fact that after the last exposure the baseline value is similar to the one recorded before the first exposure.

It is also worth noting that in the case of NO_2 detection, the fiber-optic transducer exhibited a $\Delta R/R_0$ of opposite sign with respect to that observed in the case of vapor detection. This interesting behavior has also been observed for the fiber-optic nanosensors in the standard configuration (e.g., fiber optic coated by pure SWCNTs) and could be ascribed to the electrical nature of the analyte under investigation, electron donor (vapors), or acceptor (NO_2). This aspect, however, is better discussed later.

An almost linear behavior in the sensor calibration curves (reported in Figure 20.15) was found toward most of the tested chemicals in the investigated ranges, as well as higher sensitivities (see Figure 20.16a) in case of exposure to aromatic hydrocarbons (3.2×10^{-3} and $1.3 \times 10^{-3}\,ppm^{-1}$, respectively, for xylene and toluene) than to alcohols (5×10^{-4} and $4 \times 10^{-4}\,ppm^{-1}$, respectively, for isopropanol and ethanol). In addition, the sensitivity toward NO_2 was of approx. $-1.3 \times 10^{-3}\,ppm^{-1}$.

By considering the minimum detectable $\Delta R/R_0$ achievable with the exploited interrogation system, resolutions (calculated as $\Delta R/R_{0min}/C$, where C is the analyte concentration) in the range 30–80 and 200–250 ppb have been estimated, respectively, for hydrocarbon and alcohol detection, while the minimum concentration of NO_2 that can be detected turned out to be approximately 80 ppb.

With concern to the response times (calculated as the average of the times needed for the output signal to pass from 10% to 90% of the total signal shift occurring upon analyte exposures), CdA/SWCNT composite-based sensor provided a faster response (see Figure 20.16b) in case of alcohol exposure (8 and 9 min, respectively) than in the hydrocarbon one (33 and 31 min, respectively). In particular, the ratio between the mean response time in the case of toluene and ethanol detection is approx. 3.4, while the one obtained with standard SWCNTs is approx. 15. This means that the presence of the CdA matrix not only slows the sensor response but also leads to significant differences in the adsorption dynamics of the two analytes. In addition, since the diffusivity depends upon the exploited SWCNT filler content, it is expected that variations in the SWCNT weight percentage within the CdA matrix could be able to promote further differences in the sensor response times.

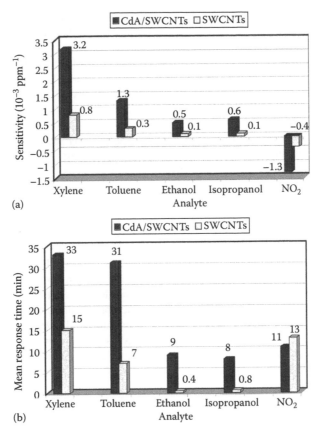

FIGURE 20.16 Comparison between sensitivities (a) and mean response times (b) of the CdA/SWCNTs and SWCNT-based fiber-optic nanosensors obtained for the five tested chemicals. (From *Sens. Actuators B*, 138(1), Consales, M., Crescitelli, A., Penza, M., Aversa, P., Veneri, P.D., Giordano, M., and Cusano, A., SWCNT nano-composite optical sensors for VOC and gas trace detection, 351–361. Copyright 2009a, from Elsevier.)

Even though the optochemical sensors based on CdA/SWCNTs exhibited quite high response times, however, it is worth emphasizing that these times are comparable with those obtained by means of many sensors based on different transducing principles (conductometric, resonator, mass-sensitive sensors, etc.) but integrating the same sensitive materials (Chopra et al. 2003, Kong et al. 2000, Lucci et al. 2005, Penza et al. 2005b).

Overall, by comparing the sensing performance of the nanocomposite-based probe with that obtained with the nanosensor coated by 10 monolayers of standard SWCNTs, it turned out that the use of the CdA/SWCNT nanocomposite coatings not only improves the robustness of the optical fiber nanosensors but also significantly enhances their sensitivity. As a matter of fact, sensor sensitivities from three to seven times higher have been observed for the investigated chemicals with respect to the counterpart fiber-optic nanosensor directly coated by SWCNTs. This can be ascribed to the higher refractive index of composite overlays in respect to those based on standard SWCNT ones.

20.5.1.1 Charge Transfer Effects on the Sensing Properties of Fiber-Optic Chemical Nanosensors Based on SWCNTs

Most of the CNT-based chemical sensors presented in literature rely on resistivity/conductivity measurements to detect changes in CNT electrical properties on exposure to various gaseous and VOC molecules. Indeed, many studies have shown that although CNTs are robust and inert structures, their electrical properties are extremely sensitive to the effects of the charge transfer induced by the analyte molecule adsorption (Kong et al. 2000, Valentini et al. 2003). In particular, it has been demonstrated that electron-accepting analytes decrease the electrical resistance of semiconducting p-type CNTs (Kong et al. 2000, Valentini et al. 2003), while the opposite effect is demonstrated for electron-donating analytes (Cantalini et al. 2003, Pena-Calva et al. 2004).

In light of this consideration, in what follows we report on the role of charge transfer on the sensing properties of SWCNT-based materials when optical detection is used. To this aim we refer to the two fiber nanosensors analyzed earlier (coated by 10 monolayers of CdA/SWCNTs and standard SWCNTs, respectively), whose sensing performance in detecting electron-donating (VOCs) and electron-accepting analytes (NO$_2$) was already reported.

To this aim, we first point out that SWCNT overlays are typically seen as a mixture of two media, graphite-like and air (with the former being the inclusion and the latter the host material) and its optical behavior can be estimated by means of an effective medium approximation (Jeon et al. 2004, 2005). Moreover, their effective index is significantly influenced by the volume fraction of air inclusion that strongly depends upon the number of SWCNT monolayers deposited as well as the nanotube distribution onto the fiber facet. However, if we consider a volume fraction of 0.5, the refractive indexes of SWCNT and CdA/SWCNT-based overlays are approx. 1.5 and 1.7, respectively. This means that it can be safely assumed that, for small relative refractive index and absorption variations, the reflectance monotonously increases with n ($S_n > 0$ in Equation 20.3) and decreases with α ($S_\alpha < 0$ in Equation 20.3).

Figures 20.13, 20.14, and 20.16 clearly show that SWCNT-based nanosensors exhibited an opposite response, in terms of $\Delta R/R_0$, when exposed to hydrocarbon and alcohol vapors (characterized by an electron donor behavior) and NO$_2$ gas (characterized by an electron-accepting nature), thus revealing that a variation of the SWCNT-based coating refractive index of opposite sign occurred in the two cases. This means that the charge transfer effects, similarly to resistive sensors, play a significant role also for optical detection.

In order to take account of the charge transfer influence, the plasma optic effect was, here, considered (Cusano et al. 2009), which allows relating the modulation of the optical properties of sensitive overlays to the changes of the free carrier concentration. Considering a classical analysis, the plasma optic effect is based on the concept that electrons in the conduction band and holes in the valence band of a semiconductor behave very much as free carriers. From simple plasma physics, a free carrier density variation ΔN (of either electrons ΔN_e or holes ΔN_h) produces a change in both the real part of the complex refractive index (Δn) and in the absorption coefficient ($\Delta \alpha$) according to the following relations (Heinrich 1990, Soref and Bennet 1987, Wooten 1972):

$$\Delta n = -\left(\frac{q^2\lambda^2}{8\pi^2c^2\varepsilon_0 nm*}\right)\cdot\left[\frac{\Delta N_e}{m_{ce^*}} + \frac{\Delta N_h}{m_{ch^*}}\right]$$

$$\Delta\alpha = -\left(\frac{q^2\lambda^2}{4\pi^2c^2\varepsilon_0 n}\right)\cdot\left[\frac{\Delta N_e}{m_{ce^*}\cdot\mu_e} + \frac{\Delta N_h}{m_{ch^*}\cdot\mu_h}\right] \tag{20.4}$$

where
  q is the electronic charge
  ε_0 is the permittivity of the free space
  λ is the operating optical wavelength
  c is the speed of light

  m_{ce^*} and m_{ch^*} are the conductivity effective masses of electrons and holes, respectively
  μ_e and μ_h are the electron and hole mobility

By assuming that charge neutrality is maintained within the device being probed, and defining a reduced mass given by $m* = (m_{ce^*}\cdot m_{ch^*})/(m_{ce^*} + m_{ch^*})$, the refractive index change can be rewritten as (Heinrich 1990)

$$\Delta n = -\frac{q^2\lambda^2\cdot\Delta N}{8\pi^2c^2\varepsilon_0 n\cdot m*} \tag{20.5}$$

where $\omega_p^2 = q^2 N/(\varepsilon_0 n^2 \cdot m*)$, is termed the plasma resonant frequency (Heinrich 1990). Equation 20.6 shows that the refractive index of a semiconducting device varies linearly with the charge–density modulation in the device itself. A more detailed and accurate analysis should involve also the effects of the scattering and losses (Wooten 1972) that, however, are not considered here. In addition, due to the fact that typically SWCNT bundles are composed of either semiconducting or metallic tubes, a more accurate model should be considered, involving a combination of the Drude term and localized Lorentzian absorption (Jeon et al. 2005). In a first-order approximation, we can say that due to the low thickness of the sensitive overlays exploited here, the fiber-film reflectance modulation due to changes of the absorption losses is small compared to those promoted by refractive index changes ($S_\alpha \ll S_n$). This means that, considering ($\Delta d_{film} = 0$) and combining Equations 20.3 and 20.5, ΔR can be approximated as follows:

$$\Delta R = S_n \cdot \Delta n = -\frac{S_n \cdot q^2 \lambda^2 \cdot \Delta N}{8\pi^2 c^2 \varepsilon_0 n \cdot m*} = -S_n^* \cdot \Delta n \qquad (20.6)$$

where $S_n^* = (S_n \cdot q^2 \lambda^2)/(8\pi^2 c^2 \varepsilon_0 n \cdot m*)$ is a constant. Equation 20.6 explains the responses of the SWCNT fiber-optic probes observed in Figures 20.13 and 20.14: in fact it turns out that when exposed to electron-accepting or electron-donating analytes, a fiber nanosensor based on CNTs can exhibit either a positive or negative reflectance change due to the variations of opposite sign occurring to ΔN. In particular, it was previously found that semiconducting SWCNTs exhibit a p-type electrical behavior (Javey et al. 2003, Johnson et al. 2007). For this reason, their interaction with electron-donating analytes (xylene and ethanol vapors) produces a diminution of the free holes density ($\Delta N_h < 0$) within the overlays as a consequence of the electron-hole recombinations, resulting in an increase of the real part of the refractive indexes ($\Delta n > 0$) and in a decrease of the absorption losses ($\Delta \alpha < 0$). On the contrary, in the case of interaction with electron-accepting analytes (NO_2), the free hole density increases ($\Delta N_h > 0$) due to the electron transfer from the sensitive nanocoatings to the analyte molecules, leading to a decrease of the film refractive index ($\Delta n < 0$) and to an increase of the absorption losses ($\Delta \alpha > 0$). This phenomenon is of significant relevance and could enable one to discriminate between contaminants of different electrical nature, on the basis of the sign of the optical sensor response (the fiber-film reflectance variation). In addition, a proper pretreatment of the SWCNTs could enable the introduction of defect sites at the sidewall of the CNTs and the consequent increase in the adsorbate binding energy and charge transfer (Robinson et al. 2006). This could result in the enhancement of sensor sensitivity to a variety of chemical substances.

In order to have a quantitative estimation of the charge density variation occurring in the SWCNT-based material upon analyte adsorption, the electrical conductivity of a SWCNT-based composite layer was measured upon NO_2 gas exposure. These measurements were in turn exploited to investigate the influence of electrical signal variations on optical responses. To this aim, a direct measure of the d.c. electrical conductance of the composite films deposited onto rough alumina (5.0 mm length, 5.0 mm width, 0.5 mm thickness) substrate with 200 μm pitch interdigitated Cr/Au (20/200 nm thick) pattern was carried out by means of two-pole probe method with an electrometer (Keithley 617). The typical time response of the electrical resistance of an LB layer of CdA/SWCNT composite upon exposure of NO_2, at room temperature, is reported in Figure 20.17.

The electrical resistance decreases when the SWCNT-based composite film is exposed to the oxidizing NO_2 gas. In particular, upon NO_2 gas adsorption, electron charge transfer occurs from SWCNT-based composite to NO_2 because of the electron-accepting power of the NO_2 molecules. Thus, the NO_2 gas depletes electrons from the SWCNT-based composite and increases the concentration of electrical holes in the p-type SWCNT-based composite, hence causing the electrical resistance to decrease. Partial desorption and unreached saturation level are observed as well, but a clear electrical response modulated by adsorption of the gas is demonstrated, revealing that a charge transfer effectively occurs between the realized LB SWCNT overlays and the analyte under investigation. Furthermore, the sensitivity of the resistive sensor based on the CdA/SWCNT overlay against NO_2, calculated as $(\Delta R^r/\Delta R_0^r)/C$, where ΔR^r is the resistance change, R_0^r is the resistance in dry air, and C is the gas concentration, was estimated to be approximately 4×10^{-3} ppm^{-1}. This relative resistance change can be rewritten as

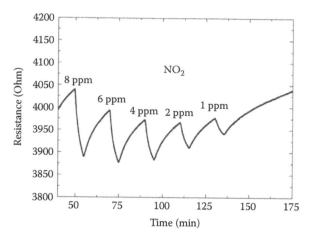

FIGURE 20.17 Electrical resistance versus time of a CdA/SWCNT thin film upon exposure to NO$_2$ gas, at room temperature. The substrate used is Cr/Au patterned alumina with 200 µm pitch interdigitated transducers. (From *Carbon*, 47, Cusano, A., Consales, M., Crescitelli, A., Penza, M., Aversa, P., Delli Veneri, P., and Giordano, M., Charge transfer effects on the sensing properties of fiber optic chemical nano-sensors based on single-walled carbon nanotubes, 782–788. Copyright 2009, from Elsevier.)

a relative conductivity change ($\Delta\sigma/\sigma$) and thus as a relative change in the free carrier density (according to Equation 20.4):

$$\frac{\Delta R^r}{R^r} = \left|\frac{\Delta\sigma}{\sigma}\right| = \left|\frac{\Delta\alpha}{\alpha}\right| = \left|\frac{\Delta N}{N}\right| \tag{20.7}$$

For this reason, it can be clearly seen that exposure of gaseous NO$_2$ at ppm level is able to induce a decrease of approx. few ‰ in the carrier density of the sensitive layer. The first consequence of this result is the decrease of the effective refractive index of the sensitive overlay according to the plasma optic formulation. Moreover, based on the dependence of the reflectance on the effective refractive index and considering the thickness and refractive index range of our overlays, a reflectance diminution is expected, which in turn is in good agreement with the experimental results obtained. In addition, the carrier density variation induced by NO$_2$ absorption and experimentally measured by the electrical tests is able to induce a refractive index change of the order of 0.1% as verified by theoretical estimations carried out taking into account the plasma optic formulation and the reported values of the SWCNT electrical properties (Bockrath et al. 2000, Wooten 1972, Xue et al. 2006, Zhou et al. 2005). This refractive index variation is consistent with the relative reflectance change revealed by the optical tests.

Overall, the results shown here, obtained by using either optical or resistive transducers, reveal that this effect, which plays a key role in CNT sensors based on resistivity/conductivity measurements, is of significant relevance also in cases where optical detection is used. As a matter of fact, according to the plasma optic effect, a free carrier density change within the SWCNT-based fiber coatings due to the analyte molecules adsorption is able to induce variations of the overlay refractive index of opposite sign, depending on the electric nature of the given contaminant (electron donor or acceptor). Finally, this mechanism could be exploited to enhance the sensing performance of the optoelectronic probes based on SWCNTs by means of proper treatments and processing of the SWCNT-based materials.

20.5.2 Improving the Discrimination Ability of SWCNT-Based Nanosensors

One of the major concerns with chemical sensors is how to improve their discrimination ability among different analytes. In fact, we are still far away from having synthetic and tunable materials that are able to mimic the tremendous molecular recognition capability of the biological receptors such as enzymes. As many other sensors based on SWCNTs (and not only), also our fiber-optic nanosensors are characterized by a poor selectivity toward a given chemical specie. This means that in the case of a multicomponent gas mixture with interfering analytes, the information provided by the sensors could be ambiguous and no straightforward information pertaining to the environment under investigation can be achieved.

Basically, there are two possible approaches in the attempt to increase the chemical sensor selectivity, and both have to be followed in synergy. The first one is a direct approach which relies on the sensitive layer functionalization in order to have a higher affinity of the material toward specific chemical species (Balasubramanian and Burghard 2005). The second one is based on the use of a hybrid system composed of multiple transducers coated by the same material (or, in the most general case, by different materials) in the form of an array. Here, sensors with a poor selective response, when considered collectively, provide unique patterns typical for each analyte. The generated response patterns are interpreted by pattern recognition algorithm for the selective detection (Zaromb and Stetter 1984).

Our attention was focused on the use of sensor arrays composed of low-selective elements combined with pattern recognition algorithms. As a matter of fact, in the following we will show the results obtained by applying the PCA on the responses of an array composed of fiber-optic nanosensors coated by different SWCNT-based overlays. In particular, we show that a suitable fiber nanosensor array is able to clearly discriminate among different analytes, either in vapor or gaseous phase.

PCA is a powerful, linear, supervised, pattern recognition technique used as a mathematical tool for analyzing, classifying, and reducing the dimensionality of numerical datasets in a multivariate problem (Gardner and Bartlett 1999). It typically decomposes the primary data matrix (made of a given number of measurements or experimental points) by projecting the multidimensional dataset onto a new coordinate base formed by the orthogonal directions with maximum variance of data. The eigenvectors of the data matrix are called principal components (PCs) and are uncorrelated among them. The PCs are ordered so that PC_1 displays the greatest amount of variance, followed by the next greatest PC_2, and so on. The magnitude of each eigenvector is expressed by its own eigenvalue, which gives a measure of the variance related to that PC. As a result of the coordinates change, a data dimensionality reduction to the most significant PCs and an elimination of the less important ones can be achieved without considerable information losses. The main features of PCA are the coordinates of the data in the new base (score plot) and the contribution to each component of the sensors (load plot). The score plot is usually used for studying the classification of the data clusters, whereas the load plot can be used for giving information on the relative importance of the sensors to each PC and their mutual correlation. In this PCA study, the array is composed of two fiber nanosensors, respectively coated by 10 monolayers of standard SWCNTs and CdA/SWCNTs composite (25 wt.%), whose performance against toluene, xylene, ethanol, isopropanol, and NO_2 has been analyzed earlier.

In addition, since we noticed a strong influence of the specific analyte specie on the dynamic behavior of the sensor output (e.g., the responses to alcohols exhibited a marked overshoot as well as a much faster increase, on exposure, than those to other VOC and NO_2), we applied the PCA both on the static and transient parameters of the response curve (as reported in Table 20.1).

Input data of primary matrix are obtained extracting, from each sensor, the parameters P1–P5 by the responses to different exposures of the five analytes. The PCA was applied to a data matrix composed by 10 columns (2 sensors × 5 parameters) and 17 rows (17 measurements × all 5 analytes). Data are further processed by the correlation matrix (centered and standardized data) to remove inadvertent weighting that arise from arbitrary units.

TABLE 20.1

Parameters Extracted from the Transient Response Curves

Parameter	Description
P1: $\Delta R/R_0$	Relative reflectance change at the equilibrium
P2: $\Delta R/R_{0\_Max}$	Maximum relative reflectance change
P3: $\Delta R/R_{0\_3\,min}$	Relative reflectance change after 3 min
P4: t_{10-90}	Response time
P5: t_{max}	Time needed to reach the maximum $\Delta R/R_0$

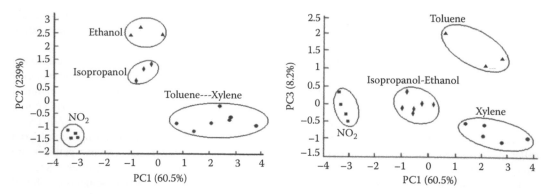

FIGURE 20.18 PCA score plot of toluene, xylene, ethanol, isopropanol, and NO_2 of the data matrix obtained from static and dynamic responses of CNT-based fiber-optic nanosensor array. (From Consales, M. et al., *Fiber Optic Chemical Nanosensors Based on Engineered Single-Walled Carbon Nanotubes: Perspectives and Challenges*, ed. C. Lethien, InTech, Rijeka, Croatia, 2009b.)

In Figure 20.18, the obtained score plots in the PC1–PC2 and PC1–PC3 planes are reported, revealing that the largest part of information has been reduced to the first PC (60.5%), which is also the most important in the discrimination of the clusters. PC2 comprises a lower amount of information (23.9%) while PC3 only the 8.2%, for a cumulative variance of 92.6%.

From score plots it can be seen that NO_2 is the most distinguishable analyte (its samples have the longer distance from those of the other analytes); this can be mainly ascribed to the fact that it is the only one for which both elements of fiber-optic sensor array exhibited negative reflectance changes. It also turned out that the score plot in the plane PC1–PC2 enables one to discriminate between NO_2, ethanol, isopropanol, and hydrocarbon in general, since toluene and xylene samples are clustered together. The same occurs for ethanol and isopropanol in the plane PC1–PC3. However, all tested chemicals (either vapors or gas) can be clearly discriminated in the PC1–PC2–PC3 space, indicating that this nanosensor array provides a high discrimination power to these species. This demonstrates that the selected features are very powerful for discrimination purposes among the different environmental pollutants in the case study.

Overall the results here reported evidence the strong potentiality of the integration of CNT-based nanoscale materials with fiber-optic technology envisaging great perspectives toward the development of high-performance SWCNT optoelectronic noses and tongues.

20.5.3 Chemical Trace Detection in Water

The amazing sensing capabilities of nanoscale SWCNT-based overlays have also been exploited for water quality monitoring applications and, in particular, for aromatic hydrocarbon trace detection at room temperature (Consales et al. 2007a, 2008, Crescitelli et al. 2008). The use of SWCNT fiber-optic nanosensors in this field was not straightforward, since it required an extensive preliminary investigation aimed to assess the stability of the sensitive overlays in aqueous environment. This investigation revealed that as the optical nanosensors are submerged in water, a strong and fast reflectance decrease occurs as a consequence of the change in the external medium refractive index (from air to water), according to Equations 20.1 and 20.2 (Consales et al. 2008, Crescitelli et al. 2008). This effect is followed by a slow variation of the reflectance itself until the equilibrium value is reached due to the interaction between CNT layer and water molecules. However, in this process, no degradation of the SWCNT-sensitive layer occurs, and once the equilibrium condition is reached (typically after a period ranging from tens of minutes to a few hours depending also on the SWCNT layer thickness), the reflectance signal becomes stable and the sensors are ready to operate.

Fiber-optic nanosensors based on standard overlays of SWCNTs were successfully exploited for the detection of toluene in water at room temperature (Consales et al. 2007a). As an example, in Figure 20.19, it is reported that the typical relative change in reflectance occurred upon its exposure to different

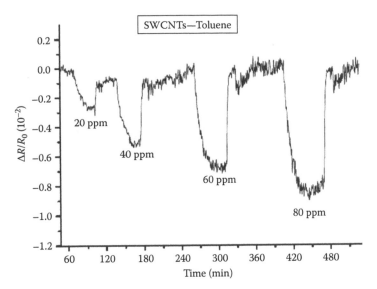

FIGURE 20.19 Typical transient response of a fiber nanosensors coated by a standard SWCNT overlay to four injections of toluene, at room temperature. (From Consales, M. et al., *Fiber Optic Chemical Nanosensors Based on Engineered Single-Walled Carbon Nanotubes: Perspectives and Challenges*, ed. C. Lethien, InTech, Rijeka, Croatia, 2009b.)

concentrations of this chemical in the range 20–80 ppm (calculated as μL/L), which reveal the capability of a SWCNT-based fiber nanosensor to detect the presence of this pollutant in water even at very low concentrations (tens of ppm).

Overall, fiber-optic nanosensors exhibited very high sensitivities, ppm resolutions, excellent desorption capability (demonstrated by the complete recoveries of the reference values of uncontaminated water), and linear operation in the investigated range (Consales et al. 2007a, 2008). Furthermore, if compared with the detection in air, the sensor sensitivities provided in the case of in-water operation resulted slightly higher, the difference being due to the dependence of the fiber-film reflectance upon the surrounding refractive index combined to different adsorption characteristics occurring in the two environments.

However, similarly to the case of detection in air, the adhesion of CNTs to the fiber end-face represents one of the major concerns to address for an actual exploitation of such sensors in practical environmental applications and for their future transfer to the market. This problem is of much more relevance in this case since particularly hard operating conditions (e.g., when the sensor is subject to a continuous water flow, like in a river or a pipeline) could even promote a detachment of the CNT-sensitive overlay from the optical fiber tip.

As already made for the in-air case, the use of CdA/SWCNTs overlays was applied also for the detection in liquid environment, as an effective method to either improve the robustness and reliability of the fiber-optic nanosensors or enhance their performance.

In Figure 20.20, it is reported that the typical transient response of a fiber-optic transducer coated by 20 monolayers of CdA/SWCNTs nanocomposite (75 wt.%) to several toluene injections in water with concentrations ranging from 20 to 100 ppm.

The optical nanosensors based on Cda/SWCNT overlays still provide high sensitivity and desorption capability, linear responses in the investigated range as well as good repeatability (e.g., they exhibit the same response amplitude when detecting analyte injections in water with same concentrations) and sub-ppm resolution.

Similar results were obtained also in the case of xylene detection measurements. The calibration curves (reporting the sensor output versus target analyte concentration) are shown in Figure 20.21. The typical higher affinity of carbon nanotube-based sensors (also showed in the in-air case) toward xylene (the sensor sensitivity is 1.0×10^{-3} ppm^{-1}) than toluene (4×10^{-4} ppm^{-1}) was confirmed (Consales et al. 2007b).

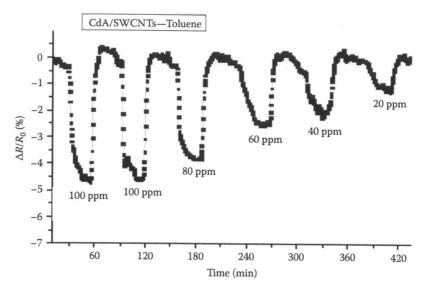

FIGURE 20.20 Typical time response of a fiber nanosensor coated by 20 monolayers of CdA/SWCNT nanocomposite overlay (75 wt.%), exposed to different concentrations of toluene in water at room temperature. (From *Sens. Actuators B*, 138(1), Consales, M., Crescitelli, A., Penza, M., Aversa, P., Veneri, P.D., Giordano, M., and Cusano, A., SWCNT nanocomposite optical sensors for VOC and gas trace detection, 351–361. Copyright 2009a, from Elsevier.)

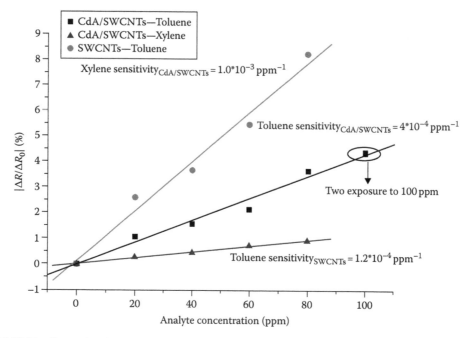

FIGURE 20.21 Comparison between the sensor characteristic curves obtained in correspondence of toluene and xylene injections in water environment. (From *Sens. Actuators B*, 138(1), Consales, M., Crescitelli, A., Penza, M., Aversa, P., Veneri, P.D., Giordano, M., and Cusano, A., SWCNT nano-composite optical sensors for VOC and gas trace detection, 351–361. Copyright 2009a, from Elsevier.)

Finally, to compare the performance of the nanocomposite-based sensor with that obtained with the SWCNT-based one, the characteristic curves of the latter against toluene in water have also been reported in Figure 20.21. The results clearly reveal that a significant enhancement in the sensor sensitivity can be obtained by the use of SWCNT composite overlays, which exhibit a threefold higher sensitivity ($4 \times 10^{-4}\,ppm^{-1}$ vs. $1.2 \times 10^{-4}\,ppm^{-1}$) to the specific pollutant (Consales et al. 2007a).

20.5.4 Hydrogen Detection at Cryogenic Temperatures

Cryogenic hydrogen is often utilized as the primary fuel source by rocket engine because of the benefits deriving from hydrogen combustion properties. Unfortunately, gaseous hydrogen has a very low flammability limit and low ignition energy, so that it becomes a very dangerous and volatile substance to work with during the rocket engine development stage. Hydrogen leakage into an oxidizing environment can easily ignite a catastrophic explosion. For this reason, the development of accurate and reliable devices for the detection of hydrogen leakage at cryogenic temperatures is of crucial importance for human safety. In 1981, Lundstrom introduced the first H_2 sensor based on thin films of palladium deposited on the top of the channel of a MOSFET device (Lundström 1981, Lundström et al. 1975). Since then this issue has been widely studied by the researchers and a number of hydrogen sensors have been proposed (Butler 1994, Chiu et al. 2009, Rivers et al. 2002, Skucha et al. 2010, Zdansky 2011). However, among the big amount of sensing devices, fiber-optic detectors have strongly attracted the attention of the scientific community in this field, especially because of their high sensitivity and lack of sparking possibilities, guaranteed by the removal of all electrical power from the test site. To date, several configurations of fiber-optic hydrogen sensors have been proposed, based on different transducing principles and materials (Buric et al. 2009, Butler 1991, Sutapun et al. 1999, Villatoro et al. 2001). The most exploited material in this field is Pd since it exhibits strong thickness and refractive index changes on H_2 molecule adsorption at room temperature. However, at cryogenic temperatures, Pd loses its excellent sensing properties (Güemes et al. 2005); thus, the main effort in this field is devoted to the research of new sensitive materials able to provide high sensitivity and good desorption features as well as fast responses at very low temperatures. Recent reports have demonstrated a high reversible adsorption of molecular hydrogen in carbon nanotubes and graphitic nanofibers (Dillon et al. 1997, Takagi et al. 2004).

On this line of argument, SWCNTs (of both closed-end and opened-end types) have also been successfully exploited as sensitive fiber overlays for the detection of gaseous H_2 at cryogenic temperatures (as low as 113 K) (Cusano et al. 2006a, Penza et al. 2004b).

In the following, the experimental results obtained by using three fiber-optic nanosensors, coated respectively by two and six monolayers of closed-end SWCNTs (namely, 2_CeSWCNTs and 6_CeSWCNTs, respectively) and two monolayers of opened-end SWCNTs (2_OeSWCNTs), are reported.

In order to test their H_2 sensing performance, the sensors were located in a cylindrical chamber, properly designed and realized in order to minimize the possible thermal variations. The chamber was, in turn, inserted within a tank containing cryogenic nitrogen (see Figure 20.22) with the aim of reaching the very low temperature of 113 K.

The temperature was monitored by a copper/copper-nickel thermocouple located close to the optical sensor head. Nitrogen gas was continuously injected in the cylindrical chamber in order to reduce the humidity content inside the test ambient before the strong temperature decreasing due to the insertion of the chamber in the cryogenic tank. To expose the SOF sensors to gaseous hydrogen with concentrations as low as 4% and 1%, nitrogen was also mixed with gaseous argon containing hydrogen.

The results obtained are shown in Figure 20.23, where the $\Delta R/R_0$ of the sensors (a) 2_CeSWCNTs, (b) 6_CeSWCNTs, and (c) 2_OeSWCNTs, exposed to decreasing concentration pulses of hydrogen at a temperature of 113 K, are reported. As evident, all the optical probes provided significant reflectance changes as a consequence of the gaseous H_2 adsorption within the sensitive nanomaterials. In particular, good sensitivity and complete recovery of the nitrogen signal after both hydrogen concentration pulses were observed for the two CeSWCNT-based sensors, whereas good sensitivity but no recovery of the steady-state value was observed for the OeSWCNT-based one.

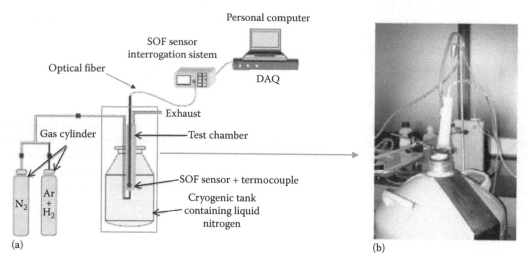

FIGURE 20.22 (a) Schematic view and (b) image of the experimental setup exploited for the H$_2$ detection at cryogenic temperatures. (From Consales, M. et al., *J. Sens.*, 2008, 29, 2008.)

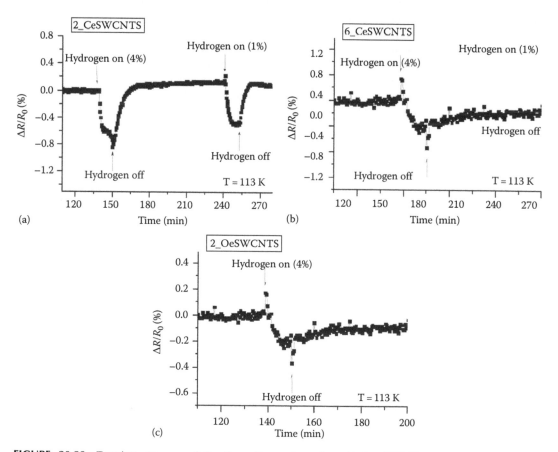

FIGURE 20.23 Transient response of the three fiber-optic probes, (a) 2 _SWCNTs, (b) 6_CeSWCNTs, and (c) 2_OeSWCNT, to decreasing concentration pulses of gaseous hydrogen, at 113 K. (From Consales, M. et al., *J. Sens.*, 2008, 29, 2008.)

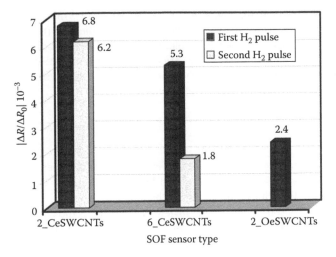

FIGURE 20.24 $\Delta R/R_0$ of the three tested nanosensors occurred on H_2 exposure. (From Consales, M. et al., *J. Sens.*, 2008, 29, 2008.)

In addition, the bar plot shown in Figure 20.24, in which the amplitude of the relative reflectance changes ($|\Delta R/R_0|$) obtained for the tested SWCNT-based probes is reported, clearly reveals that for the probe 2_CeSWCNTs the $\Delta R/R_0$ obtained for 4% and 1% of H_2 are almost similar (the ratio is approximately 1.1). In this case, a saturation of the fiber overlay sorption capability could explain the weak mismatch between the sensor responses.

Differently, for the case of the sensor 6_CeSWCNTs, a decrease of the H_2 concentration of four times leads to a relative reflectance diminution of approximately three times. In this case, however, due to the increase of nanotubes monolayers, higher saturation threshold is expected.

Furthermore, a study of the response and recovery times (t_{10-90} and t_{90-10}, respectively) has been carried out. The results obtained are reported in Figure 20.25 and reveal that the probes 2_CeSWCNTs and 6_CeSWCNTs are characterized by the same response times (approximately 4 min), but they have slightly different recovery times (approximately 9 and 11 min, respectively). This evidences that the number of carbon nanotube monolayers influences the sensor recovery features. In addition,

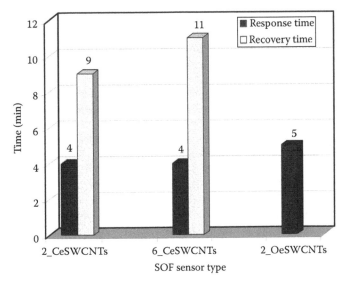

FIGURE 20.25 Mean response and recovery times for the three tested fiber nanosensors. (From Consales, M. et al., *J. Sens.*, 2008, 29, 2008.)

the probe 2_OeSWCNTs showed a response time (approximately 5 min) higher than that obtained with the CeSWCNT-based sensors. This could be attributed to the fact that while in the case of the CeSWCNTs, the H_2 molecules adsorption takes place only on the surface of the single carbon tubes and in the interstitial sites between the tubes; in the case of the counterpart OeSWCNTs, the hydrogen molecule adsorption takes place also inside the carbon tubes, thus requiring more time for the establishment of the equilibrium condition with the external environment.

20.6 Conclusions

In conclusion, in this chapter, we have reviewed the development of high-performance fiber nanosensors for environmental monitoring applications, realized by the combination of advanced nanostructured coatings based on SWCNTs with the optical fiber technology.

In particular, the chapter starts with a wide review of the state of the art of carbon nanotube-based sensors for environmental monitoring, followed by a description of the operating principle of the fiber-optic nanosensors in the FP-type reflectometric configuration. The attention is then focused on the methods so far proposed for the integration of such nanostructured materials with the optical fibers and in particular on the LB deposition, which is the technique we adopted for the fiber-optic nanosensors fabrication.

Successively, the results obtained during the last few years of research exploiting the described SWCNT-based fiber-optic nanosensors in a variety of strategic environmental monitoring applications have been reported and discussed. In particular, experimental results pertaining to VOC trace detection both in air and in water, at room temperature, and hydrogen detection at cryogenic temperatures have been presented.

The charge transfer effects on the optical properties of CNT-based overlays as a consequence of their interaction with electron-donating (hydrocarbons and alcohols) and electron-accepting analytes (NO_2) have also been discussed. Finally, we have shown how the use of standard pattern recognition method, such as the PCA, applied on both static and dynamic responses of the single elements of a fiber-optic nanosensor array, enables to enhance the typically low discrimination ability of SWCNT-based transducers, thus envisaging great perspectives for the development of advanced fiber-optic noses and tongues.

REFERENCES

Arregui, F.J., Matias, I.R., and Claus, R.O. 2003. Optical fiber gas sensors based on hydrophobic alumina thin films formed by the electrostatic self-assembly monolayer process. *IEEE Sensors Journal* 3(1):56–61.

Balasubramanian, K. and Burghard, M. 2005. Chemically functionalized carbon nanotubes. *Small* 1:180–925.

Barone, P.W., Baik, S., Heller, D.A., and Strano, M.S. 2005. Near-infrared optical sensors based on single-walled carbon nanotubes. *Nature Materials* 4(1):86–92.

Barsan, N. and U. Weimar. 2003. Understanding the fundamental principles of metal oxide based gas sensors; the example of CO sensing with SnO_2 sensors in the presence of humidity. *Journal of Physics: Condensed Matter* 15:R813.

Berruti, G., Consales, M., Cutolo, A. et al. 2011. Radiation hard humidity sensors for high energy physics applications using polymide-coated fiber bragg gratings sensors. In Paper presented at the *IEEE 2011 Conference*, Limerick, Ireland, October 28–31, pp. 1484–1487.

Bockrath, M., Hone, J., Zettl, A., McEuen, P.L., Rinzler, A.G., and Smalley, R.E. 2000. Chemical doping of individual semiconducting carbon nanotube ropes. *Physical Review B* 61:R10606.

Buric, M., Chen, T., Maklad, M., Swinehart, P.R., and Chen, K.P. 2009. Multiplexable low-temperature fiber bragg grating hydrogen sensors. *IEEE Photonics Technology Letters* 21:1594–1596.

Butler, M.A. 1991. Fiber optic sensor for hydrogen concentrations near the explosive limit. *Journal of the Electrochemical Society* 138(9):L46–L47.

Butler, M.A. 1994. Micro mirror optical-fiber hydrogen sensor. *Sensors and Actuators B* 22(2):155–163.

Byun, K.M., Yoon, S.J., Kim D. et al. 2007. Experimental study of sensitivity enhancement in surface plasmon resonance biosensors by use of periodic metallic nanowires. *Optics Letter* 32:1902–1904.

Cantalini, C., Valentini, L., Lozzi, L., Armentano, I., Kenny, J.M., and Santucci, S. 2003. NO$_2$ gas sensitivity of carbon nanotubes obtained by plasma enhanced chemical vapor deposition. *Sensor and Actuators B* 93:333–337.

Chan, M.A., Collins, S.D., and Smith, R.L. 1994. A micromachined pressure sensor with fiber-optic interferometric readout. *Sensors and Actuators A* 43(1–3):196–201.

Chen, R.J., Bangsaruntip, S., Drouvalakis, K.A. et al. 2003. Noncovalent functionalization of carbon nanotubes for highly specific electronic biosensors. *Proceedings of the National Academy of Sciences of United States of America* 100:4984–4989.

Chiu, S.Y., Huang, H.W., Liang, K.C. et al. 2009. High sensing response Pd/GaN hydrogen sensors with a porous-like mixture of Pd and SiO2. *Semiconductor Science and Technology* 24:045007.

Chopra, S., McGuire, K., Gothard, N., Rao, A.M., and Pham, A. 2003. Selective gas detection using a carbon nanotube sensor. *Applied Physics Letters* 83:2280–2282.

Cognet, L., Tsyboulski, D.A., Rocha, J.D.R., Doyle, C.D., Tour, J.M., and Weisman, R.B. 2007. Stepwise quenching of exciton fluorescence in carbon nanotubes by single molecule reactions. *Science* 316:1465–1483.

Collins, P.G., Bradley, K., Ishigami, M., and A. Zettl. 2000. Extreme oxygen sensitivity of electronic properties of carbon nanotubes. *Science* 287:1801–1804.

Consales, M., Campopiano, S., Cutolo, A. et al. 2006a. Carbon nanotubes thin films fiber optic and acoustic VOCs sensors: Performances analysis. *Sensors and Actuators B* 118(1–2):232–242.

Consales, M., Campopiano, S., Cutolo, A. et al. 2006b. Sensing properties of buffered and not buffered carbon nanotubes by fiber optic and acoustic sensors. *Measurement Science and Technology* 17:1220–1228.

Consales, M., Crescitelli, A., Campopiano, S. et al. 2007a. Chemical detection in water by single-walled carbon nanotubes-based optical fiber sensors. *IEEE Sensors Journal* 7(7):1004–1005.

Consales, M., Crescitelli, A., Penza, M. et al. 2009a. SWCNT nano-composite optical sensors for VOC and gas trace detection. *Sensors and Actuators B* 138(1):351–361.

Consales, M., Cutolo, A., Penza, M., Aversa, P., Giordano M., and Cusano, A. 2008. Fiber optic chemical nanosensors based on engineered single-walled carbon nanotubes. *Journal of Sensors* 2008:29.

Consales, M., Cutolo, A., Penza, M., Aversa, P., Giordano M., and Cusano, A. 2009b. *Fiber Optic Chemical Nanosensors Based on Engineered Single-Walled Carbon Nanotubes: Perspectives and Challenges*, ed. C. Lethien. InTech, Rijeka, Croatia.

Consales, M., Cutolo, A., Penza, M. et al. 2007b. Carbon nanotubes coated acoustic and optical vocs sensors: Towards the tailoring of the sensing performances nanotechnology. *IEEE Transactions on Nanotechnology* 6:601–612.

Costa-Fernandez, J.M., Pereiro, R., and Sanz-Medel, A. 2006. The use of luminescent quantum dots for optical sensing. *Trends in Analytical Chemistry* 25(3):207–218.

Crescitelli, A., Consales, M., Penza, M., Aversa, P., Giordano M., and Cusano, A. 2008. Toluene detection in aqueous phase by optical fiber sensors integrated with single-walled carbon nanotubes. *Open Environmental and Biological Monitoring Journal* 1:26–32.

Cusano, A., Consales, M., Crescitelli, A. et al. 2009. Charge transfer effects on the sensing properties of fiber optic chemical nano-sensors based on single-walled carbon nanotubes. *Carbon* 47:782–788.

Cusano, A., Consales, M., Cutolo, A. et al. 2006a. Optical probes based on optical fibers and single-walled carbon nanotubes for hydrogen detection at cryogenic temperatures. *Applied Physics Letters* 89(20):3.

Cusano, A., Pisco, M., Consales, M. et al. 2006b. Novel optochemical sensors based on hollow fibers and single walled carbon nanotubes. *IEEE Photonics Technology Letters* 18(22):2431–2433.

Cusano, A., Persiano, G.V., Russo, M., and Giordano, M. 2004. Novel optoelectronic sensing system for thin polymer films glass transition investigation. *IEEE Sensors Journal* 4(6):837–844.

Daghestani, H.N. and Day, B.W. 2010. Theory and applications of surface plasmon resonance, resonant mirror, resonant waveguide grating, and dual polarization interferometry biosensors. *Sensors* 10:9630–9646.

Dakin, J. and Culshaw, B. 1998. *Optical Fiber Sensors: Principle and Components*. Artech House, Boston, MA.

Day, T.M., Unwin, P.R., Wilson, N.R., and Macpherson, J.V. 2005. Electrochemical templating of metal nanoparticles and nanowires on single-walled carbon nanotube networks. *Journal of the American Chemical Society* 127(30):10639–10647.

Dillon, A.C., Jones, K.M., Bekkedahl, T. A., Kiang, C.H., Bethune, D.S., and Heben, M.J. 1997. Storage of hydrogen in singlewalled carbon nanotubes. *Nature* 386(6623):377–379.

Dresselhaus, M., Dresselhaus, G., and Eklund, P.C. 2006. *Science of Fullerenes and Carbon Nanotubes*. Academic Press, San Diego, CA.

Du, D., Wang, M., Cai, J., Qin, Y., and Zhang, A. 2010a. One-step synthesis of multiwalled carbon nanotubes-gold nanocomposites for fabricating amperometric acetylcholinesterase biosensor. *Sensors and Actuators B* 143(2):524–529.

Du, D., Ye, X.X., Cai, J., Liu, J., and Zhang, A. 2010b. Acetylcholinesterase biosensor design based on carbon nanotube-encapsulated polypyrrole and polyaniline copolymer for amperometric detection of organophosphates. *Biosensors and Bioelectronics* 25:2503–2508.

Fan, X., White, I.M., Shopova, S.I., Zhu, H., Suter, J., and Sun, Y. 2008. Sensitive optical biosensors for unlabeled targets: A review. *Analytica Chimica Acta* 620:8–26.

Fine, G.F., Cavanagh, L.M., Afonja A., and R. Binions. 2010. Metal oxide semi-conductor gas sensors in environmental monitoring. *Sensors* 10:5469–5502.

Firdoz, S., Ma, F., Yue, X.L., Dai, Z.F., Kumar A., and B. Jiang. 2010. A novel amperometric biosensor based on single walled carbon nanotubes with acetylcholine esterase for the detection of carbaryl pesticide in water. *Talanta* 83:269–273.

Gardner, J.W. and Bartlett, P.N. 1999. *Electronic Noses: Principles and Applications*. Oxford University Press, Oxford, U.K.

Gouma, P.I. 2010. *Nanomaterials for Chemical Sensors and Biotechnology*. Pan Stanford Pvt. Ltd, Singapore.

Grate, J.W. 2000. Acoustic wave microsensor arrays for vapor sensing. *Chemical Reviews* 100(7):2627–2648.

Grate, J.W. and Abraham, M.H. 1991. Solubility interactions and the design of chemically selective sorbent coatings for chemical sensors and arrays. *Sensors and Actuators B* 3(2):85–111.

Güemes, A., Pintado, J.M., Frövel, M., Olmo, E., and Obst, A. 2005. Comparison of three types of fibre optic hydrogen sensors within the frame of CryoFOS project. Paper presented at *17th International Conference on Optical Fibre Sensors*, Bruges, Belgium, May 5855:1000–1003.

Gusev, A.I. 2005. *Nanomaterials, Nanostructures, Nanotechnologies*. Fizmatlit, Moscow, Russia.

Hajjam, A., Wilson, J.C., Rahafrooz, A., and Pourkamali, S. 2010. Fabrication and characterization of resonant aerosol particle mass sensors, part II: Device fabrication and characterization. *Journal of Micromechanics and Microengineering* 20:125019.

Hartmann, J., Auge J., and Hauptmann, P. 1994. Using the quartz crystal microbalance principle for gas detection with reversible and irreversible sensors. *Sensors and Actuators B* 19(1–3):429–433.

Heinrich, H.K. 1990. Picosecond non invasive optical detection of internal electrical signals in flip-hip-mounted silicon integrated circuits. *IBM Journal of Research and Development* 34:162–172.

Heller, D.A., Jin, H., Martinez, B.M. et al. 2009. Multimodal optical sensing and analyte specificity using single-walled carbon nanotubes. *Nature Nanotechnology* 4:114–120.

Heller, D.A., Pratt, G.W., Zhang, J. et al. 2011. Peptide secondary structure modulates single-walled carbon nanotube fluorescence as a chaperone sensor for nitroaromatics. *Proceedings of the National Academy of Sciences of United States of America* 108:8544–8549.

Homola, J., Yee, S.S., and Gauglitz, G. 1999. Surface plasmon resonance sensors: Review. *Sensors and Actuators B* 54(1):3–15.

Huang X.-J. and Choi, Y.-K. 2007. Chemical sensors based on nanostructured materials. *Sensors and Actuators B* 122(2):659–671.

Huang, J., Jiang, Y., Du, X., and Bi, J. 2010. A new siloxane polymer for chemical vapor sensor. *Sensors and Actuators B: Chemical* 146(1,8):388–394.

Huang, J., Matsunaga, N., Shimanoe, K., Yamazoe, N., and Kunitake, T. 2005. Nanotubular SnO_2 templated by cellulose fibers: Synthesis and gas sensing. *Chemisty of Materials* 17:3513–3518.

Iijima, S. 1991. Helical microtubules of graphitic carbon. *Nature* 354(6348):56–58.

Jackson, D.A. 1994. Recent progress in monomode fibre-optic sensors. *Measurement Science and Technology* 5(6):621–638.

James, D., Scott, S.M., Ali, Z. and O'Hare, W.T. 2005. Chemical sensors for electronic nose systems. *Microchimica Acta* 149(1–2):1–17.

James, S.W. and Tatam, R.P. 2006. Fibre optic sensors with nano-structured coatings. *Journal of Optics A: Pure Applied Optics* 8(7):430–444.

Janata, J. 2009. *Principles of Chemical Sensors*, 2nd edn. Springer, New York.

Javey, A., Guo, J., Wang, Q., Lundstrom, M., and Dai, H. 2003. Ballistic carbon nanotube transistors. *Nature* 424:654–657.

Jeon, T.I., Kim, K.J., Kang, C. et al. 2004. Optical and electrical properties of preferentially anisotropic single-walled carbon nanotube films in terahertz region. *Journal of Applied Physics* 95(10):5736–5740.

Jeon, T.I., Son, J.H., An, K.H., Lee, Y.H., and Lee, Y.S. 2005. Terahertz absorption and dispersion of fluorine-doped single-walled carbon nanotube. *Journal of Applied Physics* 98(3):0343161–0345164.

Jeronimo, P.C.A., Araujo, A.N. and Montenegro, C.B.S.M. 2007. Optical sensors and biosensors based on sol-gel films. *Talanta* 72:13–27.

Jin, H., Heller, D.A., Kalbacova, M. et al. 2010. Detection of single-molecule H_2O_2 signaling from epidermal growth factor receptor using fluorescent single-walled carbon nanotubes. *Nature Nanotechnology* 5:302–332.

Jin, H., Heller, D.A., Kim, J.H., and Strano, M.S. 2008. Stochastic analysis of stepwise fluorescence quenching reactions on single-walled carbon nanotubes: Single molecule sensors. *Nano Letters* 8(12):4299–4304.

Johnson, K.S., Needoba, J.A, Riser, S.C., and Showers, W.J. 2007. Chemical sensor networks for the aquatic environment. *Chemical Reviews* 107(2):623–640.

Journet, C., Maser, W.K., Bernier, P. et al. 1997. Large-scale production of single-walled carbon nanotubes by the electric-arc technique. *Nature* 388:756–758.

Ju, H., Zhang, X., and Wang, J. 2011. *Nano Biosensing: Principles, Development and Application.* Springer, New York.

Kashiwagi, K. and Yamashita, S. 2010. Optical deposition of carbon nanotubes for fiber-based device fabrication. In *Frontiers in Guided Wave Optics and Optoelectronics*, ed. B. Pal, pp. 674–691. InTech, Rijeka, Croatia.

Kashiwagi, K., Yamashita, S., and Set, S.Y. 2007. Novel cost effective carbon nanotubes deposition technique using optical tweezer effect. In *Proceedings of the SPIE Vol. 6478; Photonics Packaging, Integration and Interconnects VII*, eds. A. M. Earman and R. T. Chen, pp. 6478–6415, SPIE, Bellingham, WA.

Kashiwagi, K., Yamashita, S., and Set, S.Y. 2009. In-situ monitoring of optical deposition of carbon nanotubes onto fiber end. *Optics Express* 17(7):5711–5715.

Kepley, L.J., Crooks, R.M., and Ricco, A.J. 1992. A selective SAW based organophosphonate chemical sensor employing a selfassembled, composite monolayer: A new paradigm for sensor design. *Analytical Chemistry* 64(24):3191–3193.

Kersey A.D. and Dandridge, A. 1990. Applications of fiber-optic sensors. *IEEE Transaction on Components, Hybrid and Manufacturing Technology* 13(1):137–143.

Kim, J., Heller, D.A., Jin, H. et al. 2009. The rational design of nitric oxide selectivity in single-walled carbon nanotube near infrared fluorescence sensors for biological detection. *Nature Chemistry* 1:473–481.

Kingery, W.D., Bowen, H.K., and Uhlmamr, D.R. 1976. Introduction to ceramics. J. Wiley & Sons, New York.

Kobayashi, N. 2005. Introduction to nanotechnology. BINOM. Laboratoriya Znanii, Moscow, Russia.

Kong, J., Franklin, N.R., Zhou, C. et al. 2000. Nanotube molecular wires as chemical sensors. *Science* 287:622–625.

Korotcenkov, G., Blinov, I., Brinzari, V., and J.R. Stetter. 2007. Effect of air humidity on gas response of SnO_2 thin film ozone sensors. *Sensors and Actuators B: Chemical* 122:519–526.

Korposh, S., Selyanchyn, R., and Lee, S.-W. 2010. Nano-assembled thin film gas sensors. IV. Mass-sensitive monitoring of humidity using quartz crystal microbalance (QCM) electrodes. *Sensors and Actuators B* 147(2, 3):599–606.

Kukla, A.L., Pavluchenko, A.S., Shirshov, Y.M., Konoshchuk, N.V., and Posudievsk, O.Y. 2009. Application of sensor arrays based on thin films of conducting polymers for chemical recognition of volatile organic solvents. *Sensors and Actuators B* 135(2):541–551.

Lee, S.J., Lee, J.-E., Seo, J., Jeong, I.Y., Lee, S.S., and Jung, J.H. 2007. Optical sensor based on nanomaterial for the selective detection of toxic metal ions. *Advanced Functional Materials* 17:3441–3446.

Leopold, N., Busche, S., Gauglitz, G., and Lendl, B. 2009. IR absorption and reflectometric interference spectroscopy (RIfS) combined to a new sensing approach for gas analytes absorbed into thin polymer films. *Spectrochimica Acta Part A: Molecular and Biomolecular Spectroscopy* 72(5):994–999.

Li, C.M., Dong, H., Cao, X., Luong, J.H.T., and Zhang, X. 2007. Implantable electrochemical sensors for biomedical and clinical applications: Progress, problems, and future possibilities. *Current Medicinal Chemistry* 14(8):937–951.

Lieber, C.M. 1998. One dimensional nanostructures-chemistry, physics and applications. *Solid State Communications* 107:607–616.

Lübbers, D.W. and N. Opitz. 1975. Eine neue pCO_2-bzw. pO_2-Messsonde zur Messung des pCO_2 oder pO_2 von Gasen und Flüssigkeiten. *Zeitschrift für Naturforschung C* 30(4):532–533.

Lucci, M., Regoliosi, P., Reale, A. et al. 2005. Gas sensing using single wall carbon nanotubes ordered with dielectrophoresis. *Sensors and Actuators B* 111–112:181–186.

Lundström, I. 1981. Hydrogen sensitive MOS-structures-part 1: Principles and applications. *Sensors and Actuators* 1:403–426.

Lundström, I., Shivaraman, S., Svensson, C., and Lundkvist, L. 1975. A hydrogen-sensitive MOS field-effect transistor. *Applied Physics Letters* 26(2):55–57.

Luque, G.L., Ferreyra, N.F., and Rivas, G.A. 2007. Electrochemical sensor for amino acids and albumin based on composites containing carbon nanotubes and copper microparticles. *Talanta* 71(3):1282–1287.

Macleod, H.A. 2001. *Thin Film Optical Filters*, 3rd edn. Institute of Physics, Philadelphia, PA.

Manso, J., Mena, M.L., Yanez-Sedeno, P., and Pingarron, J. 2007. Electrochemical biosensors based on colloidal gold–carbon nanotubes composite electrodes. *Journal of Electroanalytical Chemistry* 603:1–7.

Mignani, A.G., Ciaccheri, L., Cimato, A., Attilio, S., and Smith, P.R. 2005a. Spectral nephelometry for the geographic classification of Italian extra virgin olive oils. *Sensors and Actuators B* 111–112:363–369.

Mignani, A.G., Mencaglia, A.A., and Ciaccheri, L. 2005b. Fiber optic systems for colorimetry and scattered colorimetry. *Proceedings of the SPIE* 5952:89–99.

Nicholson, J.W. 2007. Optically assisted deposition of carbon nanotube saturable absorbers. In Paper presented at the *Conference on Lasers and Electro-Optics/Quantum Electronics and Laser Science Conference and Photonic Applications Systems Technologies*. Optical Society of America, Washington, DC, p. CMU6:1–2.

Nicholson, J.W., Windeler, R.S., and Di Giovanni, D.J. 2007. Optically driven deposition of single-walled carbon-nanotube saturable absorbers on optical fiber end-faces. *Optics Express* 15(15):9176–9183.

Orellana, G. 2004. Luminescent optical sensors. *Analytical and Bioanalytical Chemistry* 379(3):344–346.

Pejcic, B., Eadington, P., and Ross, A. 2007. Environmental monitoring of hydrocarbons: A chemical sensor perspective. *Environmental Science and Technology* 41:6333–6342.

Pena-Calva, A., Olmos-Dichara, A., Viniegra-Gonzalez, G., Cuervo-Lòpez, F.M., and Gòmez, J. 2004. Denitrification in presence of benzene, toluene, and m-xylene. *Applied Biochemistry and Biotechnology* 119:195–208.

Peng, N., Zhang, Q., Chow, C.L., Tan, O.K., and Marzari, N. 2009. Sensing mechanisms for carbon nanotube based NH_3 gas detection. *Nano Letters* 9:1626–1630.

Penza, M., Antolini, F., and M.V. Antisari. 2004a. Carbon nanotubes as SAW chemical sensors materials. *Sensors and Actuators B* 100(1–2):47–59.

Penza, M., Cassano, G., Aversa, P. et al. 2004b. Alcohol detection using carbon nanotubes acoustic and optical sensors. *Applied Physics Letters* 85(12):2379–2381.

Penza, M., Cassano, G., Aversa, P. et al. 2005a. Carbon nanotubes coated multi-transducing sensors for VOCs detection. *Sensors and Actuators B* 111–112:171–180.

Penza, M., Cassano, G., Aversa, P. et al. 2005b. Carbon nanotube acoustic and optical sensors for volatile organic compound detection. *Nanotechnology* 16(11):2536–2547.

Penza, M., Cassano, G., Aversa, P. et al. 2006a. Acoustic and optical VOCs sensors incorporating carbon nanotubes. *IEEE Sensors Journal* 6(4):867–874.

Penza, M., Rossi, R., Alvisi, M., and E. Serra. 2010. Metal-modified and vertically aligned carbon nanotube sensors array for landfill gas monitoring applications. *Nanotechnology* 21(10):105501.

Penza, M., Tagliente, M.A., Aversa, P., Cassano, G., and Capodieci, L. 2006b. Single-walled carbon nanotubes nanocomposite microacoustic organic vapor sensors. *Materials Science and Engineering C* 26:1165–1170.

Peterson, I.R. 1990. Langmuir–Blodgett films. *Journal of Physics D: Applied Physics* 23:379–395.

Peterson, I.R. 1992. Langmuir–Blodgett films. In *Molecular Electronics*, ed. G.J. Ashwell, 117–206. Research Studies Press, Taunton, U.K.

Petty, M.C. 1996. *Langmuir Blodgett Films: An Introduction*. Cambridge University Press, Cambridge, U.K.

Pitkethly, M.J. 2004. Nanomaterials—The driving force. *Nanotoday* 7:20–28.

Qi, P., Vermesh, O., Grecu, M. et al. 2003. Toward large arrays of multiplex functionalized carbon nanotube sensors for highly sensitive and selective molecular detection. *Nano Letters* 3(3):347–352.

Riu, J., Maroto, A., and Rius, F.X. 2006. Nanosensors in environmental analysis. *Talanta* 69(2):288–301.

Rivas, G.A., Rubianes, M.D., Pedano, M.L., Ferreyra, N.F., Luque, G., and Miscoria, S.A. 2009. *Carbon Nanotubes: A New Alternative for Electrochemical Sensors*. Nova Science Publishers, Inc., New York.

Rivers, H.K., Sikora, J.G., and Sankaran, S.N. 2002. Detection of hydrogen leakage in a composite sandwich structure at cryogenic temperature. *Journal of Spacecraft and Rockets* 39(3):452–459.

Roberts, M.E., LeMieux, M.C., and Bao, Z. 2009. Sorted and aligned single-walled carbon nanotube networks for transistor-based aqueous chemical sensors. *ACS Nano* 10:3287–3293.

Robinson, J.A., Snow, E.S., Badescu, S.C., Reinecke, T.L., and Perkins, F.K. 2006. Role of defects in single-walled carbon nanotube chemical sensors. *Nano Letters* 6:1747–1751.

Roco, M.C., Williams, R.S., and Alivasatos, P. 2000. *Nanotechnology Research Directions, Vision for Nanotechnology in the Next Decade.* Cluver Academy, Culver, IN.

Rodriguez-Pardo, L., Rodriguez, L., Gabrielli, J.F., Perrot, C., and Brendel, H. 2005. Sensitivity, noise, and resolution in QCM sensors in liquid media. *IEEE Sensors Journal* 5(6):1251–1257.

Set, S.Y., Yaguchi, H., Tanaka Y., and Jablonski, M. 2004. Ultrafast fiber pulsed lasers incorporating carbon nanotubes. *IEEE Journal on Selected Topics in Quantum Electronics* 10(1):137–146.

Shankaran, D.R., Gobi, K.V., and Miura, N. 2007. Recent advancements in surface plasmon resonance immunosensors for detection of small molecules of biomedical, food and environmental interest. *Sensors and Actuators B* 121:158–177.

Shen, C.-Y., Huang, H.-C., and Hwang, R.-C. 2008. Ammonia identification using shear horizontal surface acoustic wave sensor and quantum neural network model. *Sensors and Actuators A: Physical* 147(2, 3):464–469.

Shi, J., Zhu, Y., Zhang, X., Baeyens, W.R.G., and Garcia-Campana, A.M. 2004. Recent developments in nanomaterial optical sensors. *Trends in Analytical Chemistry* 23(5):351–360.

Siitonen, A.J., Tsyboulski, D.A., Bachilo, S.M., and Weisman, R.B. 2010a. Surfactant-dependent exciton mobility in single-walled carbon nanotubes studied by single-molecule reactions. *Nano Letters* 10(5):1595–1599.

Siitonen, A.J., Tsyboulski, D.A., Bachilo, S.M., and Weisman, R.B. 2010b. Dependence of exciton mobility on structure in single-walled carbon nanotubes. *Journal of Physical Chemistry Letters* 1(14):2189–2192.

Skucha, K., Fan, Z., Jeon, K., Javey, A., and Boser, B. 2010. Palladium/silicon nanowire Schottky barrier-based hydrogen sensors. *Sensors and Actuators B* 145:232.

Song, Y.-W., Yamashita, S., Einarsson, E., and Maruyama, S. 2007. All-fiber pulsed lasers passively mode-locked by transferable vertically aligned carbon nanotube film. *Optics Letters* 32(11):1399–1401.

Soref, R. and Bennet, B.R. 1987. Electrooptical effects in silicon. *IEEE Journal of Quantum Electronics* 23:123–129.

Steinberg, I.M., Lobnik, A., and Wolfbeis, O.S. 2003. Characterisation of an optical sensor membrane based on the metal ion indicator Pyrocatechol Violet. *Sensors and Actuators B* 90(1–3):230–235.

Stetter, J.R., Penrose, W.R., and Yao, S. 2003. Sensors, chemical sensors, electrochemical sensors and ECS. *Journal of Electrochemical Society* 150:S11–S16.

Sutapun, B., Tabib-Azar, M. and Kazemi, A. 1999. Pd-coated elasto-optic fiber optic Bragg grating sensors for multiplexed hydrogen sensing. *Sensors and Actuators B* 60(1):27–34.

Takagi, H., Hatori, H., Soneda, Y., Yoshizawa, N., and Yamada, Y. 2004. Adsorptive hydrogen storage in carbon and porous materials. *Materials Science and Engineering B* 108(1–2):143–147.

Tans, S.J., Devoret, M.H., Dai, H. et al. 1977. Individual single-wall carbon nanotubes as quantum wires. *Nature* 386:474–477.

Tansil, N.C. and Gao, Z. 2006. Nanoparticles in biomolecular detection. *Nanotoday* 1(1):28–37.

Terrones, M. 2003. Science and technology of the twenty-first century: Synthesis, properties, and applications of carbon nanotubes. *Annual Review of Materials Research* 33:419–501.

Tkac, J., Whittaker, J.W., and Ruzgas, T. 2007. The use of single walled carbon nanotubes dispersed in a chitosan matrix for preparation of a galactose biosensor. *Biosensors and Bioelectronics* 22(8):1820–1824.

Valentini, L., Armentano, I., Kenny, J.M., Cantalini, C., Lozzi, L., and Santucci, S. 2003. Sensors for sub-ppm NO_2 gas detection based on carbon nanotube thin films. *Applied Physics Letters* 82:961–963.

Valentini, L., Armentano, I., Kenny, J.M., Lozzi, L., and Santucci, S. 2004. Pulsed plasma-induced alignment of carbon nanotubes. *Material Letters* 58:470.

Varghese, O.K., Kichambre, P.D., Gong, D., Ong, K.G., Dickey, E.C., and Grimes, C.A. 2001. Gas sensing characteristics of multi-wall carbon nanotubes. *Sensor and Actuators B* 81:32–41.

Vaseashta A. and Dimova-Malinovska, D.. 2005. Nanostructured and nanoscale devices, sensors, and detectors. *Science and Technology of Advanced Materials* 6:312–318.

Villatoro, J., Diez, A., Cruz, J.L., and Andres, M. V. 2001. Highly sensitive optical hydrogen sensor using circular Pd-coated singlemode tapered fibre. *Electronics Letters* 37(16):1011–1012.

Wang, F., Gu, H., and Swager, T. M. 2008. *Journal of the American Chemical Society* 130:5392.

Wei, B.-Y., Lin, C.-S., and Lin, H.-M. 2003. Examining the gas sensing behaviors of carbon nanotubes using a piezoelectric quartz crystal microbalance. *Sensors and Materials* 15(4):177–190.

Wolfbeis, O.S. 1991. *Fiber Optic Chemical Sensors and Biosensors*, Vol. II. CRC Press, Boca Raton, FL.

Wolfbeis, O.S. and Weidgans, B.M. 2006. Fiber optic chemical sensors and biosensors: a view back. In *Optical chemical sensors (NATO Science Series II: Mathematics, Physics and Chemistry)*, ed. F. Baldini, A.N. Chester, J. Homola, and S. Martellucci, 17–46. Springer, Berlin, Germany.

Wooten, F. 1972. *Optical Properties of Solids*. Academic Press, New York.

Xue, W., Liu, Y., and Cui, T.. 2006. High-mobility transistors based on nanoassembled carbon nanotube semi-conducting layer and SiO_2 nanoparticle dielectric layer. *Applied Physics Letters* 89:163512.

Yamashita, S., Inoue, Y., Maruyama, S. et al. 2004. Saturable absorbers incorporating carbon nanotubes directly syntehsized onto substrates and fibers and their application to mode-locked fiber lasers. *Optics Letters* 29:1581–1583.

Yan, X.B., Chen, X.J., Tay, B.K., and Khor, K.A. 2007. Transparent and flexible glucose biosensor via layer-by-layer assembly of multi-wall carbon nanotubes and glucose oxidase. *Electrochemistry Communications* 9(6):1269–1275.

Ye, J. and Sheu, F.-S.. 2007. Carbon nanotube-Based Sensor. In *Nanomaterials for Biosensors*, ed. C. S. S. R. Kumar. Wiley-VCH Verlag GmbH & Co., New York

Yellampalli, S. 2011. *Carbon Nanotubes-Synthesis, Characterization, Applications*. InTech, Rijeka, Croatia.

Yeo, T.L., Sun, T., and Grattan, K.T.V. 2008. Fibre-optic sensor technologies for humidity and moisture measurement. *Sensors and Actuators A* 144(2):280–295.

Zahab, A., Spina, L., Poncharal, P., and C. Marliere. 2000. Water-vapor effect on the electrical conductivity of a single-walled carbon nanotube mat. *Physics Review B* 62(15):10000–10003.

Zare, H.R. and Nasirizadeh, N. 2007. Hematoxylin multi-wall carbon nanotubes modified glassy carbon electrode for electrocatalytic oxidation of hydrazine. *Electrochimica Acta* 52:4153–4160.

Zaromb, S. and Stetter, J.R. 1984. Theoretical basis for identification and measurement of air contaminants using an array of sensors having partially overlapping sensitivities. *Sensors and Actuators B* 6:225–243.

Zdansky, K. 2011. Highly sensitive hydrogen sensor based on graphite-InP or graphite-GaN Schottky barrier with electrophoretically deposited Pd nanoparticles. *Nanoscale Research Letters* 6:490.

Zhang, J., Hu, J., Zhu, Z.Q., Gong, H., and O'Shea, S.J. 2004. Quartz crystal microbalance coated with sol-gel-derived indium-tin oxide thin films as gas sensor for NO detection. *Colloids and Surfaces A* 236(1–3):23–30.

Zhang, T., Mubeen, S., Myung, N.V., and Deshusses, M.A. 2008. Recent progress in carbon nanotube based gas sensors. *Nanotechnology* 19:332001.

Zhou, W., Vavro, J., Nemes, N.M. et al. 2005. Charge transfer and Fermi level shift in p–doped single-walled carbon nanotubes. *Physical Review B* 71:205423.

Zudans, I., Heineman, W.R., and Seliskar, C.J. 2004. In situ measurements of chemical sensor film dynamics by spectroscopic ellipsometry. Three case studies. *Thin Solid Films* 455–456:710–715.

21

Nano-Optical Sensors for Virology

Sathish Sankar, Balaji Nandagopal, Mageshbabu Ramamurthy, and Gopalan Sridharan

CONTENTS

21.1 Introduction

Several advances have been made in the past four decades in the diagnosis of viral infections. Numerous new technologies have been introduced since 1970. The first major innovation was the countercurrent immunoelectrophoresis and its modifications which were popular for diagnosis of viral disease. The enzyme-linked immunosorbent assay (ELISA) introduced in the 1980s is still the workhorse for diagnosis of viral infections. With its use, class-specific immunoglobulins (IgG/IgM/IgA) to viruses or viral antigens can be detected. IgM detection is possible about 5–7 days post-onset of illness. From the late 1990s till date, the polymerase chain reaction (PCR) and its modifications have become widespread for viral detection. The PCR technique is useful but requires capital investment and also requires established laboratory and technical expertise. Subsequently, rapid devices based on immune-flow-through technology were introduced for viral antigen or antibody detection. These are simple to perform and useful even in point-of-care settings.

Several nanodevices based on optical sensors (optical nanosensors) have been developed for detection of viruses of medical and veterinary importance. The technology has numerous applications (Sahoo et al. 2007; Pandey et al. 2008) and has been demonstrated for efficient diagnosis of infections with human immunodeficiency virus (HIV), hepatitis B virus (HBV), hepatitis C virus (HCV), respiratory syncytial virus (RSV), herpes simplex virus (HSV), influenza virus, and other agents. The lower limit of detection (LOD) for these devices is in the range of a few femtograms per microliter and hence extremely sensitive for use directly on clinical samples.

The unusual optical, magnetic, electronic, catalytic, and mechanical properties of nanomaterials and nanoscale production of materials have great use for biomedical applications. Nanoparticles of noble metals have distinct optical properties due to their ability to support surface plasmons (plasmonic phenomonae) with localized surface plasmon resonance (LSPR). Surface-enhanced Raman spectroscopy (SERS) technologies are based on this. Both the LSPR and SERS use the principle of resonant Rayleigh scattering, refractive index sensing. SERS-based detection with the labeling of nanomaterials has formed the basis for a number of applications (Stuart et al. 2005).

QDs have a great potential as a label on a tracer molecule for optical nano methods in ligand detection. Green and orange CdTe QDs have been shown to be convenient and cheap. These serve as effective pH-sensitive fluorescent probes that could monitor the proton (H+) flux driven by ATP synthesis for detection of viruses on the basis of antibody–antigen reactions even inside living cells. Green and orange QDs labeled biosensors were shown to coexist in the detection system without interference with one another in the fluorescence assays facilitating detection of multiple agents (Deng et al. 2007; Azzazy and Mansour 2009). The development of QDs as labels on antibody (tracer molecule) has opened up a new avenue in pathogen detection. The QD-labeled antibody could be used to detect the target antigen captured on solid phase.

Yet another technique is the use of whispering gallery modes (WGMs). WGMs rely on the measurable resonances produced by light bouncing around the object's circumference, and this activity occurs inside round objects (spheres, cylinders, disks). The properties of an unknown analyte, like a virus present in a droplet of fluid, could be inferred by analyzing the resonant frequencies of the droplet's WGMs. This technique has been successfully demonstrated for influenza A viruses. When the virions bind to a microsphere cavity, a "reactive" perturbation of the resonant photon state is achieved, and this is detected by measuring signal-to-noise ratio (Vollmer et al. 2008).

Mitra et al. (2010) reported the use of nanofluidic channels in combination with optical interferometry. In this technique, scattered light from single viruses which transverse a stationary laser focus is detected with a differential heterodyne interferometer. Heterodyne detection system eliminates phase variations due to different particle trajectories, improving the recognition accuracy. The authors showed the optical detection scheme to be real-time and label-free with the ability to recognize single virus and large protein molecule. This detection system could be integrated into large nanofluidic architectures for reliable detection, sizing, and sorting of viruses. The applications are in biosensing, environmental monitoring, and quality control of viral products like vaccine particle count.

An interesting review on the development and application of plasmonic nanoprobes for biosensing and bioimaging was presented by Vo-Dinh et al. (2010). The review describes the use of plasmonics, that is, SERS gene probes, for the detection of diseases using DNA hybridization to target genes, for example, HIV gene and breast cancer genes. Also, molecular imaging was described using a hyperspectral surface-enhanced Raman imaging (HSERI) system that combines imaging capabilities with SERS detection. It is possible to identify cellular components using Raman dye-labeled silver nanoparticles in cellular systems.

Hermann et al. (2011) reported the use of tip-enhanced Raman spectroscopy (TERS) as a highly sensitive spectroscopic technique. The spectroscopic technique was used for the detection of different virus strains like Avipoxvirus and adeno-associated virus. TERS spectra obtained from different particles of the same virus strain showed variations in relative peak intensities and positions of most spectral features. The authors demonstrated the spectral variations to be higher for the larger Avipoxvirus particles than for the smaller adeno-associated virus particles.

21.2 Current Nano-Optical Technological Developments in Virus Detection

There is immense potential to develop devices (tests) for bedside diagnosis and field use. Such an approach will improve diagnostic capacity in small hospitals, at doctors' clinics, in public health laboratories, and of community physicians.

Functionalized nanoparticles covalently linked to biological molecules such as antibodies, peptides, proteins, and nucleic acids have been developed as nanoprobes for molecular detection. These functionalized NPs can provide a direct rapid method of detection of viruses with high sensitivity (Tang et al. 2007; Tripp et al. 2007). Bio-functionalized gold nanoparticles carrying antigen have been used in immunoassays. The use of gold nanoparticles for optical detection has the potential for development of sensitive detection methods. Sandwich-format fiber-optic-based platforms similar to ELISA have been developed. Here, a capture agent like an antibody is immobilized on a solid phase of a fiber-optic or capillary tube that enables capture of target molecules. A second tracer molecule (probe), an antibody conjugated to a fluorophore, facilitates target detection. The emitted light is measured by a portable instrument. Microsphere biosensor based on the resonance shifting of WGM has also been developed. The sensors could be integrated with other electronic and optical components on a semiconductor chip facilitating simple devices. Furthermore, binding of single virions is documentable by changes in the resonance frequency/wavelength of a WGM excited in a microspherical cavity.

The use of nanoparticles as tags or labels allows for the detection of infectious agents in small sample volumes directly in a very sensitive, specific, and rapid format at lower costs than current in-use technologies. Single particle sensitivity enables simple charge-based detection of macromolecules. In outbreak situations, a rapid and specific detection of the viral agent requires biosensors able to generate a quantitative signal from individual viral particles.

The application of nanowire field effect transistors has shown promise for direct real-time electrical detection of single virus particles with high selectivity. In a series of experiments with influenza virus and antibody, it was shown that nanowire arrays coated with antibodies could enable measurement of discrete conductance changes characteristic of specific binding and unbinding of influenza A but not unrelated paramyxoviruses or adenoviruses. The specificity was documented by simultaneous electrical and optical measurements using fluorescently labeled influenza A. The conductance changes corresponded to binding/unbinding of single viruses at the surface of the coated nanowire.

The first significant development in the area of single virus particle detection was achieved when silicon nanowire device arrays were synthesized by chemical vapor deposition with 20 nm gold nano-clusters as catalysts, silane as reactant, and diborane as p-type dopant (an impurity element added to a crystal lattice in low concentrations in order to alter the optical/electrical properties of the crystal) for the vapor–liquid–solid growth of boron-doped silicon nanowires. The device was used to screen different concentrations of virus suspensions (influenza type A, 10^9–10^{10} particles per mL) to determine LOD. The nanowire device arrays were exposed to the virus samples at flow rate of 0.15 mL/h in buffered solution through fluidic channels formed by 0.1-mm-thick glass cover slip for combined electrical/optical mea-surements. Electrical measurements were carried out and shown to be independent of frequency within 17 and 79 Hz range. A laser scanning confocal microscope with $DiIC_{18}$ dye excited at 532 nm was used for optical imaging. This method could detect influenza A virus particle (label-free). The binding of a single virion produced definitive changes in the resonance frequency/wavelength of a WGM excited in a microspherical cavity. On reducing the microsphere size, the magnitude of the discrete wavelength-shift signal was found to be much improved. The size and mass in the range of 5.2×10^{-16} g of a bound virus particle was determined from the optimal resonance shift (Patolsky et al. 2004).

In pioneering experiments, gold particles with diameters between 2.5 and 4.5 nm were introduced into the inner cavity of an icosahedral brome mosaic virus. The optical properties of such single gold-marked virion particles were tested in vitro for the characteristic plasmon polariton resonance. This sensitive technique shows enhancement of 10^{14}–10^{15} of Raman scattering by molecules adsorbed on rough metal surfaces and thus allows for detection of even a single molecule.

A laser coupled to a waveguide-based device has been developed. It is composed of four parallel optical channels, each coated with antibodies specific to a certain protein or virus. The application of the technology was demonstrated for the detection of HSV-1 by coating one of the waveguide channels with the appropriate herpes antibodies. The technique detects the virus over a range of concentrations ranging from as low as 10^3 to 10^7/mL. Combining the light exiting from the virus-specific channel and that from a reference channel, an interference pattern is generated. Virus binding to the antibody-coated waveguide is probed by the evanescent field of the guided light modes, causing a phase change and a change in the interference pattern, which is recognized by a monitoring device. The technology has the potential for wide application, as any antibody can be used to coat a channel for detection (Jain 2005). This sensor can be extended to any virus like HIV, severe acute respiratory syndrome (SARS) coronavirus, HBV, HCV, or the avian influenza virus (H5N1), among others (Ymeti et al. 2005). The waveguide technology has been used for the successful detection of the avian flu virus in buffer

TABLE 21.1

List of Techniques Used for the Detection of Different Viruses

Method	Development Status	Viral Agents	Reference
QD-labeled antibody	Virus-infected cell cultures	RSV	Bentzen et al. (2005)
	Clinical samples	*Toxoplasma gondii*, rubella virus, CMV, and HSV1 and 2	Yang et al. (2009)
	Clinical samples	HBV	Zhang et al. (2010)
	Clinical samples	Avian influenza subtype H5N1	Chen et al. (2010)
Surface-enhanced Raman scattering	Virus isolates	HIV-1	Hu et al. (2010)
	Infected cell cultures	Feline calicivirus (FCV)	Driskell et al. (2005)
Gold nanoparticles	Clinical nasopharyngeal swab specimens	Influenza A virus, influenza B virus, and respiratory syncytial virus A and B (RSV A/B)	Jannetto et al. (2010)
	RSV-infected Vero cells	RSV	Tripp et al. 2007
Fiber-optic-based platforms	Standard strain	Newcastle disease virus	Lee and Thompson (1995)
Whispering-gallery mode	Purified virions	Influenza A	Vollmer et al. (2008)
Young interferometer sensor	Virus isolates	HSV-1	Jain (2005)
Gold label silver stain (GLSS) coupled with multiplex asymmetric PCR	Clinical samples	HIV-1	Tang et al. (2009)

solution. Monoclonal and polyclonal capture antibodies against hemagglutinin proteins were used in the channels. The interference pattern was clearly dose dependent, and the assay had a detection limit as low as 0.0005 HAU/mL (Pipper et al. 2007).

QDs have been shown to be used to in the identification of respiratory syncytial virus (RSV) and monitor the spread of the virus inside the infected cell by labeling the F and G proteins in kinetic experiments (Bentzen et al. 2005). Research is now ongoing to develop QD mixes to simultaneously detect multiple respiratory viruses (Demidov 2004; Jain 2005). A method for the RSV detection has been developed using functionalized NPs conjugated to monoclonal antibodies for rapid and specific detection in clinical samples (Tripp et al. 2007).

The Raman spectra of viruses can be used to rapidly and readily distinguish molecular fingerprints for classifying and identifying viruses. The SERS technique is highly promising for designing new virus-detection schemes. The spectra can be collected within 1 min at a virus volume less than 5 μL. SERS establishes reproducible molecular fingerprints of several important human viruses and can be an ideal screen for bioterrorism agents as well as genotyping of viral variants that emerge in different geographical regions of the world.

Nanotechnology is still an emerging field with exciting potential for diagnosis of infectious diseases. Nanomaterials are versatile and can be engineered into biofunctionalized particles. The field of pathogen detection has begun to exploit the unique optical and magnetic properties of nanoscale materials such as fluorescent nanoparticles, metallic nanostructures, and super-paramagnetic nanoparticles for bioimaging and detection of infectious microorganisms (Tallury et al. 2010). Table 21.1 summarizing landmark developments in the application of nanotechnological methods indicates the current status of such assays.

21.3 Concept of Syndromic Diagnosis of Viral Infections

A number of diverse infectious agents cause diseases that present with similar signs and symptoms characteristic of an "infectious syndrome." This classically sets them apart from other clinical infectious conditions that are caused by an easily identifiable organism or a small number of very closely related organisms (e.g., Rickettsial species). Several syndromes may have a bacterial, viral, or other etiology. Diseases of infectious etiology are increasingly presenting with protean manifestations or with atypical manifestations. Reports suggest that infectious diseases caused by agents such as hantaviruses present with unusual manifestations. In recent years, there is a paradigm shift in the clinician's approach to diagnosis such as making a diagnosis of "acute febrile illness" (the syndromic approach), in which the clinician recognizes that there are multiple different infectious agents that could potentially cause the disease. The treatment of such conditions requires identification of the specific causative agent, and therefore relies on laboratory diagnostic procedures for specific syndromes.

The drive to develop single virus particle detection technology has to be moderated by good understanding of viral pathogenesis. A typical example is the infection with cytomegalovirus (CMV) in immunosuppressed transplant recipients. It is important to note that any virus present in the blood compartment has no bearing on pathogenesis of the infection. To a large extent, immunity controls the virus and disease is only seen when high loads of the virus are present in the blood compartment (Gimeno et al. 2008). Presently, the understanding of viral pathogenesis, diagnosis, and therapy requires consideration of many issues related to virus load in tissues and the blood compartment. The detection of even very low number of viral particles is important in certain infectious conditions but not in others. In the case of blood-borne viruses like HIV, HBV, and HCV, demonstration of even a low virus load is significant for therapy monitoring and importantly blood and blood product safety (Thibault et al. 2007; Sarrazin et al. 2008). Also, the early detection of a single particle of HSV in cerebrospinal fluid (CSF) of patients with infections of the central nervous system (CNS) is important. The detection of HSV in the CSF of CNS disorder will impact on successful therapy (Kimura et al. 2002).

21.4 Diagnostic Markers in Viral Infections

There are several considerations before designing tests to detect a viral infection. The dynamics of a viral infection have to be understood in relation to the course of viral pathogenesis. Several viruses cause only localized infection without virus in the blood (viremia). Such infections are non-systemic and include most respiratory viral infections, viral diarrheas, and majority of sexually transmitted infections of viral origin. In these infections, virus is detectable only in local secretions. Antibody production could also be seen but not of diagnostic relevance. In viral infection where there is systemic spread, viremia is established. In several situations, the infection may be initially localized with a subsequent viremic spread like HIV infection. Certain viruses may be parenterally transmitted like blood-borne viruses (HIV, HBV, and HCV) wherein virus nucleic acid and/or antigen is principally demonstrated in blood. Markers of viral infection like viral antigen or nucleic acid (RNA/DNA) or antibody to viral antigens are not usually documentable during the incubation period. Within days of infection, viral antigen and/or nucleic acid is documentable in blood in systemic viral infections. The antibody of the IgM class is also documentable by 5–7 days of manifestation of viral infection. The IgG class of antibodies appears late and often is useful to show exposure to the virus. The diagnosis of infection is established by demonstrating a rise in titer of antibodies between acute and convalescent sera samples. The timeline of viral markers is shown in Figure 21.1. These dynamics have to be factored in while designing nano-optical devices for viral diagnosis.

21.5 Viral Infectious Disease Syndromes and Nano-Optical Detection Approaches to Diagnosis

21.5.1 Viral Exanthematous Fevers and Undefined Febrile Illnesses

Some of the infectious etiologies include measles virus, rubella virus, varicella zoster virus (VZV), parvovirus B19, human herpesvirus (HHV)-6, Epstein–Barr virus (EBV), enteroviruses, and dengue virus (DENV). Bacterial infections such as scarlet fever, *Staphylococcus aureus* infections (toxic shock syndrome), meningococcemia, typhoid fever, and rickettsial infection may also present fevers with exanthema. Today, there is a widespread appreciation of the emerging and re-emerging arboviral infections because of climatic changes as a result of global warming. There was a major

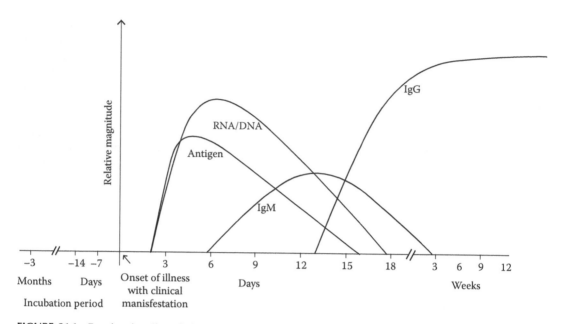

FIGURE 21.1 Putative time line of viral markers in the course of infection by agents that cause systemic infection. The *x*-axis shows a break both in the incubation period and post-onset indicating the extended time line. The *y*-axis is an approximate representation of the concentration (magnitude) of the markers in relation to each other.

multicountry epidemic of Chikungunya virus in 2005–2006 (Sudeep et al. 2008). Earlier, there was the spread of the West Nile virus (WNV) to North America in 2003–2004 (Lindsey et al. 2009). A rapid and sensitive detection method for the early diagnosis of infectious dengue virus urgently needs to be developed as this is at present a major public health problem at a global level (Vasilakis et al. 2008).

A recent study applied a circulating-flow quartz crystal microbalance (QCM) biosensing method combined with oligonucleotide-functionalized gold nanoparticles (i.e., AuNP probes) for detection of DENV. Two kinds of specific AuNP probes were linked by the target sequences onto the QCM chip to amplify the detection signal, that is, oscillatory frequency change (DeltaF) of the QCM sensor in the DNA-QCM method. The amplified target sequences from the DENV genome acted as a bridge for the layer-by-layer AuNP probes' hybridization in the method combining the amplification and detection. On AuNP size-based evaluation on the layer-by-layer hybridization, it was found that 13 nm AuNPs showed the best hybridization efficiency. The DNA-QCM biosensing method was able to detect 2 PFU/mL of DENV (Chen et al. 2009).

A novel detection system for viruses was demonstrated for WNV. The technique was based on the capture of WNV target nucleotide sequences by hybridization with complementary oligonucleotide probes. The assay was shown to have an LOD for target sequence of 10 pM. For the assay, Raman reporter tag-conjugated gold nanoparticles (GNPs) were bound to covalently linked oligonucleotide probes with magnetic removal of the unbound molecules. The SERS signature spectrum diagnostic of the reporter, 5, 5′-dithiobis (succinimidy-2-nitrobenzoate) (DSNB) was measured by laser excitation. The value of paramagnetic nanoparticle (MNP) was demonstrated for SERS detection of DNA oligonucleotides derived from the WNV genome. The short hybridization time required and reproducibility of Raman spectra acquisition make the assay simple to use for a variety of viruses (Zhang et al. 2011). The principle of this technique is shown in Figure 21.2.

21.5.2 Neurological Infections

Several viral agents that cause aseptic meningitis, encephalitis, and demyelinating disease include members of the genus enterovirus, family *Herpesviridae*, and mumps virus.

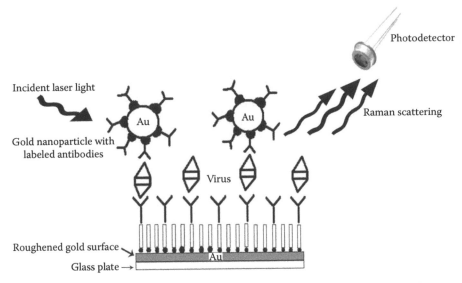

FIGURE 21.2 A drawing showing surface-enhanced-Raman-spectroscopy-based viral detection. Unique spectroscopic signatures are generated with SERS tags consisting of individual glass-encapsulated gold nanoparticles and surface-bound Raman active reporter molecules. These SERS tags are bound to a specific antibody and provide a strong, spectroscopically consistent label. Detection of Raman scattering from molecules adsorbed to rough metal surfaces enhances sensitivity, even allowing the detection of single virus particle. Superparamagnetic particles conjugated to the antibodies capture and concentrate the SERS-labeled complex at the focal point of the Raman laser using a magnetic field. The SERS readout confirms the presence or absence of the analyte.

In an early research development, the Young interferometer sensor had been applied for direct detection of viruses. The technology had been validated for HSV-1 but is broadly applicable for others as well. The detection of HSV-1 virus particles is achieved by exposing the virus-containing sample to a sensor surface coated with a specific antibody against HSV-1. The Young interferometer sensor was shown to detect HSV-1 at very low concentrations (850 particles/mL) and even directly in clinical samples (Jain 2005).

The development of a nanodevice with an ability to detect genomic material of RNA/DNA viruses in the CSF of patients with neurological disease could help in the rapid and specific diagnosis. Antiviral chemotherapy could be targeted at several *Herpesviridae* (HSV, VZV, EBV, and CMV) members early in the course of infection to prevent mortality and neurological sequelae. Nanoparticle labeled probes could be used to detect hybrids of post-amplification products anchored on capture probes coated on a matrix. Here, detection system enhances the PCR sensitivity and specificity.

21.5.3 Sexually Transmitted Infections

This group of infections is now a major problem both in developed and developing countries of Asia and Africa. Several agents cause sexually transmitted infections (STIs). The first use of SERS-active labels for primers used in PCR amplification of specific target DNA sequences has been reported. The method has the ability to combine the spectral selectivity and high sensitivity of the SERS technique with the inherent molecular specificity offered by DNA–DNA hybridization. The effectiveness of the detection scheme was demonstrated using the gag gene sequence of the HIV (Isola et al. 1998).

Wabuyele and Vo-Dinh (2005) described the use of plasmonics-based nanoprobes which comprises of a metal nanoparticle and a stem-loop DNA molecule tagged with a Raman label for the diagnosis of the HIV-1 targeting the gag gene sequence. The nanoprobe utilized the specificity and selectivity of the DNA hairpin probe sequence to detect the specific target DNA sequence. The method was shown to have spectral selectivity and high sensitivity and specificity for the diagnosis of molecular target sequences.

A novel signal amplification technology for a human papillomavirus (HPV)-DNA hybridization assay based on fluorescein diacetate (FDA) nanocrystals has been developed. The FDA nanocrystals are moderately water-insoluble precursors of fluorescein, which is dissolved and hydrolyzed by treatment with an organic solvent/hydroxide mixture. This is treated with a polymeric surfactant to create a stable, nanosized colloid with an interface for coupling streptavidin molecules. Initially, biotin labels incorporated in primers result in PCR products for the quantitative detection of the biotinylated HPV-specific DNA products, amplified in a standard PCR procedure. The amplified HPV-DNA, labeled with biotin, was hybridized with the immobilized probes of HPV 16, HPV 18, or HPV 45 in 96-well microplates. After the affinity reaction with the streptavidin FDA nanoprobe, the FDA molecules were dissolved and concomitantly converted into fluorescein by the special reagent (DMSO and 1 M NaOH in a 1:1 ratio). Florescence emission in the microwells of the plate is excited at 485 nm and recorded at 538 nm in an *fmax* fluorescence microplate reader. This approach resulted in high selectivity, short incubation times, and high sensitivity. This innovative method allows rapid detection of small amounts of target sequence in a fewer number of PCR cycles (Chan et al. 2007).

A new nanoparticle-based biobarcode amplification (BCA) assay has been developed for early and sensitive detection of HIV-1 capsid (p24) antigen. Here, anti-p24 antibody-coated microplates capture viral antigen (p24) and are linked to a detection monoclonal antibody with avidin label as the detection probe. This immune complex is detected by streptavidin-coated nanoparticle-based biobarcode DNAs for signal amplification. The signal detection uses a chip-based scanometric method. The modified BCA assay exhibited detection as low as 0.1 pg/mL and was approximately 150-fold more sensitive than conventional ELISA. In addition, the BCA assay detected HIV-1 infection 3 days earlier than ELISA in seroconversion samples and hence may serve as an alternate testing strategy to HIV RNA detection (Tang et al. 2007).

A technique for specific and rapid unlabeled detection of a virus by using a microsphere-based WGM sensor has been developed. Here, the transducer of the interaction of a whole virus with an anchored antibody is detected. It has been theoretically shown that this sensor can detect a single virion below the mass of HIV. A microfluidic device when combined with the WGM technology enables the discrimination between viruses of similar size and shape (Vollmer and Arnold 2008). The principle of WGM is demonstrated in Figure 21.3.

HPV 16/18 genotypes are transmitted by the venereal route and are associated with cervical cancer and oral squamous cell carcinoma. QDs were used in situ hybridization (ISH) to study expression of HPV16/18 in cells from oral squamous cell carcinomas by tissue microarrays. The presence of HPV16/18 high risk was detected by applying QD-ISH and, when compared to conventional ISH, was found to be more efficient (Xue et al. 2009).

A rapid method based on QDs for ToRCH-related antibodies including *Toxoplasma gondii*, rubella virus, CMV, and HSV1 and HSV2 was developed. The new assay was compared to ELISA kits for antibody detection which are considered as "gold standard." The QDs-based ToRCH microarrays have been shown to have great potential in the detection of ToRCH-related pathogens (Yang et al. 2009).

A HIV-1 DNA detection assay based on a multilayer metal–molecule–metal nanojunctions (NJs) in combination with SERS has been developed with ability to detect subattomolar concentrations in the

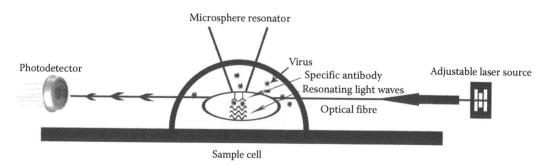

FIGURE 21.3 Schematic representation of viral detection by whispering gallery mode biosensor. The viruses are detected by discrete changes in resonance frequency of light in a microsphere cavity of the resonator by applying light from a laser source and identified with the photodetector system. Single virus particle detection is feasible either by direct determination from the magnitude of the wavelength shift or by using specific antibodies to detect unknown viruses.

range of 10^{-19} M to 10^{-23} mol. The assay was sensitive to differentiate variations down to single base mismatch discrimination. The methodology is two-step based. In the first step, unlabeled target DNA precipitated the detection molecules on the substrate forming a sandwiched structure based on the capture probe facilitating the first level amplification of target. In the second step, this sandwich is recognized by a second detection system allowing multi-metal–molecule–metal NJs between Au nanoparticles. The enhanced Raman signal of the tag molecules was due to the distance-dependent electromagnetic enhancement of SERS (Hu et al. 2010).

A visual DNA microarray capable of simultaneous detection of HIV-1 and *Treponema pallidum* has been developed. The assay used a gold label silver stain (GLSS) coupled with multiplex asymmetric PCR, where 5'-end amino-modified oligonucleotides were immobilized on a glass surface and used as the capturing probes to bind the complementary biotinylated target DNA. Gold-conjugated streptavidin was used for specific binding to biotin. The black microarray spots developed because of precipitation of silver onto nanogold particles that were bound to streptavidin. The results with this assay were comparable to real-time PCR findings. The lower LOD of the assay was 106 copies/mL of target DNA amplicons prepared by conventional PCR and multiplex PCR (Tang et al. 2009). Subsequently, the group developed a similar assay for *Ureaplasma parvum* and *Chlamydia trachomatis*. The authors used the N-terminus multiple-banded antigen of *U. parvum* and major outer membrane protein of *C. trachomatis*. Here, specific antigens were labeled with nano-gold-Staphylococcal protein A (SPA) and immobilized on a glass surface that was treated with 3-glycidoxypropyltrimethoxysilane. The bound antigens were recognized by the complementary target antibodies applied to the prepared microarray surface. The assay utilizes a "sandwich" format, wherein the nano-gold-SPA probe was used as an indicator and GLSS was applied to amplify the detection signals and produce black image on array spots, which were visible with the naked eye. The sensitivity was comparable to the fluorescent detection method. The results were found consistent with the ELISA and quantitative real-time PCR assay in clinical samples (Tang et al. 2010).

21.5.4 Hemorrhagic Fever

The agents of viral hemorrhagic fever (VHF) are all enveloped RNA viruses. The viruses are geographically restricted to the areas where their host species live. Humans are infected when they come into contact with infected hosts. However, with some viruses, after the accidental transmission from the host, humans can transmit the virus to one another. With a few exceptions, there is no cure or established drug treatment for VHFs. Because other diseases have similar clinical symptoms, specific laboratory diagnostic tests are necessary to provide the differential diagnosis especially during outbreaks. Specific diagnosis would help public health officials to institute appropriate measures to control the outbreak.

VHFs are caused by four distinct families of RNA viruses: the Arenaviridae (Lassa, Junin, Machupo, Sabia, and Guanarito), Filoviridae (Ebola and Marburg), Bunyaviridae (hantavirus), and Flaviviridae (dengue virus and Kyasanur forest disease virus). All types of VHF are characterized by fever along with bleeding disorders and can progress to high fever, shock, and death in extreme cases. Some of the VHF agents cause relatively mild illnesses, such as the Scandinavian nephropathia epidemica, while others, such as the African Ebola virus, can cause severe, life-threatening disease. It is very important to develop nanodevices which could be used in the community (fields) and hospital laboratories for rapid diagnosis specific for geographical regions. Such a diagnostic device will offer a tremendous public health benefit for early identification and institution of preventive measures.

21.5.5 Hepatitis

Hepatitis is a major problem in several parts of the globe. The virus could be transmitted by food and water, that is, enteric hepatitis viruses—hepatitis A virus (HAV) and hepatitis E Virus (HEV)—and by contaminated blood and blood products (HBV, HCV, and hepatitis D virus). The practicing physician will benefit from information on specific etiology for appropriate management with specific antiviral therapy. Over 170 million people, more than 3% of the world's population, suffer from the HCV infection, and the rate of death from liver-related mortality to HCV has increased.

A visual gene-detecting technique using gold-nanoparticle-labeled gene probes has been described. Nanoparticle-supported 3′-end-mercapto-derivatized oligonucleotide was used as the detection probe, and 5′-end-amino-derivatized oligonucleotide immobilized on glass surface was used as capturing probe. The target DNA was detected visually by sandwich hybridization using the highly sensitive "nano-amplification" and silver staining. The HBV, HCV, and HBV/HCV gene chips with gold/silver NP staining amplification method were shown to be useful in detecting these viruses in patients' samples. The detection readout was the resonance Rayleigh light scattering (RLS) spectroscopy to monitor the immobilization of gene probes on gold nanoparticle surfaces (Wang et al. 2003).

An economical and sensitive assay utilizing gold-DNA probe array was developed for detection of HAV following PCR amplification. The assay had amino-modified oligodeoxynucleotides probe (5′ position) anchored on activated glass surface to function as capture probes. Sandwich hybridization takes place between capture probes, the HAV amplicon, and gold-nanoparticle-labeled oligonucleotide probes. A silver enhancement step allowed signals to be detected by a standard flatbed scanner or just by naked eye. The assay had a lower LOD of 100 fM for HAV amplicon (Wan et al. 2005). A similar assay has been developed for HEV (Liu et al. 2006).

A novel HCV-detecting technique using a nanoparticle-supported aptamer probe was demonstrated. With the aid of nanoparticle QDs with carboxyl group as an imaging probe, and 5′-end-amine-modified RNA oligonucleotide as a capturing probe, target HCV NS3 was visually detected on chip. The QDs-based RNA aptamer for HCV NS3 showed high selectivity and specificity against other protein such as BSA. The detection limit of HCV NS3 protein was 5 ng mL^{-1} level. With a novel strategy for protein–aptamer interaction, the feasibility of applying QDs-based fluorescent detection technique to HCV viral protein assay for the development of a protein biochip was demonstrated. This scheme of QDs-mediated imaging with a target-oriented specific RNA aptamer for the detection of infectious HCV diseases provides an efficient strategy and a promising new platform for monitoring applications (Roh et al. 2010).

A quantum dots-DNA (QDs-DNA) nanosensor based on fluorescence resonance energy transfer (FRET) has been developed for the detection of the target DNA and single mismatch in HBV gene. A water-soluble CdSe/ZnS QDs was used in this assay and oligonucleotides were attached to the QDs surface to form functional QDs-DNA conjugates. Along with the addition of DNA targets and Cy5-modified signal DNAs into the QDs-DNA conjugates, sandwiched hybrids were formed. The resulting assembly brings the Cy5 fluorophore, the acceptor, and the QDs, the donor, into proximity, leading to fluorescence emission from the acceptor by means of FRET on illumination of the donor. The method was validated for the detection of synthetic 30-mer oligonucleotide targets derived from the HBV with a sensitivity of 4.0 nM by using a multilabel counter (Wang et al. 2010).

A novel microfluidic device with microbead array was developed for sensitive genotyping of HBV using QDs as labels. The device consisted of two poly(dimethylsiloxane) PDMS slabs featured with different microstructures and channel depths for the construction of a functional region comprising a chamber array and a single sampling microchannel. Highly sensitive virus DNA detection was achieved by the enhanced mass transport in the microfluidics and the rapid reaction dynamics of suspension microbead array. The device could detect 1000 copies/mL of HBV virus in clinical serum samples using in vitro transcribed RNA as the target molecules. This on-chip virus genotyping was also demonstrated to have high discrimination specificity and sensitivity using synthesized HBV DNA. This microfluidic device has advantages like rapid binding kinetics of homogeneous assays of microbead array, the liquid handling capability of microfluidics and the sensitivity of QDs for fluorescence detection. The assay provides high sensitivity and rapid DNA analysis of virus with economical reagent consumption (Zhang et al. 2010).

21.5.6 Diarrheal Diseases

Food- and waterborne viruses are a major public health concern for humans worldwide. Rapid, accurate detection of viruses is important to implement public health measures. A sandwich immunoassay format using SERS as a readout method has been reported earlier for an animal diarrheal pathogen. The assay was developed for the feline calicivirus (FCV) in infected cell y-labeled anti-FCV mAbs and was found to have a linear dynamic range of $1 \times 10^6 - 2.5 \times 10^8$ viruses/mL and an LOD of 1×10^6 viruses/mL. The performance of the assay was validated by atomic force microscopy (Driskell et al. 2005).

SERS coupled with gold SERS-active substrates was used to detect and discriminate seven food- and waterborne viruses, including norovirus, adenovirus, parvovirus, rotavirus, and coronavirus. The virus detection limit by SERS was 100 PFU/mL (Fan et al. 2010).

Driskell et al. (2010) developed an SERS-based methodology for rapid and sensitive detection of rotavirus. The SERS method was useful for direct structural characterization of viruses, and the SERS spectra for eight rotavirus strains were analyzed. It was shown that the technique could qualitatively identify and genotype the virus.

21.5.7 Respiratory Infections

These infections are a global problem leading to four million deaths in young children of developing countries of the world. Rapid diagnosis is imperative for effective management of disease.

A nanoparticle label technology with highly fluorescent europium (III)-chelate-doped nanoparticle label on high-affinity monoclonal antibodies (anti-hexon) to adenovirus has been tested. The device was highly specific with a detection limit of 5000 virus particles per milliliter of purified virus particles. The sensitivity was improved 800-fold compared to other standard methods. The nanoparticle assay showed low variation in log values and excellent linear relationship to virus concentration. The assay was assessed on nasopharyngeal samples and found to be superior to conventional methods (Valanne et al. 2005).

Agrawal et al. (2005) showed the use of dual-color QDs or FRET nanobeads labeled antibody for the rapid, sensitive and quantitative detection of RSV. The sensitivity and specificity of nanoparticles were tested on cell culture lysates of different viruses. The assay however has not been evaluated on clinical samples. The principle is demonstrated in Figure 21.4.

Vollmer et al. (2008) reported a label-free, real-time optical detection of influenza A virus particles. The binding of single virions was observed from discrete changes in the resonance frequency/wavelength of a WGM excited in a microspherical cavity. They found that the magnitude of the discrete wavelength-shift signal was adequately enhanced by reducing the microsphere size. The size and mass (approximately 5.2×10^{-16} g) of a bound virion were determined directly from the optimal resonance shift.

A semi-automated respiratory virus nucleic acid test (VRNAT) and another fully automated respiratory virus nucleic acid test SP (RVNATSP) was evaluated to detect influenza A virus, influenza B virus, and respiratory syncytial virus A and B (RSV A/B) from clinical nasopharyngeal swab specimens. Viral RNA detection is based on nucleic acid amplification followed by hybridization capture probes immobilized on a glass slide in both tests. The hybrid is detected by a gold-nanoparticle-conjugated probe. The RVNATSP sample-to-result test was shown to be capable of reliable detection of select respiratory viruses directly from clinical specimens in 3.5 h (Jannetto et al. 2010).

A fluorescent probe based on CdTe QDs covalently linked to rabbit anti-avian influenza virus H5N1 antibody was first used in a sandwich FL-linked immunosorbent assay by Chen et al. (2010) for the

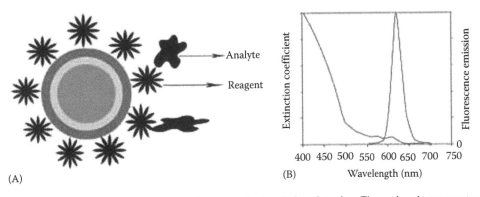

(A)   (B)   Wavelength (nm)

FIGURE 21.4 A diagrammatic representation of quantum-dot-based virus detection. The semiconductor quantum dots because of their small size and a very high luminescence make them applicable for virus detection. (A) Structure of a quantum dot with a CdTe core shell and biomolecules such as antibodies (reagent) used to capture specific viruses (analyte). (B) Emission spectra resulting from a quantum dot luminescence.

highly pathogenic AIV subtype H5N1. In this investigation, the group achieved H5N1 viral antigen detection based on these QD-antibody conjugates with an LOD of 1.5×10^{-4} μg/mL for H5N1 viral antigen. In comparison with virus isolation, the assay had a good sensitivity and specificity. Such rapid and sensitive diagnostic methods are necessary for the H5N1 surveillance. The virus is a threat to the poultry industry and poses a significant hazard to human health.

Zhang et al. (2010) described a new method to detect H5 influenza virus using QDs and magnetic beads (MBs). QDs conjugated with oligonucleotide probes were used to produce fluorescent signals. MBs which were also conjugated with probes were used to isolate and concentrate the signals. Target viral RNAs led to sandwich hybridization between the functionalized QDs and MBs. Thus, one-step hybridization facilitated the subtype determination with an LOD of 0.1 ng viral RNA.

21.5.8 Viral Opportunistic Infections, Congenital Infections, and Others

These include a group of diverse viruses which are able to cause life-threatening disease in individuals who have depressed immune responses. The patient categories include tumor bearing hosts, transplanted individuals, and those with congenital defects in the immune system.

An oligonucleotide labeled with Au nanoparticle has been created as a probe for hybridization of a single-stranded target DNA (CMV) following PCR. This is an indirect determination of the reaction of the solubilized Au^{III} ions by the sandwich-type screen-printed microband electrode (SPMBE). SPMBE makes a sensitive determination of Au^{III} in a small volume of quiescent solution enhancement during the Au^{III} mass electrodeposition time. This principle is called anodic stripping voltammetry. The LOD of this assay was 5 pM of the viral DNA (Authier et al. 2001).

There are assays available that are designed to identify the pathogen causing encephalitis in immunocompromised patients. Presently, the sensitive and specific assays which are rapid are based on PCR technology (real-time PCR assays for DNA neurotropic viruses) (Ramamurthy et al. 2011). Assays based on nano-optical methods are under development.

21.5.9 Ophthalmic Infections and Congenital Infections

Presently, a literature search does not reveal any nanotechnological developments in this field. Certain infections of the eye occur throughout the year in different parts of the globe. This is apart from some epidemic eye infections produced by EV70 and adenoviruses. Early diagnosis of infections with *Streptococcus pneomoniae*, adenoviruses, HSV1, and VZV would be extremely useful clinically to facilitate appropriate therapy. QD-labeled antibody or oligonucleotide probe may be developed either for direct antigen detection in clinical samples or unamplified pathogen DNA directly or post PCR amplification.

ToRCH agents include *Toxoplasma gondi*, rubella virus, CMV, HSV-1, and HSV-2. These agents cause morbidity and mortality among infants born to mothers infected during pregnancy. These infections are seen worldwide and, presently, the diagnosis is principally based on IgM and low avidity IgG detection by ELISA. The development of optical nanodevice on the lines mentioned above for ophthalmic infections will be a very useful advancement.

21.6 Viral Pathogenesis Studies

The study of virus pathways in living cells is vital for understanding viral effect on cell components, development, and tissue tropisms. Several techniques have emerged to facilitate this.

The shift in the plasmon polariton resonance of a single Au (gold) particle encapsulated in a virus with respect to a free particle in solution indicates a close interaction between the basic residues on the inner wall of the capsid and the negative surface charge of the particle. The authors suggested that it will be possible to use encapsulated Au particles to track changes in the viral capsid volume in a physiological environment, a development useful in unraveling virus infection events in cells (Dragnea et al. 2003).

Another report indicated a successful model of in vitro self-assembly conditions that generated infectious virions with brome virus capsid proteins around negatively charged gold nanoparticles' cores. The optical properties (elastic light scattering) and the influence of the core size of the resulting virus-like particle were described. The authors indicated the formation of a closed shell as opposed to an amorphous protein coat using different coatings on the nanoparticle core. This approach could lead to real-time monitoring of viral traffic of single virus particle and chemical sensing along the intracellular and intercellular viral pathways. This could contribute to a better understanding of the virus transport and cellular compartmentalization (Chen et al. 2005). Subsequently, Bentzen et al. (2005) also reported the use of QDs coupled with F and G proteins of respiratory syncytial virus for detection and cell trafficking of viral proteins in infected cells.

More recently, Yeh et al. (2010) described a new hybrid fluorescent nanoprobe for the real-time visualization of virus replication in living cells. The nanoprobe consisted of a nuclease-resistant molecular beacon (MB) backbone, CdSe-ZnS QDs as donors, and gold nanoparticles (Au NPs) as quenchers. A hexa-histidine-appended Tat peptide was self-assembled onto the QD surface to enable non-invasive delivery of the QD-MB-Au NP probes. Direct visualization of the fluorescent complexes formed with the newly synthesized viral RNA was possible. The QD-MB-Au NP probe thus made possible both sensitive and real-time detection of infectious viruses as well as the real-time visualization of cell-to-cell virus spread.

Chen et al. (2010) showed the usefulness of the QDs for the construction of QD-virus hybrids as an imaging probe to reveal viral infection pathways and screen antiviral agents for biological applications. The unique approach of constructing QD-virus hybrids takes advantage of the viral budding process to create enveloped virus incorporated with nanomaterials for the study of virus assembly and intracellular events in a virus infection. The study was carried out in human embryonic kidney (HEK) 293T cells transfected with three plasmids to produce lentiviruses. These allow the infected cells to express enhanced green fluorescent protein. The QD-virus hybrids while budding from the membrane surface of producer cells incorporated QDs encapsulated with alkylated chitosan (chitosan-QDs) which were pre-adsorbed via electrostatic attraction force. Lentiviruses capped with chitosan-modified QDs were thus produced by the transfected cells.

21.7 Summary

Infectious disease diagnosis is based on clinical acumen of the referring physician and laboratorians and procedures such as media-based culture for bacteria and cell culture methods for viruses, microscopy, and serology. The investigations in laboratory are directed toward established pathogens linked to the clinical conditions. Such laboratory support is limited to large hospitals especially in developing countries. Many of the technologies are not amenable to field situations or bedside diagnosis in small clinics where bulk of the patients are served. There are exciting new developments in optical nanosensors, combining immune-active reagents or bioactive molecules with electrical sensing and optical measurements opening a new dimension in viral detection. These devices will be field stable and usable even as bedside appliances. Nanotechnology-based devices for nanoscale reactions wherein the analyte is detected by optical nanosensor are a very promising field. Among the technologies that hold promise for viral diagnostics are QDs-based antigen detection or QD-DNA probe-based PCR product detection. WGM and SERS hold exciting possibilities too. In certain situations, mere virus detection may not be useful but the assay devices developed should be able to quantify the virus so that the physician can establish disease association and also monitor response to therapy. The development of assays should be followed by field testing before they are acceptable as alternatives for more cumbersome but standard technologies.

REFERENCES

Agrawal A, Tripp RA, Anderson LJ, and Nie S. Real-time detection of virus particles and viral protein expression with two-color nanoparticle probes. *J Virol*. 2005 Jul;79(13):8625–8628.

Authier L, Grossiord C, and Brossier P. Gold nanoparticle-based quantitative electrochemical detection of amplified human cytomegalovirus DNA using disposable microband electrodes. *Anal Chem*. 2001 Sep 15;73(18):4450–4456.

Azzazy HM and Mansour MM. In vitro diagnostic prospects of nanoparticles. *Clin Chim Acta.* 2009 May;403(1–2):1–8.

Bentzen EL, House F, Utley TJ, Crowe JE Jr, and Wright DW. Progression of respiratory syncytial virus infection monitored by fluorescent quantum dot probes. *Nano Lett.* 2005 Apr;5(4):591–595.

Chan CP, Tzang LC, Sin KK, Ji SL, Cheung KY, Tam TK, Yang MM, Renneberg R, and Seydack M. Biofunctional organic nanocrystals for quantitative detection of pathogen deoxyribonucleic acid. *Anal Chim Acta.* 2007;584(1):7–11.

Chen SH, Chuang YC, Lu YC, Lin HC, Yang YL, and Lin CS. A method of layer-by-layer gold nanoparticle hybridization in a quartz crystal microbalance DNA sensing system used to detect dengue virus. *Nanotechnology.* 2009 May 27;20(21):215501.

Chen C, Kwak ES, Stein B, Kao CC, and Dragnea B. Packaging of gold particles in viral capsids. *J Nanosci Nanotechnol.* 2005 Dec;5(12):2029–2033.

Chen L, Sheng Z, Zhang A, Guo X, Li J, Han H, and Jin M. Quantum-dots-based fluoroimmunoassay for the rapid and sensitive detection of avian influenza virus subtype H5N1. *Luminescence.* 2010 Nov–Dec;25(6):419–423.

Chen YH, Wang CH, Chang CW, and Peng CA. In situ formation of viruses tagged with quantum dots. *Integr Biol (Camb).* 2010 Jun;2(5–6):258–264.

Demidov VV. Nanobiosensors and molecular diagnostics: a promising partnership. *Expert Rev Mol Diagn.* 2004 May;4(3):267–268.

Deng Z, Zhang Y, Yue J, Tang F, and Wei Q. Green and orange CdTe quantum dots as effective pH-sensitive fluorescent probes for dual simultaneous and independent detection of viruses. *J Phys Chem B.* 2007 Oct 18;111(41):12024–12031.

Dragnea B, Chen C, Kwak ES, Stein B, and Kao CC. Gold nanoparticles as spectroscopic enhancers for in vitro studies on single viruses. *J Am Chem Soc.* 2003 May 28;125(21):6374–6375.

Driskell JD, Kwarta KM, Lipert RJ, Porter MD, Neill JD, and Ridpath JF. Low-level detection of viral pathogens by a surface-enhanced Raman scattering based immunoassay. *Anal Chem.* 2005 Oct 1;77(19):6147–6154.

Driskell JD, Zhu Y, Kirkwood CD, Zhao Y, Dluhy RA, and Tripp RA. Rapid and sensitive detection of rotavirus molecular signatures using surface enhanced Raman spectroscopy. *PLoS One.* 2010 Apr 19;5(4):e10222.

Fan C, Hu Z, Riley LK, Purdy GA, Mustapha A, and Lin M. Detecting food- and waterborne viruses by surface-enhanced Raman spectroscopy. *J Food Sci.* 2010 Jun;75(5):M302–M307.

Gimeno C, Solano C, Latorre JC, Hernández-Boluda JC, Clari MA, Remigia MJ, Furió S, Calabuig M, Tormo N, and Navarro D. Quantification of DNA in plasma by an automated real-time PCR assay (cytomegalovirus PCR kit) for surveillance of active cytomegalovirus infection and guidance of pre-emptive therapy for allogeneic hematopoietic stem cell transplant recipients. *J Clin Microbiol.* 2008 Oct;46(10):3311–3318.

Hermann P, Hermelink A, Lausch V, Holland G, Möller L, Bannert N, and Naumann D. Evaluation of tip-enhanced Raman spectroscopy for characterizing different virus strains. *Analyst.* 2011 Mar 21;136(6):1148–1152.

Hu J, Zheng PC, Jiang JH, Shen GL, Yu RQ, and Liu GK. Sub-attomolar HIV-1 DNA detection using surface-enhanced Raman spectroscopy. *Analyst.* 2010 May;135(5):1084–1089.

Isola NR, Stokes DL, and Vo-Dinh T. Surface-enhanced Raman gene probe for HIV detection. *Anal Chem.* 1998 Apr 1;70(7):1352–1356.

Jain KK. Nanotechnology in clinical laboratory diagnostics. *Clin Chim Acta* 2005;358:37–54.

Jannetto PJ, Buchan BW, Vaughan KA, Ledford JS, Anderson DK, Henley DC, Quigley NB, and Ledeboer NA. Real-time detection of influenza a, influenza B, and respiratory syncytial virus a and B in respiratory specimens by use of nanoparticle probes. *J Clin Microbiol.* 2010 Nov;48(11):3997–4002.

Kimura H, Ito Y, Futamura M, Ando Y, Yabuta Y, Hoshino Y, Nishiyama Y, and Morishima T. Quantitation of viral load in neonatal herpes simplex virus infection and comparison between type 1 and type 2. *J Med Virol.* 2002 Jul;67(3):349–353.

Lee, W.E. and Thompson, H.G. Detection of Newcastle disease virus using an evanescent immuno-based biosensor. *Can. J. Chem.* 1996;74:707–712.

Lindsey NP, Hayes EB, Staples JE, and Fischer M. West Nile virus disease in children, United States, 1999–2007. *Pediatrics.* 2009 Jun;123(6):e1084–e1089.

Liu HH, Cao X, Yang Y, Liu MG, and Wang YF. Array-based nano-amplification technique was applied in detection of hepatitis E virus. *J Biochem Mol Biol.* 2006 May 31;39(3):247–252.

Mitra A, Deutch B, Ignatovich F, Dyke C, and Novotny L. Nano-optofluidic detection of single viruses and nanoparticles. *ACS Nano*. 2010, 4(3):1305–1312.

Pandey P, Datta M, and Malhotra BD. Prospects of Nanomaterials in Biosensors. *Analytical Letters*. 2008;41(2):159–209.

Patolsky F, Zheng G, Hayden O, Lakadamyali M, Zhuang X, and Lieber CM. Electrical detection of single viruses. *PNAS*. 2004 Sept 28;101(39):14017–14022.

Pipper J, Inoue M, Ng LFP, Neuzil P, Zhang Y, and Novak L. Catching bird flu in a droplet. *Nat Med*. 2007;13:1259–1263.

Ramamurthy M, Alexander M, Aaron S, Kannangai R, Ravi V, Sridharan G, and Abraham AM. Comparison of a conventional PCR with real-time PCR for the detection of neurotropic viruses in CSF samples. *Indian J Med Microbiol*. 2011;29(2):102–109.

Roh C, Lee HY, Kim SE, and Jo SK. A highly sensitive and selective viral protein detection method based on RNA oligonucleotide nanoparticle. *Int J Nanomed*. 2010 May 13;5:323–329.

Sahoo SK, Parveen S, and Panda JJ. The present and future of nanotechnology in human health care. *Nanomedicine*. 2007;3(1):20–31.

Sarrazin C, Dragan A, Gärtner BC, Forman MS, Traver S, Zeuzem S, and Valsamakis A. Evaluation of an automated, highly sensitive, real-time PCR-based assay (COBAS Ampliprep/COBAS TaqMan) for quantification of HCV RNA. *J Clin Virol*. 2008 Oct;43(2):162–168.

Stuart DA, Haes AJ, Yonzon CR, Hicks EM, and Van Duyne RP. Biological applications of localised surface plasmonic phenomenae. *IEE Proc Nanobiotechnol*. 2005 Feb;152(1):13–32.

Sudeep AB, and Parashar D. Chikungunya: an overview. *J Biosci*. 2008;33(4):443–449.

Tallury P, Malhotra A, Byrne LM, and Santra S. Nanobioimaging and sensing of infectious diseases. *Adv Drug Deliv Rev*. 2010;62(4–5):424–437.

Tang J, Zhou L, Gao W, Cao X, and Wang Y. Visual DNA microarrays for simultaneous detection of human immunodeficiency virus type-1 and Treponema pallidum coupled with multiplex asymmetric polymerase chain reaction. *Diagn Microbiol Infect Dis*. 2009 Dec;65(4):372–378.

Tang J, Xu Z, Zhou L, Qin H, Wang Y, and Wang H. Rapid and simultaneous detection of Ureaplasma parvum and Chlamydia trachomatis antibodies based on visual protein microarray using gold nanoparticles and silver enhancement. *Diagn Microbiol Infect Dis*. 2010;67(2):122–128.

Tang S, Zhao J, Storhoff JJ, Norris PJ, Little RF, Yarchoan R, Stramer SL, Patno T, Domanus M, Dhar A, Mirkin CA, and Hewlett IK. Nanoparticle-based biobarcode amplification assay (BCA) for sensitive and early detection of human immunodeficiency type 1 capsid (p24) antigen. *J Acquir Immune Defic Syndr*. 2007 Oct 1;46(2):231–237.

Thibault V, Pichoud C, Mullen C, Rhoads J, Smith JB, Bitbol A, Thamm S, and Zoulim F. Characterization of a new sensitive PCR assay for quantification of viral DNA isolated from patients with hepatitis B virus infections. *J Clin Microbiol*. 2007 Dec;45(12):3948–3953.

Tripp RA, Alvarez R, Anderson B, Jones L, Weeks C, and Chen W. Bioconjugated nanoparticle detection of respiratory syncytial virus infection. *Int J Nanomed*. 2007;2(1):117–124.

Valanne A, Huopalahti S, Soukka T, Vainionpää R, Lövgren T, and Härmä H. A sensitive adenovirus immunoassay as a model for using nanoparticle label technology in virus diagnostics. *J Clin Virol*. 2005 Jul;33(3):217–223.

Vasilakis N, Durbin AP, da Rosa AP, Munoz-Jordan JL, Tesh RB, and Weaver SC. Antigenic relationships between sylvatic and endemic dengue viruses. *Am J Trop Med Hyg*. 2008 Jul;79(1):128–132.

Vo-Dinh T, Wang HN, and Scaffidi J. Plasmonic nanoprobes for SERS biosensing and bioimaging. *J Biophotonics*. 2010 Jan;3(1–2):89–102.

Vollmer F and Arnold S. Whispering-gallery-mode biosensing: Label-free detection down to single molecules. *Nat Methods*. 2008 Jul;5(7):591–596.

Vollmer F, Arnold S, and Keng D. Single virus detection from the reactive shift of a whispering-gallery mode. *Proc Natl Acad Sci USA*. 2008 Dec 30;105(52):20701–20704.

Wabuyele MB and Vo-Dinh T. Detection of human immunodeficiency virus type 1 DNA sequence using plasmonics nanoprobes. *Anal Chem*. 2005 Dec 1;77(23):7810–7815.

Wan Z, Wang Y, Li SS, Duan L, and Zhai J. Development of array-based technology for detection of HAV using gold-DNA probes. *J Biochem Mol Biol*. 2005 Jul 31;38(4):399–406.

Wang X, Lou X, Wang Y, Guo Q, Fang Z, Zhong X, Mao H, Jin Q, Wu L, Zhao H, and Zhao J. QDs-DNA nanosensor for the detection of hepatitis B virus DNA and the single-base mutants. *Biosens Bioelectron*. 2010 April 15;25(8):1934–1940.

Wang YF, Pang DW, Zhang ZL, Zheng HZ, Cao JP, and Shen JT. Visual gene diagnosis of HBV and HCV based on nanoparticle probe amplification and silver staining enhancement. *J Med Virol.* 2003 Jun;70(2):205–211.

Xue J, Chen H, Fan M, Zhu F, Diao L, Chen X, Fan L, Li P, and Xia D. Use of quantum dots to detect human papillomavirus in oral squamous cell carcinoma. *J Oral Pathol Med.* 2009 Sep;38(8):668–671.

Yang H, Guo Q, He R, Li D, Zhang X, Bao C, Hu H, and Cui D. A quick and parallel analytical method based on quantum dots labeling for ToRCH-related antibodies. *Nanoscale Res Lett.* 2009 Sep 3;4(12):1469–1474.

Yeh HY, Yates MV, Mulchandani A, and Chen W. Molecular beacon-quantum dot-Au nanoparticle hybrid nanoprobes for visualizing virus replication in living cells. *Chem Commun.* 2010 Jun 14;46(22):3914–3916.

Ymeti A, Kanger JS, Greve J, Besselink GA, Lambeck PV, Wijn R, et al. Integration of microfluidics with a four-channel integrated optical young interferometer immunosensor. *Biosens Bioelectron* 2005;20:1417–1421.

Zhang H, Harpster MH, Park HJ, Johnson PA, and Wilson WC. Surface-enhanced Raman scattering detection of DNA derived from the west nile virus genome using magnetic capture of Raman-active gold nanoparticles. *Anal Chem.* 2011 Jan 1;83(1):254–260.

Zhang H, Xu T, Li CW, and Yang M. A microfluidic device with microbead array for sensitive virus detection and genotyping using quantum dots as fluorescence labels. *Biosens Bioelectron.* 2010 Jul 15;25(11):2402–2407.

Zhang W, Wu D, Wei J, and Xiao G. A new method for the detection of the H5 influenza virus by magnetic beads capturing quantum dot fluorescent signals. *Biotechnol Lett.* 2010 Dec;32(12):1933–1937.

22

Nano-Optical Sensors for Explosive Detection

Paul B. Ruffin and Stuart (Shizhuo) Yin

CONTENTS

22.1 Introduction

To ensure the public safety, detection and identification of explosive materials and devices have become an important issue due to the increased number of explosive attacks on military and civilian targets in recent years. Although there has been tremendous progress in the field of explosive detection, a comprehensive solution to all detection issues has not yet been obtained due to the complexity of the problem. There is always a need to increase the detection sensitivity, selectivity, and standoff distance. Among the different types of detection methods, the optical method plays an important role in the explosive detection, in particular for the case of standoff detection, as many explosives have distinct spectroscopic and spatial profile signatures, which can be locally and remotely detected using optical methods. Although the focus of this chapter is nano-optical sensors for explosive detection, Section 22.2 provides a brief summary on the various types of explosive detection methods to provide an overall picture of explosive detection. Furthermore, since nano-optical sensors are basically employed for the near field/contact detection at this stage, without ignoring the importance and the urgent need of standoff detection of explosives and balancing out the overall content of this chapter, Section 22.3 gives a brief review on several popular optical methods for standoff detection of explosives. A more detailed description of nano-optical sensors for explosive detection is given in Section 22.4. It can be seen that one can further increase the detection sensitivity and selectivity by employing the state-of-the-art nano-optical technology. A conclusion is provided in Section 22.5.

22.2 Explosive Detection Methods

In recent years, many different types of explosive detection methods have been explored and developed [1]. According to the detection distance, these methods can be classified into two categories: (1) noncontact standoff detection and (2) near field and/or contact detection. Figure 22.1 summarizes typical detection technologies for each category [2].

The standoff detection methods include (1) reflection of electromagnetic wave, (2) light detection and ranging (LIDAR), (3) laser-induced breakdown spectroscopy (LIBS), (4) Raman LIDAR, (5) coherent anti-Stokes Raman spectroscopy, (6) photoacoustic, (7) terahertz spectroscopy and imaging, and (8) photothermal imaging. It can be seen that most standoff detection techniques harness optical principles because electromagnetic waves (in particular, laser beams) can propagate long (standoff) distances without much divergence.

The near field and/or contact detection techniques include (1) mass spectroscopy (MS), (2) ion mass spectroscopy, (3) electrochemical, (4) chemiluminescence, (5) microelectromechanical systems, (6) surface enhanced Raman spectroscopy (SERS), (7) Fourier transform infrared spectroscopy, (8) colorimetry, (9) thermal probe, (10) acoustic probe, etc. Since it is very challenging to position nanostructures at or near the remote target locations, nano-optical sensors are basically employed for the near field and/or contact detection (such as SERS) by taking advantages of enhanced signals by the nanostructures.

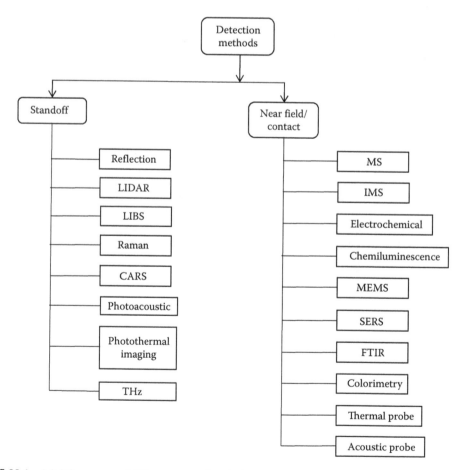

FIGURE 22.1   A brief summary of different types of explosive detection methods. (Adapted from Moore, D.S., *Sens. Imaging*, 8, 9, 2007.)

22.3 Review of Detection Methods Using Traditional Standoff Optical Sensors

This section gives a brief review on several conventional optical methods for standoff detection of explosives, including LIDAR, LIBS, photoacoustics, and photothermal imaging. Since a more detailed description on Raman technology will be provided in Section 22.4, a description of the physical mechanism of the Raman LIDAR is not included in this section.

22.3.1 LIDAR

Figure 22.2 illustrates the basic working principle of using LIDAR for the standoff detection of explosives. A series of laser pulses are sent to the suspected areas that may contain explosive materials and/or devices. These laser pulses are scattered and/or partially absorbed by the targets and the atmosphere (e.g., aerosols). A photodetector is used to detect the scattered signal, mathematically expressed as [2]

$$I_r(R,\lambda) = I_0(\lambda) \frac{A}{4\pi R^2} \beta(R,\lambda) \exp\left(-2\int_0^R \sigma(r,\lambda)dr\right), \tag{22.1}$$

where
$I_0(\lambda)$ is the spectral intensity distribution of the input laser pulses
A is the area of the photodetector
R is the standoff distance
$\beta(R,\lambda)$ is the backscattered coefficient
$\sigma(r,\lambda)$ is the extinction coefficient of the ambient atmosphere

The ranging information can be determined by measuring the transit time, as given by $R = Vt/2$. The material property of the target can be identified by directly measuring the spectral absorption [e.g., by measuring the differential absorption between two pulses with the technique, so-called, differential absorption LIDAR (DIAL)] or by analyzing the Raman shift of the target, so-called Raman LIDAR. Since the vibrational spectroscopic fingerprints of many explosive materials are within the IR spectral range, Raman LIDAR is usually a more effective method for standoff explosive detection. As an example, Figure 22.3 shows the remotely measured Raman shifts by a Raman LIDAR for explosives TATB and HMX [3]. Furthermore, with the rapid advent of compact and tunable IR laser sources (such as quantum cascaded lasers), recently, DIAL can also be used for the standoff detection of explosives.

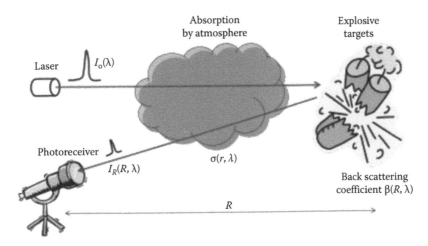

FIGURE 22.2 A conceptual illustration of standoff detection of explosives by a LIDAR. (Adapted from Liadsky, J., Introduction to LIDAR, Optech Incorporated, 2007.)

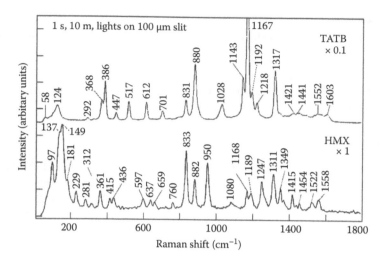

FIGURE 22.3 The remotely measured (10 m standoff distance) Raman shift of explosives TATB and HMX by a Raman LIDAR. (Reprinted from *Spectrochim. Acta Part A*, 61, Sharma, S.K., Misara, A.K., and Sharma, B., Portable remote Raman system for monitoring hydrocarbon, gas hydrates, and explosives in the environment, 2404, 2005. Copyright 2005, with permission from Elsevier.)

Also, as described in Equation 22.1, to increase the detection sensitivity and the standoff distance, a larger detection aperture, A, is preferred. However, a larger aperture usually results in a cumbersome and expensive system. This is one of the major challenges of employing LIDAR technology for long-range standoff detection of explosives. Recent efforts have been focused on how to increase the detection sensitivity by taking advantage of recently developed compact and tunable laser sources as well as integrating different types of detection technologies (e.g., combining the Raman LIDAR with LIBS) so that orthogonal signatures can be obtained.

22.3.2 LIBS

LIBS is another technology that can be used for the standoff detection of explosives. Similar to the case of LIDAR, laser pulses are used to illuminate target. However, in the case of LIBS, the intensity of laser pulses is very high (exceeding 1 GW/cm²). Such a high-intensity light field can dissociate molecules of targets and generate microplasmas [4]. The subsequent emissions from microplasmas contain emission lines of atoms existing in the original target molecules. By detecting these emission lines, the molecular compositions of the target can be determined so that it can be used for the standoff detection of explosives. Thus, LIBS technology is a kind of atomic emission spectroscopy technique.

Figure 22.4 illustrates a basic configuration of LIBS system. A high-power pulsed laser beam (e.g., from a Q-switched Nd:YAG laser) is focused on the surface of a remote target. The high light intensity at the focusing point dissociates molecules of targets and creates microplasma. The emission from the microplasma is collected by a telescopic optical system and resolved by a spectrometer. Figure 22.5 shows the detected LIBS spectrum of explosive RDX at a 20 m standoff distance [4]. It can be clearly seen that the major elements of RDX, including C_2, H, N, and O, can be detected. Thus, LIBS is a very good technique for elementary analysis.

However, LIBS technology suffers from fundamental limitations. A major limitation comes from the interference of oxygen and nitrogen in the atmosphere. To minimize the influence from the ambient atmosphere, recent efforts include the use of double-pulse and ultrafast femtosecond LIBS spectra. The double-pulse illumination can increase the detected signal and improve the reproducibility from shot to shot [5]. The ultrashort femtosecond laser pulse can minimize the influence from the ambient oxygen and nitrogen because it only deposits energy onto the target [6]. This is an ongoing research field. With the rapid advent of femtosecond laser technology, more encouraging experimental results are expected in the near future.

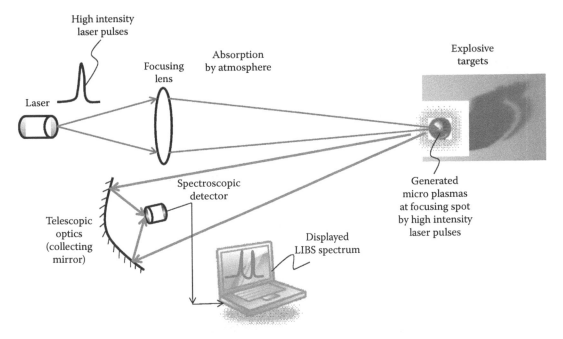

FIGURE 22.4 A basic configuration of LIBS system.

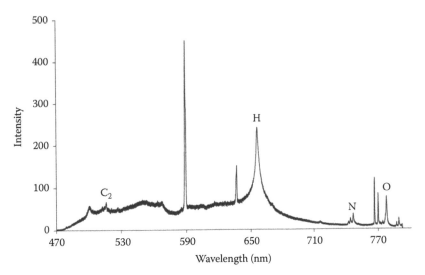

FIGURE 22.5 Remotely collected LIBS spectrum of RDX at a 20 m standoff distance. (From Munson, C.A. et al., Laser-based detection methods for explosives, Army Research Lab Report, ARL TR-4279, 2007.)

22.3.3 Photoacoustics Spectroscopy

Photoacoustic spectroscopy (PAS) is a unique technique for the identification of chemical agents. The basic working principle of PAS may be summarized as follows [7,8]: (1) the light beams are used to illuminate the target, (2) the target materials absorb the light beams and generate heat, (3) the induced thermal expansion due to heating creates pressure waves (i.e., acoustic signal), and (4) microphones are used to detect the generated acoustic signal. Since the generated acoustic waves are correlated with the optical absorption spectra of the materials, the chemical compositions of the materials can

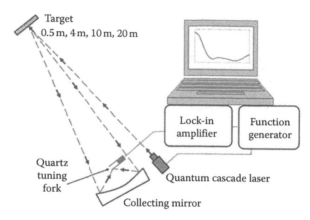

Target
0.5 m, 4 m, 10 m, 20 m

Lock-in
amplifier

Function
generator

Quartz
tuning
fork

Quantum cascade laser

Collecting mirror

FIGURE 22.6 A schematic illustration of experimental setup of standoff detection of explosives by PAS. (From Neste, C. et al., *Appl. Phys. Lett.*, 92, 234002, 2008.)

be determined by scanning the illuminating wavelengths of excited light waves. Furthermore, since the generated acoustic wave is usually weak, an acoustic cavity is usually used to amplify the acoustic wave signal via resonant effect. Although it is very effective to amplify the acoustic signal via a resonant cavity, this method cannot be directly applied for the case of standoff detection because it is difficult to position a resonant cavity at the remote location. To overcome this limitation, recently, an indirect acoustic detection method has been proposed and developed [8]. Instead of directly detecting the photogenerated acoustic wave, a pulsed laser beam is used to illuminate the target, and the scattered light from the target is detected via a quartz crystal tuning fork (QCTF). The amplitude of the vibration of the QCTF is proportional to the intensity of the scattered (and/or reflected) light beam falling on the QCTF. Since the amount of the scattered and/or reflected light is directly related to the absorption property of the target materials, the composition of the materials could be determined by measuring the vibrational amplitude of the QCTF as a function of the excited light wavelength. Figure 22.6 illustrates the configuration of the experimental setup of standoff explosive detection by PAS [8]. A tunable quantum cascaded laser with an output wavelength tuning range (9.26–9.80 μm) is focused to a remote target. A telescopic optical system (a concave mirror) is used to collect the scattered light signal and focus it on a QCTF. The vibration of QCTF is detected by a piezoelectric acoustic transducer and further amplified and processed by a lock-in amplifier-based electronic system. Figure 22.7 shows the experimental results of standoff explosive detection. One can clearly see that the measured profile of the absorption spectrum matches the reference spectrum. Further development of this work should increase the wavelength scanning range of the exciting laser source so that a more accurate detection can be achieved.

22.3.4 Photothermal Imaging

Similar to the case of PAS, the photothermal imaging is also based on the light-induced heating effect. However, instead of detecting the pressure wave generated by the thermal expansion, photothermal imaging detects the thermal emission of the heated target [9]. Again, the amount of the emission is related to the absorption property of the target. When the wavelength of the excited light beam matches the absorption line of the target material, the resonant absorption takes place, which generates the maximum heating effect. Thus, if the wavelength of the excited light matches the absorption peak of the explosive to be detected, the area containing the explosive will generate maximum thermal emission. With an IR thermal imaging detector, one can clearly see a bright spot in that area. Thus, this method can be used to detect explosives at a standoff distance with a relatively large field of view. To minimize the influence from the thermal exchange with the ambient atmosphere, a pulsed exciting wavelength light source is preferred.

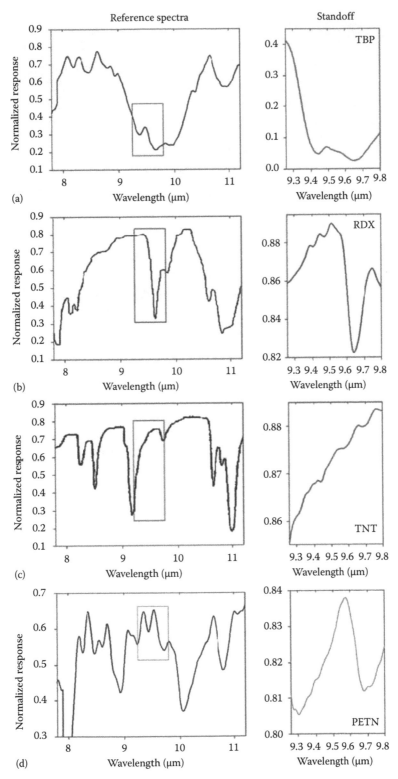

FIGURE 22.7 The experimental results of standoff detection of explosives by PAS (a) TBP detection, (b) RDX detection, (c) TNT detection, and (d) PETN detection. (From Neste, C. et al., *Appl. Phys. Lett.*, 92, 234002, 2008.)

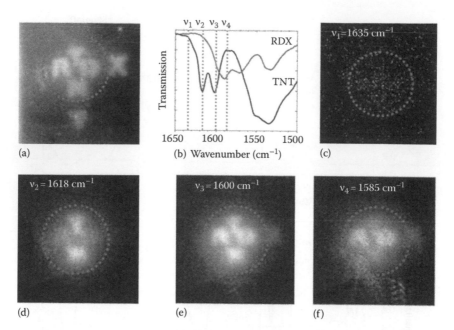

FIGURE 22.8 Experimental result of photothermal imaging. (a) Nonselective heating by a heating gun. (b) IR absorption spectra of TNT and RDX. (c) Laser heating off the resonant wavelength. (d) Laser heating with a resonant wavelength matching TNT. (e) Laser heating with a resonant wavelength matching both RDX and TNT. (f) Laser heating with a resonant wavelength matching RDX only. (From Furstenberg, R. et al., *Appl. Phys. Lett.*, 93, 224103, 2008.)

As an example, Figure 22.8a through f show the experimental result obtained in Ref. [9]. In the experiment, TNT and RDX particles are placed on a stainless steel substrate with a letter pattern "TNT" in a vertical direction and "RDX" in a transverse direction. One can clearly observe that only the targeted material can be seen by using selectively resonant thermal heating. For example, when the excited wavelength matches the resonant absorption wavelength of TNT, only the vertical "TNT" pattern can be seen, as shown in Figure 22.8d.

22.4 Nano-Optical Sensing

22.4.1 Raman Spectroscopy with Nanoengineered Materials and Structures

As mentioned in Section 22.3.1, Raman spectroscopy is an effective method for explosive detection. However, the Raman signal is much weaker $10^{-6} - 10^{-7}$ than that of the excited light intensity. Thus, a high-intensity pulsed laser (e.g., ~mJ/pulse) is usually needed in order to obtain a detectable level Raman signal, which limits the further reduction in the cost and complexity of the Raman spectroscopic system. However, with a nanoengineered structure, the Raman signal can be significantly enhanced. Although it is difficult to position the desired nanoengineered structure at the remote target location for the standoff detection, this kind of nanoengineered structure can be positioned near the target location. Thus, the Raman spectroscopy enhanced by a nanostructure can be very effective for the near field and/or contact explosive detection.

To understand how a Raman signal can be greatly enhanced by a nanostructure, first, the physical mechanism for generating Raman signal needs to be reviewed. Raman signal is generated due to the inelastic scattering of the light by molecules, which have vibrational modes. Under the small amplitude harmonic vibrational approximation, the polarizability of a molecule as a function of molecular vibrational frequency, υ_m, can be written as [10]

$$\alpha = \alpha_0 + \alpha_1 \sin(2\pi\upsilon_m t). \tag{22.2}$$

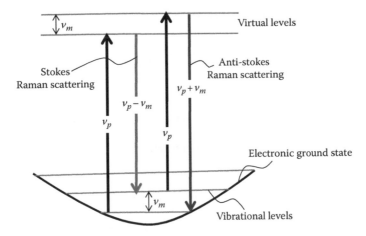

FIGURE 22.9 A graphical illustration of Rayleigh and Raman scattering processes. Interaction between the light and the material molecules.

When a light field $E = E_0 \sin(2\pi v_p t)$ where E_0 and v_p denote the magnitude of electric field and the frequency of the light field, respectively, incident on a molecule, there is a light field-induced dipole moment, given by [10]

$$p = \alpha E$$

$$= \left[\alpha_0 + \alpha_1 \sin\left(\pi v_m t\right) \right] \cdot \left[E_0 \sin(2\pi v_p t) \right]$$

$$= \alpha_0 E_0 \sin\left(2\pi v_p t\right) + \frac{1}{2}\alpha_1 E_0 \cos\left[2\pi\left(v_p - v_m\right)t \right] + \frac{1}{2}\alpha_1 E_0 \cos\left[2\pi\left(v_p + v_m\right)t \right]. \tag{22.3}$$

The first term of Equation 22.3 represents the elastic Rayleigh scattering, the second term denotes the Stokes Raman scattering that has a frequency $\upsilon_p - \upsilon_m$, and the third term represents the anti-Stokes Raman scattering that has a frequency $\upsilon_p + \upsilon_m$. Please note that Equation 22.3 is derived using a simplified classical theory model, which cannot explain the different intensities of Stokes and anti-Stokes scatterings. A more rigorous quantum theory is needed to explain the different intensities of Stokes and anti-Stokes scatterings. Figure 22.9 provides a graphic illustration of the Rayleigh and Raman scatter processes.

One can observe from Equation 22.3 that the magnitude of the Raman scattering is directly proportional to the magnitude of the electric field of the light, E_0. Since E_0 is usually much less than the internal electric field of the molecules, the Raman signal is usually very weak.

The Raman signal can be enhanced via increasing the magnitude of the excited electric field, E_0. Over 35 years ago, it was found that the intensity of Raman signal could be significantly enhanced by a rough metal surface [so-called SERS] [11]. Since then, many theoretical investigations on the enhancing mechanisms and methods have been developed. An enhancement factor up to 10^{14} has been reported [12]. In terms of enhancing mechanism, it is believed that there are basically two enhancing factors: (1) electromagnetic enhancement and (2) chemical enhancement [13]. The electromagnetic enhancement is due to the enhanced electric field caused by the light-generated plasmon near the metal surface. The chemical enhancement is due to charge transfer between the molecules and metal surface. In general, the electromagnetic enhancement is the dominant factor, and the chemical enhancement only contributes one to two enhancement factors. Thus, only a detailed discussion on electromagnetic enhancement is provided in this chapter.

A rough metal surface contains many nanostructures. For the simplicity of the theoretical investigation, one can assume that a nanostructure has a nanosphere shape, which has a diameter much smaller than the wavelength. In this case, the total electric field near the surface of the nanosphere, E_T, can be written as [13]

$$E_T = \frac{\varepsilon_1(\omega) - \varepsilon_2}{\varepsilon_1(\omega) + 2\varepsilon_2} E_{laser},\tag{22.4}$$

where

$\varepsilon_1(\omega)$, ε_2 is the complex frequency-dependent permittivity of metal and the permittivity of the ambient phase, respectively

E_{laser} denotes the electric field of the incident laser beam

Thus, at the resonant frequency [i.e., $Re(\varepsilon_1(\omega = \omega_r)) = -2\varepsilon_2$] $|E_T| \gg |E_{laser}|$. Therefore, there can be a significantly enhanced local electric field by the nanostructured metal surface, which results in an increase in the Raman signal, as described by Equation 22.3. Furthermore, the nanostructured metal surface can also serve as an antenna to control the intensity distribution of the Raman scattered signal. Therefore, the overall intensity enhancement factor can be proportional to E_T^4 (i.e., $I_R \propto E_T^4$). For example, a 10^3 increase in E_T can result in a 10^{12} increase in I_R This is why a dramatic increase in the Raman signal can be realized by employing the SERS technology.

However, at the earlier stage of SERS, the rough metal surface was a randomized nanostructured surface. It is very difficult to control the exact value of the enhancement factor, which limits its practical application because it is very challenging to conduct the consistent quantitative measurement in this case.

Fortunately, with the rapid advance of nanofabrication technology, well-controlled nanostructured metal surfaces have been fabricated in recent years. One can not only design a nanostructured surface with an optimized enhancement factor but also consistently fabricate such kind of surfaces. With the well-controlled nanostructured metal surface, the theoretically derived enhancement factor closely matches the experimental data. For example, Figure 22.10 shows the calculated enhancement factor

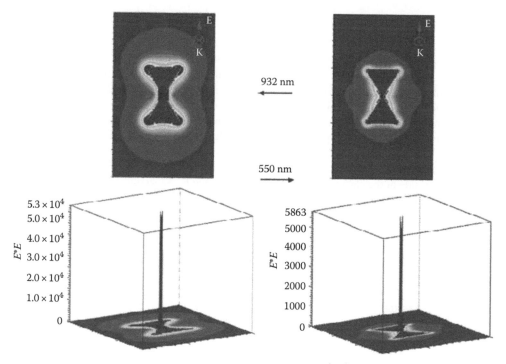

FIGURE 22.10 **(See color insert.)** The calculated enhancing factor of $\left|E_T^2\right|$ or a "tip-to-tip" prism dimer shape nanostructured silver metal surface at the resonant wavelengths of 550 and 932 nm, respectively. The prism has a 60 nm edge dimension, a 2 nm snip, and a thickness of 12 nm. (From Hao, E. and Schatz, G., *J. Chem. Phys.*, 120, 357, 2004.)

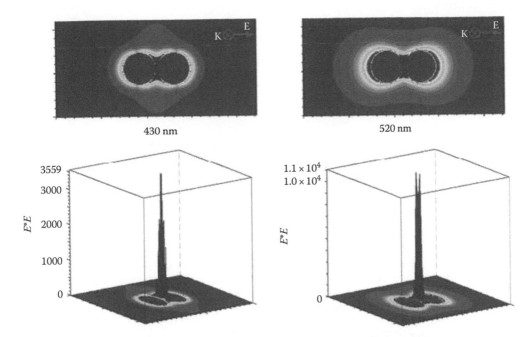

FIGURE 22.11   The calculated enhancing factor of $\left|E_T^2\right|$ or a dimer of sphere shape nanostructured silver metal surface at the resonant wavelengths of 430 and 520 nm, respectively. The dimer of the silver sphere has a diameter of 36 m and a spacing of 2 nm [14]. (From Hao, E. and Schatz, G., *J. Chem. Phys.*, 120, 357, 2004.)

for a "tip-to-tip" prism dimer nanostructured metal surface [14]. It can be seen that the enhancement factor for $\left|E_T^2\right|$ can be as high as 5.3×10^4 to the resonant wavelength 932 nm. In this case, the corresponding enhancement factor for the intensity of the Raman signal I_R is $(5.3 \times 10^4)^2 \sim 2.8 \times 10^9$ closely matches the experimentally measured enhancement factor.

The theoretical investigation also shows that the enhancement factor and resonant frequency are also largely influenced by the shape of the nanostructure. Figure 22.11 shows the calculated enhancement factor and resonant frequency for a dimer of silver nanosphere [14]. In this case, the resonant wavelengths are 430 and 520 nm, respectively. Thus, by properly designing the dimension and the shape of the nanostructure, one can obtain the required enhancement factor at the designed resonant frequency.

Nanostructure enhanced Raman spectroscopy has recently been applied to explosive detection [15]. In the experiment, first, a nanostructured metal surface, as shown in Figure 22.12, is created [15]. Then, the vapor of explosive [triacetonetriperoxide (TATP)] is exposed on this nanostructured surface. Finally, the scattered Raman signal is collected and processed. Due to the enhancing effect of this nanostructure,

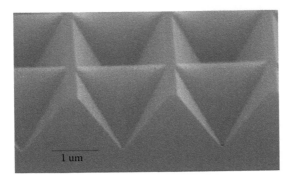

FIGURE 22.12   A picture of nanostructured surface used for the explosive detection. A triangle shape nanostructured silicon substrates coated with gold. (With kind permission from Springer Science+Business Media: *Appl. Phys. B*, Detection of explosive vapour using surface-enhanced Roman spectroscopy, 97, 2009, 723, Fang, X. and Ahmad, S.)

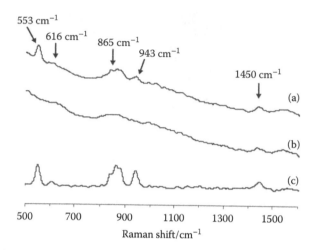

FIGURE 22.13 The experimentally measured Raman spectroscopy of explosive TATP by using a nanostructure enhanced Raman optical sensor. (a) Raman spectrum of TATP vapor, (b) Raman spectrum with TATP, and (c) Raman spectrum of TATP powder. (With kind permission from Springer Science+Business Media: *Appl. Phys. B.*, Detection of explosive vapour using surface-enhanced Roman spectroscopy, 97, 2009, 723, Fang, X. and Ahmad, S.)

a clear Raman spectrum of TATP, as illustrated in Figure 22.13 [15], is detected by using a low-power 250 mW diode laser with an output wavelength of 780 nm. Thus, by harnessing the state-of-the-art nanostructure enhanced Raman spectroscopy, one can achieve vapor level near field/contact explosive detection by using a low-power, low-cost diode laser as the exciting light source, which substantially reduces the complexity and the cost of Raman spectroscopy. Thus, a Raman nano-optical sensor can be an effective tool for the near field/contact explosive detection.

22.4.2 Nanostructure Enhanced Surface Plasmonic Sensor

It is well known that surface plasmon (SP) refers to the coherent electron oscillation at the interface between two materials (e.g., a dielectric material and a metal material) [15]. The SP excited light can propagate along the metal surface. Since the excitation of SP by light is sensitive to the incident angle and wavelength of the exciting light as well as the ambient refractive index (to be discussed in detail later), it can be harnessed for the optical sensor application. Furthermore, the performance of the optical sensor can be further enhanced by employing nanostructure and can be used for the near field/contact explosive detection.

Figure 22.14 illustrates a Kretschmann attenuated total reflection setup for SP excitation [16], which has been widely used for chemical and biological optical sensors. In this case, the reflection coefficient, r, can be shown to be [17,18]

$$r = \frac{A + B/Z_3 - Z_1\left(C + D/Z_3\right)}{A + B/Z_3 + Z_1\left(C + D/Z_3\right)} \tag{22.5}$$

where

$A = D = \cos k_{z2}d,\ B = jZ_2 \sin k_{z2}d,\ C = j\sin k_{z2}d/Z_2$

$k_{zi} = \sqrt{k_i^2 - \left(k_i \sin \theta_i\right)^2}$, i = 1, 2, 3, $k_i = n_i k_0$, $k_0 = \omega/c$ are the wave vector of space

$Z_i = k_{zi}/(\omega \varepsilon_0 \varepsilon_i)$

ε_i and μ_i are the relative permittivity and permeability for the *i*th medium, respectively

ε_0 is the permittivity of the free space and d s the thickness of the metal thin film

If we assume that the intensity of the incident light is I_0, the intensity of the reflected light is $I_R = |r^2|I_0$, which depends on both the medium property (i.e., ε_i, μ_i) and the incident angle θ_1 according to Equation 22.5. For the optical sensor application, in general, we can assume that $\mu_1 = \mu_2 = \mu_3 = 1$, and

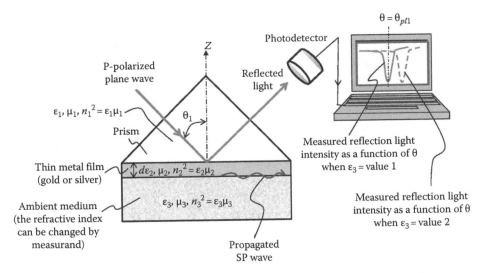

FIGURE 22.14 A conceptual illustration of Kretschmann attenuated total reflection setup for SP excitation.

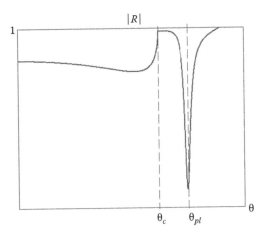

FIGURE 22.15 The calculated absolute value of the reflection coefficient |r| is a function of incident angle θ_1 of a Kretschmann attenuated total reflection setup for the case $\mu_1 = \mu_2 = \mu_3 = 1$, $\varepsilon_2 = -10 - 0.1j$, $\varepsilon_3 = 1.75$, and $\lambda = 0.6\,\mu m$ based on Equations 22.5 and 22.6. (Reproduced from Ishimaru, A., Jaruwatanakilok, S., and Kuga, Y., *Prog. Electromagn. Res. PIER*, 51, 139, 2005.)

ε_1 and ε_3 are real numbers. In this case, the complex permittivity of the metal layer ε_2 can result in a minimum reflection at certain plasmonic resonant incident angle, θ_{pl}, given by [18]

$$\sin\theta_{pl} = \sqrt{\frac{\varepsilon_2\varepsilon_3}{\varepsilon_1\left(\varepsilon_2 + \varepsilon_3\right)}}. \tag{22.6}$$

Substituting this incident angle (i.e., $\theta_1 = \theta_{pl}$) into Equation 22.5, one can obtain a minimum reflection. As an example, Figure 22.15 shows the calculated absolute value of the reflection coefficient |r| as a function of incident angle θ_1 [18]. One can clearly observe a very sharp dip at $\theta_1 = \theta_{pl}$. In this case, the propagation constant of the incident light wave matches the propagation constant of the SP wave between mediums 2 and 3 so that the energy of the incident light is coupled to the SP wave, which can be eventually absorbed by the metal layer and/or coupled out to the medium 3.

In terms of optical sensor application, ε_1 and ε_2 are fixed. According to Equation 22.6, θ_{pl} only depends on ε_3. Thus, if a measurand is related to the value of ε_3, one can measure this measurand by measuring θ_{pl}. Thus, SP resonance is a very effective tool to probe the refractive index change on the metal thin film.

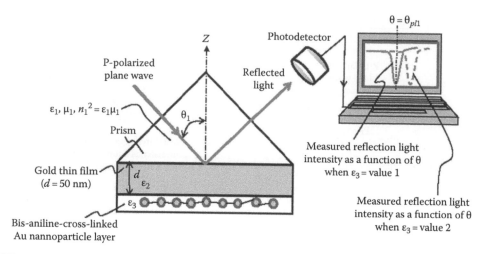

FIGURE 22.16 A schematic diagram of SP-based explosive detection by using a bis-aniline-cross-linked Au nanoparticle composite as the sensing medium. (Modified from Riskin, M. et al., *J. Am. Chem. Soc.*, 131, 7368, 2009.)

To realize the explosive detection by SP resonance, a combination of nanostructured material (e.g., gold nanoparticles) and SP resonance has recently been investigated [19]. Figure 22.16 shows the configuration of the sensing setup. A thin layer of bis-aniline-cross-linked gold (Au) nanoparticles, as illustrated in Figure 22.17 [20], is used as the sensing medium 3. Since the dimension of the Au nanoparticles is much smaller than the wavelength of the light, based on volume averaging theory, an effective dielectric constant of this composite structure can be employed, as given by [21]

$$\varepsilon_{3,eff} = (1-f)\varepsilon_{3,m} + f\varepsilon_{3,Au}, \tag{22.7}$$

where

f is the filling factor

$\varepsilon_{3,m}$ is the dielectric constant of the surrounding medium around the Au nanoparticles

$\varepsilon_{3,Au}$ is the dielectric constant of the Au nanoparticles

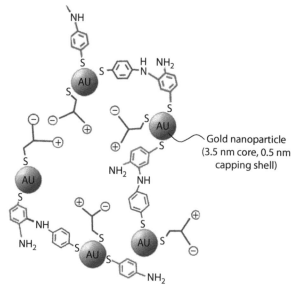

FIGURE 22.17 An illustration of bis-aniline-cross-linked Au nanoparticles composite. (From Riskin, M. et al., *Chem. Euro. J.*, 16, 7114, 2010.)

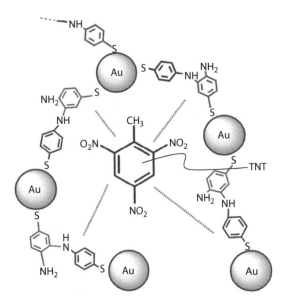

FIGURE 22.18 An illustration of interaction between the bis-aniline-cross linked Au nanoparticles composite and the explosive TNT for forming donor–acceptor complexes. (From Riskin, M. et al., *J. Am. Chem. Soc.*, 131, 7368, 2009.)

Substituting this $\varepsilon_{3,eff}$ into Equation 22.6, the SP resonant angle θ_{pl} can be determined. When the explosive trinitrotoluene (TNT) is exposed to this sensing medium, there is a π donor–acceptor interaction between the TNT and the thioaniline units and bis-aniline bridging unit, as illustrated in Figure 22.18 [19]. This results in a change of $\varepsilon_{3,m}$ that in turn varies $\varepsilon_{3,eff}$. Thus, one can observe a shift of θ_{pl} when TNT exists. The major advantage of employing Au nanocomposite for TNT detection is the high sensitivity. As shown in Figure 22.19 [19], the existence of only 200 nM of TNT can result in a noticeable shift 0.5° of θ_{pl}.

Therefore, nano-optical sensors based on SP resonance can indeed be employed for the high-sensitivity explosive detection, although it is difficult for standoff detection applications.

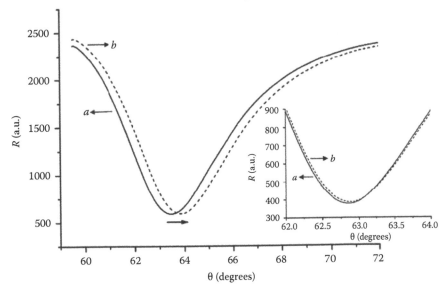

FIGURE 22.19 The experimentally measured shift of θ_{pl} when the bis-aniline-cross-linked Au nanoparticles composite is exposed to the explosive TNT. a-line: before exposure to TNT, b-line: after exposure to TNT. (From Riskin, M. et al., *J. Am. Chem. Soc.*, 131, 7368, 2009.)

22.4.3 Fluorescent Light Emission with Nanoengineered Materials and Structures

Fluorescent light emission is another method that can be used for explosive detection because certain fluorescent materials such as poly(arylene ethynylenes) can change the intensity of the fluorescent emission when they are exposed to explosives [22]. In particular, with the advent of nanoengineered fluorescent materials [such as quantum dot (QD)-based fluorophore] and the new fluorescent emission mechanism based on the interaction at the nanoscale such as Forster resonance energy transfer (FRET), a high-sensitivity and high-selectivity detection of explosives can be achieved due to the following reasons: (1) the fluorescent emission from the QD-based fluorophores has a narrow emission spectrum that increases the detection sensitivity and selectivity, (2) the central wavelength of the emitted light can be controlled by adjusting the shape and the size of the QDs that enables multichannel and multiagent detection capability that also enhances the detection sensitivity and selectivity, and (3) the FRET-based detection has a very high sensitivity (to be described in detail later).

FRET involves two types of fluorophores: a donor fluorophore and an acceptor fluorophore [23–24]. The donor fluorophore absorbs the energy of the exciting light at a frequency v_1 and transfers this energy to the acceptor fluorophore via the nonradiation energy transfer due to the dipole–dipole interaction to the acceptor fluorophore. Then, the acceptor fluorophore emits the fluorescent light at a lower frequency v_2. Figure 22.20 illustrates the FRET-based fluorescent excitation and emission process.

A unique feature of the FRET-based energy transfer is that the efficiency of the FRET energy transfer is extremely sensitive to the separation distance, r, between the donor fluorophore and the acceptor fluorophore as given by [23,24]

$$E_{FRET} = \frac{R_0^6}{R_0^6 + r^6},$$

(22.8)

where R_0 is a distance parameter, as given by [24]

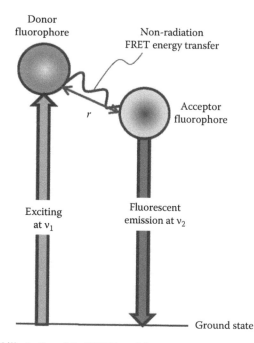

FIGURE 22.20   A conceptual illustration of the FRET-based fluorescent excitation and emission process.

$$R_0 = \left(\frac{9000(\ln 10)\kappa_p^2 Q_D}{N_A 128\pi^5 n_D^4} I \right)^{1/6},$$ (22.9)

where
  κ_p is the parameter related to the orientation of the donor and acceptor dipoles
  Q_D is the donor quantum yield of the fluorescent emission
  N_A is the Avogadro number
  N_D is the refractive index of the medium
  I is the overlap integral

Since $E_{FRET} \propto 1/r^6$ (a small change in r, e.g., sub-nm) can result in a substantial change in the efficiency of the FRET, which can in turn greatly vary the intensity of the emitted fluorescent light. This FRET mechanism has recently been employed for the high-sensitivity explosive detection [23].

  In the experiment [23], CdSe-ZnS core-shell QDs have been selected as the donor fluorophore, which has maxima emission at three wavelengths (530, 555, and 570 nm), as depicted in Figure 22.21 [23].

  Black Hole Quencher-10 (BHQ-10) dye (made by Biosearch Technologies, Novato, CA) is selected as the acceptor fluorophore. To quench the fluorescent emission from the acceptor fluorophore BHQ-10, BHQ-10 is mixed with diaminopentane (DAP) and the trinitrobenzene (TNB) to form the quencher analogue (TNB-DAP-BHQ-10), as illustrated in Figure 22.22 [23]. Thus, there will be no fluorescent emission, although the energy of the exciting light has been transferred to the TNB-DAP-BHQ-10 via FRET in this case.

FIGURE 22.21 The emission spectrum of CdSe-ZnS core-shell QDs. (From Goldman, E. et al., *J. Am. Chem. Soc.*, 127, 6744, 2005.)

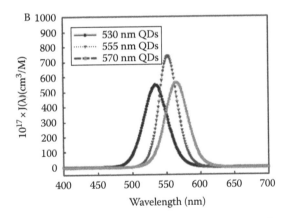

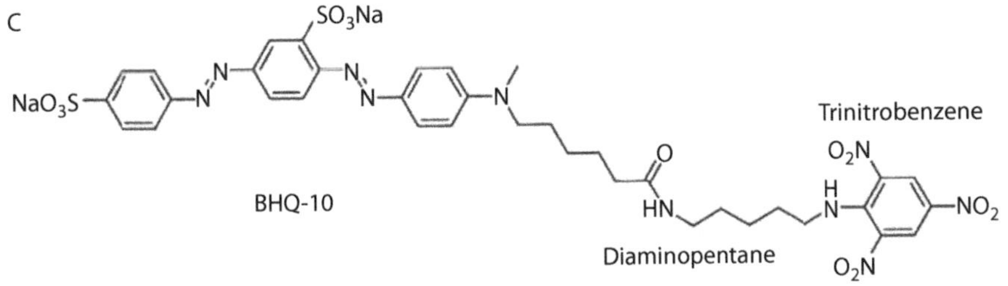

FIGURE 22.22 The chemical structure of the quencher analogue (TNB-DAP-BHQ-10). (From Goldman, E. et al., *J. Am. Chem. Soc.*, 127, 6744, 2005.)

To detect the explosive TNT, a single-chain antibody fragment TNB2-45 is mixed with QDs and TNB-DAP-BHQ-10 to form the sensing medium. When this sensing medium is exposed to the explosive TNT, the interaction between the TNT and the antibody fragment TNB2-45 results in an increase in the separation distance between the donor fluorophore OD and the quencher analogue TNB-DAP-BHQ-10, which in turn reduces the efficiency of FRET or stops the process of FRET (when the separation distance, r, is large enough). Since the energy of the excited light is not transferred from the donor fluorophore to the quencher analogue, the donor fluorophore emits the fluorescent light at the emission frequency of the donor, v_3, when it is excited, as illustrated in Figure 22.23.

Figure 22.24 shows the experimentally measured fluorescent emission from the donor QDs as a function of the concentration level of TNT [23]. One can clearly observe that the intensity of the fluorescent emission increases as the concentration level of TNT increases and a low concentration level of TNT (e.g., <1 µg/mL) can be detected.

Thus, a high-sensitivity and high-selectivity explosive detection can be realized by using the fluorescent emission based on the nanoengineered materials although it is a near field and/or contact detection.

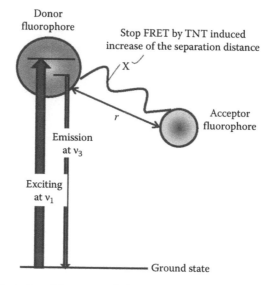

FIGURE 22.23 A graphic illustration of fluorescent mission when the sensing medium is exposed to the explosive TNT. In this case, the energy of the exciting energy will not be transferred to the acceptor fluorophore due to the increased separation between the donor and the acceptor fluorophores by the existence of TNT.

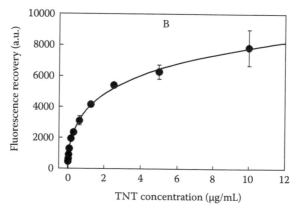

FIGURE 22.24 The experimentally measured fluorescent emission from the donor QDs as a function of the concentration level of TNT. (From Goldman, E. et al., *J. Am. Chem. Soc.*, 127, 6744, 2005.)

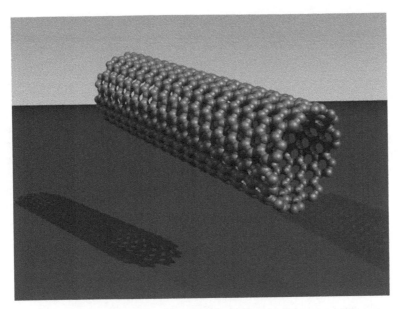

FIGURE 22.25 A graphic illustration of single-wall CNTs. (From http://en.wikipedia.org/wiki/File:Carbon_nanotube_armchair_povray.PNG.)

22.4.4 Functionalized Carbon Nanotubes

Functionalized carbon nanotubes (CNTs) can also be used as high-sensitivity nanosensors for chemical and/or explosive detection. Since CNTs contain a large amount of surface atoms (completely surface atoms in the case of single-wall CNTs, as illustrated in Figure 22.25 [25]), their electronic properties are very sensitive to their chemical environment [26]. This unique property can be harnessed for high-sensitivity sensing. Since this is a nanoelectronic sensor, instead of a nano-optical sensor, here, we just provide one example.

In this example [27], copper (Cu) nanoparticles are used to functionalize the CNTs. When this functionalized CNTs is exposed to nitroaromatic compounds (e.g., TNT), it results in a change of electrical conductivity. Figure 22.26 shows the experimentally measured peak area (μAV) as function of

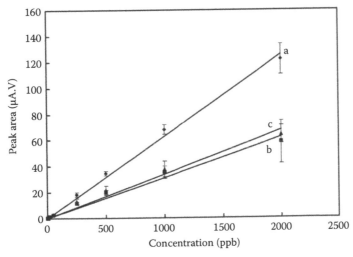

FIGURE 22.26 The experimentally measured peak area of Cu nanoparticles functionalized CNTs as a function of concentration level of TNT. (a) Peak I at −0.53 V, (b) peak II at −0.67 V, and (c) peak III at −0.77 V versus Ag/AgCl. (From Hrapovic, S. et al., *Anal. Chem.*, 78, 5504, 2006.)

concentration level of explosive TNT [27]. It clearly shows that it is possible to achieve a ppb level sensitivity. Thus, a very high-sensitivity nanosensor can be constructed based on the functionalized CNTs for the explosive detection.

22.5 Conclusion

A description of nano-optical sensors for the explosive detection has been presented in this chapter. For the purpose of comparison, a brief review of the conventional methods of explosive detection is also provided in this chapter so that one can tell the advantages and limitations of nano-optical sensors for the explosive detection. Even within the nano-optical sensors, due to the page limitation, only a few types of nanosensors and their applications to explosive detection, including (1) nanostructure enhanced Raman spectroscopic sensor, (2) nanostructure enhanced surface plasmonic sensor, (3) fluorescent light emission sensor with nanostructured material, and (4) the sensor based on functionalized CNTs, have been reviewed in this chapter. It can be seen that the performances of the sensors (such as the sensitivity, the selectivity, and the reliability) can be dramatically increased by employing the state-of-the-art nanotechnology for explosive detection. For example, by using triangular-shaped nanostructured silicon substrates coated with gold, one can consistently detect the Raman signal generated by the vapor of explosive TATP, with a low-power 250 mW diode laser as the exciting light source. By employing the nanostructure (gold nanocomposite) enhanced surface plasmonic sensor, one can detect a tiny amount (200 nM) of explosive TNT. By harnessing nanostructure enhanced fluorescent light emission, one can detect a low concentration level (<1 μg/mL) of explosive TNT. Thus, it is possible to detect the trace level of explosives by employing nano-optical sensors, which represents a significant advance in the field of explosive detection because one of the major challenges of defeating the asymmetric explosive attack is to detect the trace level explosives. However, since it is difficult to deploy the nanostructures at remote target locations, current nano-optical sensors are mainly used for the near field/contact explosive detection. In addition to further improving the enhancing factor offered by the nanostructure (e.g., by optimizing the shape, the size, and the material of the nanostructure), one future perspective is how to apply nano-optical sensors for standoff detection of explosives.

REFERENCES

1. D.S. Moore, Recent advances in trace explosive detection instrumentation, *Sens. Imaging* 8:9–38, 2007.
2. J. Liadsky, Introduction to LIDAR, Optech Incorporated, 2007.
3. S.K. Sharma, A.K. Misara, and B. Sharma, Portable remote Raman system for monitoring hydrocarbon, gas hydrates, and explosives in the environment, *Spectrochim Acta Part A*, 61A, 2404, 2005.
4. C.A. Munson, J.L. Gottfried, F.C. De Lucia, Jr., K.L. McNesby, and A.W. Miziolek, Laser-based detection methods for explosives, Army Research Lab Report, ARL-TR-4279, 2007.
5. C. Cautier, P. Fichet, D. Menut, J.-L. Lacour, D. L' Hermite, and J. Dubessy, Quantification of the intensity enhancements for the double-pulse laser induced breakdown spectroscopy in the orthogonal beam geometry, *Spectrochim Acta Part B*, 60, 265, 2005.
6. Y. Dikmelik and J.B. Spicer, Femtosecond laser – induced breakdown spectroscopy of explosives and explosive–related compounds, *Proc. SPIE*, 5794, 757, 2005.
7. C. Haisch and R. Niessner, Light and sound-photoacoustic spectroscopy, *Spectrosc. Eur.*, 14, 10, 2002.
8. C. Neste, L. Senesac, and T. Thundat, Standoff photoacoustic spectroscopy, *Appl. Phys. Lett.*, 92, 234002, 2008.
9. R. Furstenberg, C. Kendziora, J. Stepnowski, S. Stepnowski, M. Rake, M. Papantonakis, V. Nguyen, G. Hubler, and R. McGill, Standoff detection of trace explosives via resonant infrared photothermal imaging, *Appl. Phys. Lett.*, 93, 224103, 2008.
10. David W. Ball, Theory of Raman spectroscopy, *Spectroscopy*, 16, 32, 2001.
11. M. Fleischmann, P. Hendra, and A. McQuillan, Raman spectra of Pyridine absorbed at a silver electrode, *Chem. Phys. Lett.*, 26, 163, 1974.
12. K. Kneipp, H. Kneipp, I. Itzkan, R. Dasar, and S. Feld, Ultrasensitive chemical analysis by Raman spectroscopy, *Chem. Rev.*, 99, 2957, 1999.

13. A. Campion and P. Kambhampati, Surface-enhanced Raman scattering, *Chem. Soc. Rev.* 27, 241, 1998.
14. E. Hao and G. Schatz, Electromagnetic fields around silver nanoparticles and dimmers, *J. Chem. Phys.*, 120, 357, 2004.
15. X. Fang and S. Ahmad, Detection of explosive vapour using surface-enhanced Roman spectroscopy, *Appl. Phys. B.*, 97, 723, 2009.
16. E. Kretschamann, Die bestimmune Der Oberflachenrauhigkeit dunner schichten durch messing der winkelabhangigkeit der streustrahlung von oberflachenplasmaschwingungen, *Opt. Commun.*, 10, 353, 1974.
17. A. Ishimaru, *Electromagnetic Wave Propagation, Radiation, and Scattering*, Prentice Hall, London, U.K., 1991.
18. A. Ishimaru, S. Jaruwatanakilok, and Y. Kuga, Generalized surface Plasmon resonance sensors using metamaterials and negative index materials, *Prog. Electromagn. Res. PIER*, 51, 139, 2005.
19. M. Riskin, R. Vered, O. Lioubashevski, and I. Willner, Ultrasensitive surface Plasmon resonance detection of trinitrotoluene (TNT) by a Bis-aniline-cross-linked Au Nanoparticles composite, *J. Am. Chem. Soc.*, 131, 7368, 2009.
20. M. Riskin, R. Vered, M. Frasconi, N. Yavo, and I. Willner, Stereoselective and chiroselective surface Plasmon resonance analysis of amino acid by molecularly imprinted Au-nanoparticle composites, *Chem. Euro. J.*, 16, 7114, 2010.
21. A. Garahan, L. Pilon, and J. Yin, Effective optical properties of absorbing nanoporous and nanocomposite thin films, *J. Appl. Phys.*, 101, 014320, 2007.
22. J. Zheng and T. Swager, Poly(arylene ethynylene)s in chemosensing and biosensing, *Adv. Polym. Sci.*, 177, 151, 2005.
23. E. Goldman, I. Medintz, J. Whitley, A. Hayhurst, A. Clapp, H. Uyeda, J. Deschamps, M. Lassman, and H. Mattoussi, A hybrid quantum dot-antibody fragment fluorescence resonant energy transfer-based TNT sensor, *J. Am. Chem. Soc.*, 127, 6744, 2005.
24. A. Clapp, I. Medintz, and H. Mattoussi, Forster resonance energy transfer (FRET) investigations using quantum-dot fluorophores, *Chem. Phys. Chem.*, 7, 47, 2006.
25. Arneto, Carbon nanotube zigzag povary cropped.png, Wikipedia Commons, November, 27, 2007.
26. P. Hu, J. Zhang, L. Li, Z. Wang, W. O'Neill, and P. Estrela, Carbon nanostructure based field effect transistors for label-free chemical/biological sensors, *Sensors*, 10, 5133, 2010.
27. S. Hrapovic, E. Majid, Y. Liu, K. Male, and J. Luong, Metallic nanoparticles-carbon nanotube composites for electrochemical determination of explosive nitroaromatic compounds, *Anal. Chem.*, 78, 5504, 2006.

Index